St. Olaf College

APR 2 8 1998

Science Library

ELECTROSPRAY IONIZATION MASS SPECTROMETRY

ELECTROSPRAY IONIZATION MASS SPECTROMETRY

FUNDAMENTALS, INSTRUMENTATION, AND APPLICATIONS

Edited by

Richard B. Cole
Department of Chemistry, University of New Orleans, New Orleans, Louisiana

A WILEY-INTERSCIENCE PUBLICATION

JOHN WILEY & SONS, INC.

New York · Chichester · Weinheim · Brisbane · Singapore · Toronto

This text is printed on acid-free paper.

Copyright © 1997 by John Wiley & Sons, Inc.

All rights reserved. Published simultaneously in Canada.

Reproduction or translation of any part of this work beyond that permitted by Section 107 or 108 of the 1976 United States Copyright Act without the permission of the copyright owner is unlawful. Requests for permission or further information should be addressed to the Permissions Department, John Wiley & Sons, Inc., 605 Third Avenue, New York, NY 10158-0012.

Library of Congress Cataloging in Publication Data:

Electrospray ionization mass spectrometry : fundamentals, instrumentation, and applications / edited by Richard B. Cole.
    p. cm.
   "A Wiley-Interscience publication."
   Includes index.
   ISBN 0-471-14564-5 (cloth : alk. paper)
   1. Mass spectrometry. 2. Ionization. 3. Biomolecules--Analysis.
I. Cole, Richard B., 1956- .
   QP519.9.M3E44 1997
   574.19'285--dc20

Printed in the United States of America

10 9 8 7 6 5 4 3 2

CONTRIBUTORS

Mark E. Bier, Department of Chemistry, Carnegie-Mellon University, Pittsburgh Pennsylvania, USA

Andries P. Bruins, University Centre for Pharmacy, Groningen, The Netherlands

Igor V. Chernushevich, SCIEX, Concord, Ontario, Canada

Richard B. Cole, Department of Chemistry, University of New Orleans, Lakefront, New Orleans, Louisiana, USA

Pamela F. Crain, Department of Medicinal Chemistry, University of Utah, Salt Lake City, Utah, USA

Werner Ens, Department of Physics, University of Manitoba, Winnipeg, Manitoba, Canada

Christine L. Gatlin, Department of Chemistry, University of Washington, Seattle, Washington, USA

Yeunghaw Ho, Department of Chemistry, University of Alberta, Edmonton, Alberta, Canada

Paul Kebarle, Department of Chemistry, University of Alberta, Edmonton, Alberta, Canada

Barbara S. Larsen, Central Research Department, E. I. du Pont de Nemours & Company, Wilmington, Delaware, USA

David A. Laude, Department of Chemistry and Biochemistry, University of Texas at Austin, Austin, Texas, USA

Joseph A. Loo, Parke-Davis Pharmaceutical Research, Division of Warner-Lambert Company, Ann Arbor, Michigan, USA

Rachel R. Ogorzalek Loo, Department of Biological Chemistry, University of Michigan, Ann Arbor, Michigan, USA

Charles N. McEwen, Central Research Department, E. I. du Pont de Nemours & Company, Wilmington, Delaware, USA

Yoko Ohashi, Institute of Physical and Chemical Research (RIKEN), Wako, Saitama, Japan

Grace K. Poon, Drug Metabolism and Pharmacokinetics, SmithKline Beecham Pharmaceuticals, The Frythe, Welwyn, United Kingdom

Jessica M. Robinson, Department of Chemistry and Biochemistry, University of Texas at Austin, Austin, Texas, USA

Jae C. Schwartz, Finnigan Corporation, San Jose, California, USA

Joanne C. Severs, Environmental Molecular Sciences Laboratory, Pacific Northwest National Laboratory, Richland, Washington, USA

Richard D. Smith, Environmental Molecular Sciences Laboratory, Pacific Northwest National Laboratory, Richland, Washington, USA

Kenneth G. Standing, Department of Physics, University of Manitoba, Winnipeg, Manitoba, Canada

Elizabeth Stevenson, Department of Chemistry and Biochemistry, University of Texas at Austin, Austin, Texas, USA

František Tureček, Department of Chemistry, University of Washington, Seattle, Washington, USA

Gary J. Van Berkel, Chemical and Analytical Sciences Division, Oak Ridge National Laboratory, Oak Ridge, Tennessee, USA

Robert D. Voyksner, Analytical and Chemical Sciences, Research Triangle Institute, Research Triangle Park, North Carolina, USA

Guangdi Wang, Department of Chemistry, Xavier University of Louisiana, New Orleans, Louisiana, USA

CONTENTS

FOREWORD ix

PREFACE xvii

PART I FUNDAMENTAL ASPECTS OF ELECTROSPRAY IONIZATION (ESI) 1

1. On the Mechanism of Electrospray Mass Spectrometry 3
   Paul Kebarle and Yeunghaw Ho

2. The Electrolytic Nature of Electrospray 65
   Gary J. Van Berkel

3. ESI Source Design and Dynamic Range Considerations 107
   Andries P. Bruins

4. Solution, Gas-Phase, and Instrumental Parameter Influences on Charge-State Distributions in Electrospray Ionization Mass Spectrometry 137
   Guangdi Wang and Richard B. Cole

PART II ELECTROSPRAY COUPLING TO MASS ANALYZERS 175

5. Electrospray Ionization on Quadrupole and Magnetic-Sector Mass Spectrometers 177
   Charles N. McEwen and Barbara S. Larsen

6. Electrospray Ionization Time-of-Flight Mass Spectrometry 203
   Igor V. Chernushevich, Werner Ens, and Kenneth G. Standing

7. Electrospray Ionization Quadrupole Ion-Trap Mass Spectrometry 235
   Mark E. Bier and Jae C. Schwartz

8. Electrospray Ionization/Fourier Transform Ion Cyclotron Resonance Mass Spectrometry 291
   David A. Laude, Elizabeth Stevenson, and Jessica M. Robinson

PART III INTERFACING OF SOLUTION-BASED SEPARATION TECHNIQUES TO ELECTROSPRAY 321

9. Combining Liquid Chromatography with Electrospray Mass Spectrometry 323
   Robert D. Voyksner

10. **Capillary Electrophoresis–Electrospray Ionization Mass Spectrometry**   343
Joanne C. Severs and Richard D. Smith

PART IV   APPLICATIONS OF ELECTROSPRAY IONIZATION   383

11. **Electrospray Ionization Mass Spectrometry of Peptides and Proteins**   385
Joseph A. Loo and Rachel R. Ogorzalek Loo

12. **Electrospray Ionization Mass Spectrometry of Nucleic Acids and their Constituents**   421
Pamela F. Crain

13. **Electrospray Ionization Mass Spectrometry of Carbohydrates and Lipids**   459
Yoko Ohashi

14. **Drug Metabolism and Pharmacokinetics**   499
Grace K. Poon

15. **Electrospray Ionization of Inorganic and Organometallic Complexes**   527
Christine L. Gatlin and František Tureček

INDEX   571

FOREWORD

When Richard Cole asked me to write a foreword for this volume, I was highly flattered, but hardly eager. That reticence was due in part to a pile of deferred chores that grows as relentlessly as the universal entropy. A more forbidding inhibition was that in a world now teeming with electrospray (ES) users, I really didn't know what to say that hadn't already been said many times in many ways. Even so, the deadline was then far away and distant commitments always loom small, so I said yes. Tomorrow is now here and none of the nine Muses have come to my rescue. In desperation I have succumbed to banality by presenting "The Inside Story" of what has been called the electrospray revolution or "Things About Electrospray You Won't Find in the Papers!" What follows, therefore, is a highly personal account of how I became involved in the developments that led to the basis for this book.

The prologue to the story was written in 1937 at Berea College in Kentucky, where I was a senior majoring in chemistry. Jobs of any kind were scarce, so I applied for graduate study at several universities and was lucky enough to be accepted, with financial support, at both Northwestern and Yale. At first, I leaned toward Northwestern because its chemistry department was supposed to be better and its pay definitely was. But for various reasons, including a door-to-door free ride to Connecticut with my gear, I decided that fate wanted me in New Haven rather than Evanston. As a result, my chance to meet Malcolm Dole, then an Assistant Professor at Northwestern, was delayed half a century. At the time, I knew only that he had written a textbook on electrochemistry, not that he was at Northwestern. When I arrived at Yale, oddly enough as it turned out, John Zeleny was still on the Physics Faculty. Thirty years earlier, he had led the definitive pioneering studies on electrostatic dispersion of liquids into charged droplets. I was oblivious to that work for another 35 years, when I learned about Dole's electrospray technique and found that its roots were in Zeleny's investigations.

Meanwhile, I finished graduate study in 1940 and spent the next dozen years in applied research and development at Monsanto Chemical Company in Anniston, Alabama, Sharples Chemicals in Wyandotte, Michigan, and, beginning in 1945, at Experiment, Inc., a contract research and development company in Richmond, Virginia. There I became involved in combustion research for Project Bumblebee, a Navy program to develop a ramjet-powered antiaircraft missile. That experience led, in 1952, to the directorship of Project SQUID, another Navy program, administered by Princeton, on pure and applied research in "those fields of science relating to jet propulsion"—combustion, fluid flow, and heat transfer. In 1955, the Navy arranged for me to spend a year at the London Branch of the Office of Naval Research. While there I came across a 1954 paper by E. W. Becker and K. Bier describing the production of intense beams of hydrogen molecules from rarefied supersonic flows as proposed in a 1951 paper by A. Kantrowitz and J. Grey. I had been musing about using molecular beam scattering experiments to study combustion reactions, but was discouraged because the expected reaction cross sections were too small for measurement with the beam intensities available from the effusive sources of Otto Stern, Nobel Laureate and father of molecular-beam research. Moreover, the activation energies for the most interesting reactions were above 1 eV and would require source temperatures above 5000 K, much too high to be feasible. The Becker–Bier results indicated that the intensity barrier could be scaled with source gas flows that were convective as opposed to effusive. It occurred to me that if heavier reactant molecules like oxygen or chlorine were seeded into flows of hydrogen or helium, they should be accelerated to the high flow velocities achievable with those light gases. Calculations showed that, depending on their mass, such seed molecules could reach translational energies of several electron volts or more with modest source temperatures.

Back from London, I found Princeton's Department of Mechanical Engineering willing to provide a home for a research project on this idea. A proposal to NSF was fortunate enough to win support for a program that started in the fall of 1959. To make a long story short, we found that supersonic free jets from very small orifices could indeed produce intense beams of molecules with energies as high as 10 eV or more. Alas, we also found that these high translational energies were not effective in promoting chemical reaction. Meanwhile, we had learned that supersonic free jets had many other valuable features which have, ever since, played a key role in my research. Another major byproduct of the beam project was my transition from Director of Project SQUID to Professor of Mechanical Engineering, an achievement undreamt of in the philosophy of a young chemistry major at Berea College, 23 years earlier.

In 1967 I decided to accept an invitation to join Yale's newly organized Department of Engineering and Applied Science. About that time, the Bendix Corporation, producers of time-of-flight (TOF) mass spectrometers, were looking for a way to make intactions from large polymer molecules so they could sell their instruments to the burgeoning plastics industry. They interested Malcolm Dole in the problem and he came up with his now-famous idea of using Zeleny's technique to disperse dilute solutions of macromolecules as a fine spray of charged droplets into bath gas at atmospheric pressure. He reasoned that evaporation of solvent from such a droplet would increase its surface-charge density up to the Rayleigh limit, at which coulomb repulsion overcomes surface tension and the droplet breaks up into offspring droplets. Each offspring droplet would repeat that sequence, ultimately producing droplets containing only one solute molecule, which would become a free ion by retaining some droplet charge after all solvent had evaporated. This charged-residue model (CRM) for ion formation is one of the two models most often invoked to explain the formation of ES ions. The other is the ion-evaporation model (IEM) proposed some years later by Iribarne and Thomson. As discussed in the chapter by Kebarle and Ho, neither model has achieved unanimous acceptance.

Under Dole's leadership, a group at Bendix assembled an apparatus to test his idea. Sample solution was injected through a small tube or needle into a flow of dry-bath gas (nitrogen) through a cylindrical glass chamber. The needle was maintained at several kilovolts, relative to a plate constituting the end wall of the chamber. The resulting high field at the needle tip dispersed the emerging liquid into small charged droplets which drifted down the potential gradient concurrently with the flow of bath gas toward the end plate. A small orifice admitted some of the resulting mixture of ions and gas into the vacuum system as a supersonic free jet. I find it interesting that even though the Bendix people were expert in TOF techniques, a retarding potential method was chosen for mass analysis. The underlying idea was that during free-jet expansion, the ions would be accelerated to the easily calculated velocity of the nitrogen bath gas. Well downstream of the orifice, where the jet-gas density is too low to have any further effect on ion velocity, the ions passed through a set of grids on their way to a Faraday-cup electrode monitored by a sensitive electrometer. A scan of the grids' potential produced a current-voltage curve in which a dip occurred when that potential became equal to the kinetic energy of some of the arriving ions, thus preventing them from contributing to the current at the Faraday cup. The mass of the excluded ions was readily obtained from this measured value of their energy along with a value of the velocity taken as equal to that calculated for the nitrogen in the jet. Promising results for polystyrene oligomers with molecular weights (M_r's) up to 500,000 were presented in the *Journal of Chemical Physics* in 1968, the year after I arrived at Yale.

Not an avid reader of the literature, I was unaware of Dole's paper, but it stirred the interest of Professor Seymour "Sandy" Lipsky in the Medical School. A long-time mass spectrometrist, he was excited by the possibility that Dole's technique might work with large biomolecules. Sandy showed the paper to Csaba Horvath, a colleague of mine in Chemical Engineering, who had been working closely with him on the development of HPLC, a now-invaluable methodology to which they made major contributions. Dole had kindly referenced some of our Princeton papers on the acceleration of heavy molecules by light carrier gases in free-jet expansions. When Csaba saw those references, he told Sandy that I was now at Yale, so Sandy tracked me down to show me the paper. Always on the lookout for new applications of free jets, I was very much intrigued and managed to interest a new graduate student, Mike Labowsky, in repeating Dole's experiments, which by then had been confirmed in a new apparatus at Northwestern and reported in a 1970 *Journal of Chemical Physics* paper. Our vacuum system was much bigger than Dole's and had much higher pumping speeds. Moreover, we had had more experience with free jet expansions. I realized that, in his experiment, the concurrent flow of bath gas and droplets meant that solvent vapor was present in the unheated mixture of gas and ions that entered the vacuum system. It was, therefore, highly likely that adiabatic cooling during free jet-expansion resolvated the ions to an appreciable extent, thus adding to their masses by an unknown amount. As a good chemical engineer, Mike knew that desolvation would be much more effective with bath gas flowing countercurrent to the drift of droplets and ions toward the end plate. Moreover, resolvation would be avoided, because only dry bath gas would then enter the free jet. With these changes, he obtained results somewhat different from and more reproducible than Dole's.

Both Dole and we had been persuaded by the work of R. Beuhler and L. Friedman that ions as large as we hoped we were producing would not generate secondary electrons at a dynode unless they were accelerated to about half a million volts. Therefore, we did not attempt to use ion-multiplier detectors, thus depriving ourselves of the megafold gain in sensitivity that mass spectrometrists take for granted. The currents of ions we could get into the vacuum system were very small and the vibrating-reed electrometer we used to measure them was very balky. Moreover, we knew from our earlier studies that there could be substantial slip effects during acceleration of very heavy molecules by a lighter carrier gas. Indeed, we estimated that the actual velocities of Dole's macroions were probably as much as 40% less than he had assumed. For these and other reasons, we abandoned further experiments. By that time, Dole had retired from Northwestern and moved to Baylor University, where he continued his experiments. However, instead of retarding potential measurements of mass, he was using mobility measurements to characterize the ions.

In 1981, our group was joined by Masamichi "Gado" Yamashita, a young scientist I had met during a stay at the University of Tokyo. During discussions about a possible project, I suggested that it might be interesting to take another look at Dole's ES ionization. Instead of macromolecules as analytes, we would use species with molecular weights less than 400 so they could be "weighed" with a small quadrupole mass analyzer we had in the laboratory. Gado was a marvelous experimentalist and soon had converted a small "minibeam" apparatus into our first ES mass spectrometer. In an extensive set of exploratory experiments, with species small enough for our analyzer, he found that almost any solute organic molecule containing polar atoms such as O, S, N, and P would produce ions comprising the parent molecule with an anion or cation adduct. He also found that inorganic cations and anions could be produced, but generally in much lower abundances than he routinely obtained with organic solutes. As we later learned, the Aleksandrov group at the University of Leningrad (now St. Petersburg) had also begun to investigate ESI at about the same time as we did. From the fragmentary reports that we later obtained, it appeared that their emphasis was on ions from inorganic solutes.

One day in 1982 Sandy Lipsky, who had kept abreast of our work, brought VG's Brian Green to visit our lab. Brian was very interested and asked whether the technique would work with larger species. We pointed out that the upper mass limit for our analyzer was only 400 u so he arranged for VG to lend us a quadrupole with a mass range up to 1500 u. By that time, Gado had gone back to

Japan and Craig Whitehouse, who had been working in Sandy Lipsky's lab, became a graduate student in Chemical Engineering. After joining my group, he designed and built a new system incorporating the VG analyzer and a modified version of Gado's electrospray source. That new system gave beautifully clean spectra for gramicidin S and cyclosporin, two cyclic decapeptides with almost the same M_r value. A provocative difference was that most cyclosporin ions had one charge, whereas most of those from gramicidin S had two. Moreover, for some slightly larger peptides, we obtained ions with three charges. Such multiple charging was most intriguing, so we decided to explore the phenomenon further with poly(ethylene glycol)s. They were attractive as test species because their oligomers were linear polymers whose composition and structure were essentially the same no matter what their size. (Moreover, samples over an M_r range from 200 to 20,000 were available at no charge from Union Carbide!) We found that the number of charges (Na +) per oligomer increased with size, reaching at least 23 for an M_r of 20,000. Oddly enough, when we presented the data at the 1987 ASMS meeting in Denver, the only interest shown was by Charles McEwen, coauthor (with Barbara Larsen) of Chapter 5 and one of the first to recognize the virtues of ESI. He said right away that those PEG results were by far the most significant at the meeting. In contrast, a reviewer of our later paper dismissed the bands of overlapping peaks for large oligomers as spectra of dirt in the system, not of PEG oligomers!

To avoid the spectral complexity due to the wide distribution of molecular weights in PEG samples, we turned to nature to obtain samples comprising large molecules all of the same size, that is, pure proteins. We soon obtained spectra with the coherent sequences of peaks that have become the cachet of ES spectra for large molecules. The initial reaction of most mass spectrometrists to such peak multiplicity was one of horror. They were convinced that the resulting spectral complexity would make interpretation difficult, if not impossible. Moreover, the distribution of total charge among so many peaks would inevitably decrease signal/noise! They were wrong on both counts because they had reckoned without the powers of modern computers (as I still do!). One afternoon, I remarked to Matthias Mann, then a graduate student in my lab, that each of these multiple peaks really constituted an independent measure of parent ion mass. One should therefore be able to use signal-averaging methods to obtain a more accurate and reliable value of M_r. Two days later he came back with a deconvolution algorithm that allowed our little quadrupole, with a nominal mass limit of 1500 u, to determine M_r values up to 30,000 or more with an accuracy of 0.01%! At that time, M_r values from most other methods seldom had accuracies better than 5 or 10%. We presented these findings in San Francisco at the 1988 ASMS meeting and the rest, as they say, is history. There were seven ES papers at that meeting, three from Henion's lab, two from Dick Smith's, and two from ours. Six years later, in 1994, at Chicago that number had climbed to over 300 where it remained in 1995 at Atlanta. I'm told that in the archival journals covered by the Citation Index, the term "electrospray" appeared in the title and/or abstract of over 300 papers in both 1994 and

1995. I'm sure Dole never dreamed that the seed he planted would bear so much fruit. Unfortunately, germination took so long that he didn't live to see the full magnitude of the stampede that began after our results with proteins became known.

My only face-to-face encounter with Malcolm Dole was in 1985 at the San Diego ASMS Meeting. At a session in which I gave a paper, I saw him in the audience and asked him to stand up and be recognized. Even though electrospray was not yet famous, the audience was generous in its applause, for which he (and I!) were grateful. My paper included some discussion of the importance of countercurrent drying gas to keep solvent vapor out of the free-jet expansion to avoid resolvation of the ions. At the end of the session, Dole came up to say he was delighted to learn about resolvation in the free jet, a possibility of which he had been unaware. He thought such resolvation might well account for an anomaly in his drift-tube experiments that had been puzzling him for a long time—the small differences between mobilities of ES ions from large and small molecules. (I now suspect that multiple charging of the larger ions, the extent of which didn't become clear until several years later, was as much or more to blame.) After our 1989 paper on ESMS of large biomolecules appeared in *Science*, Dole wrote me a nice note thanking me for the references to his papers and asking for a reprint. I sent it right away, along with one of our paper in *Mass Spectrometry Reviews* that included a more complete and complimentary account of his work. Three months later, I received a very apologetic note saying that he had taken the reprints along on a cruise and had somehow lost them. He would be most grateful if I could find another copy of each. He enclosed a twenty dollar bill to pay for whatever costs I might incur! I returned his money with the reprints and received a very gracious note of thanks, along with a copy of his privately printed autobiography. In it he made specific mention of my asking him to stand up and take a bow at the San Diego meeting, a gesture that was obviously very meaningful to him. He also quoted, clearly with great appreciation and satisfaction, my sometime remark to the effect that his electrospray idea was "extremely ingenious." Unfortunately, I never saw or heard from Dole again. Having not been invited, I did not attend the Electrospray Workshop in November 1991 at which he was present. I was delighted to learn that he had received much well-deserved attention, so that before he passed away some months later, he could begin to realize the abundance of the harvest from the seeds he had planted.

There is a presumption in many cultures that the longer one has lived in the past, the further can he or she see into the future. That may be one reason why Editor Cole suggested that my views on electrospray's future would be welcome. However, there have been so many surprises in its past that my crystal ball sees only great risks in any attempt to predict its future. Who would have dreamed 20 years ago that by now investigators would be able to examine the behavior of biopolymer ions in the gas phase, to determine their masses with accuracies approaching parts per billion, to study the kinetics and dynamics of their inter- and intramolecular reactions and processes, and to obtain detailed information on their composition, structure, and conformation that would provide insight on these properties in solution, in vivo as well as in vitro? Who could have anticipated that living organisms could be ionized, transferred into vacuum, recovered, and found to retain their viability? Yet this *tour de force* was recently accomplished with viruses by Siuzdak and his colleagues at Scripps Institute in La Jolla, California. With these so recently unbelievable achievements of the electrospray methodology already in hand, who would dare to imply any limit on its possibilities by undertaking to guess what the future might reveal? Instead, I would simply urge investigators to exorcise their inhibitions and exercise their imaginations. May the platform of past accomplishments, so ably described in this volume, serve as a launching pad for their flights of fancy into the future. Let them dare to take off for any star whose twinkle beckons. Let them hearken to the words of Robert Browning in "Andrea del Sarto": "Ah, but a man's reach should exceed his grasp, or what's a heaven for?"

Richmond, Virginia   JOHN. B. FENN

PREFACE

This book presents a broad view of the current knowledge about the revolutionary technique of electrospray ionization—mass spectrometry (ESI–MS). For the first time, a single comprehensive volume has been compiled which recounts different perspectives of the fundamental underlying processes, the varied approaches to implementation, and the wide-ranging utility of the technique which has changed the way that researchers in physics, chemistry, and biology view the study of large molecules. The 15 contributed chapters have all been written by leading researchers in the field.

Electrospray ionization–mass spectrometry has developed at a tremendous pace since the end of the 1980s. Along with matrix-assisted laser desorption/ionization (MALDI), it has permitted new possibilities for mass-spectrometric analyses of high-molecular-weight compounds of all types, including proteins, nucleotides, and synthetic polymers. Other analytical techniques cannot provide the same level of detailed information regarding molecular weights and structures from extremely small quantities of material. Three features of electrospray ionization set it apart from other mass-spectrometric ionization techniques. The first is the truly unique ability to produce extensively multiply charged ions. This attribute allows the creation of highly charged forms of very large molecular-weight compounds which may be analyzed on virtually all types of mass spectrometers. A second distinguishing feature of ESI (shared also by "thermospray") is that samples under analysis must be introduced *in solution*. This characteristic results in a natural compatibility of ESI with many types of separation techniques, particularly those most suitable for separations of mixtures of larger molecules, such as various categories of liquid chromatography, as well as capillary electrophoresis. A third unique feature is the extreme "softness" of the ESI process which permits the preservation in the gas phase of noncovalent interactions between molecules which existed in solution, as well as the study of three-dimensional molecular conformations.

The book is divided into four main sections which can be characterized globally as fundamental and mechanistic aspects of the electrospray process, coupling electrospray to various mass analyzers, interfacing electrospray to separations techniques, and applications of electrospray to problems in biochemistry, pharmacology, and metallochemistry. Each chapter reviews the most relevant experimental work pertaining to the chapter's theme. The sequence of chapters was organized to first examine closely the physical and chemical aspects of the spray process itself, starting with a mechanistic description of the events involved in ion formation (Chapter 1). Attention then turns to electrochemical phenomena, which play an essential role in charged-droplet formation during the electrospray process (Chapter 2). Next, various designs of ion sources which permit connection of electrospray devices to mass spectrometers are depicted along with detailed consideration of the characteristics of these sprayers (Chapter 3). The fundamentals section concludes with an examination of solution, instrumental, and gas-phase considerations that influence the appearance of the electrospray mass spectrum and the distributions of multiply charged species (Chapter 4).

The second section confronts the coupling of electrospray ionization to each of the widely used mass analyzers employed in mass spectrometry. The section begins with an examination of quadrupoles and magnetic-sector instruments, which are the most popular, and which have undergone widespread development for the longest time. Different approaches to overcoming the challenge of connecting the electrospray region, where ion formation occurs at atmospheric pressure, to the high-vacuum region (the high-voltage region for magnetic-sector instruments) are detailed (Chapter 5). In the next chapter, ESI coupling to time-of-flight instruments is described, with emphasis on the advantages and performance characteristics of an "orthogonal injection" arrangement. Included are several studies of noncovalent associations of biomolecules, which are discussed from an instrumental point of view (Chapter 6). The ESI–ion-trap instrument is subsequently depicted, beginning with an overview of the fundamentals of ion trapping, and highlighted by a description of unique experiments that can be performed, such as MS^n (Chapter 7). Last, the coupling of ESI to Fourier-transform ion-cyclotron resonance instruments is described. Performance advantages such as ultrahigh resolution and accurate mass assignments, which allow direct charge-stage (hence, molecular-weight) identification of fragment ions of biomolecules, are explored in detail (Chapter 8).

Because electrospray ionization requires the introduction of samples contained in solution, it has many advantages for coupling to solution-based separations techniques. The third section of the book describes recent progress in the interfacing of these techniques to ESI, starting with the various types of liquid chromatography (Chapter 9). A separations scheme, of course, is required in many analyses of complex mixtures, including countless examples of biological, physiological, and environmental samples. The importance of coupling liquid-chromatography (LC) techniques to electrospray ionization is further emphasized when one considers that the types of compounds which are quite amenable to LC and ESI–MS (e.g., large, polar molecules) are often not suitable for gas chromatography (GC)–MS, making LC–ESI–MS quite complementary to the ever-popular GC–MS. The newer techniques falling under the heading of capillary electrophoresis (CE) have also expanded tremendously in recent years (Chapter 10). These techniques, which primarily involve separations of charged species in solution, have a certain compatibility with ESI, which produces gas-phase ions from charged droplets. The multitude of applications of "hyphenated" chromatography-ESI methodology has undergone accelerated expansion in recent years.

The final section of the book channels the explosion of applications of the ESI technique into categories delineated according to compound type. The principal groups relevant to the domain of biochemistry are addressed first. These are divided into proteins and peptides (Chapter 11), nucleic acids and constituents (Chapter 12), and carbohydrates and lipids (Chapter 13). Because of the widespread interest and utility of ESI in the domain of pharmaceutics, a separate chapter on small molecules related specifically to pharmacology and drug metabolism has been included (Chapter 14). Last, but far from least, because of the growing interest in the domains of organometallic and inorganic chemistry (which were generally considered as poorly accessible by previous mass-spectrometric ionization techniques), the new possibilities afforded by ESI in this field have been reviewed (Chapter 15).

The authors must each be commended for the effort, diligence, and punctuality shown in preparing their respective chapters. I must extend fervent gratitude to John Fenn for agreeing to share, in the Foreword, his personal retrospective view of early events (and miscellaneous frontier bushwhacking) along the winding trail leading up to this ESI–MS volume. In addition to the chapter authors, many other individuals contributed to the realization of this book. Each chapter was confidentially refereed by two leading experts in the field. My heartfelt thanks goes out to these scientists for the time they invested and for the care with which the reviews were conducted. I must thank Guangdi Wang for serving as a receptive sounding board in regard to organizational aspects of the book during its early stages. I am indebted to Jack Timberlake and Joe King of the University of New Orleans for their support of my sabbatical activities in Paris, France, and I must express my gratitude to Wenzhe Lu for keeping things running on the homefront during my absence. I ardently thank Jean-Claude Tabet for graciously hosting me during my sabbatical year at the Université Pierre et Marie Curie (Paris VI), France, where the majority of the editorial duties related to the book were performed. I remain grateful to all of my colleagues in the Laboratoire de Chimie Structurale Organique et Biologique at Paris VI for their considerable support of the project. Finally, I thank the Académie des Sciences (France) for bestowing a great honor upon me in awarding me the title of "Professeur de l'Académie" during this sabbatical year. The financial support associated with this award, generously provided by the Centre National de La Recherche Scientifique (CNRS, France) and Elf Aquitaine Inc. (France), greatly aided the timely completion of the book.

Paris, France   RICHARD B. COLE

ELECTROSPRAY IONIZATION MASS SPECTROMETRY

PART I
FUNDAMENTAL ASPECTS OF ELECTROSPRAY IONIZATION (ESI)

CHAPTER 1

On the Mechanism of Electrospray Mass Spectrometry

PAUL KEBARLE AND YEUNGHAW HO

Department of Chemistry, University of Alberta, Edmonton, Alberta, Canada

		Abstract	4
I.		Introduction	6
	A.	Electrospray as a method for the transfer of ions from solution to the gas phase	6
	B.	Electrospray, other than mass spectrometric applications	8
II.		Production of gas-phase ions by electrospray	8
	A.	The overall process	8
	B.	Production of charged droplets at the ES capillary tip—The electrophoretic mechanism	8
	C.	Electrospray as a special kind of electrolytic cell	11
	D.	Mass spectrometric evidence for the electrophoretic mechanism	12
	E.	Required electrical potentials for ES—Electrical gas discharges	13
	F.	Electrical current I due to the charged droplets—Charge and radius of droplets	15
	G.	Dependence of droplet current I on conductivity and ion concentration in the solution	17
	H.	Solvent evaporation from charged droplets leading to droplet shrinkage and Coulomb fissions	20
	I.	Mechanisms for the formation of gas-phase ions from small, highly charged droplets	25
	J.	The Iribarne–Thomson equations for ion evaporation	26
	K.	Comparison of experimental data with predictions of the ion-evaporation theory. Inclusion of surface-activity effects into ion-evaporation and charged-residue theory. Cases for which a distinction between ion-evaporation and charge-residue theory is not possible	31
	L.	Measurements of the electric field on solid charged residues—Evidence in support of the ion-evaporation theory	34
	M.	Change of solute concentration in evaporating droplets	39
	N.	Dependence of the mass spectrometrically observed analyte ion signal	

Electrospray Ionization Mass Spectrometry, Edited by Richard B. Cole.
ISBN 0-471-14564-5   © 1997 John Wiley & Sons, Inc.

	intensities on the concentration of the analyte ion and the concentrations of other electrolytes present in the electrosprayed solution	42
O.	Competition and depletion involving analyte ions—Dependence of ion signal on analyte mass	48
P.	Droplet electrospray (DES)—Significance of results to electrospray mechanism	50
Q.	Mechanism for formation of macroions	53
R.	Changes of ion charge and structure on transfer from charged droplets to the gas phase	55
III.	Conclusions	60
	References	60

Abstract

Electrospray (ES) is a method by which ions, present in a solution, can be transferred to the gas phase. The processes involved in this transfer are examined in some detail. They consist of the application of an electric field to the tip of a capillary containing a solution of electrolyte ions at concentrations $C > 10^{-6}$ mol/L. The presence of the field leads to the formation of a dipolar layer at the meniscus of the liquid at the capillary tip. The double layer is due to a partial spatial separation of the electrolyte ions. When the capillary is positive, positive electrolyte ions from the double layer are near the surface and destabilize the meniscus. A cone and a liquid jet form, and the jet emits positively charged small droplets. The droplets are charged because of an excess of positive electrolyte ions. The charged droplet stream leads to an electrical current I, which is generally less than 1 μA. The excess of unipolar ions is provided by an electrolysis process at the capillary which, when the capillary is positive, either adds positive ions to the solution or removes negative electrolyte ions from the solution. For concentrations of the original electrolyte solution greater than 10^{-5} mol/L, the number of ions added or removed by the electrolysis is much smaller than the number of electrolyte ions originally present in the solution.

The electric field required for the onset of ES increases with the surface tension of the solvent. Solvents with high surface tension (water) require high onset fields, and these fields can lead to electrical discharges which partially suppress the ES process. The onset of electrical discharges depends on the pressure and nature of the ambient gas used.

The current I due to the droplet stream, the droplet charge q, and radius R can be predicted by approximate equations derived from theoretical and semiempirical models. The current I increases with the conductivity K and the volume flow rate V_f of the solution, $I \propto (V_f K)^n$ where $n < 0.5$. The conductivity K is proportional to the concentration C of strong electrolyte in the solution. The mass spectrometrically observed total gas phase ion current, I_{ms}, is not closely coupled to the droplet current I, but depends also on the droplet radius R. Smaller droplets obtained at lower flow rates can lead to higher I_{ms}.

Solvent evaporation from the charged droplets leads to droplet shrinkage, increased coulombic repulsion, and release of coulombic strain by droplet fission. Observation of droplet fission is possible for droplets with $R > 1$ μm. The observed fission proceeds with ejection of some 20 or more offspring droplets with a radius roughly one-tenth of the parent droplet radius. The droplets carry off only 2% of the parent mass, but 15% of the parent's charge. Repeated evaporation and fission of parent and offspring droplets lead to formation of droplet populations extending down to very small droplets, $R \approx 10$ nm. The timescale for the process is in the 100 to 500 μs range, which is close to the resident time of the droplets before intake into the mass spectrometric sampling system.

There are two theories for the formation of gas-phase ions from the very small charged droplets. The equation derived by Iribarne and Thomson predicts that gas phase ion emission (called ion evaporation) occurs directly from very small droplets. This ion emission becomes the dominant process for droplets with radius $R < 10$ nm. The rival theory, of Dole and Röllgen, assumes that coulomb fission forming smaller droplets continues to occur until, ultimately, droplets containing only one excess ion are formed. Closer examination of the two models reveals that many of the observed mass spectral features can be explained with either model. Thus, both theories predict that ions which interact weakly with the solvent (low ion solvation energies, and high surface activities) will be expressed preferentially in the gas phase. Also, a true distinction between the two theories exists only in the range 1 nm $< R <$ 10 nm. For droplets where $R < 1$ nm, coulomb fission of the droplet and ion emission from the droplet become essentially indistinguishable.

Two attempts to examine quantitatively the predictions of the Iribarne–Thomson theory are described. Tang and Kebarle compare mass spectrometrically determined relative sensitivity coefficients for several ions (Li^+, Na^+, K^+, Rb^+, and Cs^+) with relative coefficients predicted by the theory. The agreement found is poor, but there is an outside chance that the thermodynamic data required for the theoretical evaluation may have larger than expected errors. In the other study, Locertales and Fernandez de la Mora determine the electric field present on solid charged residues resulting from the compete evaporation of very small droplets obtained by ES. They find that the magnitude of the observed electric field is closer to that expected on the basis of the Iribarne–Thomson theory. However, this conclusion is based on assumptions that may not be justified.

Owing to evaporation of the solvent, the concentration of solutes present in the droplets undergoes large changes. Approximate predictions for the increase of concentration of completely nonvolatile solutes, such as strong electrolytes, can be made. Predictions for the concentration changes of more volatile solutes are much more difficult. When the solutes are acids and bases, large changes of the pH of the solution can be expected.

Useful equations are available which predict the dependence of the mass spectrometrically observed intensity of a given analyte ion on the concentration of the analyte in the solution and the concentration of other electrolyte ions present in the solution. The "competition mechanism" by which other electrolytes can suppress the intensity of a given analyte ion is described.

Fenselau and co-workers have observed polyprotonated protein ions by ESMS in the positive ion mode from solutions with a basic pH (pH = 10), even though the protein is known to be deprotonated in the original solution. The observation of such unexpected ions by their group and others can be explained. It is a consequence of expected proton transfer reactions that occur as a response of the system to the change of environment from solution to gas phase.

I. INTRODUCTION

A. Electrospray as a Method for the Transfer of ions from Solution to the Gas Phase

A technique that allows the transfer of ions from solution to the gas phase, from which the ions can be subjected to mass spectrometric analysis, is of the greatest importance because close to half of the chemical and biochemical processes involve ions in solution. Electrospray is such a technique. It affords ion transfer of a wide variety of ions dissolved in a wide variety of solvents. The ions include singly and multiply charged inorganic ions such as (1) the alkali ions: Li^+, Na^+, K^+, Cs^+, Rb^+; (2) the alkaline earths: Mg^+, Ca^{2+}, Sr^{2+}, Ba^{2+}; (3) singly and doubly charged transition metal ions and their complexes with mono and polydentate ligands; (4) anions of inorganic and organic acids such as NO_3^-, Cl^-, $H_2PO_4^-$, HSO_4^-, and SO_4^{2-} (5) singly and multiply protonated organic bases such as amines, alkaloids, peptides, and proteins; and (6) singly and multiply deprotonated organic acids or organophosphates such as the nucleic acids. The solvents include practically all polar solvents, be they protic solvents like water, methanol, and ethanol or aprotic solvents like acetone, acetonitrile, and dimethylsulfoxide.

Electrospray mass spectrometry (ESMS) was introduced by Yamashita and Fenn (1) in 1984. A very similar, independent development was reported at approximately the same time by Aleksandrov and co-workers (2). Actually, electrospray as a source of gas-phase ions and their analysis by mass spectrometry was proposed much earlier (1968) by Dole (3), however Dole's experiments were too narrowly focused. They aimed at the detection of polymeric species such as the polystyrenes, which are not themselves ionized in solution, and the experimental results obtained were not convincing. Development of ESMS since 1984 has established it as a method of outstanding importance and particularly so for new biochemical applications.

The transfer of ions from solution to the gas phase is a strongly endothermic and endoergic process, because in solution the ion is strongly interacting with a number of solvent molecules which form a solvation "sphere" around the ion. The energy required to transfer a sodium ion from aqueous solution to the gas phase, estimated on the basis of thermodynamic cycles (4), is

$$Na^+(aq) \rightarrow Na^+(g)$$

$$-\Delta G°_{sol} = 98 \text{ kcal/mol}$$

$$-\Delta H°_{sol} = 106 \text{ kcal/mol} \quad (1)$$

where $\Delta G°_{sol}$ and $\Delta H°_{sol}$ stand for the reverse process, that is, the transfer of the ion from gas phase to solution. The energy is larger than the energy required to break a carbon C–C bond (~ 85 kcal/mol) and this suggests that, if the energy is supplied in one package over a short time, the act of freeing an organic ion from the solvent can also lead to fragmentation. For earlier ionization methods in which ions are transferred from solution to the gas phase, such as fast atom bombardment (FAB) and plasma desorption, which existed before the introduction of electrospray, abundant energy is supplied in a highly localized fashion over a short time. These methods lead not only to ion desolvation but also to fragmentation or even net ionization, that is, the creation of ions from neutrals. Compared to these methods, ESMS, in which the desolvation is achieved gradually by thermal energy at relatively low temperatures, is a far softer technique.

One can define the softness of the ion-transfer method as the degree to which fragmentation of the ions is avoided. From that standpoint, ESMS is the softest technique available. A more discriminating requirement would be to expect that the detected ions fully correspond to their state in solution. For example, this could mean that the structure of the gas-phase ions entering the mass analyzing region is to be exactly the same as in the solution. The experimental information available so far is equivocal. Some structural features are preserved, others are not.

To illustrate, we can consider the extensive experimental studies dealing with the "state of protonation of proteins." Early ESMS studies involving positive ions from acidic solutions indicated that the state of protonation observed with the mass spectrometer reflects the number of protonated basic sites present in the protein when in solution. Thus, the maximum number of protons observed corresponds approximately to the number of basic groups present owing to basic residues such as arginine, lysine, histidine and the terminal amino group (5–7). Furthermore, it was found that the observed charge states can provide information on the structure of the protein. Native globular proteins in which some of the basic groups are directed toward the inside of the protein, and are not accessible, showed lower degrees of protonation than the denatured species (8–14). Some of the experiments even suggested (6,7) that the distribution of the charge states observed mass spectrometrically corresponds fairly closely to the distribution at the given pH present in the solution. Later work (15) which examined solutions at different pH values showed that this is not so. The observed charge-state distribution was found to be relatively insensitive to the pH used. Furthermore, at a given acidic pH of 3.5, multiply charged negative ions were observed due to deprotonated acidic residues when operating in the negative ion ESMS mode, yet at this pH the acidic groups are not deprotonated in solution.

It is desirable to unravel problems of this type, that is, why are certain features of the state of the ions in solution retained and other features lost? The answers to such questions are clearly connected with the mechanism of electrospray and exactly what happens between the initial stage (ions in solution) and the final "stage" (mass spectral patterns observed for these ions)?

There are also, of course, many other important questions concerning the expected performance of ESMS that are directly connected with the mechanism of ESMS. An important example is the dependence of the detected ion intensity of an analyte as a function of its concentration in solution and also the related questions: (a) What are the best conditions required to detect extremely small amounts of sample? (b) Is the detected signal, when very small amounts are used, concentration dependent or mass dependent?

The overall mechanism has two major parts: (a) formation of gas-phase ions from the ions in the solution subjected to electrospray at one atmosphere and (b) the transfer of the gas-phase ions from one atmosphere to the mass-analysis section in vacuum, with attendant modification of the ions that can occur at the various pressures and electric fields of the interfaces leading to the mass analyzer. Mechanistic features concerning both of these stages are considered in this chapter.

B. Other Applications of Electrospray

Electrospray existed long before its application to mass spectrometry. It is a method of considerable importance for the electrostatic dispersion of liquids and the creation of aerosols. The interesting history and notable research advances in the field are well described in Bailey's book *Electrostatic Spraying of Liquids* (16). Much of the theory concerning the mechanism of the charged-droplet formation was developed by researchers from this group. The latest works in this area can be found in a special issue (17) of the *Journal of Aerosol Science* devoted to electrospray.

II. PRODUCTION OF GAS-PHASE IONS BY ELECTROSPRAY

A. The Overall Process

There are three major steps in the production by electrospray of gas-phase ions from electrolyte ions in solution: (1) production of charged droplets at the ES capillary tip; (2) shrinkage of the charged droplets by solvent evaporation and repeated droplet disintegrations, leading ultimately to very small highly charged droplets capable of producing gas-phase ions, and (3) the actual mechanism by which gas-phase ions are produced from the very small and highly charged droplets.

B. Production of Charged Droplets at the ES Capillary Tip— The Electrophoretic Mechanism

As shown in the schematic representation of the charged-droplet formation (Fig. 1), a voltage V_c, of 2–3 kV, is applied to the metal capillary which is typically 0.2 mm o.d. and 0.1 mm i.d. and located 1–3 cm from the counter-electrode. The counter-electrode in ESMS may be a plate with an orifice leading to the mass spectrometric sampling system or a sampling capillary, mounted on a plate, which leads to the MS. Because the electrospray capillary tip is very thin, the electric field E_c in the air at the capillary tip is very high ($E_c \approx 10^6$ V/m). The

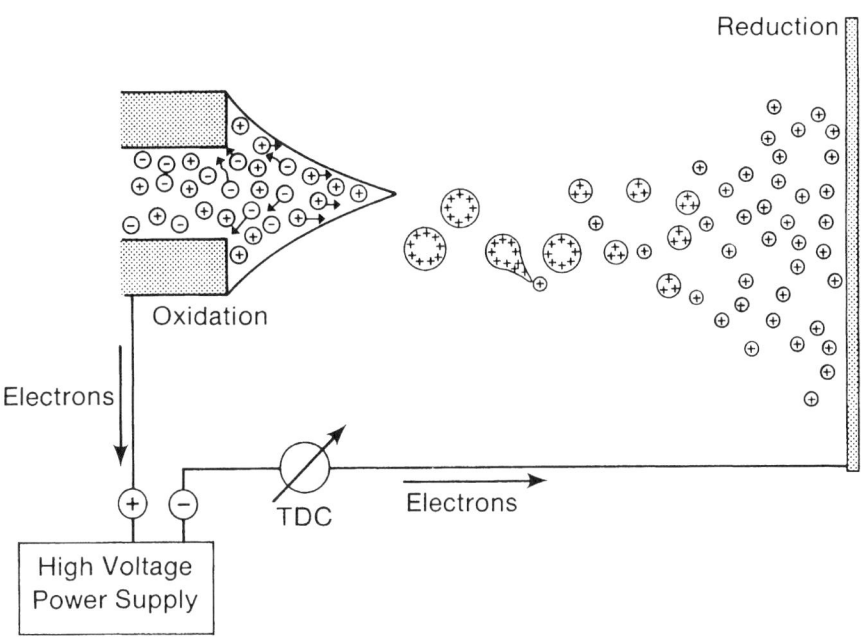

Figure 1. Schematic of major processes occurring in electrospray. Penetration of imposed electric field into liquid leads to formation of electric double layer in liquid. Enrichment of surface of liquid by positive electrolyte ions leads to destabilization of meniscus and formation of cone and jet emitting droplets with excesses of positive ions. Charged droplets shrink by evaporation and split into smaller droplets and finally gas-phase ions. Reprinted with permission from P. Kebarle and L. Tang, *Anal. Chem.*, **1993**, *65*, 972A. © 1993 American Chemical Society.

value of the field at the capillary tip, when the counterelectrode is large and planar, can be evaluated using the approximate relationship (18,19):

$$E_c = \frac{2V_c}{r_c \ln(4d/r_c)} \quad (2)$$

where V_c is the applied potential, r_c is the capillary outer radius, and d is the distance from capillary tip to the counterelectrode. For example, the combination $V_c = 2000$ V, $r_c = 10^{-4}$ m, and $d = 0.02$ m leads to $E_c \approx 6 \times 10^6$ V/m. The field E_c is proportional to V_c and the most important geometry parameter is r_c. Because E_c is essentially inversely proportional to r_c, E_c decreases slowly with the electrode separation d due to the logarithmic dependence.

A typical solution present in the capillary consists of a polar solvent in which electrolytes are soluble. For example, methanol can be used as solvent, and a simple salt like NaCl or BHCl, where B is an organic base, as the solute. Low electrolyte concentrations, 10^{-5}–10^{-3} mol/L (M), are typically used in ESMS.

For simplicity, only the positive ion mode is considered in the subsequent discussion.

When turned on, the field E_c will penetrate the solution at the capillary tip and the positive and negative electrolyte ions in the solution will move under the influence of the field until a charge distribution results which counteracts the imposed field and leads to essentially field-free conditions inside the solution. When the capillary is the positive electrode, positive ions will have drifted downfield in the solution, that is, toward the meniscus of the liquid, and negative ions will have drifted away from the surface. The mutual repulsion between the positive ions at the surface overcomes the surface tension of the liquid and the surface begins to expand, allowing the positive charges and liquid to move downfield. A cone forms, the so-called Taylor cone (20), and if the applied field is sufficiently high, a fine jet emerges from the cone tip which breaks up into small charged droplets (see Fig. 2).

The droplets are positively charged owing to the excess of positive electrolyte ions at the surface of the cone and the cone jet. Thus, if the electrolyte present in the solution is NaCl, the excess positive ions at the surface will be Na^+ ions. This mode of charging, which depends on the positive and negative ions drifting in opposite directions under the influence of the electric field, has been called the electrophoretic mechanism (19,21–24). The charged droplets produced by the cone jet drift through the air downfield, toward the counterelectrode. Solvent evaporation at constant charge leads to droplet shrinkage and an increase of the

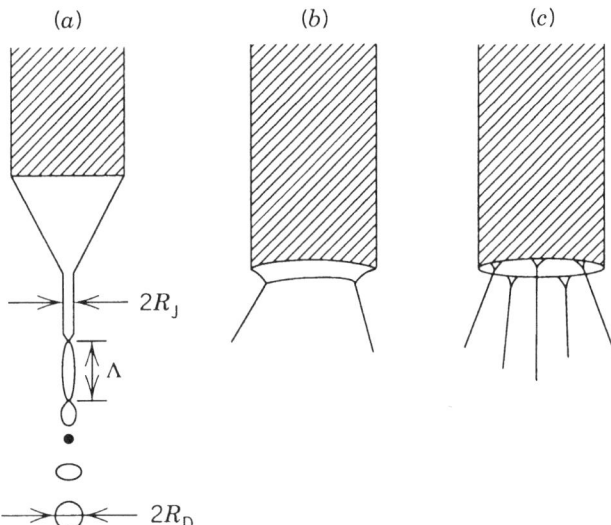

Figure 2. Some details of cone-jet mode. (*a*) Tip of cone is extended into a liquid jet. R_J = jet radius, R_D = droplet radius. For liquids of low viscosity, such as those used in ESMS, the relationship $R_D/R_J = 1.9$ is found to hold. Multijet modes are observed when voltage is increased from $a \rightarrow b \rightarrow c$. [After Cloupeau (21a).]

electric field normal to the surface of the droplets. At a given radius, the increasing repulsion overcomes the surface tension. Droplet fission occurs, followed by repeated solvent evaporation and repeated fission. The sequential fission leads ultimately to gas-phase ions by processes that are described in detail in subsequent sections.

The cone-jet mode described and illustrated in Figure 2 is only one of the possible ES modes. For a qualitative description of this and other modes, see Cloupeau (21). The cone-jet mode is most often used in ESMS. It is also the best-characterized mode in the electrospray literature (17). For recent treatments see Hendricks (21c), Smith (22), Hayati (23), and Fernandez de la Mora (24).

Taylor (20) examined the opposing effects of the electrostatic force and surface tension at the liquid surface and found that these balance when the meniscus assumes the form of a cone whose sides are straight (straight generatrix) and the half angle at the apex of the cone is $\theta = 49.3°$. This shape is expected for a static case, prior to the development of a jet at the apex. In the dynamic case, where a jet and thus a flow of charges are present, the cone apex angle and the generatrix depend on the experimental conditions. The Taylor values may be preserved when the conductivity of the solution is relatively high. At lower conductivities, the transition between the cone and the jet is more gradual, and the transition zone occurs closer to the capillary orifice (21).

C. Electrospray as a Special Kind of Electrolytic Cell

Assuming that the charge separation is electrophoretic, at a steady operation of the ES, positive droplet emission will continuously carry off positive ions. The requirement for charge balance in such a continuous-electric-current device, together with the fact that only electrons can flow through the metal wire supplying the electric potential to the electrodes (Fig. 1), leads to the supposition that the ES process must include an electrochemical conversion of ions to electrons. In other words, the ES device can be viewed as a special type of electrolytic cell (25). It is special because the ion transport does not occur through uninterrupted solution, as is normally the case in electrolysis. Part of the ion transport occurs through the gas phase, where unipolar charged droplets and later gas-phase ions are the charge carriers. A conventional electrochemical oxidation reaction should be occurring at the positive electrode, that is, at the liquid–metal interface of the capillary (Fig. 1). This reaction should be supplying positive ions to the solution, by converting atoms from the metal to positive metal ions which enter the solution and leave behind electrons (see Eq. 3). The other alternative is to remove negative ions present in the solution by an oxidation reaction, as illustrated in Eq. 4 for aqueous solutions.

$$M(s) \rightarrow M^{2+}(aq) + 2e^- \text{ (in metal)} \qquad (3)$$

$$4OH^-(aq) \rightarrow O_2(g) + 2H_2O + 4e^- \text{ (in metal)} \qquad (4)$$

One expects that the reaction with the lowest oxidation potential will dominate, which depends on the materials present in the metal electrode, the ions present in the solution and the nature of the solvent. Proof of the occurrence of an electrochemical oxidation at the metal capillary was provided by Blades et al. (25). When a Zn capillary tip was used, release of Zn^{2+} to the solution could be detected. Furthermore, the amount of Zn^{2+} released to the solution per unit time when converted to coulombs per second was found to be equal to the measured electrospray current I. Similar results were observed with stainless steel capillaries which were found to release Fe^{2+} to the solution. These quantitative results provide the strongest evidence for the electrolysis mechanism.

It should be noted that the oxidation reaction described in Eq. 3 adds ions which were not present previously in the solution. The oxidation (Eq. 4), however, provides the positive current by removing the negative counterions of positive ions that are already present in the solution. The ions added to the solution by a process like Eq. 3 are present at a very low concentration. For example, a solution of 10^{-5} M NaCl in methanol at a flow rate $V_f = 20$ μL/min led to an ES current of 1.6×10^{-7} A. The Zn^{2+} concentration produced by the Zn-tipped capillary was $[Zn^{2+}] = 2.2 \times 10^{-6}$ M. Assuming that the Na^+ ion was the analyte ion, the concentration of the ion produced by the oxidation is one-fifth of that of the analyte. The ES current I increases slowly with the total electrolyte concentration. Therefore, when the concentration of the analyte is much higher, the ratio of the concentration of the ion formed by oxidation to the concentration of analyte will be much smaller, and the presence of the electrode produced ions may not even be noticeable.

Van Berkel and co-workers (26) have shown that one can make use of the electrode reactions, at the ES capillary, to produce ions of special interest. This work is described in Chapter 2.

D. Mass Spectrometric Evidence for the Electrophoretic Mechanism

There is little explicit discussion in the pre-ESMS literature concerning the nature of the charge carriers of the droplets. Although this is a vital question to the mass spectrometrist, it is of limited interest in other applications of ES. Pfeifer and Hendricks (19) were probably the first to explicitly endorse the electrophoretic mechanism. More recently, Smith (22) and Hyati et al. (23) also endorsed the electrophoretic mechanism. The recognition came relatively late because ES was observed to occur also with apparently pure solvents to which no electrolytes had been added. It was recognized only later that these solvents invariably contained electrolyte impurities. The high electric field required for ES often leads also to the occurrence of electric gas discharges. It was probably the consideration of a possible role for gas discharges that led to the delayed recognition of the electrophoretic mechanism.

For the mass spectrometrist it is relatively easy to determine the nature of the charge carriers in the droplets. The general observation is made that, whenever an ion is present in the solution, it is also detected in the mass spectrum. This is certainly true for singly charged ions like the alkali ions and the wide variety of protonated strong organic nitrogen bases. Multiply charged ions, both of inorganic and organic origins, are sometimes not seen in their expected charge state (see Section I). However, good reasons exist to expect a change of charge state on transfer from solution to gas phase and in the interfaces leading to the mass spectrometers (see Section II.R). The situation is also very much the same for negative ions. Therefore, the mass spectrometric evidence is fully in accord with the electrophoretic mechanism, on the basis of which one expects that the charges of the droplets will be due to an excess of unipolar electrolyte ions, which originally were present in the solution accompanied by the electrolyte counterions. Because the ES capillary functions as an electrolytic half-cell (see discussion above) ions that might be produced by the electrode reaction can also be present as charge carriers in the droplets.

E. Required Electrical Potentials for ES—Electrical Gas Discharges

D.P.H. Smith (22) was able to derive a useful equation for the required electric field, E_{on}, at the capillary tip, which leads to instability of the static Taylor cone and formation of the charged jet:

$$E_{on} \approx \left(\frac{2\gamma \cos\theta}{\epsilon_0 r_c}\right)^{1/2} \tag{5}$$

This equation for the onset field, when combined with Eq. 1, leads to an equation for the potential, V_{on}, required for the onset of ES.

$$V_{on} \approx \left(\frac{r_c \gamma \cos\theta}{2\epsilon_0}\right)^{1/2} \ln(4d/r_c) \tag{6}$$

where γ is the surface tension of the solvent, ϵ_0 is the permittivity of vacuum, r_c is the radius of the capillary, and θ is the half-angle of the Taylor cone. Substituting the values $\epsilon_0 = 8.8 \times 10^{-12}$ J^{-1} C^2 m^{-1} and $\theta = 49.3$ [see Taylor (20)], one obtains:

$$V_{on} = 2 \times 10^5 (\gamma r_c)^{1/2} \ln(4d/r_c) \tag{7}$$

where γ must be substituted in Newtons per meter and r_c in meters to obtain V_{on} in volts. Table 1 shows the surface tension values for four solvents and the calculated ES onset potentials for $r_c = 0.1$ mm and $d = 40$ mm. The surface of the

TABLE 1. Onset Voltages V_{on} for Solvents with Different Surface Tension γ

Solvent	CH_3OH	CH_3CN	$(CH_3)_2SO$	H_2O
γ(N/m)	0.0226	0.030	0.043	0.073
V_{on} (kV)	2.2	2.5	3.0	4.0

solvent with the highest surface tension (H_2O) is the most difficult to stretch into a cone and jet, and this leads to the highest value for the onset potential V_{on}.

Experimental verification of Eqs. 5–7 has been provided by Smith (22) and work from our laboratory (27,28). For stable ES operation, one needs to go a few hundred volts higher than V_{on}. Use of water as the solvent can lead to the initiation of an electric discharge from the capillary tip, particularly when the capillary is negative (i.e., in the negative ion mode). The ES onset potential V_{on} is the same for both the positive and negative ion modes; however, the electric discharge onset is lower when the capillary electrode is negative (27,28).

The occurrence of an electric discharge leads to an increase of the capillary current I. Currents above 10^{-6} A are generally due to the presence of an electric discharge. A much more specific test is provided by the appearance of discharge-characteristic ions in the mass spectrum. Thus, in the positive-ion mode, the appearance of protonated solvent clusters such as $H_3O^+(H_2O)_n$ from water or $CH_3OH_2^+(CH_3OH)_n$ from methanol indicates the presence of a discharge (27). The protonated solvent ions are produced in high abundances by ES in the absence of a discharge only when the solvent has been acidified, that is, when H_3O^+ and $CH_3OH_2^+$ are present in the solution.

The presence of an electric discharge degrades the performance of ESMS, particularly so at high discharge currents. The ES ions are observed at much lower intensities than was the case prior to the discharge, and the peaks corresponding to discharge-generated ions appear with very high intensities (27,28). It is likely that the discharge reduces the electric field near the capillary tip (21), and this interferes with the charged-droplet formation.

The high potentials required for ES show that air at atmospheric pressure is not only a convenient, but also a very suitable, ambient gas for ES, particularly when solvents with high surface tension, such as water, are to be electrosprayed. The oxygen in the air has a positive electron affinity and readily captures free electrons. Initiation of gas discharges occurs when free electrons present in the gas (due to cosmic ray or background radiation) are accelerated to velocities where they can ionize the gas molecules. At pressures near atmospheric, very frequent collisions with the gas interfere with the electron acceleration process. The presence of gases that capture electrons, and convert them to atomic or molecular negative ions, suppresses the electrical breakdown. SF_6 and polychlorinated hydrocarbons also capture electrons and are more efficient discharge-suppressing gases than O_2. Both SF_6 and O_2 have been used to advantage for the suppression of discharges in ES of water (22,27,28).

Liquids with high surface tensions have high onset potentials, and cannot be electrosprayed in air even when discharge-suppressing gases have been added. This is the case for liquid metals. Liquid metals can be "electrosprayed" in vacuum, where electric discharges cannot occur. It should be noted that ES in vacuum is referred to as electrohydrodynamic (EHD) spraying and ionization in the mass-spectrometry literature (30). The term EHD is used in a much broader sense in the literature outside mass spectrometry and it includes ES of liquids at both ambient pressures and in vacuum (17).

F. Electrical current *I* due to the charged droplets—Charge and radius of droplets

The electrical current I due to the charged droplets leaving the ES capillary is easily measured (see Fig. 1) and is of interest because it provides a quantitative measure of the total number of excess positive ions that leave the capillary and could in principle, be converted to gas-phase ions.

Theoretical equations that predict the dependence of the current I, the radius R, and the charge q of the droplets were derived by Pfeifer and Hendricks (19).

$$I = [(4\pi/\epsilon)^3 (9\gamma)^2 \epsilon_0^5]^{1/7} (KE)^{3/7} (V_f)^{4/7} \tag{8}$$

$$R = \left(\frac{3\epsilon\gamma^{1/2} V_f}{4\pi\epsilon_0^{1/2} KE} \right)^{2/7} \tag{9}$$

$$q = 0.5[8(\epsilon_o \gamma R^3)^{1/2}] \tag{10}$$

where
γ = surface tension of solvent
ϵ = permittivity of solvent
ϵ_0 = permittivity of vacuum
K = conductivity of solution
E = applied electric field at capillary tip (see Eq. 2)
R = radius of droplets produced
q = charge of droplet
V_f = flow rate (volume/time)

Unfortunately, these equations were obtained on the basis of unproven assumptions. For a brief description of the assumptions and derivation see ref. 29. Nevertheless, the predictions for the current I yield values of the same order of magnitude as the experimental results. Also, predictions for the radius and charge of the droplets agree approximately with experiment, and are supported by later semiempirical derivations (22–24).

Fernandez de la Mora and Locertales (24) have proposed the following relationships, on the basis of experimental measurements of the current I, droplet sizes, droplet charges, and theoretical reasoning:

$$I = f\left(\frac{\epsilon}{\epsilon_0}\right) \left(\gamma K V_f \frac{\epsilon}{\epsilon_0} \right)^{1/2} \tag{11}$$

$$R \approx (V_f \epsilon / K)^{1/3} \tag{12}$$

$$q \approx 0.7[8\pi(\epsilon_0 \gamma R^3)^{1/2}] \tag{13}$$

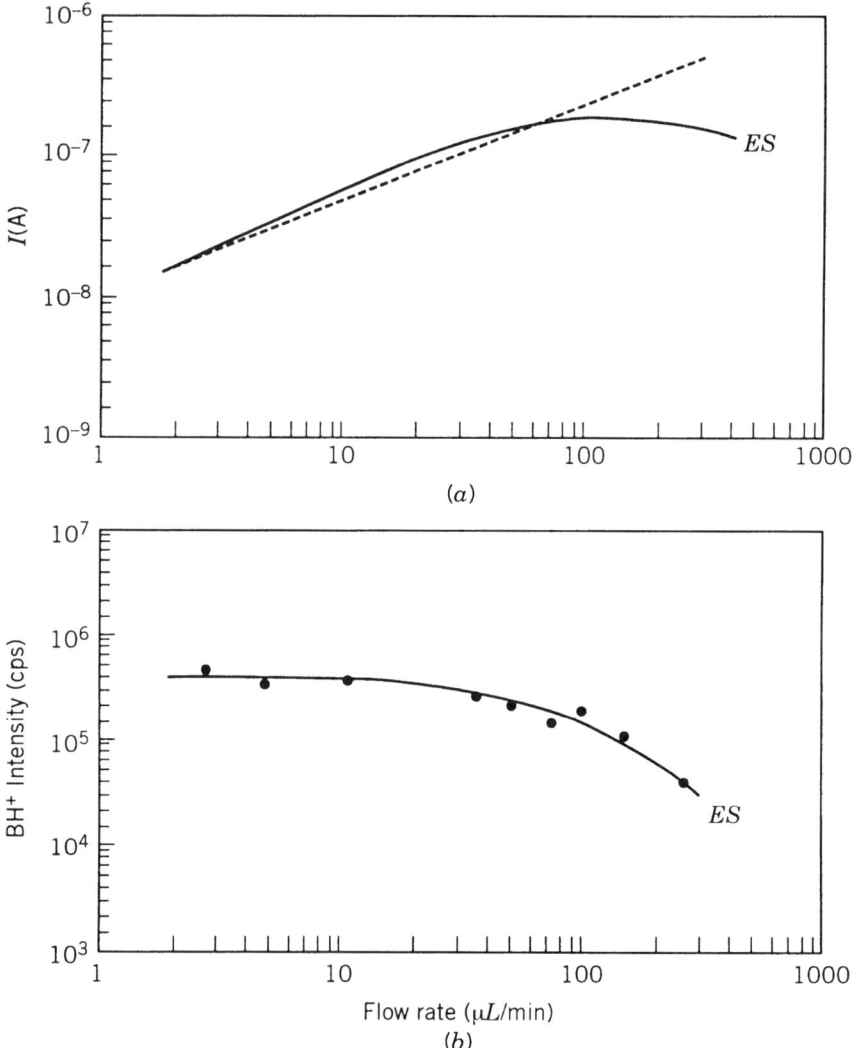

Figure 3. Capillary current and analyte ion intensity dependences on solution flow rate. (*a*) Dependence of capillary current I on solvent flow rate V_f. Solid line: cocaine. Hydrochloride at 10^{-5} M in methanol. Dashed line: slope predicted by Eq. 8. (*b*) Dependence of mass-analyzed ion intensity of the BH^+ ion of protonated cocaine on flow rate. Reprinted with permission from P. Kebarle and L. Tang, *Anal. Chem.*, **1993**, *65*, 972A. © 1993 American Chemical Society.

where $f(\epsilon/\epsilon_0)$ is a numerical function tabulated by the authors (24); the value of $f(\epsilon/\epsilon_0) \approx 18$ for liquids whose dielectric constant $\epsilon/\epsilon_0 \geq 40$. The other symbols are the same as defined above, for Eqs. 8–10. The relationships were obtained for solutions having conductivities K larger than 10^{-4} S m^{-1} (see Section II.G). For polar solvents like water and methanol, and electrolytes that dissociate essentially completely to ions, this requirement corresponds to solutions with concentrations higher than $\sim 10^{-5}$ mol/L—a concentration range that is commonly present in ESMS. The flow rates used (24) were below 1 μL/min and are thus in the lower range of flow rates used in conventional ESMS.

Although Eqs. 8 and 11 appear to be quite different, they predict a very similar dependence of the current I on the two most important experimentally variable parameters—the flow rate and the conductivity. Thus, for the flow rate, Eq. 8 predicts a dependence $I \propto V_f^{0.57}$, while Eq. 11 leads to $I \propto V_f^{0.5}$. The same dependence holds also for the conductivities K.

The dependence of the current I on the flow rate V_f observed in measurements from our laboratory (30) is shown in Figure 3a. The slope in the I versus V_f plot predicted by Eq. 8 is also shown in Figure 3a. The experimental results do not provide a straight line; nevertheless, particularly in the lower flow-rate range of 2–80 μL/min, a fair correspondence between the two slopes is observed. The results are also equally compatible with Eq. 11.

Figure 3b shows the flow-rate dependence of the observed mass-analyzed ion current of the organic analyte BH^+, which was present at a constant concentration $[BH^+] = 10^{-5}$ mol/L in the methanol solution. This plot shows that the gas-phase ion current does not correlate well with the droplet current I. Although I increases, the BH^+ intensity remains constant and even decreases at higher values of V_f. The lack of correlation between the droplet current and the ion intensity with changes of flow rate underscores the need to know more about the process by which the gas-phase ions are produced by the charged droplets.

Equations 9 and 12 predict that the droplet radius R should increase with the flow rate. The observed decrease of gas-phase ion current with flow rate, (Fig. 3b) could be associated with the increasing droplet size, that is, the yield of gas-phase ions from the charged droplets may be decreasing as the droplet size increases. Equations 9 and 12 also predict that R decreases as the conductivity K is increased. Decreases of flow rate and increases of conductivity represent a useful route for the production of very small droplets.

G. Dependence of Droplet Current *I* on Conductivity and Ion Concentration in the Solution

The dependence of the droplet current I, on the conductivity K of the solution is of special interest to the mass spectrometrist because the conductivity is proportional to the ion concentration C in solution:

$$K = \lambda_{0,m} C \tag{14}$$

Equation 14 holds at the low concentrations ($C < 10^{-1}$ mol/L) generally used in ESMS. The equivalent molar conductivity of the electrolyte $\lambda_{0,m}$ introduces the dependence on the specific nature of the electrolyte ions. The important property determining the value of $\lambda_{0,m}$ is the mobility, that is, the drift velocity of the

TABLE 2. Limiting Molar Conductivities[a] of Some Electrolytes in Methanol at 25°C

Solute	$(\lambda_{0,m})^b$	Ion	$(\ell_{0,m})^{b,c}$
HCl	190	H^+	146
$HClO_4$	214	Li^+	40
HNO_3	203	Na^+	45
LiCl	91	K^+	52
NaCl	97	Cl^-	52
KCl	104	B_R^-	56

[a] Data from Landolt, Börnstein, Zahlenwerte und Functionen; Springer Verlag; Berlin, 1960; Vol. II, pp. 366, 533, and 651.
[b] $\lambda_{0,m}$ and $\ell_{0,m}$ are given in Ω^{-1} cm² mol^{-1}. In the SI notation $\Omega^{-1}=$ S (Siemens). The SI notation conductivities $K=\lambda_{0,m}C$ are given in S m^{-1}. To obtain values of K (S m^{-1}) from $\lambda_{0,m}$ of this table, multiply by the conversion factor $=10^{-1}$ and use C in mol/L. Example: the value of K for a 10^{-3} mol/L solution of NaCl in methanol is $97 \times 10^{-1} \times 10^{-3} = 9.7 \times 10^{-3}$ S m^{-1}.
[c] Limiting molar conductivities of constituent ions where: $\lambda_{0,m}$ (MX) $= \ell_{0,m}$ (M$^+$) $+ \ell_{0,m}$ (X$^-$).

positive and negative ions under the influence of the electric field present in the solution.

Equation 14 is valid for a single electrolyte. When several electrolytes are present in the solution, K can be obtained from the sum of the conductivities:

$$K = \sum_i \lambda_{0,m\,i} C_i \qquad (15)$$

Table 2 lists the $\lambda_{0,m}$ values for several electrolytes in methanol, which is a typical solvent used in ES.

The dependence of the current I on the conductivity K for solutions of HCl and NaCl in a mixture of 60% water and 40% methanol determined from experiments in this laboratory (31) is shown in Figure 4. The current I obtained with either NaCl or HCl is different when the same concentrations are used (Fig. 4a), and the same when solutions with the same conductivities are used (Fig. 4b). This result confirms the dependence of I on the conductivity K. The slope of the linear log I versus log K plot in Figure 4 is $n = 0.22$. This leads to the relationship

$$I \propto K^n = (\lambda_{0,m}C)^n \qquad n \approx 0.22 \qquad (16)$$

This dependence is of the same form as in Eq. 8 (Hendricks) and Eq. 11 (Fernandez), but the exponents n in these equations are larger, i.e., $n = 0.57$ in Eq. 8 and $n = 0.5$ in Eq. 11.

The current I obtained with theoretically derived Eq. 8 is also shown in the figure. It is notable that the equation, which has no adjustable parameters, leads to a current which is of the same order of magnitude as the experimental result.

Fernandez de la Mora based his equations largely on experimental observations (24). The exponent $n = 0.5$ in Eq. 11 is thus supported by several measurements of the dependence of I on the conductivity K. These were made at flow

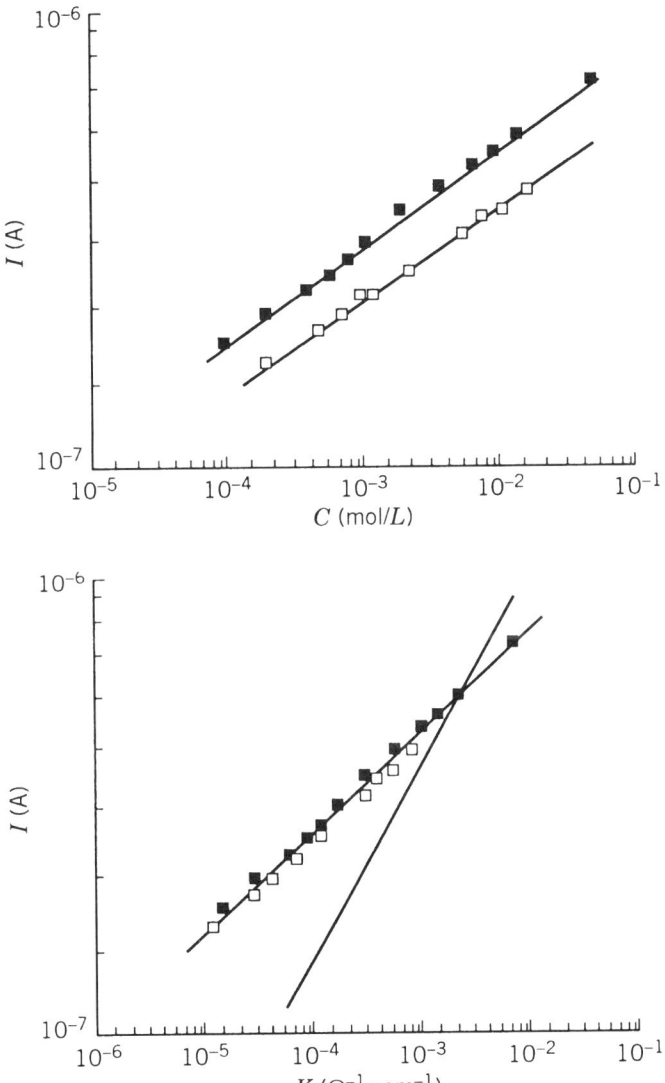

Figure 4. Dependence of capillary current I on concentration and conductivity. (a) Logarithmic plot of measured capillary current I versus concentration C of electrolytes HCl (■), and NaCl (□), in a solvent of 40% methanol and 60% water. (b) Same data plotted versus the conductivity K of the solution. This plot also shows the current I evaluated with Eq. 8 (solid line).

rates close to two orders of magnitude lower than the flow rate $V_f = 20$ μL/min used for the data represented in Figure 4. The large difference in the flow rates may be responsible for the discrepancy between the data in Figure 4, and the Fernandez equation.

As will be seen, the exact value of n in Eq. 16 is not particularly relevant to the present discussion. We will be mainly interested in the dependence of the gas-phase ion intensity of a given analyte on its concentration in the solution. When the analyte ion is the ion with by far the highest concentration in the solution, it is observed (see Section II.N) that the mass-analyzed ion intensity is approximately proportional to the droplet current I, that is, the gas-phase ion intensity has a dependence on the concentration in solution similar to that for the droplet current I.

The far more common experimental situation is that electrolytes other than the analyte electrolyte are also present in the solution. In that case, the situation is more complicated (see Section II.N), and the relationships discussed in this section prove to be of some utility.

H. Solvent Evaporation from Charged Droplets Leading to Droplet Shrinkage and Coulomb Fissions

The charged droplets produced by the spray shrink, owing to solvent evaporation while the charge remains constant. The energy required for solvent evaporation is provided by the thermal energy of the ambient air. The charge of the droplets is expected to remain constant because the emission of ions from the solution to the gas phase is highly endoergic (see Eq. 1). The shrinkage of the droplets at constant charge has been confirmed by direct observation of the droplets in special experiments. Davis *et al.* (32) trapped a given droplet at ambient pressure in a quadrupole ion trap (electrodynamic picobalance) and observed it using light-scattering techniques. Gomez and Tang (33) made "on the fly" observations of droplets emitted from an ES capillary by means of flash shadow graphs, and used Doppler anemometry to measure droplet diameters.

The decrease of droplet radius R at constant droplet charge q leads to an increase of the electrostatic repulsion of the charges at the surface until the droplets reach the Rayleigh stability limit:

$$q_{Ry} = 8\pi(\epsilon_0 \gamma R^3)^{1/2} \tag{17}$$

The Rayleigh (34) equation gives the condition at which the electrostatic repulsion becomes equal to the force due to the surface tension γ, which holds the droplet together. The charged droplet becomes unstable when its radius R and charge q satisfy Eq. 17.

It is experimentally observed that the droplets undergo fission when they are close to the Rayleigh limit. The fragmentation is generally referred to as coulombic fission. The word fission, in analogy to cell or nuclear fission, evokes a picture of the droplet dividing into two nearly equal particles. However, this is not the usual outcome of charged-droplet fission. Rayleigh (35) and Doyle (36) reported that "the liquid is thrown out in fine jets" (35) or ". . . a cloud of small droplets was ejected" (36). More detailed observations on the emission of a fine jet of droplets were provided recently by Davis et al. (32) and Gomez and

Tang (33). A tracing of a droplet undergoing fission by jet emission, as observed by Gomez and Tang (33), is shown in the insert of Figure 5. In a previous discussion of the ES mechanism (37), we called this "uneven fission," emphasizing the distinction with "even fission," where two or a few particles of nearly equal size are formed. Because of the similarity of uneven fission with the cone-jet emission of droplets from the capillary tip, a better name for uneven fission is "droplet-jet fission," and this is the name we use in the subsequent discussion.

Even fission, where the droplet fragments into two or three particles of nearly equal mass and charge, is observed much less frequently and probably only with nonpolar solvents whose conductivities are low (39). Since ESMS is normally practiced with polar solvents with higher conductivities, jet fission can be expected to be the dominant process. Droplets that have relieved the coulombic stress through jet fission will continue to evaporate solvent until they reach the Rayleigh stability limit and undergo another jet fission.

The detailed history of the evaporating and fissioning droplets depends on the initial size and charge of the droplets produced by the ES. The important

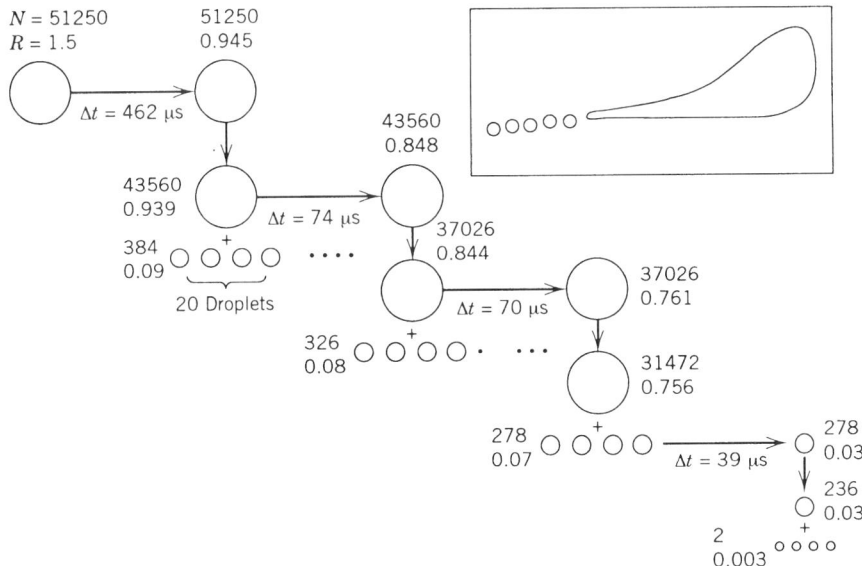

Figure 5. Schematic representation of time history of parent and offspring droplets. Droplet at top left is a typical droplet created near the capillary tip in ambient air. Solvent evaporation at constant charge leads to increasing coulombic repulsion which is released by droplet jet fission. Insert at top right illustrates droplet-jet fission as observed by Gomez and Tang (33). N = number of elementary charges on the droplet, and R = radius of droplet in micrometers. The Δt values correspond to the times required for evaporative droplet shrinkage to a size where fission occurs and were evaluated from Eq. 18b. Only the first three successive fissions of the parent droplet are shown. Reprinted with permission from P. Kebarle and L. Tang, *Anal. Chem.*, **1993**, *65*, 972A. © 1993 American Chemical Society.

parameters, which determine the radius of the droplets (see Section II.F), were the flow rate V_f and the conductivity K of the solution, (see Eqs. 9 and 12). Low flow rates and high conductivities lead to small droplets. Flow rates of a few microliters per minute and conductivities $K \approx 10^{-4}$ to 10^{-2} S m^{-1}, corresponding to electrolyte concentrations of 10^{-5} to 10^{-4} mol/L in polar solvents, lead to droplet radii of a few micrometers. It is also observed (24,38) that the charge of the droplets is not far from the Rayleigh limit. Thus, Fernandez de la Mora (24,38) cites $q_0 \approx 0.7\, q_{Ry}$ as a typical value for the initial charge on the droplets (see Eq. 13).

The time required for an initial droplet with radius R_0 and charge q_0 to reach the size R_1 that leads to the first fission can be estimated (40) with the use of expressions providing the rate of solvent evaporation from small droplets (41). When relatively volatile solvents such as methanol, water, and acetonitrile are involved, and the droplets are a few micrometers or less in diameter, the evaporation rate follows the surface evaporation limit law (41), which leads to a simple dependence (40) of the droplet radius on the time t:

$$\frac{dR}{dt} = -\frac{\alpha \bar{v}}{4\rho} \frac{p^\circ M}{R_g T} \qquad (18a)$$

$$R = R_0 - \frac{\alpha \bar{v} p^\circ M}{4\rho R_g T} t \qquad (18b)$$

where
- $\bar{v}$ = the average thermal velocity of the solvent vapor
- p° = the saturation vapor pressure of the solvent at the temperature of the droplet
- M = molar mass of the solvent molecule
- ρ = density of solvent
- R_g = gas constant
- T = temperature of droplet
- α = condensation coefficient of solvent, $\alpha \approx 0.04$ for water, ethanol, and probably methanol (41)

The time history of a parent droplet with initial radius $R_0 = 1.5\,\mu$m and charge $q_0 = 10^{14}$ C, where the solvent is methanol, the ambient temperature is 35°C, and the droplet temperature is 25°C is shown in Figure 5. The times Δt (μs) given in the figure were evaluated with Eq. 18b for methanol (p$^\circ$ = 100 torr at 25°C, $\alpha \approx 0.04$). It was assumed that jet fission occurs somewhat before the Rayleigh limit, when the charge is 0.8 q_{Ry}. This assumption is in accord with experimental observations (32,33).

Gomez and Tang (33) observed that the jet fission leads to a loss of only 2% of the parent's droplet mass and 15% of the parent's droplet charge. The radius of the offspring droplets was found to be approximately one-tenth of that of the parent droplet. Mass balance for these conditions predicts the formation of approximately twenty offspring droplets. These conditions were incorporated into the predictions shown in Figure 5. Although these predictions provide a good schematic view of the processes occurring in the droplet evolution, the numerical estimates are only very approximate, and the time dependence obtained using Eq. 18 is only an estimate. Furthermore, the Gomez and Tang (33) data for 2% loss of mass and 15% loss of charge were obtained for hexane as solvent, but are applied here to methanol.

The time required for the initial droplet ($R_0 = 1.5$ μm, $q_0 = 10^{-14}$ C) to reach the first fission is $\Delta t = 460$ μs. The initial charge assumed, $q_0 = 10^{-14}$ C, is only 0.5 of the Rayleigh charge. Had a charge which was 0.7 of the Rayleigh charge been assumed, as predicted by Eqs. 13 and 17, a considerably shorter time ($\Delta t \approx 125$ μs) would have been predicted.

Subsequent fissions of the parent droplet are also shown. These occur at progressively shorter times. The overall timescale indicated by the results in Figure 5 is hundreds of microseconds, which is not much shorter than the total residence time of the charged droplets in the atmospheric pressure region. The exact residence time depends on the type of ES interface used and probably ranges from hundreds of microseconds to a few milliseconds.

The dependence of the rate of evaporation on the vapor pressure of the solvent (Eq. 18), indicates that solvents with vapor pressures much lower than that of methanol ($p° \approx 100$ torr) will lead to acceptable ESMS sampling efficiencies only at elevated temperatures of the droplets and ambient gas (air).

A typical first-generation offspring droplet has a radius of $R \sim 0.08$ μm and a number of elementary charges $N \approx 280$ at birth (see Fig. 5). Assuming that the ions in the electrolyte present in the solvent were singly charged, N corresponds to the number of positive ions of the droplet that are not counterbalanced by negative ions present in the solution of the droplet. Such an offspring droplet reaches the instability limit in ~ 40 μs when its radius $R \sim 0.03$ μm. Assuming that the coulomb explosion of this offspring droplet is also of the jet-fission type, the resulting second-generation offspring will have $R \approx 0.003$ μm and $N \approx 2$. Droplets having radii much smaller than 1 μm are difficult to observe; therefore, the occurrence of jet fission for such droplets has not been verified by experiment.

Recently Fernandez de la Mora (39) suggested that the jet-fission process could continue down to very small droplets, and provided some criteria for the conditions present in such fissions. The treatment is based on the observed similarity between the jet-cone emission of droplets from the Taylor cone at the capillary tip (see Fig. 2), and the cone and jet emission of droplets from a small droplet (see insert of Fig. 5). The author applies and adapts expressions developed for the Taylor jet cone to the jet emission of droplets (39). Although several relationships were derived (39), only a few are mentioned here:

1. The diameter of the offspring is predicted to be approximately independent of the diameter of the parent droplet, and is given approximately by

$$d_m \approx \left[\frac{\gamma}{\rho}\left(\frac{\epsilon}{K}\right)^2\right]^{1/3} \qquad (19)$$

where γ, ρ, and K are the surface tension, density, and conductivity and ϵ is the permittivity of the solvent, that is, $\epsilon = \epsilon_r \epsilon_0$ where ϵ_r is the dielectric constant of the solvent and ϵ_0 is the permittivity of vacuum. According to this equation, smaller and smaller offspring droplets are generated not directly as the result of the decreasing diameter of the parent, but because the decrease of diameter by evaporation in the presence of an electrolyte leads to an increase in the conductivity K and thus to a decrease of d_m.

2. The charge of the droplets is expected to be below, but close to, the Rayleigh limit charge.

3. The relative mass loss of the parent droplet is much smaller than the relative charge loss.

4. The duration of the droplet-jet emission of offspring droplets is short relative to the time of solvent evaporation from the parent droplet between jet emissions.

5. The predictions are expected to be valid only for polar liquids, that is, liquids in which electrolytes are readily soluble. This includes the conventional solvents used in ESMS but not solvents like hexane.

Unfortunately, the predictions (39) could not be compared in detail with experimental results because very few experimental data are available which provide offspring droplet sizes obtained for polar solvents with known conductivities. The predictions were found (39) to agree qualitatively with the limited data available.

Because the proposed relationships [1–5] (39) have not yet been confirmed by experiment, their significance may be questioned. We believe that proof of the perceived similarity (39) between the Taylor cone-jet emission and the droplet-jet emission is not complete. Although the geometry of the disintegrations is similar, there is a significant difference in the amount of surface charging occurring in the two processes. At the capillary cone, a process of charging of the rapidly expanding surface occurs which involves ions originally present in the bulk of the solution. This charging process introduces the dependence on the bulk conductivity K. However, the fission of the charged droplets involves charges already present near the surface of the droplets and new charges are not generated. It is not clear why this process should depend on the bulk conductivity K.

Sheehan and Willoughby (42) have presented electron microscope photographs of dry residues from electrosprayed drops of a concentrated solution of cytochrome C in water–methanol. These particles have dimensions ~ 100 nm and less. Significantly, the shapes of the particles are raindrop-like with a sharp point. The strong resemblance of this shape to that of the droplets undergoing jet fissions leads to the suggestion that they are due to droplets that were undergoing jet fission, but were near saturation of the solution so that, on formation of the solid residue, the cone-jet shape was retained. This interpretation would thus present proof that jet fissions continue down to droplets with very small sizes.

However, practically all the shapes observed (42) are of the cone-jet shape, while Gomez and Tang observed only a small fraction of the droplets to be in the process of jet fission (33). It was pointed out by Fernandez de la Mora (39) that his analysis (see prediction 4 above), provides the result that the duration of jet emission is very short compared with the time of evaporation, and is thus in conflict with the observations of Sheenan and Willoughby (42). The significance of the observations (42) remains unclear.

I. Mechanisms for the Formation of Gas-Phase Ions from Small, Highly Charged Droplets

Two mechanisms have been proposed to account for the formation of gas-phase ions from very small and highly charged droplets. The first mechanism, proposed by Dole (3), depends on the formation of extremely small droplets which should contain only one ion. Solvent evaporation from such a droplet will lead to a gas-phase ion.

Dole and co-workers were the first investigators of the ESMS method (see Section I). They were interested in the determination of the molecular masses of polystyrenes and reasoned that if only one polystyrene molecule was also present in the drop, it would form an adduct with the assumed single ion and the resulting quasimolecular ions could be detected mass spectrometrically. The method was called the *charged residue method*. The presence of too many polystyrene molecules and electrolyte ions in the charged residue will complicate the analysis and, therefore, suitably dilute solutions should be used. Dole and co-workers were unsuccessful in these mass spectrometric determinations, but the *charged residue model* proposed by them survived. A more detailed consideration of, and support for, the mechanism was provided by Röllgen (43).

Iribarne and Thomson (44,45) proposed a new mechanism for the production of gas-phase ions from the charged droplets. The Iribarne theory is described in some detail in the next section. It predicts that, after the radii of the droplets decrease to a given size, direct ion emission from the droplets becomes possible and this process, which was called *ion evaporation*, becomes dominant over coulomb fission for droplets with radii $R \leq 10$ nm.

The authors (44,45) were interested in the nature of the charged species produced when very small droplets are obtained from pneumatic "atomization" of a liquid in the absence of an imposed electric field. The droplets may be charged because of statistical imbalances between positive and negative electrolyte ions present in the droplets. The charged droplets of different polarities were separated by the application of a weak electric field. Earlier [1937], on the basis of extensive mobility studies of such charged species, Chapman (46) had come to the conclusion that a few peaks in the mobility spectrum observed at the high mobility end were due to small (molecular dimensions) singly charged species, that is, gas-phase ions and not multiply charged, larger aggregates. Iribarne and Thomson, using a more definitive experimental setup, obtained similar mobility spectra and also concluded that the higher mobility peaks were due to gas-phase ions (44). In later work (45), the ions were sampled with a mass spectrometer and identified as ions originating from the solution sprayed, that is, $Na^+(H_2O)_n$ with $n = 3$–7 were observed when aqueous solutions of NaCl were sprayed.

Iribarne and Thomson do not state directly (44,45) why they did not consider Dole's charged residue mechanism (3). Possibly they were not aware of Dole's work, but they were certainly aware of the mechanism, because Chapman (46) had assumed a mechanism which is essentially the same as that proposed later by Dole. It is likely that Iribarne and Thomson chose a new mechanism because they did not believe that the charged residue mechanism was consistent with their experimental observations. The mobility measurements (44) showed a second broad maximum toward the low-mobility end. Experiments in which the concentration of the salt (NaCl) was increased led to a gradual decrease of the high-mobility peaks (gas-phase ions) and an increase of the low-mobility, broad maximum peak. Because the mobilities were measured after some "aging" of the charged particles, the low mobility group was probably due to solid $(NaCl)_n$ residues carrying some attached Na^+ ions.

The original charge on the droplets produced by statistically charged droplets was estimated to be quite low (44). The droplets, assumed to be originally 1 μm in diameter, could shrink down to ~ 10 nm without Coulomb explosions. With these conditions, the authors could also estimate the size of the solid residue that would be expected for different NaCl concentrations in the solutions used. These estimates showed that at 5×10^{-5} mol/L, solid residues of ~ 10 nm radius could be formed. Yet gas-phase ions were also observed at these and higher concentrations. The authors (44) probably came to the conclusion that gas-phase ion formation under these conditions is possible only if a considerable fraction of the charges (i.e., ions), escaped from the droplets before the formation of the solid residue. The theory for this escape process, ion evaporation, was developed in the same paper (44).

J. The Iribarne–Thomson Equations for Ion Evaporation

Iribarne and Thomson derived an equation that provided detailed predictions for the rate of ion emission from the charged droplets (44). In particular, it predicts the dependence of the rates on the chemical properties of the ions. Observed differences between the gas-phase ion intensities I_A and I_B of ions A^+ and B^+, present at equal concentrations in the sprayed solutions, were compared with predictions of the theory by Iribarne and Thomson (45) and, subsequently, other workers (25,47–50). In general, qualitative agreement between experiment and theory was found, lending support to the theory on the assumption that the experimentally observed selectivity cannot be explained by the charged residue theory of Dole (3). In section II.K we show that an expanded charged residue theory is also capable of predicting qualitatively the experimentally observed selectivities.

The Iribarne–Thomson treatment is based on transition state theory. The rate constant k_I for emission of ions from the droplets is given by

$$k_I = \frac{k_B T}{h} e^{-\Delta G^{\neq}/kT} \tag{20}$$

where k_B is the Boltzman constant, T is the temperature of the droplet, and h is Planck's constant. The free energy of activation, $\Delta G^{\neq}$, was evaluated on the basis of the model shown in Figure 6. The transition state selected by the authors resembles the products more than the initial state, that is, it is a "late" transition state. The advantage of this choice is that the energy of such a state can be expressed by a closed equation based on classical electrostatics and thermodynamics. However, in reality the transition state could occur earlier, for example, as the ion disrupts the droplet surface. The energy of such an early transition state would be much more difficult to evaluate. If a higher free-energy barrier did occur at that earlier stage, the predictions of the Iribarne–Thomson model would be invalidated.

The barrier in the Iribarne–Thomson transition state is due to the opposing electrostatic forces, namely, the repulsion of the escaping ion by the remaining charges in the droplet and the attraction between the escaping ion and the droplet, arising from the polarizability of the droplet. The attraction is larger at short distances x, between the ion and the droplet surface, but falls off faster than the repulsion as x is increased.

The equation for $\Delta G^{\neq}$ was found (44) to depend on four parameters, of which the first two are N, the number of elementary charges on the droplet, and R, the radius of the droplet. These parameters determine the value of the electric field E, which is normal to the surface of the droplet and provides the repulsive force acting on the escaping ion. The value of this field is given by the Coulomb

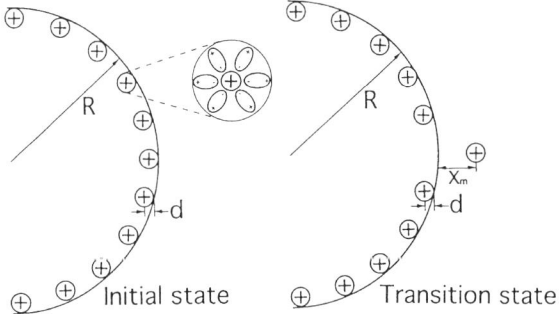

Figure 6. Iribarne–Thomson model for ion evaporation. Transition state and initial state showing droplet of radius R with an excess of N singly charged positive ions at the surface of the droplet. Ions are solvated by a solvent shell. Radius of ion plus solvent shell = d. In the transition state, one ion with a solvent shell has moved a distance x_m outside the droplet; x_m is typically 0.6 nm; $R = 8$ nm, and $N = 70$ for aqueous droplets. Adapted from [37].

equation

$$E = \frac{q}{4\pi\epsilon_0 R^2} = \frac{Ne}{4\pi\epsilon_0 R^2} \quad (21)$$

The rate constant k_I predicted by Eq. 20 increases with N and decreases with R. The other two parameters express the specific properties of the ions involved in the emission. The escaping ions leading to the lowest value of $\Delta G^{\neq}$ (Fig. 6) are not the naked electrolyte ions M^+, but ion-solvent molecule clusters $M^+(Sl)_m$ containing m solvent molecules. As mentioned in the Introduction (see Eq. 1), the transfer of a naked ion such as Na^+ from aqueous solution to the gas phase requires a large amount of energy, ~ 98 kcal/mol. However, the transfer of $Na^+(H_2O)_7$ requires only ~ 56 kcal/mol. The transfer free energies from gas phase to solution, $\Delta G°_{sol}$, for the naked alkali ions M^+ and the clusters $M^+(H_2O)_m$ are given in Table 3. These data are based on experimental thermochemical measurements available in the literature (4,51), and were evaluated in ref. 40. The strongly solvated ions Li^+ and Na^+ have large transfer energies (see Table 3), and for these ions, the Iribarne–Thomson equation predicts large activation barriers, that is, small rate constants k_I. The change of k_I with the value of the ion cluster solvation energy is illustrated by Figure 7, which depicts evaluated barriers for different values of the ion-cluster solvation energy.

The second parameter which expresses the individuality of the ions is d, the distance of the ion charge centers from the surface of the droplet (see Fig. 6). The charge centers cannot be on the droplet surface, where they would minimize their mutual repulsion, because that would partly disrupt the ion solvation shell. Strongly solvated ions like Li^+ hold on strongly to a larger number of solvent

TABLE 3. Data Used for Evaluation of Iribarne–Thomson Rate Constants and Comparison with Experiment[a]

M^+	Li^+	Na^+	K^+	Cs^+	NH_4^+	$(CH_3)_4N^+$	$(C_2H_5)_4N^+$
$-\Delta G°_{sol}(M^+)$	122	98.2	80.6	67.5	81	(54)	(49)
$-\Delta G°_{0,m}(M^+)$	74.4	56.4	36.5	23.5	—	—	—
m^b	7	7	6	5	~ 6	—	~ 0
$-\Delta G°_{sol}[M^+(H_2O)_m]$	61.2	56.5	55.8	54	55.6	(54)	(49)
$-R_{Ion}(\text{hydrated}) = d$	3.82	3.58	3.3	3.29	3.3	—	(2)
$\Delta G^{\neq}$	13.1	9.3	9.7	7.9	9.5	—	7.9
$k_I \times 10^{-5}$ s^{-1}	0.02	9.8	4.9	94	6.8	—	98
$k_I{}^c$	2×10^{-4}	0.1	0.05	1.0	0.07	—	1.0
k^d	1.6	1.6	1.0	1.0	1.3	—	~ 5
k_I (Iribarne)^e	3.5	1.25	1.5	1.0	—	—	—

[a] For the origin of all data except that in the last line, see ref. 40. All free energies are in kilocalories per mole. Radius of hydrated ions, equal to d parameter in Iribarne equation, in angstroms, 1 Å = 0.1 nm.
[b] Number of water molecules in $M^+(H_2O)_m$ which leads to lowest value of $-\Delta G°_{sol}[M^+(H_2O)_m]$.
[c] Iribarne–Thomson rate constants relative to $k_I(Cs^+) = 1$.
[d] Experimental coefficients from Table 4 expressed relative to $k(Cs^+) = 1$.
[e] Relative experimental values for k_I determined by Thomson and Iribarne (45).

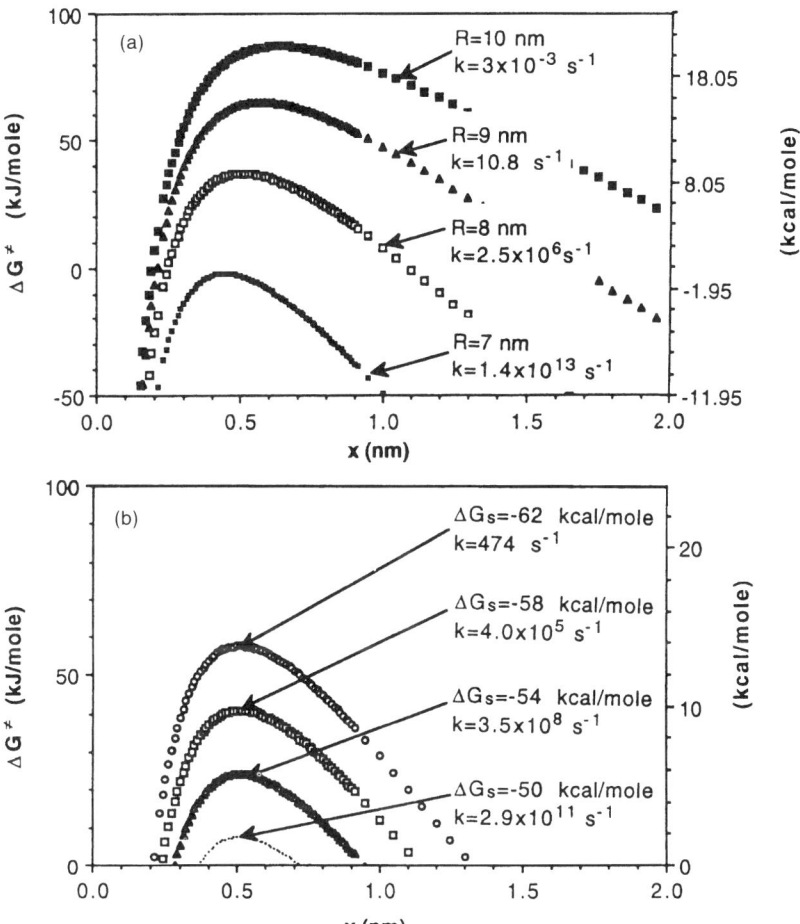

Figure 7. Iribarne–Thomson model. Dependence of free energy of escaping ion-solvent cluster on distance x between ion cluster and droplet surface. (a) The droplet radius R is treated as a parameter and $\Delta G°_{sol} = -57$ kcal/mol. (b) $\Delta G°_{sol}$ is used as a parameter and $R = 8$ nm. $N = 70$ and $d = 0.385$ nm in both figures (evaluated from Eq. 21a). Rate constants k shown in a and b were evaluated from Eq. 20. Reprinted with permission from P. Kebarle and L. Tang, *Anal. Chem.*, **1993**, *65*, 3654. © 1993 American Chemical Society.

molecules and have a larger d. Radii obtained from ion mobilities in solution (Stokes radii) provide suitable values for the parameter d (40). Such values are given in Table 3.

The expression for $\Delta G^{\ddagger}$ obtained by Iribarne and Thomson is shown below:

$$\Delta G^{\ddagger} = \left(\frac{Ne^2}{4\pi\epsilon_0(R + x_m)} - \frac{e^2}{16\pi\epsilon_0 x_m} \right) - \left(\Delta G°_{sol} + \frac{Ne^2}{4\pi\epsilon_0(R - d)} \right) \quad (21a)$$

The terms in the first bracket give the energy of the transition state while the terms in the second give the energy of the initial state (see Fig. 6). The zero level for both states corresponds to the charged drop and the ion cluster at infinite separation. The first term expresses the electrostatic potential energy due to the repulsion between the singly charged ion cluster and the charged droplet which has N elementary charges. The second term expresses the potential due to attraction between the charge of the ion and the polarizable drop, and is approximated by the electrostatic attraction resulting from the image force. The third term represents the solvation energy of the ion cluster $M^+(S1)_m$ in the neutral drop, while the last term corrects for the fact that the drop is not neutral but has N elementary charges. Thus, $Ne^2/4\pi\epsilon_0(R-d)$ corresponds to the electrostatic repulsion energy that must be overcome to bring the ion solvent molecule cluster from infinity to a distance d inside the drop. Note that this term neglects the change of permittivity which occurs when the ion cluster enters the droplet and experiences the dielectric medium of the solvent over the short distance from R to R-d. Because the dielectric constants of the solvents used are large, the potential energy change from R to R-d will be much smaller than assumed in Eq. 21a. A better approximation would be obtained by selecting a quotient equal to $4\pi\epsilon_0 R$ for the last term in Eq. 21a, that is, by neglecting the energy change for the distance d inside the droplet. The activation free energy $\Delta G^{\ddagger}$ can be obtained from Eq. 21a by treating the distance x as parameter and finding the value x_m for which $\Delta G^{\ddagger}$ is maximum.

An illustration of results obtained (40) from the Iribarne–Thomson equation is given in Figure 7. The change of $\Delta G^{\ddagger}$ and k_I for constant $\Delta G^{\circ}{}_{sol}$ ($= -57$ kcal/mol) and N ($= 70$) with the droplet radius treated as a parameter is shown in Figure 7a. These results illustrate the extremely rapid change of k_I with R. Thus, for a change from $R = 100$ Å to 70 Å, k_I increases from 3×10^{-3} s^{-1} to 1.4×10^{13} s^{-1}, that is, by 16 orders of magnitude! Iribarne assumed that ion evaporation dominates Rayleigh fission when $k_I = 10^6$ s^{-1}. This k_I value is attained when the droplets become very small, with $R = 80$ Å and $N \approx 70$ as typical values.

The results given in Figure 7b illustrate the change of $\Delta G^{\ddagger}$ and k_I when $\Delta G^{\circ}{}_{sol}$ is treated as a variable parameter and $N = 70$, $R = 80$ Å are constant. Rate constants k_I evaluated (40) for a droplet with $R = 80$ Å with $N = 70$, at a temperature T of 298K for the alkali ions, are given in Table 3. These data illustrate that the transfer energy of the cluster $M^+(H_2O)_m$, given by $-\Delta G^{\circ}{}_{sol}$ [$M^+(H_2O)_m$] in Table 3, is the decisive parameter for the value of k_I. Lithium ions, which have the largest cluster transfer energy, have the lowest k_I. The k_I for Cs$^+$ is larger by a factor of ~ 5000 relative to k_I for Li$^+$. Note that the expected monotonic rate constant increase from Li$^+$ to Cs$^+$ is not followed by the pair Na$^+$ and K$^+$, which show a smaller rate for K$^+$. This reversal is almost certainly due to experimental error in the thermochemical data (51). These data have errors in the range $\sim \pm 1$ kcal/mol, and an error of that magnitude can account for the reversal in the (Na$^+$, K$^+$) order.

Also given in Table 3 are values for the experimentally determined sensitivity coefficient k. A comparison of the theoretical and experimental results is presented in the next section.

K. Comparison of Experimental Data with Predictions of the Ion-Evaporation Theory. Inclusion of Surface-Activity Effects into Ion-Evaporation and Charged-Residue Theory. Cases for Which a Distinction Between Ion-Evaporation and Charged-Residue Theory is Not Possible

The experimental data for the relative ion evaporation rates k given in Table 3 are based on determinations of the mass spectrometrically determined ion-intensity ratio I_A/I_B when equal concentrations of electrolytes A^+X^- and B^+X^- are present in the solution that was electrosprayed. The assumption is made that

$$\frac{I_A}{I_B} = \frac{k_A}{k_B} \quad \text{when } [A^+]=[B^+] \tag{22}$$

Thomson and Iribarne (45) used this assumption and other workers adopted it also. To obtain good relative data, it is desirable that the determinations are performed under the same experimental conditions using the same apparatus. Some additional data (40) obtained under such conditions are given in Table 4.

The experimental k values for the alkali ion series Li^+ to Cs^+ (Table 3), are almost constant. A similar experimental result was obtained by Thomson and Iribarne (45) (see Table 3). However, the values calculated using experimental thermochemical data (40) predict a large increase for this series. This large numerical disagreement between the experimental and theoretical rate constants is misleading. The $\Delta G^{\neq}$ values provide a better guide to the differences involved. The experimental results, which predict nearly equal rate constants, require that the $\Delta G^{\neq}$ values be essentially the same. The calculated $\Delta G^{\neq}$ values based on the experimental thermodynamic data decrease from 13.1 kcal/mol (Li^+) to

TABLE 4. Experimentally Determined Relative Coefficients k^a

Ion[b]	k	Ion	k
Cs^+	1	$CocH^+$	10
Li^+	1.6	$Ni^{2+}(Tpy)_2$	5
Na^+	1.6	Bu_4N^+	2
K^+	1.0	Et_4N^+	5
NH_4^+	1.3	Pr_4N^+	8
$MorH^+$	3	Pen_4N^+	14
$CodH^+$	5	$C_7NH_3^+$	10
$HerH^+$	6	$C_{11}NH_3^+$	10

[a] The coefficients are defined in Eq. 35. All coefficients relative to $k_{Cs^+} = 1$. These coefficients are valid only at concentrations above 10^{-5} M. Averaged values from ref. 40.
[b] Abbreviations used: Mor = morphine, Her = heroine, Coc = cocaine, Tpy = trypyridyl, Bu = n-butyl, Et = ethyl, Pr = n-propyl, Pen = n-pentyl, $C_7 = n\text{-}C_7H_{15}$, $C_{11} = n\text{-}C_{11}H_{23}$.

7.9 kcal/mol (Cs^+). This decrease does not represent a very large difference, considering that the experimental error in the $\Delta G°_{sol}$ is ± 1 kcal/mol. However, the trend of the $\Delta G^{\ddagger}$ data to decrease is quite consistent and probably real. Therefore, the conclusion was reached (40) that, for the alkali ion series, the calculated $\Delta G^{\ddagger}$ and k_I are just barely consistent with the experimentally observed coefficients k. Note that the $\Delta G°_{sol}$ values used are for water as solvent, whereas the experimental ratios, k_A/k_B were determined in methanol.

It has generally been observed in the ESMS literature that singly charged ions which have large hydrophobic groups also have high experimental sensitivity coefficients k. Although experimental data on the solvation energies of such ions are sparse, the available values indicate low solvation energies for such species, [see the value for $(C_2H_5)_4N^+$ in Table 3]. Therefore, on the basis of the ion-evaporation theory, one expects high sensitivity coefficients for such ions, in agreement with experiment [see the relatively high k values for $(Pen)_4N^+$ and the alkyl ammonium ion $C_{11}H_{23}NH_3^+$ in Table 4. Such qualitative agreement was considered to favor the ion-evaporation theory relative to the charged-residue model (47–50,53). The charged-residue model in its initial form (3,42) did not provide a rationale for why some ionic species should have higher sensitivity coefficients than others.

In their examination of the ion-evaporation theory, Tang and Kebarle (40) pointed out that the surface activity of ions should be taken into account explicitly. Consider that ions A^+ and B^+ are present in the solution, and that the ions A^+ are more surface active than B^+. It is then to be expected that the charges at the surface of the droplet will be enriched with A^+ ions, and on that basis more A^+ ions will evaporate. The sensitivity coefficient k should, therefore, depend not only on the ion-evaporation rate constant k_I, but also on a constant K_s which expresses the bulk-to-surface ion equilibrium (40):

$$k \propto K_s k_I \quad \text{(Ion evaporation theory)} \tag{23a}$$

$$\frac{k_A}{k_B} = \frac{K_{S,A} k_{I,A}}{K_{S,B} k_{I,B}} \tag{23b}$$

For ions that have no surface activities, such as the alkali ions where $K_{S,A} = K_{S,B}$, the experimentally observed sensitivity ratio k_A/k_B, should be equal to the ratio of the Iribarne rate constants.

The realization that surface activities should be considered also has consequences for the charged-residue theory. Ions that are enriched on the surface of the droplet will be transferred preferentially to the surface of the offspring droplets because the charges on the surface of the offspring droplets originate from surface charges of the parent droplets. Because the same process repeats itself on formation of second generation offspring droplets, and so on, one might expect that the ultimate droplet, which contains only one excess ion, will carry preferentially ions that are surface active. The excess charge will not be at the surface any more, because there are no other charges to drive it toward the surface. On this basis, one expects that the experimental sensitivity coefficients should reflect the surface activities if the charged-residue theory holds (40):

$$k \propto K_s \quad \text{(Extended charged-residue theory)} \tag{23b}$$

$$\frac{k_A}{k_B} = \frac{K_{S,A}}{K_{S,B}}$$

For ion species with the same surface activity, as is the case for the alkali ions, one would expect sensitivity coefficients which are very similar, and this is in agreement with experiment (see Tables 3 and 4). The surface activities of the tetraalkylammonium ions are known (52) to increase in the order Me, Et, Pr (i.e., with increasing size of the alkyl group). The experimental sensitivity coefficients increase in the same order (see Table 4). Thus, qualitative agreement between experiment and the modified charged-residue theory (40) is present also for this series of ions. Owing to a positive correlation between the ion solvation energies and the surface activities, both ion evaporation and charged residue predict qualitatively the same direction of change for the sensitivity coefficients k. It is therefore not possible, on the basis of qualitative comparisons with experimentally determined sensitivity coefficients, to establish which theory provides a better agreement with the experimental results. Quantitative comparisons with experimental sensitivity coefficients are possible only for the alkali ions that have no surface activities, and for this system, the predictions of the Iribarne equation were not in good agreement with the experimental results (see values for k_I in Table 3).

The predictions of the ion-evaporation and charged-residue theories are distinctly different only for certain conditions. It is assumed in both theories that droplets which have radii $R > 10$ nm release coulombic stress by Rayleigh fission. According to the ion-evaporation theory, droplets $R \approx 10$ nm and smaller release coulombic stress by ion emission; according to the charged-residue theory, Rayleigh fission also continues below this size. When the droplets become very small, $R < 1$ nm, the situation changes again. Such droplets consist of some hundred solvent molecules, a few excess ionic charges (see Fig. 5), and a number of solute molecules (depending on the initial solute concentration used). Neither the Rayleigh condition (Eq. 17), nor the ion-evaporation theory (Eqs. 20 and 21), apply strictly to such a microscopic droplet. Corrections for the decreasing surface tension must be introduced and, more important, the charges on the surface must be treated as discrete charges. As solvent evaporation continues, release of coulombic stress must occur, and the emitted charges may well be individual ions solvated by several solvent molecules. This process is closer to the Iribarne–Thomson model. We consider this process to be more likely, because Rayleigh fission proceeds with a large increase of droplet surface and requires the "cooperative" repulsion of many charges. Even if the Iribarne–Thomson mechanism is quantitatively incorrect and ion evaporation does not occur as soon as predicted, the very small droplets, which have only a few ionic charges, may be expected to emit solvated gas-phase ions.

L. Measurements of the Electric Fields on Charged Residues—Evidence in Support of the Ion-Evaporation Theory

In a recent paper, Locertales and Fernandez de la Mora (54) presented experimental evidence which provides significant support for the ion-evaporation theory.

The rationale of the experiment is based on the different dependences expected for the electric field at the surface of charged droplets, which shrink by solvent evaporation, and which relieve the coulombic stress by (a) Rayleigh fission or (b) ion evaporation.

The radial field E at the surface of a drop with radius R and charge q is given by Coulomb's law (see Eq. 21):

$$E = \frac{q}{4\pi\epsilon_0 R^2} = \frac{Ne}{4\pi\epsilon_0 R^2} \quad (21)$$

The field E_{Ry} for a droplet that is at the Rayleigh limit can be determined from Eq. 21 and the Rayleigh equation (Eq. 17):

$$q_{Ry} = \pi(64\gamma\epsilon_0 R^3)^{1/2} \quad (17)$$

so one obtains:

$$E_{Ry} = \left(\frac{4\gamma}{\epsilon_0 R}\right)^{1/2} \quad (24)$$

The field E_I, for droplets that release the coulombic strain by ion evaporation, can be evaluated using the Iribarne–Thomson equations. The dependence on R is too complex to be given by a closed expression, but numerical results are readily evaluated. It is found that E increases as R decreases, but the increase is considerably slower than is the case for Rayleigh fission. A numerical example of the dependence on R of E_{Ry}, and of E_I based on data evaluated by Iribarne and Thomson (see Fig. 4 in ref. 44) is given in Table 5. The data provided are for a special droplet with an initial radius $R = 13.4$ nm and 192 elementary charges that have simultaneously reached the Rayleigh and ion-evaporation limits. After the droplet has evaporated solvent and reached the radius $R = 5$ nm, the charges and fields, illustrate that the field E_I increases less than E_{Ry}.

Obviously, if one could determine the charges and radius of the droplets

TABLE 5. Changes of Electric Field E for a Charged Evaporating Droplet Which Relieves Stress by Rayleigh Fission (E_{Ry}) or by Ion Evaporation (E_I)[a]

R	N_{Ry}	N_I	E_{Ry}	E_I	E_I/E_{Ry}
13.4	192	192	1.55	1.55	1
5.0	44	28	2.55	1.62	0.63

[a] Data for droplet radius R in nm; N is the number of elementary charges on the droplets and the electric field on droplets is given in V/nm. Numerical data was obtained from Figure 4 of Iribarne and Thomson (44).

experimentally, one could evaluate the field E of the droplets and establish whether it changes as predicted by Rayleigh fission or by ion evaporation. Unfortunately, the droplets predicted to undergo ion evaporation are very small ($R < 15$ nm) and their size cannot be determined using current methods. Furthermore, when volatile solvents are used, as is the case in ESMS, the change of radius R due to solvent evaporation is very fast. Thus, Eq. 18 predicts complete solvent evaporation within a few microseconds for such droplets. Simultaneous determinations of the charge q and radius R for such rapidly changing droplets are not possible with presently known experimental techniques. Locertales and Fernandez de la Mora (54) evaded these insurmountable problems by performing measurements of the charge and radius of charged solids resulting from the complete evaporation of droplets produced by ES. From these data, they deduced the electric fields present on the droplets just prior to formation of the solid residues and compared them with the fields predicted by the Rayleigh fission and ion-evaporation equations.

The initial droplets prepared by ES had to meet special conditions. They should have narrow distributions of size and charge. Furthermore, they should have initial diameters and charges which are close to those at onset of ion evaporation, and they should reach this onset without undergoing prior Rayleigh fissions. For the solvents used, such as formamide, droplets having an initial radius of $R \sim 20$ nm are required. The authors (54) were able to prepare such droplets, building on their previous studies (see Eq. 12 and ref. 24) on the relationships between the radius R, the conductivity K of the solution, and the flow rate V_f. The required conductivity was $K \approx 1$ S/m, corresponding to solutions of ~ 1 M concentration range. The required flow rates were very low, $V_f \sim 0.01$ μL/min. The ES produced droplets were "aged" until they reached the charged solid state, before they were subjected to mobility determinations. The mobility spectrum showed peaks for both gas-phase ions and charged solid aggregates. The mobility peak corresponding to the residues was relatively narrow, as expected for aggregates produced from monodisperse initial droplets which have not undergone Rayleigh fissions.

Using a "hypersonic impactor," an apparatus previously perfected in their laboratory, the authors (54) determined the product $R_r \rho_r$, where R_r is the radius and ρ_r is the density of the residue. From this product they evaluated R_r, assuming that the density of the residue equals the density of the solid bulk electrolyte:

$$\rho_r = \rho_{salt} \tag{25}$$

Using the same assumption, the initial droplet radius R_p was also evaluated on the basis of a mass balance between the electrolyte present in the initial droplet and that present in the residue:

$$\frac{4}{3}\pi R_p^3 M \times 1000 \text{ MW (salt)} = \frac{4}{3} R_r^3 \rho_{salt} \tag{26a}$$

$$R_p = [\rho_{salt}/\text{MW (salt)} \times 1000 \text{ M}]^{1/3} R_r \tag{26b}$$

where M is the molarity of the solution, 1000 is the conversion factor to mol/m^3, and MW is the molecular mass of the salt.

The authors (54) point out that the assumption $\rho_r = \rho_{salt}$ cannot be accepted without proof because the morphology of the residues left after evaporation of the droplets is not well understood, and consistencies ranging from dense solid spheres to snowflake particles and hollow structures can be present. Unfortunately, the authors did not attempt to examine the solids with a microscope. However, they compared the values for the radii R_p obtained from Eq. 26 with radii R_p predicted by more refined forms of Eq. 12 and found satisfactory agreement. Some representative results (54) are given in Table 6.

Table 6 also gives the ratio of the measured field E_m on the charged residue to the field E_{Ry} expected if the residue is a "droplet" containing a solid residue with a thin layer of solvent on the surface and if this system is at the Rayleigh limit. The ratio is smaller than unity, as expected when ion evaporation determined the field on the droplet (see electric field ratio in Table 5).

Viewed by themselves, the field ratios in Table 6 and the original Table 3 (54) would not be at all convincing, considering that the evaluated fields E_m and E_{Ry} are relatively close and that it is difficult to identify the charged residue when the charged droplet is still emitting ions. However, the data are strengthened when one also takes into account the mobility spectra, which showed the presence of both gas-phase ions and charged residues, as discussed above.

The authors (54) utilized the determined fields E_m and the corresponding radii

TABLE 6. Some Representative Data Involving Charged Residues[a]

Ion	Salt	d_p^b (nm)	d_r^c (nm)	N_+^d	E_m^e (V/nm)	$(E_m/E_{Ry})^f$
Li$^+$	LiCl	38.6	105	36	1.9	0.84
		27.6	7.5	18.7	1.9	0.72
$(C_3H_7)_4N^+$	$(C_3H_7)_4NBr$	32.2	18.5	82.1	1.4	0.83
		21.4	12.7	39	1.4	0.70
$(C_7H_{15})_4N^+$	$(C_4H_{15})_4NBr$	51.6	40.6	277	0.98	0.86
		43.3	30.4	202	1.02	0.81

[a] Data are from Table III of ref. 54. They represent only a small sample of the extensive data provided in the original table, which includes a wider variety of salts and solvents. Data are for formamide as solvent.
[b] $d_p = 2 \times R_p$ is the diameter of the initial droplets produced by the spray. It was evaluated with Eq. 26 and d_r, or alternatively from Eq. 12.
[c] $d_r = 2 \times R_r$ is the diameter of the solid residue, obtained from determinations with the hypersonic impactor (54), or alternatively from d_p obtained with Eqs. 12 and 26.
[d] Number of elementary charges on solid residue. Obtained with Eq. 21 where charge q was derived from mobility determinations and the known d_r.
[e] Electric field of charged residue obtained with Eq. 21 and charge q derived from mobility measurements and known d_r.
[f] E_{Ry} in ratio E_m/E_{Ry} is the electric field which would have been present on the charged residue if the charged residue was a droplet of the solvent with the same diameter and the droplet was at the Rayleigh limit. E_{Ry} is evaluated with Eq. 24 using the surface tension, γ, of the solvent. Calculations of E_{Ry} using a larger γ, as expected for a saturated solution of the salt, lead to a lower E_m/E_{Ry} ratio.

R of the residues in a rather ingenious treatment which provides additional evidence in favor of the ion-evaporation theory. The electric field is given by Eq. 21:

$$E = \frac{Ne}{4\pi\epsilon_0 R^2} \quad (21)$$

In an evaporating droplet, when ions are also evaporating, both N and R decrease; N decreased because ions leave and R decreases owing to solvent evaporation. The authors (54) argue that the field E changes rather slowly compared with the changes in N and R separately, so that for small changes of R and N it can be treated as constant. Taking the logarithm of Eq. 21, followed by differentiation with respect to time, and assuming E to be constant, one obtains:

$$\frac{d(\ell n\ N)}{dt} = \frac{2d(\ell n\ R)}{dt} \quad (27)$$

The term on the right can be evaluated using Eq. 18a, which deals with the rate of change of radius with time of an evaporating droplet. The term on the left equals the rate constant for ion evaporation k_I, taken with a negative sign:

$$-\frac{dN}{dt} = k_I N \qquad \frac{d\ \ell n\ N}{dt} = -k_I$$

Equation 27 thus provides a basis for an "experimental" determination of k_I from the droplet evaporation rate. The authors (54) also point out that the activation free energy $\Delta G^{\ddagger}$ of the theoretical expression for k_I (see Eq. 21a), can be expressed thus:

$$\Delta G^{\ddagger} = -\Delta G°_{sol} - G(E) \quad (28)$$

where $\Delta G°_{sol}$ is the solvation energy for the ion solvent cluster discussed previously (see Table 3), and $G(E)$ is the free-energy term expressing the assistance privided by the electric field E of the droplet in lowering the activation free energy.

Substituting $\Delta G^{\ddagger}$ from Eq. 28 into the rate-constant expression Eq. 20, one obtains:

$$k_I = \frac{kT}{h} e^{-[-\Delta G°_{sol} - G(E)]/kT} \quad (29)$$

Equations 18a, 27, and 29 can be combined to obtain

$$\Delta G°_{sol} + G(E) = kT\ \ell n\left(\frac{\alpha\ p°h}{2kTR}\right) \quad (30)$$

where k is the Boltzmann constant, h is Planck's constant, α is the evaporation coefficient of the solvent used, p^o is the vapor pressure, and R is the radius of the droplet.

Analytical expressions for $G(E)$ are available. The terms comprising $G(E)$ used by Iribarne and Thomson can be found in Eq. 21a. Locertales and Fernandez used the image-potential model (IPM) (54), which leads to a simpler expression:

$$G(E)_{IPM} = \left(\frac{e^3 E}{4\pi\epsilon_0}\right)^{1/2} \quad (31)$$

The authors (54) make the following uses of Eqs. 30 and 31. For charged residues obtained from aqueous solutions, and whose radii R and fields E_m were determined and whose $\Delta G°_{sol}$ values are known (see Table 3 for $\Delta G°_{sol}$), one can evaluate $G(E)$ from Eq. 30 using values of α and p^0 for water. The $G(E)$ values so obtained were close to the values predicted by Eq. 31, into which the measured E_m values were substituted (see Fig. 19 in ref. 54). This result supports the consistency of Eqs. 29–31 and thus also the validity of the ion-evaporation mechanism for these particular solutions.

However, this consistency is somewhat blemished by uncertainties concerning the value of α. The authors (54) used an evaporation coefficient $\alpha = 1$. A much lower coefficient, $\alpha = 0.04$, has been suggested in the literature (55, see p. 21). Experimental determinations of the coefficient α for volatile solvents like water are extremely difficult and the literature value (55) may be wrong. On general grounds, a low value α for water would be expected because water is a highly structured solvent and α is related to the kinetic barrier for the conversion of vapor to liquid and vice versa. Structured solvents are expected to lead to low coefficients α. The use of $\alpha = 0.04$ instead of $\alpha = 1$ in Eq. 30 will reduce significantly the consistency of the ion evaporation theory with the results described above for electrosprayed aqueous solutions.

In a second application, the solvation energies $\Delta G°_{sol}$ of tetraalkylammonium ions in formamide were obtained (54) from experimentally determined values of E and R for charged residues of the corresponding electrolytes in formamide, together with values of α and p^0 for formamide. The value of $G(E)$ was evaluated from Eq. 31, and substitution of $G(E)$, R, α, and p^0 in Eq. 30 leads to $\Delta G°_{sol}$ (see Fig. 20 and Table 6 in ref. 54). The $\Delta G°_{sol}$ values obtained by this procedure have reasonable magnitudes.

It is clear that this work (54) is an impressive effort to provide experimental evidence in a most difficult area. The results provide significant support for the ion-evaporation theory. However, the interpretation of the data also leaves some concerns:

1. The measured electric field E_m on charged residues, while lower than the Rayleigh field E_{Ry}, (see Table 6), is nevertheless close to E_{Ry}. However, errors in field measurements by a factor of 2 or more cannot be excluded. A method that

depends on conditions where E_m and E_{Ry} are expected to be of the same order of magnitude is not well suited to the task of discriminating between the two rival mechanisms because of the difficulty in obtaining accurate field measurements.

2. The charged-residue mechanism would be expected to lead to the desired result if formation of the solid occurred toward the center of the droplet and the dense solid was covered with a layer of (saturated) solution until full solidification. However, because the solvent escapes from the droplet surface, solid formation may be favored at the surface. Such solid microplatelets at the surface may have unforeseen effects on the ion evaporation from the surface and on the Rayleigh fissions of such particles.

Therefore, while recognizing the ingenuity of the approach, the mastery of the techniques, and the extensive determinations, one still cannot conclude that the ion-evaporation model has been completely proven by the work of Locertales and Fernandez de la Mora (54).

M. Change of Solute Concentration in Evaporating Droplets

The concentration of a solute that is less volatile than the solvent is expected to increase because the volume of the droplets decreases by solvent evaporation. The concentrations of solutes in the very small droplets, which are the immediate precursors of gas-phase ions, are of special interest. For example, the solute may be an acid, base, or buffer, which determines the pH of the solution that is electrosprayed. Changes in the concentration of that solute will lead to changes in pH which may influence the mass spectra observed, such as the degree of protonation of peptides or proteins.

The concentration increases of completely involatile solutes can be estimated on the basis of the assumed histories of the charged droplets, such as in the scheme presented in Figure 5. This scheme, which attempts to represent conditions in conventional ES for mass spectrometry, is quite complex because the initial droplets are relatively large and many Rayleigh fissions are interposed between the initial charged droplets and the small ion-precursor droplets. The situation would be more tractable for ES experiments leading to much smaller droplets, such as the nano ES source developed recently by Wilm and Mann (56,57) in which the initial droplets are close in size to the ion-precursor droplets. Unfortunately, extensive data obtained using this exciting technique are not yet available.

Figure 5 shows that the parent and the offspring droplets experience different increases of solute concentration. Because the offspring droplets get some 15% of the parent charge but only ~ 2% of the mass, these droplets and their own subsequent offspring evolve rapidly into highly charged, very small droplets which are the gas phase ion precursors. The increase of solute concentration in these droplets will be much smaller than that of the parent droplets, which are losing much more charge than mass. Therefore, the parent droplets are expected to be the precursors of large solid residues which are agglomerations of the solutes present in the droplets, and thus are not useful for mass spectrometric analysis.

For purposes of the following discussion, the precursor to the gas-phase ions is chosen to be a droplet with a radius $R = 0.03$ μm ($N = 278$) (see Fig. 5, lower right corner). Such a droplet could lead directly to gas phase ions by ion-evaporation if the ion evaporation mechanism is assumed to hold. Alternatively, if the charged-residue theory holds, the Rayleigh fission of this droplet will lead to 20 droplets with $R = 0.003$ μm and $N = 2$ elementary charges each. The ratio of initial volume to final volume, V_i/V_f, due to the various evaporation stages involved in the formation of that droplet, can be evaluated from the corresponding radii of the several evaporation stages leading to the droplet. Taking these into account, one obtains, for the complete change, $V_i/V_f \approx 94$. The largest ratio V_i/V_f occurs in the first stage of the evaporation where $(R_i/R_f)^3 = (1.51/0.95)^3 = 4$. Initial droplets formed closer to the Rayleigh limit will lead to a first stage ratio lower than 4. Therefore, an average ratio $V_i/V_f = 50$ will be assumed.

Completely nonvolatile solutes like NaCl and other completely dissociated electrolytes can be expected to experience a concentration increase:

$$\frac{C_f}{C_i} = \frac{V_i}{V_f} \approx 50 \tag{32}$$

The result (Eq. 32) can be used to devise a test to determine which theory—ion evaporation or charged residue—fits available mass spectral data. To test for the validity of the charged-residue theory, one can use the radius $R = 0.003$ μm of one of the 20 droplets with $N = 2$ charges to evaluate its volume. Using the known molarity M of a solute like NaCl in the initial solution and multiplying by $C_f/C_i = 50$, one can calculate the number n of neutral solute molecules in the droplet. Results obtained in this way are given in Table 7.

The table shows that somewhere within a concentration range between 10^{-3} and 10^{-2} M one expects to find one to several solute molecules in a droplet that has one or a few elementary charges. Therefore, if a single solute is present, one would expect to observe in the mass spectra ions of the type $Na^+(NaCl)_x$, where x is small, say 0–10. Experiments (30) performed under these conditions with

TABLE 7. Estimated Number n of Involatile Solute Molecules in a Dole Droplet of 0.003 μm Radius for Different Molarities M of Solute[a] in Electrosprayed Solution

M	10^{-5}	10^{-4}	10^{-3}	10^{-2}
n	3×10^{-2}	0.3	3	30

[a] Evaluated assuming that a concentration increase by a factor of 50 occurs (see Fig. 5 and text); n = number of solute molecules present in a droplet for an initial concentration of solute of M (mole/L) in solvent that is electrosprayed.

methanol as solvent provided mass spectra in which Na^+, $Na^+(H_2O)_x$, and $Na^+(MeOH)_x$ were abundant but $Na^+(NaCl)_x$ was completely absent. These observations have been cited (30,40) as evidence against the Dole charged-residue mechanism.

However, the results with NaCl in methanol (30) do not appear to be correct. More recently, Wang and Cole (58) observed the ions Cs^+ and $Cs(CsCl)^+$ in mass spectra obtained from ES of solutions of CsCl in 50:50 water–methanol mixtures. The intensity of the $Cs(CsCl)^+$ relative to the Cs^+ ions increased as the concentration of CsCl in the solution was increased. The maximum ratio of 0.35 was observed at the highest concentration, 10^{-2} M, used. No higher clusters of $Cs(CsCl)_n^+$ were observed. Using 50:50 water–acetonitrile mixtures and 10^{-1} M salt concentrations, Boyd and co-workers (59) observed abundant ion clusters. Thus, with sodium iodide, $Na(NaI)_n^+$, ions were observed in which the ion-intensity distribution envelope extended up to $n \approx 15$. Similar data were obtained with other electrolytes, such as KI, CsI, and $CsNO_3$. Ion spray was used in these experiments.

Prompted by these findings (58,59), experiments undertaken recently in this laboratory (60) led to the observation of $M(MX)_n^+$ and $X(MX)_n^-$ clusters from several MX (NaCl, NaBr, NaI, CsCl). A methanol solution, 10^{-3} M in NaCl at flow rates of 1.5 μL/min, which should lead to droplets in the 1- to 2-μm range, led to mass spectra dominated by $Na(NaCl)^+$ ions. The highest intensity was at $n = 1$, and the ion envelope extended up to $n = 12$. These results are not too far from the average $n = 3$, (Table 7), predicted with Eq. 32, and thus support the charged-residue model.

The concentration changes for solutes that are somewhat volatile (i.e., weak electrolytes and nonionic solutes) are more difficult to estimate. Under evaporation equilibrium conditions, the mole fraction Y_{St} of a solute St, in the gas phase, is given by Henry's law:

$$Y_{St} = k_{St} X_{St} \quad (33)$$

where k_{St} is Henry's constant and X_{St} is the mole fraction of the solute in solution. The following predictions may be made:

$$k_{St} > 1 \quad \text{(depletion of solute in droplet)} \quad (34)$$

$$k_{St} < 1 \quad \text{(enrichment of solute in droplet)}$$

Henry's constants are available (61) for many solutes and solvents of interest, such as the weak acids and bases used for pH control in solvents like water or methanol. Had the droplet evaporation been slow, estimates of the solute concentration changes for a given initial to final volume ratio, V_i/V_f would have been possible on the basis of liquid–vapor phase diagrams from which the Henry's constants are deduced. However, evaporation of the small ES droplets is extremely rapid and falls in the regime of "surface-controlled free molecule conditions" (41,55) (see Eq. 18). Quantitative prediction of the rates of evaporation of solvent and solute under such conditions is difficult because the available theoretical and experimental information is scarce. Because the evaporation is rapid, the solute may become entrained by the evaporating solvent. Owing to such entrainment, a highly nonvolatile solute may become less enriched in the droplet than predicted on the basis of Eq. 33 or other more detailed phase-equilibria data.

Values for the Henry's law constants of a few weak electrolytes such as ammonia and acetic acid, which are often used for pH control of solutions, are given in Table 8. These data show that the concentration of the volatile ammonia will decrease in the evaporating droplets, leading to a less basic pH than in the original solution. On the other hand, the concentrations of formic and acetic acids will increase and this will lead to more acidic pH in the evaporating droplets. Considering the very large factor by which the volume decreases, substantial changes in pH can be expected.

The most dramatic changes in pH can be expected when strong electrolyte acids like HCl or H_2SO_4 and bases like NaOH are used. For these, the concentration increases will be highest (see Eq. 32).

The high or low pH values that can result from solvent evaporation can lead to partial denaturation of proteins. Acid- or base-catalyzed denaturation occur quickly under such conditions (63), often within evolution time of the droplets (tens of microseconds).

N. Dependence of the Mass Spectrometrically Observed Analyte Ion Signal Intensities on the Concentration of the Analyte Ion and the Concentrations of Other Electrolytes Present in the Electrosprayed Solution

This section examines how the analyte ion signal intensity determined with the mass spectrometer depends on the concentration of the analyte ion in the solution that is electrosprayed and how this intensity is affected by the presence of other electrolytes. This discussion is a condensed version of previous work (40). Other electrolytes are always present as impurities in the solvent used, as coanalytes, and as buffers required in the reverse-phase liquid chromatography or capillary electrophoresis, which may have been used prior to the ESMS.

The questions concerning the expected ion intensity for a given analyte under given concentration conditions is part of the daily practice of ESMS. Trying to understand the factors determining the observed intensity leads one very rapidly into questions related to the mechanism of ES, discussed in the previous sections.

TABLE 8. Henry's Law Constants for Some Commonly Used Weak Electrolytes[a]

Compound	NH_3	HCO_2H	CH_3CO_2H
k_{St}[b]	40	~ 0.46	0.46

[a] Based on data from Clegg and Brimblecombe (61).
[b] Henry's Law constant at 25°C for aqueous solution. For definition of k_{St}, see Eq. 33.

Experimentally determined (40) ion intensities, at different concentrations of analytes in the solution that is electrosprayed, are shown in Figures 8–10. The changes in intensity of a single analyte observed in mass spectra obtained for a series of solutions with increasing analyte concentration are shown in Figure 8. A logarithmic plot was used to accommodate a wide range of ion intensities and concentrations. The plot of analyte MH^+ intensity I_A has a typical shape, which is commonly observed (40,53). A linear section with a slope ≈ 1 in the low concentration range up to $\sim 10^{-6}$ M is followed by "saturation," and even a small decrease of intensity, at the highest concentrations (10^{-3} M).

The linear section, where the intensity I_A of the analyte A is proportional to the concentration $[A^+]$, is a region useful for the quantitation of A^+ in the solution. Such a linear region is generally observed (40,49,50,53).

The key to understanding the complete analyte curve is the realization that the

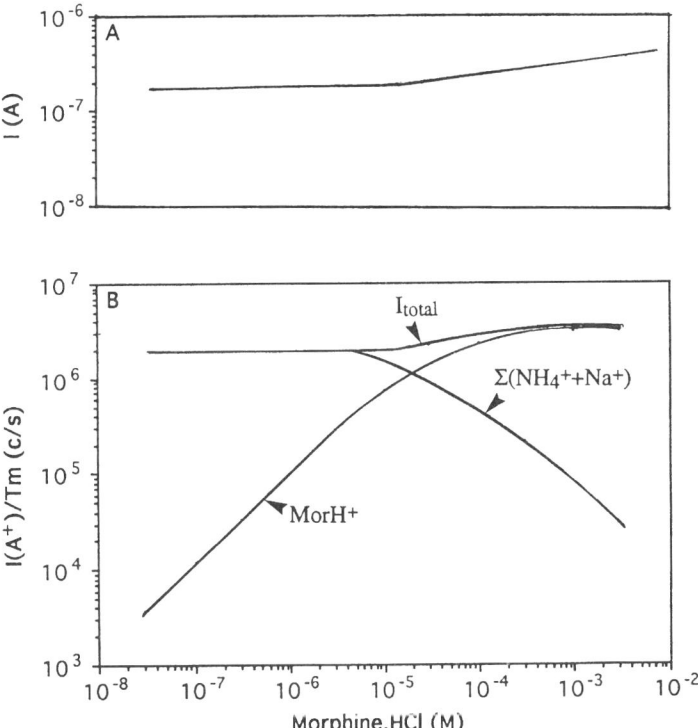

Figure 8. Dependence of capillary current I and intensity of analyte ion A^+ on analyte concentration. Capillary current I, and total ion intensity I_{total}, observed to be constant up to $\sim 10^{-5}$ M, owing to Na^+ and NH_4^+ electrolyte impurities at $\sim 3 \times 10^{-5}$ M in reagent-grade methanol. Analyte ion A^+ is protonated morphine, $MorH^+$. Ion intensities are corrected for mass-dependent transmission in the mass spectrometer by the factor T_m. Reprinted with permission from P. Kebarle and L. Tang, *Anal. Chem.*, **1993**, *65*, 372A. © 1993 American Chemical Society.

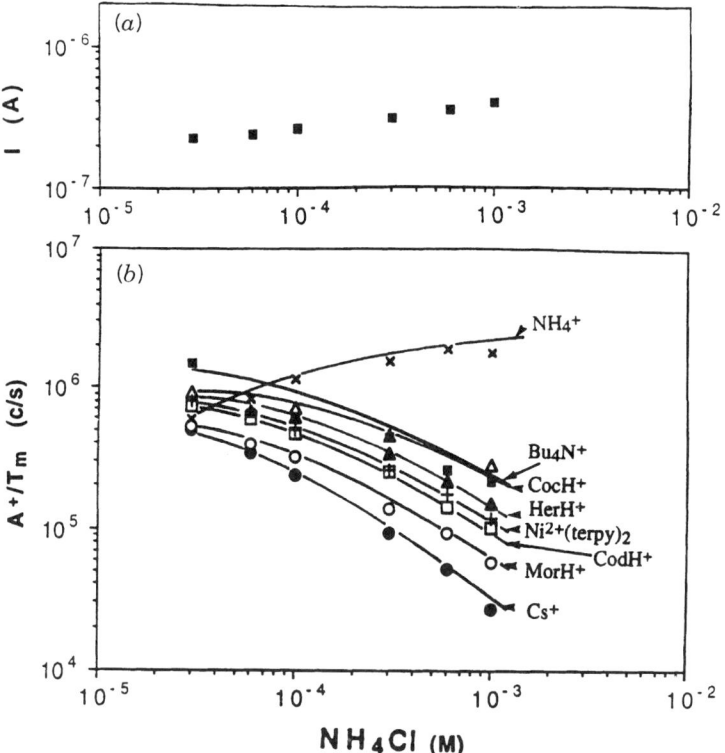

Figure 9. Effect of addition of NH$_4$Cl to mass spectrometrically observed analytes A$^+$ present at concentration [A$^+$] = 10^{-5} M. The results shown were obtained from a series of experiments in which a single analyte was present at constant concentration, and the concentration of NH$_4$Cl was increased. (*a*) Change of total capillary droplet current *I*. (*b*) Decrease of analyte ion intensities due to competition from added NH$_4^+$. Solid lines are curves predicted by Eq. 35. Reprinted with permission from P. Kebarle and L. Tang, *Anal. Chem.*, **1993**, *65*, 972A. © 1993 American Chemical Society.

solution involved is not a single electrolyte system. The solvent involved, unless special deionization procedures were applied, always contains impurity electrolyte. The impurities in the reagent grade methanol used (40) (Figs. 8 and 9) are mainly ammonium and sodium salts at a total concentration of ~ 10^{-5} M. For analyte concentrations below 10^{-5} M, ES is possible only because of the presence of an impurity electrolyte (25).

Figure 10. Ion intensities of analyte A$^+$ and B$^+$ observed at different analyte concentrations but with [A$^+$] = [B$^+$]. (*a*) Analytes K$^+$ and Cs$^+$ have the same sensitivity coefficient. (*b*) and (*c*) Analytes A$^+$ and B$^+$ have different sensitivity coefficients. The analyte with the higher sensitivity coefficient has higher intensity at high concentrations. At low concentrations, this analyte becomes depleted from charged droplets. Et = ethyl; Pen = *n*-pentyl; C$_7$NH$_3^+$ = protonated *n*-heptyl amine.

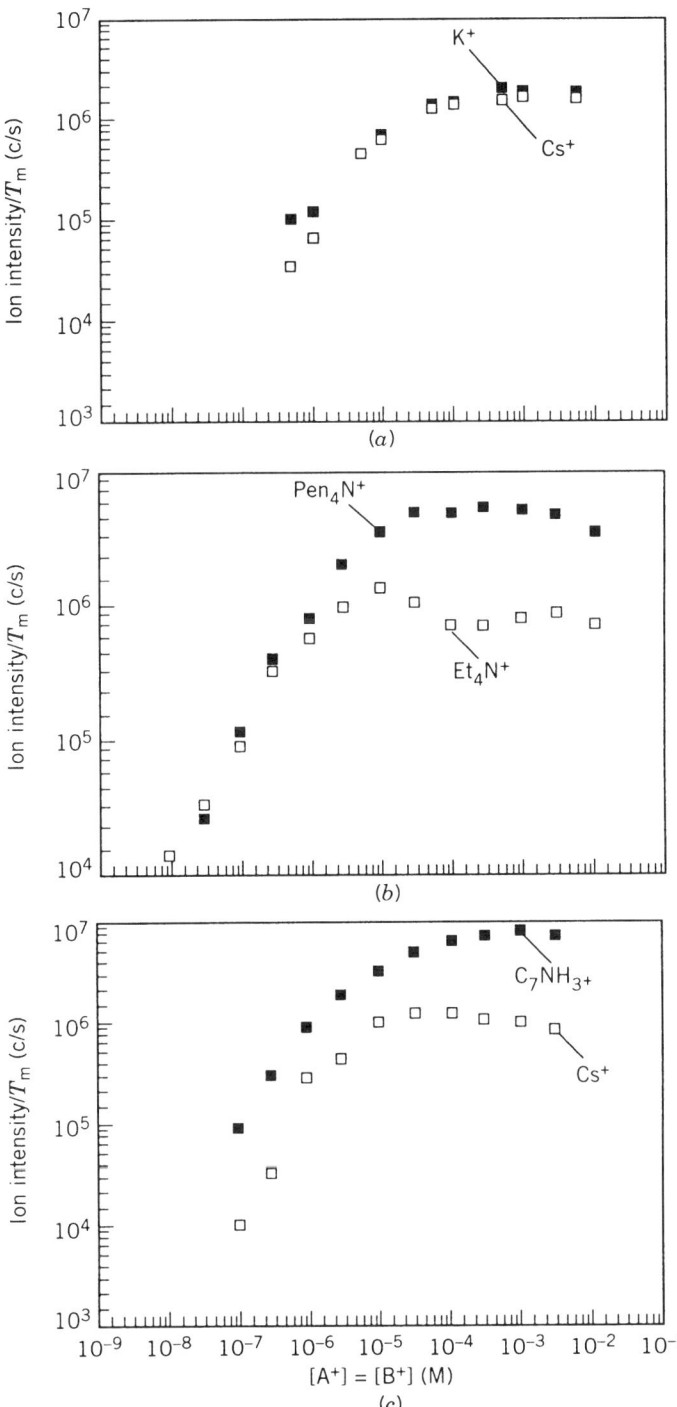

In the low-concentration range of analyte A, the capillary current I is carried by the dominant electrolyte, i.e. the impurities B, which are present at a constant concentration; I is therefore constant in this range (see Fig. 8, top). The sum of the mass-analyzed total ion intensities, $I_{tot} \approx I_A + I_B$, is also constant in this range because it is dominated by the impurities B.

Above an analyte A concentration of $\sim 10^{-5}$ M, the analyte begins to dominate and the total electrolyte concentration begins to increase. An increase of capillary current I occurs in this region, as expected from Eq. 16.

$$I \propto (\lambda_m^0 C)^n \qquad (n \approx 0.2\text{--}0.4) \tag{16}$$

Because the molar conductivity λ_m^0 for different electrolytes M^+ generally changes by less than a factor of 2 over a wide range of electrolytes (see Table 2), and the exponent n is very low, the current I is close to independent of the nature of the electrolyte (i.e., of λ_m^0). Changes of I with concentration C must be considered because C can be changed over a much wider range. However, owing to the low exponent n, these changes are still relatively small (i.e., a factor of $\approx 4\text{--}8$) for a change of C by a factor of 100.

The shape of the I_{tot} curve is similar to that of the capillary current I (Fig. 8). An approximate proportionality between the two currents is generally observed in this concentration range (25,40). For an analyte concentration above 10^{-5} M, the intensity I_B of the impurity ions B^+ decreases (Fig. 8). This decrease is a consequence of the weak dependence of the current I on the total concentration (see Eq. 16). Because the current is proportional to the total droplet charge, the addition of A barely increases the droplet charge. However, in the droplets the ions A^+ compete with the ions B^+ in the conversion process to gas-phase ions. A proportionality to concentrations of ions A^+ and B^+ in the droplets may be expected in this competition so that an increase of $[A^+]$ should lead to a decrease of production of gas-phase ions B^+, that is, to a decrease of I_B. This is exactly what is observed in Figure 8.

On the basis of such considerations, Eq. 35 has been proposed (31,40) for a two-electrolyte system, while Eq. 36 is for a three-electrolyte system:

$$I_A = fp \frac{k_A[A^+]}{k_A[A^+] + k_B[B^+]} I \tag{35}$$

$$I_A = fp \frac{k_A[A^+]}{k_A[A^+] + k_B[B^+] + k_C[C^+]} I \tag{36}$$

The equations for I_B, I_C, are analogous to those for I_A. For simplicity we will discuss the two-electrolyte equation. Here I_{A^+} is the mass spectrometrically detected ion intensity of ion A^+ corrected for the mass-dependent transmission of the mass analyzer, $[A^+]$ and $[B^+]$ are the electrolyte concentrations in the solution to be electrosprayed, and the constants k_A and k_B are sensitivity coefficients. The product fp is a factor which was assumed to be independent of the chemical nature of the ions; f is the fraction of charges on the droplets that are converted to gas-phase ions, and p is the ion-sampling efficiency, that is, the fraction of the ions detected with the mass spectrometer relative to the gas-phase ions produced by the droplets at 1 atm. The value of f depends on the quality of the droplets produced while the value of p depends on the quality of the interface.

Values of the product fp can be determined using Eq. 37:

$$I_A + I_B = fpI \qquad (37)$$

which follows from Eq. 35. Measurements of the total mass-analyzed ion current $(I_A + I_B)$ and the capillary current I lead to the product fp. When the concentration and the nature of the electrolytes were changed, fp was found to be approximately constant (40) up to a total concentration of $\sim 5 \times 10^{-4}$ M. A rough estimate of $f \approx 0.3$ has been obtained (40) for a 20-μL/m flow rate and methanol as solvent. A decrease of fp is observed above $\sim 5 \times 10^{-4}$ M (Fig. 8). It was attributed (52) to a decrease of f due to an increase of droplet size at high electrolyte concentrations. The spray was observed to contain coarse droplets at high concentrations (52). Experimental observations in other laboratories (21, 24) (see Section II.F and Eq. 12), predict not an increase but a decrease of the droplet diameter with increase of the conductivity (i.e., with increase of the concentration). However, to obtain this decrease of diameter, one must remain in the stable cone-jet domain of the flow rate and conductivity variables, and that in turn requires a decrease of flow rate at high concentrations (21, 24). If the flow rate is kept constant while the concentration is increased over a few powers of ten, as was the case for the experimental results depicted in Figure 8, the cone-jet mode may not remain stable at the imposed high flow. It was probably this instability that led to the observation of coarse droplets at high concentrations.

The value of p depends on the design of the interface between the atmospheric region and the vacuum of the mass analyzer. Values of 10^{-4} to 10^{-5} may be expected (40) when a sampling orifice is used. Such low values result from spreading of the droplets and ions due to space charge and inefficient sampling owing to the small area of the sampling orifice (see Fig. 1). Much higher values have been obtained. For example, values of p which are close to unity can be obtained with the "nano-ES" method (56,57), where the output tip of the micro-ES capillary can be placed close to or even inside the opening of the sampling capillary.

Equations 35 and 36 express the *competition* between the bulk ions to enter the gas phase. The intensity I_A (Eq. 35), depends only on the ratio k_A/k_B and not on the individual values k_A and k_B. The ratio k_A/k_B expresses the yield ratio of gas-phase ions A^+ and B^+ relative to the bulk *solution* concentrations. As noted in Section II.M, large concentration increases of the electrolytes are expected owing to the volume decrease of the droplets by evaporation. Equations 35 and 36 are valid only for strong electrolytes A, B, and C. For these, the concentrations in the droplets will increase by the same factor, and because concentration ratios are involved in these equations, this factor will cancel.

In earlier work k_A/k_B ratios were generally obtained (1,47–49) by determining the ion intensities I_A, I_B with solutions in which the concentrations were equal, that is, $[A^+] = [B^+]$. These measurements can be considered as a special case of the use of Eq. 35 which when applied to I_A and I_B, leads to

$$\frac{I_A}{I_B} = \frac{k_A[A]}{k_B[B]} \quad \text{and} \quad \frac{I_A}{I_B} = \frac{k_A}{k_B} \quad \text{for} \quad [A] = [B] \tag{38}$$

The use of Eq. 35 permits the determinations of k_A/k_B for concentration conditions where $[A] \neq [B]$.

The plots obtained in Figure 9 illustrate a situation of practical importance, that is, how the intensity of an analyte A^+ is suppressed by the presence of a second electrolyte B which is used as a buffer. In these experiments (40), the concentration of a given analyte A was kept constant, $[A^+] = 1 \times 10^{-5}$ M, while the concentration of B^+, which is always NH_4^+, is increased above 2×10^{-5} M. The solid curves in Figure 9 are fits obtained with Eq. 35. Substantial decreases of I_A are observed. For $A^+ = Bu_4N^+$, the ion with the highest k_A, the loss observed is the smallest, while Cs^+, with the lowest k_A, experiences a very large decrease, by a factor of 12, when $[NH_4^+]$ was increased from 10^{-5} to 10^{-3} M.

Equation 35 predicts that the decrease of I_A with $[B^+]$ will be greater as the value of k_B is increased. NH_4^+ has a low value of k, and therefore the decreases of I_A observed in Figure 9 are not the worst case. When the ion with the highest coefficient (Bu_4N^+) was used instead of NH_4^+, the observed decrease of the intensities of analytes Cs^+ and $MorH^+$ was much higher, by a factor of ~ 200 (40). Therefore, in actual ES, buffers with cations with low sensitivity coefficients k should be used if possible.

Recently, Smith and co-workers (53) showed that an equation of the same form as Eq. 35 can usefully be applied to ES involving effluent from capillary electrophoresis.

O. Competition and Depletion Involving Analyte Ions—Dependence of Ion Signal on Analyte Mass

More extensive experiments, particularly under conditions where the added electrolytes are at the same concentration, $[A] = [B]$, have shown (40) that a fit with Eq. 35 cannot in general be obtained over a wide concentration range such as 10^{-8}–10^{-3} M, with the same k_A/k_B ratio. The results in Figure 10 illustrate the three typical cases observed.

The ratio k_A/k_B remains constant over the complete concentration range when $k_A = k_B$. This is the case for K^+ and Cs^+ as shown in Fig. 10a; note that for the logarithmic plots used, a constant k_A/k_B corresponds to a constant vertical distance between log I_A and log I_B. The joint decrease of I_A and I_B at low $[A] = [B]$ is due to the presence of electrolyte impurity in the solvent and is analogous to the decrease discussed for Figure 8b.

When $k_A/k_B > 1$ and is approximately constant at high concentrations, a decrease to a lower constant value occurs at low concentrations. For example, the tetraalkylammonium cations Pen_4N^+ and Et_4N^+ exhibit a ratio $k_A/k_B \approx 7$ at high concentrations and $k_A/k_B \approx 1$ at low concentrations (Fig. 10b). However, the pair n-$C_7H_{15}NH_3^+$ and Cs^+ exhibit a ratio $k_A/k_B \approx 10$ at high concentrations and $k_A/k_B \approx 4$ at low concentrations (Fig. 10c).

The factors responsible for the changes noted above have been explored (40) in some detail only from the standpoint of the Iribarne–Thomson ion-evaporation theory, which was extended to also include the effect of surface activity. The high k_A/k_B ratios observed at high concentrations are due to A^+ ions having a higher Iribarne rate constant and/or higher surface activity. The observed decrease of k_A/k_B at low concentrations was attributed (40) to *depletion* of the ion A^+, whose evaporation rate is higher. At high concentrations, the number n of ions in the body of the droplet dominates over the number N of the surface charges, that is, $n \gg N$. The faster ion evaporation of A^+ from the surface does not lead to a decrease in the surface charge ratio, N_A/N_B, because A^+ ions are rapidly supplied from the droplet bulk. At low concentrations, there are many fewer ions in the body of the droplet while the surface charge number is only slightly lower (see Eq. 16). Faster evaporation of A^+ from the surface leads to a relative depletion of A^+ over B^+ and the faster emission of A^+ cannot be maintained. Therefore, the observed k_A/k_B ratio decreases at low concentrations.

The depletion phenomenon can also be expected on the basis of the charged-residue theory, as extended to include surface effects. Depletion at low concentrations occurs because the ion population on the surface of the small droplets cannot be maintained at the high ratio, $N_A/N_B = K_{SA}/K_{SB}$ (see Eq. 23b), because the number of electrolyte ions A^+ in the body of the droplets is low and cannot maintain the high surface ratio required by the surface activity constants' ratio.

The depletion phenomenon described above has two important consequences:

1. Experiments to determine coefficient ratios k_A/k_B, to be used for comparisons with predictions of the Iribarne–Thomson and charged-residue theory, should be performed only in the high-concentration range where depletion is not present. The coefficients listed in Table 3 were obtained under such conditions.

2. The depletion phenomenon is useful in the analysis of bioorganic compounds. Because the ion A^+ with the higher sensitivity coefficient k_A can become depleted from the droplets, this ion will be enriched in the gas phase and may be transferred almost completely to the gas phase. Depletion from the droplets will be expected when the ions have high-sensitivity coefficients (i.e., low solvation energies and high surface activities). Organic and bioorganic ions, which are generally the analyte ions of interest, have higher sensitivity coefficients than alkali ions or ammonia see (Table 3). Therefore, depletion of such ions from the droplets can be expected and is analytically beneficial. This effect is particularly important when small amounts of the analyte are available. In that case, a minimum amount of solution containing the analyte is injected into the ES flow and the mass spectra obtained are integrated over the flow period. The presence of other ionic species such as Na^+ at higher concentrations in the solution will

lead to charging of the droplets and depletion of the analyte from the droplets and a maximum possible ion yield from the analyte. The analyte signal will be dependent on the mass of the analyte introduced into the ES solvent flow.

P. Droplet Electrospray (DES)—Significance of Results for Electrospray Mechanism

Droplet electrospray (DES) is a method for producing charged droplets and gas-phase ions for mass spectrometric analysis (63). Neutral droplets are first produced by applying a piezoelectric buzzer to the tip of a fused-silica capillary of ~ 15 μm inner diameter. The solution is pumped through the capillary at a known volume flow rate, V_f. In the absence of vibration, a liquid filament emerges from the capillary tip and the filament remains unbroken over a considerable distance. Application of the buzzer at frequencies that are typically $\nu = 100$ kHz leads to formation of droplets which are of highly uniform size (monodisperse). A desired radius can be selected by controlling the capillary internal diameter, the flow rate, and the frequency. Thus far, controlled droplet radii in the range 6–50 μm have been produced. The number of droplets per second corresponds to the frequency, and the radius of the droplets can be obtained from the known flow rate V_f and frequency ν:

$$\nu(4/3)\pi R^3 = V_f \tag{39}$$

The droplets are essentially uncharged (excepting the weak statistical charging) because the capillary and solution are at the same electric potential as the surroundings. The droplets are electrically charged downstream of the capillary tip by placing a 50-μm platinum-rod electrode near the droplet stream. When the electrode is at a positive potential (~ 3 kV), a small-corona discharge current (~ 0.3 μA) is present in the absence of the droplet stream. When the electrode is brought within charging distance (~ 20 μm) of the droplet stream, the droplets become positively charged by gas-phase positive ions from the electric discharge. The droplet stream can be observed through a microscope with stroboscopic illumination from a light-emitting diode whose frequency was the same as the buzzer frequency. A photograph of the droplet stream is shown in Figure 11 and a higher-magnification photograph of droplets near the electrode is shown in Figure 12.

When the DES was set up in front of a mass-spectrometer sampling orifice, the gas-phase ions produced by the DES could be observed. The mass spectra obtained for solutions containing analyte concentrations of 10^{-5} M or higher were essentially the same as those obtained using conventional ESMS.

The DES method and results (63) are of significance for the mechanism in conventional ESMS (i.e., the preceding topics discussed in this chapter). As noted in Section II.C, the charging of the droplets in conventional ES is due to an electrochemical reaction at the capillary. This reaction can introduce new ions in the solution, such as iron ions when the capillary is composed of stainless steel. In

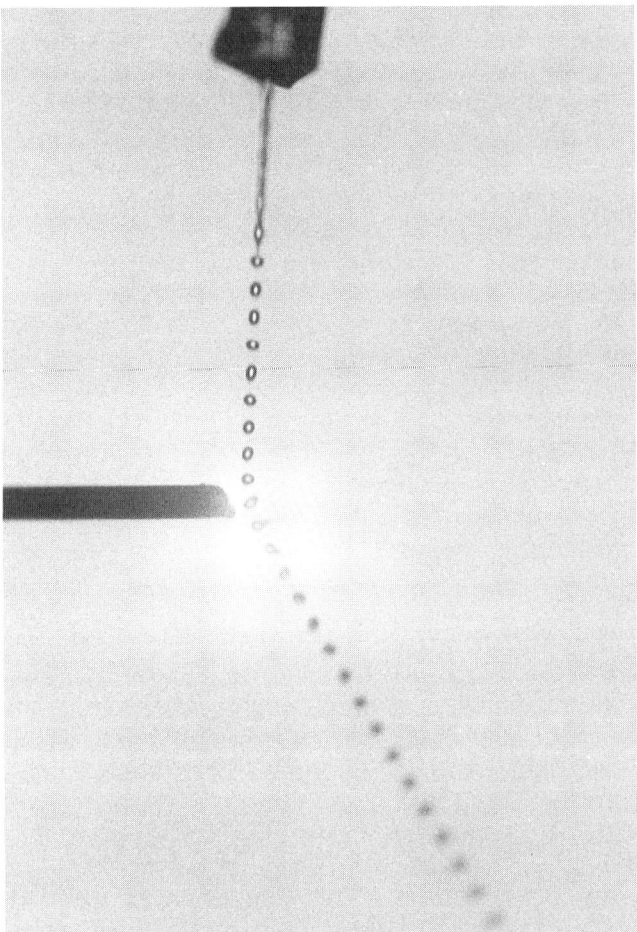

Figure 11. Droplet ES. Photograph of vibrated capillary tip showing separation of liquid jet into uncharged droplets. Droplet charging with gas-phase ions from electric discharge occurs near discharge electrode. Charged droplets produce offspring droplets by jet fission. Reprinted with permission from D. B. Hagers et al., *Anal. Chem.*, **1994**, *66*, 3944. © 1994 American Chemical Society.

DES, charging of the droplets is due to gas-phase ions produced by the electric gas discharge. These ions, NH_3^+, H_3O^+, and $CH_3OH_2^+$, are totally different from the excess ions in conventional ESMS, yet the mass spectra observed for the same analytes are essentially identical for the two modes of charging. The reason for this similarity of spectra is easily explained. The number of positive ions N required for charging the droplets is considerably smaller than the number n of electrolyte (charge-paired) positive ions present in the prepared solution. The expected ratio, n/N, can be evaluated approximately (see Eq. 1 in ref. 63) and is large for conditions where the electrolyte concentration is $M > 10^{-5}$ M. For

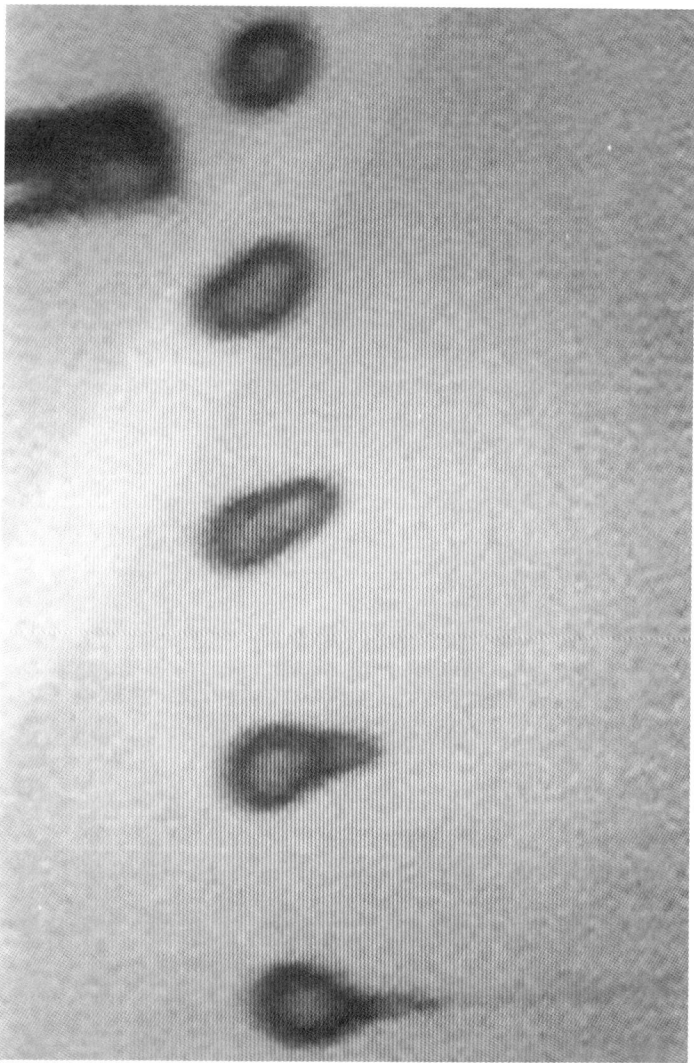

Figure 12. Close up view of droplet ES. Droplets charged at electrode are destabilized by charge and undergo jet fission. Droplet diameter = 44 μm. Vibration frequency = 93 kHz. Discharge current = 0.16 μA. Reprinted with permission from D. B. Hagers et al., *Anal. Chem.*, **1994**, *66*, 3944. © 1994 American Chemical Society.

DES, the ions created by the electric discharge will at first be present on the surface; however, ion diffusion will lead to statistical mixing with the positive electrolyte ions inside the droplet, which are far more numerous, so that the mass spectra ultimately observed will display ion intensities that are due mostly to the bulk electrolyte ions. The observed similarities between the spectra obtained with conventional ES and DES, in spite of the different modes of charging, are thus explained. The statistical mixing of surface ions and bulk ions by diffusion assumed in Section II.O, which is the basis of the "competition between bulk ions" concept proposed, is thus shown to be consistent with DES mass spectral observations, which also require a statistical mixing model.

The droplets shown in Figure 12 provide a good example of the sequence of events in the evolution of a Coulomb instability which culminates in a droplet-jet fission. The droplet fission is similar to that observed by Gomez and Tang (33) (see also Section II.H). These authors (33) photographed droplets from conventional ES using a shadowgraph technique with 25-ms flashes. The shadowgraphs were obtained with single flashes and showed droplets caught in the act of fission. The pictures of DES droplets, (Figs. 11 and 12), were obtained from shadowgraphs made using the stroboscopic technique in which a given frame represents the superposition of 3000 consecutive images and 3000 different droplets arriving in turn at the same places. The relatively sharp pictures of droplets illustrate the remarkable reproducibility of the DES process. The only images that are "smeared out" are the spray plumes, because the positions of the fine offspring droplets in the plume are not synchronized with the illumination. The shapes of the droplets observed in Figure 12 and in the work of Gomez and Tang (33) provide a good visualization of Coulomb jet fission.

Q. Mechanism for Formation of Macroions

In the previous sections we have considered only small ions. Much was learned about the formation of the charged droplets, the history of their evolution to very small and highly charged droplets that lead to gas-phase ions and the concentration changes of the solutes present in the droplets. An unequivocal decision as to whether the gas-phase ions were produced from the very small droplets by the ion evaporation or the charged-residue mechanism could not be attained. Many features of the mass spectrometric observations could be explained quite well by both mechanisms.

The status of the ion-evaporation and charged-residue theories for macroions, and particularly the analytically important large polyprotonated or polydeprotonated proteins and nucleic acids is different. The ion-evaporation theory was derived by Iribarne and Thomson (44,45) (see Section II.J) as a model for small ions. No extension of their equations for macroions is available and such an extension would require many simplifying assumptions.

A qualitative extension of the Iribarne theory which deals with multiply charged macroions has been proposed by John Fenn (64). Although an equation predicting the evaporation rate of a multiply charged large molecule was provided (see Eq. 3 in ref. 64), it is not clear whether numerical predictions can be obtained with it. In the absence of predictions which show that the rate of ion emission is of the right order of magnitude (10^6 s^{-1}), the significance of the model is difficult to establish.

Two special features must be taken into account when dealing with macroions such as proteins: (1) the large size of these entities and (2) the special nature of molecular composition of the protein *vis-à-vis* the solvent. Excepting locations that carry polar and charged residues, the protein is on the whole solvophobic in the typical polar solvents used. This solvophobicity may provide significant assistance for an escape of the protein from the droplet. When the multiply protonated protein is in the bulk of the droplet, nearby negative ions will lead to overall charge neutrality. As the protein approaches the surface, some of the counterions can be left behind such that part of the protein and its proton charges break through the droplet surface. Because the protein is solvophobic, a gradual extrusion of the charged protein can occur which will be assisted by the repulsion of the other charges on the solvent part of the surface. However, this process cannot be expected to be very fast and may not be able to compete with the rapid evaporation of solvent from the droplet, which may lead to Rayleigh fission or, if the droplet is small, to the formation of a charged residue.

The charged-residue theory does not require any significant modifications for macroions, and becomes the more natural mechanism for large macroions. A simple mechanistic picture is shown in Figure 13, depicting a droplet undergoing Rayleigh jet fission. Macromolecules with a radius $R \approx 3.5$ nm (e.g., hemoglobin), which happen to be present in the expanding cone jet, will end up in some of the microdroplets. The charged parent droplet, assumed to have a radius of $R \approx 80$ nm, leads to offspring droplets of $R \approx 8$ nm. The offspring droplet size corresponds to the size where ion evaporation of small ions becomes competitive with Rayleigh fission (44). Only several microseconds will be required for evaporation of a solvent like methanol–water to shrink the offspring droplet to the size of the macroion, $R \approx 3.5$ nm (see Eq. 18b). Small ions, which are also present as buffers or other electrolyte additives, will also evaporate if the ion evaporation theory holds. However, the macroion will probably be unable to "evaporate" out of the droplet within the same short time. In that case, the gas-phase macroion will be produced by the charged-residue mechanism. The likely validity of the charged-residue theory for large macroions has also been accepted by Fenn (64,65).

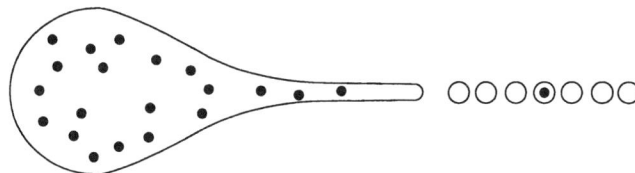

Figure 13. Droplet jet fission in the presence of macroions. Schematic representation of a droplet that has a radius of $R_D \approx 80$ nm, which leads to offspring droplets of $R \approx 8$ nm. Some of these contain one macroion each of radius $R \approx 3$ nm. Charges on all droplets are due mainly to small ion electrolytes. Only macroions are shown, as small black circles. Offspring droplets containing a macroion may lead to formation of gas-phase macroions by the charged-residue mechanism, even if the ion-evaporation mechanism holds for small ions.

R. Changes of Ion Charge and Structure on Transfer from Charged Droplets to the Gas Phase

The relationships between the intensities of detected gas-phase ions, and the corresponding concentrations of these ions in the electrosprayed solution, were presented in Sections II.N and II.O. These discussions showed that the identity of the ion was preserved in the transfer from solution to the gas phase, and that relatively simple relationships could be obtained, linking the observed gas phase ion intensities to the ion concentrations in the solution. These results may create the mistaken impression that in *all* cases there are simple relationships between the ions present in the solution and those observed in the gas phase. The ion species considered in Sections II.N and II.O were mostly relatively unreactive, singly charged ions, such as the alkali metal ions and the tetraalkylammonium ions. Furthermore, the solutions did not contain solutes that could react with these ions.

Much more complex relationships may prevail when reactive ions and solutes are present. As an example, we will consider the important class of protonated bases which yield positive ions, and deprotonated acids yielding negative ions. Such systems can engage in unexpected proton-transfer chemistry whose occurrence is associated with the unusual conditions prevailing in the transfer process from ions in the droplet to ions in the gas phase. This chemistry is essentially a consequence of the fact that solution and gas phase are two entirely different environments for ionic species.

Proton-transfer chemistry, as a factor that influences the charge-state distributions of polyprotonated peptides and proteins, is the subject of Chapter 4 by Wang and Cole. Therefore, a detailed examination of the effects on these species, of the change from solution to the gas phase environment, is not given here. However, we use one example to illustrate the effects that may occur.

The mass spectral results of Fenselau and co-workers (15) are particularly challenging when attempting to understand the nature of the process that produces the observed ions. These authors examined the positive and negative ions when a protein such as myoglobin was electrosprayed in a solution of 50 : 50 water–methanol. When the solution was made basic (pH = 10) by the addition of ammonia, they observed, in the negative-ion mode, an envelope of deprotonated negative myoglobin ions, with a charge-state distribution having a maximum of some 15 negative charges. This may be called the "expected," result because deprotonated negatively charged myoglobin ions are expected to be present in the basic solution. When the same solution was used to obtain spectra in the positive-ion mode, protonated myoglobin ions were observed with a charge distribution peaking at some 14 positive charges, even though at a pH = 10, no multiply protonated myoglobins are present in the bulk solution. On the contrary, most of the myoglobin is multiply deprotonated.

It is possible to give a simple answer to this "paradox." We will use the charged-residue model, which in our opinion affords a simple and instructive explanation. The proposed sequence of events is quite general, and can provide an explanation for both the "unexpected" and "expected" mass spectra (15). Of paramount importance are the reactions which occur during the transition from a very small droplet to a charged residue, that is, the reactions in the nascent charged residue. Three reactants are involved. The first two reactants, the multiply charged analyte ion (protein) and the corresponding counterions, are located in the disappearing interior solution. The third reactant, the ions that provide the net charge of the droplet, are located at the droplet surface. The reactions of interest occur as the solvent disappears, and as the liquid surface and surface ions come close to the protein and the counterions that surround it.

We consider here only the unexpected result, the observation of polyprotonated myoglobin ions in the positive-ion mode from the ES of a basic ammonia solution pH 10. When the droplets are formed at the capillary tip, the solution is still at pH $\approx$ 10 and the dominant ions in the solution are NH_4^+ and OH^- (we neglect the presence of methanol for simplicity). In the positive-ion mode, the charge on the droplet surface is due mainly to NH_4^+ ions. Most of the protein, which is present at a much lower concentration, is deprotonated and resides in the bulk of the solution with positive counterions (NH_4^+) nearby. As the droplets evaporate, they lose solvent and also ammonia. Because the Henry's Law constant for ammonia (see Eq. 34 and Table 8), $k_{St} \approx 40$, is larger than unity, a decrease of NH_3 concentration in the droplet is expected to occur. This has two effects: (1) the charges on the surface may partially shift from NH_4^+ to H_3O^+, assuming that the depletion of NH_3 is very severe and (2) the protein in the solution shifts to less-deprotonated states. On formation of the "final" droplet whose evaporation leads to the "charged residue," the surface ions (NH_4^+ or H_3O^+) come close to the protein. Because the solvent is disappearing, the positive and negative ions in the solution, that is, the negative groups on the protein and the NH_4^+ cations, react with one another such that neutralization results. Because of statistical fluctuations in the numbers of positive and negative charges, there are statistical fluctuations in the "neutralization," such that neutrality may reflect only the droplet-averaged condition. This is process (a) (see Eq. 40).

(a) Neutralization of bulk electrolyte ions:

$$\text{Protein} \begin{array}{c} \diagup CO_2^- \\ \diagdown NH_2 \end{array} + NH_4^+ \longrightarrow \text{Protein} \begin{array}{c} \diagup CO_2H \\ \diagdown NH_2 \end{array} + NH_3(\text{soln}) \quad (40a)$$

(b) Reaction with surface ions:

$$\text{Protein} \begin{array}{c} \diagup CO_2H \\ \diagdown NH_2 \end{array} + NH_4^+ \longrightarrow \text{Protein} \begin{array}{c} \diagup CO_2H \\ \diagdown NH_3^+ \end{array} + NH_3(g) \quad (40a)$$

Reaction b with the surface ions occurs almost simultaneously, and this protonation (Eq. 40b) leads to the protonated proteins observed in the mass spectrum. The degree of protonation is related to the number of protonating ions on the surface. Furthermore, only a small fraction of the NH_4^+ ions from the surface are expected to engage in the protonation. Coulombic repulsion by the NH_4^+ ions in the small droplet whose solvent is disappearing leads to direct escape of $NH_4^+(H_2O)_n$ clusters to the gas phase. Furthermore, as the protein becomes progressively charged owing to protonation, NH_4^+ ions from the surface become progressively repelled from the protein by coulombic forces and also end up as gas-phase NH_4^+ ions.

The protonation reaction (Eq. 40b) is expected to be exothermic in the gas phase because the basicity of the amino group attached to a carbon chain is higher than the gas-phase basicity of NH_3 (63). In solution, particularly when the concentration of NH_3 is high, the reverse reaction to Eq. 40b is expected to occur, to establish the solution equilibrium. The reverse reaction is driven by the superior solvation of NH_4^+ relative to the solvation of the protonated organic base. The occurrence of the protonation reaction (Eq. 40b), is thus a response of the system to the new environment—the gas phase.

The neutralization reaction (Eq. 40a) is of course also a response to the change from solution to the gas phase. In solution, both the positive and negative ions are stabilized by solvation, but in the absence of solvation the highly exothermic neutralization occurs.

Fenselau and co-workers (15) reported that the intensity of the protonated myoglobin observed with the basic solution (pH 10) was only approximately one tenth of the intensity observed with an acidic pH. The low intensity at the basic pH is a consequence of the direct escape of the NH_4^+ ions from the surface to the gas phase. Experiments performed in our laboratory with polyprotonated peptides in basic (ammonia) solutions have shown that much higher intensities of NH_4^+ ions relative to protonated peptide ions are observed under these conditions (66).

The higher protonated myoglobin intensities observed by Fenselau and co-workers (15) when the solution was acidic should occur because under these conditions protonated protein will be part of the droplet surface and, therefore, there will be less need for protonation by surface ions, H_3O^+, in the incipient charged residue.

It should be mentioned that Eqs. 40a and b can also be invoked to explain the observations of Fenselau and co-workers (15) when the protein escapes from the droplet by "ion evaporation." In that case, these reactions occur at the surface of the droplet and are again driven by the change from solution to the gas phase.

Other changes can also occur in response to the new environment—the gas phase. If there is some freedom of motion for the protonated basic functionality on the protein (or peptide), this group will interact with some neighboring unprotonated basic site, such as a carbonyl oxygen, and form a hydrogen-bonded cyclic structure. The formation of such cyclic structures in which the proton is dicoordinated by bonding to two basic groups was demonstrated many years ago (68). Their presence in gas-phase protonated peptides has been demonstrated more recently (69,70). Williams and co-workers (71) have modeled the conformational changes in polyprotonated peptides due to the formation of such intramolecular hydrogen bonds.

Protonated basic groups which are not stabilized by intramolecular hydrogen bonding will be solvated by one or more solvent molecules. The solvent molecules may have been retained in the transition from droplet to gas phase or acquired later from the solvent vapor present in the atmospheric pressure region of the ES ion source. It is customary to remove the solvent molecules by collision-induced dissociation (CID) in the interface region leading to the mass analyzer. This forced desolvation by CID can induce proton loss to the solvent molecules. Chait and co-workers (72) have pointed out that the degree of deprotonation of the analyte is dependent on the nature of the solvent molecules. Solvents whose molecules have high proton affinities will lead to greater deprotonation of the analyte. Thus, the charge-state distribution of the analyte is ultimately determined by the gas-phase conditions and the treatment of the analyte ions prior to mass analysis.

Gas-phase-induced changes of the charge state occur also for multiply charged ions in which the multiple charges are localized on one atom or one small group of atoms, such as multiply charged metal atoms M^{z+} or multiply charged negative ions such as SO_4^{2-} or PO_4^{3-}. Because of the high coulombic repulsions present, when the multiple charges are confined to such a small volume, some of the multiply charged ions whose existence in solution is well known cannot be observed in the gas phase. Considering the series of alkaline earths Be^{2+}, Mg^{2+}, Ca^{2+}, Sr^{2+}, and Ba^{2+}, the smallest ion, Be^{2+}, could not be observed in the gas phase (69). In general, higher-charged states can be observed when the charge is stabilized by a few or several solvent molecules (such as H_2O); Be^{2+} could not be observed even as a hydrated species. The naked Mg^{2+} ion could not be observed, but the $Mg^{2+}(H_2O)_n$ species could be observed for $n \leq 3$. Both Ca^{2+} and Sr^{2+} could be observed for $n \leq 2$, and only the largest ion Ba^{2+} could be obtained at high intensities in the naked-ion form (73).

Attempts to remove the water molecules by CID from the doubly charged alkaline earth ions, other than Ba^{2+}, led to a decrease of charge state from 2 to 1 by the reaction

$$M^{2+}(H_2O)_n \xrightarrow{CID} MOH^+(H_2O)_x + H_3O^+(H_2O)_{(n-2-x)} \quad (41)$$

at values of n indicated above. This reaction is driven not only by the Coulomb repulsion, but also by the formation of the strong bond between the metal M^+ and OH. A discussion of the thermochemistry of the process was provided (73).

Similar observations were made for several other transition metals (Mg^+, Co^{2+}, Cu^{2+}, Mn^{2+}) studied. The change of charge state was dependent also on the nature of the solvent molecules used. For example, the Cu^{2+} ion could not be observed either naked or solvated with protic solvents like water, while the use of dipolar aprotic solvents like dimethyl sulfoxide led to its observation. Collision-induced dissociation of this solvate led to the charge-state reduction reaction

$$Cu^{2+}(DMSO)_n \xrightarrow{CID} Cu^+(DMSO)_{n-1} + DMSO^+ \quad \text{at } n \approx 3 \quad (42)$$

While Eq. 41 is an intracluster proton-transfer reaction, Eq. 42 represents an intracluster charge transfer. Since the charge-transfer reaction does not involve bond formation to compensate for the bond breaking, as is the case in the proton-transfer process, it is more endothermic. Therefore, the use of the solvent DMSO does lead to a "protection" of the higher-charged state.

Triply charged transition metal ions could not be observed with any ordinary solvents. However, addition of polydentate complexing agents led to success. Thus, Co^{3+} could be observed when complexed to the hexadentate sepulchrate ligand (74). The much larger lanthanide triply charged ions, Y^{3+}, La^{3+}, Ce^{3+}, Nd^{3+}, and Sm^{3+}, could be observed in the $+3$ charged state with solvents like DMSO and DMF. Protic solvents like water led to the charge-reduced states $MOH^{2+}(H_2O)_x$ (70).

The situation is quite similar for small, multiply charged, negative ions. The naked SO_4^{2-} ion cannot be observed. It is probably unstable in the gas phase due to electron detachment (75):

$$SO_4^{2-} = SO_4^- + e \quad (43)$$

The hydrates, $SO_4^{2-}(H_2O)_n$, can be observed (75). Removal of the H_2O molecules by CID leads to H_2O loss down to $n \geq 3$. Attempts to obtain the lower hydrates lead to a reduction of the charge state:

$$SO_4^{2-}(H_2O)_3 \xrightarrow{CID} HSO_3^- + OH^-(H_2O)_2 \quad (44)$$

This reaction involves an intracluster proton transfer. Larger polysulfates, like the dithionate ion $S_2O_6^{2-}$ and peroxy disulfate, $S_2O_8^{2-}$, could be observed in the solvated and in the naked forms (75,76). So far, we have not been able to observe the phosphate ion $(HO)PO_3^{2-}$ even in its hydrated form, while doubly charged hydrated and naked polyphosphates could easily be produced (76).

The changes of ion charge and ion structure, which occur on transfer of the ions from the charged droplets to the gas phase, described above, are examples of the general sequence of relief of coulombic stress as solvent is removed. This is the last step in the overall coulombic stress relief occurring in ES. As solvent is removed from the charged droplets, coulombic stress is relieved by Rayleigh fission and possibly ion evaporation. Once the solvated gas phase ions are formed, increased coulombic stress due to removal of solvent is again relieved by emission of charges, but this time, the processes are regarded as reductions of the charge state of the ions and as chemical reactions.

III. CONCLUSIONS

Most of the processes involved in the transfer of ions from solution to the gas phase are quite well understood and familiarity with the relationships and parameters involved can be very useful to the practicing mass spectrometrist.

The exact process by which the small, highly charged droplets produce the gas-phase ions is not known. Because these droplets are so small, $R < 10$ nm, they evaporate rapidly—within a few microseconds. Thus far, no experimental method exists with which one could observe such droplets and determine whether Rayleigh fission continues, as expected on the basis of the Dole charged-residue theory, or direct emission of gas-phase ions occurs, as predicted by the Iribarne–Thomson theory.

Mass spectrometric determinations of the gas-phase ions, observed in experiments aimed at distinguishing between the predictions of the two theories, have failed to provide a clear-cut answer. This failure is not surprising. On closer examination, both theories predict qualitatively similar results. In the final stages (i.e., droplets with $R < 1$ nm), a clear distinction between Rayleigh fission and ion evaporation cannot be made.

Although a transfer of ions from solution to the gas phase is achieved with ES, the exact identity of the ion may not always be preserved. Changes in the ion, such as protonation or complexation to a neutral molecule or counterion can occur at the liquid–gas interface, that is, in the incipient gas phase. These changes occur because of the different requirements of the two media.

REFERENCES

1. (a) Yamashita, M; Fenn, J. B. *J. Phys. Chem.* **1984**, *88*, 4451; (b) Yamashita, M.; Fenn, J. B. *J. Phys. Chem.* **1984**, *88*, 4671.
2. Aleksandrov, M. L.; Gall, L. N.; Krasnov, V. N.; Nikolaev, V. I.; Pavlenko, V. A.; Shkurov, V. A. *Dokl. Akad. Nauk SSSR* **1984**, *277*, 379.
3. Dole, M.; Mack, L. L.; Hines, R. L.; Mobley, R. C.; Ferguson, L. D.; Alice, M. B. *J. Chem. Phys.* **1968**, *49*, 2240.
4. Desnoyers, J. E.; Joliceur, C. In *Modern Aspects of Electrochemistry*; Bockris, J. O. M.; Conway, B. E. Eds.; Plenum: New York, 1969; Vol. 5, p. 20.
5. Smith, R. D.; Loo, J. A.; Edmonds, C. G.; Baringa, C. J.; Udseth, H.R. *Anal. Chem.* **1990**, *62*, 882.
6. Guevremont, R.; Siu, K. W. M.; LeBlanc, J. C. Y.; Berman, S. S. *J. Am. Soc. Mass. Spectrom.* **1992**, *3*, 216.
7. Loo, J. A.; Ogozalek, R. R.; Light, J. K.; Edmonds, C. G.; Smith, R. D. *Anal. Chem.* **1992**, *64*, 81.
8. Loo, J. A.; Edmonds, C. G.; Udseth, H. R.; Smith, R. D. *Anal. Chem.* **1990**, *62*, 693.
9. Chowdhury, S. K.; Katta, V.; Chait, B. T. *J. Am. Chem. Soc.* **1991**, *112*, 9012.
10. Katta, V.; Chait, B. T. *Rapid Commun. Mass Spectrom.* **1991**, *4*, 214.

REFERENCES

11. Loo, J. A.; Udseth, H. R.; Smith, R. D. *Biomed. Environ. Mass Spectrom.* **1988**, *17*, 411.
12. Loo, J. A.; Ogorzalek, R. R.; Udseth, H. R.; Edmonds, C. G.; Smith, R. D. *Rapid Commun. Mass Spectrom.* **1991**, *5*, 101.
13. LeBlanc, J. C. Y.; Beauchemin, D.; Siu, K. W. M.; Guevremont, R.; Berman, S. S. *Org. Mass Spectrom.* **1991**, *26*, 831.
14. Ganem, B.; Li, Y.-T.; Henion, J. D. *J. Am. Chem. Soc.* **1991**, *113*, 7818.
15. Kelly, M. A.; Vestling, M. M.; Fenselau, C. C.; Smith, P. B. *Org. Mass Spectrom.* **1992**, *27*, 1143.
16. Bailey, A. G. *Electrostatic Spraying of Liquids*; Wiley: New York, 1988.
17. *Electrosprays: Theory and Applications*, Special issue *J. Aerosol Sci.* **1994**, *25*.
18. Loeb, L. B.; Kip, A. F.; Hudson, G. G.; Bennett, W. H. *Phys. Rev.* **1941**, *60*, 714.
19. Pfeifer, R. J.; Hendricks, C. D. *AIAA J.* **1968**, *6*, 496.
20. Taylor, G. I., *Proc. R Soc. London A* **1964**, *A280*, 383.
21. (a) Cloupeau, M.; Prunet-Foch, B. *J. Aerosol Sci.* **1994**, *25*, 1021. (b) Cloupeau, M. *ibid.*, p. 1143. (c) Horming, D. W.; Hendricks, C. D. *J. Appl. Phys.* **1979**, *50*, 2614.
22. Smith, D. P. H.; *IEEE Trans. Ind. Appl.* **1986**, *IA-22*, 527.
23. Hayati, I.; Bailey, A. I.; Tadros, T. F. *J. Colloid Interface Sci.* **1987**, *117*, 205.
24. Fernandez de la Mora, J.; Locertales, I. G. *J. Fluid Mech.* **1994**, *243*, 561.
25. Blades, A. T.; Ikonomou, M. G.; Kebarle, P. *Anal. Chem.* **1991**, *63*, 2109.
26. (a) Van Berkel, G. J.; McLuckey, S. A.; Glish, G. L. *Anal. Chem.* **1993**, *64*, 1586. (b) Van Berkel, G. J.; Zhou, F. *Anal. Chem.* **1995**, *67*, 2916.
27. Ikonomou, M. G.; Blades, A. T.; Kebarle, P. *J. Am. Soc. Mass Spectrom.* **1991**, *2*, 497.
28. Wampler, F. W.; Blades, A. T.; Kebarle P. *J. Am. Soc. Mass Spectrom.* **1993**, *4*, 289.
29. Juhasz, P.; Ikonomou, M. G.; Blades, A. T.; Kebarle, P. In *Methods and Mechanisms for Producing Ions from Large Molecules*; Standing, K. G.; Ens, W. Eds.; Plenum: New York, 1991.
30. Ikonomou, M. G.; Blades, A. T.; Kebarle, P. *Anal. Chem.* **1991**, *63*, 1989.
31. Tang, L.; Kebarle, P. *Anal. Chem.* **1991**, *63*, 2709.
32. (a) Taflin, D. C.; Ward, T. L.; Davis, E. J. *Langmuir* **1989**, *5*, 376. (b) Davis, E. J. *ISA Trans.* **1987**, *26*, 1.
33. Gomez, A.; Tang, K. *Phys. Fluids* **1994**, *6*, 404.
34. Rayleigh, Lord *Philos. Mag.* **1882**, *14*, 184.
35. Rayleigh, Lord *The Theory of Sound*; Dover: New York, 1945.
36. Doyle, A.; Moltett, D.R.; Vonnegut, B. *J. Colloid Sci.* **1964**, *19*, 136.
37. Kebarle, P.; Tang, L. *Anal. Chem.* **1993**, *65*, 972A.
38. Rossel, J.; Leiompart, I. G.; Fernandez de la Mora, J. *J. Aerosol Sci.* **1994**, *25*, 1093.
39. Fernandez de la Mora, J. *J. Colloid Interface Sci.* **1995**, (submitted).
40. Tang, L.; Kebarle, P. *Anal. Chem.* **1993**, *65*, 3654.
41. Davis, C. N. In *Fundamentals of Aerosol Science*; Saw, D.T. Ed.; Wiley: New York, 1978, p. 154.

42. Willoughby, R. C.; Sheenan, E. W. Studies of the physical processes in electrospray. Proceedings of the Fourth Aerosol Conference, Los Angeles, California, 1994, p. 46.
43. (a) Röllgen, F. W.; Bramer-Wegner, E.; Buttering, L. *J. Phys. Colloq.* **1984**, *45*, Supplement 12, C9-297. (b) Schmelzeisen-Redeker, G.; Buttering, L.; Röllgen, F.W. *Int. J. Mass Spectrom. Ion Processes* **1989**, *90*, 139.
44. Iribarne, J. V.; Thomson, B. A. *J. Chem. Phys.* **1976**, *64*, 2287.
45. Thomson, B. A.; Iribarne, J. V. *J. Chem. Phys.* **1979**, *71*, 4451.
46. Chapman, S. *Phys. Rev.* **1937**, *52*, 184; **1938**, *54*, 520; **1938**, *54*, 528.
47. Fenn, J. B.; Mann, M.; Meng, C. K.; Wong, S. F.; Whitehouse, C. M. *Science*, **1989**, *246*, 64.
48. Sakairi, M.; Yergey, A. L.; Siu, K. W. M.; LeBlanc, J. C. Y.; Guevremont, R.; Berman, R. S. *Anal. Sci.* **1991**, *7*, 199.
49. Rafaelli, A.; Bruins, A. P. *Rapid Commun. Mass Spectrom.* **1991**, *5*, 269.
50. Hiraoka, K. *Rapid Commun. Mass Spectrom.* **1992**, *6*, 463.
51. (a) Kebarle, P. *Annu. Rev. Phys. Chem.* **1977**, *28*, 445. (b) Keesee, R.G.; Castleman, A.W. *J. Phys. Chem. Ref. Data* **1986**, *15*, 1011.
52. Tamaki, K. *Bull. Chem. Soc. Jpn.* **1967**, *40*, 38.
53. Smith, R. D.; Wahl, J. H.; Hofstadtler, J. A. *Anal. Chem.* **1993**, *65*, 574A.
54. Loscertales, I. G.; Fernandez de la Mora, J. *J. Chem. Phys.* **1995**, *103*, 5041.
55. Pound, G. M. *J. Phys. Chem. Ref. Data*, **1972**, *1*, 135.
56. Wilm, M. S.; Mann, M. *Int. J. Mass Spectrom Ion Processes*, **1994**, *136*, 167.
57. Wilm, M. S.; Mann, M. *Anal. Chem.*, **1996**, *68*, 1.
58. Wang, G.; Cole, R. B.; *Anal. Chem.* **1994**, *66*, 3702.
59. Anaceleto, J. F.; Pleasance, S.; Boyd, R. K. *Organic Mass Spectrom.* **1992**, *27*, 660.
60. Ho, Y.; Kebarle, P. "Significance of observed clusters $M(MX)_n^+$ and $X(MX)_n^-$ to the mechanism of electrospray," to be submitted.
61. Clegg, S. L.; Brimblecombe, P.; In *Chemical Modeling of Aqueous Systems II*; Melchior, D. C.; Basset, R. L. Eds.; American Chemical Society: Washington, DC, 1990; ACS Symposium Series 416.
62. Hammes, G. D. *Principles of Chemical Kinetics*; Academic: New York, 1978; Chapter 9.
63. Hagers, D. B.; Dovichi, N. J.; Klassen, J.; Kebarle, P. *Anal. Chem.* **1994**, *66*, 3944.
64. Fenn, J. B. *J. Am. Soc. Mass Spectrom.* **1993**, *4*, 524.
65. Fenn, J. B. "How Much Pull Does an Ion Need to Escape its Droplet Prison?" Talk presented at conference on: "The Physics and Chemistry of Electrospray and Multiply Charged Ions," Sanibel Island, January 1995.
66. Ho, Y.; Kebarle P. unpublished work.
67. Lias, S. G.; Liebman, J. F.; Levin, R. D. *J. Phys. Chem. Ref. Data* **1984**, *13*, 695.
68. Yamdagni, R.; Kebarle, P. *J. Am. Chem. Soc.* **1973**, *95*, 3504; Meot-Ner (Mautner) M. *J. Am. Chem. Soc.* **1984**, *106*, 1265.
69. Zang, K.; Zimmerman, D. M.; Chung-Phillips, A.; Cassady, C. J. *J. Am. Chem. Soc.* **1993**, *115*, 10812.
70. Klassen, J. S.; Blades, A. T.; Kebarle, P. *J. Phys. Chem.* **1995**, *99*, 15509.

71. (a) Schnier, P. D.; Gross, D. S.; Williams, E. R. *J. Am. Chem. Soc.* **1995**, *117*, 6747; (b) Williams, E. R.; Gross, D. S.; Schnier, P. S.; Rodrigez-Cruz, S.; Fagerquist, C. K. *Electrostatic Forces in Multiply Protonated Gas Phase Ions.* Presented at the 43rd ASMS Conference on Mass Spectrometry and Allied Topics, Atlanta, Georgia, 1995.
72. Chait, B. T.; Chowdhury, S.; Katta, V. *On the Ionization of Peptides and Proteins by Electrospray.* Presented at the 39th ASMS Conference on Mass Spectrometry and Allied Topics, Nashville, Tennessee, 1991.
73. Blades, A. T.; Jayawera, P.; Ikonomou, M. G.; Kebarle, P. *J. Chem. Phys.* **1990**, *92*, 5900.
74. Blades, A. T.; Ikonomou, M. G.; Kebarle, P. *Int. J. Mass Spectrom. Ion Proc.* **1990**, *101*, 325; **1990**, *102*, 251.
75. Blades, A. T.; Kebarle, P. *J. Am. Chem. Soc.* **1994**, *116*, 10761.
76. Blades, A. T.; Klassen, J. S.; Kebarle, P. *J. Am. Chem. Soc.* **1995**, *117*, 10563.

CHAPTER 2

The Electrolytic Nature of Electrospray

GARY J. VAN BERKEL

Chemical and Analytical Sciences Division, Oak Ridge National Laboratory, Oak Ridge, Tennessee

	Abstract	65
I.	Introduction	66
II.	The electrolytic nature of electrospray	67
	A. An ES ion source as "an electrolytic cell of a somewhat special kind"	70
	B. Characterization of an ES ion source as a controlled-current electrolytic cell	77
III.	Analytical implications of the electrolytic nature of electrospray	86
	A. Changes in solution composition	87
	B. Chemical noise/modifications of analyte	87
	C. Electrochemical ionization	88
	1. Efficient ionization/low-level gas-phase detection	89
	2. Caveats	94
	D. Study of electrode reaction products	99
	Acknowledgments	103
	References	103

ABSTRACT

While it has been known for some time that redox reactions in the metal spray capillary of electrostatic sprayers are responsible for charge balance in these continuous-current devices, only recently has their electrolytic operation been examined thoroughly on the basis of fundamental electrochemical principles. In this review, developments in the understanding of the electrolytic nature of electrostatic sprayers, specifically the atmospheric-pressure electrospray ion source, and the analytical implications of this "built-in" electrolytic process with respect to electrospray mass spectrometry (ESMS), are examined.

Electrospray Ionization Mass Spectrometry, Edited by Richard B. Cole.
ISBN 0-471-14564-5 © 1997 John Wiley & Sons, Inc.

I. INTRODUCTION

Dispersion of a liquid into small charged droplets by an electrostatic field is a phenomenon whose observation dates back at least two centuries (1). However, the first detailed experimental studies of this phenomenon, using an electrostatic sprayer similar to that used in today's electrospray mass spectrometry (ESMS) experiments, were carried out by Zeleny in the early 1900s (2). From the time of Zeleny's work until the present, an enormous amount of research describing fundamental and applied studies of electrostatic spraying has appeared in the scientific literature (see refs. 1, 3, and 4 for recent reviews). The late 1960s and early 1970s work of Dole and co-workers (5–8) is usually credited as the first (but unsuccessful) attempt at using an atmospheric pressure electrostatic sprayer (i.e., an ES ion source) as a means to produce gas-phase ions from macromolecules in a liquid solution for analysis by mass spectrometry. The first successful demonstration that macroions could be liberated from electrically charged droplets and detected using mass spectrometry may more appropriately be attributed to the work of Iribarne and co-workers in the late 1970s (9,10). In any case, the successful, concurrent, and independent coupling of an ES ion source and a mass spectrometer by Fenn and co-workers (11–13) and Alexandrov and co-workers (14) in the early 1980s truly gave birth to the ESMS "revolution." Within a few years of Fenn's and Alexandrov's first ESMS publications, the groups of Smith (15–17) and Henion (18–21) were leading the charge by demonstrating the applicability of ESMS for the analysis of all types of nonvolatile, polar, or thermally labile compounds. A first major focus was demonstrating the applicability of ES as a liquid chromatography (LC) and capillary electrophoresis (CE) interface for MS, and later using ESMS to carry out the mass analysis and structure determination of high-molecular-weight biopolymers (22,23). Today the applications of ESMS are too numerous to catalogue easily, ranging from LCMS or CEMS applications and simple molecular-weight and structure determinations to complex studies of the solution chemistries and gas-phase structures of biopolymers. The interested reader is directed to several of the latter chapters in this book, which represent some of the most recent ESMS application overviews.

Occurring in parallel with the development of ESMS as an analytical tool (and certainly contributing to its rapid growth in utility and popularity) were numerous fundamental studies aimed at elucidating the individual steps in the ES process responsible for the liberation of gas-phase ions from the analytes in solution. A complete understanding of the relationship among (1) the various instrumental components and operational parameters of the ES device, (2) the physical and chemical nature of the solvents and the analytes in solution and charged droplets, (3) the gas-phase ions generated from the charged ES droplets, and (4) the ions ultimately observed by the mass spectrometer would surely lead to new and better analytical and fundamental applications of ESMS. Many of the more important investigations of these topics are discussed by Kebarle in Chapter 1 of this book and elsewhere (24).

Today there is generalized agreement that the ES process involves three main steps prior to mass analysis: the generation and charging of the ES droplets; droplet evaporation and the production of gas-phase ions; and secondary processes that modify the gas-phase ions in the atmosphere and the sub-atmospheric pressure-sampling regions of the mass spectrometer. The details of these individual steps and their associated analytical implications are a continued focus of study and debate. Playing an important role in the generation and charging of the ES droplets are electrolytic reactions that occur at the metal/solution interface within the metal ES capillary to maintain charge balance in this continuous-current device. In this chapter, the evolution of our understanding of the fundamental electrolytic characteristics of the ES ion source is overviewed. The analytical implications of this inherent electrolysis process for ESMS are discussed, with emphasis on utilizing the process to expand the scope of ESMS studies.

II. THE ELECTROLYTIC NATURE OF ELECTROSPRAY

Figure 1 is a schematic diagram of a typical ES ion source and the nominal electric circuit involved. The ion source is comprised of two electrodes, namely, the metal ES capillary (usually stainless steel) and the atmospheric sampling aperture plate of the mass spectrometer (also stainless steel) that are connected together (and ultimately to ground) via a high voltage supply that can output about $\pm 3-5$ kV. Under typical ESMS operating conditions, a solution containing the analyte of interest, which is normally ionic, is pumped through the ES capillary (i.e., the working electrode), held at high voltage, and sprayed toward the aperture plate (i.e., the counterelectrode). Addition of any ionic species (i.e., electrolytes) to the solution, other than the analyte (or small amounts of acids or bases to ionize the analyte), is usually avoided when possible, because their presence in solution tends to suppress the formation of gas-phase ions from the analytes of interest (25,26). However, some number of ions, either the analytes, contaminants, or deliberately added electrolytes, must be present in the solution or the ES device will not form charged droplets (27). This is because electrophoretic charge separation of these ions in solution is responsible for both formation and charging of the ES droplets. Under the influence of the applied electric field, ions of the same polarity as the voltage applied to the ES capillary migrate from the bulk liquid toward the liquid at the capillary tip, while ions of the opposite polarity migrate in the opposite direction back into the capillary. When the buildup of an excess of ions of one polarity at the surface of the liquid reaches the point that coulombic forces are sufficient to overcome the surface tension of the liquid, droplets enriched in one ion polarity are emitted from the capillary. This results in a continuous steady-state current at the counterelectrode of the same polarity as the voltage applied to the capillary (28–31). For this continuous loss of one ion polarity in the charged droplets to be maintained, the buildup of opposite charge in the capillary must be neutralized.

If this did not occur, a field in solution counter to the externally applied field would be created, thereby negating the force for charge separation, which would cause the formation of charged droplets to stop. As we now understand it, this charge-balancing process involves electrochemical oxidation/reduction of the components of the metal ES capillary and/or one or more of the species in the solution, ultimately leading to electron flow to or from the high-voltage supply, depending on the polarity applied to the capillary (30,32–38). Specifically, an oxidation reaction occurs in the ES capillary in positive-ion mode (electron flow to the voltage supply), while a reduction reaction occurs in negative-ion mode (electron flow to the capillary). The current that can be measured at the working electrode due to these reactions (i.e., the faradaic current) is equal in magnitude to the current measured at the counterelectrode (i.e., the ES current), but of the opposite polarity. Reduction/oxidation of some species at the counterelectrode (i.e., the front aperture plate of the mass spectrometer) occurs to complete the electrical circuit (and accounts for the current we measure at this electrode).

The knowledge that the fundamental operation of an electrostatic sprayer involves an electrolytic process predates their use as an ion source in ESMS. The circuit diagrams for electrostatic sprayers found in the literature from the early 1900s are drawn much the same as that shown in Figure 1. This circuit implies the occurrence of electrolytic reactions at both electrodes to complete the circuit to maintain charge balance. Therefore, one might consider these electrolytic reactions an obvious consequence of Kirchhoff's current law. Yet no mention of the details of this electrolytic process nor any implications (other than possibly charge balance) for the particular application of the spray device are easily found in the literature before the 1970s. However, in 1974, Evans and co-workers (32) discussed the electrolytic process in connection with their incorporation of an electrostatic sprayer as an organic MS ion source in what is known as electrohydrodynamic mass spectrometry (EHMS). This technique is closely related to ESMS in that it uses an electrostatic sprayer to transfer ionic species present in solution into the gas phase for mass analysis. In ESMS, the gas phase ions are formed from analytes in charged droplets at atmospheric pressure, whereas with EHMS, ions are extracted directly from solution into vacuum. The circuitry of both sprayers is analogous and, therefore, the characteristics of the charge-balancing processes that take place in the respective spray capillaries should be similar. In their initial organic EHMS studies, Evans and co-workers reported

---

Figure 1. Schematic representation of the processes that are assumed to occur in ES in a positive-ion mode. The high electric field imposed by the voltage supply causes an enrichment of positive electrolyte ions at the meniscus of the solution at the metal-capillary tip. This net charge is pulled downfield, expanding the meniscus into a cone that emits a fine mist of positively charged droplets. Solvent evaporation reduces the volume of the droplets at constant charge, leading to fission of the droplets. Charge balance is attained in the ES device by electrochemical oxidation at the positive electrode and reduction at the negative electrode. (Reproduced with permission from A. T. Blades et al., *Anal. Chem.*, **1991**, *63*, 2110. © 1991 American Chemical Society.)

II. THE ELECTROLYTIC NATURE OF ELECTROSPRAY   69

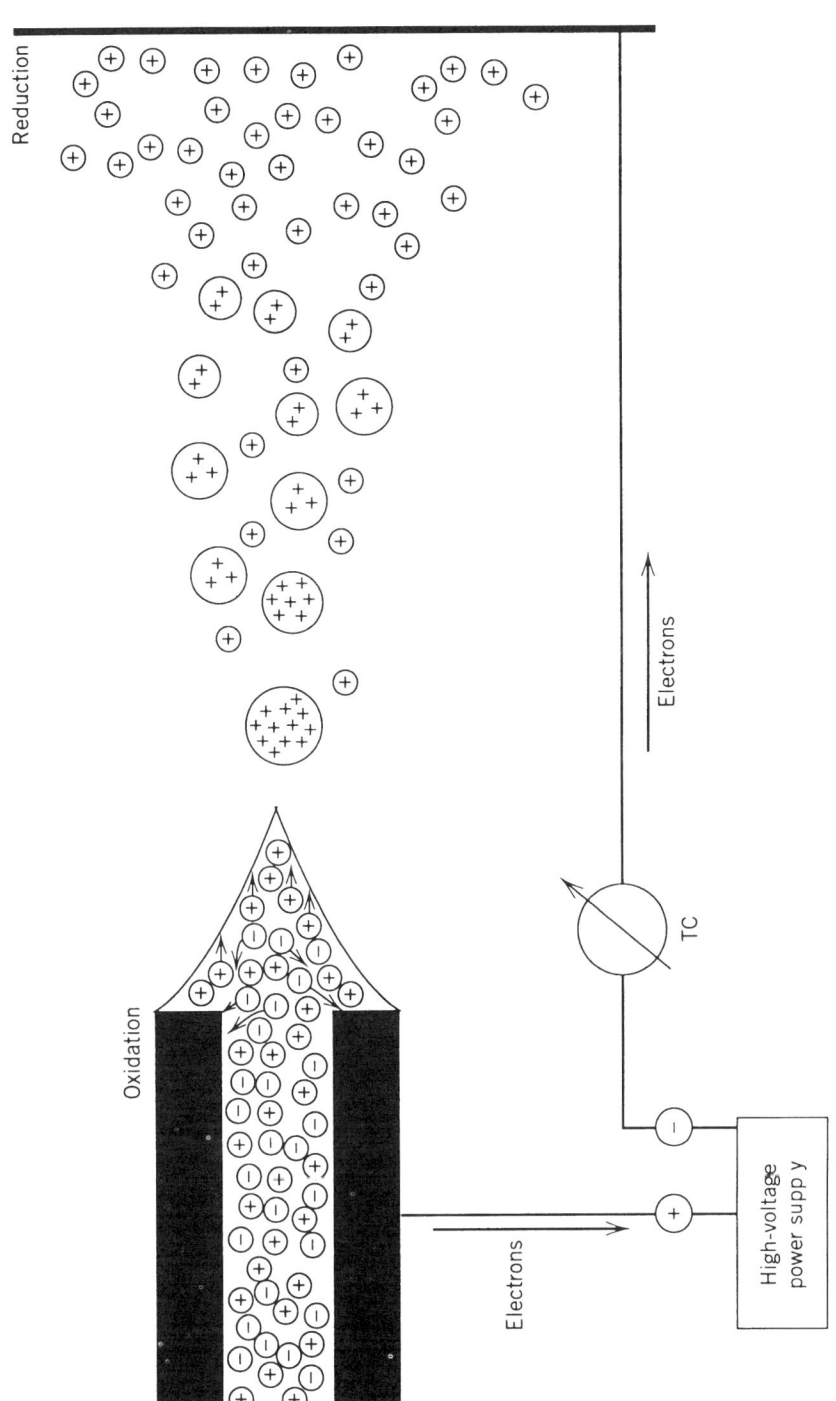

that their stainless-steel spray capillary underwent electrochemical corrosion, producing Fe^{2+} ions in solution, as proven by the observation of a series of doubly charged ions in their mass spectra that could be attributed to Fe^{2+} attachment to glycerol (G_5–G_8) and Fe^{2+} attachment to sucrose plus gycerol (Suc + G_3, Suc + G_4, and Suc_2 + G_3). They later reported that this corrosion and appearance of solvated Fe^{2+} ions could be avoided by using a pure platinum capillary in the EH ion source (33). In his review of EHMS in 1984, Cook (34) discussed further the electrolytic effect in EH and speculated regarding redox reactions of solution species (e.g., Cl^- oxidation to form Cl_2 gas) that might cause the spray instability sometimes observed in EHMS. In discussing ESMS, he also noted that "overall electroneutrality is presumably retained [in the ES ion source] by electrochemistry at the emitter surface, as in EHMS".

Although the literature shows that the electrolytic nature of electrostatic sprayers was no mystery prior to the development of ESMS, the work of Kebarle and co-workers in 1991 (35) first brought the electrolytic nature of the ES ion source to major attention in the mass spectrometry community. Like the earlier EHMS work of Evans and co-workers (32,33), the work of Kebarle's group alerted the users of ESMS that the products of the charge-balancing redox reactions formed in the metal ES capillary might be detected in the gas phase. Their work also represented the first attempt to characterize the electrolytic nature of the ES ion source in formal electrochemical terms.

A. An ES Ion Source as "An Electrolytic Cell of a Somewhat Special Kind" (35)

On the basis of the electric circuit shown for the ES ion source in Figure 1, the electrophoretic charge-separation mechanism for droplet charging and formation, and charge balance considerations, Kebarle and co-workers (35) stated succinctly the reason redox reactions must occur in the metal-capillary of the ES ion source:

> Considering the requirements for charge balance in such a continuous electric current device and the fact that only electrons can flow through the metal wire supplying the electric potential to the electrodes, one comes to the conclusion that the electrophoretic charge separation mechanism [of droplet charging and formation] requires that the [positive ion] electrospray process should involve an electrochemical conversion of ions to electrons [within the metal ES capillary]. (35)

This meant that the ES ion source could be " viewed as an electrolytic cell of a somewhat special kind ... insofar as part of the ion transport [between electrodes] does not occur through solution [as in a conventional electrolytic cell] but through the gas phase." (35) It was surmised that the redox reactions with the lowest redox potentials would predominate in this charge-balancing process, and the actual reactions occurring would depend on the particular solvent and solution composition. Furthermore, the reactions might be expected to involve neutral as well as ionic species, including the metal spray capillary [as Evans' group (32) had noted with EHMS], solvents, additives, or contaminants in the solution. Given "wet methanol" containing a chloride salt as an electrolyte, they postulated that when the ES capillary was held at a high positive voltage (i.e., positive-ion mode), the buildup of negative ions in the capillary (owing to the loss of an excess of positive ions in the droplets) might be counterbalanced by electrochemical oxidation reactions that result in the neutralization of the negative ions (Eqs. 1 and 2), the production of positive ions (Eqs. 3 and 4), or both.

$$2Cl^-(aq) = Cl_2(g) + 2e^- \quad (E^0_{red} = 1.12 \text{ V vs. SCE}) \quad (1)$$

$$4OH^-(aq) = O_2 + 2H_2O(l) + 4e^- \quad (E^0_{red} = 0.16 \text{ V vs. SCE}) \quad (2)$$

$$2H_2O(l) = O_2(g) + 4H^+(aq) + 4e^- \quad (E^0_{red} = 0.99 \text{ V vs. SCE}) \quad (3)$$

$$M(\text{capillary metal}) = M^{n+} + ne^- \quad (E^0_{red} = \text{varies with metal}) \quad (4)$$

When the needle was held at a high negative voltage (i.e., negative-ion mode), the buildup of positive ions in the capillary might be counterbalanced by electrochemical reduction of the positive ions, by the production of negative ions, or by both of these processes. No possible reactions were put forward by Kebarle's group for this mode of operation, but reduction of dissolved oxygen or protons might be included among the possible reactions. The actual potentials at which these two reactions occur would depend greatly on the nature of both the spray capillary (i.e., the working electrode material) and the solvent system (39). One should note that many of these specific charge-balancing reactions put forward by Kebarle's group are the same ones either observed (Eq. 4) or discussed by Evans and co-workers (32) and Cook (34) with regard to the electrolytic nature of the EH ion source. No mention was made of this fact by Kebarle's group; they were apparently unaware of this earlier work.

Kebarle's group chose to demonstrate the electrolytic nature of the ES ion source by "forcing" the metal comprising the ES capillary to oxidize in the hope that the metal ions so produced would be observed in the ES mass spectrum. [The observation of Fe^{2+} adduct ions in EH mass spectra provided experimental proof of the electrolytic nature of the EH ion source and the participation of the stainless-steel spray capillary in the electrolytic process (32).] This task was accomplished by selecting zinc, which is very easy to oxidize [$Zn(s) = Zn^{2+}(aq) + 2e^-$, $E^0_{red} = -1.0$ V vs. SCE], as the metal for the ES capillary tip (Fig. 2). The Zn^{2+} ions released to the solution by oxidation of the zinc tip were detected in the gas phase (Fig. 2) with the amount of Zn^{2+} observed (as determined from calibration with zinc salts, Table 1) corresponding to the amount required to maintain (i.e., charge balance) the ES current on the basis

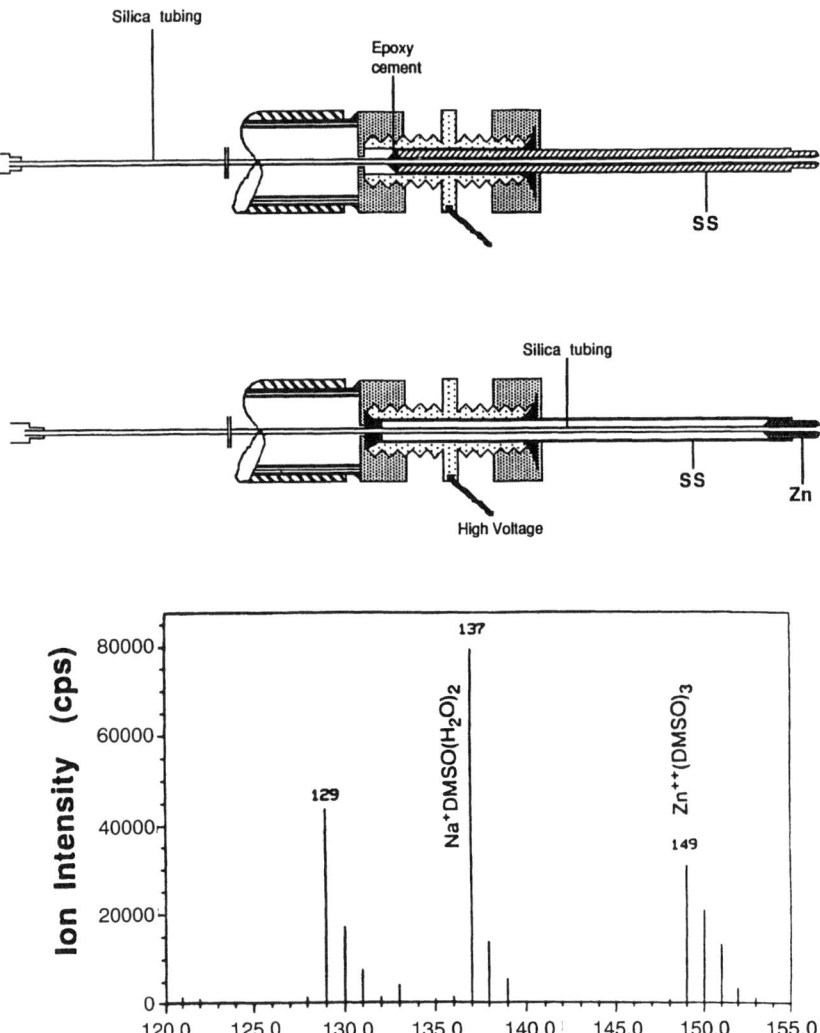

Figure 2. Stainless-steel- (SS) and zinc- (Zn) tip ES capillaries used by Kebarle's group (35) to demonstrate oxidation of the capillary material during positive-ion ESMS. The partial ES mass spectrum was obtained using the zinc-tip capillary and a spray solution containing 10^{-5} M RbCl as an electrolyte for stable spray formation and 10^{-2} M dimethyl sulfoxide (DMSO) to concentrate the Zn^{2+} gas-phase ion current in a single isotopic cluster. The ions $m/z = 149$, 150, and 151 are due to $Zn^{2+}(DMSO)_3$ isotopes. (Adapted with permission from A. T. Blades et al., *Anal. Chem.*, **1991**, *63*, 2110–2111. © 1991 American Chemical Society.)

of Faraday's law (Eq. 5). In more precise electrochemical terms, the ES current (cell current), i_{ES}, was found, as required by charge-balance considerations, to be equal in magnitude to the current due to the redox reactions occurring in the ES capillary, that is, the faradaic current at the working electrode, i_F (in this case due

TABLE 1. Determination of Zn^{2+} Concentrations for ES Involving Zinc-Tipped Capillary

	$NaCl^b$	KCl^b	$RbCl^b$	$CsCl^b$
i_{ES} (μA)[a]	0.14	0.13	0.11	0.15
$[Zn^{2+}]_{expt}{}^c$	2.2	1.7	2.0	2.4
$[Zn^{2+}]_{pr}{}^d$	2.2	1.9	1.7	2.3

Source: Adapted with permission from ref. 35. Copyright 1991, American Chemical Society.

[a] Total ES current.
[b] Supporting electrolyte, 10^{-5} M in methanol. Flow rate 20 μL/min.
[c] Zn^{2+} concentration (μM) in solution based on calibrated MS detection of Zn^{2+} (DMSO)$_3$.
[d] Zn^{2+} concentration (μM) predicted based on Faraday's law (Eq. 5) assuming oxidation of the zinc is the only redox reaction that occurs. Capillary shown in Figure 2.

only to the oxidation of the zinc capillary tip), in this electrolytic cell.

$$i_{ES} = i_F = \sum_j n_j A_j F \nu_f \qquad (5)$$

where

n_j = Number of electrons involved in the oxidation of one molecule of species j
A_j = Concentration of species j oxidized/reduced
F = Faraday constant (9.648×10^4 C/mol)
ν_f = Solution flow rate through the ES capillary

Similar, though less reproducible results, were obtained when the ES capillary was made of stainless-steel, in which case Fe^{2+} (presumably among several other metals comprising the stainless-steel, including chromium, nickel, and molybdenum), was released to the solution by oxidation of the iron in the capillary (Table 2).

Kebarle's group presented no direct evidence for the occurrence of redox reactions other than those involving the metal ES capillary material. However, when using silver as the capillary tip [Ag(s) = Ag^+(aq) + e^-, E^0_{red} = 0.56 V vs. SCE], the gas-phase signals observed for Ag^+ were less than that required to account for complete charge balance by this reaction alone. They speculated that because silver was relatively hard to oxidize, oxidation of some more easily oxidized trace-level component in the methanol solvent probably provided the remaining charge balance requirement. More direct evidence for the involvement of solution species in these redox reactions was reported at about this same time by our group at Oak Ridge National Laboratory. We found that under certain solution conditions, the molecular radical cations ($M^{+\cdot}$) of some divalent metal porphyrins [e.g., Ni^{II} octaethylporphyrin ($Ni^{II}OEP$), $Zn^{II}OEP$, and V = $O^{II}OEP$] were observed in their positive-ion ES mass spectra (40).

TABLE 2. Determination of Fe^{2+} Concentrations for ES with Stainless-Steel Capillary

Experiment	Electrolyte[a]	i_{ES} (μA)[b]	$[Fe^{2+}]_{pr}$[b]	$[Fe^{2+}]_{expt}$[d]
1	$ZnCl_2$, 10^{-4}	0.5	7.6	8.9
2	$ZnCl_2$, 10^{-4}	0.55	8.2	8.8
3[e]	$ZnCl_2$, 10^{-4}	0.53	7.9	9.3
4	KBr, 10^{-5}	0.24	3.6	5.8
5	KBr, 10^{-5}	0.22	3.3	2.9
6[e]	KBr, 10^{-5}	0.41	6.1	10.7
7	KBr, 10^{-4}	0.21	3.1	2.7
8	KBr, 10^{-4}	0.57	8.6	0.1
9	KBr, 10^{-4}	0.54	8.1	1.4
10	$NaNO_3$, 10^{-4}	0.45	6.8	3.3
11[f]	KBr, 10^{-4}	0.60	9.0	7.6

Source: Adapted with permission from ref. 35. Copyright 1991, American Chemical Society.

[a] Supporting electrolyte concentration in methanol (mol/L).
[b] Total ES current.
[c] Predicted concentration (μM) based on Eq. 5 assuming oxidation of iron in the stainless-steel capillary is the only redox reaction that occurs.
[d] Concentration (μM) based on collected electrospray in Pt dish electrode and analysis of resulting solution with ICPAES. The parts per million values obtained with ICPAES were converted to $[Fe^{2+}]$.
[e] Pt dish contained 2 mL of 5 M HCL.
[f] ES capillary was at 4-cm distance from Pt collection electrode; ES high voltage ≈7 kV. Capillary shown in Figure 2.

Molecular ions formed by loss of an electron had not been observed in ES mass spectra prior to that report. These results were interpreted to be an indication of either chemical oxidation of the porphyrins involving some species in the ES solvent system, prior to or during the spraying process, or electrochemical oxidation in the ES capillary. Both explanations seemed plausible, because chemical and electrochemical oxidation of metalloporphyrins and the stability of metalloporphyrin radical cations in solution was well documented. However, because this work was published before that of Kebarle, we were ignorant about how the electrochemistry might take place (i.e., we were also unaware of Evans' and Cook's EHMS reports and we failed on our own to recognize the consequences of charge balance in the ES ion source).

To rule out the possibility of a chemical mechanism in solution for formation of the porphyrin radical cations, or a gas-phase formation mechanism, we later used the ES ion-source configuration shown in Figure 3 (36). The stainless-steel ES capillary was placed upstream and the solution was sprayed from the end of a 30-cm length of fused-silica capillary attached to the metal capillary. A solution of the analyte of interest was pumped through the system at a rate of 2 μL/min, for several minutes, with the high voltage turned off. The high voltage was then turned on and data acquisition began. With $Ni^{II}OEP$ as the test analyte, it was found that the radical cation (m/z 590) was not observed until a time approximately equal to the calculated time (i.e., 72 s) to elute the void volume from the fused-silica/metal-capillary connection to the spray end of the silica

II. THE ELECTROLYTIC NATURE OF ELECTROSPRAY   75

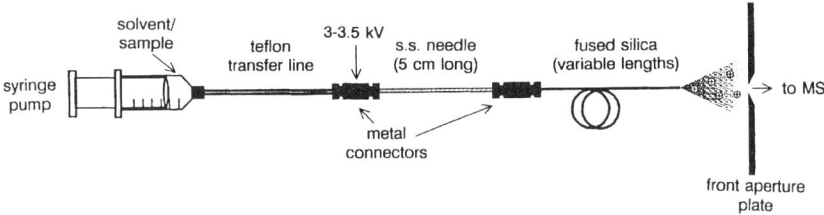

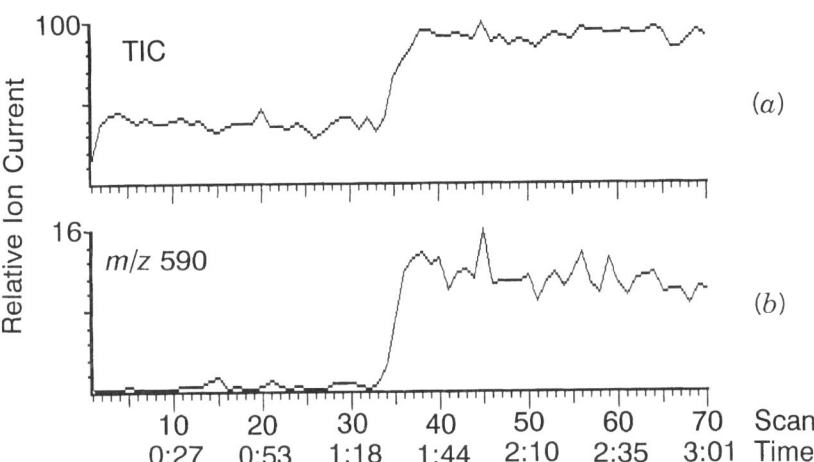

Figure 3. Schematic of the ES source geometry used by the Oak Ridge group (36) to confirm the electrolytic origin of particular radical cations observed in ESMS. The high-voltage contact to the solution is made several centimeters upstream via metal connectors and a typical ES stainless-steel capillary. The solution is sprayed from the end of a 30-cm length of fused-silica capillary. The total-ion current (TIC) profile (a) and extracted-ion current profile (b) for the radical cation of $Ni^{II}OEP$ (m/z 590) were obtained by spraying a methylene chloride/methanol/TFA (90/10/0.05% v/v/v) solution of $Ni^{II}OEP$ using this ES spray system. See text for experimental details. (Adapted with permission from G. J. Van Berkel et al., *Anal. Chem.*, **1992**, *64*, 1587–1590. © 1992 American Chemical Society.)

after the high voltage was switched on (Fig. 3b). When a salt was used as the analyte (e.g., tetrabutylammonium iodide), the expected ion signal (i.e., m/z 242) was observed immediately after the voltage was turned on. These results demonstrated that $Ni^{II}OEP$ radical cations were not in solution prior to passing through the metal ES capillary, nor would it seem that they were made by a gas-phase process. Rather, they appeared to be formed at the point of high-voltage contact to solution (i.e., within the metal ES capillary) via what we believed to be electrolytic oxidation. This interpretation was consistent with Kebarle's description of the ES ion source as an electrolytic cell (35).

Our work served to illustrate that solution species, under the appropriate operational conditions, could be involved in the charge-balancing redox reac-

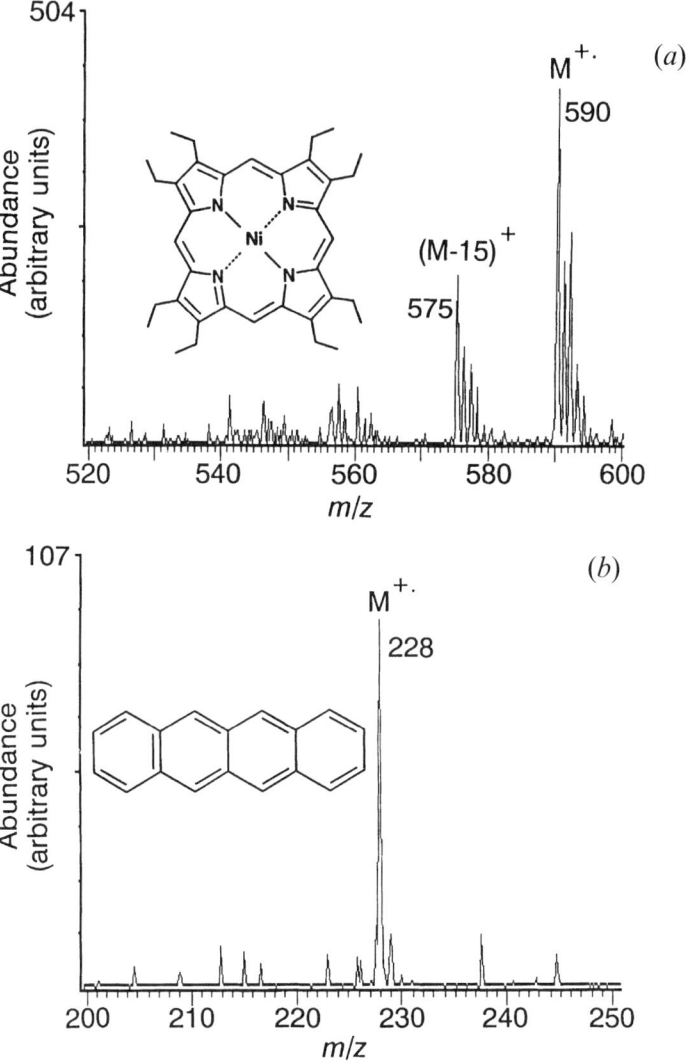

Figure 4. ES mass spectra of (a) Ni^{II}OEP sprayed from a methylene chloride/methanol (90/10 v/v) solution and (b) 2,3-benzanthracene sprayed from methlyene chloride/0.1% TFA. Spectra were acquired using a typical ES ion source in which the solution was sprayed from a stainless-steel capillary held at a high positive potential. (Adapted with permission from G. J. Van Berkel et al., *Anal. Chem.*, **1992**, *64*, 1589. © 1992 American Chemical Society.)

tions in the metal spray capillary and that the products of their reactions could be observed in the gas phase. Furthermore, this work showed that the electrolytic process could be exploited for analytical purposes. In addition to the porphyrins, we found that some polycyclic aromatic hydrocarbons (PAHs), aromatic amines, and heteroaromatics could be oxidized in the ES capillary to form radical cations and could subsequently be observed in the gas phase. Electrospray mass spectra obtained for $Ni^{II}OEP$ and 2,3-benzanthracene are shown as examples in Figure 4. A more recent study by Cole and co-workers (37) using metallocenes (see ferrocene spectrum in Fig. 5) as analytes essentially corroborated our work at Oak Ridge. Data from both of these studies showed that the nature of the analyte and its redox reaction product(s), and the ES operational conditions, including the solvents and additives, could be crucial to observing a particular ion of the analyte and defining the detection levels. For example, analytes with redox potentials less than about 0.3 V (versus SCE) were found to be almost completely oxidized by this process and detected at low levels. However, the efficiency of oxidation–detection dramatically decreased for analytes with redox potentials greater than 0.3 V. If analyte redox potentials exceeded about 1.0 V, detection even at much higher concentrations (i.e., >0.1 mM) proved difficult. Some of these results are discussed in more detail in Section III.

B. Characterization of an ES Ion Source as a Controlled-Current Electrolytic Cell

Motivated in large part by the desire to exploit the electrolytic nature of the ES ion source for analytical purposes, we continued our studies of its electrolytic

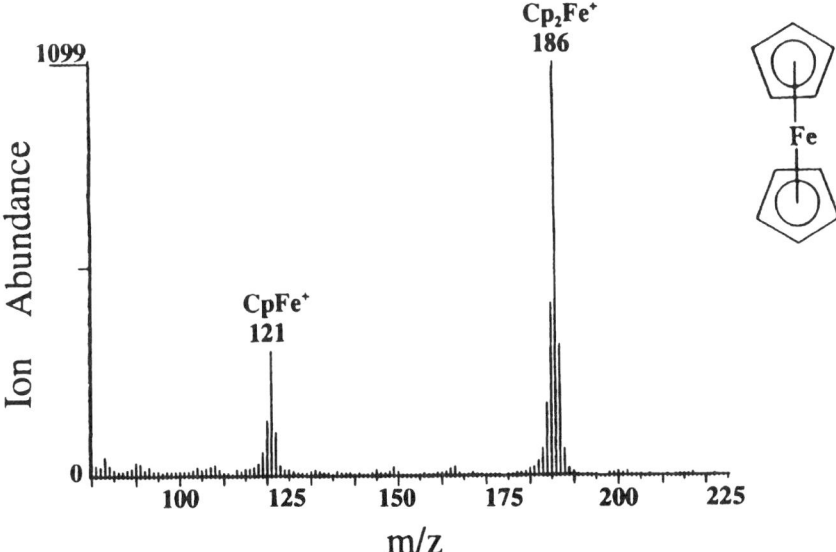

Figure 5. The ES mass spectrum of ferrocene (10^{-3} M) sprayed from acetonitrile using an ES ion source with a stainless-steel capillary. (Adapted with permission from X. Xu et al., *Anal. Chem.*, **1994**, *66*, 119–125. © 1994 American Chemical Society.)

behavior. The precise electrolytic nature of the device (i.e., the experimental parameters that determined the potential at the metal/solution interface in the ES capillary and, therefore, determined the reactions that could or could not occur) and the factors that determined to what extent specific reactions occurred were not thoroughly comprehended by us nor seemingly by the other groups studying this effect. A close reading of the papers by the Kebarle (35), Oak Ridge (36), and Cole (37) groups will show that characteristics of both a controlled-potential electrolytic (CPE) cell and a controlled-current electrolytic (CCE) cell are intermingled in describing the ES ion source. This "confusion" regarding the electrochemical details of ES stood in the way of fully exploiting the analytical benefits of the electrolytic process with regard to ESMS.

With our continued studies, we were eventually able to show experimentally what seems obvious in hindsight: the ES ion source, which is a constant or controlled-current device, operates electrolytically in an analogous fashion to a conventional CCE flow cell (38). A conventional CCE flow cell consists of a cell that houses a working electrode, a counterelectrode, and a controlled-current source (e.g., several batteries and a variable resistor of high resistance), the output of which determines the magnitude of the current flowing through the cell, that is, the cell current, i_C (41–43). The ES ion source can also be viewed as consisting of two electrodes and a controlled-current source (Fig. 6). In our view, the charged droplet formation process is the controlled-current source with the "cell current" (i.e., the ES current, i_{ES}) equal to the product of the rate at which charged droplets are formed and the average number of charges per droplet. [Although we have not fully developed this description of the ES ion source, we can say that it is more akin to describing it as a current-limited device, such as a phototube or a flame-ionization detector (44), than as a somewhat special form of a conventional electrolytic cell (35).] Altering the output of this current source (i.e., altering the magnitude of i_{ES}) requires that the rate of droplet production and/or the average number of charges per droplet be altered. This can be

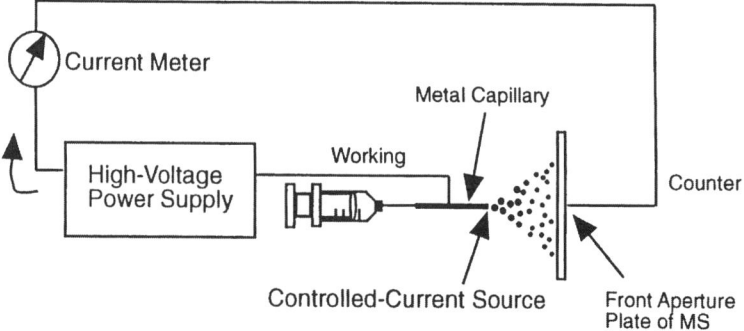

Figure 6. Schematic illustration describing an ES ion source as a CCE cell. (Adapted with permission from G. J. Van Berkel et al., *Anal. Chem.*, **1995**, *67*, 2916–2923. © 1995 American Chemical Society.)

accomplished by changing one or more of several operational parameters, as expressed in the modified Hendricks equation (24,27,45):

$$i_{ES} = H\nu_f^{\nu} \sigma_S^n E_{ES}^{\epsilon} \quad [\sigma_S = \lambda_m^0 C_E; \ E_{ES} = 2V_{ES}/r_{ES} \ln(4d/r_{ES})] \quad (6)$$

where

H = Constant, f (dielectric constant and surface tension of the solvent)
ν_f = Volumetric flow rate through ES capillary
σ_S = Specific conductivity of solution
λ_m^0 = Limiting molar conductivity of electrolyte
C_E = Concentration of electrolyte
E_{ES} = Imposed electric field at capillary tip
V_{ES} = Voltage applied to ES capillary
r_{ES} = Outer radius of the capillary
d = Distance of the capillary tip from the counter electrode

The values of the individual exponents in the equation (viz., ν, n, and ϵ) are interrelated and may vary to a degree as the individual experimental parameters are varied (24,28,36). Although this equation was originally developed on the basis of EH experiments by Pfeifer and Hendricks (28), Kebarle's group showed that it is largely valid for ES as well (24).

In direct analogy to two-electrode controlled-current electrolysis, one expects that the potential at the metal/solution interface in the ES capillary (i.e., the potential at the working electrode/solution interface, $E_{E/S}$), which ultimately determines which redox reactions can occur, will be that value for a given magnitude of i_{ES} necessary to oxidize/reduce sufficient species in the solution within the ES capillary to maintain that current (i.e., $i_{ES} = i_F$, Eq. 5). The individual species oxidized/reduced to supply i_F will do so in order of their increasing redox potentials [as in Kebarle's description (35)] until the required current is supplied. Furthermore, the extent of any reaction involving solution species will be affected by both the rate at which the species flow through the capillary (the tubular working electrode) for a given magnitude of i_{ES} (Eq. 5) and the rate of mass transfer of the species to the electrode surface. In sum, the particular charge-balancing redox reactions that actually take place, and the extent to which they take place, are governed by the magnitude of i_{ES} (which is related to the nature of the solvents, solution conductivity, and electric field at the capillary tip as expressed in Eq. 6), the respective concentrations and redox potentials of the various species in the system [including the metal(s) comprising the ES capillary], and the availability of a species for reaction at the metal–solution interface (which is determined by the rate of transport to the surface, and therefore, is related to, among other factors, capillary dimensions, solution flow rate, and species concentration and charge).

These operational characteristics are made more apparent by reference to the hypothetical $E_{E/S}$ versus i_{ES} plots in Figure 7. The solid-line curve represents a situation for positive-ion mode ES in which three electroactive species are in the

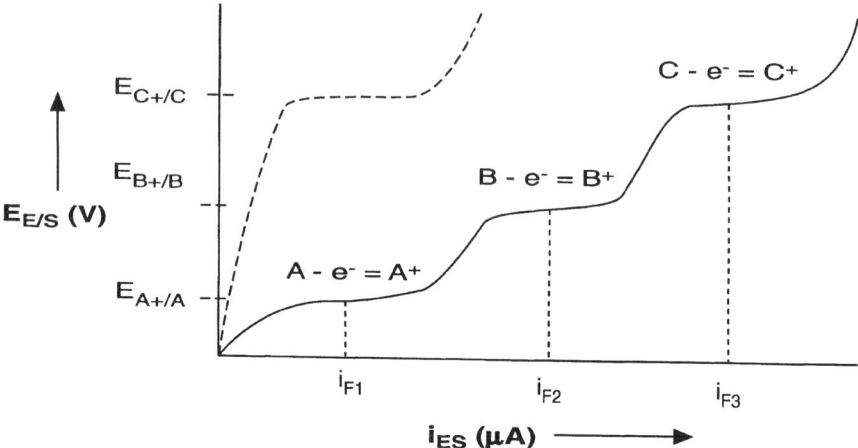

Figure 7. Schematic illustration, based on the operation of an ES ionization source as a CCE cell, showing the expected interdependence of the potential at the electrode/solution interface, $E_{E/S}$, in the ES capillary as a function of the ES current, i_{ES}, and the composition of the electroactive species in the solution. Solid line: three electroactive species (viz., A, B, and C) with electrode potentials $E_{A+/A} < E_{B+/B} < E_{C+/C}$, respectively, are present in the solution at equal concentrations. Dashed line: only the electroactive species C is present in the solution. (Reproduced with permission from G. J. Van Berkel et al., *Anal. Chem.*, **1995**, *67*, 2916–2923. © 1995 American Chemical Society.)

solution at equal concentrations (viz., A, B, and C) with redox potentials of $E_{A+/A} < E_{B+/B} < E_{C+/C}$, respectively. As the magnitude of i_{ES} increases, the value of $E_{E/S}$ increases so that a sufficient amount of these species available for reaction can be oxidized. As a result, these electroactive species are oxidized in the order of increasing electrode potential (i.e., the easiest to oxidize species is oxidized first and so on). Thus, the total faradaic current, i_{Fn}, where $n = 1$, 2, or 3, is equal to the sum of the faradaic currents, i_{FA}, i_{FB}, and i_{FC}, resulting from oxidation of the respective individual species A, B, and C. This diagram also demonstrates that changing the composition of electroactive species in solution can alter the magnitude of $E_{E/S}$ for a given i_{ES}. For example, with species A, B, and C present in the solution, and with an i_{ES} corresponding to i_{F1}, only A is oxidized and $E_{E/S}$ is equal to $E_{A+/A}$. However, if only species C is present in the solution (or the only species available for reaction), $E_{E/S}$ will be $E_{C+/C}$ because no species other than C can be oxidized more easily to maintain the required current.

We verified this controlled-current electrolysis behavior of the ES ion source using the instrumental setup shown in Figure 8 (38). This setup is similar to that shown in Figure 3 except that a UV/visible, diode-array detector, flow cell has been added in line to detect in solution the products of the redox reactions as they exit the ES capillary. Such a "spectroelectrochemical" (46) detection scheme avoids possible experimental ambiguities (or difficulties) in detection of the electrolysis products in the gas phase imposed by the spraying process (e.g., gas-phase ionization, signal suppression, or gas-phase charge-changing reactions) or

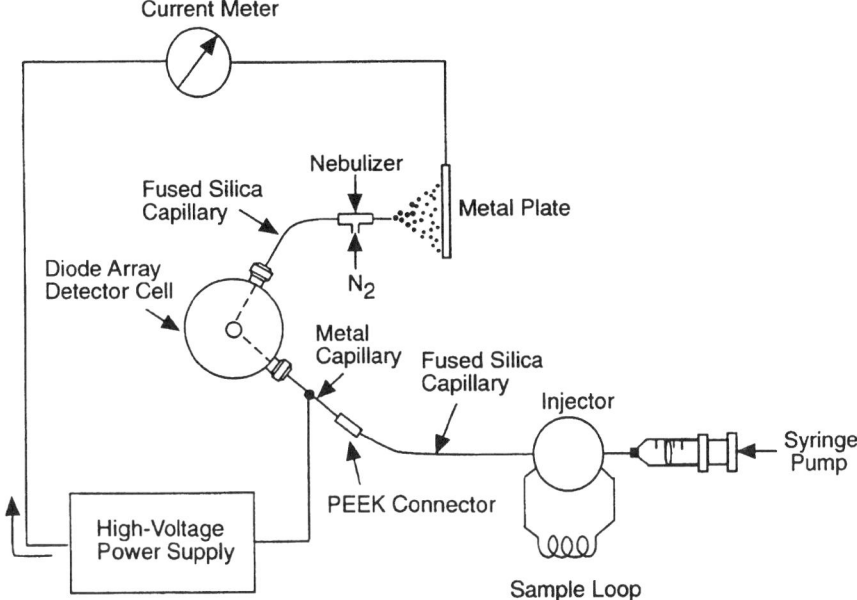

Figure 8. Schematic representation of the novel ES ionization source used to monitor in solution, by means of UV/visible spectroscopy, the extent of redox reactions that occur in the ES capillary as a function of variable experimental parameters. (Reproduced with permission from G. J. Van Berkel et al., *Anal. Chem.*, **1995**, *67*, 2916–2923. © 1995 American Chemical Society.)

subsequent mass analysis of the ions (e.g., m/z discrimination by the atmospheric sampling interface or by the mass spectrometer). In fact, this scheme requires only that the occurrence of a particular reaction can be detected by monitoring the UV/visible spectrum of the solution. The reactions under study need not produce ions that can be detected in the gas phase.

Using Ni^{II}OEP as the test analyte, we found, as expected on the basis of the discussion above, that as i_{ES} was increased by means of an increase in solution conductivity (i.e., through the addition of electrolyte to the solution, Eq. 6), the amount of porphyrin oxidized also increased. This is illustrated by the five UV/visible spectra of Ni^{II}OEP shown in Figure 9a. The spectrum recorded when no electrolyte was added to the system was that of the neutral porphyrin (λ_{max} = 386, 510, and 548 nm), while that recorded when the electrolyte concentration was 9.0 mM was that of the porphyrin radical cation (λ_{max} = 370 nm) (47,48). It was assumed that a substantial amount of porphyrin oxidation (we estimate that oxidation of as little as 0.5 μM of the porphyrin should be detectable with this system) was not observed at magnitudes of i_{ES} lower than 1.0×10^{-7} A (C_E < 0.1 mM) because these lower current levels could be maintained by oxidation of easily oxidized contaminants in the solvent system or by oxidation of iron (or other metals) in the ES capillary (i.e., electrolytic corrosion of the metal capillary). Therefore, the magnitude of $E_{E/S}$ never reached a value close to that necessary to oxidize the porphyrin [$E_{1/2} = 0.73$ V vs. SCE (36)]. However, as the concentration of electrolyte was increased and the magnitude of i_{ES} increased, the magnitude of $E_{E/S}$ increased to the value at which the porphyrin could be oxidized and contribute a portion of the required faradaic current. This same trend in the extent of porphyrin oxidation was observed when the magnitude of i_{ES} was increased (at constant electrolyte concentration) through an increase in the electric field at the capillary tip (Eq. 6), brought about by increasing the voltage, V_{ES}, applied to the capillary (Fig. 9b). In contrast, as illustrated by the data in Figure 9c, an increase in solution flow rate through the capillary resulted in a reduction in the extent of porphyrin oxidation from essentially 100% at 10 µL/min to near zero at 80 µL/min. This was interpreted to result from an increased flow rate of porphyrin through the capillary without the requisite increase in i_{ES} necessary to maintain the same extent of analyte oxidation (Eq. 5). We speculate that this change in the proportion of porphyrin oxidized might also be due, in part, to the limited rate of porphyrin mass transfer to the metal/solution interface (assumed to be mainly through molecular diffusion). At higher flow rates, the porphyrin molecules nearer the center of the capillary might not have enough time (i.e., the

---

Figure 9. (a) UV/visible spectra of Ni^{II}OEP recorded during five separate flow-injection experiments (50-µL injections) in which the voltage applied to the ES capillary ($V_{ES} = 5$ kV), the solution flow rate through the system [$\nu_f = 20$ µL/min, acetonitrile/methylene chloride (1/1 v/v)], and the concentration of Ni^{II}OEP injected [5.0 µM, acetonitrile/methylene chloride (1/1 v/v)] were kept constant and the conductivity of the solution, σ_S, varied. The concentrations of the electrolyte lithium triflate added to solutions, the value of σ_S, and values of i_{ES} measured at the apex of the eluting peaks were 0.0 mM (1.4×10^{-6} Ω^{-1} cm^{-1}, 1.2×10^{-8} A), 0.01 mM (3.7×10^{-6} Ω^{-1} cm^{-1}, 2.0×10^{-8} A), 0.1 mM (15×10^{-6} Ω^{-1} cm^{-1}, 1.0×10^{-7} A), 1.0 mM (97×10^{-6} Ω^{-1} cm^{-1}, 6.4×10^{-7} A), and 9.0 mM (880×10^{-6} Ω^{-1} cm^{-1}, 2.7×10^{-6} A). The direction of the arrows on the spectra indicate the change in the absorption peaks with increasing solution conductivity. (b) UV/visible spectra of Ni^{II}OEP recorded during an experiment in which a 5.0-µM solution of Ni^{II}OEP [acetonitrile/methylene chloride (1/1 v/v)] containing 1.0 mM lithium triflate as the electrolyte (97×10^{-6} Ω^{-1} cm^{-1}) was continuously infused through the system ($\mu_f = 20$ µL/min) and the voltage applied to the ES capillary, V_{ES}, was varied. The magnitude of V_{ES} and the corresponding magnitude of i_{ES} were 0 kV (0 A), 1.0 kV (2.7×10^{-8} A), 2.0 kV (1.45×10^{-7} A), 3.0 kV (2.75×10^{-7} A), 4.0 kV (4.4×10^{-7} A), 4.5 kV (5.3×10^{-7} A), and 5.0 kV (6.0×10^{-7} A). The direction of the arrows on the spectra indicate the change in the absorption peaks with increasing capillary voltage. (c) UV/visible spectra of Ni^{II}OEP obtained in an experiment in which a 5.0-µM solution of Ni^{II}OEP [acetonitrile/methylene chloride (1/1 v/v)] containing 1.0 mM lithium triflate as the electrolyte (97×10^{-6} Ω^{-1} cm^{-1}) was continuously infused through the ES capillary ($V_{ES} = 5$ kV) at different flow rates, ν_f. Spectra were recorded at solution flow rates of 10, 15, 30, 40, 60, and 80 µL/min. The magnitude of i_{ES} increased from 6.5×10^{-7} A at 10 µL/min to 6.9×10^{-7} A at 80 µL/min. The direction of the arrows on the spectra indicate the change in absorbance peaks with increasing flow rate. (Adapted with permission from G. J. Van Berkel et al., *Anal. Chem.*, **1995**, *67*, 2916–2923. © 1995 American Chemical Society)

electrolysis time) to diffuse to the electrode for reaction before they exit the capillary. In this case, one can assume that $E_{E/S}$ increases to a value higher so as to oxidize some other species available for reaction (e.g., the capillary metal, electrolytes, and solvents) to maintain the required current.

Only a small percentage of the total faradaic current in these experiments

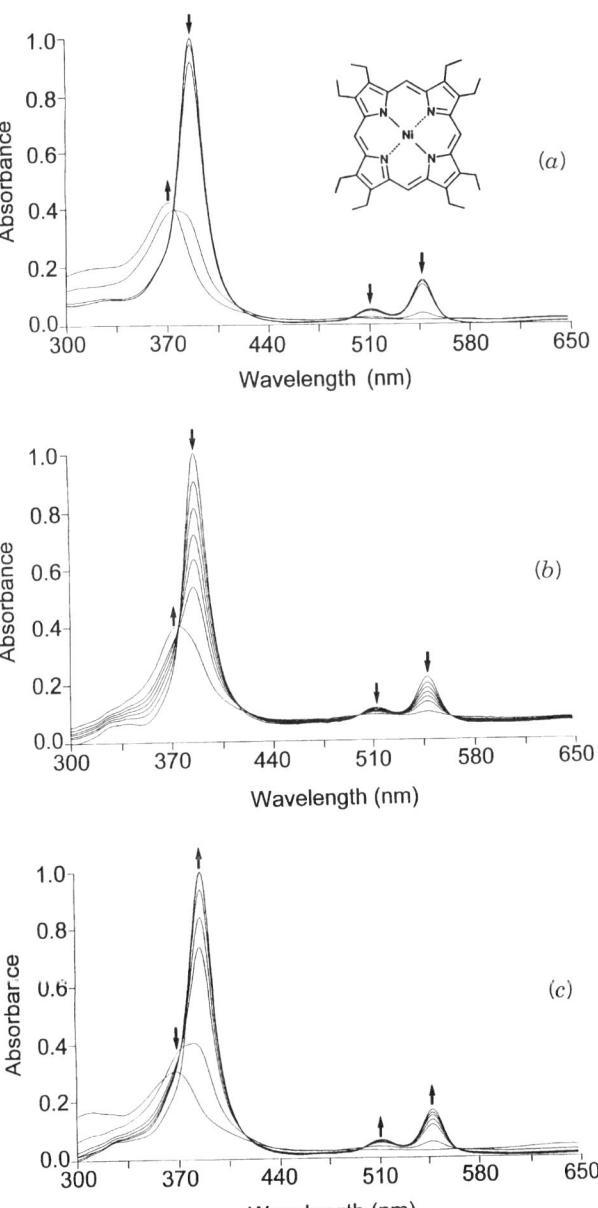

(27%, based on the data in Fig. 9a) can actually be attributed to oxidation of the porphyrin. The remainder of the faradaic current is probably supplied via oxidation of the iron (and possibly other metals) comprising the stainless-steel capillary. Typically, in the absence of passivation, there is a direct relationship between the corrosion rate of a metal (i.e., the current owing to metal oxidation) and the potential at the metal/solution interface (49). Thus, as i_{ES} is increased in the ES experiments, we might expect, in the absence of solution species easier to oxidize than the metals comprising the capillary, that the potential, $E_{E/S}$, at the metal/solution interface within the stainless-steel capillary increases, thereby increasing the corrosion rate. [Indeed, in an unpublished experiment, we found that the amount of Fe^{2+} in the solution exiting the stainless-steel capillary (detected as the 1,10-phenanthroline complex in solution using the setup in Fig. 8) increased in direct proportion to the value of i_{ES}. Moreover, the amount of Fe^{2+} detected accounted for approximately 100% of the faradaic current in the absence of any other easily oxidized species in the solution.] When $E_{E/S}$ reaches the value sufficient for porphyrin oxidation, that reaction can compete with corrosion to supply the faradaic current. For solution species that are relatively difficult to oxidize, such as $Ni^{II}OEP$, achieving the necessary value of $E_{E/S}$ required for their oxidation necessitates a relatively high i_{ES}, of which only a small fraction can ever be supplied by any reaction other than capillary corrosion. In support of this argument, we found that the fraction of porphyrin oxidized at a given i_{ES} was always greater when using a platinum ES capillary in place of stainless-steel. A platinum capillary is much more difficult to oxidize than $Ni^{II}OEP$ in a nonaqueous solvent system (42). Therefore, the potential necessary to oxidize the porphyrin is reached at a much lower value of i_{ES}, meaning that a higher fraction of total faradaic current can be supplied by porphyrin oxidation.

Further confirmation of the controlled-current electrolysis characteristics of the ES ion source was obtained by monitoring the effect that changing the composition of electroactive species in the solvent system had on the extent of porphyrin oxidation (Fig. 10). In the particular experiments performed, either

---

Figure 10. UV/visible spectra acquired in four separate flow-injection experiments (5.0 μL injections) in which the voltage applied to the ES capillary ($V_{ES} = 5$ kV), the flow rate through the system ($\nu_f = 20$ μL/min), the concentration of electrolyte in the carrier and sample solutions [acetonitrile/methylene chloride (1/1 v/v), 1.0 mM lithium triflate], and the concentration of $Ni^{II}OEP$ in the sample solutions (11 μM) were kept constant, with various amounts of other electroactive species added to the samples: (a) no other electroactive species added, (b) 30 μM ferrocene added, and (c) 28 μM anthracene added. The spectra represented by the solid lines are those of the neutral porphyrin recorded during the respective flow-injection experiments when no high voltage was applied to the capillary. The spectra represented by the dotted lines are those obtained with the high voltage turned on. Note that the fine structure on the short wavelength side of the radical cation absorption peak in (c) is due to the absorption peaks of the neutral anthracene in the solution. (Reproduced with permission from G. J. Van Berkel et al., *Anal. Chem.*, **1995**, *67*, 2916–2923. © 1995 American Chemical Society.)

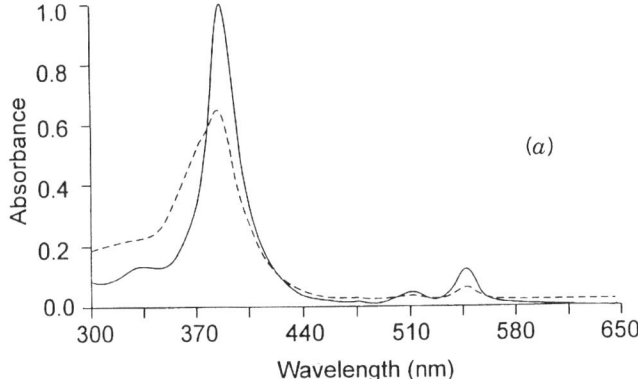

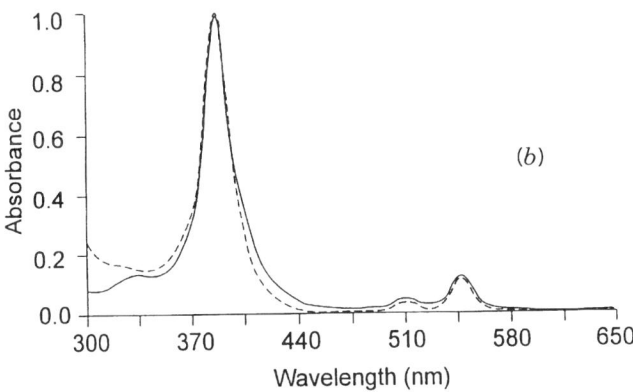

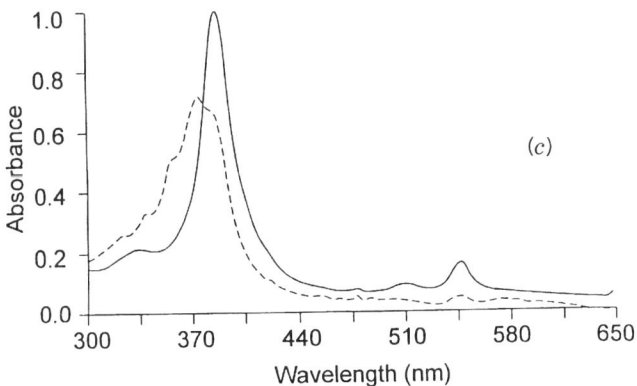

ferrocene [$E_{1/2} = 0.31$ V vs. SCE (42)] or anthracene [$E_{1/2} = 1.19$ V vs. SCE (47)] were added to the sample solution along with the Ni^{II}OEP [$E_{1/2} = 0.73$ V vs. SCE (36)]. Without the addition of the other analytes, ES conditions were such that about a 3- to 5-μM portion of the 11-μM porphyrin sample could be oxidized to the radical cation (Fig. 10a). As would be predicted from the characteristics of controlled-current electrolysis, with 30 μM ferrocene in the solution, porphyrin oxidation was not observed (Fig. 10b). In that case, a sufficient amount of the more easily oxidized ferrocene was present to supply all of the faradaic current. Therefore, $E_{E/S}$ never reached a value sufficient to oxidize the porphyrin. Also as expected, the addition of 28 μM anthracene to the sample did not affect the extent of porphyrin oxidation (Fig. 10c), because a sufficient amount of the more easily oxidized porphyrin was present to maintain the required faradaic current. Therefore, the value of $E_{E/S}$ remained below that necessary to oxidize anthracene.

Other supportive evidence for the CCE nature of the ES ion source comes from the work of Cole and co-workers (37). They found that the minimum voltage applied to the ES capillary necessary to observe the radical cations for a series of metallocenes increased in a linear fashion with the redox potential of the metallocenes. As discussed above, in the ES ion source, as the voltage applied to the ES capillary is increased, i_{ES} increases (Eq. 6). Given CCE behavior of the ES ion source, one might also expect the value of $E_{E/S}$ to increase as i_{ES} increases, thereby allowing for oxidation of the more difficult to oxidize metallocenes (see Fig. 7).

III. ANALYTICAL IMPLICATIONS OF THE ELECTROLYTIC NATURE OF ELECTROSPRAY

The electrolytic nature of an ES ion source is a fundamental part of its operation as a generator of charged droplets and, ultimately, the gas-phase ions sampled and analyzed by the mass spectrometer in ESMS. It may be considered fortunate that the electrolytic nature of ES is of little concern or consequence to the average user of ESMS. Consider, for example, that when compared to conventional electrolysis studies, the analytes that are normally the subject of analysis in ESMS (i.e., preformed ionic compounds) are "spectator" electrolytes in the solvent system. That is, under typical ES conditions, these ionic molecules aid the formation of charged droplets and the transfer of charge between the electrodes in the cell, but they do not participate in the charge-balancing redox reactions at the working electrode (i.e., the metal ES capillary). Therefore, they are unaltered by the electrolytic process. Furthermore, the redox reaction participants and products (e.g., the iron in the stainless-steel capillary and Fe^{2+}, respectively), which are the main subject of analysis in conventional electrolysis studies, normally go undetected. In any case, it is important to recognize that under some circumstances, the fact that the ES ion source operates as CCE cell can be very important. For example, as discussed below, changes in the composition of the solution entering the capillary as a result of these electrolytic processes might be an important consideration in mechanistic interpretations of the ES process. The electrolytic reactions might also result in the generation of chemical noise or an unexpected modification of the analyte. It is also important to recognize that the electrolytic nature of an ES ion source positions one to use ESMS as a means to perform on-line, controlled-current, electrochemistry–mass spectrometry. This combination may be used for the benefit of ionization–detection of specific analytes by ESMS or, from an electrochemical perspective, as a means by which to study redox reactions.

A. Changes in Solution Composition

On the basis of Faraday's law (Eq. 5), and knowledge of the particular charge-balancing redox reactions that take place in the capillary, one can determine the compositional change in the solution that takes place for a given value of i_{ES} and solution flow rate. These changes could, under certain circumstances, be quite significant, and as a result the chemical equilibria in the solution exiting the ES capillary may be quite different than those in the initial solution. This possibility is especially important to consider when attempting to discern certain mechanistic aspects of ion production in ES and to rationalize the ions ultimately observed in an ES mass spectrum. For example, consider the oxidation of water (Eq. 3) to be the major charge-balancing redox reaction to occur when spraying a weakly basic analyte from an aqueous solution whose pH measured prior to entering the capillary was 7.0. Assuming a constant value for i_{ES} of say 0.1 μA, the pH of the solution exiting the capillary for flow rates of 10, 1.0, and 0.1 μL/min would be 5.2, 4.2, and 3.2, respectively. These changes in pH are quite dramatic and would be expected to alter the ES response of a weakly basic analyte through an increase in the degree of protonation in solution.

B. Chemical Noise/Modifications of Analyte

Evans and co-workers (32,33) viewed the Fe^{2+} adduct ions observed in their EH mass spectra (resulting from the electrolytic corrosion of the stainless-steel spray capillary) as unwanted chemical noise. To avoid this problem, they switched to a platinum capillary. As Kebarle's group demonstrated (35), Fe^{2+} associated ions could be observed in ESMS if ES conditions were altered significantly to enhance their gas-phase abundance. Under more typical ESMS conditions, ions derived from electrolytically produced Fe^{2+} are not usually observed. This is due at least in part to the lower currents (and therefore, lower concentrations of Fe^{2+} produced) in ESMS than EHMS. However, Ijames et al. (50) recently reported observing ions at m/z 622 and 538 in ES mass spectra of acidified methanol/water solutions that they identified (on the basis of accurate mass measurements and tandem mass spectrometry) as the complex $Fe_3O(O_2CR)_6(L)_{0-3}$ where (O_2CR) is the acid added to the solution and L is one of several ligands corresponding to the solvent or water. Apparently, the source of the iron ions is the electrolytic corrosion of the stainless-steel spray capillary. The ES current in these experiments (although not reported) was probably relatively high owing to the high acid content and corresponding high conductivity of the solution, which would account for a high iron concentration in the solution. These iron-associated ions are nominally electrolytically derived chemical noise in the spectra, the presence of which would be eliminated by switching to a platinum capillary. Whether a "cleaner" spectrum would result depends on the electrolysis reactions and products produced within the platinum capillary.

In addition to the generation of chemical noise, the electrolytic reactions might change the analyte substantially such that the ions observed in the gas phase are of a different molecular species than that originally in the solution. One apparent example of this situation was reported by Hop et al. (51), who examined two borane salts ([(Me)$_4$N][B$_3$H$_8$] and Cs[B$_3$H$_8$]) by ESMS in the positive-ion mode. Acetonitrile solutions provided the most informative spectra; nearly all signals observed were reported to correspond to cationic cluster ions of the general formula $\{[\text{cation}^{m+}]_x[\text{anion}^{n-}]_y\}^{(mx-ny)+}$. In contrast, methanol solutions of these salts produced only $B(OCH_3)_4^-$ cluster ions under otherwise identical conditions. Off-line ^{11}B NMR analyses showed that while the borane $B_3H_8^-$ anion was present in the methanol solution entering the ES capillary, the $B(OCH_3)_4^-$ anion exited the capillary. An electrolytic methanolysis reaction in the ES capillary involving the conversion of $B_3H_8^-$ to $B(OCH_3)_4^-$ was assumed to take place, but no reaction mechanism was put forward.

C. Electrochemical Ionization

Recognition that the ES ion source operates in a fashion analogous to a CCE cell provides the information necessary to adjust experimental parameters to maximize the efficiency of the faradaic process in the ES capillary for ionization of neutral analytes in solution for subsequent gas-phase detection by the mass spectrometer. Using the ES ion source to "electrochemically ionize" an analyte provides the means to expand sensitive analysis/detection by ESMS to include certain types of neutral analytes that otherwise are ES inactive, as well as expanding the overall universality of ES as an ionization source. Those analytes found to be most amenable to electrochemical ionization and detection in ESMS form relatively stable ionic species upon electrolytic oxidation/reduction (Table 3). This is required, because the electrochemically generated ions must survive in solution from the time of their formation in the ES capillary until they are sprayed and transported into the mass spectrometer as discrete gas-phase ions. The compounds most amenable to electrochemical ionization via oxidation have relatively low potentials for oxidation and structural characteristics that aid in stabilization of the positive ion formed (often a radical cation). Compounds fitting this description are, for the most part, aromatic or highly conjugated systems that might also contain heteroatoms with lone-pair electrons and/or electron-donating groups such as –OH, –OCH$_3$, –N(CH$_3$)$_2$, and –CH$_3$ (52). Compounds most amenable to electrochemical ionization via reduction have relatively low potentials for reduction and high electron affinities. Compounds of this type are usually comprised of aromatic or highly conjugated systems substituted with electron-withdrawing groups such as halides, $-NO_2$, or $-CN$ (53). Metal-containing organics of various types are often amenable to electrochemical ionization via either oxidation or reduction (52,53).

1. Efficient Ionization/Low Level Gas-Phase Detection. Three major requirements, in addition to the nature of the analyte, are key to utilizing the CCE process in ES for efficient ionization and low-level, gas-phase detection in ESMS. The first is that the magnitude of i_{ES} must be sufficient for oxidation/reduction of the molar equivalent of all species available for reaction in the ES capillary that are as easily or more easily oxidized/reduced than the targeted analyte, including all of the analyte. This ensures that the value of $E_{E/S}$ will be that necessary to oxidize/reduce the analyte and that complete analyte electrolysis will not be limited by the magnitude of i_{ES}. To meet this requirement, the magnitude of i_{ES} required for oxidization/reduction of a given amount of analyte should be reduced by eliminating from the solvent system all species that are easier to oxidize/reduce than the analyte, which might include certain solvents, contaminants, electrolytes, other analytes, and particularly in positive-ion mode, the stainless-steel ES capillary. If necessary, the magnitude of i_{ES} can be increased by adjusting one or more of the experimental parameters that affect i_{ES}, as shown in the Hendricks equation (Eq. 6). In practice, a simple means to substantially increase i_{ES} over that current obtained under optimized ES conditions (e.g., $V_{ES} \approx 4-5\,kV$ with a fixed solvent system, capillary size, and ES source geometry) is to increase solution conductivity by addition of an electrolyte to the solvent system. The electrolytes normally employed in electrochemical experiments [e.g., alkali metal and tetraalkylammonium nitrates, perchlorates, tetrafluoroborates, hexafluorophosphates, and trifluoromethanesulfonates (triflates)] are most suitable for this purpose because they are difficult to oxidize/reduce and, therefore, do not contribute to the faradaic current (42).

The second requirement is that the analyte be available for reaction at the metal/solution interface in the ES capillary. This demands that the time for transport of the analyte to the metal/solution interface (mainly via diffusion) be short relative to the time the analyte remains within the capillary (i.e., the electrolysis time). The flow rate of the analyte through the capillary and the capillary dimensions, along with analyte concentration, will have a major affect on this availability. In general, operating at slower flow rates and using narrow-bore capillaries should enhance electrolysis efficiency by increasing the availability of the analyte for reaction.

The third requirement is that the first two steps taken to ensure efficient electrolysis avoid operational conditions that inhibit the formation of gas-phase ions from the electrolytically generated analyte ions in solution. A major obstacle in this regard is the sometimes necessary addition of an electrolyte to the analyte solution to increase the magnitude of i_{ES}. Tang and Kebarle (25,26) have shown that the mass spectrometrically detected ion current for an analyte of

TABLE 3. Analytes Reported to be Electrochemically Ionized in ES Capillary During ESMS

Compound Class	Compound[a]	Formula[b]	MW	E (V) vs. SCE[c]	Molecular Species Observed[d]	Ref.
		Oxidation (Positive-Ion Mode)				
Metalloporphyrins and chlorins	Mg^{II}OEP	$C_{36}H_{44}N_4Mg$	556	0.54	M^+	36
	Ni^{II}OEP	$C_{36}H_{44}N_4Ni$	590	0.84, 1.34	M^+, M^{2+}	36,68
	Co^{II}OEP	$C_{36}H_{44}N_4Co$	591	0.76, 0.92	M^+, M^{2+}	36,68
	Cu^{II}OEP	$C_{36}H_{44}N_4Cu$	595	0.79	M^+	36
	Zn^{II}OEP	$C_{36}H_{44}N_4Zn$	596	0.63	M^+	36
	Zn^{II}TPP	$C_{44}H_{28}N_4Zn$	676	0.77	M^+	69
	Chlorophyll *a*	$C_{55}H_{72}N_4O_5Mg$	892	0.5	M^+	70
PAHs	Anthracene	$C_{14}H_{10}$	178	1.19	M^+	36
	Pyrene	$C_{16}H_{10}$	202	1.25	M^+	36
	2,3-Benzanthracene	$C_{18}H_{12}$	228	0.98	M^+, pdts	36
	Perylene	$C_{20}H_{12}$	252	1.04	M^+	36
	9,10-Diphenylanthracene	$C_{26}H_{18}$	330	1.22	M^+	36
	Rubrene	$C_{42}H_{28}$	532	0.82	M^+	36
Aromatic amines	Tetramethyl phenylene diamine (TMPD)	$C_{10}H_{16}N_2$	164	1.00	M^+, $(M+H)^+$	36
Heteroaromatics	Phenothiazine	$C_{12}H_9N_5$	199	0.56	M^+(?), $(M+H)^+$	36
Carotenoids	β-Carotene	$C_{40}H_{56}$	536	0.75	M^+(?), M^{2+}, $(M^{2+}-H^+)^+$, pdts	64, this chapter

III. ANALYTICAL IMPLICATIONS OF THE ELECTROLYTIC NATURE OF ELECTROSPRAY

Category	Compound	Formula	MW	E	Species	Ref
Metallocenes	Decamethylferrocene	Cp_2^*Fe	326	−0.11	M^+	37
	1,1'-Dimethylferrocene	$(CH_3C_5H_4)_2Fe$	214	0.25	M^+	37
	Ferrocene	Cp_2Fe	186	0.31	M^+	37,69
	Ferrocene carboaldehyde	$CpFeC_5H_4CHO$	214	0.33	M^+	37
	Ruthenocene	Cp_2Ru	232	0.94	M^+, pdts	37
	Osmocene	Cp_2Os	322	0.82	M^+, pdts	37
Diethyldithio carbamate metal complexes	Cu^{II} diethyldithio carbamate	$[(C_2H_5)_2NCS_2]_2Cu^{II}$	359	0.73	M^+	71
Fullerenes	Buckminsterfullerene	C_{60}	720	1.6	$M^+(?)$	57

Reduction (Negative-Ion Mode)

Category	Compound	Formula	MW	E	Species	Ref
Halogens	Iodine	I_2	254	0.29	I^-, I_3^-, I_5^-	72
Quinones	Dichlorodicyano benzoquinone (DDQ)	$C_6(=O)_2(CN)_2(Cl)_2$	227	0.62	M^-	69
	Tetracyanoquino dimethane (TCNQ)	$C_6H_4[=C(CN_2)]_2$	204	0.15	M^-	69
	Benzoquinone	$C_6H_4(=O)_2$	108	−0.57	M^-	69
Fullerenes	Buckminsterfullerene	C_{60}	720	−0.36	M^-	57,69
	Fluorinated buckyball	$C_{60}F_{48}$	1632	0.79	M^-	73

[a] OEP = octaethylporphyrin, TPP = tetraphenylporphyrin.
[b] Cp^* = pentamethylcyclopentadiene, Cp = cyclopentadiene.
[c] Reduction potentials listed may be E^0, $E_{1/2}$, or peak potentials. Refer to listed references.
[d] Reference to molecular ion: as even or odd electron species omitted to avoid confusion for metal complexes with paramagnetic metals. Pdts indicates the species observed depends on solvent system employed in analysis. Question marks indicate formation by electrolytic process in ES capillary is suspect. See text for discussion.

interest, I_{A+}, may be suppressed by the presence of other ions ("foreign electrolytes") in solution that have a greater propensity (termed k) for formation of gas-phase ions, as expressed by Eq. 7:

$$I_{A+} = pf \frac{(k_{A+}C_{A+})}{(k_{A+}C_{A+}) + (k_{E+}C_{E+})} i_{ES} \qquad (7)$$

where

p = Efficiency of mass spectrometer for detecting ES-produced gas-phase ions
f = Fraction of droplet charge converted to gas-phase ions
C_{A+}, C_{E+} = Concentration of analyte and electrolyte ions present in ES solution, respectively
k_{A+}, k_{E+} = Rate constants for transfer of respective ions from droplets to the gas phase

Therefore, to minimize problems with signal suppression, it is necessary to select an electrolyte whose propensity for gas-phase ion formation, k, is small relative to that of the analyte ion, and to use it at the lowest concentration possible to provide the required i_{ES}. Lithium triflate, for example, has been shown in our work (54) to have a relatively low suppression effect, which enabled its use at concentrations up to a few millimolar to increase i_{ES} without compromising the generation of gas-phase analyte ions.

The Ni^{II}OEP data shown Figure 11 illustrate how the composition of the metal ES capillary, either stainless-steel or platinum, the addition of an appropriate electrolyte to the solution to increase i_{ES}, and the solution flow rate through the capillaries affect the observed gas-phase ion intensities for an electrolytically generated analyte ion. When no electrolyte is added to the solution (Fig. 11a), the value of i_{ES} and the gas-phase analyte ion signal for the radical cation of Ni^{II}OEP are low for both capillaries at all flow rates. On the basis of Faraday's law, the magnitude of i_{ES} necessary for complete oxidation of the porphyrin (10 μM) at each of the respective flow rates, assuming no other reactions supply i_F, are 0.08 μA (5 μL/min), 0.16 μA (10 μL/min), 0.32 μA (20 μL/min), and 0.64 μA (40 μL/min). The currents measured at each flow rate (assuming again only oxidation of the porphyrin) are sufficient to oxidize a maximum of only 5–20% of the total amount of porphyrin present, which explains the low gas-phase ion signals. With 0.1 mM of electrolyte added to the solution (Fig. 11b), the magnitude of i_{ES} with both capillaries, at all flow rates, is substantially increased. As a result, the extent of analyte electrolysis/ionization and the gas-phase ion signal levels are also increased. Note that the Ni^{II}OEP radical cation signal observed when using the platinum capillary is substantially greater than that observed when using the stainless-steel capillary owing to the supply of faradaic current by oxidation of the iron in the stainless steel. The occurrence of this oxidation reaction reduces the amount of i_F that might otherwise be supplied by oxidation of an analyte in solution. Using the ES

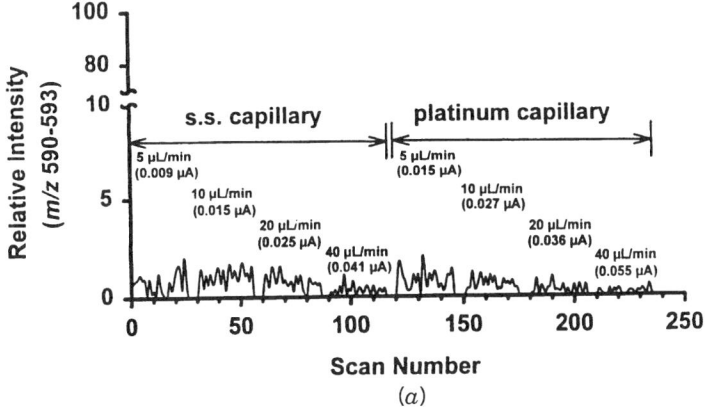

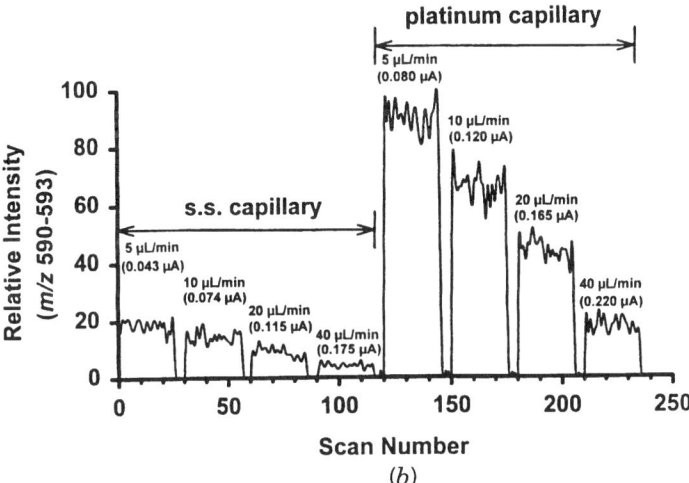

Figure 11. Ion-current intensities for the radical cation of $Ni^{II}OEP$ (m/z 590–593) measured in continuous-infusion experiments in which acetonitrile/methylene chloride (1/1 v/v) solutions of $Ni^{II}OEP$ (10 μM) containing (a) no added electrolyte and (b) 0.1 mM lithium triflate were sprayed from stainless-steel and platinum capillaries at flow rates of 5, 10, 20, and 40 μL/min. The respective flow rates and measured values of i_{ES} are also shown in the figure. The signal levels in (a) and (b) are each normalized to the maximum signal recorded in (b). (Adapted with permission from G. J. Van Berkel et al., *Anal. Chem.*, **1995**, *67*, 3643–3649. © 1995 American Chemical Society.)

capillary fabricated from platinum, which is much more difficult to oxidize than the iron in stainless steel, allows a greater fraction of i_F to be supplied by the oxidation of solution species, including the analyte. As such, more $Ni^{II}OEP$ ions are created in the solution within the platinum capillary compared to within the stainless-steel capillary, all other factors being equal. The data also show that, regardless of the capillary material, the gas-phase signal for the $Ni^{II}OEP$ radical cation decreases as flow rate increases. On the basis of Faraday's law and the measured values of i_{ES} for the platinum capillary, we calculate that 100%, 75%, 52%, and 34% of the current needed for complete analyte oxidation is available at flow rates of 5, 10, 20, and 40 μL/min, respectively. The gas-phase ion signals change in these same relative proportions as flow rate is changed. Therefore, this decrease in ion signal with increasing flow rate probably results, at least in part, because of the increased rate of $Ni^{II}OEP$ transfer through the capillary without a sufficient increase in i_{ES} to enable the same degree of analyte oxidation. Another contributing factor to reduced signal levels as flow rate increases is probably the diffusion-limited availability of the analyte for reaction at the metal/solution interface in the capillary.

When most of the key requirements for efficient electrolysis and sensitive gas-phase detection outlined above were met, we found it possible to ionize and detect neutral molecules at levels that are comparable to the lowest levels achieved for preformed ionic compounds (54). This was true even for species relatively difficult to oxidize (i.e., $E > 1.0$ V vs. SCE) when using a stainless-steel capillary (although better performance is expected with a platinum capillary), as demonstrated by the data in Figure 12. Shown in this figure are the extracted ion current profiles for the radical cation of perylene [m/z 252, $E_{1/2} = 1.04$ V vs. SCE (55)] obtained from three replicate injections of a blank solution and increasing quantities of perylene into a flowing stream of acetonitrile/methylene chloride (1/1 v/v, 20 μL/min) containing 1.0 mM of electrolyte (Fig. 12a). The detection levels of between 130 and 270 fmol are comparable to those levels that we have recorded with our instrumentation for many preformed ionic compounds under similar flow-rate and solution conditions (56). Shown in Figure 12b is the averaged, background-subtracted mass spectrum obtained from the first 270-fmol injection recorded in Figure 12a. The signal level for the radical cation is several times higher than that of the background noise, providing a clear identification of the compound.

2. Caveats. Two other important parameters to consider in regard to electrochemical ionization in ESMS are the solvent system and, for reasons that differ from those previously discussed, solution flow rate. Typical solvent systems for ESMS are comprised of various combinations of methanol, acetonitrile, and/or water along with a small amount of acidic or basic additives. Such solvent systems are chosen because of the solubility characteristics of the more common analytes (e.g., peptides and oligonucleotides), because they produce a stable spray, and because they allow for solution-phase ionization of the compounds (typically ionization via salt dissolution or acid/base chemistry). In the case of electrochemical ionization, a solvent system is chosen that allows the appropriate magnitude of i_{ES} to be generated and whose components are not more easily oxidized/reduced than the analyte. In addition, because a particular ionic species produced may be consumed by several types of rapid reactions in

III. ANALYTICAL IMPLICATIONS OF THE ELECTROLYTIC NATURE OF ELECTROSPRAY

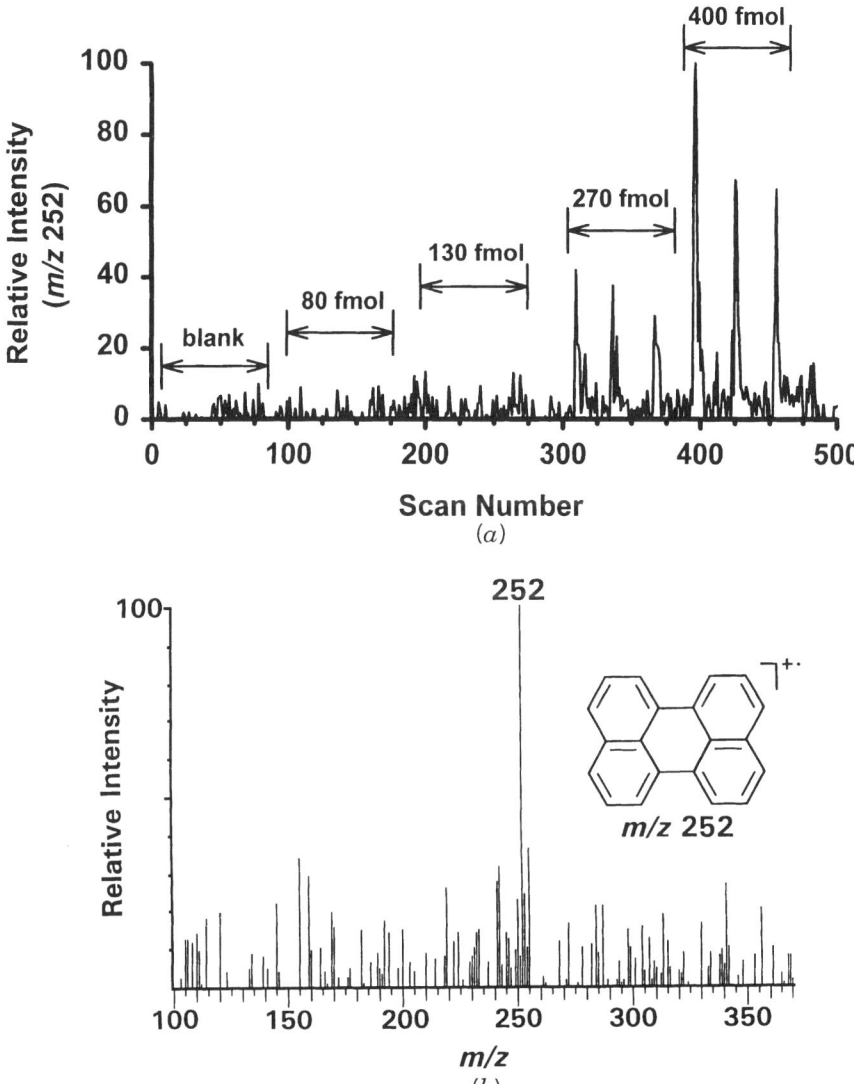

Figure 12. (a) Extracted ion current profile for the radical cation of perylene (m/z 252) obtained in a flow-injection experiment in which three replicate injections (0.5 μL) of a blank solution and analyte solutions of increasing analyte concentration (concentration shown in figure) were made into a flowing solution (20 μL/min) comprised of acetonitrile/methylene chloride (1/1 v/v) containing 1.0 mM lithium triflate. The perylene standards were prepared in a solution of the same composition as the carrier solution. (b) The averaged, background-subtracted ES mass spectrum obtained from the first 270-fmol injection of perylene as recorded in (a). (Adapted with permission from G. J. Van Berkel et al., *Anal. Chem.*, **1995**, *67*, 3643–3649. © 1995 American Chemical Society.)

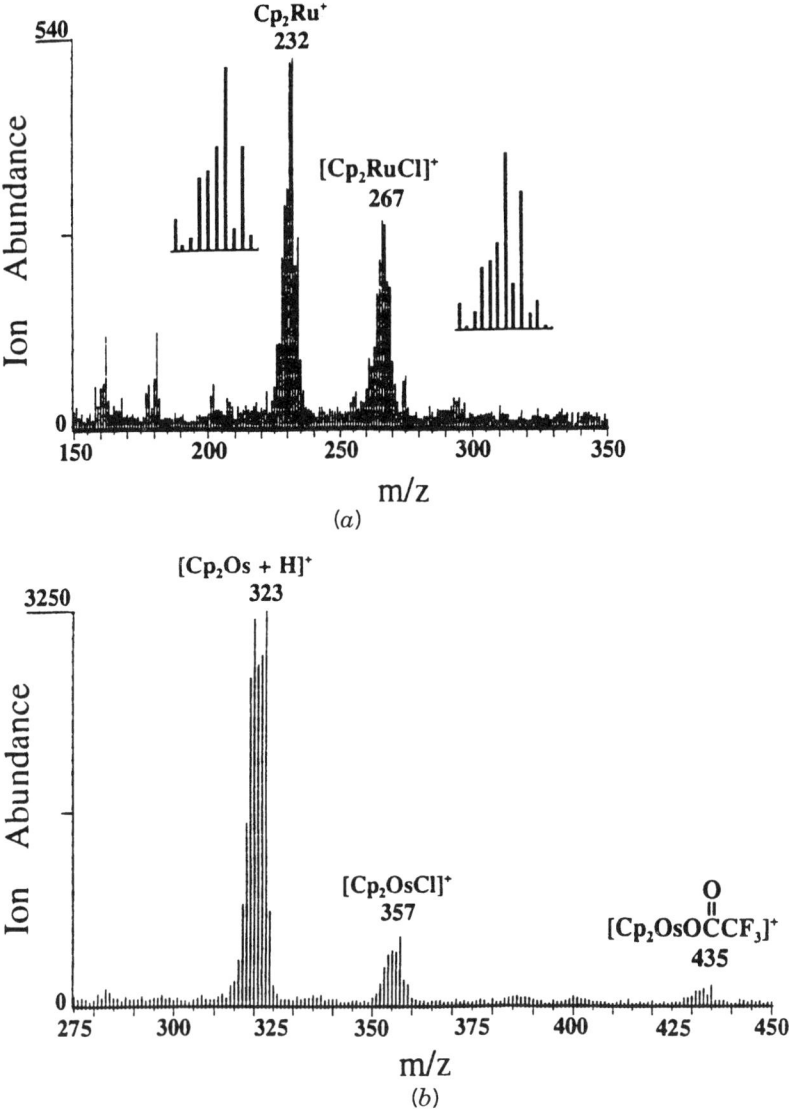

Figure 13. The ES mass spectra of (a) ruthenocene (10^{-3} M) in methylene chloride/0.05%TFA and (b) osmocene (10^{-3} M) in methylene chloride/1.0%TFA. (Reprinted with permission from X. Xu et al., *Anal. Chem.*, **1994**, *66*, 119–125. © 1994 American Chemical Society.)

solution, certain solvents/additives may have to be avoided. Particularly important with radical cations, for example, are reactions with nucleophilic solvents or solvent additives.

Cole and co-workers' (37) study of the electrochemical ionization of metallocenes in ESMS cites several examples of the occurrence of chemical follow-up reactions. For example, the oxidation products of ruthenocene (Cp_2Ru^+) and osmocene (Cp_2Os^+), which these workers were able to produce electrolytically in the ES capillary, are known to lack solution-phase stability and to be very susceptible to addition reactions with nucleophiles. The ES mass spectra of these metallocenes (Fig. 13), sprayed at 1.6 μL/min (a relatively slow flow rate) from a methylene chloride/0.5–1.0% trifluroacetic acid (TFA) solution, include peaks corresponding to the chloride-ion addition products ($[Cp_2RuCl]^+$ and $[Cp_2OsCl]^+$) and, in the case of osmocene, the trifluoroacetate addition product ($[Cp_2Os + trifluoacetate]^+$).

Electrochemists have been able to avoid or minimize these chemical follow-up reactions, thereby "stabilizing" and extending the lifetime of radical cations in solution, through the judicious choice of solvents and solvent additives (52). Protic and nucleophilic solvents (e.g., water and methanol) as well as nucleophilic solvent additives (e.g., acetate anion) are typically avoided. Aprotic, nonnucleophilic solvents such as rigorously purified and dried acetonitrile and methylene chloride are commonly employed in the electrochemical generation of radical cations. These same solvent systems have been used very successfully in our electrochemical ionization work (36,38). Chemical follow-up reactions are also exacerbated as flow rates through the ES capillary are decreased to increase electrolysis efficiency, because of the greater time the ion remains in solution following its formation. Therefore, to minimize chemical follow-up reactions for certain "unstable" ionic species, a trade-off may have to be made between a low flow rate, which offers the highest electrolysis efficiency, and a high flow rate, which minimizes the possibility that the chemical follow-up reactions will occur.

One must also be cognizant of the fact that while electrolysis of neutral molecules (or even charged analytes) in the ES capillary can be the source of ionic products observed in the gas phase, chemical-electron transfer reactions in solution and or in the gas phase might actually be the source of the ions observed. In fact, in several examples found in the literature, the electrolytic mechanism for ion formation appears to have been invoked incorrectly (see queries in Table 3). A first example stems from our own work (36). We often found that radical cation signals from certain compounds were enhanced when TFA was used as an additive in dried methylene chloride. We reasoned (on the basis of prior electrochemical reports) that TFA probably acted to stabilize the radical cations (i.e., slow the kinetics of radical cation loss by chemical follow-up reactions). We now know that in many cases this enhancement in signal was due to the increased conductivity of the solution upon addition of TFA and, therefore, a higher magnitude of i_{ES}, which in turn increased the degree of electrolysis in the capillary. Most significantly, we now know that the more easily oxidized of the compounds studied were probably chemically oxidized in that particular solvent system. We have determined that methylene chloride/ 0.1%TFA has an effective oxidation potential of 0.6–0.7 V vs. SCE (47). The radical cation of phenothiazine, for example, which we observed when spraying from this solvent system (ref. 36, Table 3), was almost certainly formed chemically rather than electrolytically.

In a more recent ESMS paper, Her and co-workers (57) reported observing the buckminsterfullerene (C_{60}) radical cation in positive-ion mode and the radical anion in negative-ion mode. Electrolytic oxidation [$E_{1/2} = 1.6$ V vs. SCE (58)] and reduction [$E_{1/2} = -0.36$ V vs. SCE (59)], respectively, of C_{60} within the metal ES capillary were deemed responsible for the formation of these ions. However, given that a stainless-steel capillary was employed and neat toluene was used as the solvent, electrolytic ionization in positive-ion mode is highly unlikely. Because of the nonpolar nature of this solvent system, the magnitude of i_{ES} would be expected to be only in the nanoampere range under true ES conditions (unfortunately, the value of i_{ES} was not reported). In positive-ion mode, because of the very positive potential for C_{60} oxidation, the small amount of faradaic current required could be supplied by reactions involving more easily oxidized contaminants in the solution or by oxidation of the capillary material. The electrolytic reduction of C_{60} to the radical anion is more easily achieved. Adding our inability to reproduce the positive-ion results in our laboratory at Oak Ridge, it appeared to us that the ions observed by Her and co-workers (57) were more likely formed via a gas-phase mechanism. Indeed, in a personal communication (60), Her acknowledged that a corona discharge at the capillary tip was the likely source for the C_{60} radical cations. As evidence, Her noted a black spot at the probe tip (caused by a discharge) and the fact that the C_{60}^+ ions were only observed beyond those voltage settings at which the low m/z background ions in their spectra [the same ions that have previously been noted to be indicative of a gas-phase discharge at the capillary tip (11,12)] increased dramatically.

It appears that Her and co-workers (57) actually performed, with an ES ion source, a form of atmospheric pressure chemical ionization (APCI) by striking a discharge at the end of the capillary tip. The C_{60} [IP = 7.6 eV (61)] might ionize directly in the discharge. Alternatively, the toluene might be ionized [IP = 8.8 eV (62)] and then charge exchange with C_{60}. A similar scheme was used by Boyd and co-workers (63) to ionize aromatic compounds, including fullerenes, by APCI. In that work, benzene was leaked into the discharge region and the radical cation of benzene formed in the discharge was used as a gas-phase, charge-transfer reagent.

An example for which both solution electron transfer and the electrolytic effect appear to be incorrectly invoked for radical cation formation in ESMS was reported by van Breemen (64). He reported that the gas-phase abundance for the molecular radical cations of certain carotenoids were enhanced by the addition of halogenated solvents to the ES solution which were thought to behave as chemical oxidants. Because low-abundance M^+ ion signals for the carotenoids were also observed without the addition of the halogenated solvents, both chemical and electrochemical formation mechanisms were claimed. Experiments in our laboratory at Oak Ridge indicated that the more likely major mechanism of ion formation was again a gas-phase process. The two halogenated solvents reported to enhance the gas-phase ion signal to the greatest degree (viz., TFA and heptafluoro-1-butanol) did not oxidize β-carotene in the solvent system that van Breemen used (viz., acetonitrile/methyl *tert*-butyl ether, 7/3 v/v). This was determined by comparison of the UV/visible spectrum of β-carotene in this solvent before and after the addition of the respective halogenated solvents [i.e., the ionic and neutral forms of β-carotene have different UV/visible spectra (47)]. The electrolytic formation mechanism can be refuted on the basis of the ions van Breemen observed in the gas phase. Cyclic-voltammetry (CV) studies of β-carotene show that it undergoes a nominal 2-electron oxidation at ~ 0.75 V and that the dication undergoes relatively rapid chemical follow-up reactions in all but the most pure and dry aprotic solvent systems (47). A major reaction product has been identified in previous studies as $(M^{2+} - H^+)^+$ formed by the loss of a proton from the dication. Thus, one would expect to see M^{2+} or $(M^{2+} - H^+)^+$ as the major gas-phase ions in the ES mass spectrum, not M^+, given an electrolytic formation mechanism for the ions. When we operated, under conditions known to maximize the electrolytic oxidation of species in solution (platinum capillary, 0.1 mM lithium triflate), these were the ions observed as illustrated by the spectra in Figure 14. In fact, the abundance of $(M^{2+} - H^+)^+$ was observed to increase at the expense of M^{2+} as flow rate decreased, as might be expected, because of the increased time for the chemical follow-up reactions to occur at the lower flow rate. One might speculate that in van Breemen's work a gas-phase discharge (or other gas-phase process) was responsible for forming the ions observed. The halogenated solvents might enhance ionization because of some change in the characteristics of the gas-phase process, owing to their presence, that favors the formation or preservation of the radical cations.

Our experience at Oak Ridge using the electrolytic nature of ES for analyte ionization has indicated that understanding the electron-transfer chemistry of the analyte of interest, in the solvent system being used, is highly desirable to avoid incorrectly interpreting the origin of certain ions observed in the ES mass spectra. The UV/visible spectra of the analyte solution aquired before and after addition of particular solvent additives may in many cases be used to confirm or to rule out the role of solution-based, chemical electron-transfer chemistry in ion formation. Cyclic voltammagrams, from the literature or acquired in-house, are extremely useful in determining if the proposed electrolytic reaction leading to the observed ion is feasible (e.g., determining if the redox potential for electron transfer is within the potential window available) and whether the ionic products of electron transfer are stable in solution for a long enough period to be detected in the gas phase.

D. Study of Redox Reactions

From an electrochemical point of view, ESMS might be viewed as a means to monitor the products of controlled-current electrolysis on line with mass spectrometry. Carrying out electrolysis in this fashion requires a very small amount of sample and provides the molecular weight, and potentially the structure, for the ionic products of the reactions. Given an experiment for

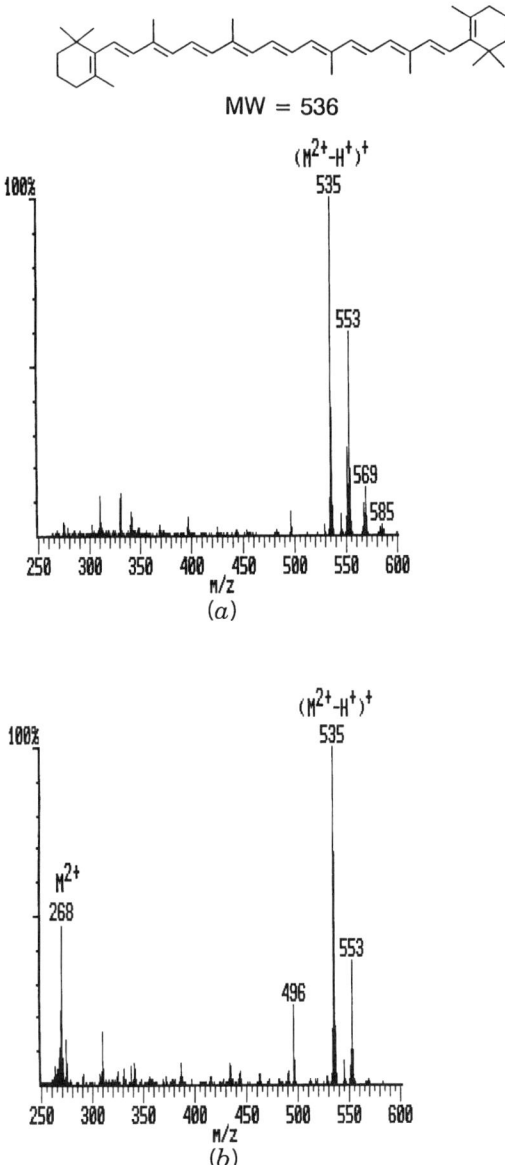

Figure 14. The ES mass spectra obtained by spraying β-carotene (11 μM) from a acetonitrile/methyl *tert*-butyl ether (70/30 v/v) solution containing 0.1 mM litium triflate at flow rates of (*a*) 5.0 μL/min and (*b*) 40 μL/min L/min using a platinum ES capillary (6.5 cm long, 0.4 mm o.d., 0.1 mm i.d.). The maximum reaction time following electrochemical ionization of β-carotene is calculated, assuming possible electrolysis along the total length of the capillary, to be approximately 10 s at 5.0 μL/min and 1.25 sec at 40 μL/min. $V_{ES} = 4$ kV.

III. ANALYTICAL IMPLICATIONS OF THE ELECTROLYTIC NATURE OF ELECTROSPRAY   101

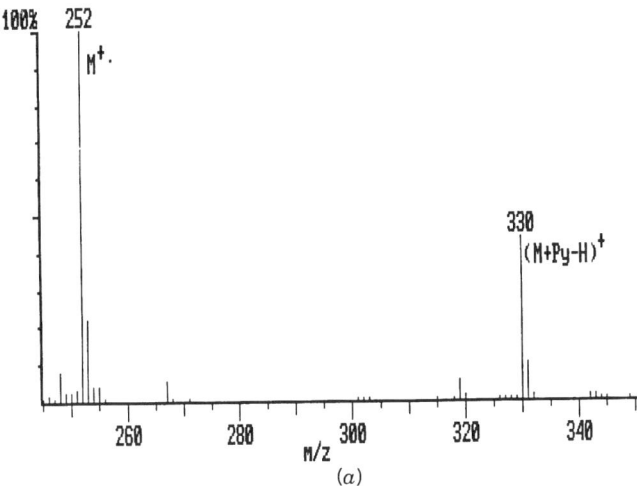

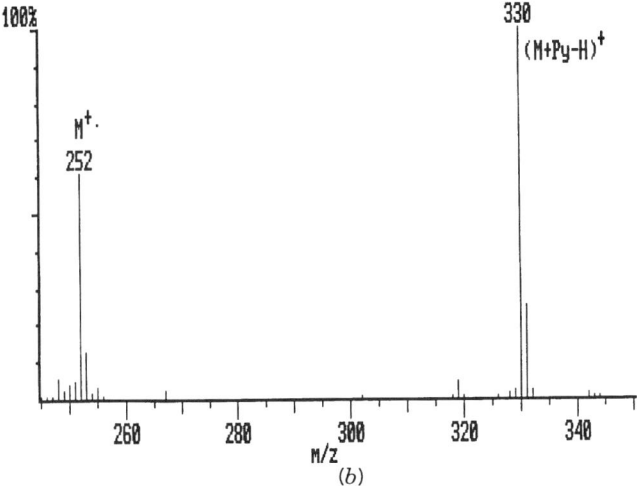

Figure 15. The ES mass spectra obtained by spraying a 27-μM perylene solution [acetonitrile/methylene chloride (1/1 v/v), 1.0 mM litium triflate] containing 0.1% pyridine by volume at flow rates of (a) 5.0 μL/min and (b) 2.5 μL/min L/min using a platinum ES capillary (6.5 cm long, 0.4 mm o.d., 0.1 mm i.d.). The maximum reaction time following electrochemical ionization of perylene is calculated, assuming possible electrolysis along the total length of the capillary, to be approximately 10 s at 5.0 μL/min and 20 sec at 2.5 μL/min. V_{ES} = 3.5 kV; py = pyridine. The reaction mechanism is illustrated in Scheme 1.

which controlled-current electrolysis is suitable, utilizing the "built-in" CCE process in ES would be expected to be much simpler than combining discrete electrochemical cells on-line with ESMS for the same purpose (65,66). To date, however, the use of ESMS to study redox reactions has not been the subject of a reviewed article in the literature. For that reason, only a brief description of some illustrative examples is presented here.

Electrochemical ionization in ESMS is in effect a use of ESMS to study the products of exhaustive controlled-current electrolysis on line with mass spectrometry. Used for this purpose, detection of the stable ionic products of a one-electron oxidation/reduction in the ES capillary is the analytical goal. Of course, electrolytic reactions involving unstable intermediates and the products of multiple electron transfers might also be studied. The data regarding unstable metallocene cations shown in Figure 13 and the β-carotene data in Figure 14 illustrate that redox reactions involving chemical follow-up reactions and multiple electron transfers might be studied with this approach. The β-carotene data further illustrate that data on the kinetics of the chemical follow-up reactions might be obtained.

It might also be possible to use ESMS to study selected nucleophilic reactions involving electrolytically generated radical cations, as illustrated by the data in Figure 15. The spectra shown were obtained by spraying a solution containing perylene and the nucleophile pyridine at about 0.1% by volume. The major ions observed are the electrolytically generated radical cation of perylene (m/z 252) and the pyridinium ion (m/z 330) formed from reaction of the radical cation with pyridine, as shown in Scheme I (67).

Again, the data show that the kinetics of the reaction might be followed by adjusting the flow rate through the capillary, which affects the amount of time available for the reaction to occur. As flow rate decreases (longer reaction time), the reaction product intensity increases at the expense of the intensity recorded for the initally formed radical cation.

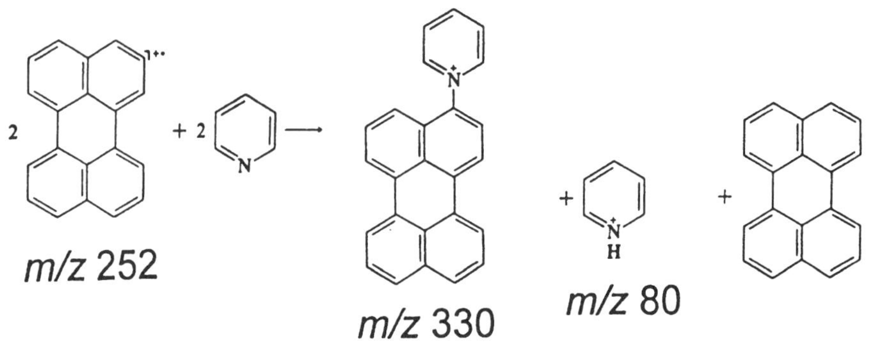

Scheme 1

ACKNOWLEDGMENTS

Special thanks are due Prof. Feimeng Zhou (University of Wisconsin-Eau Claire) who, during his postdoctoral appointment at ORNL, taught the author much electrochemistry and who was instrumental in developing our electrolytic characterization and exploitation of the ES ion source. The author also thanks Prof. Chris G. Enke (University of New Mexico) for many helpful discussions and suggestions and Prof. Kelsey D. Cook (University of Tennessee) for reprints containing the electrolytic description of the electrohydrodynamic ion source and for helpful discussions. The work described herein performed in the author's laboratory was sponsored by the Division of Chemical Sciences, Office of Basic Energy Sciences, United States Department of Energy, under contract DE-AC05-96OR22464 with Oak Ridge National Laboratory, managed by Lockheed Martin Energy Research Corp.

REFERENCES

1. Baily, A. G. *Electrostatic Spraying of Liquids*; Wiley: New York, 1988.
2. Zeleny, J. *Phys. Rev.* **1917**, *10*, 1–6.
3. Michelson, D. *Electrostatic Atomization*; Adam Hilger: New York, 1990.
4. *J. Aersol Sci.* **1994**, *25*; Special Issue: Electrosprays: Theory and Applications.
5. Dole, M.; Mach, L. L.; Hines, R. L.; Mobley, R. C.; Ferguson, L. P.; Alice, M. B. *J. Chem. Phys.* **1968**, *49*, 2240–2249.
6. Mack, L. L.; Kralik, P.; Rheude, A.; Dole, M. *J. Chem. Phys.* **1970**, *52*, 4977–4986.
7. Clegg, G. A.; Dole, M. *Bipolymers* **1971**, *10*, 821–826.
8. Teer, D.; Dole, M. *J. Polymer Sci.* **1975**, *13*, 985–995.
9. Thomson, B. A.; Iribarne, J. V. *J. Chem. Phys.* **1979**, *71*, 4451–4463.
10. Iribarne, J. V.; Dziedzic, P. J.; Thomson, B. A. *Int. J. Mass Spectrom. Ion Phys.* **1983**, *50*, 331–347.
11. Yamashita, M.; Fenn, J. B. *J. Phys. Chem.* **1984**, *88*, 4451–4459.
12. Yamashita, M.; Fenn, J. B. *J. Phys. Chem.* **1984**, *88*, 4671–4675.
13. Whitehouse, C. M.; Dreyer, R. N.; Yamashita, M.; Fenn, J. B. *Anal. Chem.* **1985**, *57*, 675–679.
14. Alexandrov, M. L.; Gall, L. N.; Krasnov, M. V.; Nikolaev, V. I., Shkurov, V. A. *Zh. Anal. Khim.* **1985**, *40*, 1272–1236.
15. Olivares, J. A.; Nguyen, N. T.; Yonker, C. R.; Smith, R. D. *Anal. Chem.* **1987**, *59*, 1232–1236.
16. Smith, R. D.; Olivares, J. A.; Nguyen, N. T.; Udseth, H. R. *Anal. Chem.* **1988**, *60*, 436–441.
17. Smith, R. D.; Barinaga, C. J.; Udseth, H. R. *Anal. Chem.* **1988**, *60*, 1948–1952.
18. Bruins, A. P.; Covey, T. R.; Henion, J. D. *Anal. Chem.* **1987**, *59*, 2642–2646.
19. Covey, T. R.; Bruins, A. P.; Henion, J. D. *Org. Mass Spectrom.* **1988**, *23*, 178–186.

20. Covey, T. R.; Bonner, R. F.; Shushan, B. I.; Henion, J. *Rapid Commun. Mass Spectrom.* **1988**, *2*, 249–256.
21. Lee, E. D.; Mück, W.; Henion, J.; Covey, T. R. *Biomed. Environ. Mass Spectrom.* **1989**, *18*, 844–850.
22. Huang, E. C.; Wachs, T.; Conboy, J. J.; Henion, J. D.; *Anal. Chem.* **1990**, *62*, 713A–725A.
23. Smith, R. D.; Loo, J. A.; Edmonds, C. G.; Barinaga, C. J.; Udseth, H. R. *Anal. Chem.* **1990**, *62*, 882–899.
24. Kebarle, P.; Tang, L. *Anal. Chem.* **1993**, *65*, 972A–986A.
25. Tang, L.; Kebarle, P. *Anal. Chem.* **1991**, *63*, 2709–2715.
26. Tang, L.; Kebarle, P. *Anal. Chem.* **1993**, *65*, 3654–3668.
27. Ikonomou, M. G.; Blades, A. T.; Kebarle, P. *Anal. Chem.* **1991**, *63*, 1989–1998.
28. Pfeifer, R. J.; Hendricks, C. D. *AIAA J.* **1968**, *6*, 496–502.
29. Smith, D. P. H. *IEEE Trans. Ind. Appl.* **1986**, *1A–22*, 527–535.
30. Hayati, I.; Bailey, A. I.; Tadros, Th. F. *J. Colloid Interface Sci.* **1987**, *117*, 205–221.
31. Hayati, I.; Bailey, A. I.; Tadros, Th. F. *J. Colloid Interface Sci.* **1987**, *117*, 222–230.
32. Simons, D. S.; Colby, B. N.; Evans, C. A., Jr. *Int. J. Mass Spectrom. Ion Phys.* **1974**, *15*, 291–302.
33. Stimpson, B. P.; Evans, C. A., Jr. *Biomed. Mass Spectrom.* **1978**, *5*, 52–63.
34. Cook, K. D. *Mass Spectrom. Rev.* **1986**, *5*, 467–519.
35. Blades, A. T.; Ikonomou, M. G.; Kebarle, P. *Anal. Chem.* **1991**, *63*, 2109–2114.
36. Van Berkel, G. J.; McLuckey, S. A.; Glish, G. L. *Anal. Chem.* **1992**, *64*, 1586–1593.
37. Xu, X.; Nolan, S. P.; Cole, R. B. *Anal. Chem.* **1994**, *66*, 119–125.
38. Van Berkel, G. J.; Zhou, F. *Anal. Chem.* **1995**, *67*, 2916–2923.
39. *Encyclopedia of Electrochemistry of the Elements*; Bard, A. J., Ed.; Marcel Dekker: New York, 1973.
40. Van Berkel, G. J.; McLuckey, S. A.; Glish, G. L. *Anal. Chem.* **1991**, *63*, 1098–1109.
41. Bockris, J. O'M.; Reddy, A. K. N. *Modern Electrochemistry*; Plenum Press: New York, 1973, Vol. 1.
42. Bard, A. J.; Faulkner, L. R. *Electrochemical Methods*; Wiley: New York, 1980.
43. Štulik, K.; Pacáková, V. *Electroanalytical Measurements in Flowing Liquids*; Ellis Horwood; Chichester, West Sussex, England, 1987.
44. Malmstadt, H. V.; Enke, C. G.; Crouch, S. R. *Electronics and Instrumentation for Scientists*; Benjamin/Cummings: Menlo Park, CA, 1981.
45. Juhasz, P.; Ikonomou, M. G.; Blades, A. T.; Kebarle, P. In *Methods and Mechanisms for Producing Ions from Large Molecules*; Standing, K. G.; Ens, W., Eds.; Plenum Press: New York, 1991, pp. 171–184.
46. Heineman, W. R. *J. Chem. Ed.* **1983**, *60*, 305–308.
47. Van Berkel, G. J.; Zhou, F. *Anal. Chem.* **1994**, *66*, 3408–3415.
48. Fuhrhop, J.-H.; Mauzerall, D. *J. Am. Chem. Soc.* **1969**, *91*, 4174–4181.
49. *Corrosion Mechanisms in Theory and Practice*; Marcus, P.; Oudar, J., Eds.; Marcel Dekker: New York, 1995.

REFERENCES

50. Ijames, C. F.; Dutky, R. C.; Fales, H. M. *J. Am. Soc. Mass Spectrom.* **1995**, *6*, 1226–1231.
51. Hop, C. E. C. A.; Saulys, D. A.; Gaines, D. F. *J. Am. Soc. Mass Spectrom.* **1995**, *6*, 860–865.
52. Bard, A. J.; Ledwith, A.; Shine, H. J. In *Advances in Physical Organic Chemistry*; Gold, V.; Bethell, D., Eds.; Academic Press: New York, 1976; Vol. 13, pp. 155–278.
53. Foster, R. *Organic Charge Transfer Complexes*; Academic Press: New York, 1969.
54. Van Berkel, G. J.; Zhou, F. *Anal. Chem.* **1995**, *67*, 3643–3649.
55. Janz, G. J.; Tomkins, R. P. T. *Nonaqueous Electrolytes Handbook*; Academic Press: New York, 1973, Vol. II.
56. McLuckey, S. A.; Van Berkel, G. J.; Glish, G. L.; Henion, J. D.; Huang, E. C. *Anal. Chem.* **1991**, *63*, 375–383.
57. Liu, T.-Y.; Shiu, L.-L.; Luh, T.-Y.; Her, G.-H. *Rapid Commun. Mass Spectrom.* **1995**, *9*, 93–96.
58. Haufler, R. E.; Conceicao, J.; Chibante, L. P. F.; Chai, Y.; Byrne, N. E.; Flanagan, S.; Haley, M. M.; O'Brien, S. C.; Pan, C.; Xiao, Z.; Billups, W. E.; Ciufolini, M. A.; Hauge, R. H.; Margrave, J. L.; Wilson, L. J.; Curl, R. F.; Smalley, R. E. *J. Phys. Chem.* **1990**, *94*, 8634.
59. Dubois, D.; Kadish, K. M.; Flanagan, S.; Wilson, L. J. *J. Am. Chem. Soc.* **1991**, *113*, 7773–7774.
60. Her, G.-R., personal communication, 1995.
61. Zimmerman, J. A.; Eyler, J. R.; Bach, S. B. H.; McElvany, S. W. *J. Chem. Phys.* **1991**, *94*, 3556–3562.
62. Lias, S. G.; Bartmess, J. E.; Liebman, J. F.; Holmes, J. L.; Levin, R. D.; Mallard, W. G. *J. Phys. Chem. Ref. Data*, Vol. 17, Suppl. 1, 1988.
63. Anacleto, J. F.; Boyd, R. K.; Pleasance, S.; Quilliam, M. A.; Howard, J. B.; Lafleur, A. L.; Makarovsky, Y. *Can. J. Chem.* **1992**, *70*, 2558–2568.
64. van Breemen, R. B. *Anal. Chem.* **1995**, *67*, 2004–2009.
65. Bond, A. M.; Colton, R.; D'Agnostino, A.; Downard, A. J.; Traeger, J. C. *Anal. Chem.* **1995**, *67*, 1691–1695.
66. Zhou, F.; Van Berkel, G. J. *Anal. Chem.* **1995**, *67*, 3643–3649.
67. Yoshida, K. *Electrooxidation in Organic Chemistry*; Wiley: New York, 1984.
68. Van Berkel, G. J.; Zhou, F. *J. Am. Soc. Mass Spectrom.* **1996**, *7*, 157–162.
69. Dupont, A.; Gisselbrecht, J-P.; Leize, E.; Wagner, L.; Van Dorsselaer, A. *Tetrahedron Lett.* **1994**, *35*, 6083–6086.
70. Van Berkel, G. J.; Quinoñes, M. A.; Quirke, J. M. E. *Energy & Fuels* **1993**, *7*, 411–419.
71. Bond, A. M.; Colton, R.; D'Agnostino, A.; Downard, A. J.; Traeger, J. C. *Anal. Chem.* **1995**, *67*, 1691–1695.
72. Hiraoka, K.; Aizawa, K.; Murata, K.; Fujimaki, S. *J. Mass Spectrom. Soc. Jpn.* **1995**, *43*, 77–83.
73. Zhou, F.; Van Berkel, G. J.; Donovan, B. T. *J. Am. Chem. Soc.* **1994**, *116*, 5455–5486.

CHAPTER 3

ESI Source Design and Dynamic Range Considerations

ANDRIES P. BRUINS

University Centre for Pharmacy, Groningen, The Netherlands

	Abstract	107
I.	Electrospray nebulization	108
II.	Electrospray construction and operation	110
	A. High-voltage connection	110
	B. Electrospray and electrical discharge	114
	C. Flow rate and sensitivity	115
	D. Position of the sprayer inside the source	116
III.	Atmospheric pressure ionization source construction	118
	A. Free jet expansion into vacuum	118
	B. Cluster ion formation	120
	1. Prevention	120
	2. Curing	122
	C. Focusing of ions at atmospheric pressure	122
	D. Vacuum system design	122
	E. Vacuum system and sensitivity	124
	F. Ion-sampling orifice	126
	G. Ion optics between sampling orfice and mass analyzer	129
IV.	Up-front collision-induced dissociation	130
V.	Mass-scale calibration	132
VI.	Dynamic range in electrospray	133
	References	135

ABSTRACT

Mass spectrometer ion sources are normally located inside a high-vacuum envelope. An ion source operating at atmospheric pressure is better suited, if not essential, for a growing number of applications. Highly polar, thermolabile, and

Electrospray Ionization Mass Spectrometry, Edited by Richard B. Cole.
ISBN 0-471-14564-5 © 1997 John Wiley & Sons, Inc.

ionic samples and biopolymers require the application of electrospray ionization (ESI) at atmospheric pressure. Electrospray ionization source design is a combination of charged-aerosol generation techniques for ion formation and atmospheric pressure source and associated vacuum technology for mass separation and detection. Limitations inherent in simple ES have been removed in mechanically assisted ESs. Vacuum-system design and ion optics are of decisive importance for sensitivity and reliability. Collision-induced dissociation just outside the source is inherent in atmospheric-pressure sources. It is easy to use, effective, and comes at no cost. Limited dynamic range at the high sample concentration end is a fundamental problem in ES. Dynamic range extension by different source design is not within reach using presently available technology.

I. ELECTROSPRAY NEBULIZATION

The nebulization of liquids by electrical forces is carried out on a very small scale in the recently developed microelectrospray accessory for mass spectrometry, and on a grand scale in the electrostatic spray painting of automobiles and the electrostatic spray deposition of pesticides on crops (1). Electrospray is the dispersion of a liquid into electrically charged droplets, and as such combines two processes: droplet formation and droplet charging. The formation of small, micron-sized droplets does not present a problem if the liquid flow rate, surface tension, and electrolyte concentration are low. An increase of one or more of these variables makes it more difficult for the electric field to produce the desired charged aerosol for mass spectrometry. The electric field strength at the sprayer tip can be increased to try and overcome the adverse effects of the aforementioned three variables, an electric field that is too high will give rise to an electrical discharge that accompanies the ES process. A discharge can be tolerated in some spray applications, but is detrimental in ESMS. Electrical discharge is particularly troublesome in the formation of negatively charged droplets.

Modifications to the simple ES system, as shown in Figure 1, are aimed at increasing the tolerance toward adverse effects of high liquid-flow rate, surface tension, and electrolyte concentration. Dilution of an aqueous solution with an organic solvent reduces surface tension. Coaxial addition of a sheath flow of methanol, acetonitrile, ethanol, isopropanol, or 2-methoxyethanol to the sample solution at the tip of the spray capillary was first used for the combination of capillary electrophoresis with ESMS and later also used for sample infusion and LC coupling with ESMS (2,3). In sheath-flow-assisted ES, the electric field alone has to disperse *and* charge the liquid in one operation.

In industrial applications of ES, the input of mechanical energy is used for the dispersion of liquids at high flow rates. Droplet charging is done by exposing the mechanical sprayer to a high electric field. For example, a rotating disk combined with an electric field created by a 100-kV power supply may be used in spray painting.

The assistance of a high-velocity gas flow or mechanical vibration is used in

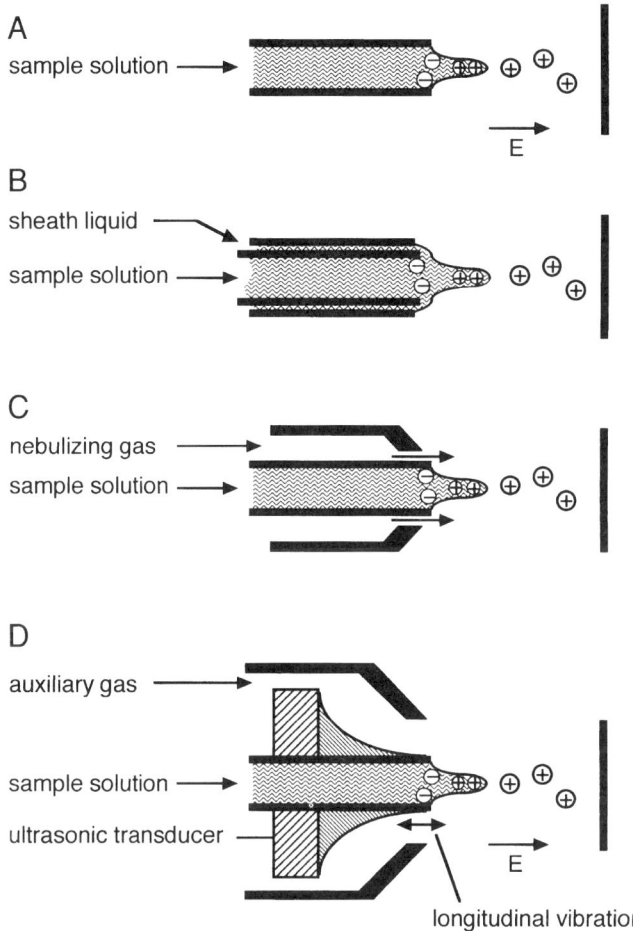

Figure 1. Aerosol formation by ES. (*a*) Simple ES; (*b*) ES with sheath flow; (*c*) ES with pneumatic assistance; (*d*) ES with ultrasonic assistance.

ESMS (4–6). In a simple approximation, the pneumatic or ultrasonic nebulizer takes care of aerosol formation, while the electric field does the droplet charging. When compared with "pure" ES, pneumatically assisted ES can handle aqueous solutions and higher flow rates without the need for critical adjustment and can be operated at a lower field strength so that electrical discharge is eliminated. Ultrasonic assistance offers the same advantages, but is a more complex and more expensive combination of mechanical and electronic devices. The trade names of some such devices are IonSpray (SCIEX) and Ultraspray (Analytica of Branford).

Charged droplets generated by ES spread out in a wide angle. A moderately fast coaxial gas flow can be used to try and keep the droplets in a narrower beam. Such pneumatic focusing can be combined with a coaxial sheath flow (3,7) and with ultrasonic nebulization (6). The focusing gas flow is not intended for pneumatic nebulization. Pneumatic focusing increases the number of ions transported into the vacuum envelope of the mass spectrometer. Pneumatically assisted ES produces a narrow beam of droplets by the very nature of the high-velocity gas flow used in the concentric nebulizer. High-flow ES in Finnigan instruments makes use of pneumatic nebulization and a separate auxiliary (focusing) gas flow that prevents wetting of the nebulizer by redeposition of droplets on the outside of the sprayer.

Pneumatically and ultrasonically assisted ES can work at liquid flow rates up to a few hundred microliters per minute. Pure, unassisted ES can be extended to very low flow rates. Microelectrospray has the advantage of extremely efficient use of a sample solution: one microliter can produce ions for about 40 min, long enough for performing a number of MS/MS experiments. A very low flow rate microsprayer is compatible with the electroosmotic flow in a capillary electrophoresis column, eliminating the need for a make-up flow.

A regular ES device is constructed from approximately 0.3 mm o.d., 0.1 mm i.d. stainless-steel tubing, or a coaxial arrangement of fused silica and stainless-steel capillaries. Stability at low flow in a microsprayer requires a miniaturized version having narrower inside and outside diameter A 10- to 50-μm i.d. fused-silica capillary is tapered on the outside by etching with hydrofluoric acid (8) or by mechanical abrasion (9). Electrical contact with the liquid stream is made by the application of silver paint (8), or by deposition of gold (9). Making contact at the tip is necessary for CE/MS (8,9). For sample introduction from a loop injector or from a syringe pump, it is sufficient to make electrical contact upstream from the tip by clamping the narrow-bore, fused-silica capillary in a metal union (10,11). Caprioli and co-workers have prepared an integrated microcolumn/microsprayer with the aim of online concentration of dilute samples (11). Wilm and Mann have constructed a microsprayer for the analysis of small volumes of peptide solutions at a flow rate of about 25 nL/min (12). A 1-mm o.d, 0.6-mm i.d. glass capillary is pulled out to an inside diameter of 1–3 μm by means of an electrode puller. The outside of the capillary is coated with gold. Approximately 1 μL of a sample solution is loaded into the capillary, and gas pressure is applied to push the liquid toward the tip.

II. ELECTROSPRAY CONSTRUCTION AND OPERATION

A. High-Voltage Connection

The tip of an ES capillary is exposed to a high electric field. In principle, the field can be generated by connecting the sprayer to a high-voltage supply and the source to ground, by grounding the sprayer and connecting the source to high voltage, and by connecting both the sprayer and the source to separate power supplies set to different voltages. Selection of either of these options depends on the mass analyzer and the system chosen for the transportation of ions from the

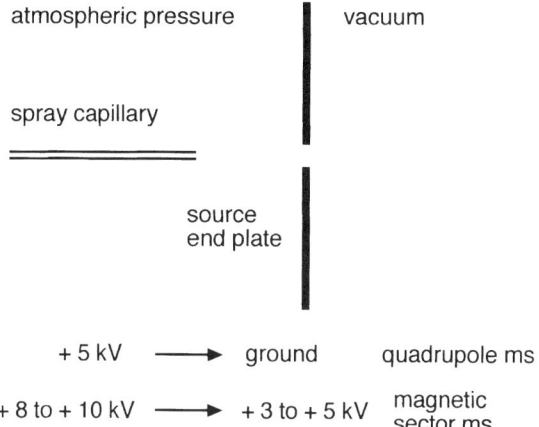

Figure 2. Voltage arrangements for ES capillary and source; no isolation between source end plate and MS accelerating voltage.

atmospheric-pressure source region into the vacuum of the mass analyzer. In a quadrupole mass spectrometer, the ion source can be at ground potential and the sprayer at up to plus or minus 5 kV (4,13). In most magnetic-sector mass spectrometers the accelerating voltage is limited to 3–5 kV during ES operation, so that the voltage on the sprayer is 8–10 kV with respect to ground (see Fig. 2).

Safety is not a major problem, since the source can be designed for protection of the operator by the use of appropriate insulation and safety interlocks. If the sprayer is at high voltage, the connection to the HPLC or infusion pump can be made from fused-silica capillary tubing. One should keep in mind that a fused-silica capillary filled with an electrolyte solution is not an insulator but a resistor, which may conduct enough to give the operator an itch if the needle or plunger of a syringe in an infusion pump is touched during ES operation. For safety reasons, one should not make or break a connection between syringe and transfer line when the ES high voltage is on. As a further safety measure, the needle of the syringe in an infusion pump can be connected to ground.

Of more practical value is the question of whether high voltage on the sprayer interferes with operation of a capillary electrophoresis system, and whether the high voltage has an influence on the transportation of a sample through the fused-silica transfer line.

Let us first consider capillary electrophoresis. Off-line CE experiments are usually conducted by applying 30 kV between anode and cathode. In CE/MS with positive ion operation, the anode is at 30 kV, while the cathode is at the high voltage of the ES interface, for example, at 5 kV. The voltage drop over the column is reduced to 25 kV, resulting in longer analysis time.

There are more electrical implications to CE/MS other than just the voltage drop over the CE column. In Figure 3A the simplest CE/MS system is drawn schematically. Current flowing from the CE power supply is carried away only to

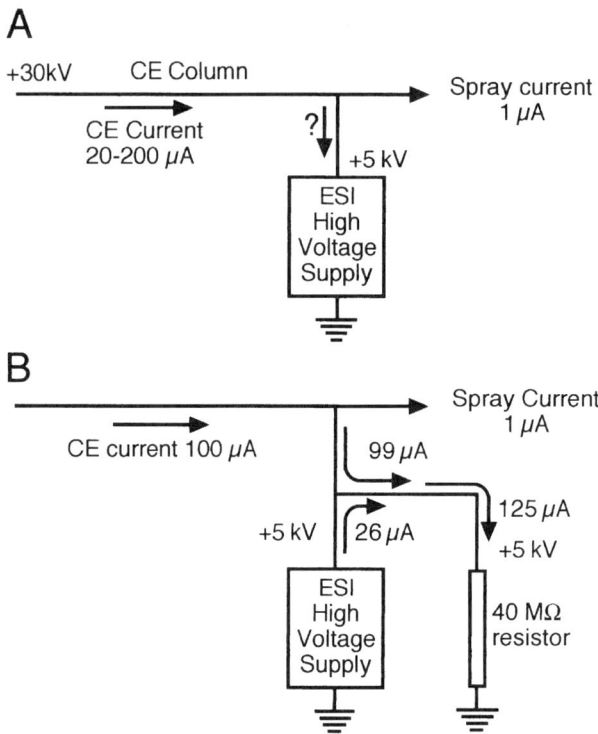

Figure 3. (*A*) Currents flowing in a simple CE/MS system using a CE high-voltage supply and an ES high-voltage supply; CE current, 100 μA; charge transportation on droplets, 1 μA; ES power supply has to sink 99 μA. (*B*): 40-MΩ load resistor connected between ES supply and ground carries 125 μA at 5 kV (Ohm's law); ES power supply has to deliver 26 μA to maintain current balance.

a minute extent as charge on sprayed droplets. Most of the current has to flow into the ES power supply, which has been designed to *supply* current, not to *sink* current. If the supply has to sink too much current, its internal feedback circuit, which stabilizes the output voltage to a preset value, cannot keep the power supply under control, resulting in an uncontrolled rise of output voltage and unstable ES. By connecting an additional load resistor, as shown in Figure 3*B*, the power supply is forced to deliver current, and output stability is maintained. The user should consult the high-voltage power supply specifications for selection of the correct resistance value. Of course these CE/MS current considerations do not apply if the spray capillary is at ground.

Less obvious and often overlooked is the possibility of electromigration in the transfer line between the electrosprayer and a syringe pump used for sample infusion. To conserve sample, the liquid flow rate may be reduced to 1 μL/min, corresponding to a linear velocity of 50 cm/min in a 50-μm i.d. capillary or 12 cm/min in a 100-μm i.d. capillary. If a positively charged sample molecule has a mobility of 10^{-3} cm^2/Vs and if a 50-cm long transfer capillary is used with 5 kV on the electrosprayer, the electrophoretic velocity of the sample is 6 cm/min in the direction from the sprayer back into the syringe. In this example, there should be no problem if a 50-μm i.d. transfer capillary is used: the liquid velocity far exceeds the opposing electrophoretic velocity. A 100-μm i.d. capillary (liquid velocity 12 cm/min at 1 μL/min) can still be used at 5 kV. The situation would be critical at 1 μL/min in a 100-μm i.d. capillary in the case of ES in a magnetic sector instrument, where the sprayer may be at $+8$ kV or higher. The narrower the internal diameter of the transfer line, the higher the liquid velocity and the smaller is the chance that electrophoretic velocity exceeds the liquid velocity. Inside the barrel of the syringe in the infusion pump, the linear liquid velocity is very low. If the syringe needle is not connected to ground, and if the plunger is pushed forward by a grounded block of metal, a voltage drop exists over the length of the liquid column inside the barrel. Electrophoretic transport of a sample may now take place inside the syringe, as shown schematically in Figure 4. The electric field inside the syringe is eliminated by connecting the syringe needle to ground.

Operation of the electrosprayer at ground potential obviates the reduction of voltage drop in CE/MS and the possibility of electromigration during infusion of a sample. The electric field necessary for the formation of positively charged droplets is generated by floating the inside wall of the ion source at negative high voltage. Ions are drawn toward the opposing wall. To accelerate ions into a mass analyzer inside the vacuum system, a more negative potential has to be applied, which would lead to a very unpractical arrangement of negative high voltages. An elegant system that eliminates the need for acceleration of positive ions toward negative high voltage is the glass capillary for the transportation of ions from the atmospheric-pressure spray region into the vacuum of the mass

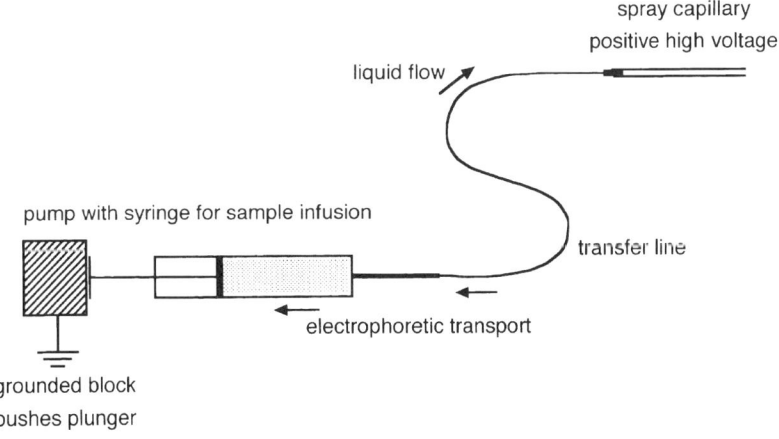

Figure 4. Possibility of electromigration of positively charged sample ions against liquid flow.

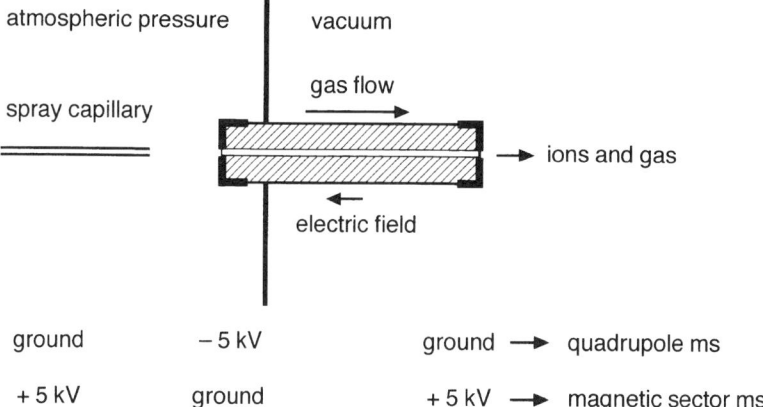

Figure 5. Voltage arrangements for ES capillary and source; isolation between source end plate and accelerating voltage by a glass capillary for ion transport.

spectrometer (see Fig. 5) (14). Ions generated in the atmospheric pressure region of the source either by ES or by atmospheric pressure chemical ionization are drawn into the vacuum region via a glass tube that is metallized at both ends. The linear velocity of the gas flow in the tube is high enough to drag ions into the vacuum even if the inlet of the tube is at a negative high voltage and the outlet is at ground potential. The grounded outlet now serves as a source of ions that can easily be focused into a quadrupole mass analyzer. The atmospheric pressure source and ion optics plus mass analyzer are electrically isolated from one another, which is particularly attractive for magnetic sector instruments. Now the outlet end of the glass capillary is at the accelerating potential of the mass spectrometer, but there is no need to have the spray chamber and the spray capillary at increasingly high voltages. In practice there is a limit to the voltage difference between inlet and outlet of the glass capillary. Apparently, it is not feasible to have the inlet at −5 kV and the spray capillary at ground, as is done in quadrupole instruments. Magnetic sector instruments that incorporate a glass capillary have the spray chamber and capillary inlet at ground, while the spray capillary is held at +5 kV, as shown in Figure 5 (15,16).

B. Electrospray and Electrical Discharge

Corona discharge between the ES capillary and its counterelectrode drastically changes the appearance of the mass spectrum. Without a discharge, the spectrum represents ions present in the solution that is dispersed into charged droplets. When a discharge takes place, the spectrum represents the products of ion molecule reactions, as is the case in atmospheric-pressure chemical ionization. At the same time, many ions present in solution can no longer be observed (13). A distinction must be made here between sample ions that exist only as ions in solution, such as quaternary ammonium ions, and ions that can be formed in solution but can also be generated by ion molecule reactions, such as amines and other reasonably volatile neutrals. Quaternary ammonium ions completely disappear when a discharge is established. Ionized neutrals may still be observed.

Electrical discharge takes place very readily if the ES needle is at a negative potential with respect to the spray chamber, since field emission of electrons from a sharp point does not require a very high field. Electrons are eliminated by flushing the space around the spray tip with an electron scavenging gas such as oxygen (17) or SF_6, or by the addition of a halogenated solvent to the liquid stream (18,19).

In a discharge, both positive and negative charge carriers are formed that will recombine with droplets having the opposite charge (20). As a result, droplet charge is neutralized and the formation of sample ions from charged droplets can no longer take place. Electrical discharge is prevented to a large extent by reduction of the electric field at the sprayer tip. Pneumatically or ultrasonically assisted ES works well at a lower electric field than pure ES.

C. Flow Rate and Sensitivity

Electrospray is a low-flow-rate technique. When the flow rate is increased for a given sample concentration, the sample ion signal does not increase. So in terms of sample concentration, sensitivity remains constant, but in terms of mass flow, the sensitivity drops when the flow rate of the sample solution is increased. This is an unusual situation in mass spectrometry because a mass spectrometer operating in the electron ionization or chemical ionization mode is a mass-flow sensitive detector.

Apparent concentration sensitivity of ES can be attributed to a decrease in droplet charging efficiency and a shift toward larger diameters in the droplet size distribution if the liquid flow rate is increased. Both effects lead to a lower ionization efficiency. The reduced efficiency of the release of ions from droplets is approximately compensated for by the introduction of more sample molecules per unit time into the interface so that the signal level at the detector remains constant.

If the liquid flow rate is increased too much, the ion signals become lower and less stable. The practical upper limit to flow rate in pure ES is 10–20 $\mu L/min$, depending on the composition of the solvent and on the use of a coaxial sheath flow (see Section I). Pneumatically assisted ES has been used up to 200 $\mu L/min$. Manufacturers have claimed good performance of ultrasonically assisted ES up to 1 mL/min. High-flow ES is always combined with the supply of heat to assist evaporation of solvents. High-flow pneumatically assisted ES (ionspray) appears to require a heated zone in the source or a heated tube for the transport of ions from the source into the vacuum in combination with a so-called liquid shield (21). In a commercial embodiment of high-flow ionspray (called Turbo-Ion-Spray), the spray plume is mixed with a hot-air flow from a heat gun in the region in front of the ion-sampling orifice of the source (22).

In ion sources based on the heated tube design by Chait and co-workers (23), as used in Finnigan mass spectrometers, the tube temperature is raised in the case of high-flow operation. At low liquid flow rate, the desolvation of charged droplets takes place mostly inside the atmospheric pressure source region, and ions are drawn into the heated tube together with solvent vapor and gas supplied to the source. At high flow rate, not only desolvated ions but also charged aerosol is drawn through the tube. Desolvation of the aerosol inside the tube generates sample ions that are transported into the mass analyzer, but nonvolatile material may be deposited inside the tube and eventually block the tube.

High-flow ES is compatible with 2-mm and wider bore columns in liquid chromatography. The need for eluent splitting is eliminated, but at the same time the load of nonvolatile contaminants on the ion source or transfer tube is much higher. All in all, it is probably better to feed a low flow rate into the sprayer and use a splitter, unless a gain in sensitivity at high flow can be substantiated, or the complexity of a splitter has to be avoided. A splitter can be set up using two capillary tubes of different length and diameter. At a given pressure drop, the flow rate through a capillary is proportional to the fourth power of the diameter and inversely proportional to the length of the capillary. A 1-m long 50-μm i.d. transfer line together with a 16-cm long, 100-μm side arm gives a split ratio of 1 : 100.

D. Position of the Sprayer Inside the Source

In the first published report of successful ES mass spectrometry (13), the spray capillary was positioned in the center of the source, in front the ion-sampling orifice of the ion source. When the experiments were repeated in Henion's group at Cornell University, the ion source of the SCIEX mass spectrometer was fully open and allowed the sprayer to be moved around during full operation. By moving to an off-axis position, the ion signal observed by the mass spectrometer was more stable, and at least as high as in the on-axis position (Fig. 6). It is desirable to take ions, but no droplets, into the mass analyzer. Charged droplets transported into the mass spectrometer may impinge on elements of the ion optics or mass analyzer and create bursts of ions that appear as spikes in the mass spectra. In the core of the aerosol generated by ES, the droplet diameter is larger than in the perimeter of the spray. Liberation of sample ions from small droplets should be more effective in the perimeter than in the core of the spray. There is more chance of incomplete droplet desolvation in the core of the spray than in the perimeter.

Off-axis positioning can be taken a step further by placing the sprayer in a diagonal position inside the source. The spray is not aimed at the sampling orifice, but at a position 1 cm beyond, in order to reduce the chance of shooting droplets into the mass spectrometer. Diagonal positioning is most effective for pneumatically assisted ES and is called "articulated ion spray." Stability is improved without loss of sensitivity.

Pure ES without pneumatic or ultrasonic assistance is more limited in freedom of positioning than the assisted sprays. The distance and voltage difference

II. ELECTROSPRAY CONSTRUCTION AND OPERATION 117

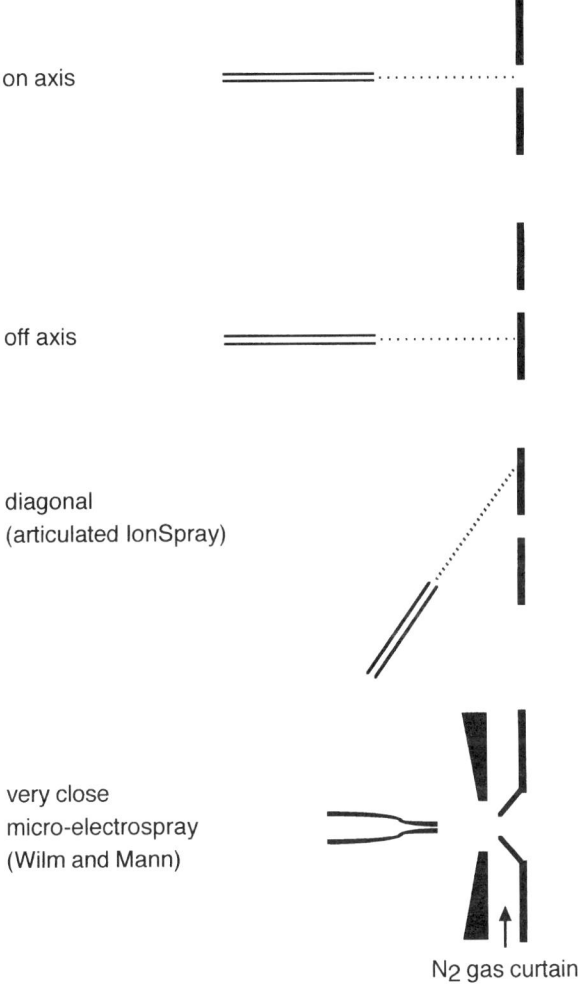

Figure 6. Arrangements for the position of the ES capillary in the ion source.

between the tip of the sprayer and the nearest part of the source determine the electric field that creates the spray and influences the performance of the spray. Optimization of position and spray voltage are interrelated.

In the case of pneumatically and ultrasonically assisted ES, the aerosol generation is determined by gas flow or ultrasonic vibration. The electric field has limited influence on the spray. Aerosol generation is independent from the position of the sprayer in the source. Optimization of sprayer position is sample dependent. The best sprayer position for pneumatically assisted ES of proteins is closer to the sampling orifice and in a narrower region than the best position for small molecules.

An important difference exists between regular and micro ES. In regular ES, the distance between the tip of the sprayer and the ion-sampling orifice of the mass spectrometer is around 1 cm, in pneumatically assisted ES it is 2 cm, or more, combined with a spray voltage of a few kilovolts, but in micro ES the capillary is positioned exactly on axis and the distance from the ion-sampling orifice is reduced to a few millimeters, with concomitant reduction of the spray voltage (12) (Fig. 6).

III. ATMOSPHERIC PRESSURE IONIZATION SOURCE CONSTRUCTION

A. Free Jet Expansion into Vacuum

Electrospray ionization is one of the ionization techniques available for an ion source operating at atmospheric pressure. Two problems are intimately related in atmospheric-pressure ionization (API) and ESMS: the transport of ions from an atmospheric pressure region into the vacuum system of the mass spectrometer, and the strong cooling of a mixture of gas and ions when expanding into vacuum. The resulting condensation of polar neutrals (notably water and solvent vapor) on analyte ions produces cluster ions having a mass far beyond the range of most mass analyzers.

$$X^+ + nH_2O \rightarrow X^+(H_2O)_n$$

A simple scheme of the expansion of a gas into a low pressure region is given in Figure 7A. The principles of the generation of molecular beams (24,25) will be explained in qualitative terms. Inside the nozzle opening and behind the nozzle the gas molecules gain a high velocity, and gas molecules with entrained ions follow straight stream lines originating approximately in the nozzle, the highest intensity of the gas flow being on the axis of the nozzle. Far away from the nozzle, the gas is pumped away, and gas molecules move at random. In the transition between directed motion and random motion, called shock waves, ions and gas molecules undergo many collisions, with scattering of the beam of ions and molecules as the result. The region inside the barrel shock wave, and between the nozzle and the Mach disk, is called the silent zone, where ions and gas move at equal speed in the same direction, and undergo strong and rapid cooling.

The location of the mach disk is important for the design of an API mass spectrometer. The distance from the nozzle is given by

$$x_M = 0.67 D_o \sqrt{\frac{P_0}{P_1}}$$

where D_o is the orifice or nozzle diameter, P_0 is the upstream (atmospheric) pressure, and P_1 is the downstream pressure in the vacuum chamber (25). For a 0.1-mm nozzle diameter and 10^{-5} Torr vacuum generated by cryogenic pump-

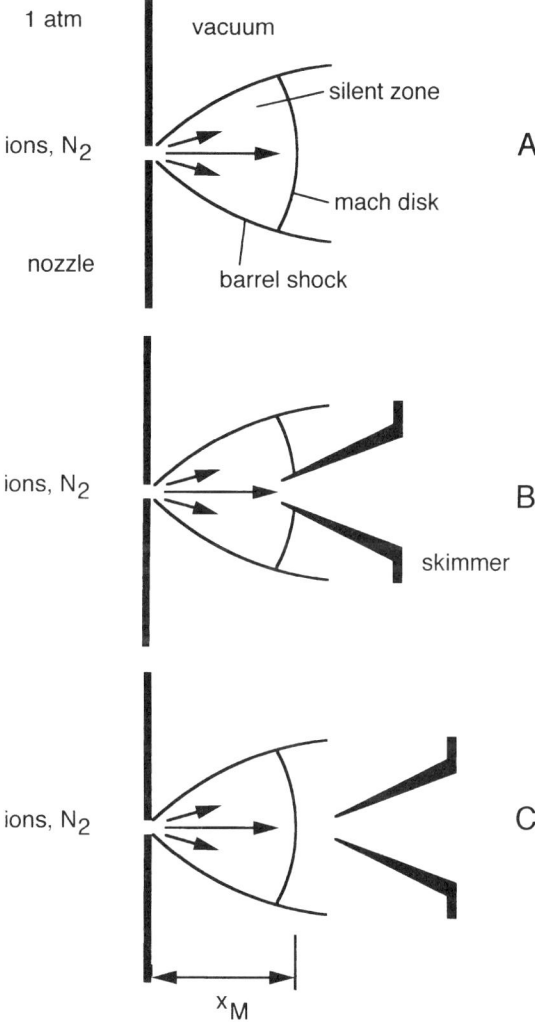

Figure 7. Free-jet expansion of gas and ions into vacuum; (*A*) basic principle; (*B*) arrangement with skimmer penetrating into the silent zone; (*C*) skimmer located more distant than the mach disk.

ing, the Mach disk is located at 670 mm, that is, beyond the dimensions of a quadrupole mass spectrometer. In an expansion through a 0.3-mm diameter orifice into a region pumped down to 1.2 Torr by a 35 m^3/h rotary pump, the Mach disk is located at 5 mm from the nozzle opening and X_M is fully determined by the pumping speed available in the area near the nozzle opening. A small pump or narrow pumping line or other obstructions will result in a small X_M. A 16-m^3/h rotary pump would have given $x_M = 3$ mm. The gas flow through a

0.1-mm i.d. (or bigger) nozzle is too large for the turbomolecular or oil-diffusion pumps of the mass analyzer. Following the practice of molecular beam machines, the central portion of the beam can be sampled into the next vacuum stage by means of a skimmer (24,25), as depicted in Figure 7B and 7C. In Figure 7B, the core of the beam is sampled from the silent zone, and the ions plus gas continue their movement in straight lines. Collection of ions from the gas flow through the skimmer should be efficient, due to the directional effect of the free jet expansion. However, since the gas flowing in the molecular beam has strongly cooled upon expansion, ions will cluster with water molecules, if present.

In Figure 7C, a sample is taken from behind the Mach disk. Ions and gas have undergone extensive scattering, and the extraction and focusing of ions is more difficult. On the other hand, the gas temperature has risen through collisions in the Mach disk, with concomitant breaking of hydrogen bonds in cluster ions. The transmission of ions and gas through the skimmer is akin to the flow of ions and gas from a regular chemical ionization source. The circumference of the skimmer opening in Figure 7B should be very sharp and free from burrs. The full angle of the skimmer cone is usually about 60°C in a molecular-beam apparatus (24,25) but can be wider in an API mass spectrometer. Fortunately, adequate transmission of ions instead of a beam of neutrals through the skimmer is subject to less-stringent conditions of mechanical quality of the skimmer.

B. Cluster Ion Formation

In any design of an API mass spectrometer, the problem of the formation of cluster ions is addressed. Polar molecules that tend to cluster with ions are water and solvent vapor present in air or generated by the evaporation of the eluate from a liquid chromatograph or electrophoresis instrument connected to the API source. All practical designs of API instruments are aimed at either prevention of clustering, or at curing the problem by breaking the clusters.

1. Prevention. If ions are allowed, but water vapor and other neutrals can be excluded from entrance into the vacuum system, formation of clusters is prevented. This can simply be achieved by forcing ions and neutrals in opposite directions by the opposing action of an electric field and a flow of dry gas, usually nitrogen, as is customary in ion-mobility spectrometry. The flow of dry gas can be restricted to the area just in front of the ion-sampling orifice, but can also be extended to the entire ion source.

Figure 8A shows the essential part of the SCIEX API source. Ions are pushed toward the interface plate by the electric field from a corona discharge needle or ES capillary. The region between the interface plate and sampling orifice plate is continuously flushed with dry nitrogen. Part of the nitrogen flow goes to the left into the ion source, pushing water, neutral contaminants, and dust away from the sampling orifice. Ions that come close to the interface plate are driven to the right, toward the sampling orifice, by a 600-V potential difference between the

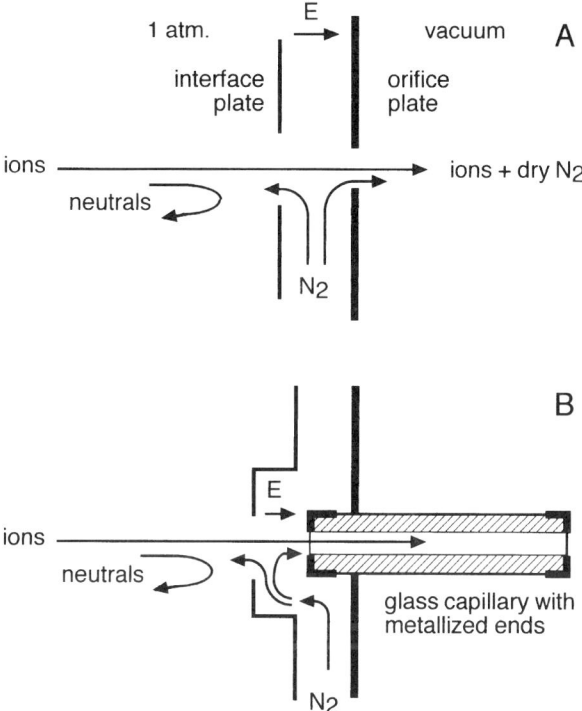

Figure 8. Prevention of cluster-ion formation and protection of sampling orifice or tube by means of: (*A*) gas curtain (SCIEX), (*B*) counter current flow of bath gas (Analytica of Branford).

interface plate and the orifice plate. The other part of the nitrogen flow goes through the sampling orifice, and carries sample ions into the vacuum of the mass analyzer. Ions have thus passed through a so-called dry nitrogen gas curtain. Clusters of ions and nitrogen are not formed, in spite of strong cooling during the expansion into the vacuum system, since nitrogen lacks the ability to form hydrogen bridges.

The countercurrent flow of nitrogen through the ES ion source developed by Fenn and co-workers (14) and built by Analytica of Branford for installation in Hewlett-Packard, Jeol, and other mass spectrometers is different in details, but serves the same purpose (see Fig. 8*B*). A curtain gas or countercurrent flow is a very elegant solution for the clustering problem. As a further benefit, the gas curtain largely avoids contamination and blockage of the sampling orifice, making the API source more reliable.

If the ion source is heated, the temperature of gas and ions remains high enough after the temperature drop resulting from the expansion, to prevent clustering.

Another way to preheat the mixture of ions and neutrals prior to expansion is via the heated transfer-tube system of Chait et al. (23). Ions and vapors pass through an approximately 20-cm long, 0.5-mm i.d. tube heated to 100–200°C.

The heated tube may also be used to help in the desolvation of ions contained in droplets. The source region itself can be at or above room temperature.

2. Curing. Ions pass through a region during the expansion into the vacuum, where the neutral gas density is falling rapidly. If ions moving in a low-density gas (pressure 10^{-3} to 1 Torr) are accelerated by an electric field in this region, either between the nozzle and the skimmer or between the nozzle and the first ion optics element, collisions take place that lead to the breaking of hydrogen bonds in cluster ions. Such a declustering by collision-induced dissociation was described in an early study on corona discharges in air (26). Ten years later, declustering by CID was adopted by Kambara (27).

In an expansion system built according to Figure 7C, ions and gas pass through the Mach disk, where initially formed cluster ions are "heated" by collisions with randomly moving background gas. Water and solvent molecules will be partially removed from clusters and the mass spectrometer will show sample ions that may still be associated with a few water or solvent molecules. A disadvantage of this simple declustering method is that the collection of ions is less efficient owing to the scatter of the ion beam in the Mach disk.

Moderate acceleration of clusters is effective and widely used for declustering. The application of too much collision energy will not only strip off solvent molecules, but also lead to fragmentation of the sample ion.

C. Focusing of Ions at Atmospheric Pressure

The very short mean free path and high collision rate at atmospheric pressure prevent classical ion focusing inside an atmospheric-pressure ion source. The geometry of the sampling orifice (flat or in a cone) and the dimensions and location of the lens or tube guiding a curtain gas (Fig. 8) are important for the ion-sampling efficiency. A correct combination of gas flow and electric field helps the guiding of ions toward the orifice. Careful attempts to drive ions to the orifice were made by Eisele (28). His aim was the sampling of very low concentrations of ions present in ambient air due to natural or other conditions. Ions were guided through a high-pressure ion optics region designed by computer simulation of electric fields. In spite of an extended electric field converging toward the sampling orifice, no increase in ion signals was observed. The conclusion was drawn that the electric field immediately in front of the sampling orifice was most important. Calculations and experiments by Sunner and co-workers (29,30) have shown that electric fields in the atmospheric pressure-ionization region are mainly dominated by space charge.

D. Vacuum System Design

Electrospray API instruments can be divided into single-stage vacuum systems and multiple-stage vacuum systems. A single-stage instrument is shown schematically in Figure 9. Ions flow from the sampling orifice through ion optics into

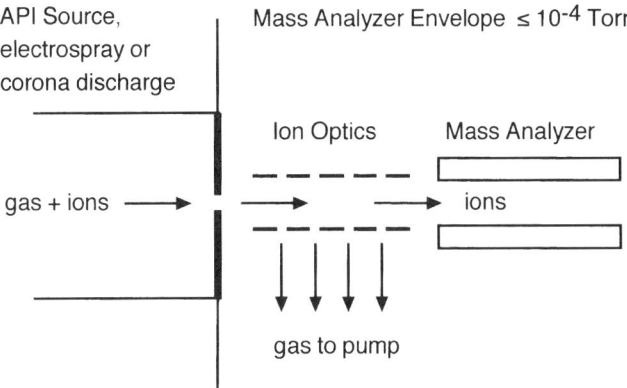

Figure 9. Electrospray (atmospheric pressure ionization) source with single stage vacuum system for the mass analyzer.

the mass analyzer. The ion optics stage can be either a stack of regular lenses or an RF only quadrupole (hexapole, octapole). The pressure tolerated by the mass analyzer sets the upper limit to the vacuum in the mass spectrometer. This pressure, together with the speed of the vacuum pump, sets the upper limit to the throughput of gas, and thus to the size of the sampling orifice. The higher the gas throughput, the more ions are transported through the sampling orifice. In older API instruments, a 10–25 μm diameter orifice compatible with a 1000-L/sec oil diffusion pump was used. A high-sensitivity, single-stage atmospheric pressure-ionization instrument built by SCIEX makes use of a 120-μm orifice protected by a gas curtain, combined with pumping by cryogenic surfaces surrounding the ion optics and quadrupole mass analyzer, rated at 100,000 L/sec.

High-speed cryogenic pumping is good from a theoretical point of view and has afforded very reliable performance in practice, but needs a big and expensive helium compressor. Manufacturers and researchers prefer a multiply staged vacuum system that can be built with less expensive vacuum pumps (Fig. 10). Ions are drawn through a sampling orifice into a first chamber, pumped to approximately 1 Torr by a rotary pump. Part of the expanding beam of gas and ions is taken into the second chamber, pumped down to 10^{-2} Torr or less. Most of the gas is pumped away in the second stage and a suitable ion-optics arrangement guides the ions into the mass analyzer region, which is pumped down to 10^{-5} Torr. The second and third stages are usually the original source housing and analyzer housing of a standard differentially pumped mass spectrometer designed for chemical ionization. This simple scheme may be elaborated in different ways, to comply with particular geometrical or electrical constraints. In recent implementations of the multiple-stage vacuum system the first or the second stage has been split in two, which increases the total number of stages to four.

Multiple-stage vacuum systems use a larger orifice, which can take more gas and ions into the first vacuum stage of the mass spectrometer. Unfortunately a

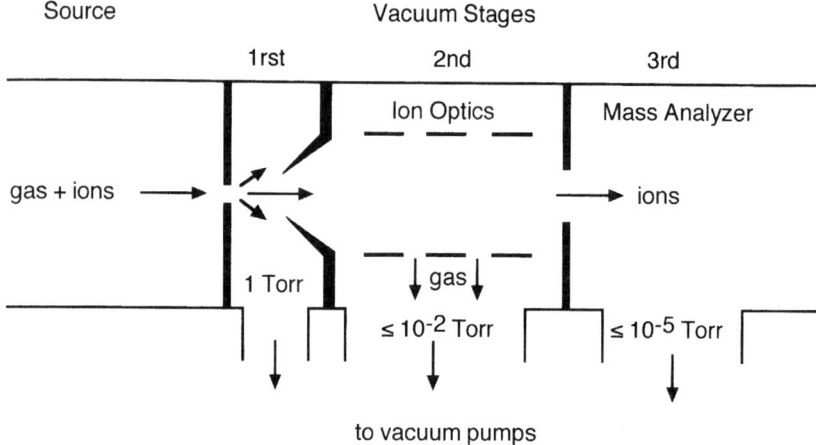

Figure 10. Vacuum system with multiple stages: 1st stage, free-jet expansion, molecular-beam stage; 2nd stage, ion optics; 3rd stage, mass analyzer.

large throughput through the sampling orifice and high sensitivity do not always go hand in hand. In a discussion about sensitivity, the first vacuum stage in Figure 10 will be called the molecular-beam stage, the second the ion-optics stage, and the third the analyzer stage.

For simplicity, we assume that no significant enrichment of heavy ions over light gas molecules takes place during passage through the first vacuum stage into the skimmer. We assume that the ion-optics region is transparent to neutrals that are removed by the first high-vacuum pump. We also assume that no ions are lost during transport through the ion-optics region. In short then, the ions to gas ratio is significantly increased in the ion-optics stage.

Typical orifice dimensions are 0.3–0.5 mm for the nozzle combined with 1.4–0.6 mm for the skimmer. The best transmission of ions is achieved when the skimmer penetrates into the silent zone of the free jet expansion so that the ion-optics stage receives a well-shaped ion beam. In attempts to build ES API sources for magnetic sector instruments, the addition of a fourth pumping stage appeared advantageous. Problems with sector instruments are arcing due to the high voltage on the ion source, electrical discharges in pumping lines, and collisions between ions and background gas in the acceleration region. Manufacturers appear to have overcome such problems. Strict adherence to molecular-beam rules may not be necessary, since the more efficient extraction of ions from an ion source in a sector instrument may offset the disadvantage of scattering of ions in intermediate pressure regions.

E. Vacuum System and Sensitivity

The goal for the design of the system for ion transport from the ES source into the mass analyzer is to introduce as many ions, but as little gas, as possible.

Ideally, ions should be separated from gas. In a simple approximation, the sensitivities of different API mass spectrometers can be judged from the throughput of gas in the relevant vacuum stage (31). In a single-stage vacuum system, the flow of ions into the mass analyzer is proportional to the gas flow through the sampling orifice. The throughput of gas into the analyzer is given by the pumping speed multiplied by the pressure in the vacuum system.

For 1000 L/sec and 10^{-5} Torr we obtain 0.01 Torr.L/sec. If the single-stage system is equipped with a 100,000 L/sec cryopump, the throughput is increased to 1 Torr.L/sec (SCIEX API 3). The sensitivity is increased by the same factor of 100.

In the discussion of multiple-stage vacuum systems, we assume for simplicity that an ion-optics stage transmits all ions toward the mass analyzer, while gas is pumped away, thereby increasing the ions to gas ratio; a molecular beam stage is supposed to transfer a portion of the gas plus ion beam, without changing the ions to gas ratio. However, recent versions of the molecular beam stage include a tube lens or focusing ring or focusing cone for increasing the ions to gas ratio as shown in Figure 11 (32). Positive ions that have spread out during the free-jet expansion are forced back toward the center through the skimmer by the application of a positive voltage on the tube, ring, or cone. A negative voltage is used for negative ions. Sensitivity is increased by a factor of 2–5, depending on the geometry. The ion-optics stage may be equipped with lenses, an RF-only quadrupole (hexapole, octapole) or other elements.

In a three-stage system, the first stage does not dramatically increase the ions to gas ratio. The ion optics stage is crucial, and with a 1000 L/sec pump and 10^{-3} Torr (approximate values for our NERMAG R3010 triple quadrupole with homemade source) a throughput of 1 Torr.L/sec can be obtained. Sensitivity roughly equal to the SCIEX API 3 instrument is obtained, at the expense, however, of some backstreaming of diffusion pump oil into the ion-optics region. In all instruments built along these lines, the pumping and pressure in the ion-optics stage determine the sensitivity. A compromise on pumping will compromise sensitivity. It has appeared that the efficiency of ion transport through an RF-only quadrupole (hexapole, octapole) increases if the pressure is raised to approximately 10^{-2} Torr (33). To obtain the same gas throughput, a smaller pump can be used. At 2.10^{-2} Torr, and 50 L/sec, the throughput is again 1 Torr.L/sec. At this high pressure, it becomes difficult to select the right vacuum pump. Diffusion pumps cannot be used at an inlet pressure above 2.10^{-3} Torr, and suffer from considerable backstreaming of diffusion pump oil. Turbomolecular pumps have reached their limits at 10^{-2} Torr, and 8.10^{-3} Torr appears a practical value for a 50 L/sec turbomolecular pump. A new development is a dual-inlet turbomolecular pump, equipped with a high-pressure port for pumping the RF-only quadrupole, and a regular low-pressure inlet port for pumping the mass analyzer (34).

The prediction of sensitivity based on vacuum considerations alone is a first approximation. The effects of the quality of the molecular-beam components, ion optics, and mass analyzer cannot be included in a simple generalization. The

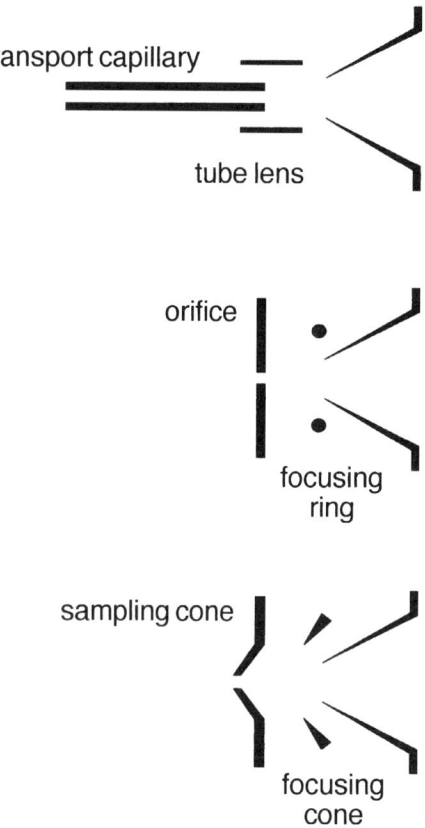

Figure 11. Tube lens (Finnigan), ring (SCIEX), or cone (Micromass) located in the molecular beam stage for forcing ions through the skimmer orifice.

atmospheric part of the ES source on each vacuum system was assumed to be equally efficient in each case. A comparison between different instrument designs would require running the same sample in different instruments under well-defined and carefully controlled conditions.

F. Ion-Sampling Orifice

Various arrangements of the sampling orifice are given in Figure 12. A hole in a disk (*a*), in the top of a cone (*b*), or in the bottom of an inverted cone (*c*) can be used equally well. A 20-cm long × 0.5 mm i.d. glass tube (*d*) or heated metal tube may also be used instead of an orifice.

The formation of molecular beams from nozzles of various shapes has been reported (35). Campargue recommends a short straight capillary with a ratio of

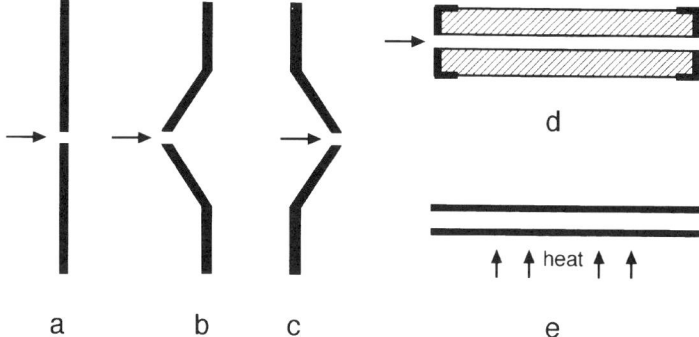

Figure 12. Sampling of ions and gas from the atmospheric-pressure ionization source via different shapes of orifices and tubes; (*a*) orifice in a flat disk (SCIEX API 100/300); (*b*) orifice in the top of a cone (Micromass, SCIEX API 3); (*c*) orifice in the bottom of a cone (Vestec); (*d*) glass tube with metallized ends (Analytica source customized for HP, Jeol, Bruker, and others), (*e*) heated metal tube (Finnigan MAT).

length to diameter equal to or greater than 2 (25). In our hands, a 0.3-mm diameter hole in a flat 1.0-mm thick stainless-steel disk and protected by a gas curtain has proven very robust and reliable in experiments with corona discharge ionization or ES during the past seven years. The operation of an API source at any desired voltage, independent of the mass analyzer, has advantages for magnetic sector instruments and for certain liquid-sample inlet systems. The transport of ions through a glass capillary instead of a metal nozzle gives the desired insulation, as discussed earlier in this chapter. The flat ends of the capillary are metal plated for efficient sampling of ions into the capillary and acceleration of ions at the exit. The gas flow drags ions through the capillary into the vacuum system, even if the electric field existing between the metallized ends opposes such a transport. Chait et al. have used a long heated metal capillary for the desolvation of ions from charged microdroplets generated by ES and introduction of the core ions into a quadrupole mass spectrometer. Another purpose served by the capillary is the transport of ions over a longer distance (e.g., 20 cm) between the API source and the ion optics and acceleration region of the mass analyzer, giving more freedom in mechanical construction.

Ion-transport efficiency through capillary tubes has been studied in detail by Lin and Sunner (36) and by Guevremont and co-workers (37). Transmitted currents were very similar for glass and metal capillaries (36). It was more difficult to obtain stable and reproducible transmitted currents with glass than with metal capillaries. Experiments were made with very long (0.6–15 m) capillaries, so that shortcomings observed are probably more pronounced than in real life in mass spectrometry, where the capillary length is 20 cm or shorter.

Contamination of the sampling orifice or tube can be troublesome. The

128 ESI SOURCE DESIGN AND DYNAMIC RANGE CONSIDERATIONS

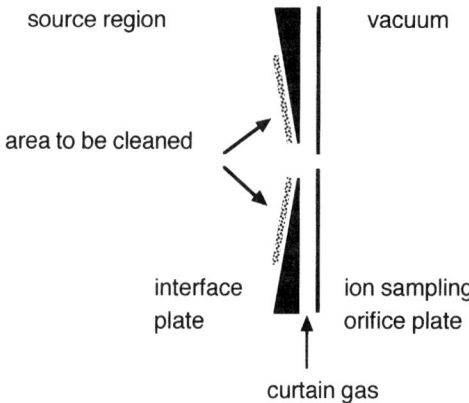

Figure 13. Area to be cleaned inside the atmospheric pressure-source region.

buildup of charge on a layer of contaminants inside an orifice or tube can effectively stop the passage of ions, while the flow of neutrals (as read from vacuum gauges) is not affected. An ES source should be designed for ease of removal and cleaning of the orifice or tube. The orifice in our instrument (0.3 mm i.d. in a 1-mm thick stainless-steel disk) can be cleaned with a sharp pin of wood and aluminium oxide powder followed by sonication in ethanol. Depending on the operating conditions, cleaning takes place after one or two months of use. Surprisingly, it has never been necessary to clean the skimmer. Tubes are less easy to clean. Blockage of a tube or orifice is prevented by the use of a gas curtain. Foolproof operation cannot be guaranteed. One should be careful and avoid spraying straight at the orifice or tube since droplets can penetrate through the gas curtain. As discussed above, off-axis or diagonal positioning not only increases stability and abundance of ion signals, but also helps to prevent contamination of the sampling orifice (nozzle) or tube. If such precautions are taken together with protection by a gas curtain, it is possible to use nonvolatile buffer components in CE/MS, or use post-column addition of a dilute sodium chloride solution in LC/MS to promote formation of $M.Na^+$ ions of polar neutral samples such as sugars. The area around the sampling orifice needs periodic cleaning (Fig. 13). Fields created by charges on contaminating layers may disturb the electric field in front of the sampling tube or orifice and prevent ions from being entrained in the gas flow into the vacuum.

Micromass instruments are equipped with a so-called pepper pot located in the atmospheric pressure region between the sprayer tip and the sampling orifice (Fig. 14). The pepper pot eliminates a line of sight between the sprayer and the sampling orifice (called sampling cone by Micromass) and reduces contamination of the orifice. Ions and small charged droplets pass through the holes in the pepper pot, while relatively big droplets impinge on the front and inside walls of

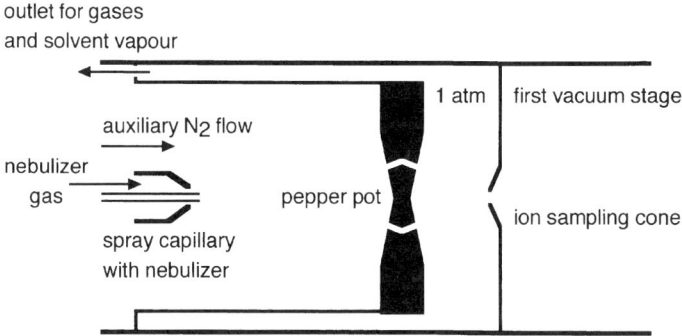

Figure 14. Pepper pot for reduction of contamination of the sampling cone in Micromass mass spectrometers.

the channels. Since the source is heated, the pepper pot also aids in the transfer of heat to droplets, solvent vapor, and auxiliary gases in the source.

G. Ion Optics Between Sampling Orifice and Mass Analyzer

In the space between the sampling orifice (or outlet of sampling tube) and skimmer, ions are transported in the free-jet expansion of gas. Part of the beam of ions and gas passes through the skimmer into the next vacuum stage. Effective focusing of ions into a very narrow beam is not possible, but ions can be forced to remain closer to the axis of the beam if a tube or ring is positioned between nozzle and skimmer (32). A positive voltage on the tube lens or ring reduces the spreading of positive ions away from the axis. The gain in sensitivity is between two- and fivefold.

Behind the skimmer the pressure is lower, but still not low enough for effective focusing with ion lenses. RF-only quadrupoles, hexapoles, and octapoles are now preferred by most manufacturers of quadrupole and magnetic-sector mass spectrometers to guide ions into the mass analyzer section. Douglas has observed that the efficiency of guiding ions is significantly increased if the pressure is raised to 10^{-2} Torr (33). Collisions between ions and background gas keep the ions on a path close to the quadrupole axis, in the same manner as ions are confined to the center of a quadrupole ion trap by collisions in 10^{-3} Torr of helium. Ions are slowed down by collisions in the RF-only quadrupole and enter the next stage with a limited energy spread of a few electron volts. Ion transmission from the RF-only multipole into the mass analyzer is easier and focusing much less dependent on mass.

The occasional penetration of charged droplets into a quadrupole mass analyzer shows up as random spikes in mass spectra. Off-axis positioning of the sprayer can reduce this aerosol noise. In Finnigan TSQ and SSQ mass spectrometers the skimmer is positioned slightly off axis. Since heavy aerosol particles follow a straight line on the axis of the ion transfer tube, they impinge on the outside wall of the skimmer. A sizable fraction of ions deviates from the center and is passed through the skimmer. Noise is eliminated, at the expense of a reduction in ion transmission.

Ion-counting detection is more immune to spikes of this nature when compared with analog current measurement, because a burst of *many* ions arriving together at the electron multiplier is recorded as *one* pulse by the ion-counting detector, while the same burst is amplified to a large voltage spike by the current-to-voltage converter of an analog detector.

IV. UP-FRONT COLLISION-INDUCED DISSOCIATION

Electrospray mass spectra are usually devoid of fragment ions. Fragments can be formed in a triple quadrupole, in an ion trap, ICR cell, or in the field-free regions of magnetic-sector mass spectrometers.

In atmospheric-pressure ionization mass spectrometry, including ES mass spectrometry, fragmentation can easily be induced in one of the higher-pressure regions of the ion passageway from the source into the mass analyzer. Acceleration of ions between nozzle (tube) and skimmer by $\Delta V(\text{N-S})$ or between skimmer and RF-only multipole by $\Delta V(\text{S-Q})$ results in collisions of ions with gas in which ions are entrained in their flow into the vacuum (Fig. 15). Sometimes this process is called "in-source CID" which is clearly a misnomer. Further

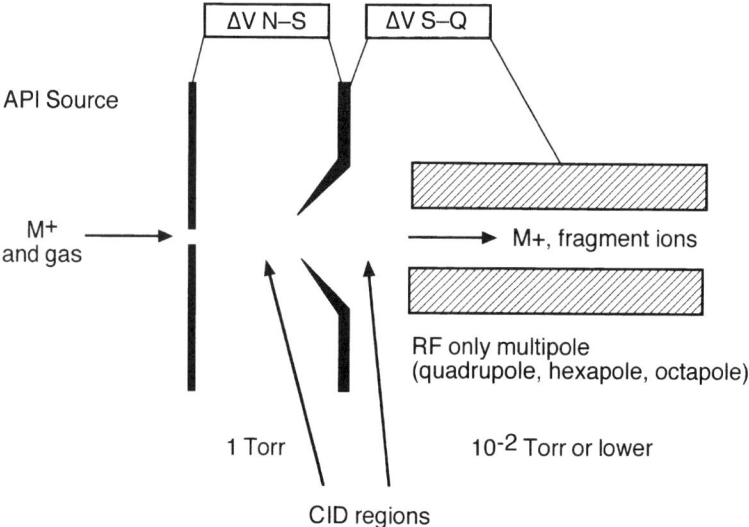

Figure 15. Collision-induced dissociation just outside the ion source.

confusion is created by the use of different names for the same process by users and manufacturers: nozzle-skimmer fragmentation, cone-voltage fragmentation, high-orifice potential fragmentation, and octapole fragmentation, to name a few.

Fragment ions produced by collision-induced dissociation are very efficiently transported into the mass analyzer. The major advantage of this poor man's MS/MS method is its simplicity: only one voltage has to be increased, no switching and adjustment of collision gas and retuning of ion optics is necessary. Of course, there is no parent ion selection possible. In LC/MS one can do one LC run without CID, and one with CID to obtain fragments of all components. An elegant application of up-front CID in LC/MS is stepping the CID (orifice) voltage during the course of each mass spectrum: a high voltage is used in the low-mass region, to generate and detect low-mass fragments, while the voltage is reduced again at, for example, m/z 200 and above, in order to collect unfragmented parent ions (38). Phosphorylated peptides were selectively detected in LC/MS of an enzymatic digest of a phosphoprotein.

Since up-front CID takes place just *outside* the ion source, it can be combined with any ionization technique that is applicable in atmospheric pressure-ion sources. In Figure 15 the accumulated effect of $\Delta V(N-S)$ and $\Delta V(S-Q)$ can lead to fragmentation. CID takes place under the condition of sufficient translational energy and gas density. In a magnetic-sector instrument, the residual gas density in the accelerating region may be high enough for CID. An additional pumping stage is used by manufacturers to eliminate unwanted CID in the accelerating region.

Mild up-front CID reduces the abundance of background ions, a distinct advantage for LC/MS or CE/MS where the contribution of background ions to the baseline noise in the total ion current trace is a serious problem. In multiply staged instruments, the transmission of ions is improved if the $\Delta V(N-S)$ is increased, in particular for big protein ions. In fact, it is not too difficult to achieve good transmission of protein ions without inducing fragmentation, because collision energy is partitioned over a large number of vibrational degrees of freedom. It is not as easy to obtain good transmission of small molecules that fragment much more easily. Manufacturers that jumped into ES with molecular-weight determination of proteins as their first goal discovered that good sensitivity for small molecules, notably pharmaceuticals and labile drug conjugates, is a different game.

In a well-designed API mass spectrometer, the abundance of a sample ion at m/z 200–300 decreases very little if $\Delta V(N-S)$ is reduced from 50 V down to a few volts. A good test compound for unwanted CID is the drug naproxen. At low collision energy, the negative-ion ES spectrum shows the $[M-H]^-$ ion at m/z 229 as the only peak. It should be easy to adjust $\Delta V(N-S)$ and $\Delta V(S-Q)$ for a spectrum free from fragment ions. Already at a moderate collision energy, where most sample ions would survive, the $[M-H]^-$ ion loses CO_2 so easily that m/z 185 becomes the base peak. In the positive-ion mode, the $M.H^+$ ion of dibutyl phthalate at m/z 279 can be used as a test ion. Fragmentation to m/z 149 is easy, but does not take place as readily as the loss of CO_2 from m/z 229 of naproxen.

naproxen

Removal of the last solvent molecule(s) from a sample ion by up-front CID may not be possible without unintended fragmentation of a labile sample ion

V. MASS-SCALE CALIBRATION

Calibration of the mass scale requires a series of ions evenly spaced throughout the mass range. Ideally, the calibrant should be easily removed from the source and leave no traces in the source or liquid-handling system. Mixtures of a protein (usually myoglobin) and some smaller peptides can be used; polyethylene glycols and polypropylene glycols are also widely used for calibration. Anacleto et al. have summarized different options, and have proposed protonated water clusters and salt clusters (39), generated by pneumatically assisted ES. Water clusters provided a calibration range up to m/z 1000 in the SCIEX API III mass spectrometer. Alkali metal halides (sodium iodide) allow calibration on cluster ions $Na^+(NaI)_n$ or $I^-(NaI)_n$ up to m/z 2400, the full mass range of this instrument.

In principle, mass-scale calibration is independent of the ionization technique. During the past seven years, we have routinely calibrated by the use of a combination of corona discharge and controlled cluster formation in the free-jet expansion. Corona discharge of the headspace of an aqueous ammonia solution creates $NH_4^+(H_2O)_n$ with $n = 0-4$ after passage through the gas curtain. If the curtain gas flow is reduced but not cut off, one can tune the gas flow to allow a controlled amount of moisture into the free-jet expansion so that a complete series of $NH_4^+(H_2O)_n$ ions is formed that extends all the way up to m/z 2000 (mass range of a NERMAG R3010) and most probably beyond this value. The calibration mass series starts at m/z 54, and one out of every five clusters is used, giving a calibration mass separation of 90 amu. The relatively low abundance of deuterium and ^{17}O, and the higher abundance of ^{18}O in the cluster ions, makes this calibrant suitable for resolution adjustment on isotope peaks.

VI. DYNAMIC RANGE IN ELECTROSPRAY

When the concentration of a sample in solution is increased, the abundance of sample ions created by ES increases until a plateau is reached. A limited dynamic range is a serious drawback in analytical applications of ES ionization mass spectrometry. The upper limit to the usable concentration range is dependent on the sample itself and on the presence of other charged material, such as electrolytes or buffers in the sample solution. Tang and Kebarle have measured the amount of charge on droplets (the spray current) and sample ion abundance as a function of concentration of a sample dissolved in reagent-grade methanol. They rationalized the upper limit as the result of interplay between droplet charging on the one hand and the competition between sample ions and background electrolyte ions leaving the charged droplets on the other hand. A close relationship between the dynamic range limit at 10^{-5} M sample concentration and the estimated 10^{-5} M background ion concentration was proposed (40). In short, the upper limit was attributed to competition of ions for insufficient charge on droplets

In our hands, the same upper limit of 10^{-5} M was found for tetrabutylammonium bromide dissolved in the highest-purity gradient-grade solvents, introduced into the ES source via repeatedly, rigorously cleaned components. The spray current keeps rising with increasing sample concentration above 10^{-5} M, while the ion signal reaches a plateau. Limitations in ion transport or in the spray process were ruled out (41). Measurement of the current arriving at the skimmer showed that the skimmer current also reaches a plateau at 10^{-5} M, while the spray current keeps rising (42) (Fig. 16). One has to conclude that above 10^{-5} M an increasing amount of charge on droplets cannot be converted into sample ions. In other words, sample ions cannot escape from droplets.

A reasonable explanation is the complete filling of droplet surface with sample ions and sample molecules at 10^{-5} M. More sample in solution would not give more sample at the droplet surface, and the upper limit to sample ion abundance is reached. This viewpoint can be supported by simple calculation and assumptions based on experiments on fission of charged droplets (43,44). Let us assume a primary droplet size of 1 μm radius and fission into droplets of 0.1 μm radius. Let us assume that sample ions and sample ion pairs all reside on the droplet surface, a reasonable assumption for surface-active quaternary ammonium salts, and an equally reasonable assumption for a good many organic compounds having a polar functional group. These small droplets have been separated from the surface of parent droplets and have the same density of sample at the surface as the parent droplets. The 0.1-μm radius droplet shrinks by evaporation down to 10 nm, and emission of ions from the surface takes place; the numbers are given in Table 1. The bottom line is that a 10-nm radius droplet has about 5 nm^2 of droplet surface available per sample ion or sample molecule, whereas the surface taken by one sample ion is of the order of 3 nm^2. The droplet surface is indeed almost full at 10^{-5} M after droplet disintegration and evaporation.

An increasing amount of charge is not converted into sample ions. This

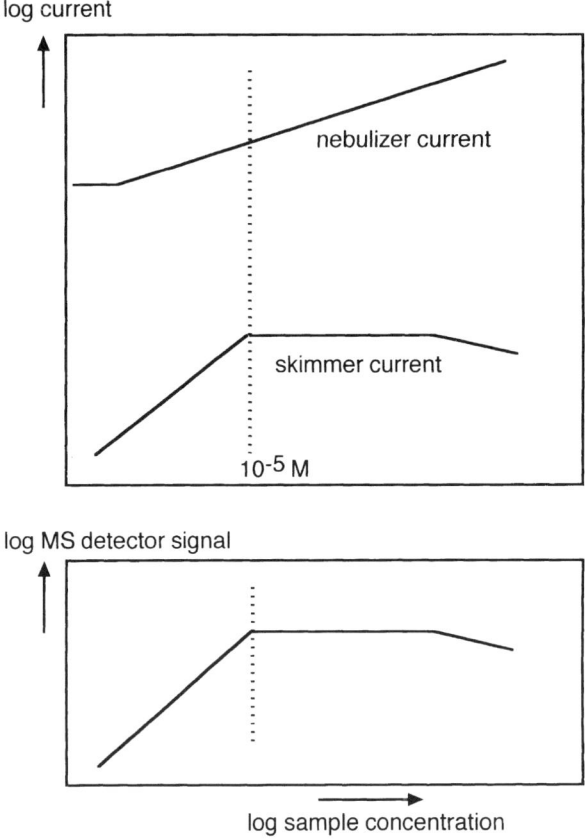

Figure 16. Nebulizer current, skimmer current, and ion signal as a function of sample concentration of tetrabutylammonium bromide dissolved in gradient-grade acetonitrile (95%) plus methanol (5%).

TABLE 1. Area Available on a Droplet Surface at 10^{-5} M

Radius of initially formed droplet	$r = 1\ \mu$m
Number of sample ions plus sample molecules	24,000
Area available per sample ion or molecule	500 nm^2
After fission to	$r = 0.1\ \mu$m
Surface density equal to parent droplet	
Area available per sample ion or molecule	500 nm^2
Number of sample ions plus sample molecules	240
After size reduction by evaporation	$r = 10$ nm
Area available per sample ion on molecule	5 nm^2
Radius of average organic ion or molecule (C–C bond length 0.15 nm)	app. 1 nm
Area taken by one ion or molecule	app. 3 nm^3

unused charge probably resides in droplets that cannot release sample ions. It is unclear at present if developments in source design will alleviate the dynamic range problem.

REFERENCES

1. Bailey, A. G. *Electrostatic Spraying of Liquids*; Electrostatics and Electrostatic Applications Series; Wiley: New York, 1988.
2. Smith, R. D.; Barinaga, C. J.; Udseth, H. R. *Anal. Chem.* **1988**, *60*, 1948–1952.
3. Mylchreest, I. C.; Hail, M. E. *United States Patent* 5,122,670; June 16, 1992.
4. Bruins, A. P.; Covey, T. R.; Henion, J. D. *Anal. Chem.* **1987**, *59*, 2642–2646.
5. Banks, J. F.; Shen, S.; Whitehouse, C. M.; Fenn, J. B. *Anal. Chem.* **1994**, *66*, 406–414.
6. Banks, J. F.; Quinn, J. P.; Whitehouse, C. M. *Anal. Chem.* **1994**, *66*, 3688–3695.
7. Hail, M. E.; Mylchreest, I. C. *United States Patent* 5,170,053; December 8, 1992.
8. Wahl, J. H.; Gale, D. C.; Smith, R. D. *J. Chromatogr.* **1993**, *659*, 217–222.
9. Kriger, M. S.; Cook, K. D.; Ramsey, R. S. *Anal. Chem.* **1995**, *67*, 385–389.
10. Gale, D. C.; Smith, R. D. *Rapid Commun. Mass Spectrom.* **1993**, *7*, 1017–1021.
11. Emmet, M. R.; Caprioli, R. M. *J. Am. Soc. Mass Spectrom.* **1994**, *5*, 605–613.
12. Wilm, M. S.; Mann, M. *Int. J. Mass Spectrom. Ion Processes* **1994**, *136*, 167–180.
13. Yamashita, M.; Fenn, J. B. *J. Phys. Chem.* **1984**, *88*, 4451–4459.
14. Whitehouse, C. M.; Dreyer, R. N.; Yamashita, M.; Fenn, J. B. *Anal. Chem.* **1985**, *57*, 675–679.
15. Larsen, B. S.; McEwen, C. N. *J. Am. Soc. Mass Spectrom.* **1991**, *2*, 205–211.
16. Cody, R. B.; Tamura, J.; Musselman, B. D. *Anal. Chem.* **1992**, *64*, 1561–1570.
17. Yamashita, M.; Fenn, J. B. *J. Phys. Chem.* **1984**, *88*, 4471–4675.
18. Cole, R. B.; Harrata, A. K. *Rapid Commun. Mass Spectrom.* **1992**, *6*, 536–539.
19. Hiraoka, K.; Kudaka, I. *Rapid Commun. Mass Spectrom.* **1992**, *6*, 265–268.
20. Whitby, K. T.; Liu, B. Y. H. in *Aerosol Science*; Davies, C.N., Ed.; Academic Press: London, 1966, Chapter III.
21. Hopfgartner, G.; Wachs, T.; Bean, K.; Henion, J.D. *Anal. Chem.* **1993**, *65*, 439–446.
22. Covey, T. R.; Anacleto, J. F. *United States Patent* 5,412,208; May 2, 1995.
23. Chowdhury, S. K.; Katta, V.; Chait, B.T. *Rapid Commun. Mass Spectrom.* **1990**, *4*, 81–87.
24. Anderson, J. B.; Andres, R. P.; Fenn, J. B. in *Adv. At. Mol. Phys.*, Bates, D. R.; Estermann, I. Eds.; Academic Press: New York, 1965, Chapter 8.
25. Campargue, R. *J. Phys. Chem.* **1984**, *88*, 4466–4474.
26. Shahin, M. M. *J. Chem. Phys.* **1966**, *45*, 2600–2605.
27. Kambara, H.; Kanomata, I. *Anal. Chem.* **1977**, *49*, 270–275.
28. Eisele, F. L. *Int. J. Mass Spectrom. Ion Processes* **1983**, *54*, 119–126.
29. Busman, M.; Sunner, J. *Int. J. Mass Spectrom. Ion Processes* **1991**, *108*, 165–178.
30. Busman, M.; Sunner, J.; Vogel, C. R. *J. Am. Soc. Mass Spectrom.* **1991**, *2*, 1–10.
31. Bruins, A. P. *Mass Spectrom. Rev.* **1991**, *10*, 53–77.

32. Mylchreest, I. C.; Hail, M. E.; Herron, J. R. *United States Patent* 5,157,260; October 20, 1992.
33. Douglas, D. J.; French, J .B. *J. Am. Soc. Mass Spectrom.* **1992**, *3*, 398–408.
34. Balzers model TMH 260–130.
35. Murphy, H. R.; Miller D. R. *J. Phys. Chem.* **1984**, *88*, 4474–4478.
36. Lin, B.; Sunner, J. *J. Am. Soc. Mass Spectrom.* **1994**, *5*, 873–885.
37. Guevremont, R.; Siu, K. W. M.; Wang, J.; LeBlanc, J. C. Y. *Proceedings of the 42nd ASMS Conference on Mass Spectrometry and Allied Topics,* May 29–June 3, 1994, Chicago, IL, p. 999.
38. Carr, S. A.; Huddleston, M. J.; Bean, M. F. *Protein Sci.* **1993**, *2*, 183–196.
39. Anacleto, J. F.; Pleasance, S.; Boyd, R. K. *Org. Mass Spectrom.* **1992**, *27*, 660–666.
40. Tang, L.; Kebarle, P. *Anal. Chem.* **1993**, *65*, 3654–3668.
41. Kostiainen, R.; Bruins, A. P. *Rapid Commun. Mass Spectrom.* **1994**, *8*, 549–558.
42. Zook, D. R.; Bruins, A. P. *Proceedings of the 42nd ASMS Conference on Mass Spectrometry and Allied topics, May 29–June 3, 1994,* Chicago, IL, p. 1015; *Int. J. Mass Spectrom. Ion Processes,* **1997**, in press.
43. Gomez, A.; Tang, K. *Phys. Fluids* **1994**, *6*, 404–414.
44. Kebarle, P.; Tang, L. *Anal. Chem.* **1993**, *65*, 972A–986A.

CHAPTER 4

Solution, Gas-Phase, and Instrumental Parameter Influences on Charge-State Distributions in Electrospray Ionization Mass Spectrometry

GUANGDI WANG

Department of Chemistry, Xavier University of Louisiana, New Orleans, Louisiana

and

RICHARD B. COLE

Department of Chemistry, University of New Orleans, Lakefront, New Orleans, Louisiana

	Abstract	138
I.	Background and overview	138
II.	Role of solution equilibria and solution-phase chemistry	140
	A. Effect of solution pH	142
	B. Solvent effects	146
	C. Effect of analyte concentration	148
	D. Effect of counterions and electrolytes present in solution	153
	E. Effect of analyte structures/conformations	156
III.	Effect of instrumental conditions	161
	A. Instrument geometry	161
	B. Countercurrent gas and cooling gas	162
	C. Skimmer voltage and in-source collision-induced dissociation	164
IV.	Gas-phase modifications	165
	A. Ion solvation and the observation of adduct species	165
	B. Proton-transfer reactions	167
	C. Intramolecular coulombic repulsion	168
	D. Defining the maximum obtainable charge state	170
V.	Conclusion	171
	References	172

Electrospray Ionization Mass Spectrometry, Edited by Richard B. Cole.
ISBN 0-471-14564-5 © 1997 John Wiley & Sons, Inc.

ABSTRACT

In the domain of mass spectrometry, the observation of ionic species which exhibit extensive multiple charging is a phenomenon which is unique to the electrospray ionization (ESI) technique. In virtually all reported mass spectra displaying highly charged species, a distribution of charge states is observed. The shape and placement of this distribution along the m/z axis of the mass spectrum depends upon many experimental factors. This chapter presents a compilation of results from laboratory work that offer insight into the complex series of processes that ultimately determines the appearance of the charge state distribution in the ESI mass spectrum. The chapter begins with an examination of the solution phase chemical processes which can exert an effect on the appearance of the mass spectrum prior to the analyte's arrival at the mass spectrometer. Initial solution considerations include hydronium ion (pH) and other electrolyte concentrations, solvent polarity, as well as analyte concentration and conformation. Confronted next are influences attributable to instrumental factors related to the design of the ESI mass spectrometer. The effect of specific operating conditions, including temperature and gas flow, upon the charge state distributions are examined. Lastly, processes which occur subsequent to gas-phase ion production but prior to ion detection are addressed. These include desolvation, proton transfer reactions, and the destabilizing effect of coulombic repulsion between neighboring charged sites on extensively charged molecules.

I. BACKGROUND AND OVERVIEW

The inherently unique characteristic of extensive multiple charging distinguishes electrospray ionization (ESI) mass spectrometry from all other ionization techniques which currently exist. In fact, one of the earliest efforts made by Dole et al. (1,2) was to retain multiple charges on the droplets presumably containing only one macromolecule, such that upon solvent evaporation, multiply charged single macroions could be formed. This portrayal of the mechanism responsible for multiple charging of free gas-phase ions has been referred to as the charged-residue model (CRM). Although an alternative interpretation of the generation of gas-phase ions was proposed in the late 1970s by Thomson and Iribarne (3–6), their ion-evaporation model (IEM) did not address the issue of multiple charging. It was not until the latter portion of the 1980s, when Fenn and co-workers first reported the mass spectrometric observation of doubly charged peptides (7) and multiply charged protein ions (8,9), that the potential of the ESI technique in bioanalytical applications began to become widely appreciated. Subsequently, the mechanism of formation of multiply charged analyte ions was intensively investigated (10–34).

As noted in a review by Fenn et al. (10), Dole's CRM seemed unable to explain the production of singly charged macroions in ESI experiments. This is because the increasing charge/mass ratios of the smaller daughter droplets would most likely yield multiply charged ions. Furthermore, the flux of analyte molecules in the electrosprayed solution was much greater than the flux of unit charges; hence, the ultimate droplet could contain more than one molecule even if it contained only one charge (10). In extending the Iribarne–Thomson ion-evaporation model to rationalize the phenomenon of multiple charging, Fenn and co-workers (10) suggested that the charge-state distributions of PEGs and proteins reflect the ion desorption rate which depends on droplet charge density. As the surface charge density increases during the course of droplet evaporation, ions of lower charge states start to desorb; further evaporation leads to the desorption of higher charge state ions at a faster rate, hence a higher ion current. Continued increase in charge density at the droplet surface eventually results in a decreasing flux of ions of even higher charge states, because the majority of ions have desorbed prior to collecting the additional charges characteristic of the most abundant observed charge states. This scenario largely rationalizes the commonly observed "envelope" of charge-state distributions of large molecules (e.g., proteins and polynucleotides). In a later publication (11), Fenn further refined a model elaborating the mechanism of formation of multiply charged ions in the ESI process. His treatment contends that the spacing of charges on a desorbed ion is closely related to the spacing of charges on the droplet surface at the moment that the ion desorbs. The charge state of a desorbed protein ion is thus determined by the charge density of the droplet surface and the size and configuration of the protein molecule (i.e., the number of charges on the droplet surface that it can span), rather than the number of basic residues contained in the protein molecule.

There is general agreement regarding the importance of protonation or deprotonation processes that yield positively or negatively charged species, respectively, in ESI–MS. In positive-ion experiments, the number of basic sites on the analyte molecule is believed to intervene in determining the maximum number of protons which may be attached to this analyte. Proteins and peptides are by far the most commonly studied species in ESI mass spectrometry. The number of basic residues in a protein was reported by Loo et al. [12,15] and Covey et al. [13] to be directly related to the charge state of the analyte species observed in ESI mass spectra, that is, the upper limit to the number of charges carried by proteins in ESI–MS was found to be determined by the number of basic residues contained in the protein. For several proteins containing multiple cysteine–cysteine disulfide bonds, the observed maximum charge states were considerably lower than those predicted by the discussed relation above (15). The disparity was attributed to the inaccessability of some basic sites which were "buried" in the tighter conformation which was imposed by the disulfide bonds, that is, the oxidized form of the cysteine sulfhydryl group. This argument was supported by experimental observation of a shift in protein charge states toward higher values upon reduction of cysteine–cysteine bonds to the sulfhydryl form (15). Other examples showing incongruity between the maximum number of charges in ESI mass spectra and the number of basic sites available on analyte proteins were catalogued in a review by Smith and co-workers (34).

Numerous cases have now been reported in which protein molecules undergoing conformational changes from folded native structures, to extended denatured ones, provoke shifts in ESI charge states toward higher values. These changes in conformation may be induced by altering the solution pH (16,29), introducing organic solvent (17), heating (18,19), or by other denaturing reagents (20). These observations appear to support the notion that the ESI mass spectra of peptides and proteins are highly influenced by their solution phase conformation and acid–base equilibria. Indeed, based on aqueous titration data of selected proteins, Guevremont et al. (21) concluded that the shape of ion abundance profiles observed in ESI-MS arose from the distribution of "preformed" ions in aqueous solution phase. On the other hand, Loo et al. (14) and Kelly et al. (31) showed that positively charged protein ions could be desorbed from solutions of high pH, wherein these proteins were known to bear overall negative charges. Likewise, negative ion species were observed in ESI mass spectra even when the solution pH was low enough to ensure that solution equilibria dictated a positive charge. Discrepancies were also noted between solution-phase equilibria and gas-phase observations for smaller molecules (peptides) whose charge states in ESI–MS varied quite weakly in response to widely differing solution pH conditions (25). In this case, these smaller peptides exhibited little or no conformational changes in response to varying pH.

It is now generally believed that solution-phase equilibrium considerations contribute to, but do not determine, the appearance of ESI mass spectra of multiply charged species. Factors such as instrumental conditions (14,32), solvent effects (17,23,24,35), types of anions present (22,28), analyte concentration (11,15,25,27,36), analyte conformation (16–18,22), and gas-phase modifications (33,37,38) must also be considered in accounting for the detected analyte charge-state distributions. This chapter presents an overview of results related to the underlying factors which determine the appearance of the distribution of charge states observed in ESI mass spectra. It is hoped that a somewhat clearer picture might emerge, even though different reports relevant to the multiple charging process occurring in ESI, at times, seem contradictory.

II. ROLE OF SOLUTION EQUILIBRIA AND SOLUTION-PHASE CHEMISTRY

Multiply charged ions observed in ESI mass spectra are most often formed from either neutral polar molecules onto which cations (most commonly H^+, Na^+, K^+, and NH_4^+) become attached (positive-ion mode) or removed (negative-ion mode), or, from ionic species bearing a permanent charge site. Adduct formation between neutral analytes and solution anions is also possible (39). Because analytes observed by ESI–MS are initially dissolved in various kinds and/or mixtures of solvents, a comparison between how analytes exist in solution and how they are detected as gas-phase ions is certainly appropriate.

It may be helpful to start with the simplest case, namely, when analytes form singly charged cations in the positive-ion ESI mode. For solvated ionic species in solution such as alkali-metal cations, quaternary ammonium ions, or protonated organic bases, the detection of their gas phase ionic counterparts may appear rather straightforward. Nevertheless, phenomena inherent to the desorption process which enable the conversion of solution-phase ions into gas-phase ions can lead to differences in sensitivity for individual ionic species. For example, according to ion evaporation theory, strongly solvated ions having large transfer free energies ($-\Delta G_{sol}$) from gas phase to solution are predicted to have large activation barriers ($\Delta G^{\neq}$) to escape from electrosprayed droplets, implying low rate constants (k) for desorption (11,40,41). Experimentally, qualitative agreement has been reached with this theoretical description, as ions predicted to have the highest rate constants for desorption were indeed observed to appear in the highest abundances in the ESI mass spectra of two-component mixtures (40,41). In such binary mixtures, the signal intensity of a given singly charged analyte (A^+) in the presence of a competing singly charged species (B^+) has been described mathematically as

$$I_{A+,ms} = \frac{pfk_{A+}[A^+]}{k_{A+}[A^+] + k_{B+}[B^+]} I \quad (1)$$

where $I_{A+,ms}$ is the ion current of analyte A^+ detected by the mass spectrometer, p is a constant expressing the efficiency of the mass spectrometer for detecting gas-phase ions produced in the ESI process, f is the fraction of droplet charge converted to gas-phase ions, $[A^+]$ and $[B^+]$ are the electrolyte concentrations initially present in the electrosprayed solution, k_{A+} and k_{B+} are the rate constants expressing the rate of transfer of the respective ions from charged droplets to the gas phase, and I is the total ES current leaving the ESI capillary. For a two-component system, Eq. 1 describes the factors that influence the detectability of a specified analyte in solution.

Clearly, the presence of excess charges on the droplets (polarity determined by the ES voltage) is the principal driving force for conversion of solution species into gas-phase ions. In a conventional ES device, at least a portion of the excess charges on the droplets may arise from the removal of the counterions associated with the analyte (e.g., if present in a salt form) via electrochemical processes. Thus, upon solvent evaporation, the analyte ions formed in this manner can be converted into gas-phase ions. Recent results (42) obtained on a droplet electrospray (DES) device (where charge is acquired on originally neutral droplets via a gas discharge process) show that the nonanalyte excess charges added to droplets lead to virtually identical mass spectra for analyte species when compared to conventional ESI mass spectra. In the DES device, no opportunity for removal of counterions of analyte species exists. This implies that the desorption of gas-phase ions from an initially neutral droplet will have to be initiated by converting excess charges originating from the discharge process into ion-paired species (discharge ion plus counterion of charged analyte), thus liberating an analyte ion. The process would have to be repeated n times to produce an analyte ion bearing n charges.

Logically, one can first examine the solution conditions for analyte species capable of forming multiply charged ions in the gas phase. In general, analytes such as proteins, peptides, oligonucleotides, and other molecules containing multiple basic or acidic sites exist in solution with a distribution of charge states as multiply charged positive or negative ions. The exact form of the distribution is highly pH dependent. It is surely this notion of a distribution in the molecules' charged states that appear both in the solution phase and in the gas phase which has led to numerous efforts to correlate solution chemistry to ESI mass spectral appearance. On the other hand, analytes such as polyethylene glycol (PEG), one of the most thoroughly studied macromolecules in early ESI–MS practice by Fenn and co-workers (43,44), present a somewhat different situation. For PEG, comparison of charging between the solution phase and the gas phase is more difficult because protonation of analyte via Bronsted-type acid–base equilibria is virtually negligible owing to the very weak basicity of PEG. In contrast, much less is known concerning the extent of Na^+ attachment in electrostatically neutral solutions (i.e., before the solution is electrosprayed) containing such analytes.

A. Effect of Solution pH

Variation of solution pH can substantially change the acid–base equilibrium of an analyte species, thus altering its degree of positive or negative charging via protonation/deprotonation. Chowdhury et al. (16) first showed that the lowering of solution pH beyond a critical threshold gave rise to a dramatic, discontinuous jump in observed ESI charge states of protein molecules toward higher values, as shown in Figure 1. This discontinuity was attributed to a shift in protein conformation from the native to the denatured state incurred by the pH changes. The work of Guevremont et al. (21) compared the solution distribution of charges carried by protein molecules and the charge states of the proteins detected during ESI–MS experiments. While more acidic solutions certainly produce a higher degree of solution protonation of proteins in a given conformation, the unfolding of the tertiary structure of proteins at low pH must also contribute to the shift in charge-state distributions observed in ESI–MS. In other words, variation of solution pH invokes a compounded effect of: (1)

Figure 1. Positive ion ESI mass spectra of 10^{-5} M bovine cytochrome c obtained with different acetic acid concentrations in aqueous protein solutions. Solution conditions: (*a*) 4% acetic acid (pH = 2.6); (*b*) 0.2% acetic acid (pH = 3.0); and (*c*) no acid (pH = 5.2). The labels on the peaks, $n+$, indicate the number of protons, n, attached to the protein molecule. The observation of two distinct distributions of charge states was correlated to a change in protein conformation from a folded (higher pH) to an unfolded (lower pH) state. Reprinted with permission from S. K. Chowdhury et al., *J. Amer. Chem. Soc.*, **1990**, *112*, 9012–9013. © 1990 American Chemical Society.

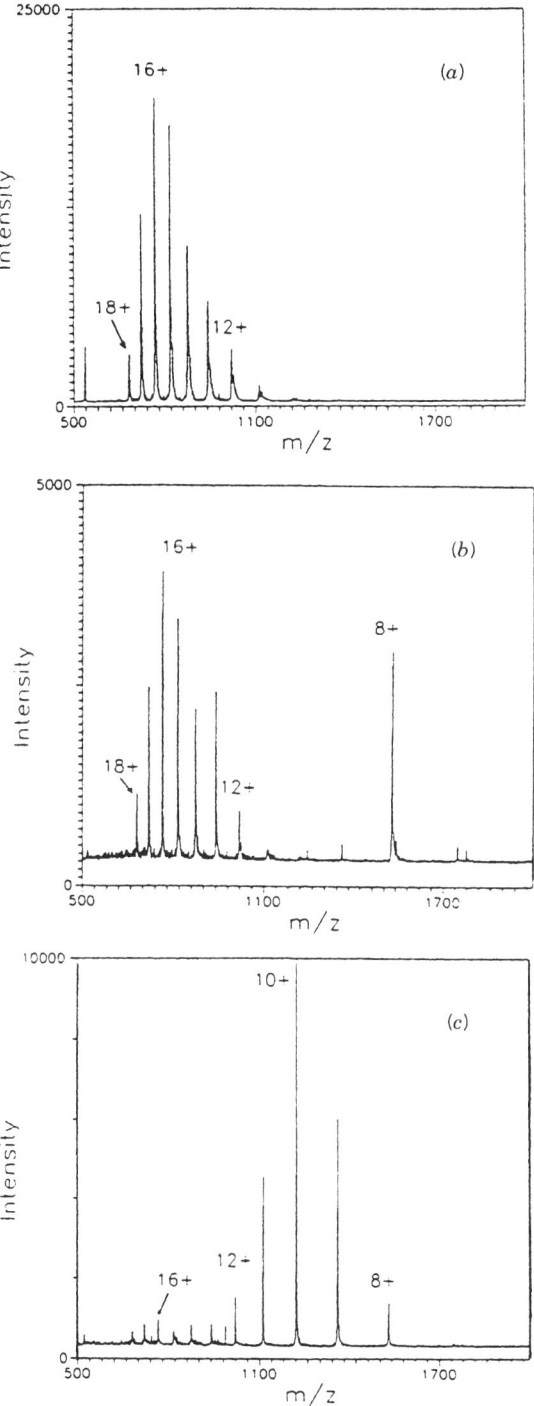

altering acid–base equilibria, while (2) simultaneously perturbing the three-dimensional conformation of the protein molecule. In contrast to the reported qualitative matching (21) concerning protein charge states in solution and in ESI mass spectra, Loo et al. (14) and Kelly et al. (31) reported results showing the appearance of multiply protonated proteins in positive-ion ESI mass spectra obtained from highly basic initial solutions, where analyte ions are known to exist in anionic form. In addition, multiply deprotonated molecules were obtained in the negative ion mode from highly acidic starting solutions (31) (i.e., proteins existed initially in cationic form). In the context of acid–base equilibria, protein conformational differences between the extremes of pH (i.e., highly acidic and highly basic) were not specifically discussed. Nevertheless the question was raised as to what degree the solution pH influences the appearance of ESI mass spectra.

In a later report (25), the small peptides bradykinin and gramicidin S ($m_r \sim 1000$ Da) were employed as test compounds to examine the effect of pH on the resultant charge states observed in ESI mass spectra. The use of such small peptides circumvented the complication of conformational changes, hence, the effect of pH alone on analyte charge distributions could be more clearly revealed. It was found that a solution hydronium ion concentration variation of 7 orders of magnitude resulted in only minor changes in the charge states observed in ESI mass spectra (Fig. 2). The degree of peptide protonation observed in the gas phase was highly disproportionate to the degree of protonation in solution. This study provided further evidence of the discrepancy between solution-phase equilibria and gas-phase charge states, leading to the conclusion that the ESI mass spectra are far from a direct reflection of the solution-phase charge pattern. Most notably, when acidic sites on analyte molecules are almost completely deprotonated, multiply protonated gas-phase ions can still be observed (14,25,31), and conversely, positively charged protein ions in solution can readily be converted to negatively charged gas-phase ions.

In rationalizing these phenomena, several factors have been considered. First, the formation of multiply charged gas-phase ions may be inherent to the electrospray process itself, for example, the excess charges near the droplet surface may provide the principal driving force, rendering the formation of multiply charged species rather independent of the solution pH (11). Secondly, the hydronium ion concentration in droplets may undergo changes during the course of the ES process. Indeed, Gatlin and Turecek (45) suggested that in positive-ion ESI, acidity may be enhanced by 10^3- to 10^4-fold in the outer thin layer of the droplets. Similarly, droplets formed in the negative-ion mode (bearing excess negative charge) are likely to be enriched in OH^-, and thus may undergo a considerable pH increase as compared to the bulk solution. This argument does not exclude the possibility that observed gas-phase ions originate from their preformed counterparts in droplets if one considers that the solution species have changed their distribution profile in response to changing conditions during the course of droplet evaporation.

The discrepancy between solution-phase equilibria based upon initial pH

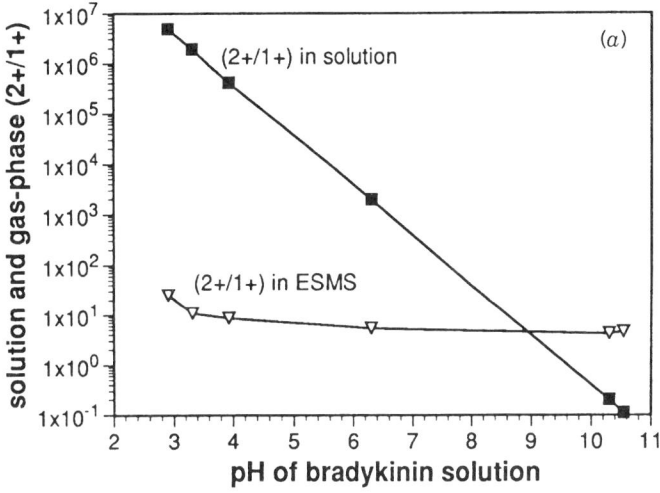

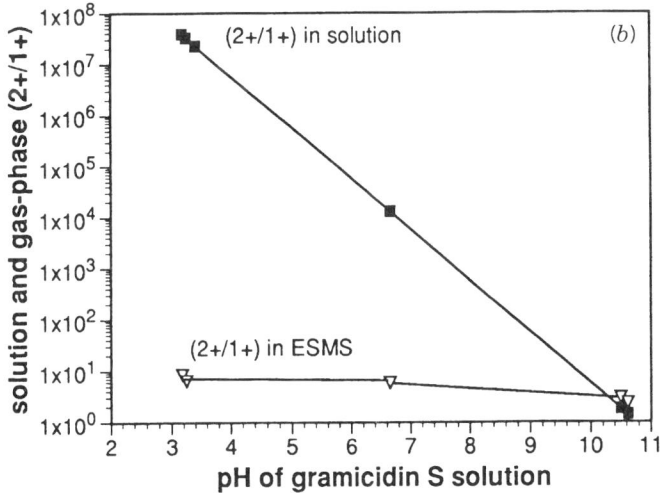

Figure 2. Comparison of calculated solution phase protonation versus gas-phase charge states observed in ESI mass spectra. Plots show the ratios of doubly to singly charged species versus solution pH (solid squares = solution-phase calculation; inverted hollow triangles = ESI–MS data) for (*a*) bradykinin and (*b*) gramicidin S. The ESI charge states undergo only minor changes in response to major variations in solution pH and calculated solution protonation. The data used to generate these plots was taken from ref. 25.

conditions, and observed charge state distributions in ESI mass spectra has been considered further in terms of ion pairing and coulombic repulsion effects (25). The possibility was brought forth that attractive forces between multiply charged species and nearby counterions may increase at progressively later moments in the droplet lifetime, as fewer solvent molecules are available for solvation of charged species. Increased ion pairing between attached protons and counterions may lower the level of attached protons observed in ESI mass spectra relative to what existed in the initial solution. Consideration of the increased coulombic repulsion between charge sites on protonated gas-phase molecules versus their solution-phase counterparts would also lower the tendency to observe high levels of charging in the gas phase compared to solution (25). Lastly, the production of protonated peptides and proteins in the presence of small, neutral nitrogen bases has also been examined (46). It was proposed that protonated polypeptides may be desorbed into the gas phase with the nitrogen bases attached. These complexes could then dissociate in the gas phase with a partitioning of the protons between the nitrogen base and the polypeptide, thus lowering the charge states of the polypeptides in ESI mass spectra relative to the initial solution condition (46).

A limited number of studies have investigated the effect of hydronium ion (or sodium ion) concentration on the distribution of charge states of other types of polymers such as oligodeoxynucleotides (47) or polyethylene glycol polymers (43,44). The maximum charge states of oligonucleotides appearing in negative-ion ESI mass spectra were reported to diminish with decreasing pH (47). The precise pH dependence (i.e., the specific pH value corresponding to a shift in the maximum charge state) varied according to the specific heterocyclic bases present in tested oligonucleotides. Results indicated a correlation of charge state observed in ESI–MS with the basicity (pK_a) of the nucleoside bases (47). The shift in charge states appeared rather gradual and continuous compared to previously discussed protein results.

In the case of PEGs, the issue of pH is not so relevant. However, solution phase chelation of sodium ions by electrons located on oxygen atoms of adjacent repeat units of the PEG backbone, or on separate PEG molecules, may also contribute to the multiple charging of PEG oligomers. Fenn and co-workers (43,44) studied the relationship between the maximum number of charges observed in ESI and the size of the PEG oligomer. They presumed that the maximum number of charges on PEG is reached when the energy associated with charge attachment equals the electrostatic repulsion of the centermost charge. This treatment gives the upper limit of multiple charging of the free gas-phase ions.

In summary, the initial solution pH is known to exert an important influence on solution-phase equilibria of charge attachment to analyte molecules, but it has a tempered influence on the observed ESI mass spectra of multiply charged ions. Other factors that play roles in determining analyte charge-state distributions are discussed in the following sections.

B. Solvent Effects

Early studies of electrohydrodynamic processes revealed that solvent properties can influence spray characteristics in a variety of ways (48–50), notably the dependences of the "onset potential" (i.e., the minimum potential required to form the characteristic Taylor cone) on surface tension, the spray current on conductivity, and the droplet size on viscosity. A mass spectrometrist's interest, however, lies in the role that solvent plays on the quality and appearance of the ESI mass spectrum.

Two detailed negative-ion ESI–MS studies investigated the dependence of charge-state distributions on a variety of solvent types (23,24). Despite the difficulty in varying one solvent parameter without simultaneously altering others, obtained results led to the conclusion that for analytes bearing permanent charge sites, the dielectric constant of the solvent medium plays a predominant role in determining the charge-state distributions observed in ESI mass spectra. Notably, solvents of higher dielectric constant shifted the charge-state distributions toward higher values, as shown in Figure 3. The mass spectra appearing in Figure 3a–e are listed in descending order of polarity of the chlorinated solvent employed, as measured by the dielectric constant. The analyte phospholipid, cardiolipin, is characterized by two ionizable phosphate groups wherein singly or doubly charged anions may be formed by removal of one or two counterions, respectively. The lipid portion of the molecule exists as a mixture of forms of varying hydrocarbon chain length ($-CH_2-$), thus giving rise to a distribution of observed m/z values separated by 14 m/z units for singly charged species and 7 m/z units for the doubly charged variety.

The influence of solvent polarity on analyte charge states can be viewed as a contribution from solution chemistry to the gas-phase ions generated in ESI. When dissolved in solution, a neutral compound (AB_2) may dissociate as follows:

$$AB_2 \rightleftharpoons AB^- + B^+ \quad (2)$$

$$AB^- \rightleftharpoons A^{2-} + B^+ \quad (3)$$

where AB^- and A^{2-} represent, respectively, the singly and doubly deprotonated/decationized forms of the analyte phospholipids observed in negative-ion ESI–MS (23, 24), and B^+ is the counterion (e.g., H^+, Na^+, etc.). The dissociation equilibria above (Eqs. 2 and 3) shift to the right in solvents of higher polarity, whereas solvents of lower polarity drive these equilibria to the left. The charge states observed in ESI mass spectra (Fig. 3) reflect this trend in solution. It thus appears that more polar solvents, which can better stabilize multiply charged species in solution, favor the formation of gas-phase ions of higher charge states (23,24).

The validity of this argument was further investigated in positive-ion ESI–MS using diquaternary ammonium salts as test analytes (28). Similarly, these neutral analytes (CD_2) undergo dissociation in dielectric media as

$$CD_2 \rightleftharpoons CD^+ + D^- \quad (4)$$

$$CD^+ \rightleftharpoons C^{2+} + D^- \quad (5)$$

where C^{2+} represents the diquaternary ammonium dication and D^- is the counterion (Cl^- or CF_3COO^-, etc.). Both C^{2+} and the monocation CD^+ are readily detected in positive-ion ESI–MS. The diquaternary ammonium salts are directly analogous to the analytes employed in the negative ion ESI–MS study discussed previously in that the singly charged ions result from the dissociation of one counterion while the doubly charged analyte ions represent species wherein two counterions have departed from the neutral species. Indeed, results obtained for diquaternary ammonium ions in positive-ion ESI–MS (28) are in good agreement with those from negative-ion ESI–MS experiments (23,24). The abundance of the doubly charged ion relative to that of the singly charged ion decreased markedly as solvent polarity was progressively decreased. This trend was observed for a series of *n*-alcohols of decreasing polarity (increasing hydrocarbon chain length) and for a series of low-molecular-weight chlorinated solvents of varying polarity (and chlorination).

These observations may be revealing in terms of the mechanism of ion formation in the ES process. Solvents of higher polarity can offer increased stabilization to charge separation in solution (23,24), which in turn facilitates the "electrophoretic" process of charged-droplet formation (50–53) at the ES capillary exit. In the electrophoretic process, negatively charged ions migrate toward the positively charged ES capillary (positive-ion mode) where they may undergo oxidation. Analogously, in the negative-ion mode, positively charged species migrate to the capillary held at high negative voltage where reduction processes occur. This removal of counterions from positively charged analytes when the ESI capillary is held at a positive potential, or conversely, from negatively charged analytes when the ESI capillary is held at a negative potential, can contribute to the formation of droplets bearing excess charges of one polarity. It is from these droplets that multiply charged gas-phase ions are formed and detected by the mass spectrometer. It is thus reasonable to conclude that solvents with higher polarity have enabled a greater degree of charge separation and ion solvation in the bulk solution as well as in the ES process, thus, leading to gas-phase ions of higher charge states (23,24).

C. Effect of Analyte Concentration

The relationship between signal intensity and original analyte solution concentration has been studied extensively (11,40,54,55). A typical relation between ESI–MS response and concentration is characterized by two distinct regions.

Figure 3. Negative-ion ESI mass spectra of the disodium salt of cardiolipin dissolved in 10% chloroform and 90% (*a*) methanol, (*b*) dichloromethane, (*c*) 1,2-dichloroethane, (*d*) chloroform, and (*e*) carbon tetrachloride. The solvents are listed in order of decreasing polarity as measured by the dielectric constant. The relative abundances of doubly charged species progressively decrease with decreasing dielectric constant of the solvent. Reprinted by permission of Elsevier Science, Inc. from R. B. Cole and A. K. Harrata, *J. Am. Soc. Mass Spectrom.*, **93**, *4*, 546–556. © 1993 American Society for Mass Spectrometry.

II. ROLE OF SOLUTION EQUILIBRIA AND SOLUTION-PHASE CHEMISTRY 149

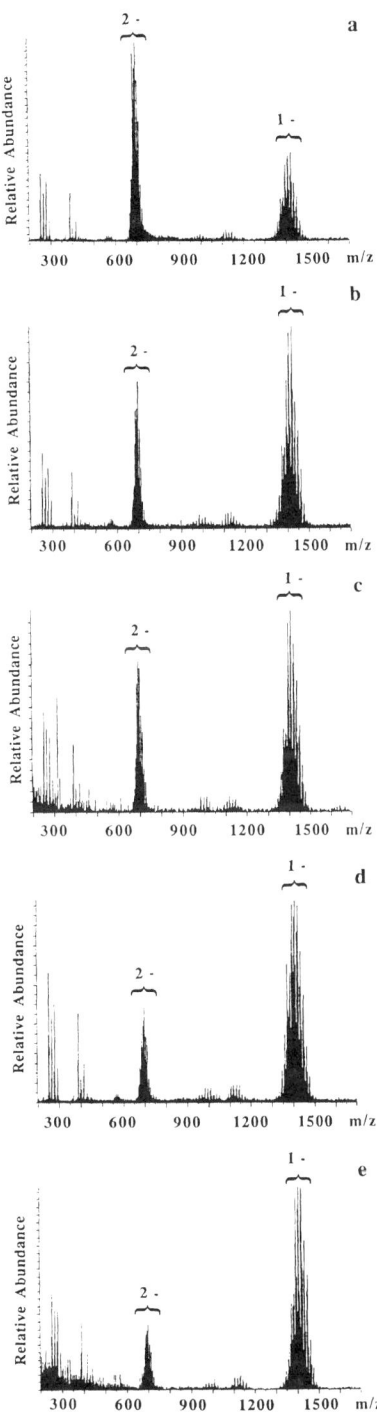

The first region, often referred to as the linear dynamic range, represents a concentration range from the lower limit of detection up to about 10^{-5} M, where the signal response increases linearly with increasing concentration. In a second region encompassing progressively higher concentrations, the signal intensity levels off and finally decreases as concentration is further raised.

For a given multiply charged analyte, ions of a discrete charge state exhibit similar dependences on concentration in that the abundance of ions of a particular m/z value has a "linear dynamic range" followed by a "level-off" region as a function of increasing concentration. However, the rate of increase in the linear range, and the concentration where transition into the "level-off" region occurs, are variable for ions of different charge state, such that the resultant overall charge-state distribution of the analyte is shifted to progressively lower values as the concentration is raised (27). The interpretation of the effect of analyte concentration on the charge-state distribution observed in ESI–MS hinges upon an understanding of the mechanism of ion formation in the ES process. At elevated concentrations, the decrease in the relative abundances of ions bearing a higher number of charges has been hypothesized by Chowdhury et al. (56) to be a consequence of an increase in the competition for the limited number of available charges on the droplets. Similarly, Smith et al. (57) proposed that a different "saturation" limit exists for ions of different charge states, which could be responsible for the onset of the decrease in abundances of highly charged ions at lower concentrations than those bearing fewer charges.

A more detailed rationale was described by Fenn (11) as a part of his ion-evaporation model depicting a stepwise process of multiply charged ion formation. According to that model, the spacing of charges on a desorbed ion is determined by the spacing of charges on the surface of the droplet at the moment when that ion is converted into the gas phase. As a charged droplet evaporates and its surface area is reduced, the surface charge density increases, which leads to a greater number of charges within reach of binding sites on a molecule. This implies that ions desorbed at different moments over the course of droplet evaporation will experience different surface charge densities, thus acquiring a different number of charges. Raising the concentration of analyte results in an increased number of analyte molecules in droplets, hence an increased rate of desorption of lower charge-state ions at earlier stages of droplet evaporation when the lower surface charge density is conducive to the analyte molecules desorbing with less charges. The consequence is that most of the droplets' charges are depleted by the evaporation of ions of lower charge states, leaving relatively few ions of higher charge states at later moments of droplet evaporation, because the supply of droplet charges has been reduced.

With a somewhat different approach, the roles of counterions and analyte concentration on observed charge-state distributions in positive- and negative-ion ESI–MS were examined using analyte species originally present in the salt form (27). Even though increasing analyte concentrations were accompanied by increases in the charge associated with each analyte *in solution*, the charge states of all analytes observed in ESI mass spectra were shifted without exception toward lower values in positive- and negative-ion ESI–MS as a function of increasing analyte concentration. In elucidating the underlying mechanism, the ratio (N/N_0), that is, the ratio of the total number of excess charges (N) to the total number of analyte molecules (N_0) in all droplets was introduced. The total number of excess charges (N), corresponding to the number of elemental charges leaving the ES needle, was estimated by measuring the current arriving at the counterelectrode and dividing by the elemental charge, I/e (Eq. 6, given in units of charges per second). The total number of analyte molecules entering sprayed droplets (N_0) was deduced from the product of the flow rate and the analyte solution concentration, ACU (Eq. 6, given in units of molecules per second).

$$\frac{N}{N_0} = \frac{I/e}{ACU} \quad \frac{(\text{charges/sec})}{(\text{molecules/sec})} \qquad (6)$$

where I represents the electrospray current (C/s), e is the elemental charge (1.602×10^{-19} C/charge), A is Avogadro's number (6.023×10^{23} molecules/mol), C is the analyte concentration (mol/L), and U is the solution flow rate into the mass spectrometer (L/s). The N/N_0 ratio thus indicates the number of excess charges available per analyte molecule present. The value of N/N_0 was compared with the value of the average analyte charge state as the concentration was raised. Because the former value decreased concurrently with observed analyte charge states, it was suggested that a diminishing N/N_0, indicative of an increasing number of ion pairs of multiply charged analyte species and counterions in solution, was the underlying factor leading to the detection of gas-phase ions of lowered charge states.

This rationale is illustrated in Figures 4 and 5, where a portion of the outer layer of a charged droplet is shown schematically. This droplet portion (Fig. 4, top) contains five dibasic analyte molecules represented as ellipses bearing two protons (shown as two "plus" charges) indicative of the overwhelmingly preferred solution-phase 2+ analyte charge state. Taking into account the presence of four counterions (labeled A^-) and two other cations (smallest circles bearing positive charges), this droplet portion contains eight excess positive charges per five analyte molecules (i.e., $N/N_0 = 8/5$ for this droplet portion). In the process of droplet evaporation and deformation, available A^- counterions may stay close to protonated sites on analyte molecules. As droplet evaporation proceeds, the two counterions may either stay attached to one diprotonated analyte ion which will end up as a residual neutral (Fig. 4, path a, doubly charged ellipse bearing two A^- counterions), or, as depicted in path b, they may be charge neutralized by two protons (forming H^+,A^- ion pairs) which have departed from two separate analyte molecules and thus yield two monoprotonated gas-phase analyte ions (singly charged ellipses) which were originally absent in the droplet. Because the latter result better resembles the analyte charge-state distribution observed in ESI mass spectra, the process depicted in path b is proposed to be chiefly responsible for the disparity between solution equilibria of protonation and gas-phase charge-state distributions.

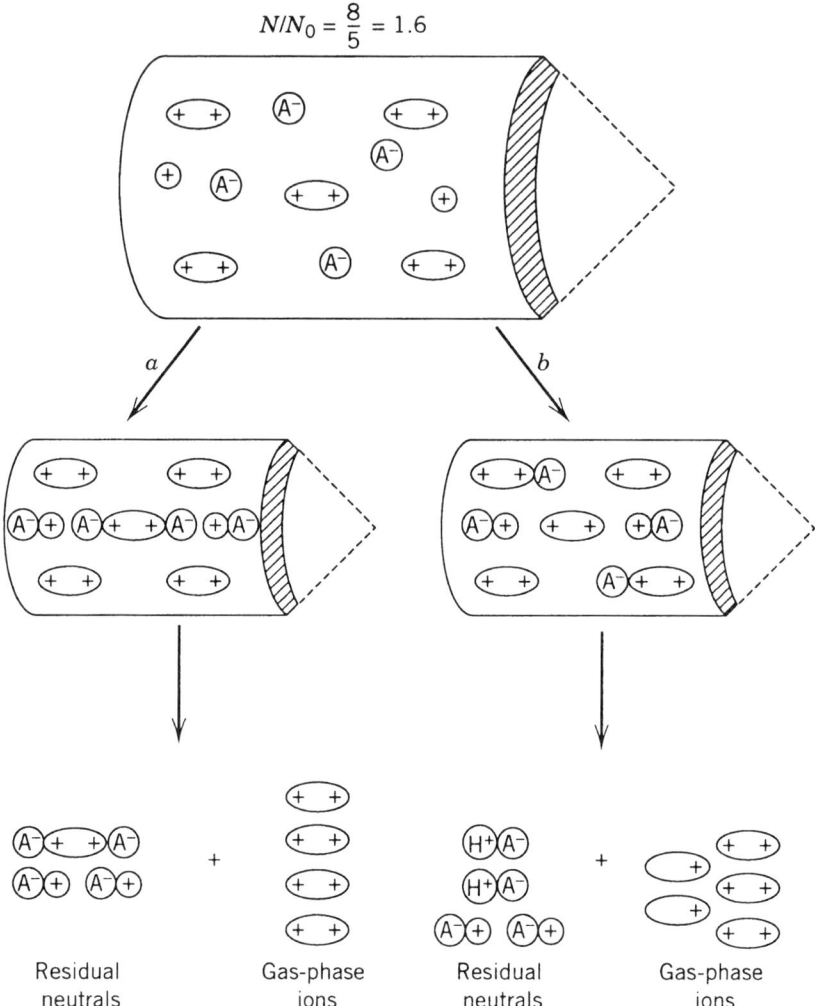

Figure 4. Schematic representation of the outer portion of a charged droplet containing an excess of positive charges. Doubly protonated analyte molecules (doubly charged ellipses) were initially introduced in the salt form with two anions (circles labeled A^-) attached. Droplet charging occurred by electrophoretic removal of anions and electrochemical production of cations (smallest circles bearing + charges). The ellipses are arranged in horizontal fashion for visual clarity. The anions are likely to be located just below the near-surface layer where excess positive charges reside. The middle panels show a later stage of the droplet lifetime where solvent has evaporated to decrease the droplet size, and the probability of forming contact ion pairs has increased. The diagram depicts the relation between N/N_0 (top) and the charge states of ESI-generated gas-phase ions (bottom). Pathway *b* corresponds more closely to experimental observations. Reprinted with permission from G. Wang and R. B. Cole, *Anal. Chem.*, **1995**, *67*, 2896. © 1995 American Chemical Society.

If the number of analyte molecules in the droplet portion increases from 5 to 8 as shown in Figure 5, because of the weak dependence of the electrospray current I on the solution conductivity, σ, ($I = H\sigma^n$, $n = \sim 0.2$–0.3) (40,41,58), where H is a proportionality constant, the number of excess charges increases only from about 8 to 10. In moving from the lower concentration situation (Fig. 4) to that of higher concentration (Fig. 5), the N/N_0 value would thus be lowered from $(8/5) = 1.6$ to $(10/8) = 1.25$, and there will be a higher number of A^- present as counterions in the droplet. Spreading the counterions over as many diprotonated analyte ions as possible will ultimately result in a lowering of the overall charge associated with the analyte as more protonated sites are charge-neutralized by A^- counterions. Singly charged analyte ions that are thereby formed (with one of the two protonated sites bearing a A^- counterion) presumably desorb into the gas phase accompanied by the dissociation of a neutral HA molecule to form the mass spectrometrically observed singly charged MH^+. The driving force for the departure of the HA molecule may be thermal, or it may be the result of collision-induced dissociation (CID). At higher concentrations (Fig. 5), the proportion of molecules undergoing this process is raised relative to the lower-concentration case (Fig. 4); hence, the average ESI detected charge state is reduced.

The scenario described above is supported by the observation of singly charged ions of the form $(MH_2^{2+} + Cl^-)^+$ which arise from the attachment of one Cl^- counterion to the doubly charged analyte cation (27). When analyte species are doubly protonated diamines (gramicidin S and 4,4'-bipiperidine), the intensity of the peak corresponding to $(MH_2^{2+} + Cl^-)^+$ is weak, owing to the apparent ease with which the HCl molecule departs in the gas phase, thereby yielding MH^+ as the predominant singly charged species (27). However, when the analyte is a diquaternary ammonium dication (i.e., (M^{2+} + 2counterion$^-$), neutral form) whose charges are "permanently" anchored on the molecule unless removed by fragmentation, a counterion (e.g., Cl^- or CF_3COO^-) must remain attached to one of the positive charges of the diquaternary ammonium ion in order to generate a singly charged ion (28). The abundance of $(M^{2+} + Cl^-)^+$ was observed to increase monotonically with increasing diquaternary ammonium chloride concentration, providing additional evidence that a decreasing N/N_0 ratio, reflecting this higher concentration, induces the shift in observed charge states toward lower values.

D. Effect of Counterions and Electrolytes Present in Solution

The effect of counterions present in solution on the appearance of analyte ions in the ESI mass spectrum may be viewed as having two components—the nature of, and the concentration of, counterions. The nature of the counterion was reported by Mirza and Chait (22) to influence the analyte charge state. They found that the charge states of proteins and peptides were shifted to lower values because of the neutralization of the positive charges by counterions present in solution. Furthermore, the magnitude of the shift in charge-state distributions of

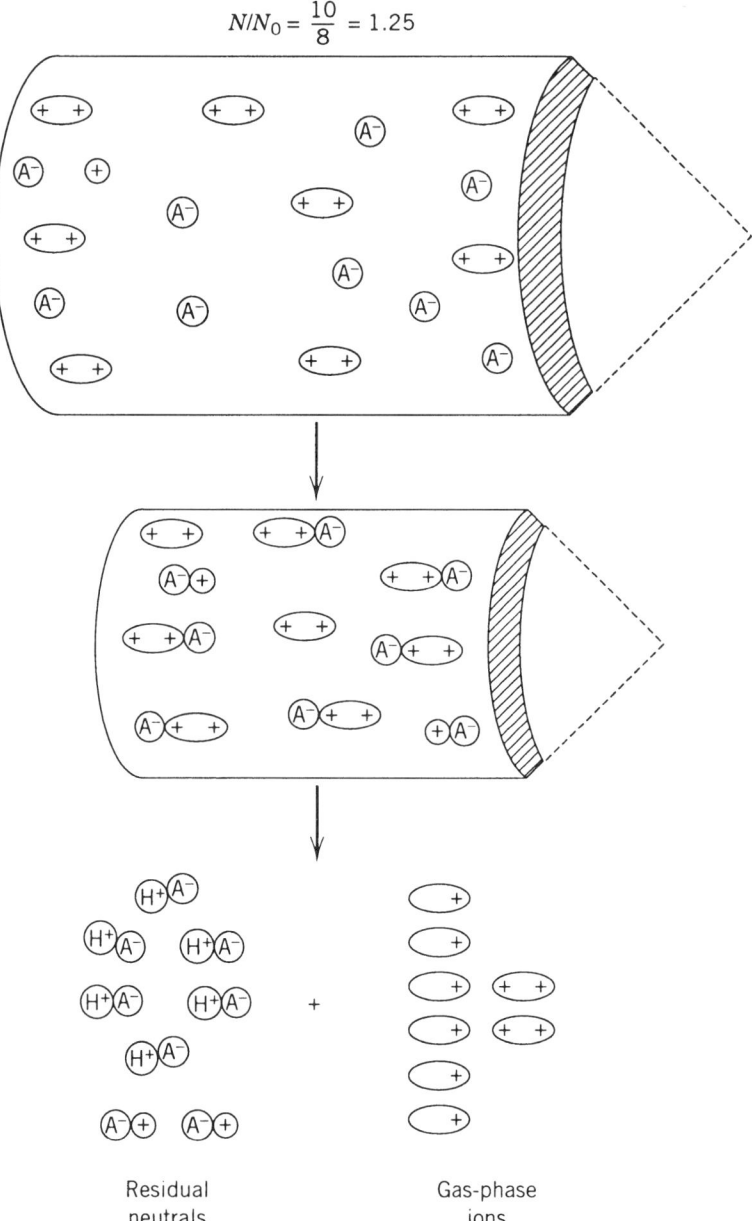

Figure 5. Schematic representation of the outer portion of a charged droplet containing a higher concentration of the same dibasic analyte, originally introduced in the salt form, as was shown in Figure 4. The symbols employed are explained in Figure 4. The observed increase in the proportion of singly charged ES-generated ions at higher concentration is attributed largely to the reduced N/N_0 value. Reprinted with permission from G. Wang and R. B. Cole, *Anal. Chem.*, 1995, **67**, 2897. © 1995 American Chemical Society.

proteins and peptides depends on the nature of the anionic species in the order $CCl_3COO^- > CF_3COO^- > CH_3COO^- \approx Cl^-$, regardless of the form in which the anions were introduced. This ranking coincides with the electroselectivity of anions toward anion exchange resins (22). This trend also holds true for compounds that exist in the salt form, such as the chloride salts of quaternary ammonium cations (28). In Figure 6, the ESI mass spectrum of a diquaternary ammonium salt (diquat 1, chloride salt, Fig. 6a) is compared with the ESI mass spectrum of an equimolar mixture of diquat 1 chloride salt and cesium trifluoroacetate (Fig. 6b). Note that there are actually two equivalents of Cl^- ions present for each equivalent of CF_3COO^- when the diquat 1 chloride salt and cesium trifluoroacetate are present in equimolar quantities. Clearly, CF_3COO^- attaches more efficiently to the doubly charged diquat 1 ion than Cl^-, and thus yields a higher abundance of singly charged ions, leaving all other parameters unchanged. When external electrolyte (cesium trifluoroacetate) is used as the source of CF_3COO^-, the singly charged diquat 1 initially introduced in the chloride salt form is represented in the ESI mass spectrum (Fig. 6b) predominantly by $(M-2Cl+CF_3COO)^+$ at m/z 770, although the peak due to $(M-Cl)^+$ at m/z 691 is still discernable. In the case where the original counterion Cl^- has been removed, and CF_3COO^- is now present as the sole counterion of the diquaternary ammonium dication (Fig. 6c), the only singly charged ion which appears is $(M-CF_3COO)^+$ at m/z 770, and its abundance relative to the doubly charged species is higher than that of the sum of the abundances of the singly charged ions in the mixture of chloride and trifluoracetate salts (Fig. 6b). These results imply that counterions with higher affinities for cationic species tend to shift charge-state distributions observed in ESI–MS to a greater extent toward lower values.

When counterions are introduced into analyte solutions by addition of external electrolytes, the effect on analyte charge-state distributions appears to be analyte dependent. For large molecules such as proteins, the addition of external electrolytes may exert little effect on the overall charge-state distributions observed in ESI–MS (26). This is illustrated in Figure 7 in which the appearance of the charge-state profile of the protein myoglobin is nearly constant despite a dramatic decrease in the total myoglobin signal intensity due to four orders of magnitude increase in CsCl concentration. Note the progressive increase in Cs^+ and $Cs(CsCl)^+$ signal intensities in descending from Figure 7a to Figure 7d. The near constancy of analyte charge states with increasing CsCl concentration is also observed for lysozyme (26) and some tested diquaternary ammonium compounds (28).

Increasing concentrations of other types of electrolytes, such as ammonium acetate, caused a slight shift in charge states of lysozyme toward lower values (26). The basicity of ammonium acetate likely favored increasing removal of charge-bearing protons from lysozyme as the concentration of this electrolyte was raised. Furthermore, when analytes are either smaller in size, or have a low efficiency of conversion from solution into gas-phase ions *relative* to the electrolyte ions, the charge states have been observed to be even more variable with increasing electrolyte concentrations. For example, the addition of CsCl caused the charge states of 4,4'-bipiperidine to be lowered significantly, and those of gramicidin S to be lowered slightly, whereas the charge states of proteins and diquaternary ammonium cations are largely unaffected by the presence of increasing amounts of CsCl (26,28). Moreover, even though the average charge state of diquat 1 is shifted considerably to lower values by increasing the concentration of cesium trifluoroacetate (28), that of lysozyme is not. The exact reason behind these differences is not well understood. However, it appears that the magnitude of the effect(s) of counterion concentration is analyte dependent. It is possible that, given a common anion (e.g., Cl^-), the desorption rates of positively charged analyte ions relative to that of Cs^+ may influence the extent of anion interaction with the analyte; hence, the ESI charge-state distributions. For example, when a two-component equimolar solution of CsCl and protein (e.g., myoglobin or lysozyme) was subjected to ES conditions, the protein exhibited much higher peak intensities than that of Cs^+, indicating that the efficiency of desorption of these multiply charged protein ions was considerably larger than that for Cs^+. The charge-state distributions of these proteins were virtually unaffected by increasing concentrations of the salt (26). In another case where the ESI mass spectrum of an equimolar mixture of CsCl and 4,4'-bipiperidine was obtained, the intensity of 4,4'-bipiperidine was only moderately higher than that of Cs^+, indicating a smaller difference between the desorption ionization efficiency of the analyte and that of Cs^+. The distribution of charge states of 4,4'-bipiperidine shifted to lower values in the presence of increasing quantities of the salt (59). In other words, a larger difference between the desorption rate constant of analyte ions and that of Cs^+ may lead to a smaller degree of variation in analyte charge state with varying electrolyte concentration.

E. Effect of Analyte Structures/Conformations

Variation of the experimental conditions under which analytes are subjected to ESI can influence the charge-state distributions observed in the mass spectra, as will be explored in Section III. Given the same solvent and instrumental conditions, however, the molecular structure and the three-dimensional conformation of the molecule can significantly influence the number of charges the analyte can carry in the gas phase. Molecules of similar size, but with functional groups differing in the ability to retain charge, may give rise to different charge-state distributions in ESI mass spectra. Again, the effect of analyte structure should be considered in both the solution phase and the gas phase.

Figure 6. Positive-ion ESI mass spectra of a 10^{-4} M diquaternary ammonium salt in 1:1 methanol-water using (a) Cl^- as the orignal counterion; (b) Cl^- as the original counterion, but containing 10^{-4} M cesium trifluoroacetate; (c) with CF_3COO^- as the original counterion. The increased binding ability of CF_3COO^- promotes production of singly charged species relative to Cl^-.

II. ROLE OF SOLUTION EQUILIBRIA AND SOLUTION-PHASE CHEMISTRY   **157**

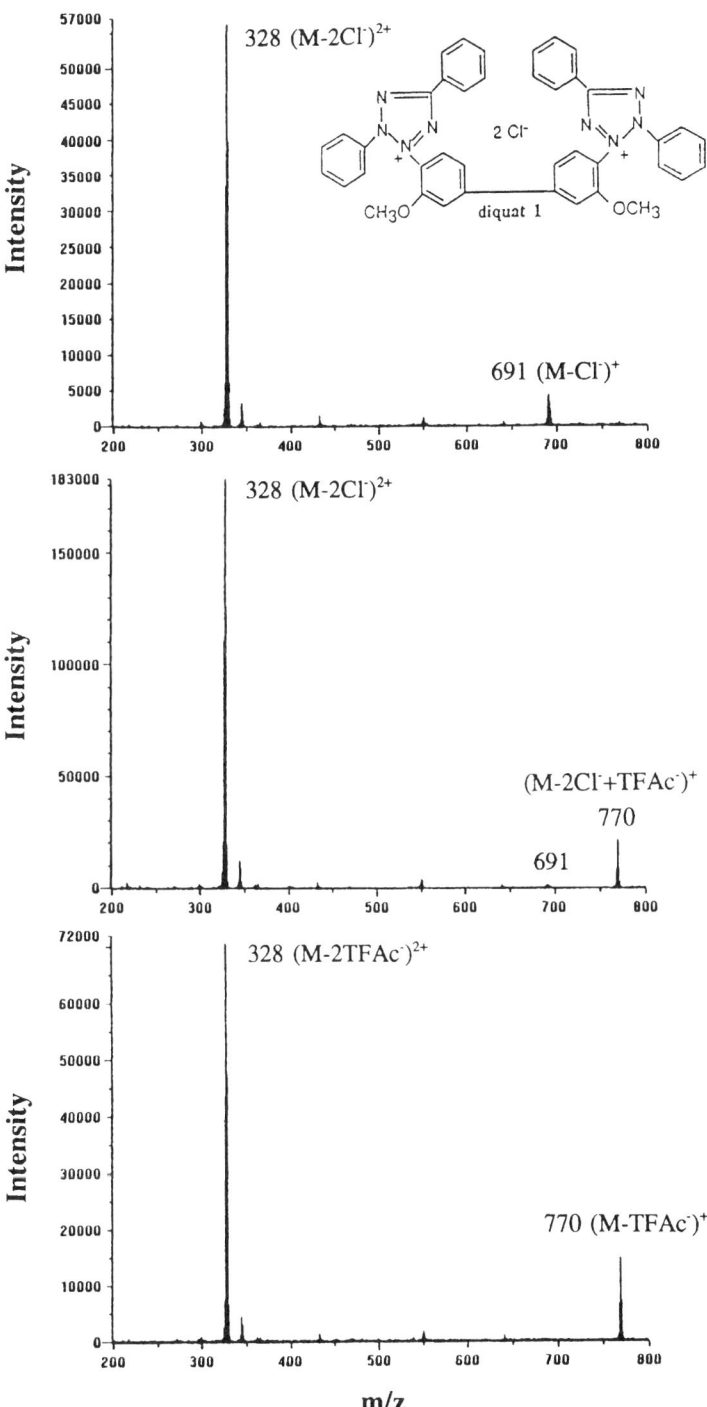

In a situation where protons constitute the charge excess, although it has been shown that the degree of solution protonation (25,31,45) may be poorly reflected in ESI mass spectra, the availability of basic sites (basic enough to be protonated in a given solution) sets an upper limit to the maximum number of charges that can attach to an analyte molecule. This is well illustrated in the ESI mass spectra of two small molecules, 4,4'-bipyridyl and 4,4'-bipiperidine, which have nearly the same size and conformation such that the coulombic repulsion between the intramolecular protonated sites is approximately the same (the effect of coulombic repulsion is discussed in Section IV.C). Although singly charged species were observed for both compounds, only 4,4'-bipiperidine exhibited an abundant doubly charged ion in its ESI mass spectra (59). The pK_{a2} for 4,4'-bipyridyl is extremely small in aqueous solution, making it virtually impossible to protonate the second nitrogen.

The structural effect is also evident in proteins when tertiary structure (conformation) is altered. In a tightly folded native conformation, proteins offer less access to ionizable sites as demonstrated by the lower degree of solution-phase hydrogen–deuterium exchange relative to the denatured conformation (60–62). A tightly folded conformation thus yields lower charge states in ESI mass spectra. If the proteins adopt a more extended, or denatured form, the previously "buried" basic sites become available for protonation, thereby giving rise to a shift in charge-state distributions to higher values (lower m/z values). Various means of denaturing proteins have been employed, including lowering the solution pH, as first demonstrated by Chowdhury et al. (16). This seminal example of the effect of protein conformation on the charge states observed in ESI mass spectra was shown in Figure 1 (16). Bovine cytochrome c undergoes conformational changes when the solution pH is changed from 5.2 to 2.6. The most intense peak shown in the mass spectrum (Fig. 1c), obtained at a solution pH of 5.2, corresponds to the native protein carrying 10 positive charges. As the solution pH is lowered to 3.0, the protein is partially denatured, which is reflected in the ESI mass spectrum (Fig. 1b) as having two discrete charge-state distributions centered at 16+ and at 8+. Further decreasing the solution pH to 2.6 forces the protein to completely unfold, and the resultant mass spectrum yields only one distribution of charge states, which is centered at 16+ (Fig. 1a).

Another method of denaturing proteins involves reducing the disulfide bridges which are present in cysteine-containing proteins, as first demonstrated by Loo et al. (15). The ESI–MS of several proteins containing various numbers of disulfide linkages was investigated with and without reaction with 1,4-dithiothreitol (DTT), a reagent that reduces disulfide bonds to sulfhydryl

---

Figure 7. Positive-ion ESI mass spectra of a 5×10^{-6} M horse heart myoglobin in 1:1 methanol-water plus 0.1% acetic acid (v/v) with CsCl added in the following concentrations: (a) 10^{-6} M, (b) 10^{-4} M, (c) 10^{-3} M, and (d) 10^{-2} M. Despite the progressive decrease in myoglobin signal intensity with augmenting salt concentration, the distribution of myoglobin charge states is essentially unperturbed. Reprinted with permission from G. Wang and R. B. Cole, *Anal. Chem.*, 1994, **66**, 3705. © 1994 American Chemical Society.

II. ROLE OF SOLUTION EQUILIBRIA AND SOLUTION-PHASE CHEMISTRY   **159**

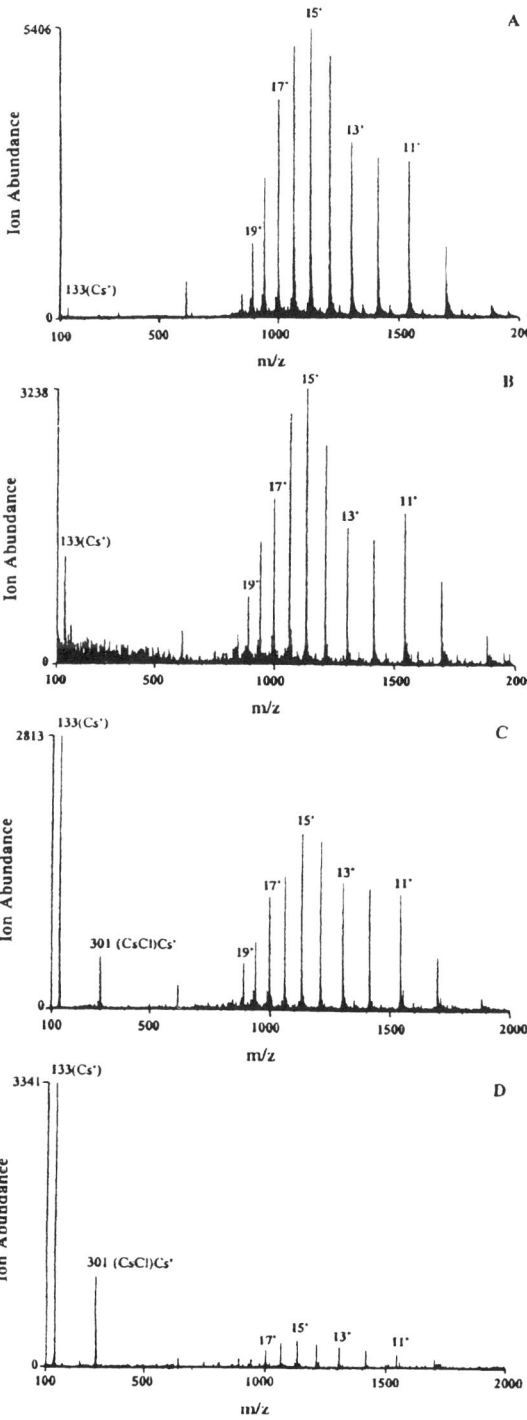

groups (15). All proteins exhibited an increase in their ESI–MS charge states upon addition of DTT. For example, hen egg white lysozyme shows a maximum charge state of 14 (Fig. 8A), five charges less than the total number of basic residues (i.e., Arg, Lys, and His) present on the protein molecule. However, upon treatment with DTT, the ESI mass spectrum of lysozyme (Fig. 8B) shows a maximum of 20 charges that have been attached to the protein. This charge-state shift has been attributed to the reduction of 4 cysteine–cysteine linkages, thus allowing the protein to unfold into a more extended conformation which can accommodate more charges.

Le Blanc et al. (18) investigated the effect of heat on the conformation of globular proteins, and the resultant change in the ESI charge-state distributions. These globular proteins were found to exhibit increased charge states with

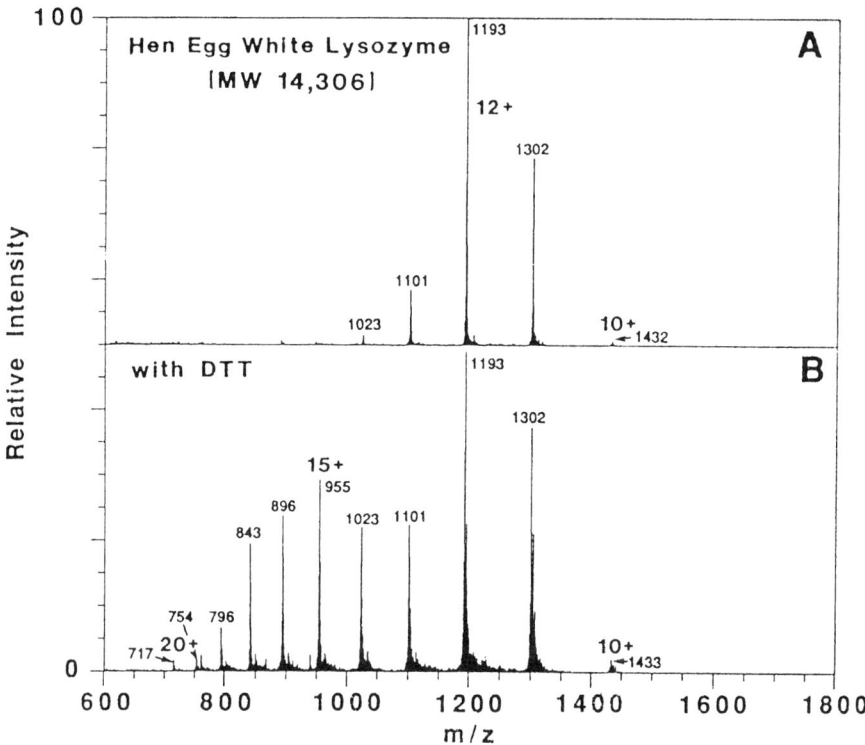

Figure 8. (A) Positive ion ESI mass spectrum of hen egg white lysozyme in 5% glacial acetic acid; and (B) ESI mass spectrum of the same solution upon addition of dithiothreitol. The marked shift in charge states toward higher values in the lower mass spectrum was attributed to disulfide cleavage reactions which allow the protein to adopt a more extended conformation, thus permitting the protein to accommodate a higher number of charges. Source: Reprinted with permission from J. A. Loo et al., *Anal. Chem.*, 1990, **62**, 695. © 1990 American Chemical Society.

increasing temperature which was attributed to a thermal denaturation effect (18). Similar results were also obtained when the ES needle temperature was raised up to 98°C by a heating device designed by Mirza and Chait (19). Use of organic solvents represents another means of inducing conformation changes in proteins. Loo et al. (17) reported that addition of various amounts of organic solvents such as methanol, acetonitrile, isopropanol, and acetone to protein solutions caused shifts in charge-state distributions in ESI–MS. The minimum amount of organic solvent required to induce the appearance of discrete charge-state distributions in regions of the ESI mass spectra other than those observed in the absence of organic solvent, were largely consistent with previous studies of protein conformational changes induced by alterations in solvent composition, as measured by other instrumental techniques such as circular dichroism and nuclear magnetic resonance (17).

III. EFFECT OF INSTRUMENTAL CONDITIONS

Even after closely following a set protocol for sample solution preparation, ESI mass spectra obtained under different circumstances can have widely varying appearances. This becomes obvious when comparing mass spectra acquired in different laboratories, using different types of instrumentation. Differences can also be observed when comparing data obtained on two different instruments, perhaps of the same model constructed by the same manufacturer, when certain adjustable parameters are not duplicated. Moreover, ESI mass spectra acquired on the same instrument at different time periods can exhibit some degree of variability, even when an attempt is made to faithfully reproduce adjustable parameters, if other specific conditions (e.g., pressure or temperature in a given source region, or instrument contamination) are less well controlled.

A. Instrument Geometry

A good example of how different source geometries can affect the appearance of the ESI mass spectrum is given in a comparison of data from cytochrome c obtained on the same mass spectrometer which had been equipped with two different style ESI sources (33). In an arrangement employing a countercurrent (nitrogen) drying gas, a higher level of protonation was observed than in an arrangement employing a heated capillary inlet with no drying gas. Presumably more proton abstraction is allowed in the latter configuration owing to the higher level of solvent species present in the absence of drying gas. The same authors made the observation that larger distances between the spray capillary and the sampling orifice led to discrimination against higher charge states. The degree of desolvation and the possibility for varying levels of proton transfer and CID of multiply charged analyte molecules, which depend upon the specific instrument geometry and daily operating conditions, will surely influence the appearance of the charge-state distribution. The results above (33) imply that construction of an ESI source, wherein the amount of solvent species entering the mass spectrometer is diminished, can result in the production of more highly charged ions.

B. Countercurrent Gas and Cooling Gas

In extending concepts of ion evaporation theory, Fenn (11) has proposed that the observed distribution of charge states in an ESI mass spectrum should be influenced by the evaporation rate of the charged droplet. Fenn's model depicts the charged droplet with excess charges located on the outer surface of the droplet, wherein the distance between charges is rather uniform. As analyte species approach the droplet surface during the course of droplet evaporation, they come into contact with the excess charges located on the droplet surface. Over the course of the shrinking droplet's lifetime, the surface charge density increases and the coulombic repulsion between adjacent charges becomes more severe. When the field acting on the surface charges becomes sufficient to overcome the activation energy barrier that an ion must surmount to escape from the droplet, ions may lift into the gas phase, depleting the surface of excess charges. Fenn argues (11) that the specific surface-charge density of the droplet at the moment of ion evaporation will exert an important influence upon the charge state of an ion desorbed at a certain instant in time. More specifically, ions desorbed in the early stages of evaporation (where the surface-charge densities are relatively low) will carry fewer charges than those desorbing at later stages, when the surface charge density has increased.

It was further reasoned (11) that if the evaporation rate could be slowed, the length of time that the droplet spent in the low surface-charge density regime would be prolonged. This could lead to an increased proportion of lower charge-state ions in the total population of desorbed ions, that is, the charge-state distribution would shift toward lower values. To test this hypothesis, after acquiring the mass spectrum shown at the top of Figure 9, the flow rate of the heated counterflow bath gas was increased. This increase had the effect of increasing the droplet evaporation rate, resulting in a decrease in the time that the droplet existed in the lower surface-charge density regime. As the mass spectrum at the bottom of Figure 9 illustrates, the distribution of charges clearly shifted to higher charge states (lower m/z values) when the drying gas flow rate was higher, and the evaporation rate was faster.

Figure 9. Effect of droplet evaporation rate on the charge-state distribution of positively charged ESI-generated ions of cytochrome c in 1 : 1 methanol-water at a concentration of 0.8 μmol/L. The flow rate of drying gas in the lower spectrum is about six times that in the upper spectrum. The flow rate for cooling gas was the same for both mass spectra. There is a clear shift to higher charge states (lower mass-to-charge ratio values) at the higher flow of drying gas for which the evaporation rate is faster. Reprinted by permission of Elsevier Science, Inc. from J. B. Fenn, *J. Am. Soc. Mass Spectrom.*, 1993, **4**, 533. © 1993 American Society for Mass Spectrometry.

III. EFFECT OF INSTRUMENTAL CONDITIONS   **163**

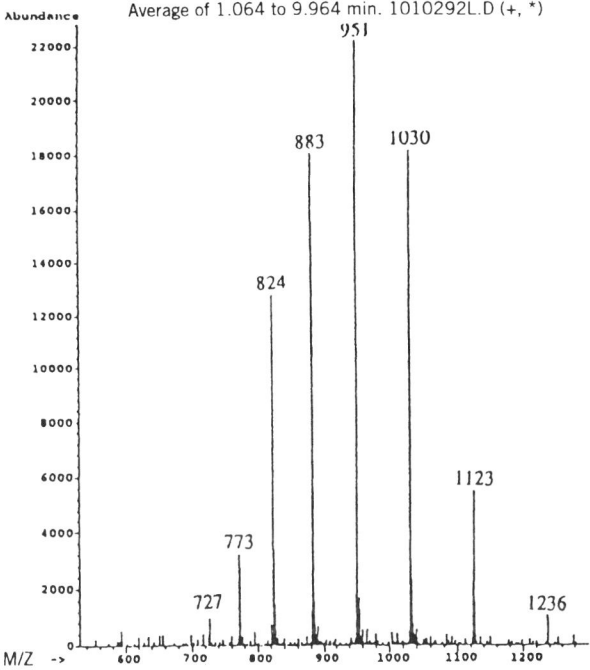

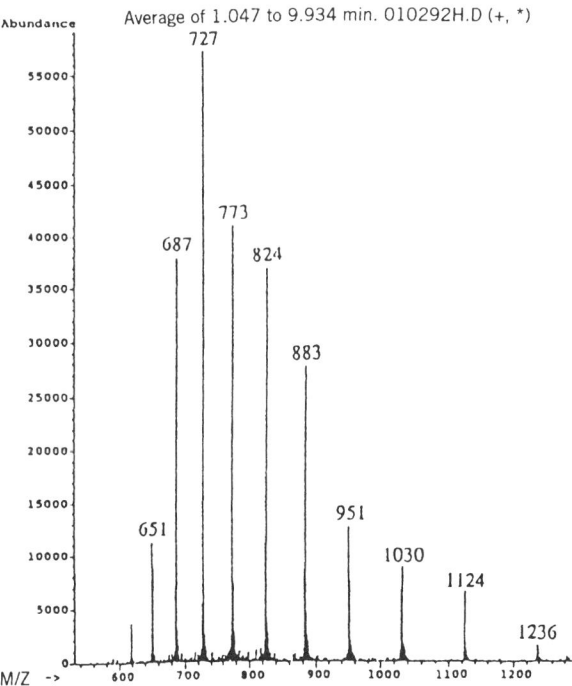

164 SOLUTION, GAS-PHASE, AND INSTRUMENTAL PARAMETER INFLUENCES

In addition to altering the flow rate of the heated bath gas, the bath gas temperature was also varied in order to exert another type of control on the evaporation rate (11). As a result, the charge-state distribution was reported to shift toward higher values with increasing temperatures, although the applicable range of temperatures yielding acceptable signals was rather narrow. This result is consistent with the idea that increasing the evaporation rate will reduce the time available for desorption of ions during stages of low-droplet charge density, leading to fewer ions of lower charge state. Lastly, the flow rate of a concurrent (same direction as solution flow) "cooling gas" was also varied (11) while holding other parameters constant. Increasing the flow rate of this concurrent cooling gas had the effect of slowing down the evaporation process (resulting in larger droplet radii) and decreasing the time before "ion sampling," that is, the time period of acceptance of ions into the mass spectrometer orifice was reduced. Over the range of cooling gas-flow rates employed, increased cooling gas-flow rates led to a steady shift in the charge state of the most abundant ion toward lower values. These results are consistent with a description of events wherein ions desorbed during earlier stages of droplet evaporation carry with them lower numbers of charges, relative to those desorbed at later stages.

C. Skimmer Voltage and In-Source Collision-Induced Dissociation

The effect of varying the "skimmer" voltage, sometimes called the "cone" voltage (i.e., the potential applied to the conical metal guide whose orifice separates the ESI source into two distinct pressure regions), to provoke fragmentations has been exploited in a large number of ESI–MS studies. The voltage applied to the skimmer serves as both an electrostatic lens to guide ions more efficiently toward the detector, and as a means to invoke energetic collisions with residual gases. The latter capability is most efficiently achieved when the proper balance of pressure considerations is found in a particular region of the ion source (presumably the region just after the skimmer). Collisions are most efficient when the pressure is sufficiently low to allow a significant mean free path of travel before a collision occurs, and sufficiently high to allow a significant number of collisions which can produce daughter ions in high abundances. Increasing the skimmer voltage presumably augments the translational energy of ions exiting the skimmer orifice. With the proper balance between mean free path of movement and collisional cross section, fragmentations are increased, although the maximum amount of energy transferable to analyte ions in this manner appears to be significantly below that attainable in the collision cell of a triple quadrupole mass spectrometer.

Ashton et al. (32) documented the effect of varying the skimmer voltage on charge-state distributions. They showed a progressive shift in charge states toward lower values with successive increases in the skimmer voltage. Although the details were not expressly spelled out in their manuscript, it seems that the lowering of charge states must be attributable to the dislodging of protons from their stable sites at the moment of collision, most likely via gas-phase transfer to solvent molecules or other solution components, or residual gases. Even under conditions where in-source CID is minimized by maintaining the skimmer voltage at the lowest value which affords acceptable lensing properties (i.e., the minimal skimmer voltage where signal abundance remains adequate), it is still possible that significant CID may be occurring between desorption and detection. This question is a rather difficult one to probe, of course, and effects of this type can be minimized on a given instrument, but perhaps never totally removed.

IV. GAS PHASE MODIFICATIONS

The final section of the chapter is devoted to developing an understanding of how processes that occur after ions have desorbed from solution into the gas phase can influence the appearance of the ESI mass spectrum. This section builds upon the discussion concerning CID (clearly a gas-phase process) as a function of skimmer potential that was introduced in Section III.C. It is generally contended that gas-phase ions desorbed from solution are initially highly solvated. As the ion proceeds into progressively lower-pressure regions of the mass spectrometer, desolvation, rearrangements, and other gas-phase reactions, including charged-particle transfer (e.g., proton transfer) may occur. Several factors can promote such reactions, including thermal considerations, collisional activation, and coulombic forces encountered by highly charged molecules.

A. Ion Solvation and the Observation of Adduct Species

Ions formed via ES are generally believed to exist in the ion source as highly solvated species, prior to entry into the mass analyzer. Adduct ions formed between analytes and solvent constituents (e.g., acetic-acid adducts of multiply protonated peptides) that existed at zero nozzle-skimmer voltage bias were eliminated when this voltage was raised to 100 V (63). Another example showed a reduction in "tailing" of multiply charged peaks toward higher m/z values when the nozzle-skimmer voltage was increased. The existence of this "tail" in the mass spectrum was attributed to noncovalent solvent association with desorbed ions. This type of association became less visible with increasing in-source CID at higher nozzle-skimmer voltages. The same study (63) reported a diminution of the relative contribution of adduct species to the observed ion current with increasing bath gas temperature, as more thermal energy was apparently available for dissociation. Furthermore, higher charge-state species were observed to engage in a higher degree of solvation with neutral acetic-acid molecules. This phenomenon was attributed to a higher degree of stabilization via solvation for the more highly charged species.

In-source CID at increasing skimmer voltages decreases the degree of solvation by neutral species present in the ESI solvent and in the atmospheric-pressure region of the ESI source. In addition, at even higher skimmer voltages, fragmentation processes typical of multiple-collision conditions begin to appear, as has been documented in many instances in both positive-ion (63) and negative-ion (64) work. For collisional activation of large, multiply charged proteins, species of higher charge state are more susceptible to decomposition because of their higher inherent reactivity (to be discussed in Section IV.C) and because they have a higher translational energy than lower charge-state ions at the same acceleration voltage. This higher kinetic energy results in a higher collision energy with residual gas species, rendering collisions more energetic and efficient. Figure 10 shows the effect of varying the nozzle-skimmer voltage difference on the ESI mass spectrum of horse heart myoglobin (63). In progressing from Figure 10A–D, the nozzle-skimmer voltage difference was raised from 85 to 285 V. As the voltage difference progressively increased, the highest charge-state ions disappeared (e.g., those ions below m/z 700), and the charge-state distribution shifted toward lower values. In Figures 10C and 10D, poorly resolved fragment ions, presumably covering a wide range of fragment charge states and masses, contribute to a progressively growing background signal. This study (63) also reported that when species formed via in-source CID were mass selected for subsequent tandem mass spectrometry (MS/MS) experiments,

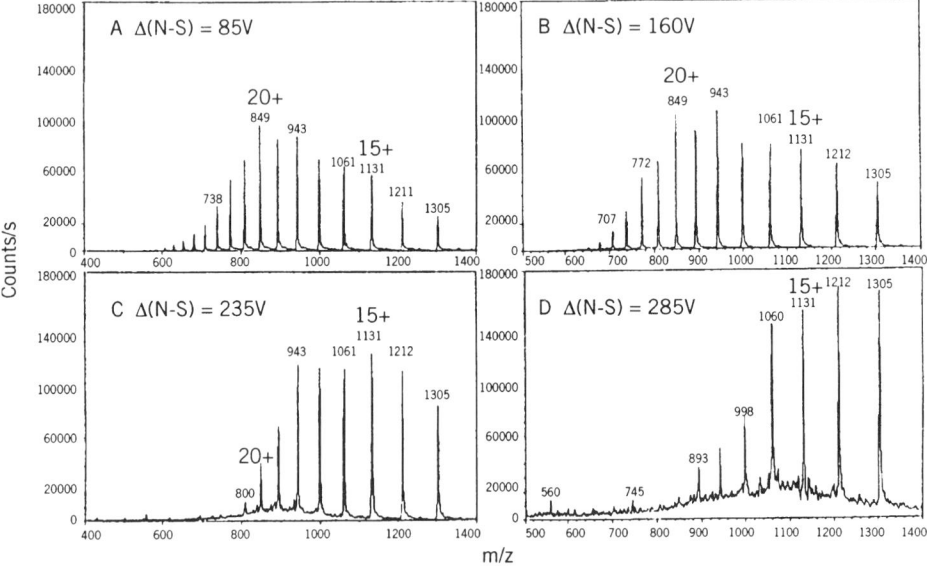

Figure 10. Positive-ion ESI mass spectra of horse heart myoglobin at various values of nozzle-skimmer bias, Δ(N-S). The major peaks represent protonated molecules of various charge states. As Δ(N-S) is raised, peaks of the highest charge states (lowest m/z value) disappear. In panels C and D, poorly resolved fragment ions produce a rising background signal. Reprinted by permission of Elsevier Science, Inc. from R. D. Smith et al., *J. Am. Soc. Mass Spectrom.*, 1990, **1**, 59. © 1990 American Society for Mass Spectrometry.

ensuing decompositions could be invoked with a reduced amount of additional energy transfer, relative to "cold" ions, because the in-source CID products were already excited ("heated") by energy uptake during the in-source collision process. The CID mass spectra of very large species may still show a relatively limited number of products owing to the large amount of internal energy required to invoke fragmentation. The CID spectra obtained from highly charged precursors, of course, can be complicated immensely by the numerous possible fragment ions and charge states appearing over a narrow m/z range.

B. Proton-Transfer Reactions

Proton-transfer reactions, whether controlled in a reaction cell, or taking place in the ES source shortly after ion formation, will clearly influence the appearance of charge-state distributions in ESI mass spectra. This effect was studied by McLuckey et al. (38) in experiments where a gas-phase base (e.g., dimethylamine) was reacted with a multiply charged protein to diminish the degree of multiple charging of the protein by proton abstraction. Deprotonation rate constants were demonstrated to increase with increasing charge state of the reactive ion. Follow up MS/MS experiments (performed on a quadrupole ion trap) (65) using 1,6-diaminohexane as the reactive base, demonstrated how proton-transfer reactions from multiply charged proteins to the reagent base, accompanied by a reduction in charge state, can be used to identify the charge state (and thus the mass) of unknown peaks. A similar scheme employing clustering reactions with the same reactive base was used to identify the charge states of multiply charged CID product ions. It was noted that the most reactive ions were those bearing the highest number of charges. Moreover, after a certain number of proton transfers to the base, the remaining analyte-bound protons became unreactive. This observation was rationalized on the basis of a reduced coulombic repulsion between neighboring charges, and an increased basicity of the remaining proton-bearing sites on analyte molecules.

Shortly afterwards, Winger et al. (33) demonstrated that the addition of gaseous water (acting as a base) could radically shift the distribution of charge states toward lower values. The relative pressure of the reactive gas was observed to influence the extent of proton transfer more than other parameters. Increasing pressures led to lower analyte charge states, but also simultaneously reduced the signal intensity. As a control, the experiments above were repeated under the same conditions, except that Ar replaced the H_2O vapor at the same pressure; no significant shift in the charge-state distribution was observed when the possibility for transfer to a nucleophilic species was eliminated. In related studies, Ogorzalek Loo et al. (66) reported a decreasing reactivity of multiply charged protein ions with decreasing charge state to be a general phenomenon for all proteins studied. Differences in reactivity (proton transfer to diethylamine) were more apparent between species of lower charge state, as those of the highest charge states all reacted quite rapidly with the basic reagent. Cassady et al. (67) reported findings which echoed the increased reactivity (i.e., higher rate constants for proton transfer) of higher charge states of a given multiply protonated molecule. Moreover, when amines of progressively higher gas-phase basicity were employed as reactant gases, proton transfer rates were shown to increase.

C. Intramolecular Coulombic Repulsion

Studies on the influence of electrostatic forces on the propensity for multiply charged analytes to retain charges have been conducted mostly using proteins as model analytes (33,38,65–69,71,73,78). Proton affinities (PAs) of component amino-acid residues contained in multiply protonated proteins differ from corresponding individual singly charged amino-acid residues because of the combined effects of intramolecular interactions, and changes in barriers to proton-transfer reactions caused by the presence of nearby charges. Gas-phase intramolecular interactions, such as self-solvation of charge sites, will stabilize the charge at a given amino-acid residue relative to an analogous site on the monomer of the same amino acid. On the other hand, in the gas phase, dielectric shielding of charged sites by the solvent is lost, resulting in significantly increased coulombic repulsion between charge sites. This can lower the barrier for departure of protons, or other forms of charge removal.

"Apparent proton affinity" values (and apparent gas-phase basicity values) for various sites on a protein thus depend upon the degree of solvation and dielectric shielding provided by the accessible, polarizable portions of the molecule itself, and upon the combined coulomb energy which all other charge sites exert upon a given charge site. Relative to the affinity for a binding site on a single neutral amino acid, self-solvation will increase the proton affinity for the same amino acid contained in a polypeptide chain, while the presence of nearby charges will lower the apparent affinity for that same site.

Busman et al. (68) verified the notion that ions of higher charge states are indeed destabilized relative to their less-charged counterparts by comparing differences in activation energies for unimolecular dissociation reactions of multiply protonated melitten molecules bearing different numbers of charges. Activation energies were found to be consistently lower for more highly charged species. Winger et al. (33) suggested that entropic factors can also enhance the propensity for proton transfer under seemingly unfavorable conditions. It was argued that highly charged gas-phase ions are conformationally constrained in their structure and torsional motion due to the high level of coulombic repulsion. Proton transfer to diminish the coulombic repulsion can reduce this strain, thus resulting in a gain in entropy which can help to overcome a positive enthalpy of reaction.

Ogorzalek Loo et al. (66) found that lower charge states of the disulfide reduced forms of three examined proteins were not as reactive as their equally charged counterparts in the nonreduced (native) form. Increased coulombic forces arising from the more highly constrained conformation of the native forms (comprising several disulfide bridges) were proposed to contribute to this increased propensity toward proton transfer. In a follow-up study (69), lower charge-state protein ions were observed to be less sensitive to the compactness of the protein (disulfide intact vs. disulfide reduced) as compared to higher charge-state ions. This conclusion was drawn after probing rates of proton transfer to triethylamine, trimethylamine, and 1,6-hexanediamine from various multiply charged proteins. Reactions with trimethylamine were found to proceed much faster than expected based purely on proton-affinity considerations, a finding which was attributed to coulombic repulsion considerations.

Employing a series of specifically constructed arginine-containing peptides, Downard and Biemann (70) investigated charging behavior as a function of peptide size and basic residue (arginine) arrangement. While small peptides of the series were observed to protonate fully, the average number of charges carried per arginine residue was observed to decrease for larger peptides with the same repeating primary structure. Interestingly, no significant increase in the extent of charging was found as the distance between potential charge-bearing sites was increased. It was suggested that the role of coulombic repulsion effects in the ESI charging process may have been overstated, based on the observation that the proximity of potential charge-bearing sites did not appear to substantially affect charging behavior (70).

Williams and co-workers (71,72) have offered an experimental approach to quantifying coulomb energy in multiply protonated molecules. Coulomb energies were calculated from differences in apparent gas-phase basicities of individual charge states of multiply charged peptides (72), proteins (71,73), and diaminoalkanes (74). Their calculations imply that the effects of polarizability, dipole moment, and impact parameter are small compared to Coulomb energy differences calculated for a homologous series of diprotonated diaminoalkanes. After estimating apparent gas-phase basicities using a bracketing approach (75–77) in a Fourier transform, ion-cyclotron resonance instrument, the magnitude of the coulomb energy was derived from proton-transfer reactions of isolated multiply protonated cytochrome c ions using an estimated dielectric polarizability term for gas-phase multiply protonated ions (71). For the 21+ charge state of cytochrome c (the maximum reported in the literature), the minimum Coulomb energy was found to be 24 eV considering a linear structure. Calculated Coulomb energy increased exponentially with charge state, while gas-phase basicity values dropped progressively. Results indicated that considerable scrambling of protons occurs in the gas phase, and that the conformation of the protein tends toward the denatured state after desorption into the gas phase (71). The latter results corroborate those of Ogorzalek Loo and Smith (78) who did not find significant differences in gas-phase reactivity between native and denatured forms of three evaluated proteins. In considering the possibility of proton scrambling in the gas phase, Cassidy et al. (67) propose that a multiply protonated protein in a given charge state may have different protonation sites with very similar basicities; these different structures would thus yield similar reaction rates.

Williams and co-workers (71) attribute an augmented stability (reduced reactivity) of very low charge-state species (e.g., 3+ of cytochrome c) to increased intramolecular interactions relative to the least basic of the individual amino acids believed to be carrying the charge. They contend that the maximum charge state that can be observed in the ESI mass spectrum depends upon the relative apparent gas-phase basicity of this most highly charged ion, which will be just below that of the gas-phase basicity of the components of the solvent used in the ESI process.

D. Defining the Maximum Obtainable Charge State

The pioneering experiments of Fenn and co-workers, investigating charge attachment to polyethylene glycols (43,44), led to an early model to define the maximum charge state obtainable in ESI. They proposed that the capacity of a PEG molecule to retain charge reaches its upper limit when, because of coulomb repulsion by other charges, the electrostatic potential energy of the centermost charge equals the energy that binds it to its site (43). In other words, if the electrostatic potential energy surpasses the binding energy, this charge will not remain attached. In the case of polypeptides, the early observations of Loo et al. (12) and Covey et al. (13) led to the proposition of the existence of a direct relation between the upper limit to the number of charges that a protein may carry in the gas phase and the number of basic residues contained in that protein. Building upon this rough approximation, a model for calculation of the maximum charge state of multiply protonated peptides and proteins has been presented by Williams and co-workers (73). Apparent gas-phase basicities of multiply charged ions from 13 proteins were calculated. These values were then compared to the gas-phase basicity of the solvent employed, while reasoning that it is energetically and kinetically favorable for an ion with an apparent gas-phase basicity inferior to that of the solvent to undergo proton transfer to the solvent. This procedure allowed calculation of a maximum charge state for a given protein, equal to that of the highest charge state ion whose calculated apparent gas-phase basicity is below the gas-phase basicity of the solvent. Obtained maximum charge state values compared favorably with experimental values reported in the literature, pointing to the conclusion that the maximum charge state for proteins is dictated by gas-phase reactivity with residual solvent. However, for peptides comprised of few basic amino acids, the maximum charge state was found to correlate more closely with the solution charge state based upon a calculated number of amino acids that are protonated in solution (73). A protein bearing the calculated maximum number of charges is postulated to be formed from a solvent-protein cluster which could conceivably carry more charge than that found on the protein in bulk solution, and also more than that of a "naked" gas-phase protein ion, due to charge stabilization afforded by the solvent dielectric. As the final solvent molecules depart, partitioning of the charge between analyte and solvent is proposed to be governed by the gas-phase chemistry of the species comprising the final cluster.

V. CONCLUSION

Between the time that an initially neutral sample solution is pumped into the ES capillary, and the moment that ions are displayed on the mass spectrometer's readout device, a complex series of events is unleashed during the ESI process. The initial solvent conditions clearly define the starting condition for analyte molecules. Before the ES process is triggered, any ions that are "preformed" in solution will have counterions in the nearby vicinity. As this solution comes under the influence of the ES high voltage, changes begin to occur. Redox products are necessarily formed at the metal–solution interface and they are likely to be contained in the released liquid. Thereafter, the solution contains a charge imbalance of one polarity. The liquid stream is transformed into charged droplets whereby excess charge is localized near the surface layer. Uneven fission of these initially formed droplets increases the charge–mass ratio of produced offspring droplets from which detected ions are ultimately desorbed. As the solvent departs from charged droplets, counterion interactions assume increased importance, such as the possibility to form "contact" ion pairs between charged analyte molecules and available counterions. This increasing propensity to form ion-paired analyte species at progressively later stages of the ESI process can exert a significant effect upon how much charge will be retained by the analyte in the gas phase.

The ensemble of results reported in this chapter, illuminating various aspects of the charging process in ESI, point to the formation of a final charged cluster comprised of solvent and analyte molecules, along with possible electrochemical reaction products and other additives. The dissociation of this cluster leads to a partitioning of charge between solvent components and the remaining analyte molecule(s). At the end of this process, the maximum obtainable analyte charge state is likely to reflect the gas-phase chemistry of the species in intimate contact with the available charge. It is not reasonable to expect that analyte ions could be produced bearing more charge than was present in the final charged solvent–analyte cluster. Other species exhibiting acid–base behavior (including other analyte molecules) present in a solvent–analyte cluster can surely compete for available charge, and all equilibria involving charge-bearing species can contribute to the equation which dictates how much charge is ultimately carried by a given analyte molecule. Further complicating the description of events are thermochemical considerations such as free jet expansion of the sprayed solution and kinetically controlled collisional "heating" of charged species which take place in the midst of often ill-defined thermal and pressure gradients within the mass spectrometer. The grey area between highly developed models describing neutral solutions, and relatively recently elaborated models of gas-phase ion behavior, is the essence of the intrigue of the ESI process. If the current level of fascination in comprehending all aspects of the ESI technique continues, the detailed description of the process promises to undergo evolution and refinement for many years to come.

REFERENCES

1. Dole, M.; Mack, L. L.; Hines, R. L.; Mobley, R. C.; Ferguson, L. D.; Alice, M. B. *J. Chem. Phys.* **1968**, *49*, 2240–2249.
2. Dole, M.; Hines, R. L.; Mack, L. L.; Mobley, R. C.; Ferguson, L. D.; Alice, M. B. *Macromol.* **1968**, *1*, 96.
3. Iribarne, J. V.; Thomson, B. A. *J. Chem. Phys.* **1976**, *64*, 2287–2294.
4. Thomson, B. A.; Iribarne, J. V. *J. Chem. Phys.* **1979**, *71*, 4451–4463.
5. Thomson, B. A.; Iribarne, J. V.; Dziedzic, P. J. *Anal. Chem.* **1982**, *54*, 2219–2224.
6. Iribarne, J. V.; Dziedzic, P. J.; Thomson, B. A. *Int. J. Mass Spectrom. Ion Phys.* **1983**, *50*, 331–334.
7. Whitehouse, C. M.; Dreyer, R. N.; Yamashita, M.; Fenn, J. B. *Anal. Chem.* **1985**, *57*, 675–679.
8. Meng, C. K.; Mann, M.; Fenn, J. B. *Z. Phys. D.* **1988**, *10*, 361–368.
9. Mann, M.; Meng, C. K.; Fenn, J. B. *Proceedings of the 41st ASMS Conference on Mass Spectrometry and Allied Topics*, San Francisco, CA, **1988**, 1207–1208.
10. Fenn, J. B.; Mann, M.; Meng, C. K.; Wong. S. K.; Whitehouse, C. *Mass Spectrom. Rev.* **1990**, *9*, 37–70.
11. Fenn, J. B. *J. Am. Soc. Mass Spectrom.* **1993**, *4*, 524–535.
12. Loo, J. A.; Udseth, H. R.; Smith, R. D. *Biomed. Environ. Mass Spectrom.* **1988**, *17*, 411–414.
13. Covey, T. R.; Bonner, R. F.; Shushan, B. I.; Henion, J. *Rapid Commun. Mass Spectrom.* **1988**, *2*, 249–256.
14. Loo, J. A.; Udseth, H. R.; Smith, R. D. *Rapid Commun. Mass Spectrom.* **1988**, *2*, 207–210.
15. Loo, J. A.; Edmonds, C. G.; Udseth, H. R.; Smith, R. D. *Anal. Chem.* **1990**, *62*, 693–698.
16. Chowdhury, S. K.; Katta, V.; Chait, B. T. *J. Am. Chem. Soc.* **1990**, *112*, 9012–9013.
17. Loo, J. A.; Ogorzalek Loo, R. R.; Udseth, H. R.; Edmonds, C. G.; Smith, R. D. *Rapid Commun. Mass Spectrom.* **1991**, *5*, 101–105.
18. Le Blanc, J. C. Y.; Beuchemin, D.; Siu, K. W. M.; Guevremont, R.; Berman, S. S. *Org. Mass Spectrom.* **1991**, *26*, 831–839.
19. Mirza, U. A.; Cohen, S. L.; Chait, B. T. *Anal. Chem.* **1993**, *65*, 1–6.
20. Vorm, O.; Chait, B. T.; Roepstorff, P. *Proceedings of the 41st ASMS Conference on Mass Spectrometry and Allied Topics*, San Francisco, CA, **1993**, pp. 621a–621b.
21. Guevremont, R.; Siu, K. W. M.; Le Blanc, J. C. Y.; Berman, S. S. *J. Am. Soc. Mass Spectrom.* **1992**, *3*, 216–224.
22. Mirza, U. A.; Chait, B. T. *Anal. Chem.* **1994**, *66*, 2898–2904.
23. Cole, R. B.; Harrata, A. K. *Rapid Commun. Mass Spectrom.* **1992**, *6*, 536–539.
24. Cole, R. B.; Harrata, A. K. *J. Am. Soc. Mass Spectrom.* **1993**, *4*, 546–556.
25. Wang, G.; Cole, R. B. *Organic Mass Spectrom.* **1994**, *29*, 419–427.
26. Wang, G.; Cole, R. B. *Anal. Chem.* **1994**, *66*, 3702–3708.
27. Wang, G.; Cole, R. B. *Anal. Chem.* **1995**, *67*, 2892–2900.
28. Wang, G.; Cole, R. B., *J. Am. Soc. Mass Spectrom.* **1996**, *10*, 1050–1058.

29. Feng, R.; Konishi, Y. *J. Am. Soc. Mass Spectrom.* **1993**, *4*, 638–645.
30. Winger, B. E.; Light-Wahl, K. J.; Ogorzalek Loo, R. R.; Udseth, H. R.; Smith. R. D. *J. Am. Soc. Mass Spectrom.* **1993**, *4*, 536–545.
31. Kelly, M. A.; Vestling, M. M.; Fenselau, C. C. *Org. Mass Spectrom.* **1992**, *27*, 1143–1147.
32. Ashton, D. S.; Beddell, C. R.; Cooper, D. J.; Green, B. N.; Oliver, R. W. A. *Org. Mass Spectrom.* **1993**, *28*, 721–728.
33. Winger, B. E.; Light-Wahl, K. J.; Smith, R. D. *J. Am. Soc. Mass Spectrom.* **1992**, *3*, 624–630.
34. Smith, R. D.; Loo, J. A.; Ogorzalek Loo, R. R.; Busman, M.; Udseth, H. R. *Mass Spectrom. Rev.*, **1991**, *10*, 359–451.
35. Edmonds, C. G.; Loo, J. A.; Barinaga, C. J.; Udseth, H. R.; Smith, R. D. *J. Chromatography* **1989**, *474*, 21–37.
36. Chowdhury, S.; Katta, V.; Chait, B. T. *Rapid Commun. Mass Spectrom.* **1990**, *4*, 81–87.
37. Hunter, A. P.; Severs, J. C.; Harris, F. M.; Games, D. E. *Rapid Commun. Mass Spectrom.* **1994**, *8*, 417–422.
38. McLuckey, S. A.; Van Berkel, G. J.; Glish, G. L. *J. Am. Chem. Soc.* **1990**, *112*, 5668–5670.
39. Cheng, X.; Gao, Q.; Smith, R.D.; Simanek, E. E.; Mammen, M.; Whitesides, G. M. *Rapid Commun. Mass Spectrom.* **1995**, *9*, 312–316.
40. Kebarle, P.; Tang, L. *Anal. Chem.* **1993**, *65*, 972A–986A.
41. Tang, L.; Kebarle, P. *Anal. Chem.* **1993**, *65*, 3654–3668.
42. Hager, D. B.; Dovichi, N. J.; Klassen, J.; Kebarle, P. *Anal. Chem.* **1994**, *66*, 3944–3949.
43. Wong, S. F.; Meng, C. K.; Fenn, J. B. *J. Phys. Chem.* **1988**, *92*, 546–550.
44. Nohmi, T.; Fenn, J. B. *J. Am. Chem. Soc.* **1992**, *114*, 3241–3246.
45. Gatlin, C. L.; Turecek, F. *Anal. Chem.* **1994**, *66*, 712–718.
46. Le Blanc, J. C. Y.; Wang, J.; Guevremont, R.; Siu, K. W. M. *Org. Mass Spectrom.* **1994**, *29*, 587–593.
47. Tong, X.; Henion, J.; Ganem, B. *J. Mass Spectrom.*, **1995**, *30*, 867–871.
48. Rayleigh, J. W. S. *Philos. Mag.* **1882**, *14*, 184.
49. Taylor, G. I. *Proc. R. Soc.* **1964**, *A280*, 383–397.
50. Smith, D. P. H. *IEEE Trans. Ind. Appl.* **1986**, *IA-22*, 527–535.
51. Hayati, I.; Bailey, A. I.; Tadros, Th. F. *Nature* **1986**, *319*, 41–43.
52. Blades, A. T.; Ikonomou, M. G.; Kebarle, P. *Anal. Chem.* **1991**, *63*, 2109–2114.
53. Pfeifer, R. J; Hendricks, C. D. *AIAA J.* **1968**, *6*, 496–502.
54. Sunner, J.; Nicol, G.; Kebarle, P. *Anal. Chem.* **1988**, *60*, 1300–1307.
55. Kostiainen, R.; Bruins, A. P. *Rapid Commun. Mass Spectrom.* **1994**, *8*, 549–558.
56. Chowdhury, S. K.; Katta, V.; Chait, B. T. *Rapid Commun. Mass Spectrom.* **1990**, *4*, 81–87.
57. Smith, R. D.; Loo, J. A.; Edmonds, C. G.; Barinaga, C. J.; Udseth, H. R. *Anal. Chem.* **1990**, *62*, 882–899.

58. Tang, L.; Kebarle, P. *Anal. Chem.* **1991**, *63*, 2709–2715.
59. Wang, G.; Cole, R. B. unpublished results.
60. Katta, V.; Chait, B. T. *Rapid Commun. Mass Spectrom.* **1991**, *5*, 214–217.
61. Katta, V.; Chait, B. T. *J. Am. Chem. Soc.* **1993**, *115*, 6317–6321.
62. Suckau, D.; Shi, Y.; Beu, S. C.; Senko, M. W.; Quinn, J. P.; Wampler, F. W.; McLafferty F. W. *Proc. Natl. Acad. Sci. USA* **1993**, *90*, 790–793.
63. Smith, R. D.; Loo, J. A.; Barinaga, C. J.; Edmonds, C. G.; Udseth, H. R. *J. Am. Soc. Mass Spectrom.* 1990, *1*, 53–65.
64. Harrata, A. K.; Domelsmith, L. N.; Cole, R. B. *Biol. Mass Spectrom.*, **1993**, *22*, 59–67.
65. McLuckey, S. A.; Glish, G. L.; Van Berkel, G. J. *Anal. Chem.*, **1991**, *63*, 1971–1978.
66. Ogorzalek Loo, R. R.; Loo, J. A.; Udseth, H. R.; Fulton, J. L.; Smith, R. D. *Rapid Commun. Mass Spectrom.* **1992**, *6*, 159–165.
67. Cassady, C. J.; Wronka, J.; Kruppa, G. H.; Laukien, F. H. *Rapid Commun. Mass Spectrom.*, **1994**, *8*, 394–400.
68. Busman, M.; Rockwood, A. L.; Smith, R. D. *J. Phys. Chem.*, **1992**, *96*, 2397–2400.
69. Ogorzalek Loo, R. R.; Winger, B. E.; Smith, R. D. *J. Am. Soc. Mass Spectrom.*, **1994**, *5*, 1064–1071.
70. Downard, K. M.; Biemann, K. *Int. J. Mass Spectrom. Ion Proc.*, **1995**, *148*, 191–202.
71. Schnier, P. D.; Gross, D. S.; Williams, E. R. *J. Am. Chem. Soc.*, **1995**, *117*, 6747–6757.
72. Gross, D. S.; Williams, E. R. *J. Am. Chem. Soc.*, **1995**, *117*, 883–890.
73. Schnier, P. D.; Gross, D. S.; Williams, E. R. *J. Am. Soc. Mass Spectrom.*, **1995**, *6*, 1086–1097.
74. Gross, D. S.; Rodriguez-Cruz, S. E.; Bock, S.; Williams, E. R. *J. Phys. Chem.*, **1995**, *99*, 4034–4038.
75. DeFrees, D. J.; McIver, R. T.; Hehre, W.J. *J. Am. Chem. Soc.*, **1980**, *102*, 3334–3338.
76. Gorman, G. S.; Amster, I. J. *J. Am. Chem. Soc.*, **1993**, *115*, 5729–5735.
77. Wu, J.; LeBrilla, C. B. *J. Am. Chem. Soc.*, **1993**, *115*, 3270–3275.
78. Ogorzalek Loo, R. R.; Smith, R. D. *J. Am. Soc. Mass Spectrom.*, **1994**, *5*, 207–220.

PART II
ELECTROSPRAY COUPLING TO MASS ANALYZERS

CHAPTER 5

Electrospray Ionization on Quadrupole and Magnetic-Sector Mass Spectrometers

CHARLES N. McEWEN AND BARBARA S. LARSEN

Central Research Department, E. I. du Pont de Nemours & Company, Wilmington, Delaware

	Abstract	177
I.	History	178
II.	ESI source designs for quadrupole mass spectrometers	179
III.	ESI source designs for magnetic-sector mass spectrometers	181
IV.	Data treatment	182
V.	Mass accuracy	186
VI.	ESI and fragment ion information	188
VII.	LC coupling to ESI–MS	190
VIII.	LC/MS Applications	191
IX.	Trace analysis using LC–ESI/MS/MS	194
X.	Interesting ESI applications	197
	References	198

ABSTRACT

Electrospray ionization (ESI) mass spectrometry was first demonstrated and commercialized on quadrupole mass spectrometers and has been shown to be well suited to this type of instrumentation. Electrospray ionization interfaced to a triple quadrupole mass spectrometer, especially in conjunction with liquid chromatography, is an extremely powerful analytical method for structure analysis of low- or high-molecular-weight compounds. Magnetic-sector mass spectrometers, while less commonly interfaced to ESI, can provide better mass accuracy, higher resolution, and with an array detector, higher sensitivity. These gains come at the expense of a more demanding vacuum requirement for the magnetic-sector ion source.

Electrospray Ionization Mass Spectrometry, Edited by Richard B. Cole.
ISBN 0-471-14564-5 © 1997 John Wiley & Sons, Inc.

I. HISTORY

Electrospray ionization (ESI) coupled to quadrupole mass analyzers has revolutionized mass spectrometry. The phenomenal success of this technique is primarily the result of multiple charging which makes it possible to accurately measure the molecular weights of high-mass compounds with excellent sensitivity using commonly available instrumentation such as quadrupole and magnetic-sector mass spectrometers having limited mass-to-charge (m/z) ranges. The ease of interfacing ESI to liquid separation methods has also enhanced its utility.

The ES process dates to experiments by Zeleny (1). However, it was Malcolm Dole's experiments in the late 1960s and early 1970s that suggested ESI could be used to mass analyze large molecules (2). The possibility of using mass spectrometry or plasma chromatography were discussed by Dole et al. (3). Analyses of electrosprayed macromolecules, mass analyzed by retarding potentials, produced data that were interpreted as resulting from singly and multiply charged ions. Dole eventually concluded that accurate molecular weights of polymers could not be determined by electrospray ionization (4).

Yamashita and Fenn (5) reported in 1984 on the successful interfacing of ESI to a quadrupole mass spectrometer. They were able to demonstrate that nonvolatile materials could be ionized without fragmentation using either positive or negative ionization. Dole and co-workers (6), in a paper published the same year, reported on the mobility of polystyrene and lysozyme ions in nitrogen gas using an ion-drift spectrometer. For lysozyme, they reported that ions having 1–3 positive charges were observed. Previously, using plasma chromatography, Dole reported observing 1–4 charges for polystyrene with a molecular weight of 97,000 (4). Also in 1984, Aleksandrov et al. (7) reported on the interfacing of ESI to a magnetic sector mass spectrometer and on the detection of molecular ions for low mass compounds. By increasing the voltage between the aperture plates separating the atmospheric pressure region and the high-vacuum region, fragmentation for a dipeptide was induced. In 1987, Bruins et al. (8) published data obtained by interfacing ESI to an atmospheric pressure quadrupole mass spectrometer with nitrogen gas assisting nebulization to achieve higher liquid flow rates. The term "ion spray" was applied to distinguish the higher flow rate nebulization technique from the low-flow ESI.

Electrospray ionization did not catch the attention of the wider mass-spectrometry community until Meng et al. (9) reported that high-mass proteins were observed in ESI as multiply charged ions and could therefore be mass analyzed using quadrupole mass spectrometers (10,11). Work began immediately in several laboratories to reproduce these results. Loo et al. (12) interfaced ESI to a quadrupole instrument with such speed that they had confirmed the Fenn group's results within months and extended the work to proteins of even higher mass. Interfacing ESI to a magnetic-sector instrument was technically more difficult and had previously been demontrated only on low-mass molecules (7,13). Allen and Lewis (14), using a reduced-pressure ES probe reported observing gramicidin S and bradykinin on a single-focusing magnetic-sector instrument. Meng et al. (15,16) reported on the first experiments interfacing ESI to a high-performance, double-focusing, magnetic-sector mass spectrometer. Larsen and McEwen (17) reported improvements in the ESI source design which made it possible to resolve the molecular ion of insulin and observe proteins with molecular weights greater than 40,000 D. Gallagher and Chapman (18) also reported on an ESI source design for a magnetic-sector mass spectrometer which allowed them to observe the multiply charged molecular ions for bovine albumin (MW 66300).

Reviews have been published on the history of the ES process (19) as well as on ESMS (20, 21).

II. ESI SOURCE DESIGNS FOR QUADRUPOLE MASS SPECTROMETERS

The ES process occurs at atmospheric pressure, but the ions that are produced must be analyzed in the high vacuum of the mass spectrometer. The Fenn–Whitehouse source design shown in Figure 1 overcomes this problem by using a glass capillary which allows ions formed at atmospheric pressure to enter into the vacuum region of the mass spectrometer while limiting the gas load. The laminar flow characteristics in the capillary prevents ions from being neutralized at the surface. The glass capillary had two distinct advantages: (1) the aperture can be enlarged to reduce the potential for clogging without increasing the gas load and (2) ions can be swept up a potential gradient greater than 10 kV (22). The second

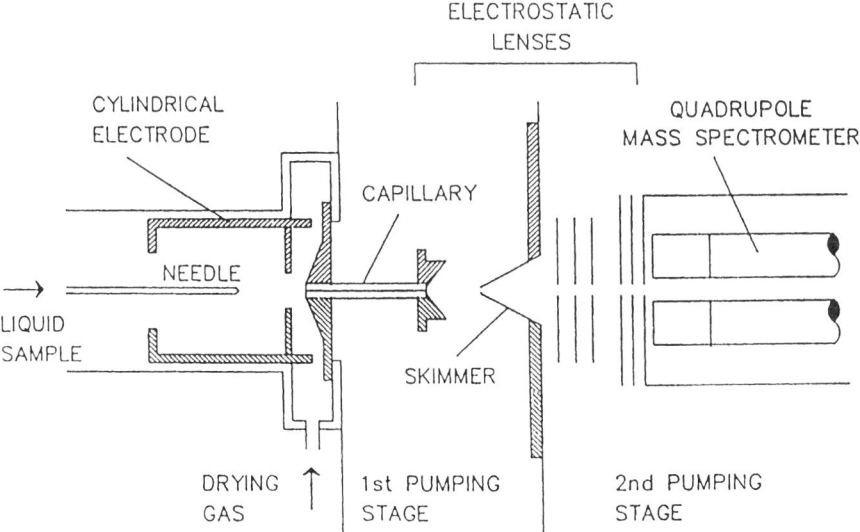

Figure 1. Fenn–Whitehouse ES ion-source design. Taken from J. B. Fenn et al., *Mass Spectrom. Rev.*, **1990**, *9*, 37. © 1990 John Wiley & Sons, Inc. Reprinted by permission of John Wiley & Sons, Inc.

feature makes it possible to apply voltages to the metal-coated ends of the glass capillary independent of each other. In this way, the ES needle can be operated at ground potential with the entrance end of the glass capillary at ± 4000 V and the exit (vacuum) end near ground potential.

The ions exiting the glass capillary expand nearly isentropically into the first vacuum stage. The result is a supersonic free jet with the associated aerodynamic shock waves. A nozzle-skimmer design provides a method for differential pumping that reduces the pressure in the analyzer region and allows sampling of the supersonic free jet without disrupting the molecular flow (23,24).

In the orignial Fenn–Whitehouse source design, an attempt was made to focus ions in the atmospheric pressure region and also in the vacuum region using electrostatic lenses. The lenses in the vacuum region were placed after the skimmer and were also used to steer the beam. Modifications of this source design include a "tube" lens which is placed between the capillary exit and the first skimmer. In addition, the electrostatic lens assembly in the source vacuum region can be replaced by hexapole or octapole lenses for better transmission. Chowdhury et al. (25) discovered that the glass capillary can be replaced by a metal capillary. Heating the metal capillary is effective in desolvating the ESI ions so that use of a heated countercurrent gas is not necessary. Addition of a second skimmer to further reduce pressure in the source region has been shown to be valuable for magnetic-sector instruments because of the need to obtain better vacuum conditions in the region of high-voltage ion acceleration. The atmospheric sampling method originally used on the atmospheric-pressure quadrupole mass spectrometer was a thin walled aperture with "curtain gas" and cyropumping on the vacuum side of the orifice. This method for sampling the ES plume has proven to be very reliable. Another source design which uses a small aperture with a skimmer and dual pumping instead of cyropumping was reported by Jarvis et al. (26).

Bruins et al. (8) introduced pneumatically assisted ES which they referred to as "ion spray". By incorporation of a liquid shield between the ESI needle and the vacuum orifice, Henion's group extended ion spray to flows up to 2 mL/min (27). A variation of pneumatically assisted ES is thermally assisted ES, which provides higher flow rates by heating the ESI needle (28). The advantage of these techniques is that flow rates can be increased from the microliters per minute flow rate of ESI to flow rates compatible with the 4.6-mm columns commonly used in liquid chromatography, thus simplifying the interface between the LC and ESI source. However, the ion currents sampled from ESI are concentration dependent so that signal does not increase with higher flow rates. In fact, it is very difficult to keep the ion current from decreasing as the flow rate increases. The high flow rate also presents a possible health problem unless the aerosols produced are efficiently trapped.

Because ESI is concentration sensitive and not flow sensitive, it has been recognized that operating at low flow rates improves the detection limit of the technique. Hunt et al. (29) achieved extremely high sensitivity by operating ESI interfaced to a quadrupole mass spectrometer at microliter flow rates. An ESI source designed to operate at low nanoliter flow rates was reported by Wilm and Mann (30). Nanoflow ESI was found to be more tolerant of 100% water and small concentrations of trifluoroacetic acid. The technique in its simplest form involves spraying the liquid from a gold-coated borosilicate glass capillary that has been drawn out to an inner tip diameter of 1–3 μm. Liquid is drawn up into the tube by capillary action and drawn out by the application of sufficient voltage to initiate the ES process. Because only a few nanoliters of solution are consumed per minute, the amount of sample needed for an analysis is extremely low.

III. ESI SOURCE DESIGNS FOR MAGNETIC-SECTOR MASS SPECTROMETERS

Early attempts to interface ESI sources to magnetic-sector instruments were only successful for low mass ions (3,13). In attempting to interface ESI to a high-performance and thus high accelerating voltage magnetic-sector instrument, Meng et al. (15) discovered low sensitivity even with peptides such as bradykinin. On the other hand, very stable aromatic dye structures could be sprayed with good sensitivity and good resolution. The Fenn–Whitehouse (Analytica) source was designed with the largest aperture capillary and skimmers that the vacuum would allow in order to pass the maximum number of ions into the mass analyzer. Meng et al. (15) reasoned that the peptide ions were not surviving the high acceleration because of collisional fragmentation at the high pressures of the experiment. By reducing the inner diameter of the glass capillary and the skimmer aperture they improved the ion source pressure by approximately an order of magnitude. The lower pressure in the ion-acceleration region greatly improved the sensitivity for peptides and small proteins.

Further reductions in the inner diameter of the glass capillary and increased pumping improved transmission of high-mass ions in the magnetic-sector instrument, allowing Larsen et al. (17) to observe a full-scan spectrum of lysozyme at 15 fmole consumed. Introduction of dual skimmers with an additional stage of pumping further improved the high mass capabilities of ESI on magnetic-sector instruments. Gallagher et al. (18) added a small diffusion pump to the second pumping stage to further reduce the ion source pressure. The success of ESI on magnetic-sector instruments for the analysis of high-mass and labile molecules was demonstrated by Loo (31) in work with large subunit protein complexes. Loo was able to observe the molecular ion of a tetrameric complex of rabbit muscle pyruvate kinase ($M_r \sim 232,000$) on a Finnigan MAT 900Q mass spectrometer using a PATRIC detector.

While pressure reduction was one major ingredient in making ESI work on a kilovolt accelerating-voltage magnetic-sector instrument, the most difficult hurdle to overcome was operating at high voltage across a pressure drop from atmosphere to approximately 10^{-6} Torr. Because gas is highly conducting at moderate pressures, the pumps on the first stage of pumping need to be isolated from ground potential. One method of electrically isolating a rotary vacuum pump is to use an isolation transformer which can withstand the ion source potential. Another method is to use a belt-driven pump in which the vacuum pump, mounted on wood or plastic, is electrically isolated from the motor by an insulating pulley and/or drive belt. Standard drive belts can be used if a Delrin® or other electrically insulating pulley is used on either the motor or pump. Electrical isolation of the pumps means that they will operate at source potential, thus adding an additional hazard to magnetic-sector instruments. Shielding is needed to prevent exposure to any high-voltage equipment.

An obvious disadvantage of magnetic-sector mass spectrometers is the need to operate at high ion source voltages, which makes it necessary to better isolate the transmitted ions from the surrounding gas to minimize collisions during ion acceleration and to prevent electrical discharge between the high voltage and ground potential. However, for commercial instruments, this disadvantage is only discernable to the user by the higher cost of the ion source. A properly designed ion source should operate in a similar manner to one interfaced to a quadrupole instrument. In our experience, similar ESI source designs give similar sensitivity on magnetic sector and quadrupole mass spectrometers with the exception that greater sensitivity can be obtained with a quadrupole instrument by degradation of the resolution, and on a magnetic-sector instrument, by use of an array detector (32). The magnetic-sector instrument can operate at higher resolution, achieve more accurate mass measurement, cover a wider m/z range, and provide high voltage collision induced fragmentation.

IV. DATA TREATMENT

The ability of ESI to mass analyze large molecules (MW = 150,000–200,000) catapulted the technique to prominence. However, ESI is proving to be useful for the analysis of compounds with molecular weights below 1000 (5,33). For molecules with molecular weights below 500 Da, ESI is usually characterized by singly charged ions. Compounds of higher molecular weight have an increased abundance of multiply charged ions. Above a molecular weight of approximately 2000 Da, singly charged ions are usually in low abundance or are not observed. Thus, the manner in which data must be handled changes in going from low- to high-mass materials.

For low-mass compounds, the most abundant peaks observed are most likely the $[M+H]^+$ ions in the positive-ion mode and $[M-H]^-$ ions in the negative ion mode. Doubly and, rarely, triply charged ions as well as dimers may also be observed. Doubly charged positive ions will be observed at a mass that is $(M+2C^+)/2$ where M is the molecular weight and C^+ is the ionizing cation (H^+, Na^+, etc.). The dimer will be observed at $2M+C^+$. The mass-to-charge ratios for the $[M+H]^+:[M+2H]^{2+}$ and $[2M+H]^+:[M+H]^+$ ions are equivalent. Therefore, a doubly charged/singly charged ion pair cannot be distinguished from a singly charged/dimer ion pair based solely on the mass-to-charge ratio. However, if the instrument has sufficient resolution, the isotope peaks can be used to determine if the ion is singly or doubly charged (34). A singly charged ion will have ^{13}C isotope peaks that are spaced at 1 m/z unit intervals, while doubly charged ions will have these isotopes spaced at $1/2$ m/z unit intervals. If the instrument resolution is insufficient to resolve isotopes, there may be additional clues as to whether an ion is multiply charged. For example, Figure 2 (top) shows $[M+H]^+$ and the $[M+Na]^+$ ions for gramicidin S which are separated by 22 m/z units. The doubly charged ions (Fig. 2, bottom) are separated by 11 m/z units.

For negative-ion ES, the most common ions observed are the loss of one or more protons. Distinguishing between the relationships of singly charged, $[M-H]^-$, the doubly charged, $[(M-2H)/2]^=$, and the dimer, $[2M-H]^-$ requires the same considerations as discussed above for the positive-ion case.

With high-mass ions, a series of multiply charged molecular ions are observed (11). As shown in Figure 3 for ubiquitin, ESI mass spectra consist of a series of molecular ions that differ only in the number of charges on the ions. The molecular weight for positive ions is determined by multiplying the measured m/z ratio minus the mass of the cation (C^+) times the number of charges (n) on that ion.

$$\mathrm{MW} = n(m/z - C^+) \tag{1}$$

The unknowns in determining the molecular weight are the number of charges and the mass of the ionizing cation. For proteins, ionization is usually by protonation, but for compounds with mostly oxygen functionality, such as polysaccharides or polyethylene glycols, ionization is usually by Na^+ or other alkali-ion addition.

The charge on any ion in Figure 3 can be determined by the isotope pattern, where the charge is the inverse of the mass-to-charge spacing of nearest-neighbor isotopes expressed in fractions of a mass-to-charge unit. Therefore, for isotope spacings of $1/2$ m/z unit intervals, the charge is 2, and for spacings of $1/10$ m/z unit intervals, the charge is 10. However, determining charge in this manner requires that the instrument be operated at a sufficiently high resolution to resolve the ^{13}C isotope peaks.

Because most instruments do not have sufficient resolution to separate isotopes, an alternate method uses the mass-to-charge ratio of two adjacent multiply charged ions. For an ion of n_j elemental charges with a measured mass-to-charge ratio of m_j and the adjacent peak of higher m/z, m_k, with n_k charges where $n_k = n_j - 1$ and $M =$ molecular weight, then $M = n_j(m_j - x)$ where $x = 1$ for H^+ (or 23 for Na^+, etc.). Solving two simultaneous equations:

$$M = n_k(m_k - x) = (n_j - 1)(m_k - x) \tag{2}$$

Transposing,

$$(n_j - 1)(m_k - x) = n_j(m_j - x) \quad \text{and} \quad n_j(m_k - x) - n_j(m_j - x) = m_k - x \tag{3}$$

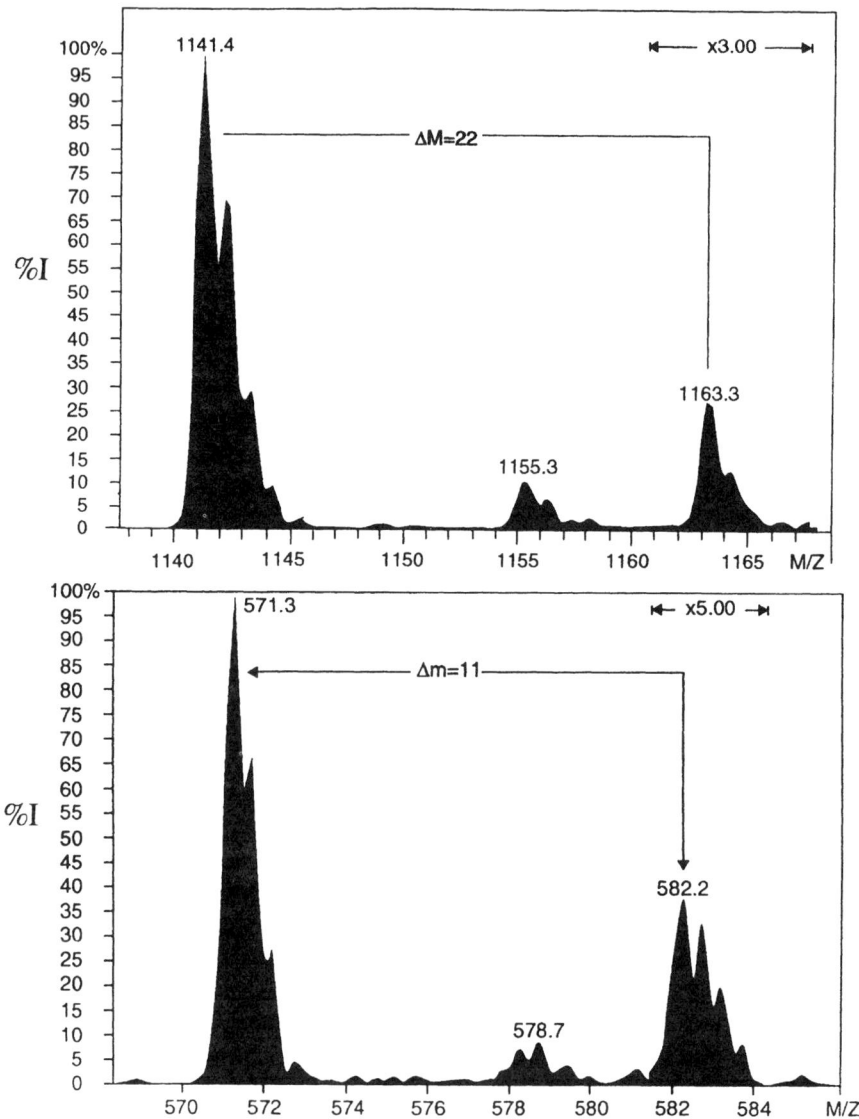

Figure 2. Portion of ESI mass spectrum of gramacidin S showing the m/z difference for $[M+H]^+$ and $[M+Na]^+$ (top) and $[M+2H]^{2+}$ and $[M+H+Na]^{2+}$ (bottom).

Therefore,

$$n_j = \frac{m_k - x}{m_k - m_j} \qquad n_k = \frac{m_j - x}{m_k - m_j} \qquad (4)$$

The charge on each peak in Figure 3 can be determined by Eq. 4. In fact,

IV. DATA TREATMENT

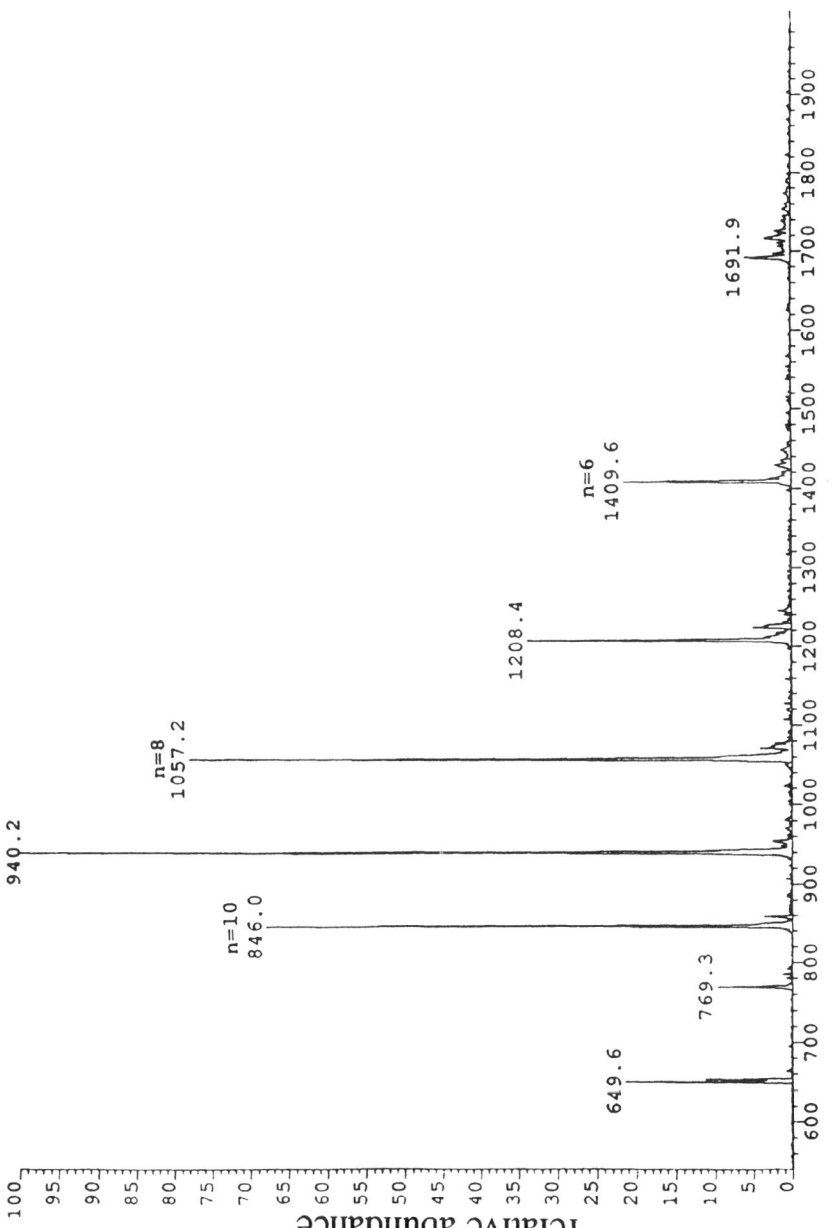

Figure 3. Magnetic-sector ESI spectrum of ubiquitin showing the charge (n) on the molecular ions.

determining the charge on any one peak determines the charge on the others because the charge on each adjacent peak will increase by 1, moving to lower m/z, and decrease by 1, moving to higher m/z.

Another means for deconvoluting a multiply charged spectrum is through computer algorithms where the adduct ion mass must be a known so that the multiply charged peaks can be summed into the singly charged representation. The first algorithm for transforming multiply charged spectra was published by Mann et al. (35). Zhou et al. (36) refined the algorithim for the first commercial deconvolution program. Labowsky et al. (37) have expanded the algorithm from a two-dimensional analysis to a "three-dimensional surface analysis" where the mass of the adduct ions are included in the function. Further statistical packages such as *MAX ENT* have afforded higher-resolution solutions from the data (38).

V. MASS ACCURACY

Electrospray data is usually acquired by averaging a number of scans together and smoothing to provide good peak intensity and shape. Except for the very low mass compounds, the ions detected will be multiply charged. The error for the molecular-weight measurment is determined by the error of the measured mass to charge multiplied by the number of charges on the peak. For example, an error on the measured mass to charge of 0.1 for an ion with 10 charges leads to an error of 1.0 Da for the molecular-weight determination. Fortunately, for mass-measurement purposes, ESI spectra have several molecular ion peaks that differ in mass to charge only by the number of charges the ion possesses. Thus, random errors in determining the m/z of multiply charged ions, such as those produced by ion statistics, will tend to cancel when several values of the molecular weight from the same spectrum are averaged. For this reason, it is possible with careful measurement to obtain average mass measurement accuracies of 100 ppm or better on quadrupole mass spectrometers (39,40) and 10 ppm or better on magnetic-sector instruments to at least 20,000 Da (41–45).

Several factors influence mass-measurement accuracy in both quadrupole and magnetic-sector instruments. Unresolved overlapping peaks are always a problem for accurate mass measurements. Purification can reduce the problem of overlapping signals if they are caused by impurities. Resolution can be especially important in removing mass contributions from overlapping peaks (41,44). Figure 4 shows the effect of resolution on a mixture of the +10 ion for lysozyme and oxidized lysozyme. To obtain high sensitivity for the high-mass compounds, quadrupole instruments are typically operated with limited resolution (approximately 300). When the resolution is 300, the oxidation product, 16 u higher in mass than the molecular ion, is unresolved (41). Furthermore, using either a centroid measurement or peak top measurement will produce a significant mass error which, as pointed out above, will be multiplied by the

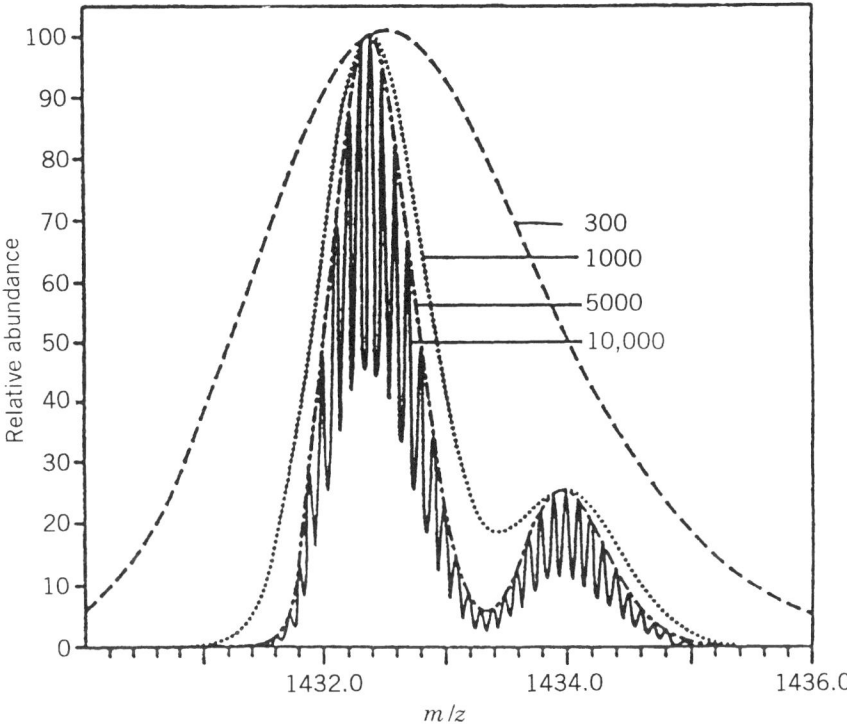

Figure 4. The effect of resolution on the +10 charge state molecular ion of lysozyme. Taken from C. N. McEwen and B. S. Larsen, *Rapid Commun. Mass Spectrom.*, **1992**, 6, 173. © 1992 John Wiley & Sons, Inc. Reprinted by permission of John Wiley & Sons, Inc.

number of charges on the ion being measured to determine the molecular weight. This kind of error is not compensated by averaging measurements of the multiply charged molecular ion peaks.

Unlike quadrupole instruments, double-focusing magnetic-sector instruments can achieve 1000 resolution with little loss in sensitivity. The oxidation product can be resolved at this resolution (see Fig. 4), but on close inspection some error in mass measurement will be obtained by either centroid or peak-top measurements. Increasing the resolution to 2000, which causes a moderate loss in sensitivity for double-focusing magnetic-sector instruments, is sufficient in this example to provide little peak broadening beyond the isotope contributions. The resolution necessary to obtain good mass data is dependant on m and on Δm where m is the mass being measured and Δm is the mass difference in the overlapping peaks. Higher m and lower Δm requires increased resolution. In addition, if Δm is less than the width of the isotope cluster for the peaks, complete resolution cannot be achieved.

The mass being measured for the example shown in Figure 4 is the average chemical mass unless much higher resolution is achieved to resolve the individual isotopes. The average chemical mass depends on the isotope ratio for each element that is present. A small error can be introduced if high-mass compounds are obtained from sources with, for example, different ratios of ^{13}C. Some magnetic-sector instruments are capable of resolving isotope peaks for reasonably high-mass compounds (44). However, in addition to the loss in sensitivity that is sacrificed in achieving high resolution, identifying the monoisotopic peak can be a significant problem (46).

Resolution and mass accuracy have been demonstrated on magnetic-sector instruments (41–45). Using an internal reference for accurate measurement (Fig. 5) requires that each multiply charged peak be visually inspected to insure the highest mass accuracy (41). Peaks representing ions present in low abundances may not contain sufficient detected ions to produce a symmetrical peak. These peaks should be eliminated along with any peaks unresolved from the internal reference peaks.

VI. ESI AND FRAGMENT ION INFORMATION

A great advantage of ESI is its ability to provide very soft ionization. Conditions can be achieved in which only molecular ions are observed. This advantage, however, becomes a disadvantage when fragment information is needed to characterize or confirm structures. With MS/MS instrumentation it is possible to select a molecular ion, collisionally fragment that ion, and mass measure the fragment ions. The method is valuable as a means of characterizing unknown structures or confirming the presence of a known compound. The MS/MS strategy is especially useful for mixtures where selecting a molecular ion from a number of ions present in the mass spectrum provides additional specificity. MS/MS in combination with ESI has been demonstrated to be a useful method for sequencing peptides (47), especially doubly charged ions generated from a tryptic digest (48,49). The tryptic peptide ions have basic sites which carry the charge at the ends of the peptide so that fragment ions are singly charged. Sequence information can readily be obtained from the fragmentation pattern.

Fragment ions can also be generated in the ESI ion source by forcing the ions exiting the atmosphere–vacuum interface through a potential gradient (7). This is done by imposing a voltage difference between the atmosphere–vacuum aperture and the first skimmer. In this region, the pressure is in a regime where multiple collisions between ions and neutrals will occur. The applied voltage increases the energy imparted to the ion during collisions with neutral molecules. The degree of fragmentation is determined by the energetics of the collision process which is dependent on the voltage difference between the aperture and the skimmer and the gas pressure in that region.

With in-source fragmentation, it is best to have reasonably pure compounds, because no mass selection of ions is possible before fragmentation. For mixtures, a LC can be used to separate the individual compounds. In-source fragmentation has been applied to peptide sequencing and to the identification of small

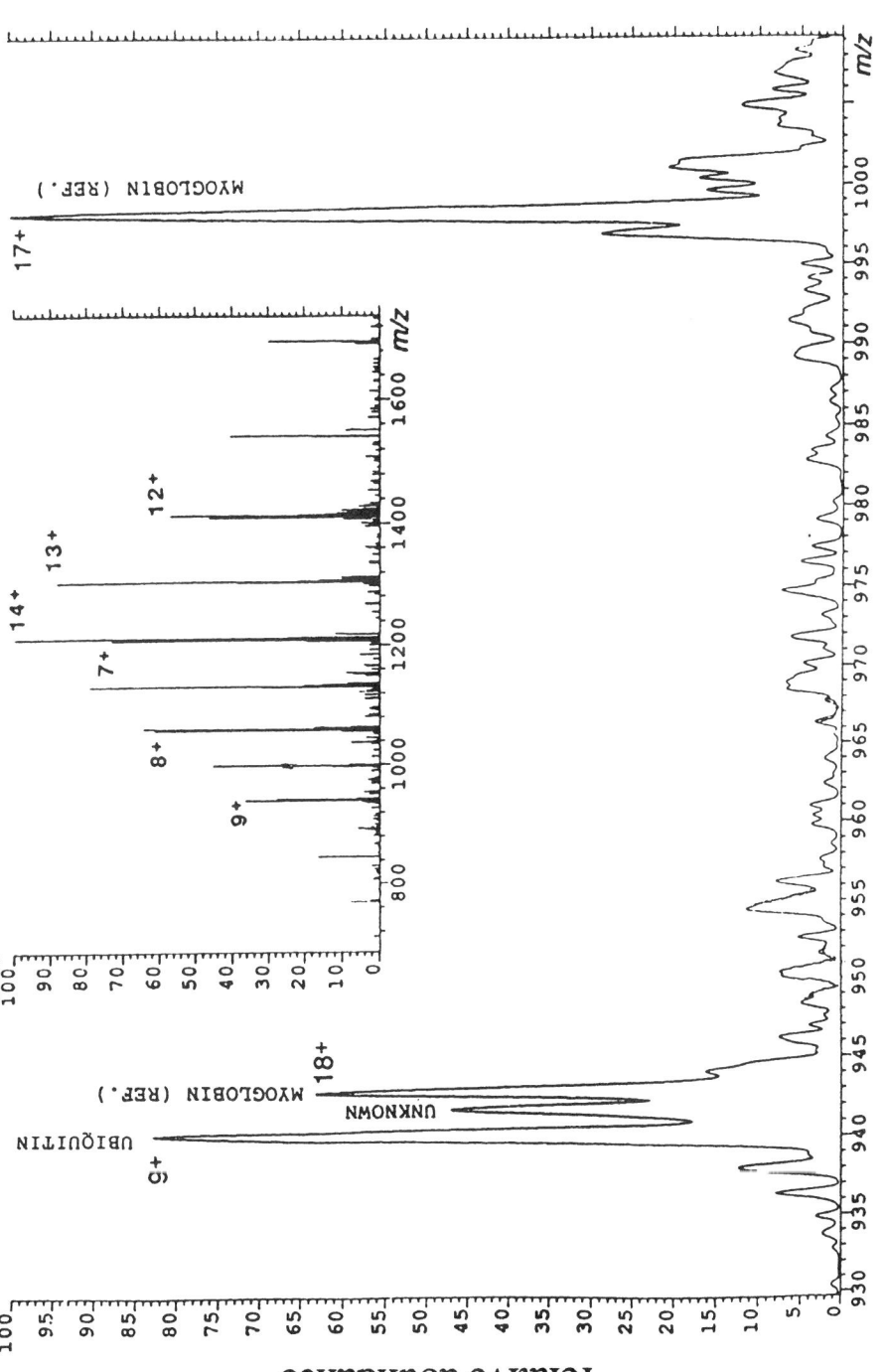

Figure 5. Demonstration of peak interference in accurate mass measurement on a double-focusing magnetic-sector mass spectrometer using the multiply charged ions from myoglobin as an internal mass reference. Taken from C. N. McEwen and B. S. Larsen, *Rapid Commun. Mass Spectrom.*, **1992**, 6, 173. © 1992 John Wiley & Sons, Inc. Reprinted by permission of John Wiley & Sons, Inc.

molecules (34,50,51). For the peptides, singly and doubly charged fragment ions were distinguished by the m/z separation of ^{13}C isotopes as described above (34). Proteins have also been shown to produce fragment ions by in-source fragmentation (52,53). With this technique, only a single-sector mass spectrometer is necessary. It also has the added advantage that selection of an abundant molecular ion versus abundant fragmentation is simply a matter of switching a voltage.

A further advantage of in-source fragmentation is achieved with MS/MS instruments. Fragment ions generated from in-source collisions can be mass selected and caused to undergo further fragmentation. This is equivalent to an MS^3 experiment and provides additional structural information for the identification of unknown structures.

VII. LC COUPLING TO ESI–MS

Because ESI is an atmospheric-pressure ionization process, coupling it with a liquid chromatograph requires no additional vacuum considerations. Whitehouse et al. (54) first interfaced LC to an ESI mass spectrometer. Bruins et al. (8) used pneumatically assisted nebulization to make ESI compatible with flows from conventional 4.6-mm LC columns. Most commercial ESI sources are compatible with flow rates from at least a 1.0-mm column, and some can operate up to flow rates compatible with standard 4.6-mm columns. However, ESI is concentration dependent, and, as noted previously, using higher flow rates does not increase the observed ion current. In practice, introducing the flow from a 1.0-, 2.2-, or 4.6-mm column directly into the ESI atmospheric-pressure enclosure reduces the ion signal to less than what is observed when the effluent from the column is split and only a few microliters per minute enter the ESI source. Using the full flow from a column has the advantage of eliminating the splitter, which can be a source of problems. The disadvantages are some loss in sensitivity and the release of large quantities of solvent and aerosols into the atmosphere. For safe operation, a closed system for the ESI process should be used to prevent solvent vapors and aerosols from entering the lab atmosphere.

Implementing a splitting arrangement has the advantage of operating under the highest ESI sensitivity and the remaining flow from the splitter can be sent to other detectors, collected for other uses, or discarded. By varying the lengths and/or inner diameter of the tubing that connects the splitter to the ESI source and to the UV detector, it is possible to balance the flow so that the sample enters both detectors simultaneously. The simplest splitter design is to use a low-dead volume tee with different sized tubing to restrict the flow to the mass spectrometer while feeding the majority of the flow to the UV detector. The flow is adjusted by cutting the tubing length. Alternatively, commercial splitters are available which are reliable for interfacing an LC to ESI.

The best method for increasing the sensitivity of the mass spectrometer coupled to an LC is to use columns that operate at low flow rates (see Table 1).

TABLE 1. Parameters to Consider in ESI–LC/MS

	Column Diameter (mm)	Injection Volume (μL)	Flow Rate (μL/min)	Split Ratio (MS:UV)
Conventional HPLC	4.6	100	1000	1:200
Microbore HPLC	1–2.1	5	50	1:10
Packed FS micro-HPLC	0.2–0.6	0.5	5	No split
Open-tubular HPLC	0.05–0.2	0.001	<1	Makeup flow

At equal sample concentrations, the smaller-diameter columns, which operate at lower flow rates, result in a better ratio of signal response to analyte consumed than larger-diameter columns. Thus, a capillary column operating at 2 μL/min is 50 times more efficient for ESI than a 1-mm column operating at 100 μL/min. Any increase in signal produced by sharper LC peaks will also be reflected in signal intensity provide that chromatographic resolution is maintained by the LC–ESI coupling. The practical application of nanoflow technology is just beginning to be realized. The lower flow rates that are attained with capillary-zone electrophoresis (CE) also result in high sensitivity; CE has been interfaced to ESI on magnetic-sector (55) and quadrupole (56,57) mass spectrometers.

Of special note is the voltage that is applied to the ESI needle. By using a glass capillary to transfer ions from atmospheric pressure to the first vacuum region as described previously, it is possible to operate the ESI needle at ground potential. Any source without this design will operate the needle at a voltage of between 2 and 5 kV above the potential of the skimmer (for positive ions). Obviously, any current flow through the solution from the ESI needle to the LC is not desirable. The resistance of the solution changes depending on its composition, so that it is possible to conduct current through the LC–ESI coupling. The smaller the inner diameter of the LC–ESI coupling, the higher the resistance to current flow. This then favors low flow rates and small-diameter capillaries, but can be especially important for magnetic-sector instruments that do not use the glass-capillary approach to reduce the ESI needle potential. For example, an ESI source with the accelerating voltage set at 6 kV, without the glass capillary, will require that the voltage on the atmospheric pressure–vacuum orifice be 6 kV. For positive-ion sampling, the ESI needle will need to be at 8–11 kV!

VIII. LC/MS APPLICATIONS

Applications of LC–ESI/MS cover the range of low- to high-mass compounds. An example of a separation of a matrix of low mass compounds is given in Figure 6, which shows the UV chromatographic trace (top) and the total ion

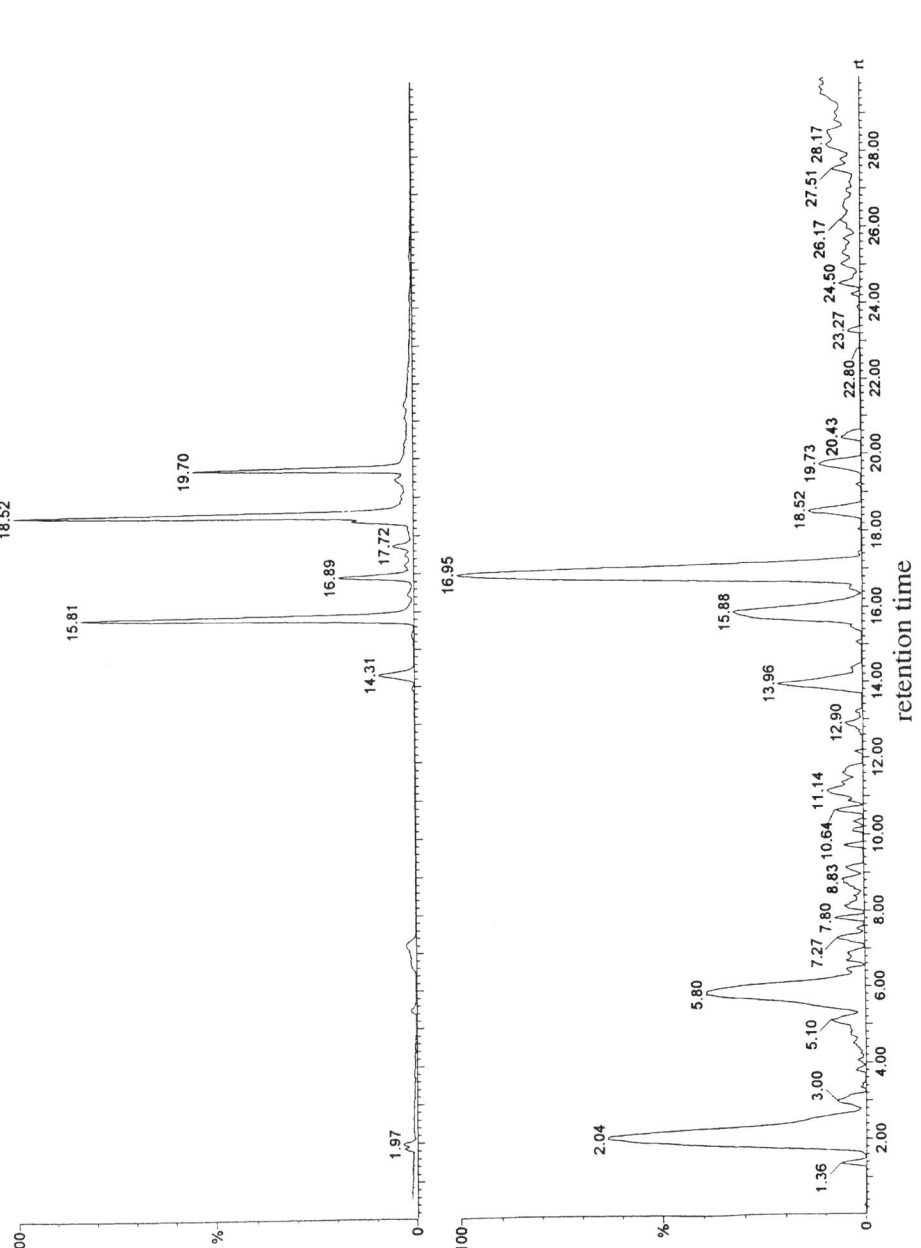

Figure 6. A UV trace of an LC separation of an unknown mixture (top) and total ESI ion current chromatogram (bottom) showing differences in selectivity.

VIII. LC/MS APPLICATIONS   **193**

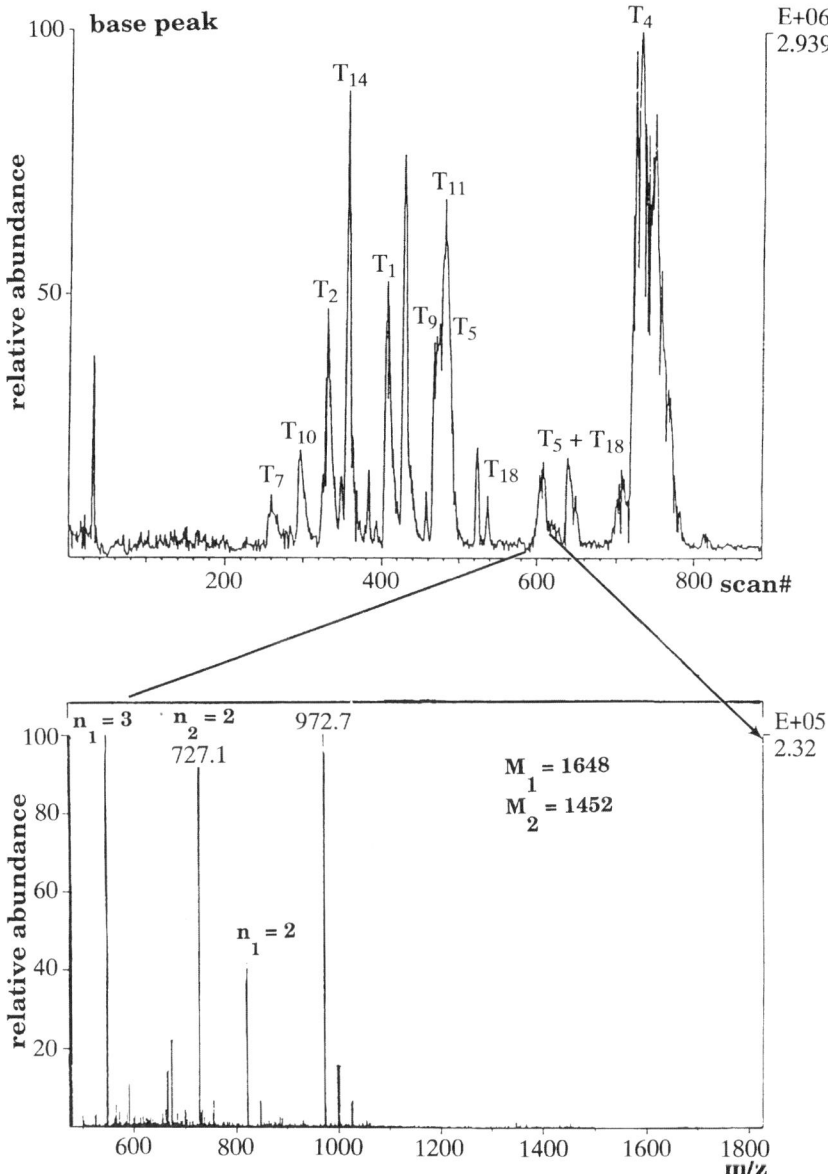

Figure 7. Base peak chromatogram of 10 pmoles of a tryptic digest of β-lactoglobin (top) and the mass spectrum of a nontryptic peak (bottom).

current chromatogram obtained from the mass spectrometer (bottom). Clearly, the two plots are not identical and demonstrate that the two detectors have different sensitivities to different structural types. Compounds that are not easy to see in the total ion chromatogram can easily be observed by plotting the $[M + H]^+$ ion current.

LC–ESI/MS is especially valuable for moderate mass compounds such as peptides produced by tryptic digestion of proteins. The LC chromatogram of 10 pmole of a tryptic digest of β-lactoglobin is shown in Figure 7 (top), and the ESI mass spectrum of a nontryptic cleavage showing at least three components is displayed at the bottom of the figure. Although many of the peptides could be identified by infusing the digest into the ESI source, more are identified after an LC separation step. Additionally, isolated peptides can be collisionally fragmented either in the ion source or using MS/MS instrumentation to provide sequence information.

For further information on ESI interfaced with chromatographic separations methods, see Chapters 9 and 10.

IX. TRACE ANALYSES USING LC–ESI/MS/MS

An extremely powerful method for the analysis of mixtures is to combine liquid chromatographic or capillary-zone electrophoresis separations with ESI/MS/MS instrumentation (29,58). This methodology makes it possible to identify compounds in complex matrices at extremely low levels with high specificity. When working near the limit of detection of this technology, it is extremely important to use blanks to assure that cross contamination does not occur during the preparation of standard solutions for quantitation.

For trace analysis, even with a prior separation step, when working with complex matrices, chemical background noise in a mass spectrum can mask the presence of the compounds of interest. However, with MS/MS instrumentation, the additional mass analyzer adds specificity and reduces the limits of detection by removing all background ions except those near in mass to the ions of interest. Fragmentation of the mass-selected ions produces a fragment-ion spectrum void of chemical noise and thus the true instrument detection limit is achieved. For work requiring the highest sensitivity, the instrument is generally set to observe the three most intense fragment ions so that the time associated with scanning the entire mass range is reduced. The compound of interest is considered to be detected if all three fragment ions are observed in the approximate ratio of their expected abundance. Not to observe an abundant fragment ion when others of equal or lesser expected abundance are observed is considered a negative finding.

There are several advantages to using ESI for trace analyses. The most significant is the ability of ESI to ionize labile and/or nonvolatile materials with little or no fragmentation and with high sensitivity. In addition, ESI is a selective ionization method which eliminates some background chemical noise. Hydrocarbons, for example, are not ionized by ESI. An added advantage is the ease of interfacing ESI with liquid separation methods.

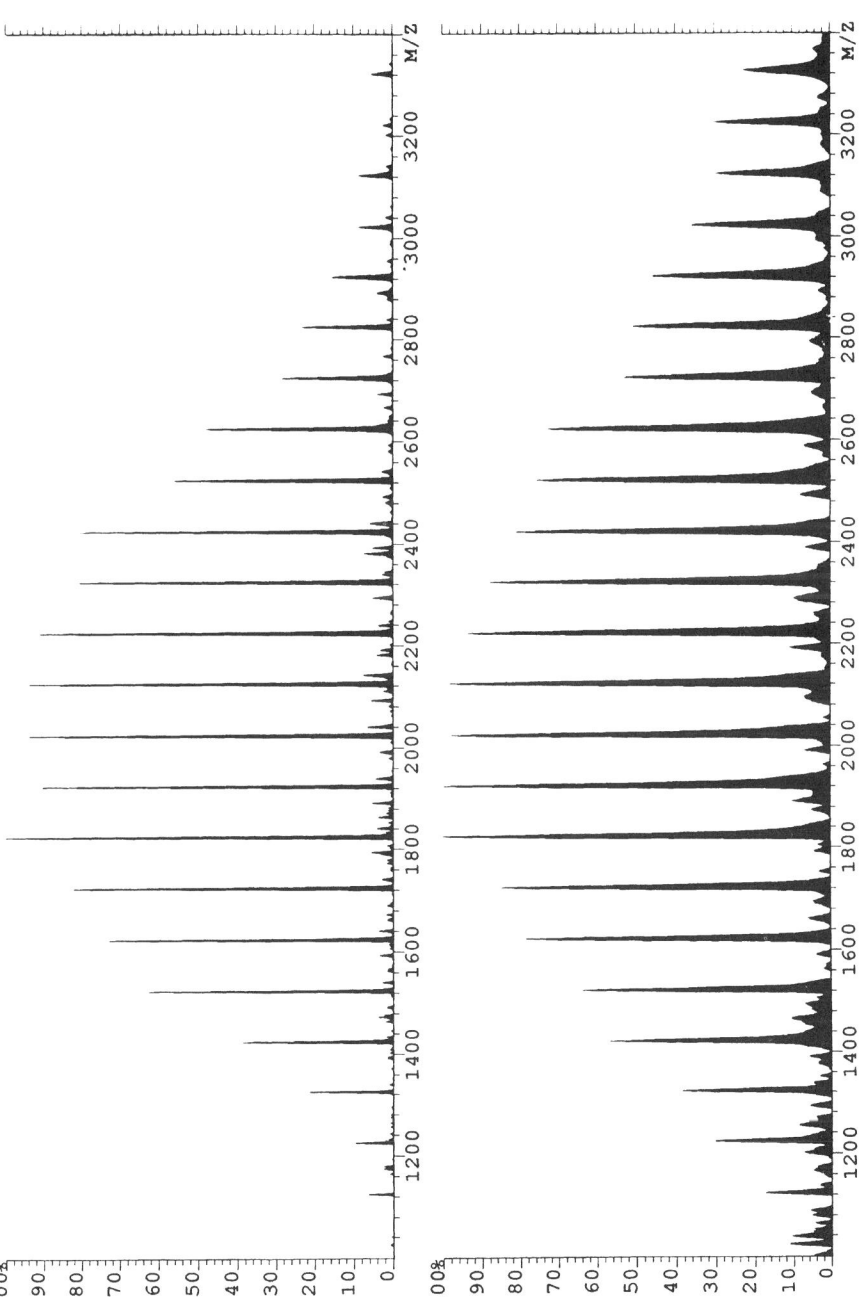

Figure 8. The transformed (singly charged) ESI mass spectrum of a PMMA 2400 standard (top) and the matrix-assisted laser desorption–ionization mass spectrum (bottom).

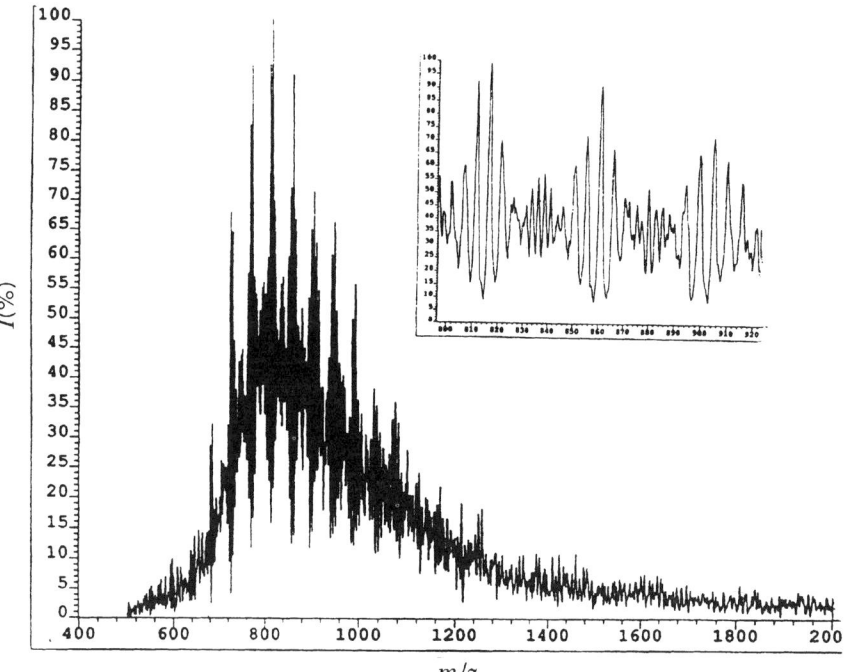
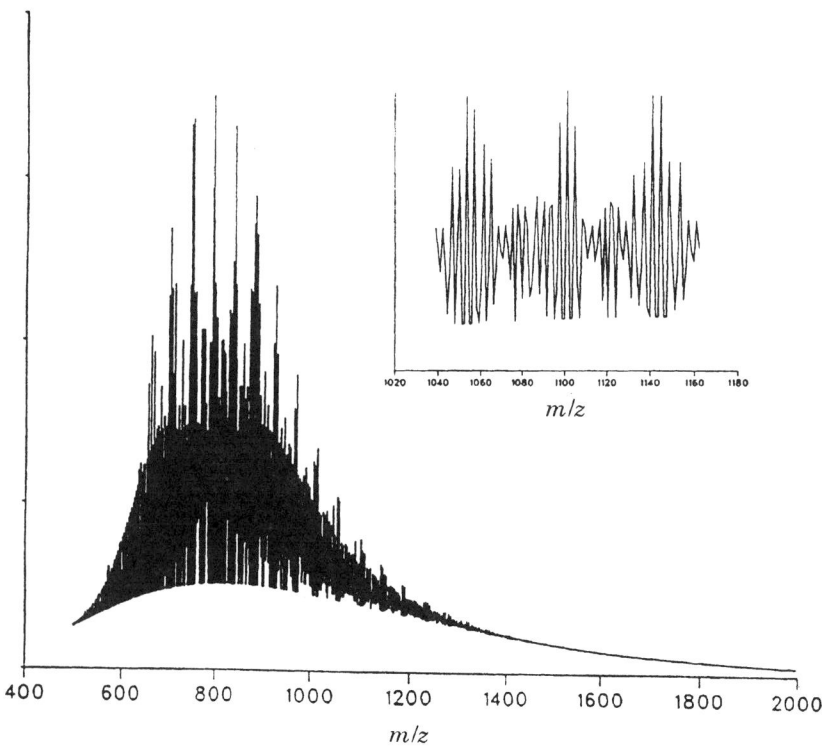

X. INTERESTING ESI APPLICATIONS*

There have been far too many noteworthy applications of ESI on quadrupole and magnetic-sector mass spectrometers to do more than mention a few in this chapter. Obviously, the entire area of protein-molecular-weight and structure analysis has been especially fruitful (59). Additionally, the detection of non-covalent receptor–ligand complexes by ESI mass spectrometry opens a new area of inquiry. Ganem et al. (60,61) were the first to demonstrate that noncovalent complexes could be observed using ESI under conditions in which the complexes are known to exist in solution. These researchers were able to observe immuno-suppressant agents (molecular weights 804 and 913 Da) bound to an immuno-suppressive binding protein (molecular weight 11812 Da). However, under conditions in which the protein was denatured, no binding was observed. These authors also observed the enzyme (hen egg white lysozyme) time-course hydrolysis of a hexasaccharide of N-acetylglucosamine by monitoring the noncovalent complexes (60). Numerous other studies have resulted from this work that demonstrate the possible applications of this technology (31,62–67), as well as the pitfalls (68).

Protein folding is an intramolecular example of noncovalent interaction and it was Katta and Chait (69) who demonstrated that conformational changes in proteins could be followed by measuring hydrogen–deuterium exchange over time using ESI. These authors probed the conformational changes in bovine ubiquitin induced by the addition of methanol to aqueous acetic acid solutions of the protein. In combination with nuclear magnetic resonance (NMR) spectroscopy, ESI was used to distinguish mechanisms of protein folding in hen lysozyme (70). Conformational changes induced by site-directed mutagenesis have also been studied using this methodology (71,72). ESI was also used to observe hydrogen-exchange protection in the GroEL bound state of α-lactalbumin (73).

The synthesis and screening of combinatorial libraries allows large numbers of molecules to be tested in assays for biological activity. Although the use of noncovalent bonding to identify compounds from a library that bind most tightly with a receptor may be possible using ESI (74–76), a more common use of ESI, often in combination with atmospheric pressure chemical ionization (APCI), is in the identification or confirmation of the components of synthetic combinatorial libraries (77–81). Both ESI and APCI are atmospheric-pressure ionization techniques and use common source designs. The two techniques complement one another very well.

Polymer analysis is an area that was of interest to Dole in his early work with ESI (2,4,6), and was also pursued by Fenn (20). Kallos et al. (82) used ESI to

---

* See also Chapters 11–15.

Figure 9. The ESI mass spectrum of a poly(ethyleneglycol) 12600 standard (top) and computer simulation of the multiply charged spectrum (bottom). Insets are expansion of a narrow mass window.

determine the molecular weight and polydispersity of polyamidoamine starburst dendrimers having narrow polydispersity. Prokai and Simonsick (83) produced narrow polydisperse fractions by coupling size-exclusion chromatography (SEC) with ESI–MS for octylphenoxypoly(ethoxy)ethanol oligomers. Mass analysis of the SEC fractions using ESI–MS allows calibration of the SEC chromatogram for a specific polymer. Hunt et al. (84) examined a series of polyester oligomers in paint resins. They were able to determine average molecular weights, polydispersity, the distribution of end groups, and the average frequency of branching. McEwen et al. (85) demonstrated that equimolar-isolated PMMA oligomers incorporating 25 and 50 monomer units gave equal ion currents by ESI, provided peak compression is accounted for when transforming the spectrum to the singly charged representation. Peak compression is accounted for by multiplying the area of a doubly charged peak by 2, a triply charged peak by 3, and so forth, to obtain a quantitative representation of the number of ions represented by each peak.

Figure 8 is a comparison of the transformed ESI mass spectrum of PMMA 2400 (top) and the matrix-assisted laser desorption/ionization (MALDI) mass spectrum of the same material (bottom). Although present computer programs can transform polymer spectra of low-mass polymers, higher-mass polymers become much too complex as a result of multiple charging of each of the oligomers in the distribution. This is seen in Figure 9 (top), which is the ESI mass spectrum of a narrow-polydispersity polyethylene glycol (PEG) with $M_w = 12{,}600$. All ions appear at a mass-to-charge ratio between approximately 600 and 1700. Figure 9 (bottom) is the computer simulation of a multiply charged spectrum of a polymer distribution that would be expected for PEG 12,600. The inserts in Figure 9 (top and bottom) show the similarity of the actual and simulated spectrum of PEG over a narrow mass range. These data suggest that a sophisticated computer-transform program could provide at least compositional information from moderate-mass polymer spectra.

A striking example of the power of ESI mass spectrometry is the work from Hunt's laboratory in which the sequences of peptide antigens bound to the surface of class I MCH molecules were determined using LC-ESI/MS/MS (29,86). The number of different peptides presented by the cells may exceed 1000. Of these, 200 are observed by mass spectrometry with a lower detection limit estimated at 30 fmole. Peptides were sequenced from this mixture at the 100- to 300-fmole level.

REFERENCES

1. Zeleny, J., *Phys. Rev.* **1917**, *10*, 1.
2. Dole, M.; Hines, R. L.; Mack, L. L.; Mobley, R. C.; Ferguson, L. D.; Alice, M. B. *Macromolucles* **1968**, *1*, 96 and *J. Chem. Phys.* **1968**, *49*, 2240. Cleg G. A.; Dole, M. *Biopolymers* **1971**, *10*, 821.
3. Dole, M.; Cox, H. L.; Gieniec, J. *Adv. Chem. Ser.* **1973**, *125*, 73.

4. Dole, M.; Gupta, C. V.; Mack, L. L.; Makamae, K. *Polym. Prep. Am. Chem. Soc. Div. Polym. Chem.* **1977**, *18*, 188.
5. Yamashita, M.; Fenn, J. B. *Iyo Masu Kenkyukai Koenshu* **1984**, *9*, 203.
6. Gieniec, J.; Mack, L. L.; Nakamae, K.; Gupta, C.; Kumar, V.; Dole, M. *Biomed. Mass Spectrom.* **1984**, *11*, 259.
7. Aleksandrov, M. L.; Gall, L. N.; Krasnov, N. V.; Nikolaev, V. I.; Paulenko, V. A.; and Shkurou, V. A. *Dokl. Akad. Nauk. SSSR* **1984**, *277*, 379.
8. Bruins, A. P.; Covey, T. R.; Henion, J. D. *Anal. Chem.* **1987**, *59*, 2642.
9. Meng, C-K.; Mann, M.; Fenn, J. B. *Proceedings of the 36th ASMS Conf. on Mass Spectrom. and Allied Topics* **1988**, p.771.
10. Meng, C. K.; Mann, M.; Fenn, J. B. *Z. Phys. D: At., Mol. Clusters* **1988**, *10*, 361.
11. Fenn, J. B.; Mann, M.; Meng, C.-K.; Wong, S.-F.; Whitehouse, C. M. *Science* **1989**, *246*, 64.
12. Loo, J. A.; Udseth H. R.; Smith, R. D. *Biomed. Environ. Mass Spectrom.* **1988**, *17*, 411.
13. Two ESI sources were built in 1984 by C. M. Whitehouse of the Fenn group at Yale that were based on the Fenn–Whitehouse quadrupole design, but were made to fit a VG-ZAB mass spectrometer. VG (Fisons) and these authors at DuPont received the sources, but were only able to observe low-mass ions.
14. Allen, M. H.; Lewis, I. A. S. *Rapid Commun. Mass Spectrom.* **1989**, *3*, 255.
15. Meng, C-K.; McEwen, C. N.; Larsen, B. S.; Whitehouse, C. M.; Fenn, J. B. *Proceedings of the 2nd International Symposium on Mass Spectrometry in the Health and Life Sciences*, San Francisco, CA, 1989.
16. Meng, C-K.; McEwen, C. N.; Larsen, B. S. *Rapid Commun. Mass Spectrom.* **1990**, *4*, 147.
17. Larsen, B. S.; McEwen, C. N. *J. Am. Soc. Mass Spectrom.* **1991**, *2*, 205.
18. Gallagher, R. T.; Chapman, J. R. *Rapid Commun. Mass Spectrom.* **1990**, *4*, 369.
19. Hamdan, M.; Curcuruto, O. *Int. J. Mass Spectrom. Ion Processes* **1991**, *108*, 93.
20. Fenn, J. B.; Mann, M.; Meng, C.-K.; Wong, S.-F.; Whitehouse, C. *Mass Spectrom. Rev.* **1990**, *9*, 37.
21. Mann, M. *Org. Mass Spectrom.* **1990**, *25*, 575. Meng, C. K.; Fenn, J. B. *Am. Biotechnol. Lab.* **1990**, *8*, 54.
22. Whitehouse, C. M.; Yamashita, M.; Meng, C.-K.; Fenn, J. B. *Proceedings of the 14th Int. Symp. on Rarefied Gas Dynamics*, July 16–20 1984, Tsukuba Science City, Japan.
23. Anderson, J. B.; Andres, R. P.; Fenn, J. B. *Adv. Chem. Phys.* **1966**, *10*, 275 and *Adv. At. Mol. Phys.* **1965**, *1*, 345.
24. The nozzle skimmer arrangement was first used by Dole for sampling ESI-generated ions (3). Interestingly, Dole references earlier work by Fenn in support of using a conical skimmer to sample a molecular beam (23).
25. Chowdhury, S. W.; Katta, V.; Chait, B. T. *Rapid Commun. Mass Spectrom.* **1990**, *4*, 81.
26. Jarvis, S.; Bateman, R. H.; Green, B. N. *Proceedings of the 37th Ann. Conf. on Mass Spectrom. and Allied Topics*, May 21–26 1989, p. 1017.
27. Hopfgartner, G.; Wachs, T.; Bean, K.; Henion, J. *Anal. Chem.* **1993**, *65*, 439.

28. Lee, E. D.; Henion, J. D. *Rapid Commun. Mass Spectrom.* **1992**, *6*, 727.
29. Hunt, D. F.; Henderson, R. A.; Shabanowitz, J.; Sakagouchi, K.; Michel, H.; Sevilir, N.; Cox, A. L.; Appella, E.; Engelhard, V. H. *Science* **1992**, *255*, 1261.
30. Wilm, M. W.; Mann, M. *Int. J. Mass Spectrom. Ion Processes* **1994**, *136*, 167.
31. Loo, J. A. *J. Mass Spectrom.* **1995**, *30*, 180.
32. Cody, R. B.; Tamura, J.; Finch, J. W.; Musselmann, D. B. *J. Am. Soc. Mass Spectrom.* **1994**, *5*, 194.
33. Voyksner, R. B.; Pack, T. *Rapid Commun. Mass Spectrom.* **1991**, *5*, 263.
34. Meng, C. K.; McEwen, C. N.; Larsen, B. S. *Rapid Commun. Mass Spectrom.* **1990**, *4*, 151.
35. Mann, M.; Meng, C. K.; Fenn, J. B. *Anal. Chem.* **1989**, *61*, 1702.
36. Zhou, J. X. G.; Jardine, I. *Proceedings of the 38th ASMS Ann. Conf. on Mass Spectrom. and Allied Topics*, June 3–8, 1990, p134.
37. Labowsky, M.; Whitehouse, C. M. *Rapid Commun. Mass Spectrom.* **1993**, *7*, 71.
38. Ferrige, A. G.; Seddon, M. J.; Green, B. N.; Jarvis, S. A.; Skilling, J. *Rapid Commun. Mass Spectrom.* **1992**, *6*, 707.
39. Feng, R.; Konishi, Y.; Bell, A. W. *J. Am. Soc. Mass Spectrom.* **1991**, *2*, 387.
40. Chait, B. T.; Kent S. B. H. *Science* **1992**, *257*, 1885.
41. McEwen, C. N.; Larsen, B. S. *Rapid Commun. Mass Spectrom.* **1992**, *6*, 173.
42. Chapman, J. R.; Gallagher, R. T.; Mann, M. *Biochem. Soc. Trans.* **1991**, *19*, 940.
43. Perkins, J. R.; Tomer, K. B. *Anal. Chem.* **1994**, *66*, 2835.
44. Cody, R. B.; Tamura, J.; Musselman, B. D. *Anal. Chem.*, **1992**, *64*, 1561. Dobberstein, P.; Schroeder, E. *Rapid Commun. Mass Spectrom.* **1993**, *7*, 861. Jiang, L.; Moini, M. *J. Am. Soc. Mass Spectrom.* **1995**, *6*, 1256.
45. Starrett, A. M.; DiDonato, G. C. *Rapid Commun. Mass Spectrom.* **1993**, *7*, 12.
46. Yergey, J.; Heller, D.; Hansen, G.; Cotter, R. J.; Fenselau, C. *Anal. Chem.* **1983**, *55*, 353.
47. Baringa, C. J.; Edmonds, C. G.; Udseth, H. R.; Smith, R. D. *Rapid Commun. Mass Spectrom.* **1989**, *3*, 60.
48. Lee, E. D.; Henion, J. D.; Covey, T. R. *J. Macrocolumn Sep.* **1989**, *1*, 14.
49. Covey, T. R.; Huang, E. C.; Henion, J. D. *Anal. Chem.* **1991**, *63*, 1193.
50. Katta, V.; Chowdhury, S. K.; Chait, B. T. *Anal. Chem.* **1991**, *63*, 174.
51. Voyksner, R. D.; Pack, T. *Rapid Commun. Mass Spectrom.* **1991**, *5*, 263.
52. Loo, J. A.; Udseth, H. R.; Smith, R. D. *Rapid Commun. Mass Spectrom.* **1988**, *2*, 207.
53. Loo, J. A.; Edmonds, C. G.; Udseth, H. R.; Smith, R. D. *Anal. Chim. Acta.* **1990**, *241*, 167.
54. Whitehouse, C. M.; Dreyer, R. N.; Yamashita, M.; Fenn, J. B. *Anal. Chem.* **1985**, *57*, 675.
55. Perkins, J. R.; Tomer, K. B. *Anal. Chem.* **1994**, *66*, 2835.
56. Lee, E. D.; Muck, W.; Henion, J. D.; Covey, T. R. *J Chromatogr.* **1988**, *179*, 404.
57. Smith, R. D.; Loo, J. A.; Edmonds, C. G.; Baringa, C. J.; Udseth, H. R. *J. Chromatogr.* **1990**, *516*, 157.
58. Hail, M.; Lewis, S.; Jardine, I. *J. Microcolumn Sep.* **1990**, *2*, 1040.

59. Siuzdak, G. *Proc. Natl. Acad. Sci.* **1994**, *91*, 11290.
60. Ganem, B.; Li, Y-T.; Henion, J. D. *J. Am. Chem. Soc.* **1991**, *113*, 6294.
61. Ganem, B.; Li, Y-T.; Henion, J. D. *J. Am. Chem. Soc.* **1991**, *113*, 7818.
62. Feng, R.; Castelhano, A. L.; Billedeau, R.; Yuan, Z. *J. Am. Soc. Mass Spectrom.* **1995**, *6*, 1105.
63. Baca, M.; Kent, B. H.; *J. Am. Chem. Soc.* **1992**, *114*, 3992.
64. Ganguly, A. K.; Pramanik, B. N.; Tsarbopoulos, A.; Covey, T. R.; Huang, E.; Fuhrman, S. A. *J. Am. Chem. Soc.* **1992**, *114*, 6559.
65. Ogorzalek Loo, R. R.; Goodlett, D. R.; Smith, R. D.; Loo, J. A. *J. Am. Chem. Soc.* **1993**, *115*, 4391.
66. Li, Y-T.; Hsieh, Y.-L.; Henion, J. D.; Senko, M. W.; McLafferty, F. W.; Ganem, B. *J. Am. Chem. Soc.* **1993**, *115*, 8409.
67. Aplin, R. T.; Baldwin, J. E.; Schofield, C. J.; Waley, S. G. *FEBS Lett*, **1990**, *277*, 212.
68. Smith, R. D.; Light-Wahl, K. J. *Biol. Mass Spectrom.* **1993**, *22*, 493.
69. Katta, V.; Chait, B.T. *Rapid Commun. Mass Spectrom.* **1991**, *5*, 214.
70. Miranker, A.; Robinson, C. V.; Radford, S. E.; Aplin, R. T.; Dobson, C. M. *Science* **1993**, *262*, 896.
71. Zhang, Z.; Smith, D. L. *Protein Sci.* **1993**, *2*, 522.
72. Halgand, F.; Jaquinod, M.; Caffrey, M.; Fitch, J.; Cusanovich, M.; and Forest, E. *Proceedings from the 43rd ASMS Conf. on Mass Spectrom. and Allied Topics*, May 21–26, 1995, p. 1257.
73. Robinson, C. V.; Gross, M.; Eyles, S. J.; Ewbank, J. J.; Mayhew, M.; Hartl, F. U.; Dobson, C. M.; and Radford, S. E. *Nature* **1994**, *372*, 645.
74. Cheng, X.; Chen, R.; Bruce, J. E.; Schwartz, B. L.; Anderson, G. A.; Hofstadler, S. A.; Gale, D. C.; Smith, R. D.; Gao, J.; Sigal, G. B. *J. Am. Chem. Soc.* **1995**, *117*, 8859.
75. Quesnel, A.; Casrouge, A.; Kourilsky, P.; Abastado, J. P.; Trudelle, Y. *Peptide Res.* **1995**, *8*, 44.
76. Bruce, J. E.; Anderson, G. A.; Chen, R.; Cheng, X.; Gale, D. C.; Hofstadler, S. A.; Schwartz, B. L.; Smith, R. D. *Rapid Commun. Mass Spectrom.* **1995**, *9*, 644.
77. Metzger, J. W.; Stevanovic, S.; Bruenjes, J.; Wiesmueller, K. H.; Jung, G. *Methods* **1994**, *6*, 425.
78. Carell, T.; Winter, E. A.; Sutherland, A. J.; Rebek, J., Jr.; Dunayenskiy, Y. M.; Vouros, P. *Chem. Biol.* **1995**, *2*, 171.
79. Dunavevskiv, Y. M.; Vouros, P.; Carell, T.; Winter, E. A.; Rebek, J., Jr. *Proceedings of the 43rd ASMS Conf. on Mass Spectrom. and Allied Topics*, May 21–26 1995, p. 493.
80. Fitch, W. L.; Shah, N.; Holmes, C. P.; Look, G. C. *Proceeding of the 43rd ASMS Conf. on Mass Spectrom. and Allied Topics*, May 21–26 1995, p. 486.
81. Wagner, D. S.; Brown, B. B.; Geysen, H. M. *Proceeding of the 43rd ASMS Conf. on Mass Spectrom. and Allied Topics*, May 21–26 1995, p. 488.
82. Kallos, G. J.; Tomalia, D. A.; Hedstrand, D. M.; Lewis, S. *Rapid Commun. Mass Spectrom.* **1991**, *5*, 383.
83. Prokai, L.; Simonsick, W. J., Jr. *Rapid Commun. Mass Spectrom.* **1993**, *7*, 853.
84. Hunt, S. M.; Binns, M. R.; Sheil, M. M. *J. Appl. Polym. Sci.* **1995**, *56*, 1589.

85. McEwen, C. N.; Simonsick, W. J., Jr.; Larsen, B. S.; Ute, K.; Hatada, K. *J. Am. Soc Mass Spectrom.* **1995**, *6*, 906.
86. Henderson, R. A.; Michel, H.; Sakaguchi, K.; Shabanowitz, J.; Appella, E.; Hunt, D. F.; Engelhard, V. H. *Science* **1992**, *255*, 1264.

CHAPTER 6

Electrospray Ionization Time-of-Flight Mass Spectrometry

IGOR V. CHERNUSHEVICH,* WERNER ENS, AND KENNETH G. STANDING

Department of Physics, University of Manitoba, Winnipeg, Manitoba, Canada

	Abstract	204
I.	Introduction	204
II.	Experimental method	207
	A. Time-of-flight mass spectrometer (TOF III)	208
	B. Electrospray ion source	208
	C. Orthogonal injection	210
	D. Chemicals	212
III.	Instrument performance	212
	A. Mass resolution	212
	B. Accuracy of mass determination	214
	C. Sensitivity and dynamic range	215
	D. Secondary electron-resolved mass spectrometry	217
	E. Mass/charge (m/z) range	218
IV.	Coupling ES–TOF MS with separation techniques	219
	A. Capillary electrophoresis at Analytica of Branford	219
	B. On-line LC/LC/MS at PerSeptive Biosystems	221
V.	Studies of noncovalent associations of biomolecules	222
	A. Complexes of enzymes	224
	B. Noncovalent interactions in soybean agglutinin	226
	C. Immunoconjugate F(ab')2-CPG2	227
	D. Catalase HP II	229
VI.	Conclusion	230
	Acknowledgments	231
	References	231

---

* Present address: SCIEX, Concord, Ont., L4K 4V8, Canada.
Electrospray Ionization Mass Spectrometry, Edited by Richard B. Cole.
ISBN 0-471-14564-5 © 1997 John Wiley & Sons, Inc.

Abstract

The principles and practice of electrospray ionization time-of-flight mass spectrometry (ES–TOF MS) with orthogonal injection are discussed, as well as the most recent results obtained on the instrument built at the University of Manitoba and on those developed by several other groups. The method combines electrospray ionization with the advantages of TOF MS, such as unlimited mass range, moderate resolution (R_{FWHM} up to 10,000), fast response, and the ability to register all ions simultaneously without scanning. These features are illustrated by the coupling of ES–TOF instruments to capillary electrophoresis and to liquid chromatography, and by several studies of noncovalent associations, which require both high sensitivity and high m/z range (m/z up to 16,000), both of which are provided by this instrument.

I. INTRODUCTION

There are several reasons for the renaissance of the time-of-flight mass spectrometry technique (TOF MS) during the last decade. First, the well-known advantages of TOF MS include a mass range limited only by the ion detector, high ion transmission, and relatively low cost. The ability to register ions of all masses at the same time provides a sensitivity over the full spectrum equal to that in the single ion-monitoring regime. Second, several approaches have been developed for improving the mass resolution of TOF MS, which originally was less than several hundred. The most common method is the application of the electrostatic mirror originated by Mamyrin et al. (1,2), with which, for example, a resolution $m/\Delta m_{FWHM}$ up to 35,000 for cesium ions has been reported by Bergmann et al. (3). Third, TOF MS is a natural choice for pulsed-ionization techniques such as plasma desorption (PDMS) and laser desorption (LD), because it provides a complete spectrum for each event. However, the most important reason for the recent interest is probably the introduction of matrix-assisted laser desorption (MALDI) in 1987–1988 by Karas and Hillenkamp (4,5). Together with electrospray (ES), MALDI has revolutionized the mass spectrometry of biomolecules, particularly large ones.

Electrospray ionization was first coupled with quadrupole (6–8) and magnetic-sector mass spectrometers (9), and is now recognized as a powerful analytical method (10–13), as described in this volume. A major advantage of TOF spectrometers is their "unlimited" mass range, as noted above. This would not seem to offer much help to ES, because one of the advantages of that technique is the modest range in m/z of the ions usually produced (≤ 2000), which makes them suitable for analysis in quadrupole filters and easily detectable by conventional electron multipliers (14). Fenn et al. (10,11) have presented arguments that the range of m/z normally produced by ES sources is restricted to these values. However, it is difficult even to check this prediction experimentally in most quadrupole instruments because of their rapid decrease in efficiency as m/z increases; examples in which a much larger m/z range is necessary have

I. INTRODUCTION

recently been demonstrated (15,16), especially when ions are studied from near-physiological conditions (e.g., see refs. 17-20).

Although the marriage of MALDI and TOF MS is quite natural, there are difficulties in coupling an inherently continuous ES source to a TOF instrument. Like other continuous ionization sources, ES cannot be coupled to TOF mass analyzers in the most straightforward manner (injection along the spectrometer axis) without a tremendous loss in sensitivity. An example of interfacing a continuous chemical-ionization source to the TOF mass analyzer was described by Pinkston et al. (21); the need to extract short ion packets with a small energy spread meant that only 0.0025% of the total ion current contributed to the recorded spectra. Boyle et al. (22) tried to improve the situation in their first ES–TOF instrument by introducing a storage region where low-energy ions could be accumulated between the extraction pulses. The duty cycle was increased by a factor ~ 60, but the resolution did not exceed 100.

Fortunately, TOF instruments can tolerate a relatively large spatial or velocity spread in a plane perpendicular to the spectrometer axis (the z axis), as witnessed by the large sources (≤ 1 cm diameter) typically used in fission-fragment desorption (PDMS). This tolerance can be exploited by injecting ES ions into the TOF instrument perpendicular to the axis (23-33), that is "orthogonal injection" (see Figs. 1 and 2). The history of orthogonal injection in TOF MS dates back to the 1950s (34); it was used to study ions formed in the interaction of a pulsed-ionizing electron beam with a cold molecular beam. In the 1960s the Bendix Corporation produced an orthogonal acceleration TOF MS instrument with an atmospheric-pressure plasma ion source (35). Orthogonal injection has also been used with other types of ion sources: ions from flames (36), LSIMS (37,38), a cluster-ion source (39), electron-impact ionization (40,41), atmospheric-pressure ionization (24,42) and inductively coupled plasma (43). This geometry provides a high-efficiency interface for transferring ions from a continuous beam to a pulsed mode. Another advantage is the small velocity spread in the z direction that is usually observed, making high resolution easier to obtain.

The first ES TOF instrument using orthogonal injection was constructed by Dodonov et al. (23). It is shown in Figure 1, together with a spectrum of insulin. Note that this spectrum of multiply charged ions was reported in 1987, about the same time that Fenn recorded his first ES spectra of proteins (14). Subsequent work with the spectrometer has included peptide analysis (26) and studies of the kinetics of tryptic digestion of melittin (44). A careful account of the limitations on its performance has been presented (25,45).

Several other ES–TOF spectrometers have been constructed more recently (27-33). The existing ES–TOF instruments with orthogonal injection have already achieved high sensitivity [in the femtomole range (26,27,31,32)] and reasonable resolution [$M/\Delta M_{\text{FWHM}} \sim 2000$ (26), ~ 3000 (46), and ~ 5000 (31)]. The ES–TOF has been coupled recently to liquid chromatographs (32,33) and capillary electrophoresis (46,47), utilizing the ability of TOF MS to record full spectra without scanning on a millisecond time scale.

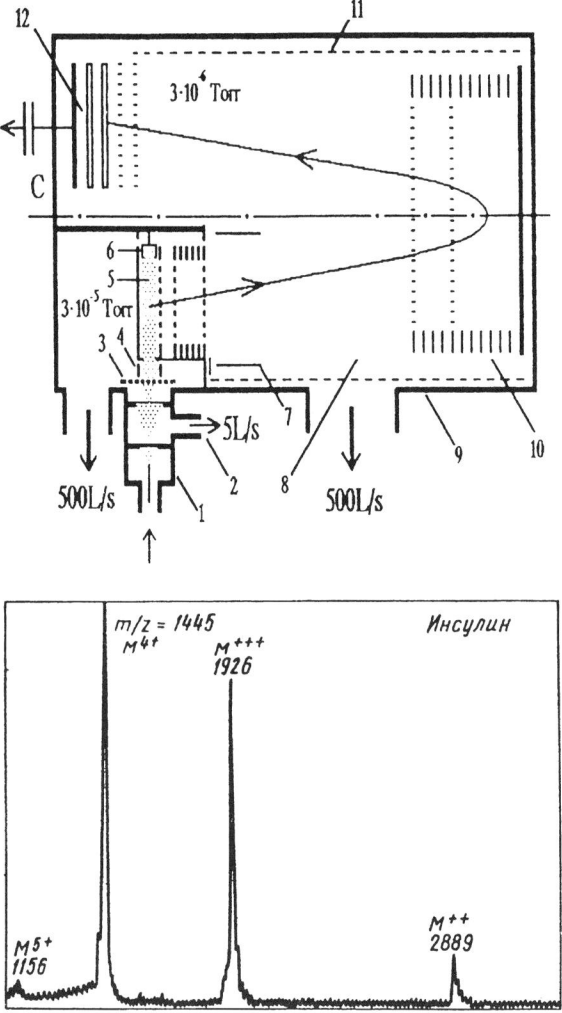

Figure 1. Top: Schematic diagram of the reflecting TOF mass spectrometer with orthogonally injected ES ions developed by Dodonov et al. Reprinted with permission from D. A. Mirgorodskaya et al., *Anal. Chem.*, **1994**, *66*, 99–107. © 1994 American Chemical Society. Bottom: the first spectrum of multiply charged ions of bovine insulin obtained with the instrument (23).

An alternative approach for coupling an ES source to TOF MS has been developed by Lubman et al. (48–50). An on-axis configuration of an ES source, an ion trap, and a TOF mass analyzer was used in this instrument. The ion trap accumulates ions between the extraction pulses, increasing the duty cycle significantly. Resolution of ~ 3000 and sensitivity in the femtomole range were measured. An advantage of the configuration is the ability to perform MS/MS

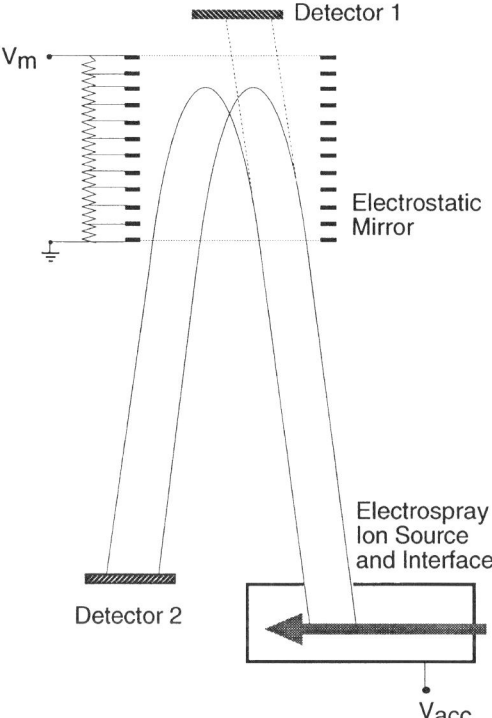

Figure 2. Schematic diagram of the reflecting TOF mass spectrometer (TOF III) developed at the University of Manitoba.

measurements with collision-induced dissociation of ions in the ion trap. However, although this hybrid instrument works well with light ions (e.g. peptides), there appear to be some difficulties when examining heavier ions (50).

In our laboratory we have coupled a reflecting time-of-flight spectrometer to an ES ion source using orthogonal injection. Our early evaluation of the properties of this instrument has been reported (31). In this chapter, we discuss the principles and practice of ES–TOF mass spectrometry with orthogonal injection using our own machine as an example. We examine as well the most recent results obtained on this instrument and on those developed by several other groups.

II. EXPERIMENTAL METHOD

The instrument built at the University of Manitoba (TOF III) is shown in Figure 2. It consists of a reflecting time-of-flight mass spectrometer with its axis vertical and an ES ion source floated at the acceleration potential V_{acc}. The source continuously injects a slow (~ 10 eV/charge) horizontal ion beam into the spectrometer, where it is formed into vertically accelerated ion packages by an orthogonal injection system.

A. Time-of-Flight Mass Spectrometer (TOF III)

Apart from the ion source, the instrument is almost identical to our spectrometer TOF II (51,52), used for measurements in secondary ion mass spectrometry (SIMS) and MALDI. The vertical stainless-steel vacuum chamber has a diameter of 25 cm and a length ~ 1.2 m. It contains a single-stage electrostatic mirror similar to the one in TOF II (51), in which resolving powers $M/\Delta M_{FWHM}$ up to $\sim 13,000$ have been obtained for alkali halide ions, and up to ~ 5000 for organic ions of mass $< \sim 3000$ u, when the secondary ions were produced by pulsed Cs^+ ion bombardment (51,52). The mirror consists of 30 rings; it has an inner diameter of 12 cm and a length of 33 cm. The diameter of the entrance window of the mirror is 8 cm, somewhat larger than in TOF II, because of the large dimensions of the ion-source region in ES. The potential of the entrance window is defined by a 90% transmission grid.

The orthogonal injection system and the detector for reflected ions are mounted symmetrically on the bottom flange of the chamber with 11 cm between centers. A detector for undeviated particles is mounted on the top flange. The spectrometer may be run either in the reflecting mode, with an electric field in the mirror, or in the linear mode, with the mirror voltage off. In the first case, reflected ions are observed in the lower detector; in the second case, all ions are observed in the upper detector. The total flight time corresponds to an effective path length of ~ 2.8 m in the reflecting mode and ~ 1.5 m in the linear mode. Each detector consists of two 40-mm diameter microchannel plates in a chevron configuration. In both cases, the detector signal passes to a preamplifier and a constant-fraction discriminator with a 10-mV threshold. It is then registered in a single ion-counting mode by an Orsay time-to-digital converter (TDC), model CTN-M2, connected to an Atari TT computer (53). The TDC can record up to 255 stops for each extraction pulse, but only one stop per extraction pulse for a given m/z species. For high counting rates, this could cause distortion. However, in the experiments described here, the total counting rate is in most cases less than one ion per injection pulse (on average), so distortion of this kind is negligible.

B. Electrospray Ion Source

Spraying is performed in an open chamber with a countercurrent of dry heated (70°C) nitrogen as a curtain gas (Fig. 3). Either normal ES, with a solution flow rate of 0.25 μL/min, or nanospray (54), with a flow rate ~ 20 nL/min, is used. The atmosphere-vacuum interface constructed for the instrument is similar to the one designed by Chait et al. (55). A sample of the aerosol products produced enters the spectrometer through a three-stage differential pumping system. The inlet of the first stage is a stainless-steel capillary (2) 12 cm long with an inner

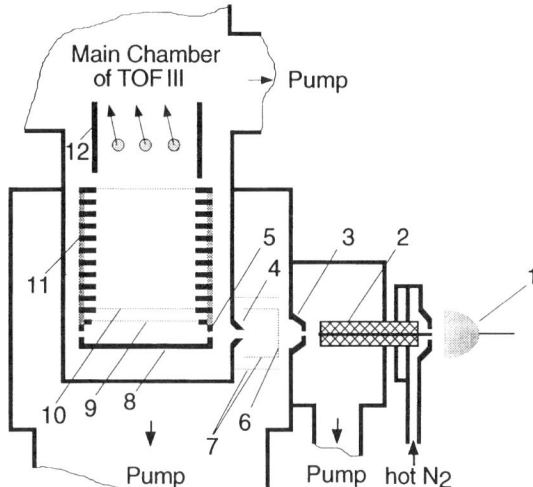

Figure 3. Schematic diagram of the ES ion source and the modulator for orthogonal injection; 1, electrospray source; 2, heated stainless-steel capillary; 3, skimmer 1; 4, skimmer 2; 5, slit; 6, focusing grid; 7, cylindrical grids; 8, plate (extraction electrode); 9 and 10, grids; 11, accelerating column; 12, deflection plates.

diameter 0.5 mm; it is pumped by a 6-L/s mechanical pump to a pressure ~ 3 Torr. The capillary is heated to $\leq 120°C$. The second stage follows a 0.15-mm diameter aperture in a cone skimmer (3) with a 3-mm diameter flat top, and is evacuated by a 450-L/s turbo pump to $\sim 10^{-5}$ Torr. The third stage, entered through a 2 mm × 6 mm horizontal slit (cut in the flat tip of a second skimmer) is connected to the mass spectrometer chamber, which is pumped by a 1000-L/s cryopump to a pressure between 2 and 3×10^{-7} Torr.

The ES source as a whole is elevated to the DC acceleration potential, normally $+4\,kV$. The relatively high pressure in the first stage inhibits discharges to the grounded mechanical pump; a 1-m length of plastic hose is sufficient to hold up to 10 kV at the 3-Torr operating pressure. The region at gas pressures between 1 and 0.01 Torr, where discharges are likely, is screened from the electric field inside the cone skimmer. The second stage runs at $\sim 10^{-5}$ Torr, so discharges to the grounded turbopump are also inhibited.

Electrospray sources for quadrupole spectrometers commonly place the skimmer inside the Mach disk (stagnated shock wave) of the free jet issuing from the nozzle. The undisturbed core of the jet then diverges very slowly and is able to carry ions into the quadrupole without the assistance of a strong electric field. In contrast to this arrangement, our skimmer is placed beyond the Mach disk, 8 mm from the nozzle, so the stream of gas passing through the skimmer has a large divergence. As a result, ions passing through the skimmer suffer fewer collisions with the gas; such collisions add undesirable momentum in the vertical direction.

The potentials applied to the source components gradually decrease (with one exception) from the spraying capillary to the final skimmer. Typical values are as follows: 3.0–3.5 kV between the ES capillary and capillary 2 to provide good conditions for spraying, with the total spray current from 0.1 to 0.3 μA; 50–350 V between capillary 2 and the skimmer, to provide controllable breakup of clusters and fragmentation of the ions; 5–10 V between the first and second skimmer and the injection electrodes, which normally are at V_{acc} (between the extraction pulses). A grid (6) between the two skimmers floated at -30 to -50 V with respect to V_{acc} provides some focusing of the ion beam without significant broadening of the velocity distribution. Possible penetration of the electrostatic field from parts of the system at ground potential is prevented by careful screening using three cylindrical grids.

C. Orthogonal Injection

The third stage of the ES source is contained in an aluminum, 127-mm cube mounted on the bottom of the spectrometer so that the ion beam is perpendicular to the vertical axis of the spectrometer, as shown in Figure 2. As mentioned above, the ions are accelerated by 5–10 V applied to the first skimmer, and a pair of horizontal slits (second skimmer 4 and slit 5 in Fig. 3) 2 mm wide and 20 mm apart restricts the vertical-beam dimension between the deflection electrodes to ≤ 4 mm and the vertical-velocity spread to $\leq 1/5$ of the horizontal-velocity spread.

The ions are injected into the spectrometer by a pulse applied to one of the horizontal extraction electrodes (plate 8). Between extraction pulses both electrodes (plate 8 and grid 9) are at the DC accelerating potential ($V_{acc} = 4$ kV), allowing the slow electrosprayed ions to fill the 25-mm long storage gap. The ratio of the filling time (determined by the velocity of the electrosprayed ions) and the time between pulses (determined by the flight times of the ions with the largest m/z value to be observed) defines the duty cycle. The velocity of the ions entering the injection gap depends on their mass because, in addition to the uniform velocity gained in the gas jet, their energy is increased by a constant amount (~ 5 eV) per charge. As this horizontal velocity remains constant during the whole flight time of the ions (in the absence of further deflection in the field-free region), the duty cycle is determined by the ratio of the length of the storage gap (25 mm) and the distance between the centers of the storage gap and the detector (11 cm). Thus, the duty cycle is $\sim 20\%$ for the ions with the largest observed m/z value, and is less for other ions.

The extraction pulse applied to the plate (8) has an amplitude of 420 V, a risetime of ~ 50 ns, and a duration of ~ 25 μs. The ion bunch is ejected through a 90% transmission grid (9) into the DC acceleration region, and then across ~ 4 kV through two more 90% transmission grids (10, 11) into the spectrometer. Electric-field penetration through grid (9) can cause ions to leak out of the storage gap during the storage phase. To prevent this, the intermediate grid (10) is held slightly above the acceleration potential ($+9$ V) between pulses; it is pulsed to ~ 400 V below V_{acc} during extraction. It is worthwhile mentioning that the risetime of the pulses does not affect the resolution significantly: ion peaks with a 3-ns width were recorded when the risetime exceeded 100 ns. The presence of a finite risetime results only in a small shift of the mass spectra.

The lengths of the pulsed and DC acceleration regions are 6, 6, and 54 mm, and the corresponding voltages are chosen so as to give a nearly uniform electric field over the three regions during the pulse. Under these conditions, ions originating at different vertical positions in the region between the plates are focused onto a horizontal plane after a free-flight path approximately twice as long as the acceleration region, that is, a plane about 19 cm above their original position (56). This plane then serves as a virtual "object plane" for the spectrometer; the axial velocity spread of the ions in the object plane is then corrected to first order by the electrostatic mirror (51). For more detailed calculations of resolution in an ES–TOF mass spectrometer, see the papers by Dodonov et al. (25,45).

The ion beam has to be directed a few degrees away from the axis of the instrument after leaving the acceleration region so that the reflected ions will strike the detector. If the initial component of ion velocity does not meet this condition, the ions must be deflected. However, the angle of deflection should be kept to a minimum for optimum resolution (25,57), because the ion packet is no longer parallel to the plane of the detector after deflection. In the first approximation, the slope of the ion packet after deflection is equal to the deflection angle of the ion trajectory, so the contribution of the deflection to the width of the peak is proportional to the applied deflection voltage.

A pair of deflection plates (12) (7 cm long, 5 cm wide, and 6 cm apart) with symmetrical applied potentials gave adequate steering of the beam without significant deterioration of the resolution (for deflection voltages ≤ 40 V). However, the best resolution in our instrument was obtained with nearly zero deflection voltages when the initial energy was reduced to 5–8 eV, consistent with the argument above. This leads to $\sim 50\%$ decrease in sensitivity, because low-energy ions are more difficult to focus into a parallel beam.

For a vertical acceleration potential of 4 kV, the instrument will accept ions with horizontal velocities in the storage region corresponding to a range of energies between 2.5 and 10 eV per charge. Light ES ions already have an energy near this range, so little additional deflection is required; usually less than 70 V is applied to the deflection plates. Because part of the energy of the ES ions comes from the gas jet, the energy, and therefore, the optimum deflection voltage has a mass dependence. For ions with m/z above a few thousand, a deflection voltage between 100 and 300 V was used. This effect imposes some limitations on the range of m/z values that can be detected simultaneously without mass discrimination, and reduces the resolution for these ions, as discussed above.

It is possible to overcome these limitations by collisional damping of ions in an rf-quadrupole ion guide at relatively high pressure (0.01–1 Torr), as introduced by Douglas and French (58) and Xu et al. (59) for quadrupole mass spectrometers. Our initial results obtained with a short rf-quadrupole placed in an additional pumping stage after the first skimmer showed that the m/z discrimination is significantly reduced (60). However, most of the measurements discussed below were done before the rf-quadrupole was installed in the instrument, so we do not discuss it in detail here.

The description above applies to the instrument built at the University of Manitoba, which is representative of other ES–TOF instruments mentioned in the Introduction (23–33). These are based on the same principles and differ from ours mainly in the design of the ion source and the atmosphere–vacuum interface, the type of electrostatic mirror used (if any), the method of recording mass spectra (transient recorder or TDC), voltages, and geometrical parameters. Some of the characteristic features of these mass spectrometers are outlined in the next section, in which the results obtained in the corresponding instrument are discussed.

D. Chemicals

Solutions of peptides and proteins with concentrations from 10^{-6} to 10^{-5} M were prepared from deionized water (50–100%) and reagent-grade methanol (0–50%). Glacial acetic acid (99.9985% purity, Alfa Æsar) was added to solutions in studies of peptides and proteins from denaturing conditions; aqueous solutions of ammonium acetate (99.999% purity, Aldrich Chemical Co.) or ammonium bicarbonate (Fisher Scientific Co.) were used in studies of higher-order structures. The compounds measured were obtained from Sigma Chemical Co., unless otherwise indicated.

III. INSTRUMENT PERFORMANCE

A. Mass Resolution

A direct comparison of the resolution in the linear and reflecting modes of our instrument has been carried out under constant-ion-source conditions (31). The time focusing for each mode was optimized independently. In the linear mode, the best resolution was obtained when the ions were focused onto an object plane at the detector position; this required a small extraction pulse (~ 180 V). The resolving power in this case never exceeded 500. However, no attempts were made to optimize the source geometry for the operation in the linear mode; higher values of resolution ($R \geq 1000$) have recently been measured by Dresch et al. (46,47) in a short linear instrument.

For reflecting-mode operation, the extraction pulse was increased to 420 V to bring the object plane close to the extraction region. Under this condition, the optimum setting of the mirror voltage (4.6 kV) was found to be close to the full accelerating voltage, so the ions spent enough time in the mirror to obtain full velocity correction (51). As a result, the resolving power was greatly improved (to $M/\Delta M_{\text{FWHM}} > 5000$). Consequently, all the experiments discussed below were done in the reflecting mode.

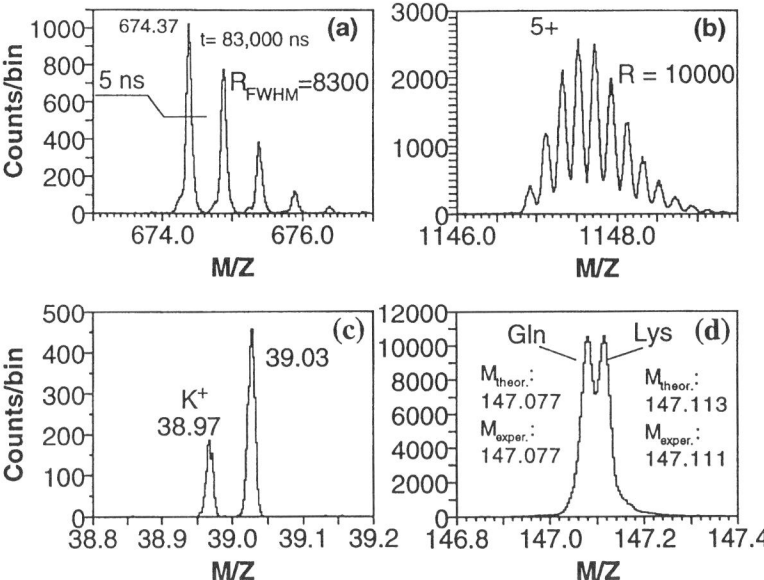

Figure 4. Resolution in the reflecting mode: (*a*) doubly charged ion of substance P (1346.73 Da); (*b*) segment of the mass spectrum of bovine insulin (5732.6 Da); recorded with the rf-quadrupole ion guide as described in ref. 60; (*c*) background ions with $m/z = 39$; (*d*) partially resolved peaks of the amino acids glutamine and lysine having the same nominal mass.

Several spectra shown in Figures 4–6 demonstrate the resolution measured in the reflecting mode of our instrument. The largest values of resolution between 8000 and 10,000 were obtained for peptides with mass 1000–6000 Da (Fig. 4*a* and *b*). In these cases, small or zero voltage was applied to the deflection plates as discussed above, and 50–70% of the ion current to the detector was thus sacrificed. However, the typical resolution was ~4000, even when conditions were optimized for maximum sensitivity. This resolution is sufficient to distinguish separate isotopic peaks for most peptides, and hence to determine their individual charge states. It is particularly useful in interpreting complicated spectra from mixtures such as tryptic digests or spectra resulting from fragmentation of multiply charged ions. For example, Figure 5*a* shows the spectrum of a tryptic digest of a recombinant protein IclR of mass 29,608 Da, prepared by H. W. Duckworth and L. Donald of our Chemistry Department. Trypsin cleaves at arginine and lysine, and a fragment normally contains only one of these basic residues; but in this case the protein contained 13 histidines and the product masses ranged up to nearly 4000 Da. Thus, there was a reasonable probability of obtaining multiple charges on the fragments (up to 6+ as it turned out). Figure 5*b* shows an expanded view of a small section of the spectrum, where the unit mass resolution enables the charge on each fragment to be determined.

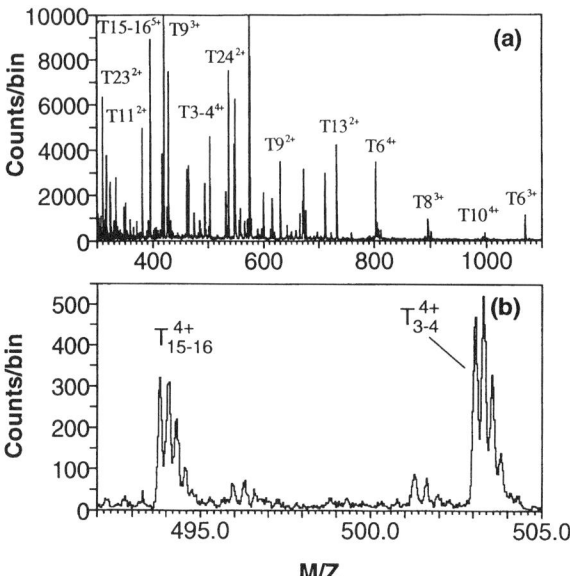

Figure 5. (*a*) The *m/z* spectrum of a tryptic digest of the IclR recombinant protein (29,608 Da) and (*b*) the expansion of a small *m/z* region of this spectrum.

For smaller ions the resolution decreases, but it remains sufficiently high to resolve two or more peaks within the same nominal mass, as demonstrated in Figures 4c for $m/z = 39$ and in Figure 4*d* for a pair of amino acids, glutamine and lysine. Peaks in this mass range are often used as the first peak for mass calibration, and misinterpretation of an unresolved peak (e.g., K^+) may lead to a large error in mass assignments. This is another advantage of TOF mass spectrometers over quadrupoles, where the resolution usually increases linearly with mass in this range.

For large masses ($M > 6000$), the effective resolution is determined by the width of the unresolved isotopic distribution. In cytochrome c (12,360 Da), for example (not shown), we have measured an apparent resolving power ~1500. Larger proteins could in principle give higher resolution owing to the smaller relative width of the isotopic pattern, but the observed values are usually limited either by the presence of unresolved adducts or by the heterogeneity of the proteins. The best results in our instrument have been achieved for human apotransferrin (Calbiochem, San Diego, CA), MW = 79,550 Da, which was sufficiently homogeneous and pure to give a resolution $m/\Delta m_{FWHM}$ close to 2000 (Fig. 6*a*).

B. Accuracy of Mass Determination

Mass calibration of ESI TOF spectra can be obtained with the monoisotopic peaks of the doubly and singly charged ions from Substance P (1346.73 Da). The

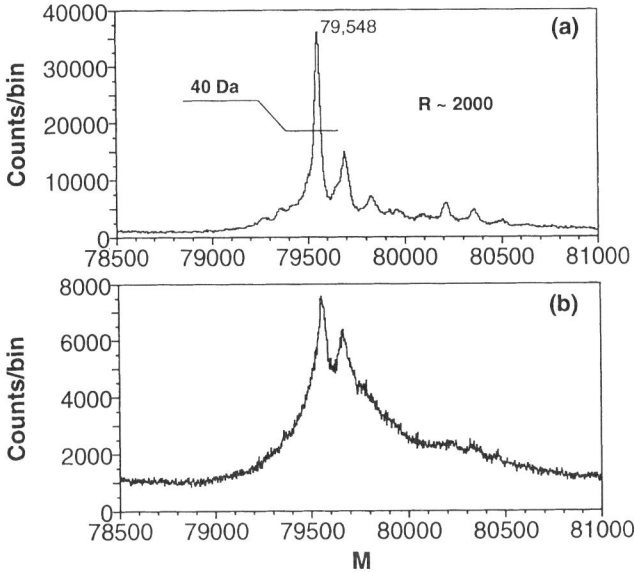

Figure 6. (*a*) Deconvoluted mass spectrum of human apotransferrin obtained from charge states 30–36; (*b*) mass spectrum recorded in the same conditions as *a* but at a higher pressure (1×10^{-6} Torr instead of 0.3×10^{-6} Torr) in the main chamber of the spectrometer.

good separation of the isotopic peaks for this compound provides high accuracy; the mass accuracy for small ions ($M < 2000$) is 10–20 ppm with the internal calibration. In some cases, this allows one to confirm the elemental composition, or to distinguish between two peaks of the same nominal mass as shown in Figure 4*c* and *d*. For large ions ($10,000 < M < 80,000$) and external calibration, the accuracy of mass determination is ~ 100 ppm. Mass calibration of the ions of Substance P was quite accurate, even when the ions of interest were far from the m/z interval determined by the two reference peaks: the molecular weight of human apotransferrin determined from the spectrum shown in Figure 6*a* differs from that determined by Feng et al. (61) by only 8 Da.

C. Sensitivity and Dynamic Range

The geometry of our ES–TOF instrument was chosen as a compromise between better sensitivity and better resolution, with emphasis on the latter: the pair of 2-mm-wide slits and the large distance between the skimmer orifice and the storage region (~ 80 mm) reduce the velocity and spatial spread of ions, as well as the ion current to the drift region. However, the instrument sensitivity was still comparable to that for most quadrupole instruments, or perhaps

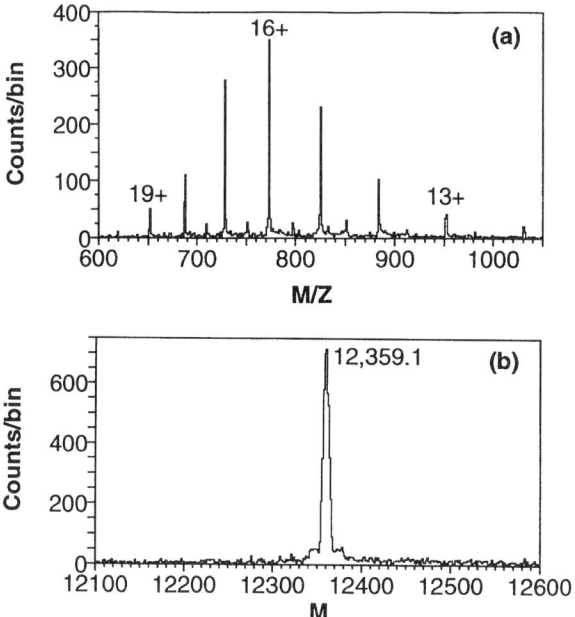

Figure 7. (a) The m/z spectrum and (b) deconvoluted spectrum recorded for 5×10^{-7} M solution of cytochrome c (12,360 Da) at a flow rate of 0.25 μL/min; 2 fmole were consumed in 1 s recording time.

somewhat better. Figure 7a shows a spectrum of cytochrome c (12,360 Da) recorded from a water–methanol solution at a protein concentration 5×10^{-7} M. The spectrum was recorded for 1 s, during which 2 fmol of protein were consumed. This high sensitivity is a result of the high duty cycle of pulse formation ($\sim 20\%$), nearly complete transmission of the pulsed beam through the TOF spectrometer, and simultaneous detection of all ions (i.e., no scanning). Figure 7b shows the deconvolution of the same spectrum of cytochrome c made from the charge states $z = 14$–18, allowing determination of the mass with an accuracy ± 1 Da.

The dynamic range of the instrument is closely related to the sensitivity. Figure 8a shows the singly charged molecular ion of methionine enkephalin (573.22 Da), where the spectrum was recorded for 1 min and 3 pmol were consumed. The first four isotopic peaks are visible in the spectrum. When the vertical scale is expanded by a factor of 150, as shown in Figure 8b, the fifth and sixth isotopic peaks appear, where the latter has an intensity ~ 2000 times lower than the first isotopic peak (i.e., it corresponds to 1.5 fmol). This large dynamic range is a benefit of the single-ion-counting capability of the recording system.

Recently, we increased the sensitivity still further by the introduction of the "nanospray" technique (54), as discussed below.

III. INSTRUMENT PERFORMANCE 217

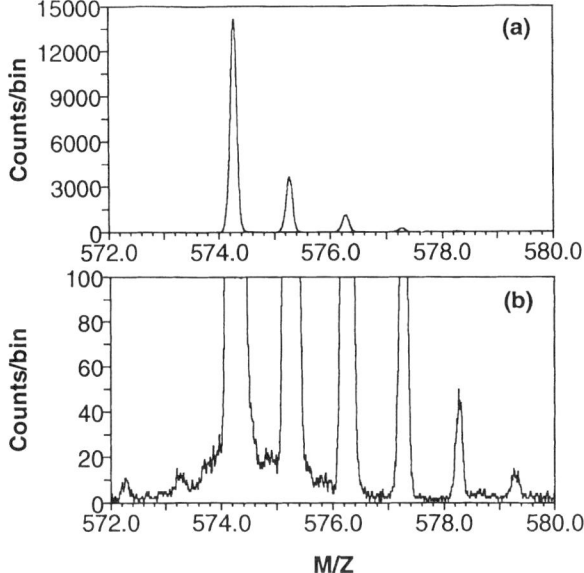

Figure 8. The m/z spectrum of methionine enkephalin (573.22 Da) accumulated for 1 min. The molecular ion is shown (*a*) on a normal scale and (*b*) with the vertical scale expanded by a factor of 150.

D. Secondary Electron-Resolved Mass Spectrometry

The ion-counting technique gives us one more possibility that can be very useful for biochemical applications. It is known that the secondary electron yield of ions hitting a detector increases with mass for a given velocity (62). Thus, ions with different masses but the same mass-to-charge ratio (i.e., the same velocity) can be resolved by changing the threshold of the discriminator. The method has already been used in Uppsala and called "secondary-electron resolved-mass spectrometry" (63,64), and has been reinvented independently by several other groups (65–67). We have found that the method is useful both for reducing low-mass chemical noise in the spectra of proteins (66) and for reducing the "fragmentation" noise in the case of collision-induced dissociation of proteins in the vacuum chamber (68). Mass spectra of a mixture of methionine enkephalin, leucine enkephalin, bovine insulin, and albumin recorded at two different thresholds of the constant-fraction discriminator are shown in Figure 9*a* and *b*. Nearly all ions hitting the detector are recorded with the 10-mV threshold (Fig. 9*a*). At a higher threshold (200 mV), the intensity of the peptide peaks changes dramatically, while 70% of the protein ions are still recorded (Fig. 9*b*). This feature is especially valuable when the protein sample under investigation contains low-mass impurities, as demonstrated in ref. 66 for spectra of an old sample of the above mentioned IclR-protein. The initial spectrum was spoiled badly by the presence of chemical noise, including both low-mass impurities and

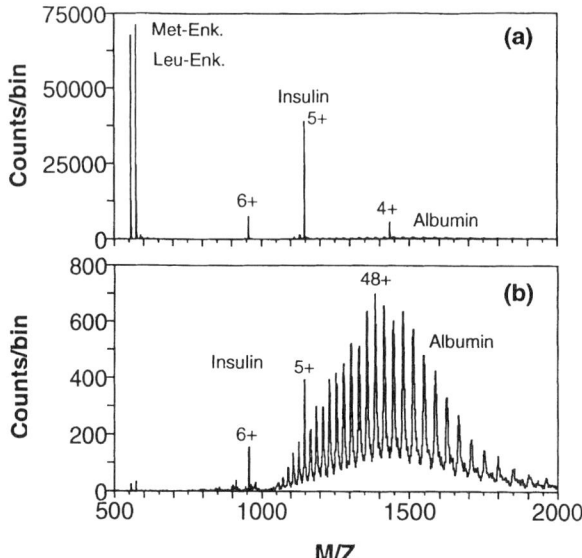

Figure 9. Mass spectra of a mixture of methionine enkephalin, leucine enkephalin, bovine insulin, and albumin recorded at two different thresholds of the constant-fraction discriminator: (*a*), 10 mV; (*b*), 200 mV.

protein fragments which fell into the same m/z range as ions of the protein itself. To avoid recording these low-mass ions, the spectrum was also recorded with a higher threshold. Increasing the threshold to 200 mV led to a tremendous loss of counts, but improved significantly the quality of the mass spectrum. In this technique, spectra with low and high thresholds are recorded simultaneously using two discriminators, so separate measurements are not necessary. Note that single ion counting is necessary for the procedure to be fully effective.

E. Mass/Charge (*m*/*z*) Range

In principle, the m/z range of a TOF spectrometer is unlimited; the only restrictions arise in ion production and detection. As noted in the Introduction, the ions produced by ES sources usually lie within the m/z range of most quadrupole spectrometers (≤ 2400). However, an increasing number of exceptions to this rule are appearing; many examples are given in Section V for solutions under near-physiological conditions. Even under denaturing conditions, the m/z ratio can sometimes be large; for example, Figure 10*a* shows the m/z spectrum of a monoclonal antibody anti-(human α_1-acid glycoprotein) of molecular weight close to 150,000 Da in 7% acetic-acid solution. The protein solution was filtered by centrifuging through an Ultrafree 30,000 NMWL filter (Millipore) prior to spraying. The spectrum has a maximum at about m/z 3000, and ions (including those of dimers) are observed up to $m/z \sim 7000$. Deconvolu-

IV. COUPLING ES–TOF MS WITH SEPARATION TECHNIQUES

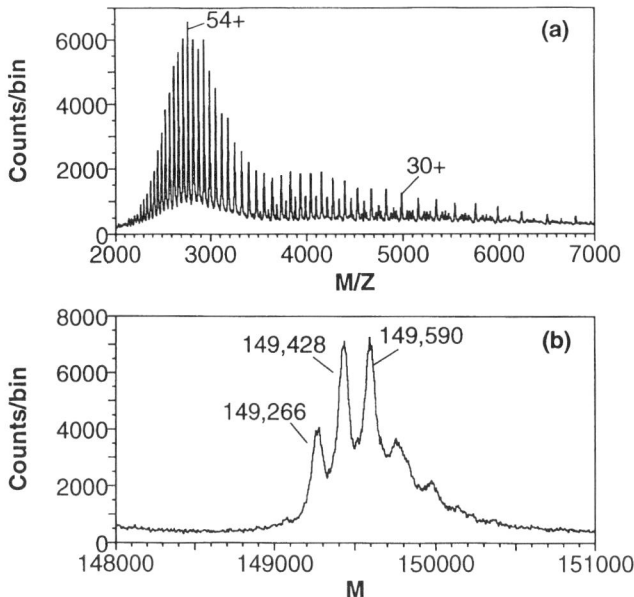

Figure 10. The (*a*) m/z spectrum and (*b*) deconvoluted spectrum of murine anti-(human α_1-acid glycoprotein) monoclonal antibody recorded from a 5×10^{-6} M solution in water with 7% acetic acid. Deconvolution was obtained from charge states 45–60.

tion of the spectrum (Fig. 10*b*) shows several partially resolved peaks with a mass difference of 162 Da, corresponding to the different degrees of glycosylation.

IV. COUPLING ES–TOF MS WITH SEPARATION TECHNIQUES

Separation systems such as liquid chromatography (LC) and capillary-zone electrophoresis (CZE) can be coupled on line with ESI sources to allow mass spectrometric detection and analysis of individual species in a mixture, as described in this volume. However, to achieve the full benefits of such combinations, the mass analyzer should have high sensitivity, adequate mass resolution, and a short characteristic scan time. Time-of-flight instruments seem to offer most of these desirable features. The results obtained by two other groups are discussed in this section: Boyle et al. from Analytica of Branford and Verentchikov et al. from PerSeptive Biosystems.

A. Capillary Electrophoresis at Analytica of Branford

Capillary electrophoresis (CE) allows a complete separation to be performed in a few minutes with no loss in component resolution. Because of the small sample volumes required (a few nanoliters or less), CE is quickly becoming the analytical tool of choice for applications which are truly sample limited. Single-constituent effluence peak widths of one second or less have been demonstrated by optical-detection methods. Conventional scanning-mass spectrometers as detectors for CE are reaching their performance limits whenever a spectral mass range exceeding a few hundred Daltons has to be scanned at a rate exceeding one spectrum per second. The TOF instrument developed at Analytica of Branford (28) is capable of recording more than 10 full spectra per second with minimal dead time, each spectrum being a sum of multiple (several thousands) single scans. It was designed as a short linear instrument of the Wiley–McLaren type (56), with a drift path of 37 cm. Mass resolution of $R_{FWHM} = 1000$ has been achieved in this short linear configuration, and $R_{FWHM} = 3000$ when equipped with an ion mirror and an extended drift path of 130 cm. The CE column was interfaced recently to the ES ion source (46,47) by means of the sheath-flow mechanism introduced by Smith et al. (69); in this setup, a sheath liquid makes the electrical contact to the CE buffer solution and at the same time provides for a stable ES process.

Figure 11 gives a comparison of total ion currents (TIC) acquired for a separation of standard peptides with two instruments: a quadrupole running at its smallest possible scan time of 2 s/spectrum, and the TOF instrument with an integration time of 0.125 s/spectrum. With the quadrupole, the first three peaks

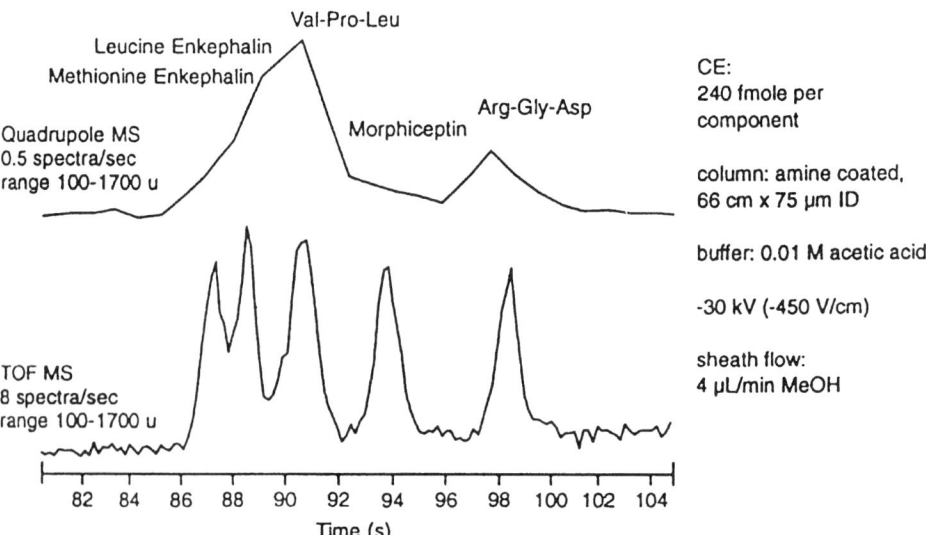

Figure 11. Comparison of the TICs from a peptide separation using the TOF and quadrupole instruments. (Presented at the 43rd ASMS Conference on Mass Spectrometry and Allied Topics, Atlanta, May 21–26, 1995 by T. Dresch et al., see ref. 46. Reprinted by permission.)

appear to have comigrated, whereas the morphiceptin peak is missing entirely, because it elutes completely while the mass spectrometer is slowly scanning unrelated regions of the m/z spectrum. The effects of the liquid-sheath flow rate, electric field strength, and integration time on sensitivity and peak shape have been examined in this instrument for separation of peptide and protein mixtures (47). A sensitivity limit of 10 fmole has been established for leucine enkephalin.

B. On-line LC/LC/MS at PerSeptive Biosystems

On-line LC/LC/MS experiments with in-column tryptic digestion, reverse-phase chromatographic separation, and orthogonal-injection TOF mass spectrometry have been carried out recently (33). The LC/LC was done on an "Integral" analytical workstation (PerSeptive Biosystems), which can automatically control multicolumn experiments. A sample of 500 pmole of horse-heart myoglobin was taken from the autosampler and inserted into the column with the immobilized trypsin. After a 30-min residence time, the sample was eluted and captured onto a dilution-capture column. Following a brief desalting, the sample was captured onto a 1-mm C18 Vydac column. The separation was performed at a flow rate of 50 μL/min and a gradient from solvent A (0.1% TFA in water) to solvent B (80% acetonitrile, 0.1% TFA in water) was reached in 75 min. The eluent was passed directly into the ion-spray source coupled to a breadboard TOF mass spectrometer with an effective drift path of 1.5 m. The data-acquisition TDC board (Diamond Schmidt) provided 2.5 ns resolution at a pulsing rate of 10 kHz. Each spectrum represents an average of 1000 transients.

Figure 12a shows a segment of a total-ion chromatogram recovered from mass spectra in the m/z range from 450 to 1500. An expansion of the peak of the T10 + T11 fragment is shown in Figure 12b. The acquisition rate of 1 spectrum/s considerably exceeds the normal requirements for LC separations for a typical peak width from 10–15 s: the rule "10 spectra per chromatographic peak" applicable to scanning instruments is not valid here because a full spectrum is acquired every 100 μs, thus eliminating mass discrimination. Increasing the acquisition rate decreases the number of ion counts in each spectrum; however, even a single spectrum (Fig. 12c) provides good mass accuracy ($\sim$ 30 ppm for all peptides) and a clear isotopic pattern (Fig. 12d). To recover an isotopic pattern and good mass accuracy for a smaller sample, one has to average all spectra across the corresponding chromatographic peak. The minimum necessary amount of myoglobin was calculated as $\sim$ 10 pmol.

There may be a substantial difference between the sensitivities in the LC/MS and sample-infusion experiments owing to the large flow rates in LC. The infusion experiments illustrated by Figure 13 were carried out on a similar, but $\sim$ 10 times more sensitive instrument, with a 1.1-m effective drift path. The microspray source was driven by a syringe pump to maintain a constant flow rate of 60 nL/min. Figure 13a shows a spectrum of a 10^{-6} M neurotensin solution recorded for 0.05 s, (i.e., consuming only 50 mol of neurotensin). The expansion for a doubly charged ion reveals a clear isotopic pattern.

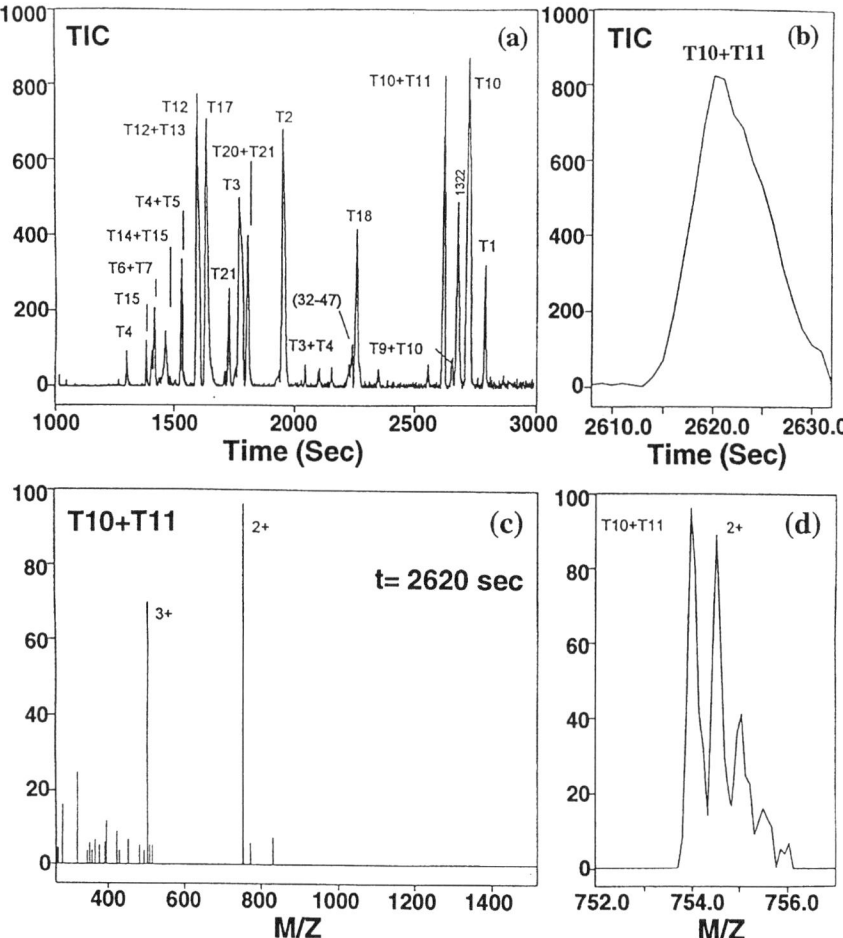

Figure 12. (*a*) Total ion chromatogram of the myoglobin tryptic peptides recovered from ESI mass spectra; (*b*) an expansion of the T10 + T11 fragment peak; (*c*) ESI mass spectrum recorded at $t = 2620$ s of the chromatogram shown in *a*; *d* an expansion for a doubly charged ion. (Presented at the 43rd ASMS Conference on Mass Spectrometry and Allied Topics, Atlanta, May 21–26, 1995 by A. Verentchikov et al., see ref. 33. Reprinted by permission. © 1995 ASMS.)

V. STUDIES OF NONCOVALENT ASSOCIATIONS OF BIOMOLECULES

Most ESI–MS studies with proteins have been carried out in acidic solutions. Such solutions, particularly in combination with organic solvents, denature biopolymers, which then acquire sufficient charge to be detected within the limited m/z range of most commercial quadrupole mass spectrometers. This environment is usually sufficiently unfriendly so as to destroy any higher-order

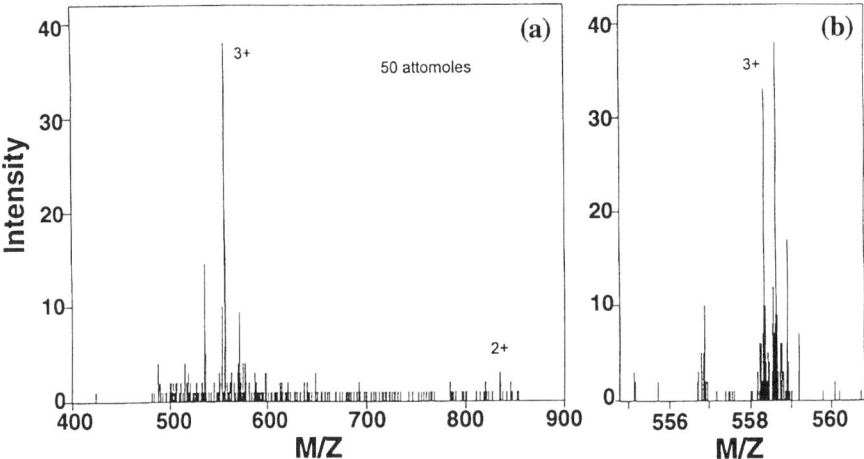

Figure 13. (*a*) The ESI mass spectrum of a 10^{-6}-M neurotensin (1673 Da) solution recorded for 0.05 s ; (*b*) an expansion for a doubly charged ion.

structure, which must be examined closer to physiological conditions. However, it has been shown recently that careful control of these conditions, combined with the gentle nature of the ES technique, enable study of noncovalent interactions (e.g., see refs. 18,19,70). In these studies, the charge acquired by the complex may be much lower than the charge observed under denaturing conditions, resulting in m/z values beyond the range of most quadrupole instruments or ion traps. The simple example of carbonic anhydrase shown in Figure 14 illustrates the difference in charge-state distributions obtained from (*a*) acidic and (*b*) near-neutral solutions. Note that in the latter case carbonic anhydrase was recorded together with its cofactor Zn, while the Zn has been expelled in the acidic solution.

Thus, the unlimited m/z range of the TOF spectrometer together, with the mild ionization of ES, provide a powerful tool for observation of the noncova-

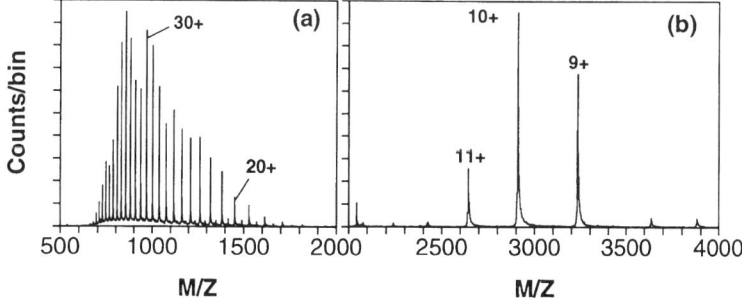

Figure 14. The ESI mass spectra of carbonic anhydrase II recorded from (*a*) a solution containing 2% acetic acid (pH ~ 2); and (*b*) a 6-mM solution of ammonium acetate (pH ~ 6).

lent associations of biomolecules. About 30 of these interesting entities have been studied in our instrument, in collaboration with other groups, and some results have been published (20,71–74). Most of our collaborations have arisen from the inability of other instruments and techniques to solve specific biochemical problems. In the following sections, we give a brief overview of some of these studies, discussing them from the instrumental rather than the biochemical point of view.

A. Complexes of Enzymes

It is well known that naturally occurring protein inhibitors of proteinases form complexes with the enzyme at which they are targeted. Investigations of such complexes may give insight into the mechanism of inhibitor action. Pepsin and its inhibitor pepstatin form a complex for pH values in the range 2 to 4. Porcine pepsin itself is difficult to observe by ESI–MS in the positive mode because it has only 4 basic amino-acid residues and 42 acidic ones. Nevertheless, charge states up to $z = 11$ are present in its spectra, which implies that not only Arg, Lys, or His residues are protonated in this molecule, but also some others, probably Gln. The declustering voltage ΔU has little effect on the spectra of pepsin itself, but the situation changes when pepstatin is added to the same solution (Fig. 15). A weakly bound complex is then formed between the two molecules (Fig. 15a), which is dissociated easily by increasing ΔU from 40 to 70 V (Fig. 15b).

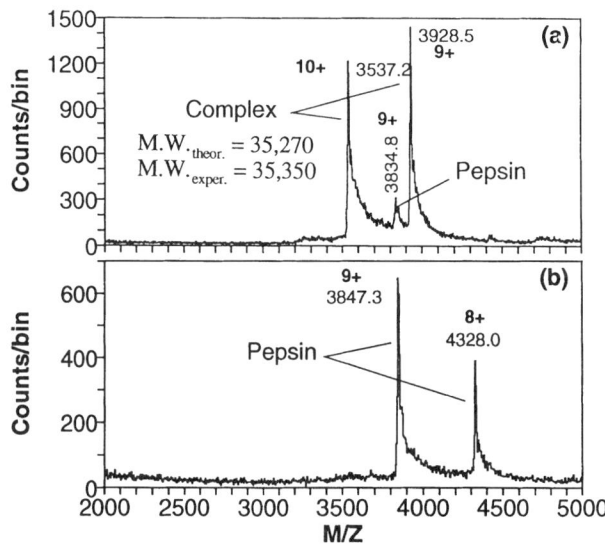

Figure 15. The ESI mass spectra of a noncovalent complex formed between porcine pepsin (34,584 Da) and its inhibitor pepstatin (685.5 Da) recorded at (a) $\Delta U = 40$ V and (b) $\Delta U = 70$ V between capillary ("nozzle") and skimmer; concentrations were 2×10^{-5} M for pepstatin and 6×10^{-6} M for pepsin; pH ~ 3.5.

Another enzyme studied, 4-oxalocrotonate tautomerase (4OT), was prepared by total chemical synthesis at the Scripps Research Institute (71,72). This enzyme has a monomer mass of 6810 Da. Preliminary gel-perfusion measurements (75) indicated that the enzyme was a pentamer in solution. However, the recently solved X-ray structure (76) showed the enzyme as a hexamer in the crystalline state; it was not clear whether this describes the actual situation in solution. The ESI mass spectrum of 4OT (Fig. 16a) shows the presence of monomer, *hexamer*, and some dimer ions from near-neutral solution (pH ~ 7.5), establishing the oligomerization state in solution as that seen by X-ray crystallography. The hexamer disappeared from the solution when its pH was reduced to ~ 4 (not shown). In Figure 16, spectra of 4OT are shown as a function of the declustering voltage ΔU. It is interesting that hexamer ions disappear at $\Delta U = 150$ V, but the pentamer is relatively stable up to 300 V. This is difficult to reconcile with the X-ray studies that picture the hexamer as a trimer of dimers, but we should keep in mind that structure and magic numbers may possibly be different in vacuum from those in the solids. The mirror-image D form (72) and several mutants of this enzyme (71) have also been studied in our instrument.

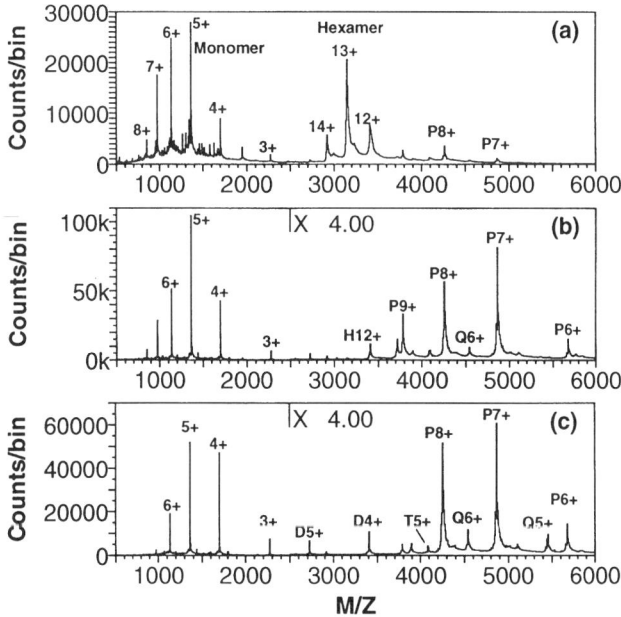

Figure 16. The ESI mass spectra of the enzyme 4-oxalocrotonate tautomerase (4OT, monomer MW = 6810.5 Da) recorded from 5-mM aqueous solution of ammonium bicarbonate (pH 7.5) at different voltages ΔU: (*a*), 120 V; (*b*), 180 V; (*c*), 250V. H, Hexamer; P, pentamer; Q, tetramer; T, trimer; and D, dimer. Synthesized by M. Fitzgerald and S. B. H. Kent, Scripps Research Institute, La Jolla, CA (71,72).

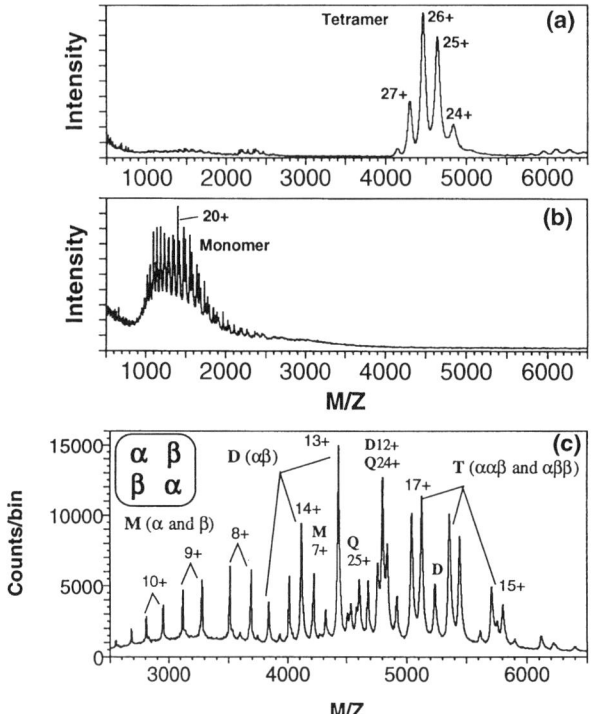

Figure 17. The ESI mass spectra of native SBA with different concentrations of acetonitrile in solution: (*a*), 0%; (*b*), 30%. Concentration of protein—2×10^{-6} M in 0.02% formic acid in water; capillary-skimmer potential $\Delta U = 60$ V. (*c*) The ESI mass spectrum of SBA-1 at high capillary-skimmer potential ($\Delta U = 160$ V). M, monomers; D, dimers; T, trimers; and Q, tetramers. Inset: spatial arrangement of the subunits α and β in SBA-1. Samples prepared by X. J. Tang, C. F. Brewer and S. Saha, Albert Einstein College of Medicine, New York (20).

B. Noncovalent Interactions in Soybean Agglutinin

Soybean agglutinin (SBA) is a noncovalently bonded tetrameric glycoprotein with a molecular weight of $\sim 116,000$ Da, which has recently been examined in a triple-quadrupole mass spectrometer (77). During the study, it was found that no protein ions could be detected for aqueous SBA. However, intense signals corresponding to the SBA monomers were observed for SBA solutions containing 25% or more acetonitrile. It was suspected that this dependence on the acetonitrile content was caused by a change of the quaternary structure of SBA to give ions beyond the m/z range of the quadrupole instrument. To prove this, SBA was studied in our ES–TOF MS (20). Figure 17*a* and *b* show the spectra obtained from the aqeous SBA solution titrated with 0% and 30% acetonitrile. Since the ESI–MS conditions were kept the same and only tetramers were observed in pure water, these data provide strong evidence that the SBA tetramer species observed arise from specific noncovalent association of the subunits in solution.

To probe the quaternary structures of the tetramers, SBA-I was chosen for further study because it is a relatively simple tetramer consisting of only two different subunits in equal abundance, called α and β. Figure 17c illustrates the mass spectrum obtained from aqueous SBA-I by increasing the capillary-skimmer potential to 160 V. The ions then observed are daughters of the multiply charged tetramers, the dominant ions in the spectrum of Figure 17a. Several important pieces of information can be extracted from the fragment-ion spectrum of the tetramers. For example, it was found that nearly all the dimer ions consist of one α and one β subunit; there were almost no dimer ions consisting of pure α or pure β subunits. This result suggests that the α–β interactions are much stronger than α–α or β–β interactions. The spatial arrangement of the subunits in SBA-I is depicted in the inset of Figure 17c. It is not clear why the interactions between the intact subunit and the C-terminal truncated subunits are stronger than the interactions between two identical subunits, but this is consistent with the observation that all tetrameric isomers found in native SBA consist of both intact and C-terminal truncated subunits. Initial experiments with the refolded SBA-I tetramers (73) showed that new complexes are unlikely to be generated in the refolding process. Only a small fraction of $\alpha\beta_3$ was formed, and no $\alpha_3\beta$ was detected after refolding. Moreover, the decay rate of $\alpha\beta_3$ was found to be much larger than that of normal SBA-I ($\alpha_2\beta_2$).

C. Immunoconjugate F(ab′)$_2$-CPG2

Another complex studied in our ES–TOF instrument was a site-specific immunoconjugate F(ab′)$_2$-CPG2, prepared from an F(ab′)$_2$-like fragment of the monoclonal anti-CEA murine IgG1 A5B7 and a mutant of the dimeric enzyme carboxypeptidase G2 (74). The ions of the complex were recorded with $m/z > 6000$ and they were the dominant ions in the spectra (Fig. 18). The mass determined experimentally is in fairly good agreement with that predicted by theory. The complex was observed at pH ~ 6, and it falls apart at pH < 4.5.

As we go to larger masses, a question arises: what limitations we may expect for larger complexes? Dodonov et al. (25) noticed that the resolution for higher masses decreased in their instrument, which operated at the relatively high vacuum-chamber pressure of 3×10^{-6} Torr; ions of compounds with $m > 30,000$ Da could never be observed. We have recently studied the influence of ion-molecule collisions in the vacuum chamber of the instrument on the mass spectra of proteins (68). This influence is illustrated in Figure 6 by two spectra of human transferrin recorded at different pressures in the vacuum chamber; both resolution and sensitivity deteriorate at higher pressure. We have determined that this is a result of collision-induced dissociation of ions in the field-free region. For higher masses, we might expect an even larger effect.

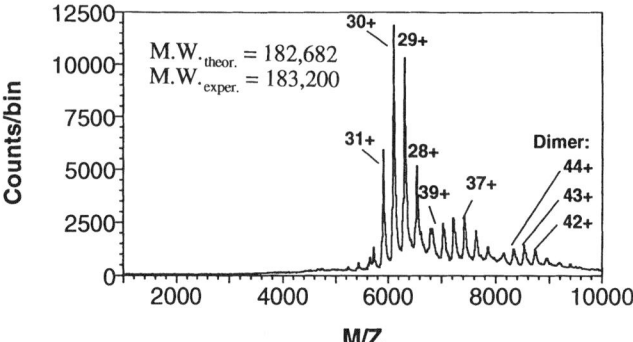

Figure 18. The ESI mass spectrum of immunoconjugate F(ab′)2-CPG2; concentration, 2×10^{-6} M; $\Delta U = 200$ V. Sample prepared by R. C. Werlen and M. Lankinen, Centre Medical Universitaire, Geneva, Switzerland (74).

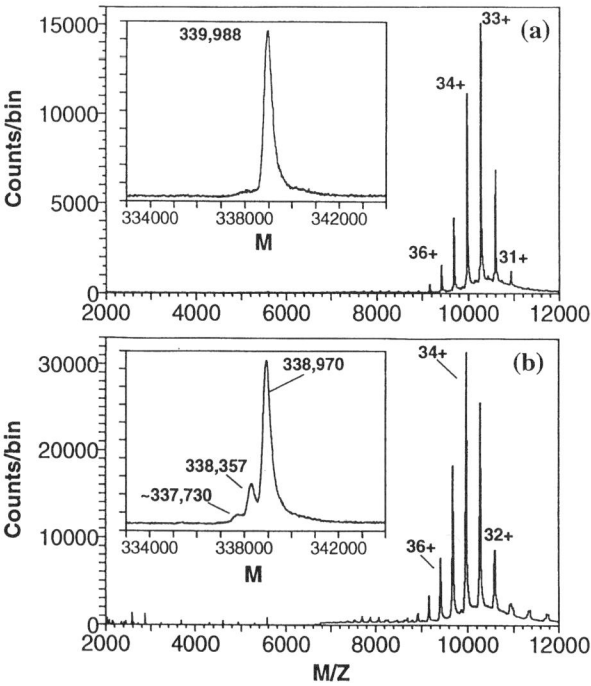

Figure 19. (*a*) The ESI mass spectrum and deconvoluted spectrum (inset) of catalase HP II (339,100 Da) recorded at increased pressure in the main chamber of the ESI–TOF instrument (1.3×10^{-6} Torr) and $\Delta U = 250$ V; (*b*) ESI mass spectrum and deconvoluted spectrum (inset) of catalase HP II recorded at $\Delta U = 350$ V. Sample prepared by P. C. Loewen, Department of Microbiology, University of Manitoba, Winnipeg, Canada.

D. Catalase HP II

The heaviest complex we have studied is catalase HP II from *E. coli*, a tetrameric protein with mass 339,100. To our surprise, the spectrum of catalase (Fig. 19a) did not change when the pressure in the vacuum chamber was increased: both resolution and sensitivity remained the same. This observation implies that the cross section for collision-induced dissociation is not entirely determined by the dimensions of the ion, and it gives us some hope of recording spectra of even higher masses. It is also worthwhile to emphasize that the charge-state distribution of catalase is centered around $m/z \sim 10,000$. The narrow peaks of HP II allow determination of the mass with an accuracy of 0.03%. The tetramer of HP II is known to be very stable: it can be broken into monomers only by boiling with SDS. The results of our attempts to break it by collision-induced dissociation are shown in Figure 19b. The few additional peaks which appeared at the highest declustering voltage ($\Delta U = 350$ V) are those corresponding to the loss of one or two hemes.

We have chosen the HP II tetrameric complex as a test compound to investigate the advantages of the "nanospray" process (54). This technique was shown to be very efficient; ion current intensities similar to those in a normal spray can be obtained with nanospray at a flow rate of less than 25 nL/min. However, our initial attempts to apply nanospray for studying ions from near-neutral pH were totally unsuccessful: only broad humps were observed in all spectra. Such humps usually appear when the concentration of ionic impurities in a solution is high. Thus, we suspected that a high concentration of sodium ions might be present in the solutions owing to their contact with the glass surface and the high surface-to-volume ratio in the microcapillary. It turned out that washing the capillary with ~ 5 μL of 5 mM ammonium acetate was sufficient to remove most of the sodium. A segment of an HP II spectrum recorded in the same conditions as those in Figure 19b, but with nanospray, is presented in Figure 20. A 3-μL sample with protein concentration $\sim 10^{-5}$ M was loaded into a gold-plated capillary with ~ 2 μm i.d. at the tip. This amount is enough to spray

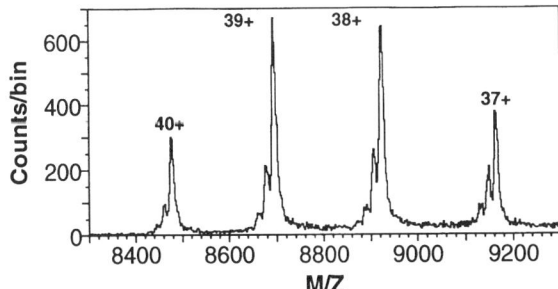

Figure 20. The ESI mass spectrum of catalase HP II recorded with nanospray; 10^{-5} M solution in 5 mM ammonium acetate; 3 μL (30 pmole) injected; $\Delta U = 350$ V.

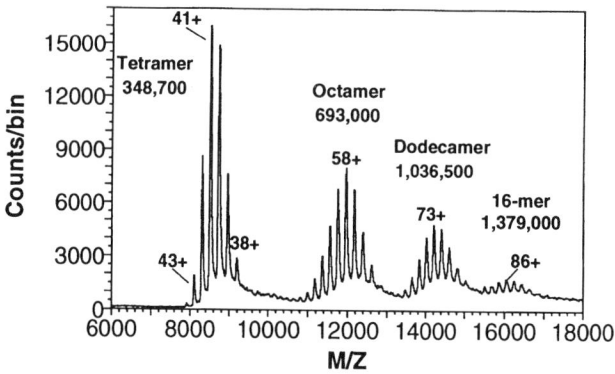

Figure 21. The ESI mass spectrum of clusters of catalase HP II recorded at $\Delta U = 100$ V with nanospray.

solution continuously for many hours. The spectrum shown was recorded in 10 s, and less than 50 fmol were consumed.

At low values of the declustering potential, clusters of the tetramer appear in the spectrum (Fig. 21), presumably owing to less specific association of the tetramers. The mass of some of these clusters is more than 1 MDa and the m/z reaches 17,000.

VI. CONCLUSION

The high resolution, sensitivity, and mass accuracy of ES–TOF mass spectrometers may find practical applications in many areas. Although the geometry of our ES–TOF instrument was chosen by compromising between better sensitivity and better resolution, with emphasis on the latter, the instrument attains a femtomole level of sensitivity. Unimolecular decay processes are almost entirely completed in the ES source before the ions are injected into the TOF spectrometer, so the observed spectra are very clean, even when there is appreciable excitation of the ions. High resolution (up to 10,000) is particularly useful in interpretation of complicated spectra from mixtures such as tryptic digests or spectra resulting from fragmentation of multiply charged ions. The mass accuracy for small ions is sufficient in some cases to confirm elemental composition or to distinguish between two peaks of the same nominal mass.

The high m/z range and sensitivity of the instrument, together with the mild ionization conditions of ES, are ideal for investigating noncovalent associations of biomolecules. Most of our difficulties in these studies arise not from the instrumental performance, but from the preparation of samples: purification and choice of solution conditions. The presence of impurities (salts, acids, etc.) was especially critical when the solutions under investigation were at pH 6–8: a large number of unresolved or partially resolved adducts is then observed in the spectra. Unfortunately, many of these adducts cannot be removed either by mechanical filtering of solutions or by increasing the declustering potential, because the complex itself may be too unstable to survive declustering. However, some water and/or buffer molecules may become an integral part of a complex, so the complex cannot exist without them. Therefore, the accuracy of mass determination for noncovalent associations is always poorer than that for proteins alone, and the experimentally determined mass may be larger by several hundred daltons than the calculated value. Application of the nanospray technique in studies of noncovalent associations provides higher sensitivity and more stable spraying of almost any solutions that are important in practical applications.

ACKNOWLEDGMENTS

This work was supported by grants from the U.S. National Institutes of Health (GM-30605) and from the Natural Sciences and Engineering Research Council of Canada. We are grateful to Anatoli Verentchikov, who participated in this work at its early stages. We thank Andrew Krutchinsky for assistance in creation of the rf-quadrupole ion guide, with which some of the data presented here were obtained. We are indebted to Matthias Mann and Matthias Wilm who shared their nanospray experience and capillaries with us and to Alexander Dodonov who brought this knowledge to our laboratory under the NATO linkage grant program. We thank Thomas Dresch and Anatoli Verentchikov for revealing the results of their research before its publication. We are also grateful to Victor Spicer for his technical assistance.

REFERENCES

1. Karataev, V. I.; Mamyrin, B. A.; Shmikk, D. V. *Sov. Phys.-Tech. Phys.* **1972**, *16*, 1177–1179.
2. Mamyrin, B. A.; Karataev, V. I.; Shmikk, D. V.; Zagulin, V. A. *Sov. Phys. JETP* **1973**, *37*, 45–48.
3. Bergmann, T.; Martin, T. P.; Schaber, H. *Rev. Sci. Instrum.* **1989**, *60*, 792–793.
4. Karas, M.; Hillenkamp, F. *Anal. Chem.* **1988**, *60*, 2299–2301.
5. Karas, M.; Bachmann, D.; Bahr, U.; Hillenkamp, F. *Int. J. Mass Spectrom. Ion Proc.* **1987**, *78*, 53–68; see also Tanaka, K.; Waki, H.; Ido, Y.; Akita, S.; Yoshida, Y.; Yoshida, T. *Rapid Commun. Mass Spectrom.* **1988**, *2*, 151–153.
6. Iribarne, J. V.; Dziedzic, P. J.; Thomson, B. A. *Int. J. Mass Spectrom. Ion Phys.* **1983**, *50*, 331–347.
7. Yamashita, M; Fenn, J. B. *J. Chem. Phys.* **1984**, *88*, 4451–4459.
8. Kambara, H. *Anal. Chem.* **1982**, *54*, 143–146.
9. Aleksandrov, M. L.; Gall, L. N.; Krasnov, N. V.; Nikolayev, V. I.; Pavlenko, V. A.;

Shkurov, V. A. *Dokl. Akad. Nauk* **1984**, *277*, 379–383 (*Dokl. Phys. Chem.* **1985,** *277*, 572–576).

10. Fenn, J. B.; Mann, M.; Meng, C. K.; Wong, S. F. *Mass Spectrom. Rev.* **1990**, *9*, 37–70.
11. Mann, M.; Fenn, J. B. in *Mass Spectrometry, Clinical and Biomedical Applications*, Desiderio, D. Ed.; Plenum: New York, 1992; Vol. I; pp. 1–35.
12. Smith, R. D.; Loo, J. A.; Loo, R. R. O.; Busman, M.; Udseth, H. R. *Mass Spectrom. Rev.* **1991**, *10*, 359–452.
13. Smith, R. D.; Loo, J. A.; Edmonds, C. G. in *Mass Spectrometry, Clinical and Biomedical Applications*, Desiderio, D. Ed.; Plenum: New York, 1992; Vol. I, p. 37–98.
14. Meng, C. K.; Mann, M.; Fenn, J. B. Z. *Phys.* **1988**, *D10*, 361–368.
15. Feng, R.; Konishi, Y. *Anal. Chem.* **1992**, *64*, 2090–2095; **1993**, *65*, 645–649.
16. Mirza, U. A.; Cohen, S. L.; Chait, B. T. *Anal. Chem.* **1993**, *65*, 1–6.
17. Winger, B. E.; Light-Wahl, K. J.; Ogorzalek Loo, R. R.; Udseth, H. R.; Smith, R. D. *J. Am. Soc. Mass Spectrom.* **1993**, *4*, 536–545.
18. Loo, J. A.; Ogorzalek Loo, R. R.; Andrews, P. C. *Organic Mass Spectrom.* **1993**, *28*, 1640–1649.
19. Light-Wahl, K. J.; Schwartz, B. L.; Smith, R. D. *J. Am. Chem. Soc.* **1994**, *116*, 5271–5278.
20. Tang, X. J.; Brewer, C. F.; Saha, S.; Chernushevich, I. V.; Standing, K. G.; Ens, W. *Rapid Commun. Mass Spectrom.* **1994**, *8*, 750–754.
21. Pinkston, J. D.; Rabb, M.; Watson, I. Th.; Allison, I. *Rev. Sci. Instrum.* **1986**, *57*, 583–592.
22. Boyle, J. G.; Whitehouse, C. M.; Fenn, J. B. *Rapid Commun. Mass Spectrom.* **1991**, *5*, 400–405.
23. Dodonov, A. F.; Chernushevich, I. V.; Dodonova, T. F.; Raznikov, V. V.; Tal'roze, V. L. USSR Patent #1681340A1 (Feb. 1987).
24. Dodonov, A. F.; Chernushevich, I. V.; Laiko, V. V. *12 Int. Mass Spectrom. Conference*, Amsterdam, August 1991, Extended Abstracts, p. 153.
25. Dodonov, A. F.; Chernushevich, I. V.; Laiko, V. V. In *Time-of-Flight Mass Spectrometry*; Ed. Cotter, R. J. American Chemical Society: Washington, DC, 1994; Symposium Series 549, pp. 108–123.
26. Mirgorodskaya, O. A.; Shevchenko, A. A.; Chernushevich, I. V.; Dodonov, A. F.; Miroshnikov, A. I. *Anal. Chem.* **1994**, *66*, 99–107.
27. Boyle, J. G.; Whitehouse, C. M. *Anal. Chem.* **1992**, *64*, 2084–2089.
28. Dresch, T.; Gulcicek, E.; Banks, J. F.; Whitehouse, C. M.; Fenn, J.; Boyle, J. G. *Proceedings of the 41st ASMS Conference on Mass Spectrometry and Allied Topics*; 1993, San Francisco, CA p. 16.
29. Clayton, E.; Bateman, R. H. *Rapid Commun. Mass Spectrom.* **1992**, *6*, 719–720.
30. Bateman, R. H.; Green, M. R.; Scott, G. *Proceedings of the 42nd ASMS Conference on Mass Spectrometry and Allied Topics*; 1994, Chicago, IL p. 1034.
31. Verentchikov, A. N.; Ens, W.; Standing, K. G. *Anal. Chem.* **1994**, *66*, 126–133; see also Verentchikov, A. N.; Ens, W.; Standing, K. G. *Proceedings of the 41st ASMS Conference on Mass Spectrometry and Allied Topics*; 1993, San Francisco, CA p. 4.

32. Verentchikov, A. N.; Blakley, C.; Martin, S.; Vestal, M. *Proceedings of the 42nd ASMS Conference on Mass Spectrometry and Allied Topics*; 1994, Chicago, IL p. 773.
33. Verentchikov, A. N.; Hsieh, F.; Tomany, M.; Gabeler, S.; Vestal, M.; Martin, S. *Proceedings of the 43rd ASMS Conference on Mass Spectrometry and Allied Topics;* 1995, Atlanta, GA p. 613.
34. Benson, S. W.; Grossman, J. J.; Elkin, H. S. US Patent No. 2,938,116, May 24, 1960 (filed April 2, 1956).
35. O'Halloran, G. J.; Fluegge, R. A.; Betts, J. F.; Everett, W. L. Technical Documentary Report, No. ASD-TDR-62–644, Parts I & II, prepared under contracts Nos. AF33(616)-8374 and AF33(657)-11018 by the Bendix Corporation, Research Laboratory Division, Southfield, MI, 1964.
36. Gerhard, Ph.; Loffler, S.; Homann, K. H. *Chem. Phys. Lett.* **1987**, *137*, 306–310.
37. Olthoff, J. K.; Lys, I. A.; Cotter, R. J. *Rapid Commun. Mass Spectrom.* **1988**, *2*, 171–175.
38. Emary, W. B.; Lys, I.; Cotter, R. J.; Simpson, R.; Hoffman, A. *Anal. Chem.* **1990**, *62*, 1319–1324.
39. Kennedy, R. A.; Kung, C.-Y.; Miller, J. P. In *Ion and Cluster Ion Spectroscopy and Stucture*; Maier, J. P. Ed.; Elsevier: Amsterdam, 1989; pp. 213–239.
40. Dawson, J. H. J.; Guilhaus, M. *Rapid Commun. Mass Spectrom.* **1989**, *3*, 155–159.
41. Coles, J.; Guilhaus, M. *Trends Anal. Chem.* **1993**, *12*, 203–213.
42. Sin, C. H.; Lee, E. D.; Lee, M. L. *Anal. Chem.* **1991**, *63*, 2897–2900.
43. Myers, D. P.; Li, G; Mahoney, P. P.; Hieftje, G. M. *J. Am. Soc. Mass Spectrom.* **1995**, *6*, 400–410 and 411–420.
44. Mirgorodskaya, E. P.; Mirgorodskaya, O. A.; Dobretsov, S. V.; Shevchenko, A. A.; Dodonov, A. F.; Kozlovskiy, V. I.; Raznikov, V. V. *Anal. Chem.* **1995**, *67*, 2864–2869.
45. Laiko, V. V.; Dodonov, A. F. *Rapid Commun. Mass Spectrom.* **1994**, *8*, 720–726.
46. Andrien, B. A.; Banks, J. F.; Boyle, J. G.; Dresch, T.; Haren, P. J.; Whitehouse, C. M. *Proceedings of the 43rd ASMS Conference on Mass Spectrometry and Allied Topics*; 1995, Atlanta, GA p. 999.
47. Banks, J. F.; Dresch, T. *Anal. Chem.* **996**, *68*, 1480–1485.
48. Michael, S. M.; Chien, B. M.; Lubman, D. M. *Anal. Chem.* **1993**, *65*, 2614–2620.
49. Chien, B. M.; Lubman, D. M. *Anal. Chem.* **1994**, *66*, 1630–1636.
50. Qian, M. G.; Lubman, D. M. *Anal. Chem.* **1995**, *67*, 234A–242A.
51. Tang, X.; Beavis, R.; Ens, W.; Lafortune, F.; Schueler, B.; Standing, K. G. *Int. J. Mass Spectrom. Ion Processes* **1988**, *85*, 43–67.
52. Tang, X. Ph.D. Thesis, University of Manitoba, 1991.
53. Ens, W.; Standing, K. G.; Verentchikov, A. In *Proceedings of the International Conference on Instrumentation for Time-of-Flight Mass Spectrometry*; Blanar G. J.; Cotter R. J.; Eds.; LeCroy Corp.: Chestnut Ridge, NY, 1992, pp. 137–144.
54. Wilm, M. S.; Mann, M. *Int. J. Mass Spectrom. Ion Processes* **1994**, *136*, 167–180.
55. Chowdhury, S. K.; Katta, V.; Chait, B. T. *Rapid Commun. Mass Spectrom.* **1990**, *4*, 81–87.
56. Wiley, W. C.; McLaren, I. H. *Rev. Sci. Instrum.* **1955**, *26*, 1150–1157.

57. Guilhaus, M. *J. Am. Soc. Mass Spectrom.* **1994**, *5*, 588–595.
58. Douglas, D. J.; French, J. B. *J. Am. Soc. Mass Spectrom.* **1992**, *3*, 398–408.
59. Xu, H. J.; Wada, M.; Tanaka, J.; Kawakami, H.; Katayama, I.; Ohtani, S. *Nucl. Instr. Meth.* **1993**, *A333*, 274–281.
60. Krutchinsky, A. N.; Chernushevich, I. V.; Spicer, V.; Ens, W.; Standing, K. G. *Proceedings of the 43rd ASMS Conference on Mass Spectrometry and Allied Topics*; 1995, Atlanta, GA p. 126.
61. Feng, R.; Konishi, Y.; Bell, A.W. *J. Am. Soc. Mass Spectrom.* **1991**, *2*, 387–401.
62. Beuhler, R. J.; Friedman, L. *Nucl. Instr. Methods* **1980**, *170*, 309–315.
63. Axelsson, J.; Reimann, C. T.; Hakansson, P.; Sundqvist, B. U. R.; Demirev, P. *Proceedings of 41st ASMS Conference on Mass Spectrometry and Allied Topics*, May 31–June 4, 1993, San Francisco, CA, p. 757.
64. Axelsson, J.; Reimann, C. T.; Sundqvist, B. U. R. *Int. J. Mass Spectrom. Ion Processes* **1994**, *133*, 141–155.
65. Bondarenko, P. V.; Grant, P. G.; Macfarlane, R. D. *Int. J. Mass Spectrom. Ion Processes* **1994**, *131*, 181–192.
66. Standing, K. G.; Chernushevich, I. V.; Ens, W.; Verentchikov, A. N. *Proceedings of 42nd ASMS Conference on Mass Spectrometry and Allied Topics*, May 29–June 3, 1994, Chicago, IL, p. 1148.
67. Loo, J. A.; Pesch, R. *Anal. Chem.* **1994**, *66*, 3659–3663.
68. Chernushevich, I. V.; Verentchikov, A. N.; Ens, W.; Standing, K. G. *J. Am. Soc. Mass Spectrom.* **1996**, *7*, 342–349.
69. Smith, R. D.; Barinaga, C. J.; Udseth, H. R. *Anal. Chem.* **1988**, *60*, 1948–1952.
70. Przybylski, M.; Glocker, M. O. *Angew. Chem. Int. Ed. Engl.* **1996**, *35*, 806–826.
71. Fitzgerald, M. C.; Chernushevich, I. V.; Standing, K. G.; Whitman, C. P.; Kent, S. B. H. *Proc. Natl. Acad. Sci. USA*, **1996**, *93*, 6851–6856.
72. Fitzgerald, M. C.; Chernushevich, I. V.; Standing, K. G.; Kent, S. B. H.; Whitman, C. P. *J. Am. Chem. Soc.* **1995**, *117*, 11075–80.
73. Tang, X.-J.; Saha, S.; Brewer, C. F.; Chernushevich, I. V.; Standing, K. G. *Proceedings of the 43rd ASMS Conference on Mass Spectrometry and Allied Topics;* 1995, Atlanta, GA p. 1259.
74. Werlen, R. C.; Lankinen, M.; Smith, A.; Chernushevich, I. V.; Standing, K. G.; Blakey, D. C.; Shuttleworth, H.; Melton, R. G.; Offord, R. E.; Rose, K. *Tumor Targeting* **1995**, *1*, 251–258.
75. Chen, L. H.; Kenyon, G. L.; Curtin, F.; Harayama, S.; Bembenek, M. E.; Hajipour, G.; Whitman, C. P. *J. Biol. Chem.* **1992**, *267*, 17716–17721.
76. Roper, D. I.; Subramanya, H. S.; Shingler, V.; Wigley, D. B. *J. Mol. Biol.* **1994**, *243*, 799–801.
77. Mandal, D. K.; Nieves, E.; Bhattacharyya, L.; Orr, G. A.; Roboz, J.; Yu, Q.-T.; Brewer, C. F. *Eur. J. Biochem.* **1994**, *221*, 547–553.

CHAPTER 7

Electrospray-Ionization Quadrupole Ion-Trap Mass Spectrometry

MARK E. BIER* and JAE C. SCHWARTZ
Finnigan Corporation, San Jose, California

	Abstract	236
I.	Introduction	236
II.	Brief fundamentals of quadrupole ion traps	238
	A. The QIT geometry and quadrupole field	238
	B. The ion-trap stability diagram and the resonant frequencies of ions	240
	C. The quadrupole ion-trap scan function and timing diagrams	241
	D. The effects of scan rate	243
	E. The effects of space charge	243
	F. The effects of helium-damping gas	244
III.	Electrospray–quadrupole ion-trap instrumentation	244
	A. Coupling the electrospray ion source to the quadrupole ion trap	244
	B. Trapping ions in the analyzer	248
	C. Electrospray and ion–molecule reactions in an ion trap	248
	D. Mass range of the quadrupole ion trap	249
	E. Detection of ejected ions	249
IV.	Capabilities of the electrospray–quadrupole ion trap	250
	A. Operation in the full-scan MS mode	250
	B. Operation in the SIM and the high-resolution scan modes	253
	1. Ion-isolation techniques	254
	2. High resolution of electrosprayed ions in a quadrupole ion trap	258
	C. Operation with multiple stages of MS: MS^n and SRM^n	258
	D. Additional methods of ion excitation in a quadrupole ion trap	261
	1. Photoinduced dissociation	261
V.	Applications of the ES quadrupole ion trap	262
	A. Analyses of peptides	262
	B. Analyses of proteins	265

---

*Present address: Department of Chemistry, Carnegie Mellon University, Mellon Institute, Pittsburgh, Pennsylvania.

Electrospray Ionization Mass Spectrometry, Edited by Richard B. Cole.
ISBN 0-471-14564-5 © 1997 John Wiley & Sons, Inc.

    C. Analyses of oligonucleotides   267
    D. Charge-state determination by ion reactions   272
        1. Charge-state determination by ion–molecule reactions   273
        2. Charge-state determination by ion–ion reactions   273
    E. Capillary electrophoresis ES–QIT   277
    F. Detection limits determined on an ES–QIT   279
    G. ES–QIT data-dependent scans   279
VI. Summary   281
    Acknowledgments   281
    References   282

ABSTRACT

Electrospray (ES) ionization coupled to a quadrupole ion trap (QIT) mass spectrometer is a powerful analytical tool. Electrospray enables the intact flight of polar, thermally labile molecules such as drugs, DNA, RNA, sugars, peptides, and proteins, while the compact QIT has the mass range, resolution, and scan modes (e.g., MS^n) essential for analysis. The details of coupling an ES source to a QIT are discussed and examples of different scan-mode capabilities are given. Molecular-weight determinations of multiply charged ions are possible through a spectral-deconvolution method or by using a high-resolution scan. Proton transfer ion–molecule and unique ion–ion reactions may also be used for molecular-weight determinations. Liquid chromatographs and capillary electrophoresis instruments have been directly interfaced to the ES–QIT. Ion structure and genealogy are elucidated by the use of MS^n-scans. Detection limits are at the low-femtomole levels for full-scan MS and in the low-femtomole to high-attomole levels for selective-ion monitoring (SIM), selective-reaction monitoring (SRM), and MS/MS analyses.

I. INTRODUCTION

Electrospray (ES) ionization of "macroions" was pioneered almost 30 years ago through the efforts of Malcolm Dole and colleagues (1). Dole's group electrosprayed polystyrene molecules with molecular weights of 51 and 411 kDa, and later, electrosprayed the proteins lysozyme (MW = 14.4 kDa) and zein (MW = 38 kDa) (2–4). The simple energy analyzer that they used was able to detect dimers and trimers of the 51-kDa polystyrene and the 14.4-kDa lysozyme by measuring stopping potentials for these ions. They found that the larger polystyrene macroion (411 kDa) carried up to five charges. In 1973, Dole et al. discussed the use of a linear quadrupole rod and a time-of-flight mass spectrometer to measure the mass-to-charge ratios of these ions, but it was not until Fenn and co-workers at Yale University (5) and Aleksandrov and co-workers in Russia (6) coupled an ES source to a mass spectrometer in 1984 that the field of

ES mass spectrometry began. In 1988, mass spectrometry took a leap forward when Fenn's group (7) discovered that electrosprayed proteins may indeed carry multiple charges, from 2 to greater than 100, and they further showed that a distribution of these multiply charged ions could be used to calculate accurate molecular weights.

Wolfgang Paul and H. Steinwedel at the University of Bonn publicly described a quadrupole ion-trap (QIT) mass spectrometer in 1953 (8,9) and three years later were issued a German patent (10) that included the three-electrode arrangement of the QIT. In 1989, Paul shared the Nobel Prize in Physics (11) with H. G. Dehmelt at the University of Washington and N. F. Ramsey at Harvard University. The award was given jointly to Paul for his work on radio frequency (rf) ion traps and to Dehmelt and Ramsey for their work on atomic precision spectroscopy. Since the time of its conception to the present day, a major thrust of research activity has involved these quadrupole field devices. The popularity of the QIT is attributed to the birth of the instrument as a scanning mass spectrometer, made possible by the invention of the mass selective instability scan by Stafford, Kelley, and Stephens (12,13) and the commercial introduction of an ion-trap mass spectrometer in 1983 by Finnigan Corporation. Also, often overlooked, is the fact that sophisticated computer control has been absolutely instrumental in QIT development and operation.

Over the past 15 years, a multitude of performance enhancements has been developed for the QIT. These advancements have been both fortuitous and by design. The analyzer performance enhancements include the use of helium dampening gas (12), variable ionization times (14), resonance ejection (15), rf/dc isolation (16), MS^n scans (15,17,18), ion injection (19), precursor and neutral loss scans (20,21), mass-range extension (22–24), high resolution (25–28), and tailored waveforms (TWF) (29–33). In addition to these improvements, the ion trap is also used as an electrostatic vessel for carrying out gas-phase ion–molecule reactions (34,35). Many types of ion sources and inlets have been coupled to the QIT. They include chemical ionization (36,37), liquid secondary ion-mass spectrometry (LSIMS) (19), thermospray (38, 39), glow discharge (40), membrane (41), particle beam (39,42), matrix-assisted laser desorption ionization (MALDI) (43–47), atmospheric pressure chemical ionization (APCI) (48), and ES ionization. The union of ES and the QIT, in particular, has formed an extremely powerful and exciting analytical instrument (49–51).

The remainder of this chapter focuses on ion trap technology developed to analyze ions formed by an ES source. Our discussion starts with a brief section on some fundamentals of the QIT and is followed by ES–QIT instrumentation, capabilities, and applications. Several QIT review papers have been published and are highly recommended. They include general reviews (52–56), one on high-pressure ion sources coupled to ion traps (57), and two specifically focused on ES coupled to ion traps (58,59), as discussed here. Although this chapter covers certain aspects of the ES source, Chapters 1–4 of this book are recommended to those interested in learning specifically about the ES source and this unique ionization process in detail.

II. BRIEF FUNDAMENTALS OF QUADRUPOLE ION TRAPS

Some theory is discussed here to provide a basic understanding of the QIT analyzer. A thorough coverage of QIT theory can be found in books by March and Hughes (60) and Dawson (61), and a three-volume book edited by March and Todd (62).

A. The QIT Geometry and Quadrupole Field

The QIT is a three-electrode device (see Fig. 1). The basic shape of the device can be imagined by rotating the hyperbolic electrode cross sections in space around the z-axis (ion axis). Two of the electrodes formed are identical and are called *end caps*, while the third electrode is donut shaped and is called the *ring* electrode. The ring electrode is sandwiched between the two end caps at a precise distance typically maintained by ceramic or quartz spacers. In Figure 1 the QIT is shown configured to store ions from an ES source. The holes in the end caps are for *ion injection* into the device from the ES source and *ion ejection* out of the device to a detector. A small negative dc offset is applied to all three electrodes to inject positive electrosprayed ions.

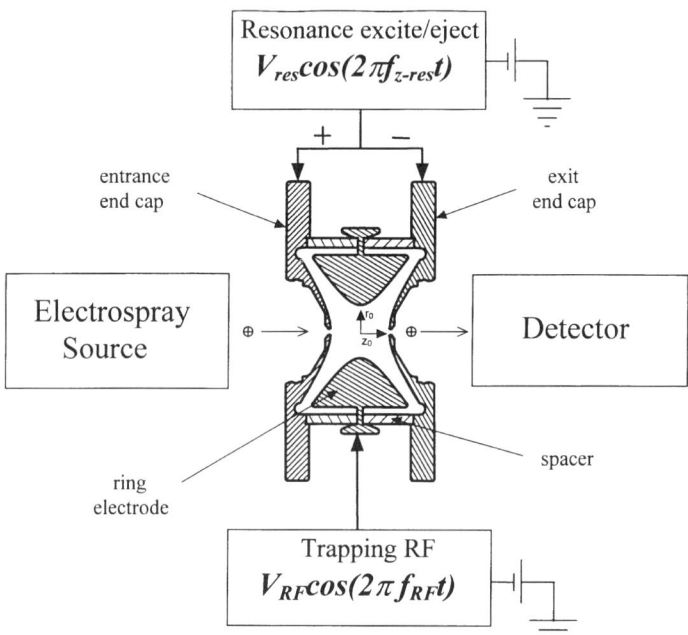

Figure 1. A three-dimensional QIT consisting of two hyperbolic end caps and a central hyperbolic ring electrode positioned between an ES source and a detector. The trapping rf voltage is applied to the ring electrode and the supplementary resonance excitation–ejection rf voltage is applied in dipolar fashion to the end-cap electrodes.

II. BRIEF FUNDAMENTALS OF QUADRUPOLE ION TRAPS

A three-dimensional quadrupole field is formed by the three electrodes when a suitable rf voltage $[V_{rf}\cos(2\pi f_{rf}t)]$ is applied to the ring electrode as shown in Figure 1. The quadrupole field traps ions by continuously forcing them toward the center of the device. The force on an ion is linearly proportional to the distance the ion is from the center of the QIT. Ions stored inside the trap follow trajectories described by the second-order Mathieu differential equation. There is some error in the predicted ion trajectory, however, because ion motion is effected by space-charge repulsion, the imperfect quadrupole trapping field, and collisions with gases (e.g., helium) which are unaccounted for in the differential equation. Solutions to the differential equation are in terms of the Mathieu parameters a_z and q_z as shown in Eqs. 1 and 2 for the three-dimensional ion-trap case (60):

$$a_z = -2a_r = \frac{-16eU}{m(r_0^2 + 2z_0^2)\Omega^2} \tag{1}$$

$$q_z = -2q_r = \frac{-8eV}{m(r_0^2 + 2z_0^2)\Omega^2} \tag{2}$$

In Eqs. 1 and 2, r represents the radial direction, z represents the axial direction, U is the dc amplitude, V is the rf amplitude, e is the charge on an ion, m is the mass of an ion, r_0 is the inner radius of the ring electrode, z_0 is the axial distance from the center of the device to the nearest point on one of the end cap electrodes, and $\Omega = 2\pi f_{rf}$ where f_{rf} is the frequency of the main rf voltage.

One specific, but often cited, geometry of the ion trap is defined by Eq. 3. Solving for z_0^2 in Eq. 3 and substituting the result into Eqs. 1 and 2 above, simplifies these Mathieu parameter equations.

$$r_0^2 = 2z_0^2 \tag{3}$$

In the early 1980s, however, researchers at Finnigan Corporation observed that ion traps built with the specific geometry of Eq. 3 demonstrated poor mass accuracy. Mass assignments were shifted and the errors were compound dependent. Although not well understood, Syka (63) has suggested that these mass shifts are due to a difference in ion radial distributions that occur because of the different ion collision cross sections, the high pressure of helium in the trap, and the effects of higher-order fields (64) that are caused by field imperfections resulting from the entrance and exit holes. Mass shifts were reduced to negligible deviations by spacing the end caps outward by a factor of $0.11z_0$ (65). Syka has proposed that such an alteration of the geometry improves the homogeneity of the quadrupole field near the center of the trap where the ions are stored. A "stretched" ion-trap geometry requires accounting for the z_0 characteristic dimension using the complete form of Eqs. 1 and 2.

B. The Ion-Trap Stability Diagram and the Resonant Frequencies of Ions

Plotting and overlapping the solutions to the Mathieu equation in (a,q) space for the r and z dimensions forms the QIT stability diagram. A portion of the stability diagram including the solutions where $a_z = 0$ is shown in Figure 2. The Mathieu parameters, a and q, are indicative of the stability and motion of an ion in the three-dimensional quadrupole field.

The operating or scan line represents the (a_z,q_z) points in a stability diagram that ions pass through until they are scanned out of the trap. For the basic mass-selective instability scan, the operating line starts at $(a_z,q_z) = (0,0)$ and passes through $(a_z,q_z) = (0,0.908)$ as seen in the diagram. Since q_z is inversely proportional to mass (m), high mass-to-charge ratio ions have a lower q_z value than low mass-to-charge ratio ions, as shown by the different-size circles on the diagram. In this case, m_9^+ has a larger mass-to-charge ratio than m_8^+. By increasing V, the rf amplitude applied to the ring electrode, ions positioned along the operating line shown in Figure 2 move to higher q_z values. At the edge of the stability diagram at $q_{z-\text{edge}} = 0.908$, instability ensues in the z dimension and the ions leave the QIT through the holes in the end caps. The lowest mass-to-charge ratio ion, becomes unstable first, followed sequentially by the next higher mass-to-charge ratio ion, and so on.

Ions of a specific mass-to-charge ratio have a fundamental frequency of motion unique to their q_z value and this unique frequency is often used to resonate these ions. The frequencies of ion motion in the z dimension are defined by Eq. 4 where f_h is a frequency component of the ion motion, f_{rf} is the frequency

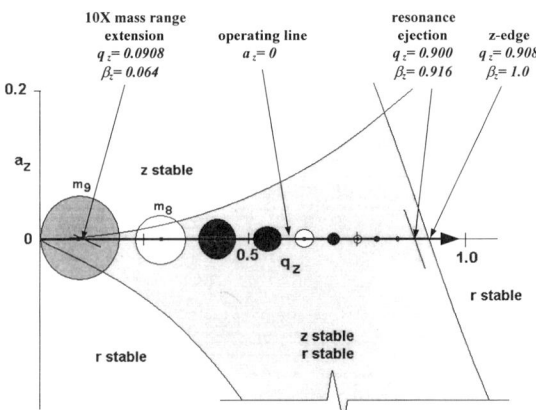

Figure 2. A portion of the QIT stability diagram (shaded region) showing the mass-selective instability operating line at $a_z = 0$. The circles represent ion q_z values on the operating line for various ions with different mass-to-charge ratios (e.g., where $m_9^+ > m_8^+$). The mass-selective instability scan is supplemented with resonance ejection at $q_z = 0.90$ for a normal scan or at $q_z = 0.0908$ to extend the mass range by a factor of 10.

of the rf applied to the ring electrode, and β_z is a parameter used to relate the frequency of an ion to its position on the stability diagram (61).

$$f_h = (hf_{rf} \pm \beta_z f_{rf}/2) \qquad \text{where } h = \{0, \pm1, \pm2, \pm3, ..\} \qquad (4)$$

By knowing the q_z of an ion, one can calculate the β_z value (60,61) and thus the resonant frequencies of an ion following Eq. 4. At low β_z, the frequency components ($|h| > 0$) have a negligible effect on ion motion, and unless large voltages are applied to the end caps, these higher-order frequencies are usually insignificant and thus, not utilized. In general, the fundamental resonant frequency of an ion is defined as the frequency where $h = 0$ (i.e., $f_{z\text{-res}} = f_0 = \beta_z f_{rf}/2$).

A dipole signal [$V_{res}\cos(2\pi f_{z\text{-res}} t)$] is applied to the end caps as shown in Figure 1, for *resonance excitation* and/or *resonance ejection*. For either of these processes to occur efficiently, however, the frequency used must equal the fundamental resonant frequency of the ion, $f_{z\text{-res}}$. For example, during the highly practiced mass-selective instability scan supplemented with resonance ejection (see Figure 2), ions are ejected from the trap axially as they are scanned into resonance at a frequency that corresponds to $q_{z\text{-eject}} = 0.900$, $\beta_{z\text{-eject}} = 0.916$. Resolution, peak height, and sensitivity are greatly improved when using the mass-selective instability scan mode supplemented with resonance ejection.

C. The Quadrupole Ion-Trap Scan Function and Timing Diagrams

Unlike triple-quadrupole mass spectrometers, where each operation on the ion beam is separated in space (i.e., either in Q1, Q2, or Q3), the QIT operates on the ions over a period of time, but within the same analyzer. For example, an ion-trap MS–MS scan consists of time periods for ion injection, isolation, excitation, and mass analysis. These steps are diagramed in Figure 3, which shows a plot of the rf amplitude on the ring electrode and is referred to as an *ion-trap scan function*. To avoid the effects of space charge resulting from too many ions, a *prescan* precedes the *analytical scan*. A prescan is rapid and is used to determine the proper analytical ion injection time. Based on the prescan ion-count measurement, the data system calculates a suitable ion injection time that does not inject an overabundance of ions that would cause the deleterious effects of space charge. For example, if the ion signal level doubles, as measured by a prescan, the ion-injection period during the analytical scan is shortened to half of the time used in the previous analytical scan and the data system scales up the recorded counts by a factor of 2. The prescan may include many of the steps found in the analytical scan, but the prescan ion-injection time is typically much shorter in duration (≤ 10 ms) and the prescan mass-analysis step is rapid, because only an ion-current measurement is required. Use of variable ion gate times has extended the linear dynamic range of the ion trap to five orders of magnitude in GC–MS applications (14,66).

During the *ion-injection* step (step 1), ions are gated electrostatically from an

Ion Trap Scan Function

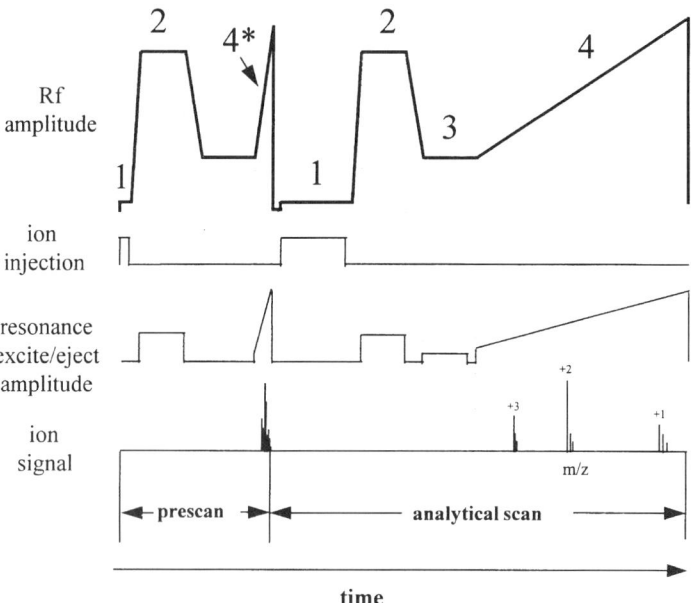

Figure 3. A simplified scan function for the quadrupole ion trap showing the prescan and the analytical scan which makes up one microscan. The four steps of the QIT operation: ion injection (1), isolation (2), excitation (3), and mass analysis (4) are shown in the scan function. Step 1 injects ions into the trap for storage. Steps 2 and 3 are shown by the amplitude of the dipolar resonance excitation–ejection signal for isolation and excitation. The ion signal is acquired during the analytical mass-analysis segment of the scan function.

external ES source into the QIT for a specific time determined by the prescan. The ions are trapped by the quadrupole field by maintaining a suitable rf voltage on the ring electrode and by using a partial pressure of helium in the trapping chamber of about 1 mTorr. For a MS–MS scan, an *isolation* step (step 2) must be performed to select a precursor ion(s) using one of a variety of isolation techniques that are explained in more detail in the sections to follow. After isolation, energy is deposited into the ion during an *excitation* step (step 3). Excitation of the ions is accomplished typically, through collisions with gases, but other means such as photon absorption, which deposits enough internal energy to cause dissociation, can be used. Finally, the last step (step 4) of every scan function involves *mass analysis*. During the mass-analysis step, ions are ejected sequentially out of the QIT chamber through the holes in the end caps. The ions that pass through the exit end cap impinge on a post acceleration dynode–electron multiplier detector. These four steps, ion injection, isolation, excitation, and mass analysis, typically require 0.001–1000 ms, 5–30 ms, 5–30 ms, and 10–400 ms, respectively. The results of the scan are then sent to the data system and the scan function can be repeated. One complete scan function ending in mass analysis has been termed a *microscan*.

D. The Effects of Scan Rate

Scan rate (Da/sec) refers to the rate of ejection of ions out of the trap during the mass-analysis step. Scan rates were first increased during the development of the prescan for variable ionization times (14). Later, at Purdue University, the scan rate was increased when the mass range was extended (24). Rapid scan rates can greatly improve detection limits because the ion current increases (i.e., the signal height increases) and the resulting shorter scan functions allow for improved signal averaging (67). The greater speed of ion ejection, however, causes mass peaks to broaden on the mass scale even though the peaks narrow in time. The decreased resolution occurs when using these high-scan rates because the ion undergoes a fewer number of resonance cycles before ejection. Rapid scan rates are less sensitive to the effects of space charge and, hence, a greater number of ions can be trapped without causing mass shifts or further degradation in mass resolution (67). An example of the use of a rapid scan rate is given in the application section on peptides.

Alternatively, the scan rate can be decreased (26,25). A decreased scan rate allows for an increase in the number of resonant cycles of an ion before ejection occurs and, hence, the resolution is greatly improved. Because this scan rate is reduced for the high-resolution scan, the mass range is kept small to maintain a high sampling rate across a chromatographic peak. The high resolution scan is also more susceptible to the effects of space charge than fast scans and, as a result, the number of ions trapped for this scan is reduced. Each scan rate, whether fast or slow, requires a proper calibration of the resonance-ejection amplitude versus-mass-to-charge ratio.

E. The Effects of Space Charge

Space-charge effects are caused by the distortion of an electrostatic field, namely the applied trapping field, owing to the presence of the electrostatic fields from one or more ions. The greater the number of ions stored inside a QIT, the greater the effects of space charge. This perturbation of the applied field forces ions to follow nonideal trajectories and is a fundamental limitation for all QITs. Most notably, space charge causes a degradation in resolution, a reduction in peak height, and a shift in mass assignments (68,69). At severe space-charge conditions, the mass peaks are further broadened and reduced in peak height to the point where they flatten into the baseline. Beyond the extreme limit of space charge, the ion density becomes so large that additional ions injected into the quadrupole field may not be trapped at all, or previously trapped ions may be displaced. Ion displacement should occur in descending order of mass-to-charge ratio based on storage forces at various q_z values (61).

In a simplified analogy, the QIT acts like a bucket. At "1/10 full," a bucket works well at carrying water and water may be poured out of the bucket with little or no spillage. However, filling a bucket to the brim makes it difficult to carry or to pour water. And, of course, any attempt to fill a bucket above the brim immediately results in spillage of water out of the bucket. To avoid the effects of space charge, the number of ions in the QIT chamber are regulated to the "1/10 level" to achieve optimum mass-to-charge analysis. This control occurs by calculating the optimum analytical ion injection time as determined by the prescan to achieve the "1/10 level" (see Section II.E).

F. The Effects of Helium-Damping Gas

In the early 1980s, researchers at Finnigan Corporation were interfacing the first gas chromatograph (GC) to a QIT. Because this QIT utilized internal ionization, the GC column was inserted directly into the QIT chamber. Serendipitously, they found that the resolution and sensitivity of the instrument was greatly improved when approximately 1 mTorr of helium was present in the QIT chamber (12,70). This was especially evident for ions with higher mass-to-charge ratios. The improvement results from a collisional dampening effect of the ions by the low-molecular-weight helium atoms. The ions are kinetically cooled to the center of the trap and occupy an area less than 2 mm in diameter as measured by tomography experiments (71). Ions are damped kinetically over a period of a few milliseconds, which is pressure dependent, and this allows them to be ejected from the ion trap in dense ion packets during the mass-analysis step.

Successful trapping of injected ions into a QIT from an external source such as an ES ion source is extremely dependent on the helium pressure. Without helium, or another suitable gas, only a negligible number of ions is trapped. At a helium pressure of 1 mTorr, ions are readily trapped. At helium pressures greater than 1 mTorr, trapping efficiency is improved; however, the mass resolution is reduced if the pressure is too high. Because of these pressure constraints, pulsing helium into the ion-trap chamber only during the ion-injection step has been considered by several researchers. Typically, the helium pressure in the QIT for the work shown in this chapter ranged between 1 and 4 mTorr.

III. ELECTROSPRAY–QUADRUPOLE ION-TRAP INSTRUMENTATION

A. Coupling the Electrospray Ion Source to the Quadrupole Ion Trap

A critical problem encountered when coupling an ESI source to a mass spectrometer is that the analyte is dissolved in a liquid at atmospheric pressure, but the mass spectrometer functions at low pressures in a vacuum chamber. About eight orders of magnitude lie between the atmospheric pressure of the ES ion source and the ambient pressure of a QIT. Despite these significant pressure differences, several laboratories successfully coupled the ES source to a QIT mass spectrometer (49–72). Van Berkel, Glish, and McLuckey at Oak Ridge National Laboratory were the first to couple an ES ion source to a QIT in 1990 (49) followed by Schwartz and Jardine at Finnigan Corporation (50). In 1992, Mordehai and co-workers at Cornell University coupled their own ES source to the first bench-top QIT (Saturn II^{TM}, Varian Instruments, Palo Alto, CA) for use in LC and CE studies (51). Bruker-Franzen of Germany has also recently introduced an ES–QIT instrument (ESQUIRE^{TM}). The Oak Ridge ES–QIT instrument is briefly reviewed here, is described in more detail in the literature (49), and was used to collect much of the data in the applications section (Section V). Recently, Finnigan Corporation introduced an ES–QIT mass spectrometer (LCQ^{TM}) (73,74), which has been under development during the past six years, and is used as an example here and, for a detailed description of QIT operation, in the next section.

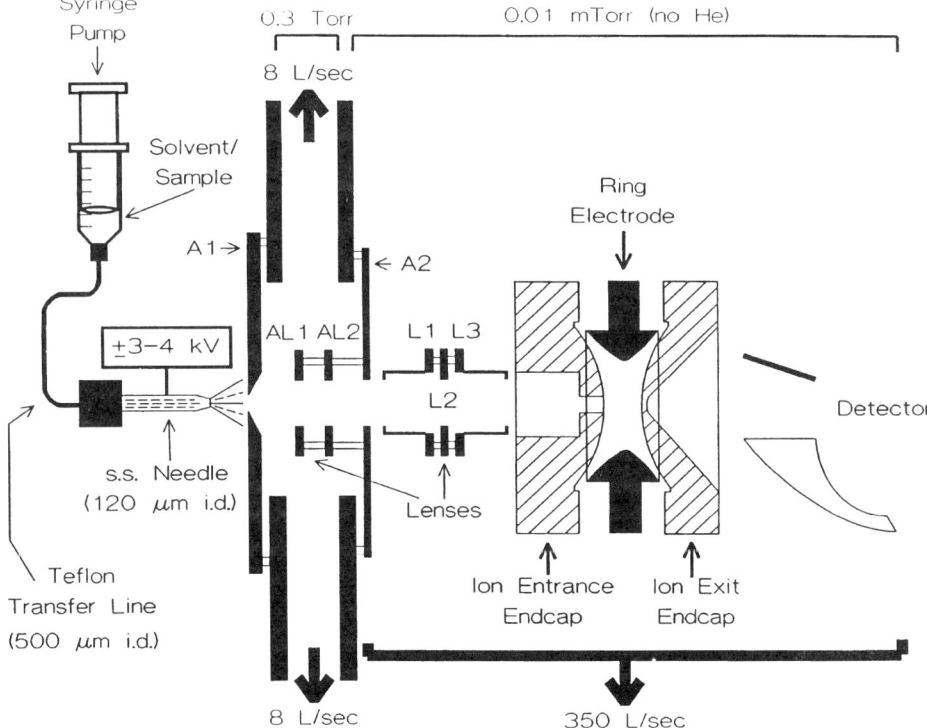

Figure 4. Cross-sectional view of the first ES–QIT mass spectrometer. Reprinted with permission from G. L. Van Berkel et al., *Anal. Chem.*, **1990**, *63*, 1284. © 1990 American Chemical Society.

The design developed by Oak Ridge to couple an ES ionization source to a QIT is shown in Figure 4. They modified an atmospheric glow-discharge ionization source (40) for use as an ES source by adding two lens elements. Sample solutions were sprayed from a 120-μm i.d. "electrospray" needle positioned at the entrance orifice. The orifice (100 μm) in lens element A1 samples the electrosprayed liquid from the needle and introduces the analyte into the first of two differentially pumped chambers. Chamber one is maintained at a pressure of 0.3 Torr while chamber two, the QIT chamber, is maintained at a pressure of 0.01 mTorr (uncorrected). The ions are electrostatically focused by a three-element lens stack (L1, L2, L3) onto a hole in the entrance end cap. Ion packets are injected into the QIT at the appropriate time by pulsing one of two semicircular plates (L2). A potential difference of 300 V between the two half plates deflects the ion beam away from the QIT entrance orifice when not injecting ions. Ions ejected from the QIT were detected with a dynode–electron multiplier detector. Finnigan ITMS^{TM} electronics were used to control the QIT.

Most ES interfaces desolvate solvent-ion clusters by gaseous collisions and/or by heat. The ES source developed by Sciex Corporation (Toronto, Canada) used a nitrogen "curtain" near the orifice to break up ES solvent clusters (75). The Finnigan ES source utilizes a heated metal capillary developed by Chait's group at Rockefeller University (76). The Fenn–Whitehouse ES source (Analytica of Branford, Branford, CT) used a countercurrent flow of hot nitrogen as a drying gas (77) and Henion's group at Cornell used a heated lens (51). The ES source constructed by Oak Ridge was not heated, and was prone to produce highly solvated biomolecular ions that were then trapped. To assist in desolvating these ions, they extended the trapping time and collisionally activated the solvated ions by resonance excitation prior to mass analysis (49). The early Oak Ridge ES source lead to an interesting study of protonated water and methanol clusters (78). They found that highly clustered solvent $(H_2O)_n H^{+n}$ and $(CH_3OH)_n H^{+n}$ ions underwent rapid desolvation when $n > 6$. The Oak Ridge group has also pointed out that a softer ES source and ion-transport process is desirable when studying weakly bound complexes by ES–MS–MS in a QIT (57).

The Finnigan ES–QIT mass spectrometer is shown in Figure 5. The vacuum system consists of a rough pump and a dual-port, turbomolecular pump to maintain operating pressures P_2 through P_4 at 1 Torr, 1×10^{-3} Torr, and 2.5×10^{-5} Torr (uncorrected), respectively. The solution is electrosprayed from a stainless-steel needle at approximately 4.5 kV, in the positive-ion mode, and is sampled by the capillary, which is typically heated to a temperature of 200°C. The capillary (400 μm i.d., 11.5 cm long) helps complete the desolvation process and serves as the nozzle for the supersonic expansion of the gas into the next chamber. The supersonic free jet of solvent molecules exiting the heated capillary forms a barrel shock which includes a mach disk downstream (79). A metal skimmer is positioned inside the barrel shock and is "attached" to the mach disk for optimum ion transmission. Ions are transmitted through the skimmer using the tube lens as a gating element. Positive ions are pulsed through the skimmer by applying 0 to +200 V to the tube lens while −150 V stops transmission of the

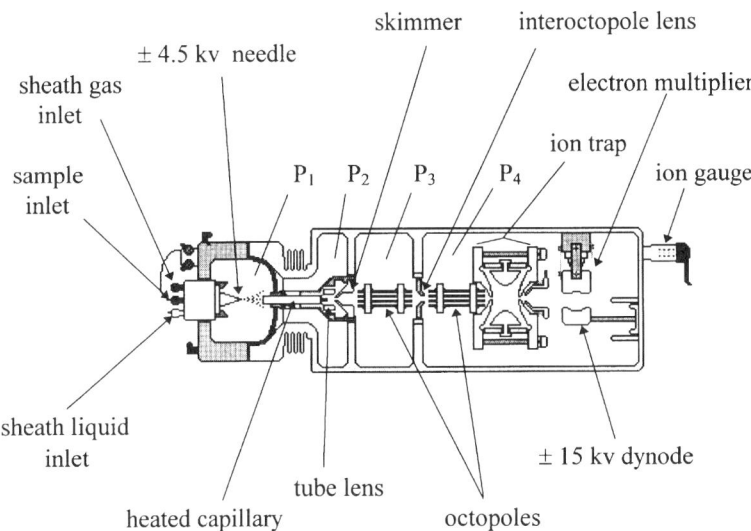

Figure 5. Electrospray quadrupole ion-trap mass spectrometer (LCQ^{TM}) developed by Finnigan Corporation featuring a heated capillary and dual-octopole ion guides to inject ions directly into the trapping chamber. $P_1 \approx 760$ Torr, $P_2 \approx 1$ Torr, $P_3 \approx 1 \times 10^{-3}$ Torr, $P_4 \approx 2.5 \times 10^{-5}$ Torr, and $P_{trap} \approx 1-2 \times 10^{-3}$ Torr.

ions through the skimmer. The sampled ions are collected by the first rf octopole and are transmitted to the second rf octopole through an interoctopole lens. Each octopole is 5 cm long ($r_0 = 3.3$ mm) and both are operated at 2.5 MHz and 400 V_{pp}.

Octopole ion guides were used for several reasons. First, they efficiently transmit ions through a region of relatively high pressure where ion scattering would otherwise occur (80–82). It has been shown that low-mass ions are scattered easily by gases in static lens systems, whereas the rf octopole reduces these effects and improves transmission as was found with rf collision cells in triple-stage quadrupole instruments (83). Second the octopoles also allow for efficient differential pumping and, although each can have a separate dc offset potential for optimum ion transmission, only one rf supply is required. Unique to this design is the use of an octopole to transmit ions directly to the ion-trap chamber by placement of the second octopole inside the entrance end cap of the QIT (see Figure 5).

The interoctopole lens has a 2.5 mm i.d. aperture and serves three functions: (*i*) to transmit ions during the ion injection period, (*ii*) to act as the conductance limit for differential pumping, and (*iii*) to act as a potential barrier against the transmission of large, charged particles formed by the ES source. Many users of ES sources have noted the appearance of large, random noise spikes in the mass spectrum. Our studies have shown that these spikes were attributed to large multiply charged particles that apparently strike the detector and result in noise spikes of high intensity in the mass spectrum. To stop these particles, several hundred positive volts are applied either to the exit lens (lens after the exit end cap) or to the interoctopole lens after the ion-injection period, but before the mass-analysis period. This applied voltage dramatically reduces the particle noise during the subsequent mass-analysis part of the scan function (74). We note that the reduction of background ES noise has also been studied by Ramsey et al. using resonance-excitation techniques (84).

B. Trapping Ions in the Analyzer

For all ES ion-trap systems to date, ions have been injected axially through the entrance end cap aperture. It has been shown that ions can be injected and efficiently trapped directly from an octopole ion guide, despite having a significant transverse ion-kinetic energy (74). To efficiently trap the axially injected ions, it is crucial to have an appropriate rf amplitude applied to the ring electrode and to have the partial pressure of helium in the trap at a minimum of 1–2 mTorr. The injected ions collide with helium and the ion radial and axial trajectories are reduced while the trapping field continuously forces the ions toward the center of the device.

Louris et al. (19) originally plotted the intensity of injected ions of different mass-to-charge ratios from perfluorotributylamine versus the rf amplitude on the ring electrode. He observed that different mass-to-charge ratio ions had different optimum rf amplitudes. Since then, it has been determined that to a first approximation, a linear relationship exists between the optimum trapping rf level and mass-to-charge ratio (85) and the slopes of these lines increase with ion-injection energy (74). The higher rf amplitudes are believed to reflect the need for stronger rf trapping fields at the higher ion-kinetic energies. Ion-injection energies in the work described here are typically between 4 and 20 eV.

C. Electrospray and Ion–Molecule Reactions in an Ion Trap

One of the limitations of ion traps that use internal electron ionization (EI) is that, given the long storage times (e.g., 10 ms and greater), some ions react with neutral molecules (86). These ion–molecule reactions are dependant on the analyte reactivity, the partial pressure of the neutral reactant (e.g., water, methanol, nitrogen, or the analyte neutral), and duration of the reaction time. Little reactivity of the protonated ions are observed in ES–QIT because they are even-electron ions. In general, even-electron ions are less reactive than odd-electron ions formed by EI sources. In addition, ions formed by ES have already been exposed to the high partial pressures of reactants from solvents such as water, methanol, and acetonitrile, or acids like acetic or trifluoroacetic in the capillary–skimmer region. As a result, an ion–molecule reaction should have already taken place in the ES interface prior to trapping. Product ions formed from resonance excitation in the trap chamber, however, are not exposed to the high pressure of gases found in the ES source. Usually, these product ions are also the less reactive even-electron ions, but the relatively long QIT storage times increase the probability that an ion–molecule reaction may occur. Routine ES–QIT analysis has shown that unwanted ion–molecule reactions are indeed uncommon and that when they do occur, they often are structurally informative (48). Desirable ion–molecule reactions are discussed in Section V.

D. Mass Range of the Quadrupole Ion Trap

The mass range of the first commercial QIT was 650 Da. Since ES sources are often used to create ions greater than m/z 650, a method was needed to extend the mass range of the QIT. In 1990, the mass range was extended to over 70,000 Da by decreasing the $q_{z\text{-eject}}$ point at which ions are resonantly ejected from the trapping chamber (24). For example, if all other parameters remain constant, by decreasing $q_{z\text{-eject}}$ by a factor of 10, from $q_{z\text{-edge}} = 0.908$ to $q_{z\text{-eject}} = 0.0908$, the maximum mass range increases from 650 to 6500 Da as defined by Eq. 5 and indicated in Figure 2. For example, now $m_9^+ = 5000$ can be ejected resonantly, as shown in this illustration:

$$(m/z)_{\max} = \frac{8V_{\max}}{(r_0^2 + 2z_0^2)\Omega^2 q_{z\text{-eject}}} \tag{5}$$

In Eq. 5, $V_{\max}$ is the maximum amplitude of the rf and $q_{z\text{-eject}}$ is the point on the stability diagram where the ions are resonantly ejected. By using a lower $q_{z\text{-eject}}$, however, the scan rate (Da s^{-1}) increases and resolution is reduced. In this example, the scan rate increases to 55,550 Da s^{-1}. This rapid scan rate will also result in a fewer number of points acquired across a mass peak unless a higher data-acquisition rate is used (see Section II.D). A solution to this problem would be to reduce the scan rate to 5555 Da s^{-1} and employ an adjustable mass range to avoid long scan times.

When the $q_{z\text{-eject}}$ point is reduced for mass range extension, artifact peaks or *ghost peaks* can be observed (65,87). Ghost peaks are the result of ions that have fallen between the $q_{z\text{-eject}}$ point and $q_{z\text{-edge}} = 0.908$ of the stability diagram at some time before or during the mass-analysis scan. During the mass-selective instability scan, these ions are ejected at the $q_{z\text{-edge}}$ rather than at the resonance point and appear as broad under-resolved peaks. To avoid ghost peaks, rather than working at a low $q_{z\text{-eject}}$ value to extend the mass range, one could alternatively lower the fundamental rf frequency, reduce r_0, and/or increase the $V_{\max}$ as indicated by Eq. 5. The rf amplitude ($V_{\max}$) has a practical working limit of approximately 8.5 kV$_{0-p}$, above which arcing may occur. The Finnigan LCQ^{TM} achieves a mass range of 2000 D by using a V_{0-p} of 8.5 kV, a r_0 of 7 mm, and a f_{rf} of 760 kHz, and by reducing $q_{z\text{-eject}}$ to less than $q_{z\text{-edge}} = 0.908$ (73,74).

E. Detection of Ejected Ions

Ions are damped to the center of the trap by helium prior to the mass-selective instability scan supplemented with resonance ejection so that they may be ejected in dense packets. Ion ejection at a high $q_{z\text{-eject}}$ value will cause the kinetic energies of high mass ions to reach several kiloelectronvolts (74,88). The kinetic energy is imparted to the ions by the high rf amplitude used during mass analysis. Despite these high ion-axial ejection energies, an off-axis ± 15 kV dynode can deflect and focus these high mass ions onto its surface for detection. When higher helium pressures are used to trap injected ions more efficiently, the detector may be differentially pumped to avoid the detrimental effects of ion scattering or arcing (89).

IV. CAPABILITIES OF THE ELECTROSPRAY–QUADRUPOLE ION TRAP

This section discusses the capabilities of the ES–QIT and exhibits representative examples of spectra. The discussion starts with an example of a MS scan function and evolves this analytical scan to cover the SIM, high-resolution, SRM^n, and MS^n scan modes.

A. Operation in the Full-Scan MS Mode

As discussed previously, the operation of the QIT is frequently described using scan functions. Figure 6 shows the rf scan function used to generate a full-scan MS and includes the waveforms from other dynamic devices used in the LCQ^{TM}. The prescan, although essential for automated operation, has been removed from all scan-function figures in this section for clarity. Several features of the scan function are noteworthy. For example, the ion-injection period has been divided into four segments of increasing amplitude. The trapping of injected ions is highly dependent on the level of the rf amplitude applied to the ring electrode. The tube lens is also adjusted to higher potentials during each successive ion-injection period to improve the transmission of higher mass-to-charge ratio ions. This segmentation of the ion-injection period allows for a more uniform capture of ions across the mass range. Figure 6 also shows the use of an optional tailored waveform (TWF) which can be applied to the end caps during ion injection. In this case, the TWFs consist of a "sum of sines" and are used to isolate a mass range of interest during the ion-injection periods (31,32,90). The inset shows the frequency domain spectrum [fast Fourier transform (FFT)] of the second sum-of-sines TWF revealing the discrete frequencies used to eject all ions except those within the notch. For full-scan MS, the notch can be quite wide. Tailored waveforms are discussed in more detail in the SIM section that follows. The resonance ejection voltage is increased linearly with mass-to-charge ratio for optimum resolution throughout the mass range and for a linear mass calibration (61,91). The interoctopole lens permits ion transmission during the ion injection periods and stops the transmission of the supposed large charged particles during mass analysis, as stated earlier. The rf amplitude applied to the octopoles is also turned off after ion injection to ensure that a higher order frequency (2.5 MHz) is not induced onto the entrance end cap during ion storage and to

Scan Function for Full Scan MS

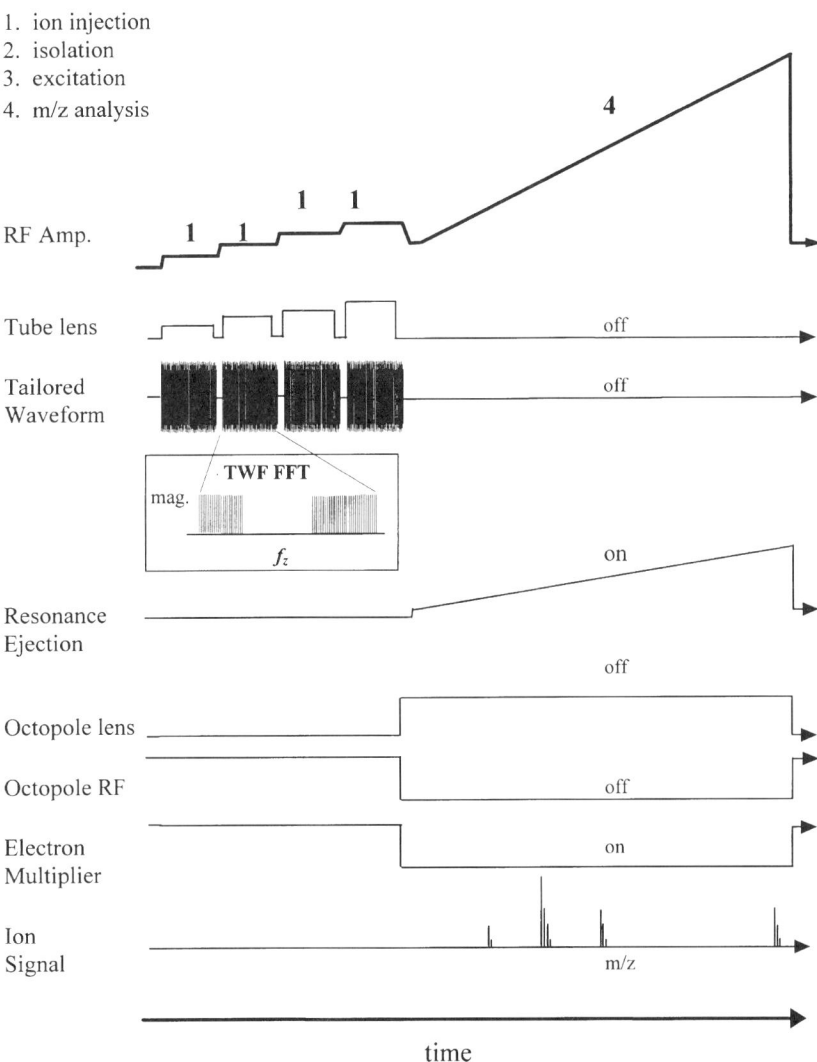

Figure 6. Scan function for the full scan MS mode of operation using an ES–QIT mass spectrometer. Only the ion injection (1) and the mass analysis (4) steps are required. An optional sum-of-sines TWF is used during ion injection to eject unwanted ions from the trap. A frequency domain spectrum [fast Fourier transform (FFT)] of the sum-of-sines TWF reveals the notch, as shown in the inset. The tube-lens ion gate, the interoctopole element, and the octopole rf voltage are turned on to maximize ion transmission during ion injection. The resonance ejection amplitude is increased linearly for optimum resolution and linearity during mass analysis. The multiplier is turned on during mass analysis (4) to detect the ion signal.

reduce charged particle noise (92). The electron multiplier is turned off during the ion injection period to protect it from the ion current due to untrapped ions. The last plot in Figure 6 shows an example of a hypothetical ion signal from this scan.

An example of a full-scan ES mass spectrum using a Finnigan LCQ^{TM} is shown in Figure 7. The sample contained a mixture of Ultramark 1621 [a mixture of fluorinated phosphazenes (PCR, Gainesville, FL)], the tetrapeptide MRFA (MW = 523.3), and caffeine (MW = 194.1). The sample mixture was dissolved in a 1% acetic-acid solution of 50:25:25 acetonitrile:methanol:H_2O and infused at 3 $\mu L/min$ into the ES ion source. The positive-ion spectrum shows the $(M+H)^+$ ion from caffeine at m/z 195.2, the $(M+H)^+$ from MRFA at m/z 524.3, the doubly charged ion from MRFA $(M+2H)^{+2}$ at m/z 262.6, and the singly protonated fluorinated phosphazenes $(M_n+H)^+$ observed between m/z 800 and 2000. Moini reported Ultramark 1621 to be an exceptional mixture to calibrate an ES or APCI mass spectrometers because the molecular ions are

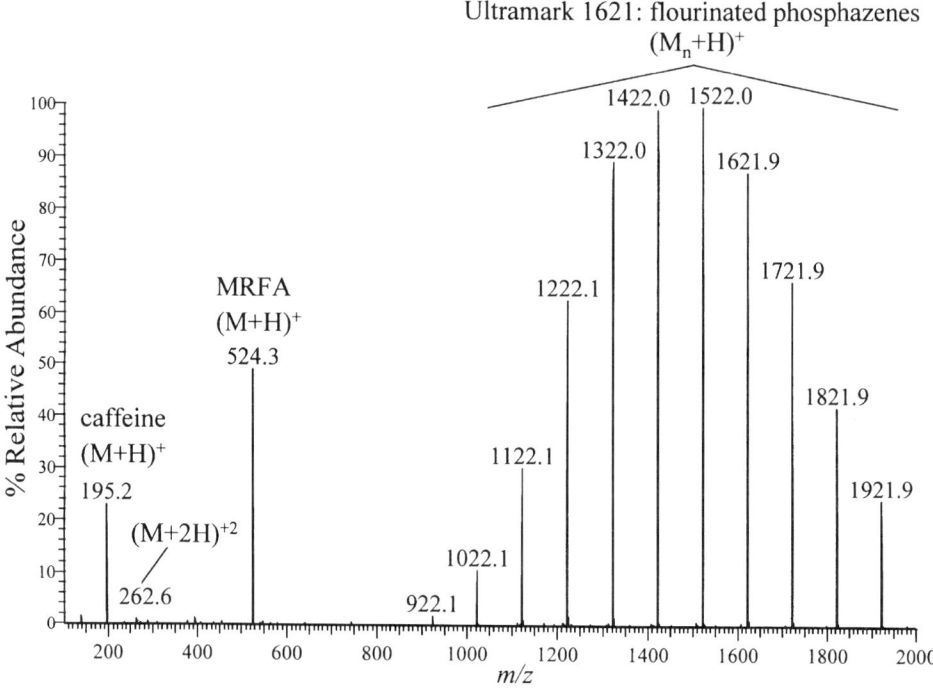

Figure 7. Electrospray full-scan spectrum of Ultramark 1621 [a mixture of fluorinated phosphazenes (PCR, Gainesville, FL)], the tetrapeptide MRFA, and caffeine made up in a solution of 1% acetic acid in a 50:25:25 solution of acetonitrile, methanol, and H_2O. The positive-ion spectrum shows the $(M+H)^+$ ion from caffeine at m/z 195.2, the $(M+H)^+$ from MRFA at m/z 524.3, the doubly charged ion from MRFA $(M+2H)^{+2}$ at m/z 262.6, and the mixture of protonated fluorinated phosphazenes, $(M+H)^+$, observed between m/z 800 and 2000. From M. E. Bier, Finnigan Corporation, 1996.

IV. CAPABILITIES OF THE ELECTROSPRAY–QUADRUPOLE ION TRAP   253

readily formed in both positive and negative ion modes, singly charged, and evenly spaced every 100 Da over a large mass range (93).

A full-scan mass spectrum can also be used to analyze the product ions formed from precursor ions that are activated by collisions in the ES interface. In this mode of operation, the precursor ion presumably collides vigorously with solvent molecules and dissociates, provided there is a large potential gradient in the tube lens–skimmer region (ca.1 Torr). Figure 8 shows the tube lens collision-activated dissociation (CAD) spectrum obtained from angiotensin I ($MW_{avg} = 1296.5$). The intense B, A, Y, sodiated, and doubly charged ions are identified. Tube-lens CAD MS can be complimentary to MS–MS because a shorter scan time is used and the $q_{z\text{-inject}}$ value may be set lower to trap the low-mass fragments. In contrast, a relatively high $q_{z\text{-excite}}$ value of 0.25 is often required in MS–MS scans and some low-mass fragments are lost if they have a $q_z > 0.908$ (see Section IV.C). The major disadvantage of the tube-lens CAD experiment is that it has no selectivity, and therefore, may result in a product spectrum consisting of a mixture of various background analyte fragments. A more selective and definitive experiment uses the ES–QIT MS^n scan mode described later.

B. Operation in the SIM and the High-Resolution Scan Modes

As with ion-beam mass spectrometers, quadrupole ion traps have improved ion signal-to-noise ratios in analyses by SIM versus the full-scan mode (94). The

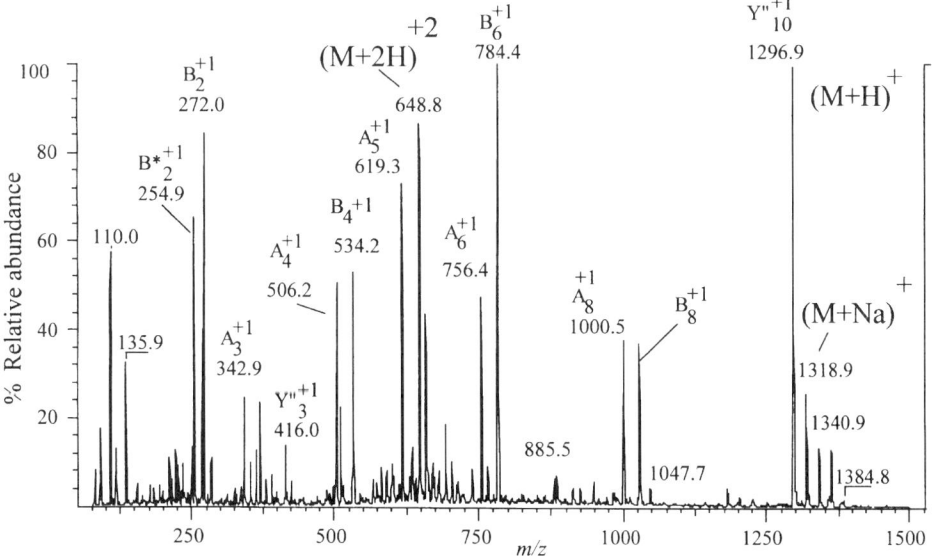

Figure 8. Electrospray tube-lens–skimmer CAD spectrum of angiotensin I (MW = 1295.7 Da) acquired using a prototype Finnigan LCQ™. From M. E. Bier, Finnigan Corporation, 1995.

longer a mass spectrometer spends time detecting a selected ion, the lower the detection limit is for that ion. Quadrupole rod instruments have an advantage over the QIT in the SIM scan mode because for quadrupole instruments the duty cycle can increase from less than 0.1% for one ion in a scan, from m/z 100–2000, to nearly 100% in a SIM scan [duty cycle = (ion collection or detection time/total scan time) $\times$ 100%]. However, the QIT has a duty cycle advantage in the full-scan mode of operation over linear quadrupole instruments. For low-level ion signals in the QIT, the duty cycles can reach levels > 50% in the full-scan mode and can increase to levels > 90% for the SIM-scan mode. The duty cycle is sample amount dependent in a QIT, unlike in the linear quadrupole case.

A scan function for SIM is shown in Figure 9. Although using the isolation TWF during injection is optional, and is not shown in this figure, a TWF isolation step after the ion-injection period is required in the SIM-scan mode. This high q_z isolation step is absent from the scan function shown in Figure 6, but provides a significant advantage when operating a QIT in the SIM mode. In this case, the trap is filled beyond the "1/10 level" with a variety of ions, including the ion of interest (refer to Section II.E). After a TWF has been applied, unwanted ions are ejected and leave behind a greater abundance of the isolated ion at the "1/10 level". In effect, the selected ion is concentrated in the QIT.

1. Ion-Isolation Techniques. Over the years several methods of isolating ions in a QIT have been implemented. Methods include rf/dc isolation (18,95), forward and reverse rf resonance-ejection isolation (96–98), and various forms of TWF isolation (21,31,90,100,99–104). The rf/dc isolation methods position the ion of interest near the boundaries of the stability diagram for isolation. In this case, the parameter a_z is set to a nonzero value. The rf resonance-ejection method sweeps the main rf amplitude and/or the resonance-ejection frequency applied to the end caps in both the forward and/or reverse directions to eject all but the ion of interest. This technique has been shown to yield high-resolution isolations and can be used to analyze multiply charged ions (97).

Recently, a sum-of-sines TWF consisting of many discrete frequencies has been applied to the end caps in dipolar fashion to isolate a narrow mass-to-charge ratio window (Figure 9). Ideally, one would like to apply a continuous band of frequencies at all β_z values except at $\beta_{z\text{-isolate}}$ [refer to Eq. (4)] for the ion of interest; in practice, a sum-of-sines approach (102) is simpler to implement. Sine waves are added together so that each discrete frequency is applied at the desired power for efficient resonant ejection across the entire mass range. The waveform has a notch in the frequency domain at $f_{z\text{-res}}$, as shown in the boxed inset of Figure 9, which allows these ions to remain stable. The discrete frequencies are calculated, typically, at spacings of 250, 500, or 1000 Hz (90,105) or at every integer mass (100). A high $q_{z\text{-isolate}}$ (e.g., 0.8–0.9) for the isolation step provides the optimum resolution because of the high degree of frequency dispersion of different masses at these values.

The sum-of-sines TWF approach is somewhat similar to the stored-waveform, inverse Fourier transform (SWIFT) TWF, introduced by Marshall

Scan Functions for SIM and High Resolution

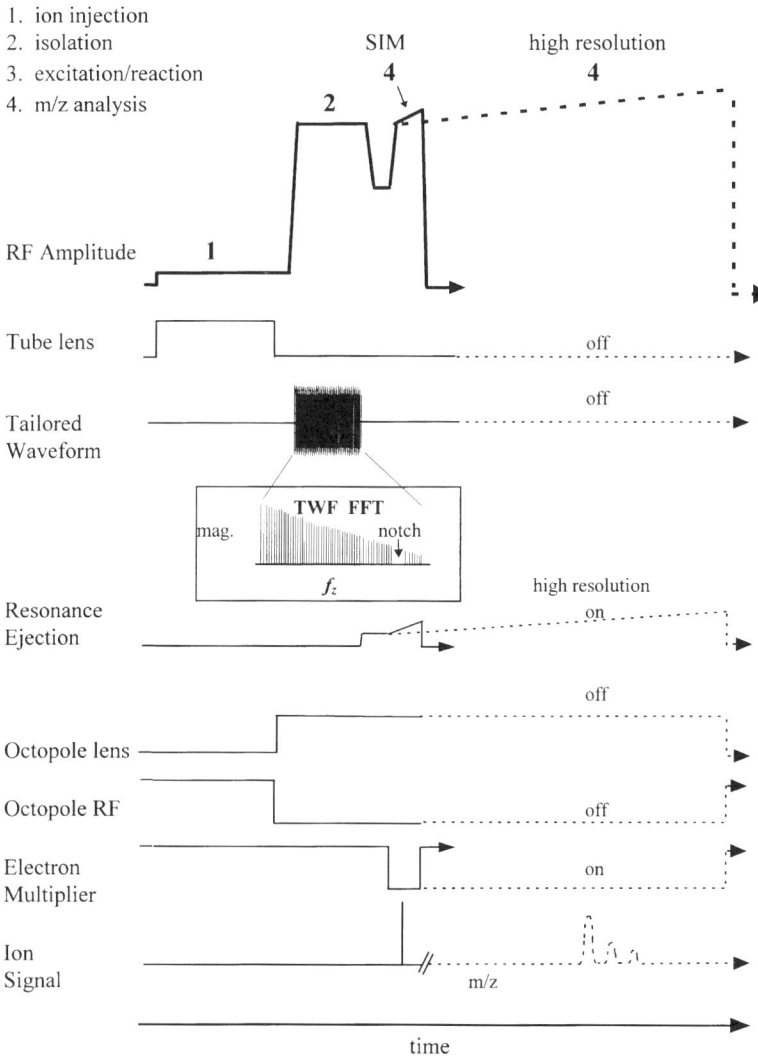

Figure 9. Scan functions for the SIM and high-resolution MS modes of operation for the ES–QIT. Only the ion-injection (1), ion-isolation (2), and mass-analysis (4) steps are required. In this example, a sum-of-sines TWF is used during high q_z isolation step to eject unwanted ions from the trap. The optional ion-injection TWF has been turned off. A frequency domain spectrum (FFT) of the sum-of-sines TWF reveals the notch in frequencies applied as shown in the inset. The tube-lens ion gate, the interoctopole element, and the octopole rf voltage are turned on to maximize ion transmission during ion injection. The resonance-ejection amplitude is increased linearly during mass analysis for optimum resolution. The multiplier is turned on during the mass-analysis segment (4) to collect the ion signal.

(106,107). Both create a defined time-domain waveform consisting of many frequencies. The SWIFT–TWF, however, is calculated by taking the inverse Fourier transform of a specified excitation or ejection frequency–domain spectrum. This synthesized time-domain spectrum can be applied to the end caps of the QIT for dipolar excitation similar to the sum-of-sines TWF discussed above. Guan and Marshall (108) as well as Julian (109) have discussed the use of SWIFT–TWF in the QIT in detail.

A SIM analysis of reserpine, m/z 609, is shown in Figure 10 using a flow-injection analysis (FIA). The 1-μL injections were made into a 200-μL/min flow rate consisting of a solvent system of 1% acetic acid in 50:50 methanol:water. In this experiment, the ion-injection step did not include a TWF isolation step, but a TWF was used in isolation step (2), as shown in Figure 9. A detection limit of 8 fmol (5 pg) with a signal-to-noise ratio of 7 was measured for this set of data.

Isolation of electrosprayed ions may also occur outside of the QIT. Jonscher and Yates have built a hybrid linear quadrupole–QIT mass spectrometer. The linear quadrupole is used to filter out all but a 10-Da mass window of ions which contains the analyte mass-to-charge ratio of interest. The 10-Da mass window is then injected into the QIT. Preliminary results show a reduction in space-charge effects for the analysis of peptides by using this prefiltering technique (110).

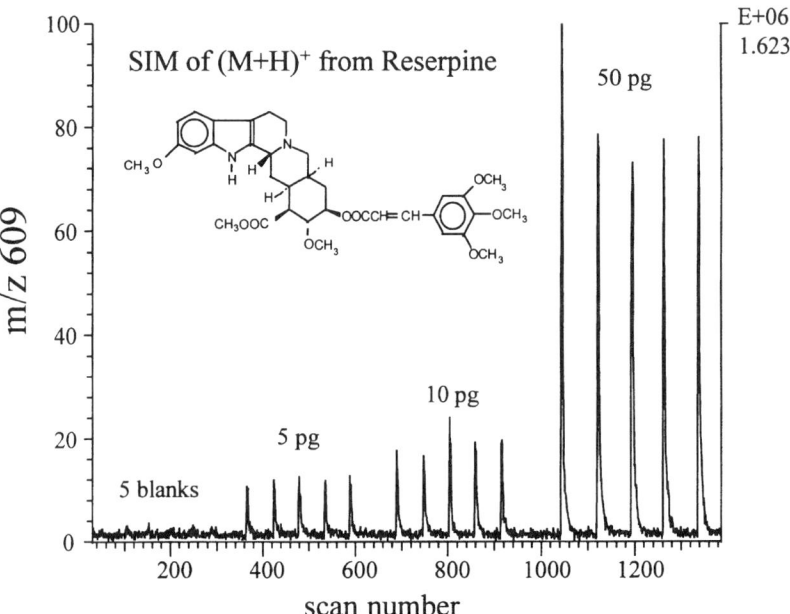

Figure 10. Flow-injection SIM analysis of reserpine (MW = 608). Five 1-μL injections each of 0, 5, 10, and 50 pg were made into a flow rate of 200 μL/min of 1% acetic acid in 50:50 methanol:water. From M. E. Bier; Finnigan Corporation, 1995.

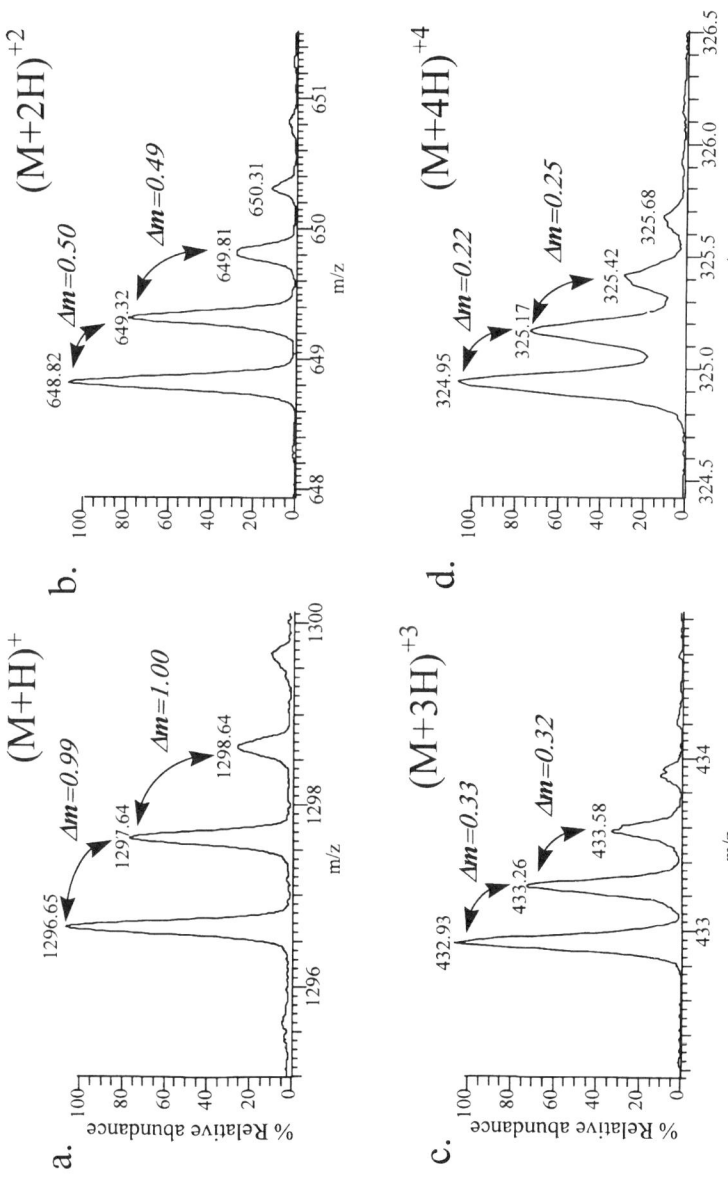

Figure 11. High-resolution charge-state determination of four different multiply charged isotope distributions from electrosprayed angiotensin I, MW = 1295.7 Da. The charge state was automatically determined to be (a) + 1, (b) + 2, (c) + 3, and (d) + 4. From J. C. Schwartz, Finnigan Corporation, 1996.

2. High Resolution of Electrosprayed Ions in a Quadrupole Ion Trap.

Molecular-weight calculations of electrosprayed ions require the determination of the ion charge state. High-resolution mass analysis ($m/\Delta m \geq 10,000$ at m/z 2000) achieved by reducing the scan rate of the QIT (Da s^{-1}) and supplementing this scan with an appropriate resonance-ejection voltage (25,26) can be used to ascertain the charge state of ions. This high resolution scan is shown in the mass analysis part of the SIM scan function in Figure 9. The mass resolution is improved because the ion spends more time in resonance during the ejection process (61). High resolution in an ES–QIT has been reported at levels greater than 1×10^6 (111,112); however, this has not become routine and the mass accuracy has not been shown to be commensurate (113). At a resolution of 8000, the ES–QIT can easily determine a charge state of up to ± 4 at m/z 2000 by measuring the difference in the mass-to-charge ratio of two adjacent isotope peaks ($\Delta m/z$). For example, if $\Delta m/z = 1$, the charge state is 1, if $\Delta m/z = 0.5$, the charge state is 2, if $\Delta m/z = 0.33$, the charge state is 3, and so on. Figure 11 shows an example of charge-state determination from four different multiply charged isotope distributions of electrosprayed angiotensin I. Charge states much greater than 4 require even higher mass resolution. For example, a mass resolution of approximately 25,000 would be required to resolve isotopes of a charge state of 25 at m/z 1000 for a 25-kDa protein. This resolving power has been demonstrated in FTMS (114,115); however, higher-resolution scans of multiply charged ions in a QIT have been limited to < 7 charges because these scans have lower S/N ratios, the scan rates are slow (e.g., 200 ds^{-1}), and the ions are more susceptible to space charge and fragmentation effects in this scan mode.

C. Operation with Multiple Stages of MS: MS^n and SRM^n

In MS–MS (MS^2 = MS^n where $n = 2$), there are two stages of mass spectrometry. The first stage of "mass spectrometry" occurs during the isolation step of an ion or ions, while the second stage occurs during the mass-analysis step. The scan function for MS^2 (Figure 12) requires one more step than the SIM-scan function, namely, the excitation or activation step. This step could alternatively be used as an ion–molecule reaction step. A resonance excitation period is typically 5–30 msec in duration and requires an excitation voltage of less than 5 V$_{pp}$ measured differentially, end cap to end cap. The resonance excitation frequency is calculated using Eq. 4 for a specified $q_{z\text{-excite}}$ value. During the resonance-excitation step, ions undergo CAD in the ion trapping chamber where the resonant precursor ion repeatedly collides with helium buffer gas. A heavier collision gas partner can be mixed with the helium for improved internal-energy deposition (116–118). Given that enough internal energy is deposited by this multiple-collision process, the precursor ion fragments. In 1987, Louris studied energy deposition and CAD efficiency in a QIT for the resonance-excitation process (119). For the systems studied, tetraethylsilane, n-butylbenzene, and nitrobenzene, the CAD process was nearly 100% efficient when the internal energy deposited was less than 2 eV.

IV. CAPABILITIES OF THE ELECTROSPRAY–QUADRUPOLE ION TRAP

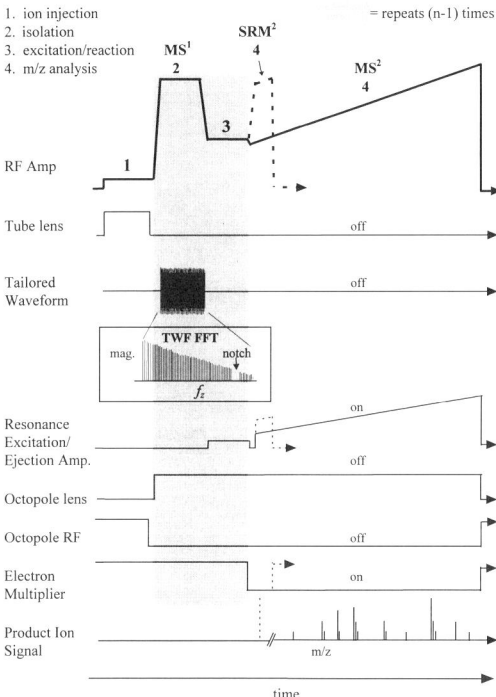

Figure 12. Scan functions for the SRM^n and MS^n modes of operation for the ES–QIT mass spectrometer. All four operations, ion injection (1), ion isolation (2), excitation (3), and mass analysis (4) are executed. In this example, a sum-of-sines TWF is used to isolate only the precursor ion(s) of interest. A frequency-domain spectrum (FFT) of the sum-of-sines TWF reveals the notch in frequencies applied as shown in the inset. The tube-lens ion gate, the interoctopole element, and the octopole rf are turned on to optimize ion transmission during ion injection. Resonance excitation (step 3) dissociates the precursor ion. The multiplier is turned on and the resonance-ejection amplitude is increased linearly during the mass analysis segment (4), allowing detection of the product-ion spectrum.

Typically, the $q_{z\text{-excite}}$ value is chosen between 0.2 and 0.3 for resonance excitation. This allows the precursor ion to gain enough internal energy to fragment yet remain trapped in a strong rf field during resonance excitation. It also allows for a majority of the product ions to fall within the stability boundary at $q_z = 0.908$. For example, if a doubly charged precursor ion [$(M+2H)^{+2}$, having m/z 500, (MW = 998)] is resonantly activated at $q_{z\text{-excite}} = 0.2$, product ions with $q_z > 0.908$ are not trapped. In this case, no product ion of $m/z < (q_{z\text{-excite}}/q_{z\text{-edge}}) \times$ (precursor m/z) = $(0.2/0.908) \times 500$ Da = 110 Da will be stored. This *low-mass cutoff* is a fundamental limitation of this type of CAD in a QIT.

The loss of product-ion information below the low-mass cutoff can be overcome by using additional isolation and excitation steps on lower mass

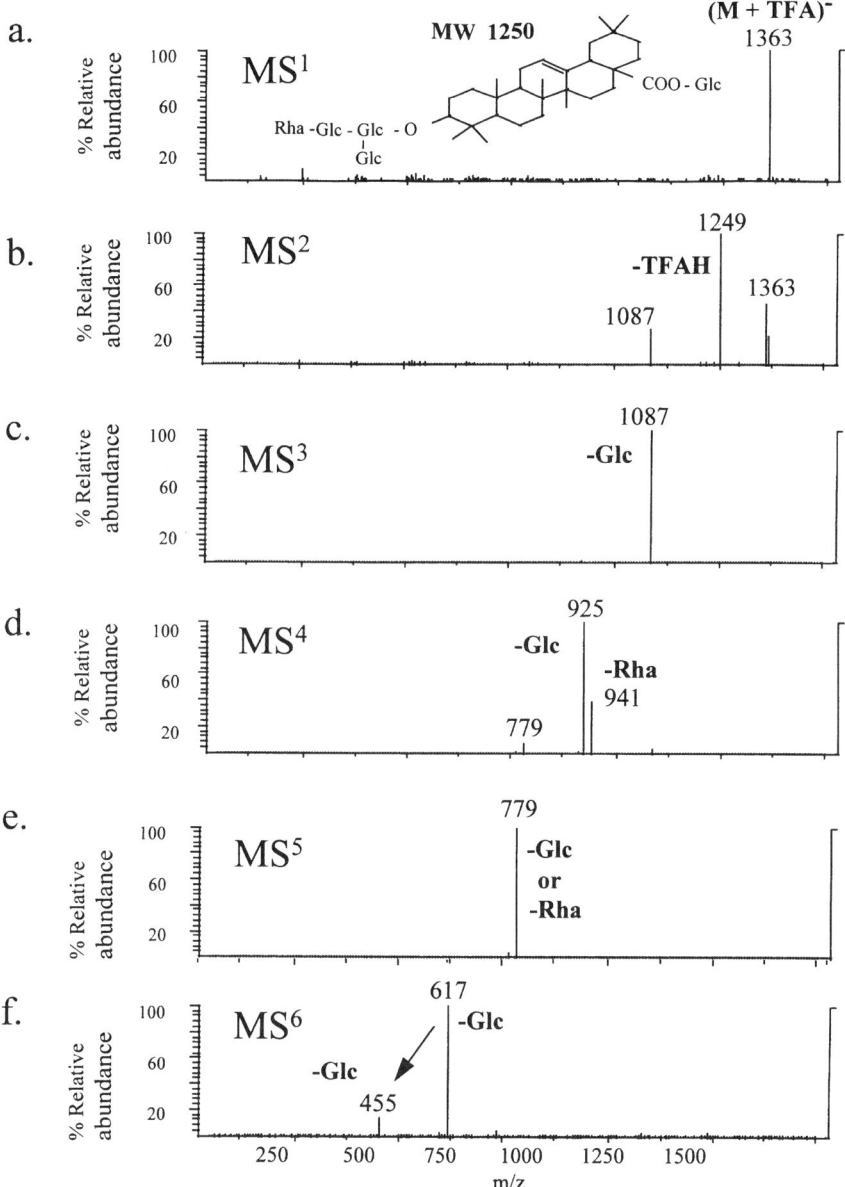

Figure 13. The MS^6 of oleanolic acid glycoconjugate (MW = 1250). (*a*)–(*f*) Six stages of MS resulting in the loss of the trifluoroacetic-acid adduct and either a glucose or rhamnose group as noted. From A. Land, Finnigan Corporation, 1995. Sample provided by the Universite de Lausaune, Lausaune, Switzerland.

fragments (e.g., MS^n scans). In fact, these additional MS^n scans provide a wealth of information regarding ion genealogy and structure not obtainable from MS^2 scans (120). The power of the MS^n scan is demonstrated in Figure 13a–f. The results of six stages of MS (five stages of resonance CAD) are shown for oleanolic acid glycoconjugate (MW = 1250). Each stage of MS shows a loss of one or two fragments, allowing for a straightforward genealogical interpretation. Loss of the trifluoroacetic acid adduct (TFAH) was followed by losses of either a glucose (Glc) or rhamnose (Rha) group (121). The MS^n-scans, however, do have limitations in that the ion current available for the mass-analysis step is diminished as n increases, and scan functions can become long in duration because SRM^n- or MS^n-scans require ($n-1$) additional isolation and excitation steps.

The mass-analysis step after isolation and excitation may cover one ion of interest or a narrow range of ions for a SRM^n scan with no significant loss in duty cycle (74). This was demonstrated in a FIA experiment where the detection limit of reserpine using a (50-Da product mass range) SRM scan was 8 fmol (5 pg) with a signal-to-noise ratio of 20. This detection limit was obtained by injecting 1 μL of 5 pg/μL reserpine into a 200-μL/min flow of 1% acetic acid in 50:50 methanol:water and summing three product ions—397$^+$, 436$^+$, and 448$^+$.

D. Additional Methods of Ion Excitation in a Quadrupole Ion Trap

Other ion-trap CAD schemes were developed to increase energy deposition and/or the degree of fragmentation. For example, a characteristic of using single-frequency resonance excitation is that when the product ions are formed, they are not activated, because they have a resonance frequency that is different from the applied frequency [i.e., they are stored at a different β_z, see Eq. (4)]. Broadband excitation is a variation of the resonance-excitation method that can be used to activate these product ions. In this case, a broad band of frequencies is applied to the end caps to excite the precursor ion as well as any product ions formed from the first CAD (122). A low-frequency activation technique has also been studied in ion traps. By applying a relatively low-frequency square wave to the end caps, ions can also be activated (123). This activation process is somewhat similar to the nonresonance collision process in the multipole-collision cell used in a triple-quadrupole mass spectrometer, and the process has shown high energy deposition. Qin and Chait have also shown encouraging preliminary results using a red-shifted, off-resonance excitation technique (124). They applied a large excitation amplitude (21 V$_{pp}$) off-resonance and observed efficient fragmentation of the singly charged protonated peptide, Substance P, into structurally significant B$_3$ through B$_{11}$ product ions.

1. Photoinduced Dissociation. Almost all of the past ion-activation analyses with the QIT has used CAD excitation; however, photoinduced dissociation (PID) has shown promise as a means of obtaining high-internal energy deposition and has been demonstrated in ion-cyclotron resonance (ICR) (125–127) as well as in the QIT (128–130). Louris et al. used a fiber optic to introduce light from a Nd:YAG laser into the QIT chamber to photodissociate protonated benzaldehyde, butylbenzene, and perfluoropropylene (131). The spectra collected in this study suggested that PID deposits more energy than CAD in a QIT. More recently, Stephenson et al. (132–134) have built an ES–QIT which incorporated PID. This instrument is unique because it includes mirrored light optics on the ring electrode. Their design focuses the laser light into a 0.3-cm^2 aperture of the ring electrode and allows for eight passes of the laser beam. The longer path length improves the PID efficiencies. A comparison was made between CAD and PID product spectra acquired in a QIT from the trisaccharide raffinose using a pulsed CO_2 laser with a maximum energy of 1.1 J. Both the CAD and the PID spectra showed the loss of a single monosaccharide at m/z 343 from the protonated molecular ion at m/z 505. But the PID spectrum showed more extensive fragmentation, leading to the formation of the protonated monosaccharide ion at m/z 164. The CAD and PID data were collected with the precursor ion at $q_{z\text{-excite}} = 0.1$ (135), but dissociation was more extensive in the PID experiment, apparently because of the higher internal energy deposited. Energy deposition at low $q_{z\text{-excite}}$ resonance CAD is limited, because ion ejection of the precursor ion will result before dissociation if the resonance-excitation amplitude is excessive. Photoinduced dissociation, however, appears to offer higher activation at lower $q_{z\text{-excite}}$ values, which allows lower mass-to-charge ratio fragments to be observed. Extensive fragmentation from photodissociation has also been demonstrated in the activation of angiotensin I. In addition to the amount of energy deposited, PID offers a narrow energy distribution of activation, which should allow for more selective fragmentation. A drawback to the PID technique is that it requires the added expense and maintenance for the laser and light optics and the ring electrode must be modified. A limitation of the PID process is that the energy deposition is dependent on the molecular structure. The latter problem is becoming less restrictive, however, because more powerful lasers with wider wavelengths of excitation are now available (132).

V. APPLICATIONS OF THE ES–QUADRUPOLE ION TRAP

A. Analyses of Peptides

It is of interest to biochemists to determine the molecular weight and amino-acid sequence of peptides. A molecular weight can be determined by using a high-resolution scan or by calculating a deconvolved spectrum from the multiply charged ion distributions. An example of a high-resolution molecular-weight determination of a quadruply charged peptide, interleukin-8 (rat), is shown in Figure 14a and b. The isotope peaks are separated by 0.25 Da, which clearly indicates a +4 charge state. The exact molecular weight of the C^{12} isotope of interleukin-8 was measured to be 7840.38 ± 0.14 Da using the mass-to-charge ratios of all the isotopes. The theoretical exact mass of the C^{12} isotope of

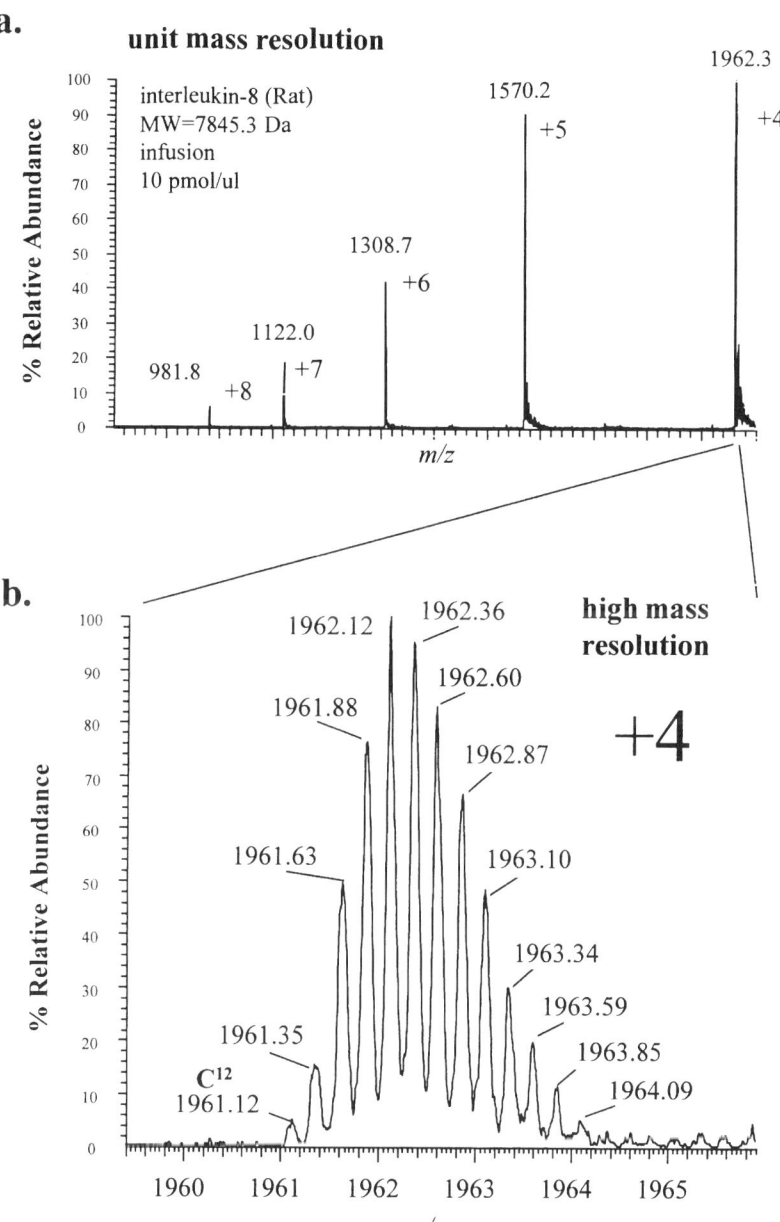

Figure 14. High-resolution molecular-weight determination of interleukin-8 (rat) +4 ion. The spectrum in (*a*) shows several of the multiply charged interleukin-8 ions while (*b*) shows the high-resolution spectrum of the +4 charge state used to determine the C^{12} species MW = 7840.38 (theoretical MW = 7840.09 Da). From M. Sanders, Finnigan Corporation, 1995.

interleukin-8 is 7840.09 Da. This data indicated that four of the −SH groups were oxidized to two disulfide bonds. The second method of a peptide molecular-weight determination on a Finnigan LCQ^{TM} utilized an algorithm to deconvolve the multiply charged peak distribution (136). A 1-μL injection (20 pmol/μL) of bovine ubiquitin (MW$_{avg}$ = 8564.9 Da) gave a deconvolved average molecular weight of 8565.0 Da. Remarkably, the high-resolution method determined the monoisotopic molecular weight of interleukin-8 to within 0.3 Da (0.004%), and the deconvolution method determined the average molecular weight of bovine ubiquitin to within 0.1 Da (0.002%). These results are tabulated in the lower half of Table 1 with other peptide and protein examples.

TABLE 1. Average and Monoisotopic Molecular Weights of Proteins and Peptides by ES–QIT

	Theoretical MW$_{avg}$ (Da)	ES–QIT Mw$_{avg}$ (Da)	Mass Accuracy (%)
Protein			
Lysozyme (chicken egg)	14306.2	14305.0	0.006[a]
Carbonic anhydrase (bovine erythrocytes)	29024.6	29025.2	0.002[a]
Cytochrome c (bovine)	12230.9	12231.5	0.005[a]
β-Lactoglobulin A (bovine milk)	18363.3	18364.5	0.006[a]
Apomyoglobin (equine)	16951.5	16951.9	0.002[a]
Hemoglobin-alpha chain (bovine)	15053.2	15053.7	0.004[a]
Hemoglobin-beta chain (bovine)	15953.3	15053.5	0.001[a]
Peptide			
Ubiquitin (bovine)	8564.9	8565.0	0.002[a]
Interleukin-8 (rat)	7840.09[b]	7840.38[b]	0.004[c]
Angiotensin I	1295.68[b]	1295.63[b]	0.004[c]
Calciseptine	7031.2[b]	7030.8[b]	0.006[c,d]
Bradykinin	1059.56[b]	1059.62[b]	0.006[c]
Cytochrome c Peptides			
TGQAPGFTYTDANK	1469.68	1469.66	0.001[c]
KTEREDLIAYLK	1477.82	1477.86	0.003[c]
GITWKEETLMEYLENPKK	2208.11	2208.19	0.004[c]

Source: All data was collected using a Finnigan LCQ^{TM}.
[a] Deconvolution method.
[b] C^{12} MW.
[c] High-resolution method.
[d] From ref. 141.
Data acquired by D. Gale, T. Vasconcellos, M. Sanders, M. Sweeney, T. Chaudhary, J. Schwartz, and A. Land, Finnigan Corporation, 1996.

A mixture of peptides can be sequenced by first separating them using LC and then analyzing the individual peptides by ES–QIT MS^n. The first on-line coupling of a 1-mm i.d. LC column to an ES–QIT showed useful ion currents at the 2.5-pmol level of injected tryptic peptides and at the 300-fmol level for human serum albumin (MW$_{avg}$ = 66 kDa) (137). Since then several labs have coupled capillary LC to a ES–QIT for peptide sequencing (72,74) using the short-column capillary LC technique adapted from Kennedy and Jorgenson (138). In using this method, a capillary column (100 μm i.d. and 15 cm long) was packed with 10-μm beads of POROS R2^{TM} (Perseptive Biosystems, Framingham, MA) and inserted directly into the ES needle as prescribed. This configuration reduces dead volume and thus avoids peak broadening. A standard mixture of 5 angiotensins (Michrom BioResources, Inc., Auburn, CA) was hydrostatically loaded onto the head of the column and was separated within a 15-min gradient elution. The flow rate (200 μL/min) from a Michrom micro-HPLC pump (Auburn, CA) was split 200:1 and delivered to the capillary column. Figure 15 shows the MS–MS product-ion spectrum of the isolated doubly charged ion (442.3^{+2}), and its isotopes, from 10 fmol of the peptide RVYVHPI (MW$_{exact}$ = 882.6 Da) acquired on a LCQ^{TM}. At the 10-fmol level, ions can readily be assigned to be the Y$_2''^{+1}$, B$_5^{+2}$, A$_7^{+2}$, B$_4^{+1}$, A$_5^{+1}$, B$_5^{+1}$, and B$_5^{+1}$ product ions. Multiply charged peptides fragment readily at low-resonance excitation voltages compared to singly charged peptides and this allows for the use of lower $q_{z-excite}$ values (139). As mentioned earlier, the lower $q_{z-excite}$ values also permit the storage of the lower mass-to-charge ratio fragment ions. Similar data to that shown in Figure 15a has been collected using triple quadrupole mass spectrometers (140); however, a decrease in the resolution of the quadrupoles to give 2–3 Da wide peaks was required to achieve the necessary 10X improvement in signal-to-noise level. The ES–QIT MS–MS data shown in Figure 15a was collected at unit-mass resolution.

Bier et al. have shown that the detection limit of peptides for sequencing can be improved by operating the QIT using rapid scan rates (67). Figure 15b shows a MS–MS spectrum of 550 attomoles of RVYVHPI analyzed in a QIT at 16X the normal scan rate of 5555 Da s^{-1}. The faster scan rate of 88,880 Da s^{-1} improved the signal height by approximately 6X, and additional scans were collected for improved signal averaging. The mass resolution, however, was decreased and mass-to-charge ratio assignments were less accurate owing to the reduced number of points acquired across a mass peak (Section II.D).

B. Analyses of Proteins

One of the most powerful applications of ES mass spectrometry is the determination of the molecular weights of proteins. Deconvolution algorithms are often used to calculate the molecular weight of proteins utilizing the mass-to-charge ratios of all of the multiply charged mass peaks present in the ES mass spectrum. This multiplicative method of using all the charge states for the molecular-weight determination has a high degree of accuracy. Figure 16 shows the ES mass

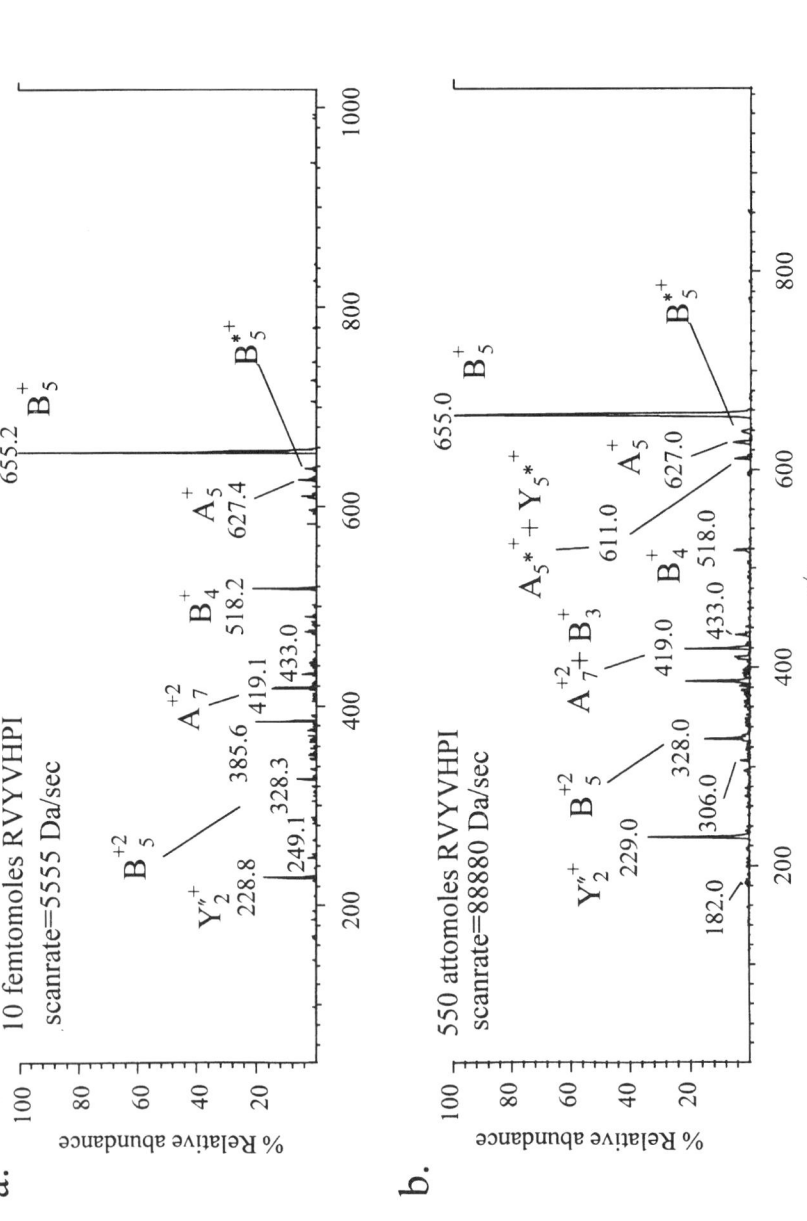

Figure 15. Capillary LC ES–MS–MS spectra of the doubly charged peptide RVYVHPI (MW$_{exact}$ = 882.6 Da) at m/z 442.3 loaded on a 100-μm i.d. capillary column. Spectrum (*a*) shows the results at 10 fmol using the normal scan rate of 5555 Da s^{-1} while (*b*) shows approximately the same S/N ratio at 550 attomoles using a scan rate of 88,880 Da s^{-1}, but with reduced mass resolution. From refs. 67 and 74.

spectra of three proteins, cytochrome c (bovine), hemoglobin (bovine), and apomyoglobin (equine) that were acquired using a Finnigan LCQ^{TM}. A deconvolution algorithm was applied to each multiply charged spectrum and the deconvolved spectra are shown in the insets of Figure 16a–c. The mass accuracy for these three proteins, along with the additional mass accuracy data for other proteins, is shown in Table 1 and was found to be better than 0.01%.

Multiply charged myoglobin, which can retain the prosthetic heme group (holomyoglobin) has been analyzed by CAD in a QIT (142). Isolation of the +8 charge state between m/z 2190–2310, followed by resonance excitation, yielded the heme group at m/z 616 and the +7 multiply charged state of apomyoglobin as shown in Figure 17. In the absence of resonance excitation, myoglobin was stored for 1 s at a pressure of 1 mTorr helium with no dissociation to heme and apomyoglobin species. This study suggests that other biologically important complexes bound noncovalently may be studied with a QIT using the soft-ionization process of ES.

C. Analyses of Oligonucleotides

Although there are hundreds of QIT instruments in use in the world today, few have been configured with negative-ion formation and detection capabilities (143–148). With the advent of matrix-assisted laser desorption ionization (MALDI) and ES ionization, there has been a renewed interest in biomolecular ions that are negatively charged in solution, but could not be ionized intact by most other methods.

Oligonucleotides are one class of biomolecules that are readily ionized by ES in the negative-ion mode because of the ability of the phosphodiester linkages and phosphate groups to stabilize a negative charge (149). Sodium and other metal counterions are commonly found bound to these sites, forming salts with the formula $(M-nNa^+)^{-n}$ and $(M-nNa^+ + mH^+)^{(m-n)}$ (150). However, Figure 18 shows an ES mass spectrum of 5'-dTAGTCTAG-3' with little evidence of bound alkali counterions. In this case, the oligonucleotide was desalted prior to analysis by an ammonium acetate–ethanol precipitation. The desalted oligonucleotide was infused in a 50:50 water:isopropanol solution at 3 μL/min. The charge state and, therefore, the molecular weight may be determined by the spacings of the -2 anion isotopes, as shown in the high-resolution inset.

Tandem mass spectrometry of multiply charged oligonucleotide anions by ES was demonstrated in a QIT by McLuckey et al. (150,151). These authors proposed a nomenclature for different oligonucleotide fragment types analogous to the one developed for peptides (152). Sequencing of small oligonucleotides was possible using MS–MS, but MS^n was recommended for larger oligomers. An example of the product ions observed from collisional activation of an oligonucleotide is shown in Figure 19 (153). The oligomer (M), where M = 5'-d(TGCATCGT)-3', was electrosprayed to form $(M-7H^+)^{-7}$, isolated, and then resonantly excited for 20–40 ms to form product ions. The major route of decomposition involved the loss of an adenine anion (A$^-$), as was seen with other

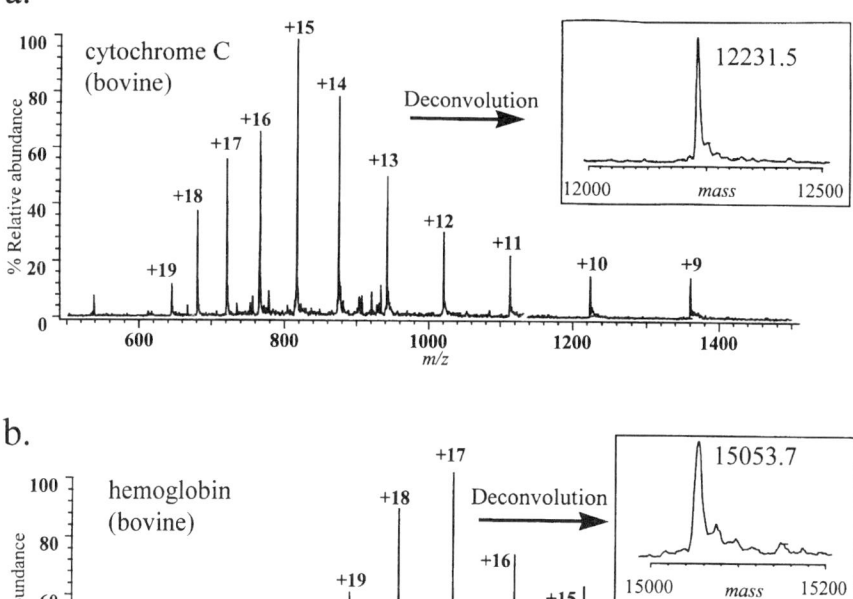
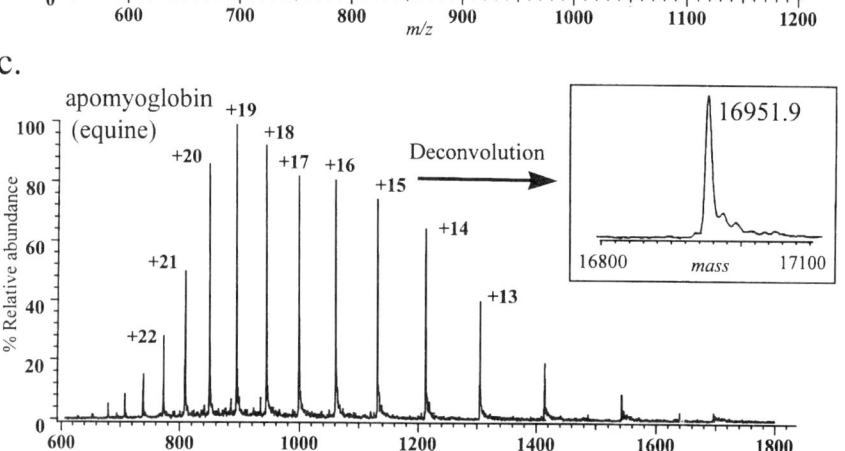

Figure 16. The ES mass spectrum and molecular-weight deconvolution of three proteins: (*a*) cytochrome c, (*b*) hemoglobin (bovine), and (*c*) apomyoglobin (equine) using a Finnigan LCQ^{TM}. A deconvolution algorithm was applied to each multiply charged spectrum and the deconvolved spectra are shown in the three insets indicating the MW$_{avg}$. From T. Vasconcellos; Finnigan Corporaton, 1996.

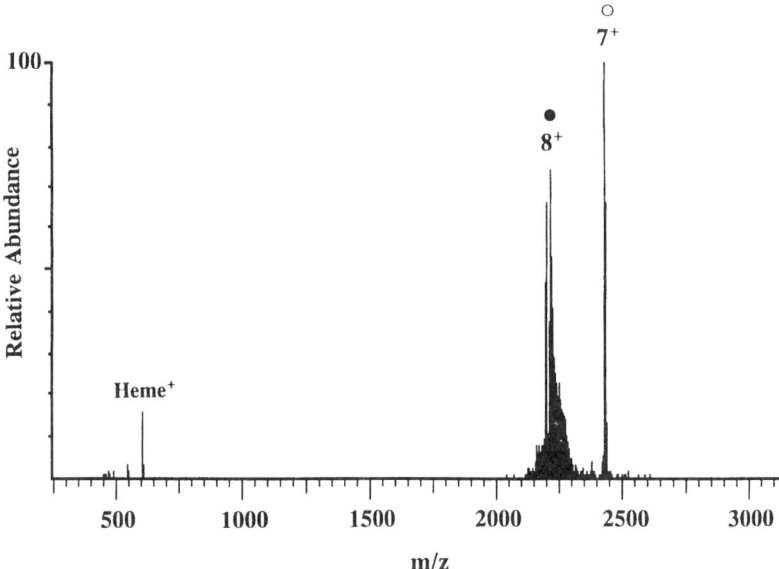

Figure 17. The MS–MS spectrum of the +8 charge state of myoglobin. The spectrum shows the loss of the heme group at m/z 616 from myoglobin. Reprinted with permission of Elsevier Science Inc. from "Gaseous Myoglobin Ions Stored at Greater than 300°K" by S. A. McLuckey, and R. S. Ramsey, *Journal of the American Society for Mass Spectrometry*, **1994**, 5, 324–327. © 1994 American Society for Mass Spectrometry (142).

adenine-containing oligonucleotides in their work. Complementary fragment ions can be located in the MS–MS spectrum to aid in fragment identification. For example, $A^-/(M-7H^+-A^-)^{-6}$ are complementary ions produced from the activated precursor, while $w_4^{-4}/[a_4-B_4(A)]^{-2}$ and $w_4^{-3}/[a_4-B_4(A)]^{-3}$ are complementary ions of activated $(M-7H^+-A^-)^{-6}$. Recent ES work where large-duplex DNA was electrosprayed into a QIT shows an exciting potential application for diagnosing genetic diseases. Doktycz and co-workers have shown ES mass spectra in a QIT from 72- and 75-base single-stranded oligonucleotides and a duplex DNA consisting of complementary strands with 72 base pairs (154). Their ES spectrum was acquired from a 20-μM duplex DNA solution containing 10 mM ammonium acetate. The average charge on this duplex DNA was determined to be approximately −31. Although analytes electrosprayed with high salt concentrations in aqueous solutions show reduced sensitivity, their data still demonstrated the utility of an ES–QIT instrument for the analysis of large, double-stranded DNA. Oligonucleotides up to 76 residues have also previously been electrosprayed and analyzed by Smith et al. using a quadrupole mass spectrometer (155).

Duplex DNA has also been collisionally activated in a QIT. Although the precursor ion was isolated at low resolution, Figure 20 shows the product

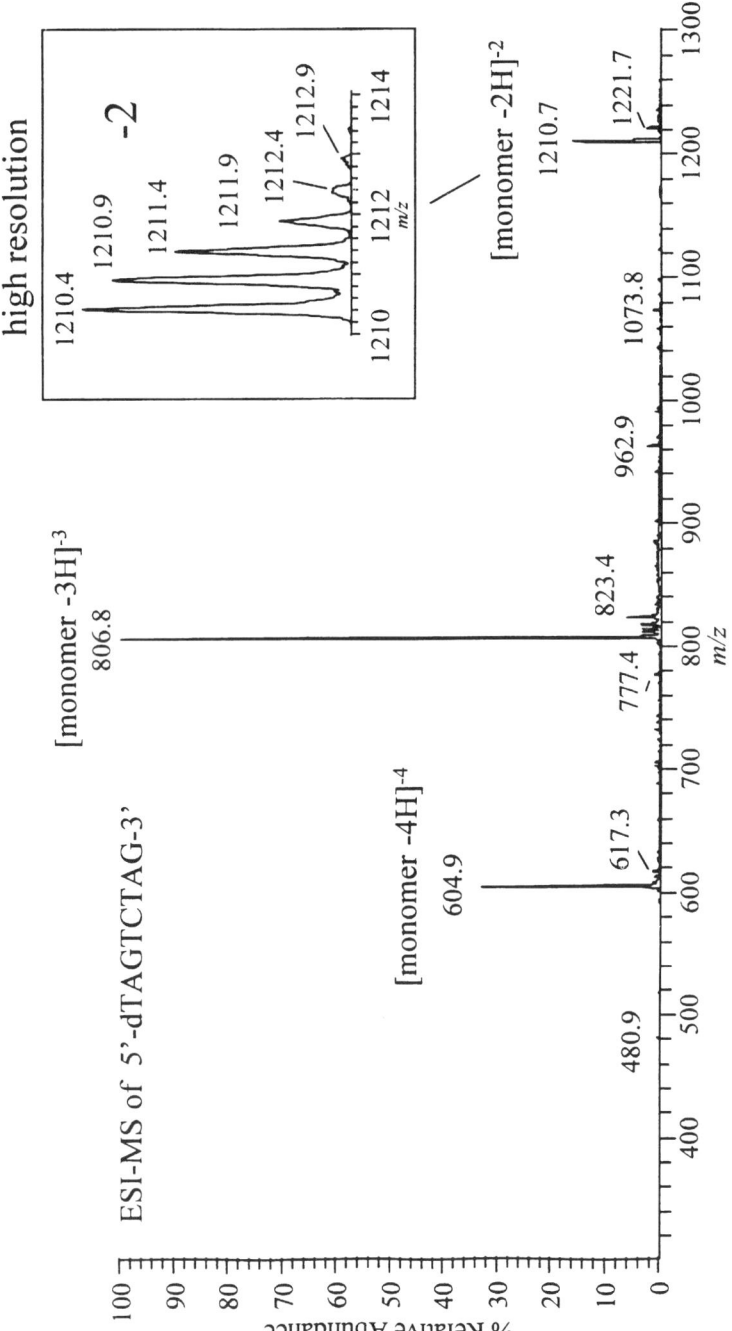

Figure 18. The ES–QIT MS of 5'-TAGTCTAG-3' under normal resolution conditions showing only minor alkali ion addition due to the use of a desalting procedure prior to analysis. The inset shows the high-resolution mass spectrum of the -2 anion. From D. C. Gale, Finnigan Corporation, 1996. Sample provided by R. Griffey, ISIS Pharmaceuticals, Inc.

V. APPLICATIONS OF THE ES–QUADRUPOLE ION TRAP 271

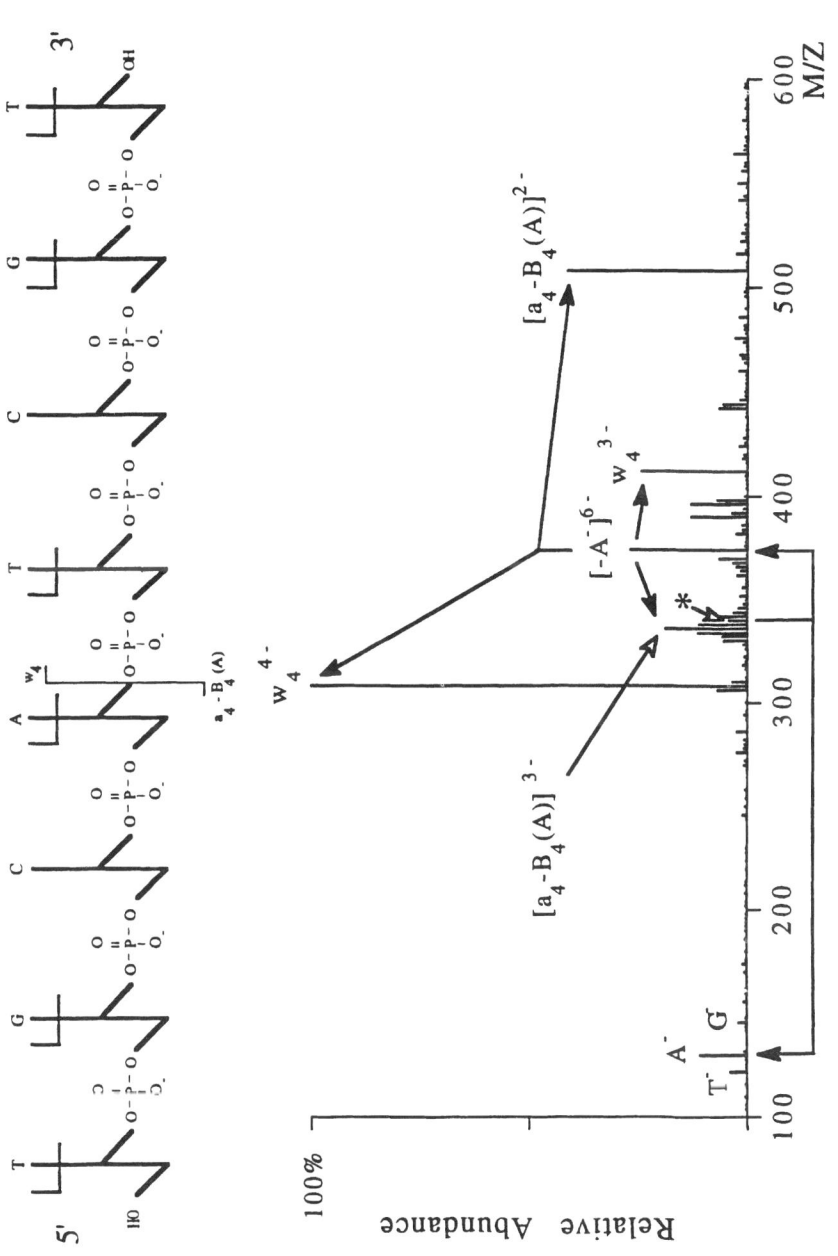

Figure 19. The ES–QIT MS–MS spectrum of $(M-7H^+)^{-7}$ ion, where $M = 5'$-d(TGCATCGT)-$3'$. The asterisk indicates the mass-to-charge ratio of the precursor ion. Closed arrowheads indicate the genealogy of some complementary pairs of fragments. Reprinted with permission from S. A. McLuckey and S. Habibi-Goudarzi, *J. Amer. Chem. Soc.* **1993**, *115*, 12085. © 1993 American Chemical Society.

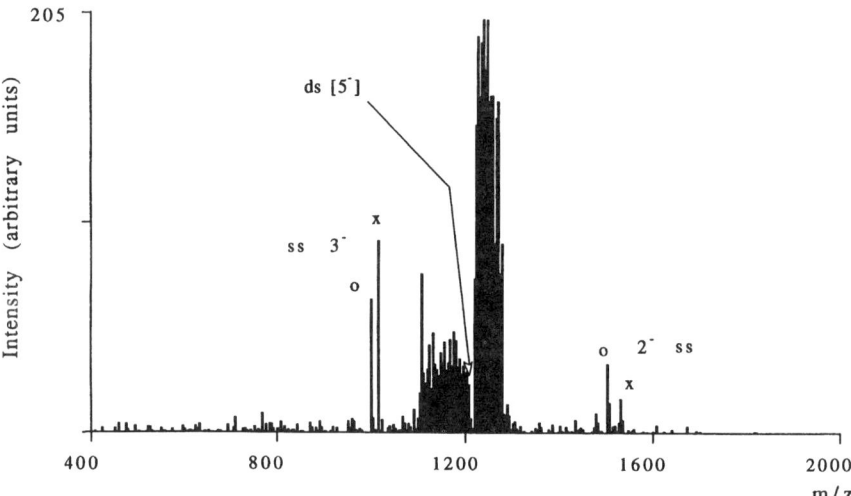

Figure 20. The MS–MS product spectrum from the resonance excitation of the electrosprayed duplex decamer consisting of 5′-d(ACA TTC TGG C)-3′ and 5′-d(GCC AGA ATG T)-3′. The labels x and o represent the single-strand fragments from the excitation of the −5 duplex, respectively. Reprinted with permission from M. J. Doktycz et al., *Anal. Chem.*, **1994**, *66*, 3416. © 1994 American Chemical Society.

spectrum from the resonance excitation of the selected duplex decamer consisting of 5′-d(ACATTCTGGC)-3′ and 5′-d(GCCAGAATGT)-3′ (156). The labels x and o represent the single-strand fragments from the excitation of the −5 duplex, respectively. This data demonstrated that the peak resonantly excited in the large mass window isolated is indeed a duplex ion by noting the complementary product ions (o^{-3} and x^{-2}) and (x^{-3} and o^{-2}) formed from the dissociation of the −5 precursor duplex DNA. This data suggests that the QIT may be ideal at elucidating structural changes on DNA, such as sites of alkylation, by using MS^n.

D. Charge-State Determination by Ion Reactions

As mentioned earlier, at unit mass resolution, there is uncertainty in determining charge states much greater than 2 by isotope separation. In addition to the high-resolution method demonstrated in the peptide section above, the next few sections give examples of charge-state determinations by the use of ion–molecule reactions and ion–ion reactions even at high (> 7) charge states.

1. Charge-State Determination by Ion–Molecule Reactions. McLuckey and Goeringer (157) showed that ion traps could be used to determine the charge state of both highly charged positive or negative macromolecules by using ion–molecule reactions. Macromolecules have broad isotopic distributions which increase with mass. Real samples are also never pure, both in the mixture of counterions that can be present and in the macromolecular heterogeneity. This can result in an overlap of adjacent mass-to-charge ratio distributions. By isolating a narrow window of the broad overlapping distributions, and then allowing a gas-phase reagent to react with these ions, the charge state can readily be determined. This is demonstrated in Figure 21a, where *E. coli t*-RNA was electrosprayed into a QIT and a narrow mass range was selected from the anions. The narrow mass range was next reacted with trifluoroacetic acid for 200 ms. The trifluoroacetic acid transfers protons to the isolated distribution and two higher mass to charge ratio distributions are observed and are shown in Figure 21b. Assuming that the newly formed product distributions at higher mass-to-charge ratio differ in mass by 1 negative charge and 1 Da, owing to the addition of a proton [i.e., $m/z = (m+1)/(z+1)$], the charge of the isolated precursor can be determined.

Similarly, this ion–molecule reaction procedure can be used to determine the charge state of CAD products (158). The $(M + 4H)^{+4}$ cation of melittin was mass selected and resonantly excited to produce the Y_{13}^{+3} at m/z 542. Next, this ion was isolated for a MS^3 reaction step with the strong base 1,6-diaminohexane. The charge state of this ion was determined by the newly observed product ion formed from the single proton-transfer reaction producing Y_{13}^{+2} at m/z 812.

A known clustering reaction can also be used to determine charge states (158). By reacting 1,6-diaminohexane (MW = 116) with the $(M + 4H)^{+4}$ ion of bovine insulin, several cluster species incorporating one, two, and three bases are formed as shown in Figure 22a. Figure 22b shows the MS^3 spectrum where the precursor ion $(M + 4H)^{+4}$ undergoes CAD and the product ion m/z 1430 is isolated and undergoes a reaction with 1,6-diaminohexane. Again one, two, and three bases are attached to this product ion. Given the assumption that two adjacent peaks differ by one base molecule, the charge state (n) for both species is determined to be +4. To make this determination, the amount of increase in m/z of the adjacent peak was divided into m_B, the molecular weight of the base molecule, as shown by Eq. 7:

$$n = m_B/|m/z_1 - m/z_2| \qquad (7)$$

2. Charge-State Determination by Ion–Ion Reactions. Although there has been considerable effort directed at studying ion–molecule reactions in a QIT, there have been few studies involving reactions between positive and negative ions. Quadrupole ion traps and Fourier transform ion-cyclotron resonance-mass spectrometers are unique in that both positive and negative ions can be stored simultaneously (159–162) at least prior to reactive annihilation. The polarity of the ions may also be selected in a QIT by adjusting the a_z value to an

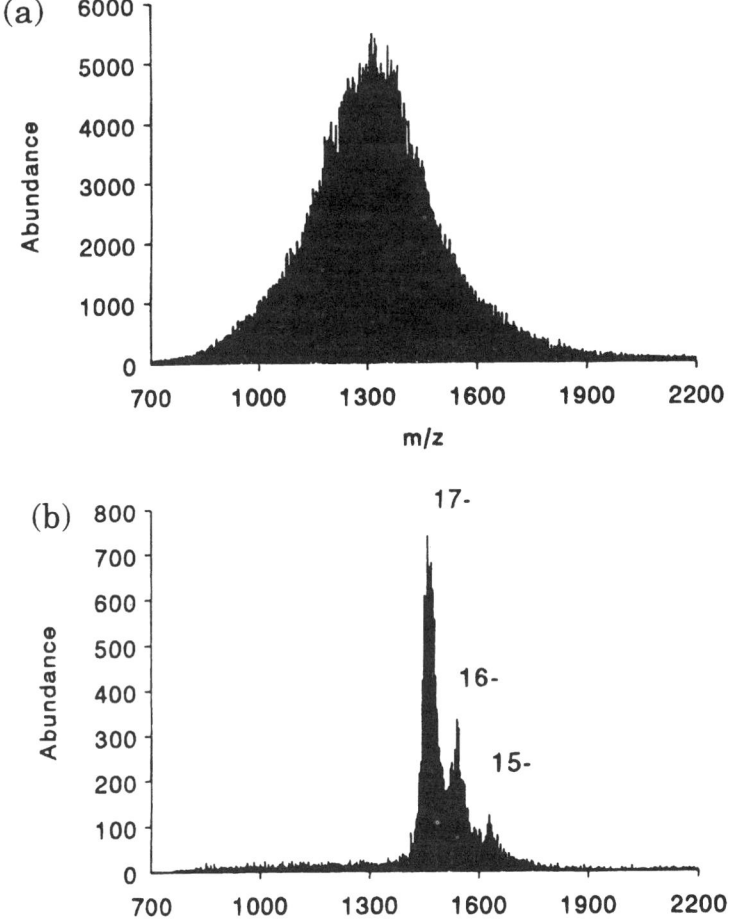

Figure 21. (*a*) Electrospray mass spectrum of anions of *E. coli* *t*-RNA, strain W. (*b*) Product spectrum resulting from isolating a narrow mass range of the ions shown in the spectrum shown in (*a*) and reacting them with trifluoroacetic acid for 200 ms. The trifluoroacetic acid transfers protons to the isolated distribution of ions and two higher mass-to-charge ratio distributions are observed. From these two new product distributions, the mass and charge of the isolated precursor ions can be determined. Reprinted with permission from S. A. McLuckey and D. E. Goeringer, *Anal. Chem.*, **1993**, *67*, 2493. © 1995 American Chemical Society.

appropriate point in the stability diagram. The parameter a_z is set to a calculated nonzero value by applying a dc potential to the ring electrode at some time prior to the mass-selective instability scan.

Herron, Goeringer, and McLuckey (163) have electrosprayed and isolated triply and doubly charged anions from the single-stranded deoxynucleotide 5′-d(AAAA)-3 and allowed the anions to undergo ion–ion proton-transfer reactions with protonated pyridine cations. Pyridine vapor was introduced

V. APPLICATIONS OF THE ES–QUADRUPOLE ION TRAP   275

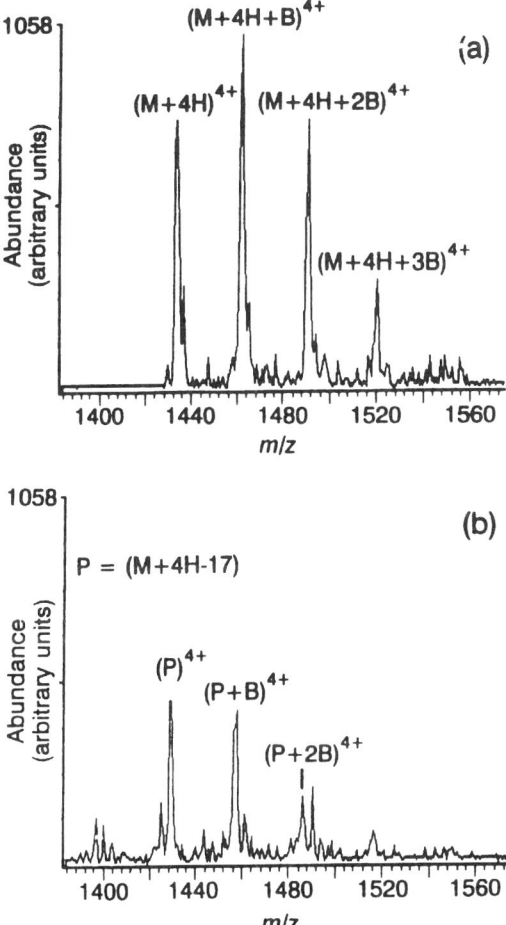

Figure 22. (*a*) The MS–MS spectrum of bovine insulin $(M+H)^{+4}$ where the precursor ions were allowed to react with 1,6-diaminohexane. (*b*) The MS^3 spectrum where the isolated precursor ion undergoes CAD and then the isolated product ions at m/z 1430 react with the base. The charge state was determined to be +4 in both cases. Reprinted with permission from S. A. McLuckey et al., *Anal. Chem.*, **1991**, *63*, 1975. © 1991 American Chemical Society.

into the vacuum chamber at $1-3 \times 10^{-7}$ Torr (uncorrected) and the protonated cations were produced by internally ionizing pyridine by electron ionization using a radially injected electron beam. This pressure of pyridine was sufficient to cause complete protonation of the molecular ion through self-chemical ionization (self-CI) (164–166) in 10 ms. Next, the single-stranded DNA anions were injected into the QIT and simultaneously stored with the pyridine cations for up to 1 s during which time the ion–ion proton-transfer reactions occurred.

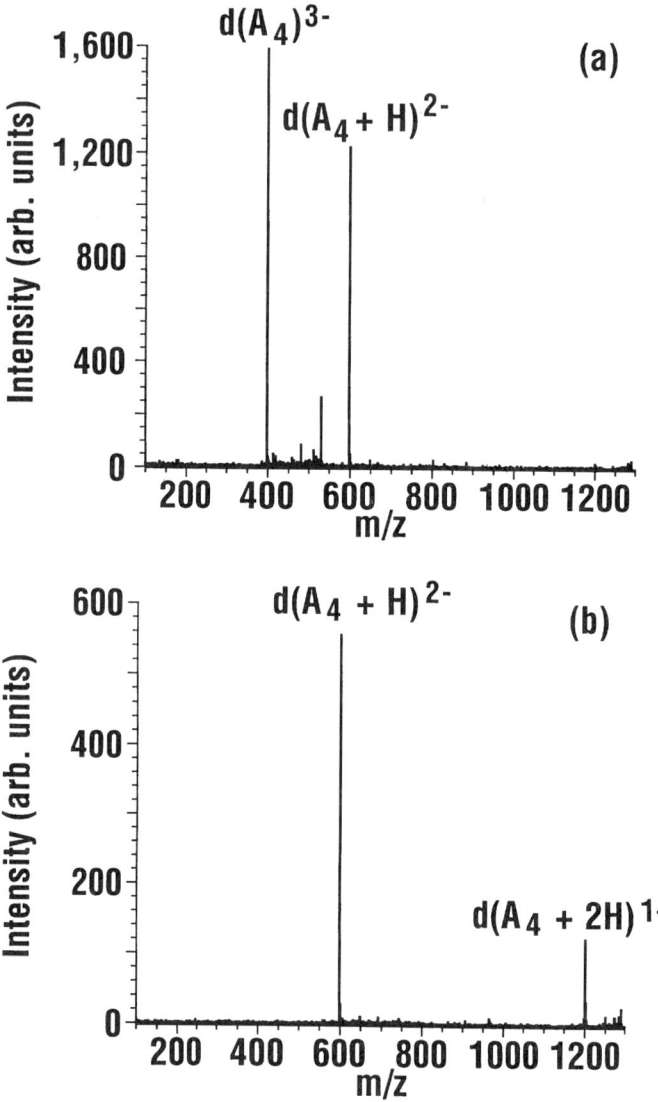

Figure 23. The MS–MS product ions formed from the reaction of isolated d(A$_4$)$^{-3}$ with protonated pyridine. (*a*) The d(A$_4$)$^{-3}$ ion undergoes ion–ion proton transfer with the protonated pyridine to form the doubly charged species d(A$_4$ + H)$^{-2}$. (*b*) The result from reacting the doubly charged product anion d(A$_4$ + H)$^{-2}$ from the first reaction with the protonated pyridine to form the singly charged anion d(A$_4$ + 2H)$^{-1}$. Reprinted by permissiom of Elsevier Science Inc. from "Ion–Ion Reactions in the Gas Phase: Proton Transfer Reactions of Protonated Pyridine with Multiply Charged Oligonucleotide Anions," by W. J. Herron, D. E. Goeringer, and S. A. McLuckey, *Journal of the American Society for Mass Spectrometry*, **1995**, *6*, 529–532. © 1995 American Society for Mass Spectrometry (163).

Figure 23a shows the product ions formed from the reaction of isolated $d(A_4)^{-3}$ with protonated pyridine. The $d(A_4)^{-3}$ ion undergoes ion–ion proton transfer with the protonated pyridine to form the doubly charged species $d(A_4+H)^{-2}$.

Figure 23b shows the result from further reacting the doubly charged product anion $d(A_4+H)^{-2}$ from the first reaction with protonated pyridine to form the singly charged anion $d(A_4+2H)^{-1}$. Surprisingly, they reported very little fragmentation for what is a very exothermic process. More recently, the Oak Ridge group has shown ion charge-state determinations in MS^3 experiments where pyridine reacts with product anions of 5′-d(AAAA)-3′ and the A-chain of bovine insulin (167). In these examples, a reaction time of 200 ms allows for the proton transfer reaction to proceed. The isolated product ions produced by CAD were further reacted with protonated pyridine to produce protonated MS^3 product ions. Again, by assuming a change in mass of 1 Da, the charge state was readily obtained.

One of the limitations of all three of the charge-state determination techniques discussed above is that the long reaction times (200 ms) necessary, lengthen the overall scan time. As a result, fewer scans would be acquired across an eluting peak from an LC or CE column. Reducing the mass range to two or three isotope distributions, however, can help shorten the scan time. An additional limitation of the ion–molecule charge state determination method is that it relies on the reactivity of the ion with a given neutral and all ions do not react to the same degree, if at all. Also, the neutral reactant is always present in the QIT chamber in the ion–molecule experiment, unless a pulsed valve is used to introduce the gas only during the reaction period. The ion–ion reaction method is far more universal in reactivity and is more easily controlled (167), but it does require the ionization and capturing of both positive and negative ions.

E. Capillary Electrophoresis ES–QIT

Capillary electrophoresis (CE) is an ion-mobility separation technique that is rapid and efficient at resolving biomolecules such as peptides and proteins. Chapter 10 covers the interfacing of this technique to ES and a review of CE–MS is available for further reading (168). Coupling such a technique to a QIT stems from the fact that this analyzer has proven valuable at analyzing biomolecules from an ES source. Several laboratories have coupled CE to an ES–QIT (169–172) and have achieved impressive results. Attomole detection limits have been achieved, although often with concentrated samples. An early example of CE–MS from our laboratory, using an experimental ES–QIT (169), separated a mixture of three peptides at the 190-fmol level. Two of the peptides, renin substrate (MW = 1759 Da) and substance P (MW = 1349 Da), are shown separated in Figure 24 at a high-CE resolution ($N > 100,000$) due the use of an aminopropyl-silated fused-silica column. This column was recommended by Moseley et al. (173) so that the separation could take place at an acidic pH for optimum ES ionization and still provide narrow CE peak widths.

Henion et al. reported the first use of CE–ES–QIT to quantitatively deter-

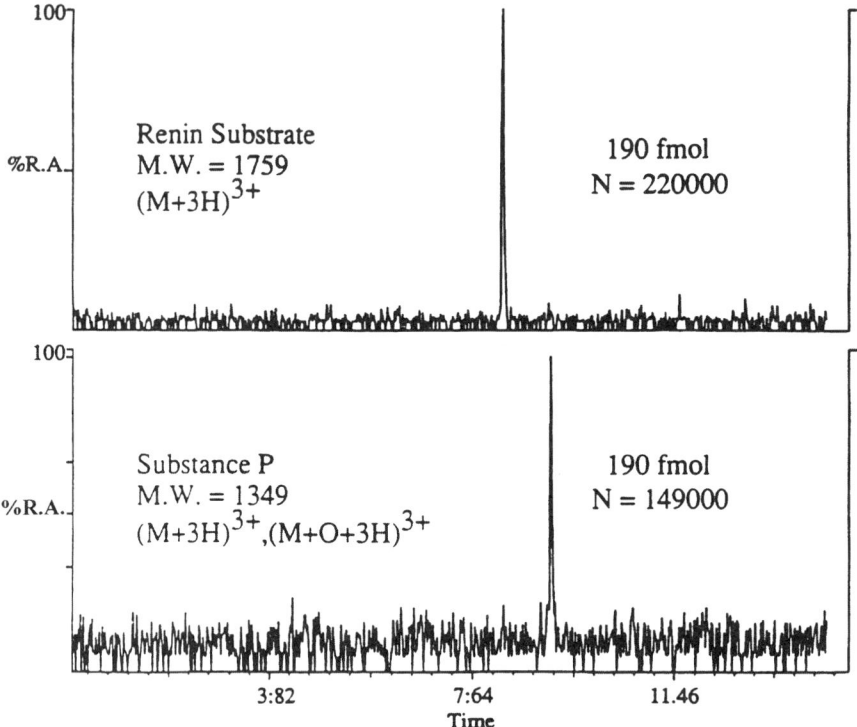

Figure 24. The CE–ES–QIT analysis of the peptides renin substrate (MW = 1759 Da) and substance P (MW = 1349 Da), separated at a CE resolution where $N > 100,0000$. The separation occurred in an aminopropyl-silated, fused-silica column at acidic pH to achieve optimum ES ionization conditions. From ref. 169.

mine the amounts of the isoquinoline alkaloids berberine and palmatine in the bark of *Phellodendron wilsonii* (171). Dried *P. wilsonii* tree bark is used in China and other countries to treat dysentery and jaundice (174). Relative migration times and ES source CAD was used to identify these compounds. Signal-to-noise ratios greater than 10 were demonstrated for mass chromatograms of a synthetic mixture of these compounds at the 400–500 attomole level. At the picogram level, the calibration curves for berberine and palmatine were linear, with correlation coefficients of 0.998 and 0.999, respectively. Berberine and palmatine were determined in the bark at 13.95 μg/mg and 0.234 μg/mg, respectively.

The determination of mid-attomole detection limits for the peptide leucine-enkephalin in total-ion electropherograms using 20-μm i.d. capillaries without using a liquid sheath has been acheived by Ramsey et al. (175). The 20-μm i.d. capillaries were tapered down to 3 μm i.d. to form the ES tip and coated with gold to make the electrical connection. A 650-attomole injection resulted in a S/N ratio of 6:1 in the total ion electropherogram (TIE) and 25:1 in the reconstructed mass electopherogram of $(M + H)^+ = 556$. Overall, the nonliquid sheath-capillary ES interface showed a gain in signal of 3 to 4 over the sheath-flow ES source.

F. Detection Limits Determined on an ES–QIT

Detection limits of analytes measured in ES–QIT mass spectrometers are quite noteworthy. Table 2 lists detection limits for various molecules analyzed by LC, CE, and FIA coupled to an ES–QIT. The lowest limits of detection are measured when an isolation step (e.g., TWF) is used to accumulate and/or isolate only the ion(s) of interest. For example, in Table 2, the detection limit for reserpine in SIM and SRM scan modes is better than in the full-scan mode. Although QIT full MS scans have a higher duty cycle over ion-beam instruments, detection limits can be reduced in the full-scan MS mode because of matrix effects. In this latter case, at low analyte amounts, the QIT is filled up primarily with matrix ions, which leaves little space to store the ions of interest. Detection limits are outstanding in the full-scan MS–MS mode because the duty cycle is high and the effects of the matrix are reduced during isolation.

G. ES–QIT Data-Dependent Scans

Data-dependent scans are a type of scan which utilizes algorithms that have been written to make real-time unassisted decisions to control the acquisition mode or

TABLE 2. Detection Limits of Various Analytes by ES–QIT[a]

Analyte	Introduction Technique	Scan Type (m/z)	Ion(s) (m/z)	Detection Limit	S/N (Ref.)
Octaethylporphyrin	FIA	300–600	535^+	18 fmol	4 (176)
RVYVHPI	μLC	MS–MS	442^{+2}	10 fmol	20 (74)
	μLC	Fast-MS–MS	442^{+2}	550 amol	20 (67)
Bradykinin	μCE	100–1300	TIE	180 fmol	20 (84)
Leu-enkephalin	μCE	100–1200	556^+	650 amol	25 (175)
	μCE	100–1200	TIE	650 amol	6 (175)
Renin substrate	μCE	100–2000	587.3^{+3}	190 fmol	15 (169)
Substance P	μCE	100–2000	$450.7^{+3}, 456.0^{+3}$	190 fmol	8 (169)
T-T (sodiated)	FIA	100–350	275^+	730 fmol	4 (177)
Berberine	LC	150–400	336^+	19 fmol	3 (178)
	μCE	140–400	TIE	51 fmol	20 (173)
	μCE	140–400	336^+	90–130 amol	3 (173)
Magnoflorine	μCE	140–400	TIE	37 fmol	15 (173)
	μCE	140–400	343^+	90 amol	3 (173)
Reserpine	FIA	SIM	609^+	8 fmol	7
	FIA	SRM	$397^+, 436^+, 448^+$	8 fmol	20
	FIA	200–700	609^+	82 fmol	10

[a] TIE = total ion electropherogram; μLC = capillary LC (100 μm i.d.); flow rate $\approx$ 1 μL/min; FIA = flow-injection analysis (200 μL/min), LC = 2 mm i.d. column zorbax; μCE = 20–25 μm i.d. column; 50 μm i.d. column (175); μCE = aminopropyl–silated column (169); T–T = thymine–thymine.

system parameters. Essentially, the data system executes the next scan type based on previously collected data (179–181). For example, an ES–QIT system can be programmed to implement the following experiment: (1) acquire a full-scan mass spectrum, (2) select the base peak from the mass spectrum if it meets a preset signal-threshold requirement, (3) determine the charge state and, thus, ion mass to charge ratio by a high-resolution scan, and (4) acquire a MS–MS scan with the appropriate product mass range for that ion. Data-dependent scans provide system automation that can greatly increase the amount of information gained from a single LC analysis. The QIT has the advantage over other mass spectro-

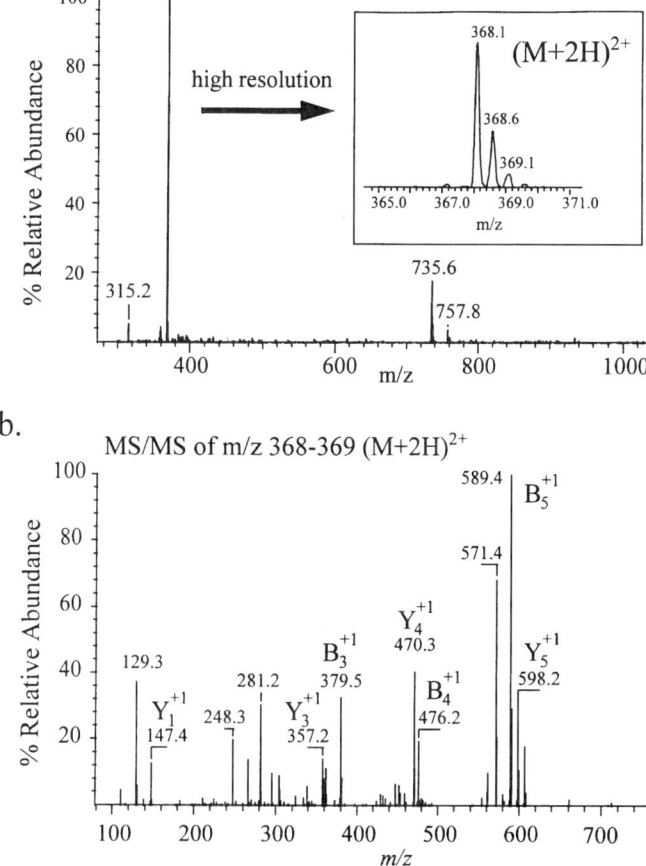

Figure 25. An example of data-dependent scanning during an LC separation of a myoglobin tryptic digest. (*a*) A full-scan MS of a component of the digest eluting from the column. This scan is followed by a charge-state determination of the largest peak in the MS spectrum by a high-resolution scan shown in the inset. A MS–MS scan of the ion is shown in (*b*) providing product ion information for automated protein identification. From A. Land, Finnigan Corporation, 1995.

meters because it can immediately switch between a full-scan MS and a CAD product scan in the presence of helium collision gas with no loss in signal, CAD efficiency, or time to switch gases on or off. Ion-beam mass spectrometers such as triple-stage quadrupoles, would have to reduce the gas pressure in the collision cell to achieve similar full-scan MS results. The initial work using data-dependent scans on a QIT has identified proteins associated with hypertrophy of cardiomyocytes (182), analyzed digests from proteins separated by two-dimensional polyacrylamide gel electrophoresis (183), identified proteins (184), and identified phosphorylation sites of peptides by a MS–high resolution–MS^2–MS^3 sequence of scans (185,186). Figure 25a,b shows an example of the dependent scan discussed above, acquired during the LC elution of a myoglobin tryptic digest. In Figure 25a, an MS scan of 10 pmol of an unknown digest constituent is shown. The charge state and thus fragment mass is determined for m/z 368.5, as shown in the inset by a high-resolution scan. The charge state was determined to be +2 and the mass range was set for the MS–MS scan shown in Figure 25b. This peptide was searched for and identified in a protein database by a MS–MS cross-correlation algorithm (SEQUEST) (187–189) and found to be the subsequence HKIPIK (MW = 734) from horse-heart myoglobin. Protein identification using MS^3 and MS^4 scans are also now being explored with successful results (190).

VI. SUMMARY

Over 40 years of experimentation have passed since Wolfgang Paul's invention of the quadrupole ion trap. The QIT has truly evolved over that time into a remarkable, high-performance mass spectrometer. The marriage of ES to a QIT has made the coupling of LC and CE separation techniques routine and the analyst can now readily employ an assortment of scan types (e.g., MS^n, SIM, SRM^n) for real-time detailed structural and/or quantitative analyses. High-resolution scans used for charge-state determinations are now routine for up to four charges at m/z 2000 and even higher-resolution scans are expected in the future. Detection limits determined with the ES–QIT are exemplary, especially in the full-scan MS–MS mode. Finally, the QIT can be an extremely useful electrostatic reaction vessel for electrosprayed ion–molecule and ion–ion reactions. This area of study will no doubt see future growth and may lead to some enlightening biochemical discoveries. With these new capabilities in the hands of many more scientists, new niches for the modern Paul ion trap will be revealed and the traditional ones should flourish.

ACKNOWLEDGMENTS

We thank Dr. Pamela B. Nakajima, Dr. Jon W. Amy, and the Reviewers for helpful comments. We thank Dr. Scott McLuckey and Dr. James L. Stephenson of Oak Ridge National Laboratory for generously supplying us with preprints. We thank Mr. Kevin Wheeler, Dr. Adrian Land, Dr. Mark Sanders, Mr. Matt Sweeney, Mr. Tom Vasconcellos, Dr. David Gale, Dr. Tanuja Chaudhary, and Dr. Ian Jardine of Finnigan Corporation for some of the peptide and protein data and helpful comments. Finally, we would like to thank the Finnigan LCQ^{TM} development team.

REFERENCES

1. Dole, M.; Mach, L. L.; Hines, R. L., Mobley, R. C.; Ferguson, L. D.; Alice, M. B. *J. Chem. Phys.* **1968**, *49*, 2240.
2. Clegg, G. A.; Dole, M. *Biopolymers* **1971**, *10*, 821–826.
3. Dole, M.; Cox, H. L.; Gieniec, J. *Advances in Chemistry Series, No. 125, Polymer Molecular Weight Methods*; Mason, E.A. Ed.; Electrospray Mass Spectroscopy, Publisher, 1973, Chapter 7, p. 73.
4. Gieniec, J.; Mack, L. L.; Nakamae, K.; Gupta, C.; Kumar, V.; Dole, M. *Biomed. Mass Spectrom.* **1984**, *11*, 259–268.
5. Fenn, J. B.; Mann, M.; Meng, C. K.; Wong, S. F. *Mass Spectrom. Rev.* **1990**, *9*, 37.
6. Aleksandrov, M. L.; Gall, L. N.; Krasnow, V. N.; Nikolaev, V. I.; Palenko, V. A.; Shkurov, V. A.; Baram, G. I.; Gracher, M. A.; Knorre, V. D.; Kusner, Y. S. *Bioorg. Khim.* **1984**, *10*, 710.
7. Meng, C. K.; Mann, M.; Fenn, J. B. *Z. Phys. D.* **1988**, *2*, 95.
8. Paul, W.; Steinwedel, Z. *Z. Naturforsch.* **1953**, *8a*, 448–450.
9. Paul, W.; Steinwedel, H. *Z. Naturforsch. A*, **1956**, *104*, 672.
10. Paul, W.; Steinwedel, H. *German Patent 944900*, **1956**.
11. Paul, W. *Angew. Chem. Int. Ed. Engl.* **1990**, *29*, 739.
12. Stafford, G. C.; Kelley, P.; Stephens, D. R. U.S. Patent 4540884, 1985.
13. Stafford, G. C.; Kelley, P. E.; Syka, J. E. P.; Reynolds, W. E., Todd J. F. J. *Int. J. Mass Spectrom. Ion Proc.* **1984**, *60*, 85–98.
14. Stafford, G. C., Taylor, D. M.; Bradshaw, S. C. U.S. Patent 5107109, 1992.
15. Syka, J. E. P.; Louris, J. N., Kelley, P. E.; Stafford, G. C.; Reynolds, W. E. U.S. Patent 4736101, 1988.
16. Weber-Grabau, M. U.S. Patent 4818869, 1989.
17. Louris, J. N.; Cooks, R. G.; Syka, J. E. P.; Kelley, P. E.; Stafford, G. C., Todd, J. F. J. *Anal. Chem.* **1987**, *59*, 1677–1685.
18. Louris, J. N.; Brodbelt-Lustig, J. S.; Cooks, R. G.; Glish, G. L.; Van Berkel, G. J.; McLuckey, S. A. *Int. J. Mass Spectrom Ion Proc.* **1990**, *96*, 117–137.
19. Louris, J. N.; Amy, J. W.; Ridley, T. Y.; Cooks, R. G. *Int. J. Mass Spectrom. Ion Proc.* **1989**, *88*, 97–111.
20. Johnson, J. V., Pedder, R. E.; Yost, R. A. *Int. J. Mass Spectrom. Ion Proc.* **1991**, *106*, 197–212.
21. Johnson, J. V.; Pedder, R. E.; Yost, R. A.; Story, M. S. U.S. Patent 5075547, 1991.

22. Kaiser, R. E.; Cooks, R. G., Stafford, G. C.; Syka, J. E. P.; Hemberger, P. H. *Int. J. Mass Spectrom Ion Proc.* **1991**, *106*, 79–115.
23. Kaiser, R. E.; Cooks, R. G.; Moss J.; Hemberger, P. H. *Rapid Commun. Mass Spec.*, **1989**, *3*, 50–53.
24. Kaiser, R. E. Jr.; Louris, J. N.; Amy, J. W.; Cooks R. G.; *Rapid Commun. Mass Spec.*, **1989**, *3*, 225–229.
25. Schwartz, J. C.; Syka J. E. P.; Jardine I. *J. Am. Soc. Mass Spectrom.* **1991**, *2*, 198–204.
26. Kaiser, R. E. Ph.D. Thesis, Purdue University, 1990, pp. 120–123.
27. Londry, F. A.; Wells, G. J.; March, R. E. *Rapid Commun. Mass Spec.* **1993**, *7*, 43–45.
28. Goeringer, D. E.; McLuckey, S. A.; Glish, G. L. Proceedings of the 39th Annual Conference of Mass Spectrometry and Allied Topics, Nashville, TN, 1991, pp. 532–533.
29. Kelley, P. E. U.S. Patent No. 5134286, 1992.
30. Kelley, P. E. U.S. Patent No. 5206507, 1993.
31. Hoekman, D. J.; Kelley, P. E. U.S. Patent 5,256,875, 1993.
32. Goeringer, D. E.; Asano, K. G.; McLuckey, S. A.; Hoekman, D.; Stiller, S. W. *Anal. Chem.* 1994, **66**, 313–318.
33. Guan, S.; Marshall, A. G. *Anal. Chem.* **1993**, *65*, 1288–1294.
34. Brodbelt, J. S.; Cooks, R. G. *Anal. Chem. Acta* **1988**, *206*, 239–251.
35. McLuckey, S. A.; Van Berkel, G. J.; Glish, G. L. *J. Am. Chem. Soc.* **1990**, *112*, 5668.
36. Louris, J. N.; Syka, J. E. P.; Kelley, P. E. U.S. Patent 4,686,367, 1987.
37. Weber-Grabau, M.; Bradshaw, S. C.; Syka, J. E. P. U.S. Patent 4,771,172, 1988.
38. Kaiser, R. E., Jr.; Williams, J. D.; Schwartz, J.C.; Lammert, S.A.; Cooks, R. G.; *Proceedings of the 37th Annual Conference of Mass Spectrometry and Allied Topics*, Miami, FL, 1989, pp. 369–370.
39. Bier, M. E.; Hartford, R. E. Herron J. R.; Stafford, G. C. *Proceedings of the 39th ASMS Conference on Mass Spectrometry and Allied Topics*, Nashville, TN, 1991, pp. 538–539.
40. McLuckey, S. A.; Glish G. L.; Asano, K. G. *Anal. Chim. Acta.* **1989**, *225*, 25–35.
41. Lister, A. K.; Wood, K. V.; Cooks, R. G.; Noon, K. R. *Biomed. Environ. Mass Spectrom.* **1989**, *18*, 1063.
42. Bier, M. E.; Winkler, P. C.; Herron, J. R. *J. Am. Soc. Mass Spectrom.* **1993**, *4*, 38–46.
43. Louris, J. N.; Amy, J. W.; Ridley, T. Y.; Cooks, R. G. *Int. J. Mass Spectrom. Ion Processes* **1989**, *88*, 97–111.
44. Heller, D. N.; Lys, I.; Cotter, R.; Uy, O. M. *Anal. Chem.* **1989**, *61*, 1083–108.
45. Cox, K. A.; Williams, J. D.; Cooks, R. G.; Kaiser, R. E. *Biol. Mass Spectrom.* **1992**, *21*, 226.
46. Yates, J., *Rapid Commun. Mass Spectrom.* **1993**, *7*, 20–26.
47. Schwartz, J. C.; Bier, M. E. *Rapid Commun Mass Spectrom.* **1993**, *7*, 27–32.
48. Taylor, L. C. E.; Singh, R.; Cahng, S. Y.; Johnson, R. L.; Schwartz, J. S. *Rapid Commun. Mass Spectrom.* **1995**, *9*, 902–910.
49. Van Berkel, G. J., Glish, G. L.; McLuckey, S. A. *Anal. Chem.* **1990**, *63*, 1284–1295.

50. Schwartz, J. C.; Jardine, I. *Proc. of the 38th ASMS Conf. on Mass Spectrom. and Allied Topics*, Tucson, AZ 1990, pp. 16–17.
51. Mordehai, A. V.; Hopfgartner, G.; Huggins, T. G.; Henion, J. D. *Rapid Commun Mass Spectrom.* **1992**, *6*, 508–516.
52. March, R. E. *Int. J. Mass Spectrom. Ion Procs.* **1992**, *118/119*, 71–135.
53. McLuckey, S. A.; Van Berkel, G. J.; Goeringer, D. E.; Glish, G. L. *Anal. Chem.* **1994**, *66*, 689A–696A.
54. Todd, J. F. J. *Mass Spectrom. Rev.* **1991**, *10*, 3.
55. Todd, J. F. J.; Penman, A. D. *Int. J. Mass Spectrom. Ion Process.* **1991**, *106*, 1.
56. Cooks, R. G.; Glish, G. L.; McLuckey, S. A.; Kaiser, R. E. *Chem. Eng. News*, **1991**, *69*, 26.
57. McLuckey, S. A.; Van Berkel, G. J.; Goeringer, D. E.; Glish, G. L. *Anal. Chem.* **1994**, *66*, 737A–743A.
58. McLuckey, S. A.; Van Berkel, G. J.; Glish, G. L.; Schwartz, J. C. In *Practical Aspects of Ion Trap Mass Spectrometry*; March, R. E.; Todd, J. F. J. Ed.; CRC Series Modern Mass Spectrometry: New York, 1995; Vol II, 1995; pp. 89–141.
59. Schwartz, J.C.; Jardine, I. In *High Resolution Separation and Analysis of Biological Macromolecules, Part A: Fundamentals*; Karger, B.; Hancock, W. Eds.; Spectrum Pub. Services: York, PA; 1996, Vol. 270.
60. March, R. E.; Hughes, R. J. *Quadrupole Storage Mass Spectrometry*; Winefordner, Ed.; Wiley: New York, 1989, Vol. 102.
61. *Quadrupole Mass Spectrometry and its Applications*; Dawson, P.H. Ed.; Elsevier: Amsterdam, 1976 (republished by the American Instutute of Physics, Woodbury, NY, 1995).
62. *Practical Aspects of Ion Trap Mass Spectrometry*; March, R. E.; Todd, J. F. J. Eds.; CRC Press; Boca Raton, FL, 1995, Vols. 1–3.
63. *Practical Aspects of Ion Trap Mass Spectrometry*; March, R. E.; Todd, J. F. J. Eds.; CRC Press; Boca Raton, FL, 1995; Vol. 1, Ch. 1.
64. Wang, Y.; Franzen, J. *J. Int. J. Mass Spectrom. Ion Procs.* **1992**, *112*, 167–178.
65. Louris, J.; Schwartz, J.; Stafford, G.; Syka, J.; Taylor, D. *Procs. of the 40th ASMS Conf. on Mass Spectrometry and Allied Topics*, Washington, DC May 31–June 5, 1992, 1003–1004.
66. Yost, R. A.; McClennen, W.; Snyder, A. P. *Proceedings of the 35th Annual Conference of Mass Spectrometry and Allied Topics*, Denver, CO, 1987, pp. 789–790.
67. Bier, M. E.; Schwartz, J. C.; Zhou, J.; Syka J. E. P.; Taylor D.; Land A.; James, M.; Fies, B.; *Procs. of the 43rd ASMS Conference on Mass Spectrometry and Allied Topics*, Atlanta, GA, May 21–26, 1995, 988.
68. Cleven, C. D.; Cox, K. A.; Cooks, R. G.; Bier, M. E. *Rapid Commun Mass Spectrom.* **1994**, *8*, 451–454.
69. Cox, K. A.; Cleven, C. D.; Cooks, R. G. *Int. J. Mass Spectrom. Ion Procs.* **1995**, *144*, 47–65.
70. Stafford, G. C.; Kelley, P. E.; Syka, J. E. P.; Reynolds, W. E.; Todd, J. F. J. *Int. J. Mass Spectrom. Ion Phys.* **1984**, *60*, 85–98.
71. Williams, J. D.; Cooks, R. G.; Syka, J. E. P.; Hemberger, P. H.; Nogar, N. S. *J. Am. Soc. Mass Spectrom.* **1993**, *4*, 792–797.

72. Yates, N.; Kottmeier, D.; Shabanowitz, J.; Hunt, D. *Procs. of the 42nd ASMS Conference on Mass Spectrometry and Allied Topics*, Chicago, IL, May 29–June 3, 1994, p. 212.
73. Schwartz, J. C.; Bier, M. E.; Taylor, D. M.; Zhou, J.; Syka, J. E. P.; James, M. S.; Stafford, G. C. *Procs. of the 43rd Conference on Mass Spectrometry and Allied Topics*, Atlanta, GA, May 21–26, 1995, p. 1114.
74. Bier, M. E.; Schwartz, J. C.; Zhou, J.; Taylor, D. M.; Syka, J. E. P.; James, M. S.; Fies, B.; Stafford, G. C. *Procs. of the 43rd Conference on Mass Spectrometry and Allied Topics*, Atlanta, GA, May 21–26, 1995, p. 1117.
75. Bruins, A. P.; Covey, T. R.; Henion, J. D. *Anal. Chem.* **1987**, *59*, 2642–2646.
76. Chowdhury, S. K.; Katta V.; Chait B. T. *J. Am. Soc. Mass Spectrom.* **1990**, *1*, 382–388.
77. Whitehouse, C. M.; Dreyer, R. N.; Yamashita, M.; Fenn, J. B. *Anal. Chem.* **1985**, *57*, 675–679.
78. McLuckey, S. A.; Glish, G. L.; Asano, K. G.; Bartmess, J. E. *Int. J. Mass Spectrom. Ion Procs.* **1991**, *109*, 171–186.
79. Ashkenas, H.; Sherman, F. S. In *Rarefied Gas Dynamics*, De Leeuw, J., Ed.; Academic: New York, 1966; Vol. 2, p. 84.
80. Teloy, E.; Gerlich, D. *Chem. Phys.* **1974**, *4*, 417.
81. Tosi, P.; Fontana, G.; Longano, S.; Bassi, D. *Int. J. Mass Spectrom. Ion Proc.* **1989**, *93*, 95–105.
82. Syka, J. E. P.; Szabo I. *Procs. of the 36th ASMS Conf. on Mass Spectrometry and Allied Topics*, June 5–10, San Francisco, CA, 1988, pp. 1328–1329.
83. Yost, R. A.; Enke, C. G. *Anal. Chem.* **1979**, *51*, 1251A–1264A.
84. Ramsey, R. S.; Goeringer, D. E.; McLuckey S. A. *Anal. Chem.* **1993**, *65*, 3521–3524.
85. Kaiser, R. Ph.D. Thesis, Purdue University 1990, 120–123.
86. Eichelberger, J. W.; Budde, W. L.; Slivon, L. E. *Anal. Chem.*, **1987**, *59*, 2730–2732.
87. Schwartz, J. C.; Louris, J. N. U.S. Patent 5285063, 1994.
88. Reiser, H.-P.; Kaiser, R. E.; Savickas, P. J.; Cooks, R. G. *Int. J. Mass Spectrom. Ion Procs.* **1991**, *106*, 237–247.
89. Mordehai, A. V.; Henion, J. *Rapid Commun. Mass Spectrom.* **1993**, *7*, 205–209.
90. Louris, J. N.; Taylor, D. M. *U.S. Patent* 5,324,939, 1994.
91. Doroshenko, V. M.; Cotter, R. J. *Rapid Commun. Mass Spectrom.* **1994**, *8*, 766–776.
92. Bier, M. E. U.S. Patent Pending.
93. Moini, M. *Rapid Commun. Mass Spectrom.* **1994**, *8*, 711–714.
94. Wells, G.; Hunston, C. *Anal. Chem.* **1995**, *67*, 3650–3655.
95. Yates, N. A.; Yost, R. A.; Bradshaw, S. C.; Tucker, D. B. *Proceedings of the 39th Annual Conference of Mass Spectrometry and Allied Topics*, Nashville, TN, 1991, pp. 1489–1490.
96. Kaiser, R. E.; Cooks, R. G.; Syka, J. E. P.; Stafford, G. C. *Rapid Commun. Mass Spectrom.*, **1990**, *4*, 30–33.
97. Schwartz, J. C.; Jardine, I: *Rapid Commun. Mass Spectrom.*, **1992**, *6*, 313–317.
98. McLuckey, S. A.; Goeringer, D. E.; Glish, G. L. *J. Am. Soc. Mass Spectrom.* **1991**, *2*, 11–21.

99. Doroshenko, V. M.; Cotter R. J. *Procs. of the 43rd Conference on Mass Spectrometry and Allied Topics,* Atlanta, GA, May 21–26, 1995, p. 1102.
100. Shaffer, B. A.; Karnicky, J.; Buttrill, S. E. *Procs. of the 41st Conference on Mass Spectrometry and Allied Topics,* San Francisco, CA, May 30–June 4, 1993, p. 802.
101. Goeringer, D. E.; Asano, K. G.; McLuckey, S. A. *Anal. Chem.* **1994**, *66*, 313–318.
102. Taylor, D.; Schwartz, J.; Zhou, J.; James, M.; Bier, M.; Korsak, A.; Stafford, G. *Procs. of the 43rd Conference on Mass Spectrometry and Allied Topics,* Atlanta, GA, May 21–26, 1995, p. 1102.
103. Schubert, M.; Nagel, M.; Wang, Y.; Franzen, J. *Procs. of the 43rd Conference on Mass Spectrometry and Allied Topics,* Atlanta, GA, May 21–26, 1995, 1106.
104. Wells, G.; Huston, C. *J. Am. Soc. Mass Spectrom.* **1995**, *6*, 928–935.
105. Goeringer, D. E.; Asano, K. G.; McLuckey, S. A.; Hoekman, D.; Stiller, S. W. *Anal. Chem.* **1994**, *66*, 313–318.
106. Marshall, A. G.; Wang, T.-C.L.; Ricca, T. L. *J. Am. Chem. Soc.* **1985**, *107*, 7893–7897.
107. Chen, L.; Wang, T. C. L.; Ricca, T.; Marshall, A. G. *Anal. Chem.* **1987**, *59*, 449–454.
108. Guan, S.; Marshall, A. G. *Anal. Chem.* **1993**, *65*, 1288–1294.
109. Julian, R. K.; Cox, K.; Cooks, R. G. *Procs. of the 40th Conference on Mass Spectrometry and Allied Topics,* Washington, D.C., 1992, pp. 943–944.
110. Jonscher, K. R.; Yates J. R. III *Procs. of the 43rd Conference on Mass Spectrometry and Allied Topics,* Atlanta, GA, May 21–26, 1995, p. 990.
111. Williams J. D.; Cox, K.; Morand, K. L.; Julian R. K.; Julian R. K.; Kaiser, R. E. *Procs of the 39th ASMS Conf. Mass Spectrom. and Allied Topics,* Nashville, TN, 1991, pp. 1481–1482.
112. Londry, F. A.; Wells, G. J.; March, R. E. *Rapid Commun Mass Spectrom.* **1993**, *7*, 43–45.
113. Williams, J. D.; Cooks, R. G. *Rapid Commun. Mass Spectrom.* **1992**, *6*, 424–527.
114. Beu, S. C.; Senko, M. W.; Quinn, J. P.; Wampler, III, F. M.; McLafferty, F. W. *J. Am. Soc. Mass Spectrom.* **1993**, *4*, 557–565.
115. Winger, B. E.; Hofstadler, S. A.; Bruce, J. E.; Udseth, H. R.; Smith, R. D. *J. Am. Soc. Mass Spectrom.* **1993**, *4*, 566–577.
116. Morand, K. L.; Cox, K. A.; Cooks, R. G. *Rapid Commun Mass Spectrom.* **1992**, *6*, 520–523.
117. McLuckey, S. A.; Glish, G. L.; Asano, K. G. *Anal. Chim. Acta* **1989**, *22*, 25–35.
118. Doroshenko, V. M.; Cotter, R. J. *Anal. Chem.* **1996**; *68*, 463–472.
119. Louris, J. N.; Cooks, R. G.; Syka, J. E. P.; Kelley, P. E.; Stafford, G. C.; Todd, J. F J. *Anal. Chem.* **1987**, *59*, 1677–1685.
120. Strife, R. J.; Schwartz, J.; Bier, M.; Zhou, J. *Procs. of the 43rd Conference on Mass Spectrometry and Allied Topics,* Atlanta, GA, May 21–26, 1995, p. 160.
121. Wolfender, J.-L.; Rodringuez, S.; Hostettmann, K.; Winfried, W.-R. *J. Mass Spectrom. Rapid Comm. Mass Spectrom.* **1995**.
122. McLuckey, S. A.; Goeringer, D. E.; Glish, G. L. *Anal. Chem.* **1992**, *64*, 1455–1460.
123. Wang, M.; Wells, G. *Proceedings of the 41st Annual Conference of Mass Spectrometry and Allied Topics,* San Francisco, CA, **1993**, p. 463.

124. Qin J.; Chait, B. T. *Procs. of the 43rd Conference on Mass Spectrometry and Allied Topics*, Atlanta, GA, May 21–26, 1995, p. 1100.

125. Williams, E. R.; Furlong, J. J. P.; McLafferty, F. W. *J. Am. Soc. Mass Spectrom.* **1990**, *1*, 288–294.

126. Little, D. P.; Speir, P. J.; Senko, M. W.; O'Connor, P. B.; McLafferty, F. W. *Anal. Chem.* **1994**, *66*, 2809–2815.

127. Castro, J. A.; Nuwaysir, L. M.; Ijames, C. F.; Williams, C. L. *Anal. Chem.* **1992**, *64*, 2238–2243.

128. Hughes, R. J.; March, R. E.; Young A. B. *Int. J. Mass Spectrom. Ion Phys.* **1982**, *42*, 255–263.

129. Hughes, R. J.; March, R. E; Young A. B. *Can. J. Chem.* **1983**; *61*, 824.

130. Ensberg, E. S.; Jefferts, K. B. *Astrophys. J.* **1975**, *195*, L89.

131. Louris, J. N.; Brodbelt, J. S.; Cooks, R. G. *Int. J. Mass Spectrom. Ion Procs.* **1987**, *75*, 345–352.

132. Stephenson, J. L., Jr., Booth, M. M.; Boue, S. M.; Eyler, J. R.; Yost, R. A. *The Analysis of Biomolecules Using Electrospray Ionization/Ion Trap Mass Spectrometry and Laser Photodissociation*; Snyder, P., Ed.; American Chemical Society: Washington, DC, 1996; Chapter XX.

133. Stephenson, J. L.; Booth, M. M.; Shalosky, J. A.; Eyler, J. R.; Yost, R. A. *J. Am. Soc. Mass Spectrom.* **1994**, *5*, 886–893.

134. Boue, S. M., Stephenson, J. L.; Yost R. A. *Procs. of the 43rd ASMS Conference on Mass Spectrometry and Allied Topics*, Atlanta, GA, May 21–26, 1995, p. 408.

135. Stephenson, J. L. Personal communication.

136. BIOMASS™ deconvolution program, Finnigan Corp., **1991**.

137. McLuckey, S. A.; Van Berkel, G. J.; Glish, G. L.; Huang, E. C.; Henion, J. D. *Anal. Chem.* **1991**, *63*, 375–383.

138. Kennedy, R. T.; Jorgenson, J. W. *Anal. Chem.* **1989**, *61*, 1128–1135.

139. Schwartz, J. C.; Jardine, I.; Edmonds, C. G. *Procs. of the 40th ASMS Conference on Mass Spectrometry and Allied Topics*, Washington, DC 1992, pp. 709–710.

140. Cox, A. L.; Skipper, J.; Henderson, R. A.; Darrow, T. L.; Schabanowitz, J.; Engelhard, D. F.; Hunt, D. F.; Slingluff, C. L. *Science*, **1994**, *274*, 716.

141. Land, A.; Sanders, M.; Schwartz, J.; Gale, D.; Chaudhary, T. *Procs. of the 44th ASMS Conference on Mass Spectrometry and Allied Topics*, Portland, OR, May 12–17, 1996.

142. McLuckey, S. A.; Ramsey, R. S. *J. Am. Soc. Mass Spectrom.* **1994**, *5*, 324–327.

143. McLuckey, S. A.; Glish, G. L., Kelley, P. E. *Anal. Chem.* **1987**, *59*, 1670–1674.

144. Berberich, D. W.; Yost, R. A. *J. Am. Soc. Mass Spectrom.* **1994**, *5*, 757–764.

145. March, R. E.; Hughes, R. *Practical Organic Mass Spectrometry*; Wiley: New York, 1989.

146. Harrison, A. G. *Chemical Ionization Mass Spectrometry*, 2ed.; CRC Press: Boca Raton, FL; 1992, p. 1.

147. Mather, R. E.; Todd, J. F. J. *Int. J. Mass Spectrom. Ion Phys.* **1980**, *33*, 159.

148. Schermann, J. P.; Major, F. G. *Appl. Phys.* **1978**, *16*, 225.

149. Covey, T. R.; Bonner, R. F.; Shushan, B. I., Henion, J. *Rapid Commun. Mass Spectrom.* **1988**, *2*, 249–256.
150. McLuckey, S. A.; Van Berkel, G. J. V.; Glish, G. L. *J. Am. Soc. Mass Spectrom.* **1992**, *3*, 60–70.
151. McLuckey, S. A.; Habibi-Goudarzi S. *J. Am. Soc. Mass Spectrom.* **1994**, *5*, 740.
152. Roepstorff, P.; Fohlman, *J. Biomed. Mass Spectrom.* **1984**, *11*, 601.
153. McLuckey, S. A.; Habibi-Goudarzi, S. *J. Am. Chem. Soc.* **1993**, *115*, 12085.
154. Doktycz, M. J.; Hurst, G. B.; Habibi-Goudarzi, S.; McLuckey, S. A.; Tang, K.; Chen, C. H.; Uziel, M.; Jacobson, K. B.; Woychik, R. P.; Buchanan, M. V. *Anal. Biochem.* **1995**, *230*, 205–214.
155. Smith, R. D.; Loo, J. A.; Edmonds, C. G.; Barinaga, C. J.; Udseth, H. R. *Anal. Chem.*, **1990**, *62*, 882–899.
156. Doktycz, M. J.; Habibi-Goudarzi, S.; McLuckey, S. A. *Anal. Chem.* **1994**, *66*, 3416–3422.
157. McLuckey, S. A.; Goeringer, D. E. *Anal. Chem.* **1995**, *67*, 2493–2487.
158. McLuckey, S. A.; Glish, G. L.; Van Berkel, G. J. *Anal. Chem.* **1991**, *63*, 1971–1978.
159. Schermann, J. P.; Major, F. G. *Appl. Phys.* **1978**, *16*, 225.
160. Mather, R. E.; Todd, J. F. J. *Int. J. Mass Spectrom. Ion Phys.* **1980**, *33*, 159.
161. Williams, J. D.; Cooks, R. G. *Rapid Commun. Mass Spectrom.* **1993**, *7*, 380–382.
162. Gorshkov, M. V.; Guan, S.; Marshall, A. G. *Rapid Commun. Mass Spectrom.* **1992**, *6*, 166.
163. Herron, W. J.; Goeringer, D. E.; McLuckey, S. A. *J. Am. Soc. Mass Spectrom.* **1995**, *6*, 529.
164. McLuckey, S. A.; Glish, G. L.; Asano, K. G.; Van Berkel, G. J. *Anal. Chem.* **1988**, *60*, 2312–2314.
165. Pannell, L. K.; Pu Q. L.; Fales, H. M.; Mason, R. T.; Stephenson, J. L. *Anal. Chem.* **1989**, *61*, 2500–2503.
166. Ratnayake, W. M. N.; Timmins, A.; Ohshima, T.; Ackman, R. G. *Lipids.* **1986**, *21*, 518–524.
167. Herron, W. J.; Goeringer, D. E.; McLuckey, S. A. Submitted to *Anal. Chem.* **1995**.
168. Hofstadler, S. A.; Wahl, J. H.; Bruce, J. E.; Smith, R. D. *J. Am Chem. Soc.* **1993**, *115*, 6983.
169. Schwartz, J. C.; Jardine, I. *Proceedings of the 40th ASMS Conference on Mass Spectrometry and Allied Topics*, Washington, DC, 1992, pp. 707–708.
170. Ramsey, R. S.; Goeringer, D. E.; McLuckey, S.A. *Anal. Chem.* **1993**, *65*, 3521–3524.
171. Henion, J. D.; Mordehai, A. V.; Cai J. *Anal. Chem.* **1994**, *66*, 2103–2109.
172. Ramsey, R. S.; McLuckey, S. A. *J. Am. Soc. Mass Spectrom.* **1994**, *5*, 324.
173. Moseley, M. A.; Jorgenson, J. W.; Shabanowitz, J.; Hunt, D. F.; Tomer, K. B. *J. Am. Soc. Mass Spectrom.*, **1992**, *3*, 289–300.
174. Li, C. P. *Chinese Herbal Medicine*; U.S. Department of Health, Education, and Welfare, Public Health Service, National Institutes of Health: Washington, DC, 1974.
175. Ramsey, R. S.; McLuckey, S. A. *J. Microcolumn Separations*, in press.
176. Van Berkel, G. J.; McLuckey, S. A.; Glish, G. L. *Anal. Chem.* **1991**, *63*, 1098–1109.

177. Ramsey, R. S.; Van Berkel, G. J.; McLuckey, S. A.; Glish, G. L. *Biological Mass Spectrom.* **1992**, *21*, 347–352.
178. Lim, H. K.; Wu, W-N.; Mordehai, A.; Henion J. D. *Procs. of the 41st Conference on Mass Spectrometry and Allied Topics*, San Francisco, CA, May 31–June 4, 1993, p. 51.
179. Stahl, D. C.; Martino, P. A.; Swiderek, K. M.; Davis, M. T.; Lee T. D. *Proc. of the 40th ASMS Conference on Mass Spectrometry and Allied Topics*, Washington DC, May 31–June 5, 1992, pp. 1801–1802.
180. Mylchreest, I.; Campbell, C.; Wheeler, K.; Wakefield, M. *Procs. of the 43rd Conference on Mass Spectrometry and Allied Topics*, Atlanta, GA, May 21–26, 1995, p. 436.
181. Wakefield, M. R.; Sanders, M.; Josephs, J. L.; Mylchreest, I. *Procs. of the 43rd Conference on Mass Spectrometry and Allied Topics*, Atlanta, GA, May 21–26, 1995, p. 173.
182. Arnott, D.; King, K.; Bier, M.; Land A.; Stults, J. *Procs. of the 43rd ASMS Conference on Mass Spectrometry and Allied Topics*, Atlanta, GA, May 21–26, 1995, p. 31.
183. Kanai, M.; Seta, K.; Nakayama, H.; Isobe, T.; Land, A. P.; Bier, M. E. *Procs. of the 43rd Conference on Mass Spectrometry and Allied Topics*, Atlanta, GA, May 21–26, p. 616. 1995.
184. Yates, J. R.; Eng, J.; Schieltz, D.; Link, A. *Procs. of the 43rd Conference on Mass Spectrometry and Allied Topics*, Atlanta, GA, May 21–26, 1995, p. 325.
185. Gillece-Castro, B. L.; Arnott, D. P.; Bier, M. E.; Land, A. P.; Stults, J. T. *Procs. of the 43rd Conference on Mass Spectrometry and Allied Topics*, Atlanta, GA, May 21–26, 1995, p. 302.
186. Gillece-Castro, B. L.; Arnott, D. P.; Bier, M. E.; Land, A. P.; Stults, J. T. (submitted to *J. Amer. Soc. Mass Spectrom.*, 1996).
187. Eng, J. K.; McCormack, A. L.; Yates, J. R. *J. Amer. Soc. Mass Spectrom.* **1994**, *5*, 976–989.
188. Yates, J. R.; Eng, J. K.; McCormack, A. L.; Schieltz, D. *Anal. Chem.* **1995**, *67*, 1426–1436.
189. Eng, J. K.; McCormack, A. L.; Yates, J. R. *J. Am. Soc. Mass Spectrom.* **1994**, *5*, 976–989.
190. Eng. J. K.; Yates, J. R.; Schieltz, D. M. *Procs. of the 43rd Conference on Mass Spectrometry and Allied Topics*, Atlanta, GA, May 21–26, 1995, p. 641.

CHAPTER 8

Electrospray Ionization/Fourier Transform Ion Cyclotron Resonance Mass Spectrometry

DAVID A. LAUDE, ELIZABETH STEVENSON, and JESSICA M. ROBINSON
Department of Chemistry and Biochemistry, University of Texas at Austin, Austin, Texas

	Abstract	292
I.	Introduction	292
II.	ESI/FTICR interface considerations	293
	A. Interface concerns	293
	1. Pressure gradient	293
	2. Magnetic-field gradient	294
	B. Approaches to instrument design	294
	1. External source configurations	295
	2. Magnetic-field focusing ESI/FTICR	296
	C. Separations techniques and ESI/FTICR	297
	1. Capillary electrophoresis/ESI/FTICR	299
	2. Liquid chromatography/ESI/FTICR	299
III.	Experimental methodologies	301
	A. Kinetic-energy dependent-ion trapping in ESI/FTICR	301
	B. Quadrupolar excitation and remeasurement in ESI/FTICR	303
	C. Ion dissociation in ESI/FTICR	306
	1. Collision-induced dissociation (CID) in ESI/FTICR	306
	2. Sustained off-resonance irradiation (SORI) in ESI/FTICR	307
	3. Infrared multiphoton dissociation (IRMPD) in ESI/FTICR	307
	4. Surface-induced dissociation (SID) in ESI/FTICR	309
IV.	Chemical information derived from FTICR detection of electrosprayed ions	309
	A. Ultrahigh-resolution detection in ESI/FTICR	309
	B. Accurate mass assignment in ESI/FTICR	311
	C. Reaction chemistry in ESI/FTICR	311
	D. Gas-phase conformation studies in ESI/FTICR	312
	E. Single-ion detection in ESI/FTICR	313

Electrospray Ionization Mass Spectrometry, Edited by Richard B. Cole.
ISBN 0-471-14564-5 © 1997 John Wiley & Sons, Inc.

V. Conclusions 315
Acknowledgments 316
References 316

ABSTRACT

The development of Fourier transform ion cyclotron resonance mass spectrometry for high-performance analysis of electrosprayed biomolecular ions is described. Design considerations in constructing a successful interface between source and spectrometer are addressed and successful ESI/FTICR configurations are described. The extraordinary analytical figures of merit achieved with these instruments are noted, including routine mass resolving power in excess of 10^6 at 1000 Da/z permitting isotope resolution within charge states of proteins in the 10- to 100-kDa range, mass accuracy with errors at the 100 ppb level, single-ion detection, and a mass range extending to 10^8 Da. Other features described include near-unit ion-dissociation efficiency for intermediate-sized proteins, nondestructive remeasurement of a single-ion packet, techniques for exploring gas-phase conformation and approaches to the interface for separation techniques.

I. INTRODUCTION

Fourier transform ion cyclotron resonance mass spectrometry (FTICR), developed in the mid-1970s by Comisarow and Marshall (1), was quickly touted by many as the mass analyzer to replace sector instruments in high-performance applications. Unfortunately, the next 15 years of development for FTICR were notable in many ways for an inability to deliver promised performance in realistic sampling environments. Considerable advances were made in the ability to deliver large biomolecular ions from sources such as Cs^+-secondary ion-mass spectrometry (SIMS) to the low-pressure FTICR trapped ion cell (2). However, the extraordinary high-resolution performance for which FTICR is noted under pristine conditions could not be replicated under nonideal conditions (3). The growing recognition that superior FTICR mass-resolution performance at low mass could not be sustained at very high mass also raised serious questions as to the analytical viability of FTICR for techniques such as matrix-assisted laser desorption ionization (MALDI) (4) when compared to time-of-flight (TOF) mass analyzers. At the beginning of the 1990s, FTICR continued to be regarded by the general mass-spectrometry community as a curiosity, capable of generating extraordinary mass spectra, but without any obvious application that could exploit this performance.

The suspect perspective of FTICR described above has changed remarkably in the last five years and FTICR today enjoys a popularity and promise to rival the predictions of the late 1970s. The single most important reason for this change in perspective is clearly the successful coupling of electrospray ionization (ESI) to FTICR. The number of ESI/FTICR instruments is today growing exponentially—as few as three ESI/FTICR instruments were operational in 1993 (5-7) as compared to about 20 now in use worldwide. There are many reasons for this rapid increase in popularity. For example, the typical mass-to-charge range of ESI ions is between 1000 and 2000 Da/z and coincides with the FTICR mass range for which clear domination exists in comparison to other mass analyzers in producing high resolution and accurate mass spectra. This high mass-resolution capability of FTICR satisfies an important ESI requirement for high resolving power; ES spectra are typically more complex than those achieved with other high mass sources because of the folding back of multiply charged peaks into a small mass-to-charge region. A second area of compatibility of ESI and FTICR is tandem mass spectrometry. Large, multiply charged biomolecules formed by ESI are more susceptible to fragmentation using traditional ion-dissociation techniques, but the spectra are of little value without the ability to assign charge states to product ions. However, FTICR provides the necessary information through the resolution of isotopic peaks within individual charge states and as a consequence ESI/FTICR is now showing promise in generating structural information for biomolecules in the 10- to 100-kDa range from a single mass spectrum. Finally, FTICR is the ideal technique to study the reactivity and higher-order structure of electrosprayed biomolecular ions. In contrast with the microsecond to millisecond timescale for most mass analyzers, the trapped ion cell of an FTICR provides an attractive environment in which to probe these complex ions from seconds to hours.

This chapter presents a complete review of the ESI/FTICR instrumentation and methodology developed through 1995. The material to be presented is divided into three categories which in many ways parallel the areas of continuing growth of the technique. A discussion of interface considerations in the evolution of instrument design and status of current instruments is provided. The adaptation of existing and new FTICR methodologies for ESI analysis is then described. Finally, a description of current performance standards and achievements with the ESI/FTICR experiment is presented.

II. ESI/FTICR INTERFACE CONSIDERATIONS

A. Interface Concerns

1. Pressure Gradient. It is well known that FTICR performance varies inversely with pressure. Fourier transform ion cyclotron resonance detection takes place on the tens of millisecond to second timescale and during this time collisions with background neutrals alter the trajectory of ions and consequently degrade the analytical measurement. As a rule, the trapped ion-cell pressure must be in the low 10^{-8} to 10^{-9} Torr range to achieve the level of performance that distinguishes FTICR from other mass analyzers (8). This ultrahigh-vacuum pressure range is not immediately compatible with an ESI source that operates at atmospheric pressure and delivers a liquid spray to the mass analyzer. As a result, there has been a considerable investment of effort in the interface design.

The efficient delivery of a well-behaved ion beam from the ESI source to the FTICR trapped ion cell across a pressure gradient from atmosphere to 10^{-9} Torr analyzer pressure places a great burden on pumping and ion-optical requirements. Methods for reducing the gas load include differential pumping across small orifices and the use of shuttering devices to minimize the introduction of neutrals to the mass analyzer. Presently ESI/FTICR spectrometers employ 3–5 stages of differential pumping to achieve the desired pressure reduction. This elaborate use of differential pumping in turn complicates the alignment of the ion beam in order to avoid significant attenuation of the ES ion current at the trapped ion cell.

2. Magnetic-Field Gradient. FTICR detection of electrosprayed ions is currently performed in superconducting solenoidal magnets with field strengths ranging from 3.0 to 9.4 T. Because commercial ion sources are not generally designed to operate in strong magnetic fields or to be housed in the small volume of a superconducting magnet bore, these sources are most often positioned a considerable distance from the FTICR magnet. Thus, an additional interface design consideration must be addressed in the delivery of the ES ion beam into the trapped ion cell across a tremendous magnetic-field gradient (9). One problem that arises when ES ions approach the magnetic field gradient with a significant radial component of motion is that a magnetic moment is established which increases as the ion penetrates into the magnet bore (10). As a result, the radial amplitude of the ion trajectory increases as the axial component of ion motion decreases. If the initial ratio of perpendicular to parallel ion velocities is sufficiently large, the forward ion motion will come to a stop and the ions will reverse direction without ever entering the trapped ion cell. This phenomenon is known as the magnetic mirror effect. The problem is especially severe for the ESI experiment because of the relatively broad spatial and kinetic energy distributions of the ion beam. Some type of focusing optics becomes necessary to direct a substantial fraction of an electrosprayed ion beam along suitable ion trajectories and into the trapped ion cell.

Successful solutions to the magnetic mirror effect include the injection of high-kinetic-energy ions through electrostatic lenses and low-energy injection using rf ion guides. An alternate solution is simply to position an ES source in the magnetic field adjacent to the FTICR trapped ion cell.

B. Approaches to Instrument Design

Given the twin difficulties of overcoming high magnetic-field and pressure gradients in designing an effective ESI/FTICR spectrometer, the resulting interface configurations are necessarily intricate. Brief descriptions of the various designs are presented below.

1. External Source Configurations. The first ESI/FTICR experiment was performed by McLafferty and co-workers on the University of Virginia external-source FTICR built by Hunt and Shabanowitz (5). The use of two quadrupole lenses to increase ion-injection efficiency was borrowed directly from the original FTICR external source designed by McIver and Hunt for Cs^+-SIMS (2). In the double-quadrupole design, the first set of rods provided focusing for ions exiting the ESI source. A second set of quadrupole rods then delivered the ions through the fringing magnetic field to the trapped ion cell. Even though pressure problems limited the performance of the instrument, the merit of the interface design was demonstrated and mass resolution values exceeding 100,000 were sufficient to resolve the isotopic cluster of individual charge states.

McLafferty was soon to employ quadrupole focusing in the construction of a dedicated 2.8 T ESI/FTICR instrument at Cornell University with which routine high-resolution ESI mass spectra were obtained by FTICR (11). High-resolution and high-mass accuracy FTICR spectra of proteins (12) and ion-dissociation spectra (13) were obtained with this instrument, but sensitivity problems necessitated a complete redesign of the instrument, which is shown in Figure 1 (14,15). This external-source configuration, which remains in active use, includes a 6.2-T magnet, five stages of differential pumping to achieve a base pressure of 2×10^{-9} Torr, three quadrupole lenses for ion focusing, an open-geometry trapped-ion-cell configuration (16), and laser photodissociation capabilities. The current Cornell instrument was the first to demonstrate high-resolution and

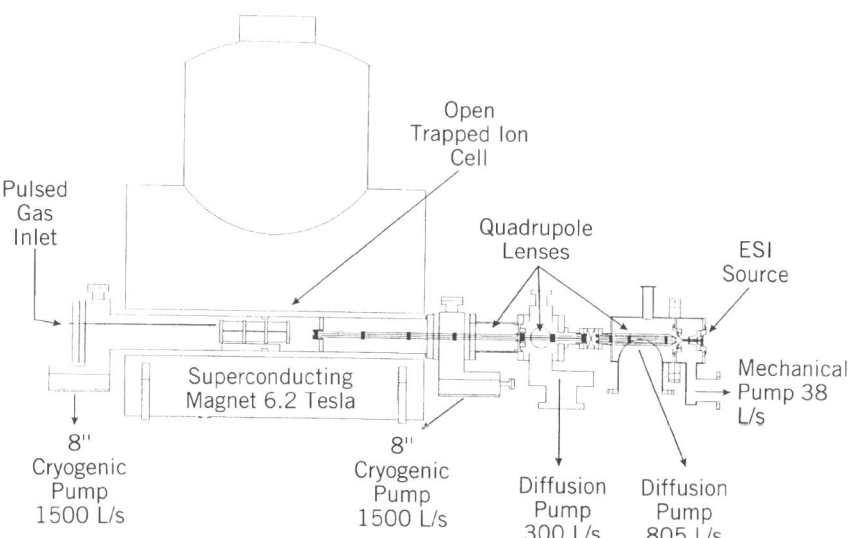

Figure 1. Schematic of the Cornell ESI/FTICR which employs electrostatic focusing and a 6.2-T superconducting magnet. (Reprinted by permission of Elsevier Science, Inc. from S. C. Beu et al., *J. Am. Soc. Mass Spectrom.*, **1993**, *4*, 558. © 1993 American Society for Mass Spectrometry.)

high-mass-accuracy data for proteins beyond 10 kDa, to use H/D exchange experiments to investigate gas-phase ion conformation and to perform efficient ion dissociation of multiply charged ions by various techniques including sustained off-resonance irradiation (17) and infrared photodissociation (18,19).

A second ESI/FTICR system based upon a 7.0-T superconducting magnet and quadrupole ion optics was constructed in Richard Smith's laboratory at Battelle Northwest in 1993 (7). This instrument, which utilizes two quadrupoles, including a second 1 m length set of rods to guide the ions through the fringing magnetic field, has achieved an exceptional level of performance by using extraordinary efforts to minimize the neutral pressure in the analyzer cell. Two shutters are installed on the system to reduce the flow of background neutrals and six stages of differential pumping, including a final cryopanel stage in the bore of the magnet, are employed. System pressure during ion detection is below 10^{-9} Torr. Exciting innovations, such as high-performance capillary electrophoresis/mass spectrometry and the detection of a single electrosprayed DNA fragment ion with a molecular weight exceeding 10^8 Da were achieved with this instrument.

A second approach to FTICR external-ion source design, developed in Europe in the 1980s by Wanczek and co-workers (20), would become the basis for the Bruker Spectrospin instrument that is sold today. This instrument employs electrostatic focusing in a design intended to deliver ions from any type of source to the trapped ion cell. An ion pulse exiting the source is accelerated to several thousand volts in order to successfully traverse the strong magnetic field and is subsequently decelerated for trapping once inside the magnetic field (21). The external source ESI/FTICR offered by Extrel FTMS likewise uses electrostatic lenses to deliver ions to the trapped ion cell, but differs in that low acceleration voltages are used and ions are continuously injected rather than pulsed into the trapped ion cell (22). Both commercial spectrometers permit high-performance resolution detection of proteins in the 10- to 20-kDa range but do not achieve the level of performance of either the Cornell or Battelle Northwest systems. However, since Bruker and Extrel are the only current vendors of ESI/FTICR, the use of electrostatic focusing optics is likely to find increasing presence in the mass-spectrometry community.

Brief mention is made here of some alternative approaches to external focusing ESI/FTICR instruments that are currently in the developmental stage. At the National High Magnetic Field Laboratory (NHMFL), Marshall and co-workers are employing octupole guides to focus ions through the fringing magnetic field (23). At Oak Ridge National Laboratory, Buchanan and co-workers are employing a wire ion guide to focus electrosprayed ions into the trapped ion cell (24). Eyler and co-workers are designing an ESI/FTICR interface to accommodate an inductively coupled plasma (25).

2. Magnetic-Field Focusing ESI/FTICR. In 1993, Laude and co-workers constructed the second operational ESI/FTICR spectrometer (6). In a substantial departure from the external source designs just described, the University of

Texas instrument was based upon an interface design that positioned the ES source directly in the strong magnetic field, adjacent to the trapped ion cell. The current version of this magnetic-field focusing ESI/FTICR is depicted in Figure 2 (26). The motivation for forming the electrospray in a radially homogeneous magnetic field is that ions dispersed from the skimmer nozzle with a large radial displacement are constrained by the field into cyclotron orbits. Even ions with radial kinetic energies on the order of hundreds of electrovolts will assume orbits of only a few millimeters and hence be accessible for subsequent FTICR detection. The potential analytical advantage of the design is a sensitivity gain because an increased fraction of the ions leaving the source will be properly aligned with the magnetic field and be directed to the trapped ion cell. Placement of the source in the magnet bore also obviates the need for ion-focusing optics.

Placement of the ES source in the strong magnetic field creates a new problem—an inaccessibility to the magnet bore for adequate differential pumping. With the external-source design in which ESI occurs outside the magnet bore, a linear series of pumping stages separated by conductance limits provides between 10^2 and 10^3 pressure differentials until the desired base pressure is achieved. The luxury of unlimited axial displacement is not available when the ES source is positioned in the magnet bore. For typical superconducting magnet bore lengths, at most 20–30 cm may separate the analyzer-trapped ion cell and ESI source regions. As shown in Figure 2, a concentric-tube network of differentially pumped vacuum chambers coupled with a mechanical shutter permit system base pressure in the 10^{-9}-Torr range to be achieved without losing the sensitivity advantage due to restricted conductance limits (27).

Among other applications, the spectrometer was used to develop optimized trapping protocols for ESI/FTICR and remeasurement pulse sequences for repetitive detection of a single ion packet. Also, by eliminating the requirement for ion-focusing optics in the interface, very narrow ES kinetic-energy distributions could be generated. This feature is exploited in the resolution of higher-order structural differences in proteins.

The use of magnetic-field focusing ESI/FTICR is now being evaluated in spectrometers being constructed in Eyler's group at the University of Florida at Gainesville (28) and in Smith's group at Batelle Northwest (29).

C. Separations Techniques and ESI/FTICR

The coupling of modern electrophoretic and chromatographic separation techniques with mass spectrometry is of increasing interest to the biochemical and biomedical communities. The success of both high-performance liquid chromatography/mass spectrometry (HPLC/MS) and capillary electrophoresis/mass spectrometry (CE/MS) has been hastened by the development of ESI, ionization which provides a convenient interface between source and detector. The implementation of HPLC/FTICR and CE/FTICR are particularly attractive because of the fast duty cycle and sensitivity of the FTICR experiment, the large information content of the high-resolution mass spectrum and the

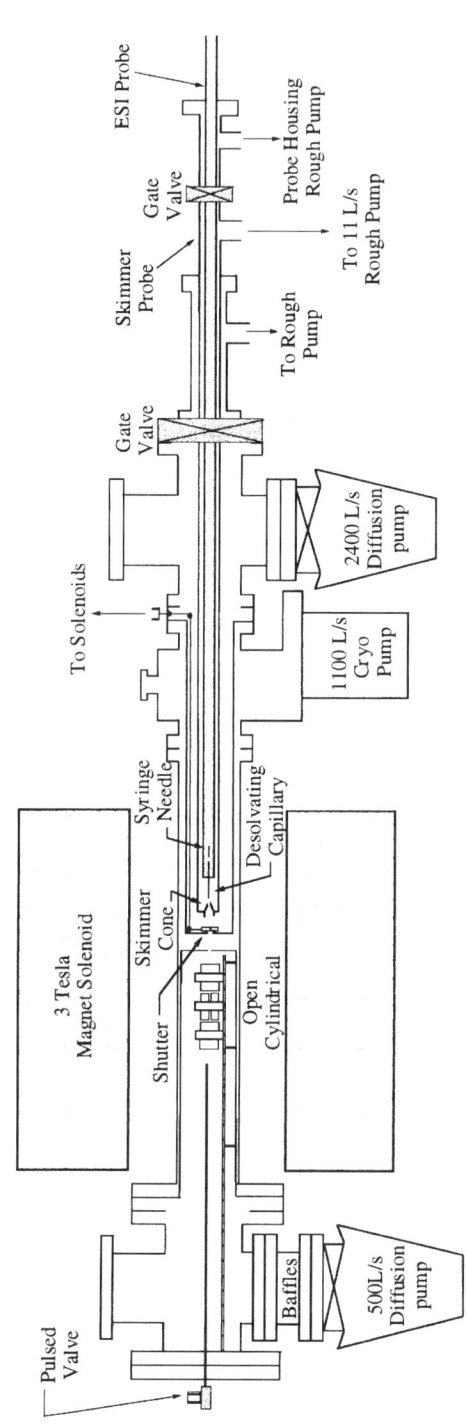

Figure 2. Schematic of the Texas ESI/FTICR which employs magnetic-field focusing and a 3.0-T superconducting magnet. (Reprinted with permission from Z. Guan et al., *Rev. Sci. Instrum.*, **1995**, *66*, 4508. © 1995 American Institute of Physics.)

facility with which tandem mass spectrometry might be implemented. Unfortunately, the coupling of separation techniques with FTICR has historically been slowed by the restrictive pressure requirements for FTICR detection—a 1000-fold lower pressure is needed in the FTICR mass-analyzer region compared to most mass analyzers. Fortunately, with the successful development of an effective ES interface for FTICR, the general problem of coupling high gas-load sources with FTICR detection is solved and the extension to CE and HPLC should be a fairly straightforward process.

1. Capillary Electrophoresis/ESI/FTICR. Workers at Batelle Northwest developed the first successful CE/FTICR interface and have acquired the first on-line high-resolution mass spectra of CE samples (30–33). In their first demonstration of the feasibility of the CE/ESI/FTICR interface, a mixture of six proteins was separated and then detected on line by FTICR (30). Sufficient mass-spectral resolving power was achieved to observe the individual isotopes in charge states for proteins as large as carbonic anhydrase (MW 28,802 Da). The actual amount of sample consumed per FTICR scan was about 20 attomoles.

An interesting modification of the instrument which greatly reduced the data storage requirements and increased mass resolution was the direct monitoring of ES ion current at a shutter assembly which was isolated from the vacuum chamber (31). When a solute zone arrives at the shutter region, the shutter opens, allowing electrosprayed ions into the mass-analyzer region. When the shutter closes, the trapped ion cell pressure reduces sufficiently to allow mass resolution as high as 164,000 to be obtained for carbonic anhydrase. In later work with the CE/FTICR instrument, a tandem mass spectrometry experiment was performed in which electrosprayed CE solute bands were subjected to sustained off-resonance irradiation (SORI) to produce dissociation spectra for proteins as large as equine apomyoglobin (32) with resolving power of 50,000. A total ion electropherogram and product ion spectrum from this work are shown in Figure 3.

Smith and co-workers applied CE/FTICR to the challenging task of directly analyzing cellular proteins (33). In an initial study, 5–10 human erythrocyte cells were drawn into the capillary and then lysed upon exposure to the electrophoresis buffer. About 4.5 fmol of the α and β chains of hemoglobin were detected in spectra with resolution of about 45,000. Future efforts to reduce the direct CE/ESI/FTICR analysis to single cells will focus on improved FTICR sensitivity achieved by using quadrupolar excitation (34) and by prefocusing CE bands prior to detection.

2. Liquid Chromatography/ESI/FTICR. Research in HPLC/FTICR is not as well developed as the CE/FTICR experiment. To date, a single manuscript on the subject of LC/ESI/FTICR by the Bruker Spectrospin instrument group demonstrates that HPLC and ESI may be effectively coupled to FTICR through the standard Bruker Spectrospin external source (21). This source employs high-voltage electrostatic focusing optics to deliver ions to the cell where a gated

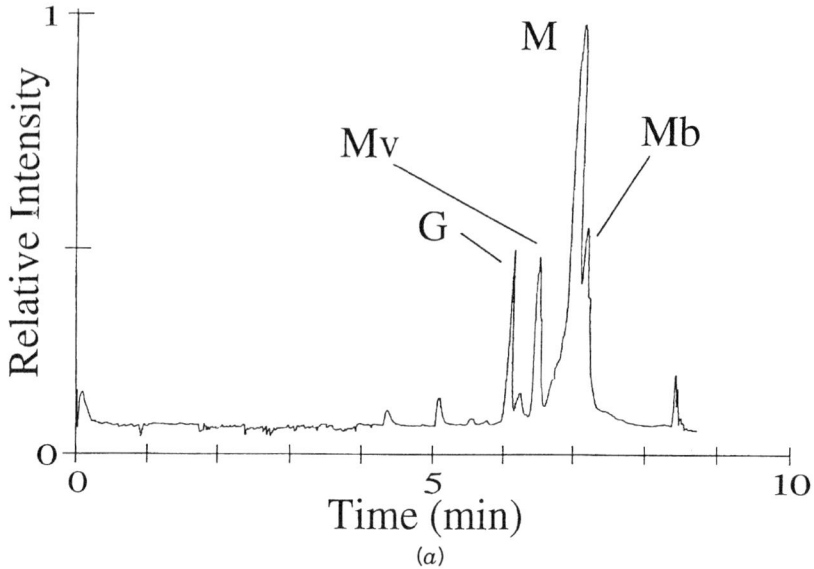

G-I|G|A|V|L|K|V|L|T-T|G|L|P-A-L-I-S-W-I-K-R-K-R-Q-Q-NH_3^+

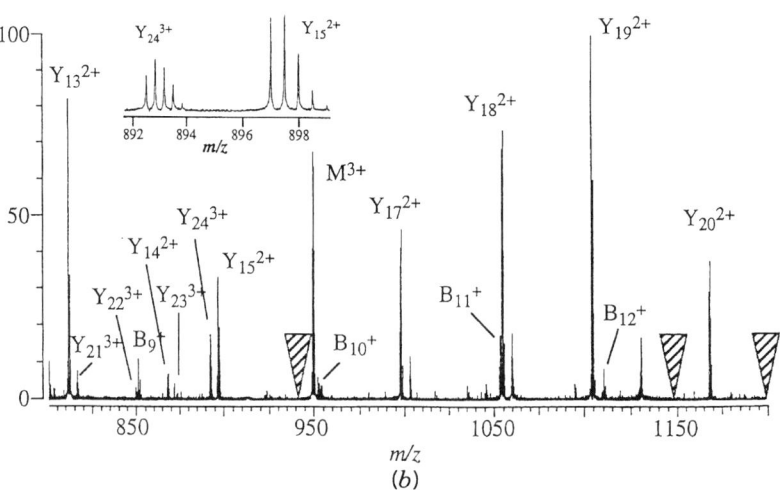

Figure 3. (a) Total-ion electropherogram from CE/ESI/FTICR detection of a mixture of melittin, gramicidin S, and equine apomyoglobin. (b) Melittin product ion spectrum from a tandem mass spectrometry experiment performed during a CE/ESI/FTICR experiment. The mass resolution achieved was in excess of 50,000. (Reprinted by permission of Elsevier Science, Inc. from S. A. Hoffstadler et al., *J. Am. Soc. Mass Spectrom.*, **1994**, *5*, 898. © 1994 American Society for Mass Spectrometry.)

trapping-pulse sequence is employed to trap ES ions for detection. In the results presented, a five-component mixture of small peptides was separated by a reverse-phase gradient. A mass spectral resolution of about 35,000 was sufficient to resolve isotopic peaks within charge states.

In a first step toward implementing on-line ESI/FTICR detection, McLafferty and co-workers developed a microinjection loop for introducing samples to their spectrometer (35). A 50-fmol sample of ubiquitin injected onto the loop yielded a complete high-resolution mass spectrum. When gated trapping was used on a smaller sample, a 5-amol sample was consumed to generate a spectrum with signal-to-noise (S/N) ratio of 23 : 1 and mass resolution of 30,000. A further advance from the Cornell group was the implementation of microscale ESI/FTICR using drawn fused-silica capillary glass and low nanoliter per minute flow rates (36). This permitted a reduction in ES sample volume to the 200 pL level with a total sample loading of 216 amol to produce an ESI/FTICR mass spectrum of cytochrome c with S/N of 60 : 1 and a resolving power exceeding 10^5.

III. EXPERIMENTAL METHODOLOGIES

A. Kinetic-Energy Dependent-Ion Trapping in ESI/FTICR

An examination of the FTICR trapped-ion cell reveals that ion motion in the radial dimension is confined by the magnetic field. However ion motion along magnetic field lines in the z dimension must be constrained by the imposition of an electrostatic potential well. Because ion injection from external sources occurs along this access, it is essential that the axial kinetic energy of electrosprayed ions be determined so that the appropriate potential can be applied to the trap plates. If the kinetic energy of electrosprayed ions is incompatible with the trapping well, ions directed at the trapped ion cell are not trapped for detection. This feature of the FTICR experiment distinguishes it from other forms of mass analysis which typically operate successfully over a wide range of source kinetic energies. This increased complexity of the FTICR experiment is not necessarily a limitation—if ES source kinetic energies are effectively managed, the trapping electrodes can be employed as kinetic-energy filters in various analytical capacities. These are described below.

Traditionally, trapping procedures for externally formed FTICR ions can be divided into two categories based upon whether the trap-plate potential is altered during the injection process—gated trapping (37)—or held at a constant value throughout the injection process—accumulated trapping (38).

In the first successful ESI/FTICR experiments, McLafferty and co-workers employed a gated-trapping procedure in which the trap potentials were initially grounded to allow ES ions into the trapped ion cell (5,11,12). Voltage was then applied to the trap plates to establish a potential well in which ions with kinetic energy below the potential well depth were retained and detected. In principle, this gated trapping procedure can be highly efficient for a packet of monoenergetic ions. However, the ES source typically puts out a continuous beam of ions with a broad kinetic energy. Consequently, the duty cycle for McLafferty's experiment was extremely poor and spectra exhibited low S/N.

Laude and co-workers optimized the conditions for accumulated trapping in ESI/FTICR as an alternative to the gated-trapping experiment (39). They found that if the trap potential is closely matched to the average kinetic energy per unit charge of an electrosprayed ion a substantial number of ions are trapped by increasing the injection time, and consequently the experimental duty cycle, until the desired trapped-ion population is achieved. Using accumulated trapping, ESI/FTICR spectra with S/N exceeding several hundred to one from a single injection event became common place. Later McLafferty and co-workers at Cornell verified that the primary ion-cooling mechanism is collisional and began the now common practice of using a pulsed valve to introduce a transient high-pressure collision gas during the injection period (14,15). Smith and co-workers at Batelle Northwest then demonstrated that accumulated trapping efficiency could be enhanced by inserting an auxiliary electrode to increase the effective pathlength for ion collisions in the trapped ion cell (40).

As a first example that the FTICR accumulated trapping procedure could be used as a kinetic energy selective filter, Laude and co-workers at the University of Texas demonstrated that individual components in electrosprayed protein mixtures could be selectively trapped through appropriate adjustment of source and trapping conditions (41). For example, as shown in Figure 4, the trap potential is tuned to select for one of three proteins in a mixture: lysozyme, albumin, and cytochrome c. The other proteins are not detected because they

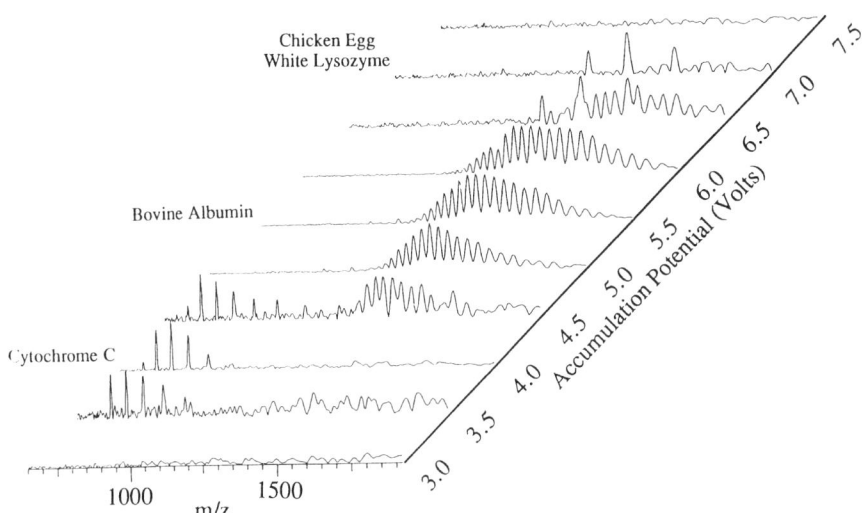

Figure 4. Equimolar three-component mixture of cytochrome-c, bovine albumin, and chicken egg white lysozyme separated using a fixed skimmer potential of 14 V at variable accumulation potentials. (Reprinted with permission from S. A. Hofstadler et al., *Anal. Chem.*, **1993**, *65*, 315. © 1993 American Chemical Society.)

either possess insufficient kinetic energy to overcome the potential barrier at the first trap electrode or too much kinetic energy to be cooled by collisions and trapped for detection. Laude and co-workers found that the magnetic-field focusing interface had the potential to generate electrosprayed ion beams with extremely narrow kinetic energy distributions (42). For example, trapping profiles indicated that a kinetic energy distribution as small as a few hundredths of an electrovolt could be defined. The obvious disadvantage of using the magnetic-field focusing source under such selective trapping conditions is the inability to select experimental parameters that trap such a narrow distribution of ions. To overcome this problem, Laude and co-workers developed a technique for modulating the skimmer potential and thereby expanding the kinetic-energy distribution of electrosprayed ions exiting the source (43). They found that an additional advantage of this energy-modulation procedure was that the kinetic-energy dependent mass discrimination in the ES charge envelope of FTICR spectra was minimized.

B. Quadrupolar Excitation and Remeasurement in ESI/FTICR

There are many conditions necessary to generate high-performance FTICR spectra. Although detection at low pressure is most often cited, of equal importance is the generation of a well-behaved ion cloud appropriately positioned in the trapped ion cell. It may be possible to trap spatially unfocussed ion beams with broad kinetic energy spreads in the FTICR trapped ion cell, but ion lifetimes in the cell are reduced and general analytical figures of merit, including sensitivity and mass resolution, are degraded. Even under the best conditions, most ionization techniques, including ESI, will deliver a less than ideal ion packet to the FTICR. Consequently, considerable tuning time may be necessary to find the narrow range of trapping, excitation, and detection conditions which will generate an acceptable spectrum. Clearly, it is important to develop methodologies for improving the reliability of FTICR detection and reducing the time spent in optimizing experimental parameters.

One of the most important causes of nonideal behavior in the FTICR cell is the existence of magnetron motion. Magnetron motion results whenever an ion introduced to the trapped ion cell has a guiding center of motion that is radially disposed from the center line of the cell (10). One of the deleterious effects of magnetron motion is the radial drift of ions at high pressure and high ion density in the presence of a strong radial trapping field. This ultimately leads to radial-ion ejection from the trapped ion cell, but even under less extreme conditions, in the decreased coherence of the ion cloud leads to poorer mass resolution and decreased sensitivity. The initiation of magnetron motion for injected ES ions has been demonstrated by Laude and co-workers (44) and is certainly a hindrance to performance optimization in even the best-designed ESI/FTICR interfaces.

An exciting new development in FTICR is the introduction of quadrupolar excitation (QE) (34), a technique borrowed from the physicists, in which an ion's unwanted magnetron motion is converted into cyclotron motion. The value of quadrupolar excitation is that when properly applied to any poorly behaved ensemble of ions, it yields a packet of ions spatially focused in the center of the trapped ion cell. Marshall and co-workers demonstrated that QE leads to improved sensitivity, mass resolution, and dynamic range in a variety of FTICR experiments (34).

Smith and co-workers were the first to demonstrate the advantage of QE for ESI/FTICR (45). They used QE in a selective-ion-accumulation (SIA) mode to allow a single charge state to be isolated and accumulated in the trapped ion cell. The advantage of this procedure was an enormous increase in FTICR experiment dynamic range. They later applied QE in the selective accumulation of electrosprayed noncovalent complexes in the trapped ion cell (46). They speculated that the SIA event evidently provided a gentler cooling environment for stabilizing the weakly bonded species; for example, the myoglobin–heme complex was detected with a much higher relative abundance when SIA was employed. In the same work, Smith and co-workers also demonstrated a SIA/tandem mass-spectrometry pulse sequence used to generate the dissociation products of the noncovalent adducts.

Quadrupole excitation can also be employed as an augmentation to simple collisional cooling in a type of experiment called remeasurement (47). The FTICR technique is unique among mass analyzers in that the image-current detection process is nondestructive. Consequently, it should be possible to recapture the ions at the end of the FTICR experiment for further chemical and physical evaluation. McLafferty and co-workers were the first to demonstrate the feasibility of the remeasurement experiment in FTICR (48). They found that simple collisional cooling of very large ions was effective in returning the ion packet to the center of the trapped ion cell. A maximum remeasurement efficiency of 98% was obtained for gramicidin-Dubos molecular ion at 1904 Da/z introduced to the trapped ion cell from a plasma-desorption source. They speculated that more massive ions, for example, those introduced by ESI, could be detected with even higher remeasurement efficiency.

Work by Laude and co-workers verified the facility with which the remeasurement experiment could be applied to ESI/FTICR (49). They demonstrated that for proteins beyond 10 kDa, the same packet of electrosprayed ions could be remeasured hundreds of times with unit remeasurement efficiency. For example, 250 consecutive scans of the same ion population of bovine albumin dimer yielded the expected 16-fold improvement in S/N when coadded. Laude and co-workers demonstrated that if space-charge conditions in the trapped ion cell are minimized, electrosprayed spectra can be remeasured with sufficient resolution to distinguish the isotopes within an individual charge state (50). Laude and co-workers also evaluated the general analytical utility of QE for improving the remeasurement efficiency of electrosprayed proteins (51). It was found that QE greatly extended the regions of the trapped ion cell within which remeasurement could be achieved. For example, unit remeasurement efficiency was achieved over 52% of the cell radius, compared to 18% of the cell radius when simple collisional cooling was employed. The remeasurement technique was extended to the more difficult case of small proteins as 100 coadded scans of a single melittin ion packet yielded the expected tenfold increase in S/N when QE was employed for cooling purposes. In a related paper, Laude and co-workers demonstrated that the original QE experiment performed as a four-electrode experiment could be simplified and performed with similar performance as a two-electrode experiment (52).

Finally, in an interesting application of the remeasurement experiment, Laude and co-workers demonstrated that a real-time monitoring of a gas-phase, ion-molecule reaction could be performed in which the reactivity of the same ion

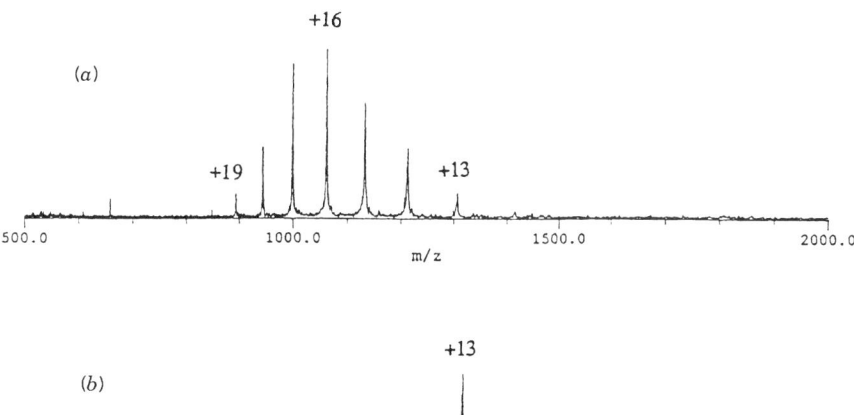

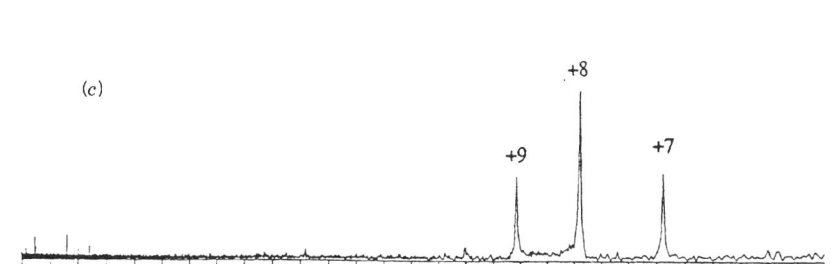

Figure 5. Spectra of horse heart myoglobin (*a*) at zero reaction time, (*b*) after a 10-s reaction time, and (*c*) after a 60-min reaction time. Each spectrum was acquired from the same ion population with a continuous 1.5×10^{-8} Torr background of diethylamine. (Reprinted with permission from Z. Guan et al., *Anal. Chem.*, **1995**, *67*, 1456. © 1995 American Chemical Society.)

packet could be detected continuously by FTICR over a detection period of minutes to hours (53). For example, as shown in Figure 5, the reaction chemistry of a single packet of electrosprayed horse apomyoglobin ions with a background of diethlyamine was monitored by FTICR every 2 s for a 60-s time period. The competitive reactions of adduct formation and charge stripping are evident. This type of experiment was performed on the same ion packet for a full hour without a significant reduction in the population of ions in the trapped ion cell.

C. Ion Dissociation in ESI/FTICR

1. Collision-Induced Dissociation (CID) in ESI/FTICR. One extremely attractive feature of ESI is the possibility that the increased coulombic forces present in multiply charged ions will sufficiently destabilize the ion to permit more efficient ion dissociation. This potential has been realized and ESI now enjoys a particular advantage over other ionization sources in promoting efficient dissociation of extremely large proteins.

Once again, in the tandem mass-spectrometry experiment, FTICR detection is ideally matched with ESI for a variety of reasons. The most obvious benefit is the superior resolution of the FTICR experiment. Given that ESI already generates a more complicated precursor ion spectrum than typically observed with other high mass-ion sources, it is likely that product spectra will be extremely complex and will require high mass resolution to extract the desired analytical information. This problem is exacerbated in the ES experiment because information about product-ion charge state cannot usually be extracted from a low-resolution spectrum. Fortunately, high-resolution spectra generated by FTICR routinely provide isotopic spacings within a charge state. With this information, charge-state assignment within product spectra is easily accomplished. Consequently, FTICR is best suited among all mass analyzers to exploit the facility with which ion dissociation is accomplished in ESI.

The pioneering work in the area of ES ion dissociation occurred in McLafferty's group (13–15). In these early experiments, the method of choice for ion dissociation was nozzle-skimmer dissociation, a technique developed by Smith and co-workers for dissociation of electrosprayed ions first applied in quadrupole mass analyzers (54). In the best nozzle-skimmer ESI/FTICR work (55), McLafferty and co-workers demonstrated that high-resolution ion-dissociation spectra could be obtained for carbonic anhydrase with 80% efficiency; they were able to observe over 100 isotopic clusters, most of which were due to cleavage near proline sites in the molecule. They also demonstrated that ESI/FTICR tandem mass spectrometry could be performed with sufficient resolution and mass accuracy to distinguish variations in protein sequence associated with substitution of a single amino acid unit.

Nozzle-skimmer dissociation techniques have the disadvantage that precursor ion selection is not accomplished prior to ion dissociation. This is a particular problem in ES given the large number of ions in the ES charge envelope. Actually, the problem is mitigated in FTICR more so than with other mass analyzers because the high resolving power of the FTICR simplifies mass and charge assignment in product spectra. Nevertheless, there has been a substantial effort to develop alternative ion-dissociation methodologies for ESI/FTICR. These include various forms of collisional dissociation, photodissociation, and surface dissociation, each of which was first demonstrated and evaluated at Cornell.

2. Sustained Off-Resonance Irradiation (SORI) in ESI/FTICR. The conventional FTICR collision-induced dissociation experiment first applied to ES ions by McLafferty and co-workers was ineffective, even for small peptides (15). They instead turned to the use of lower-energy collisional techniques such as sustained off-resonance irradiation (SORI), a technique first demonstrated by Jacobsen and co-workers for dissociation of small ions (17). The advantage of the lower-energy collisional techniques is that internal energy is allowed to accumulate in the molecule during a lengthy excitation process. It is not necessary for ions to assume a large radial trajectory prior to collision, as occurs with the higher-energy CID experiment. The low-energy experiment is a particular benefit to FTICR because, as described earlier, optimum detection occurs for ions that remain in a small coherent packet. Thus, SORI is the most attractive of various low-energy techniques because it is simple to implement without the need for additional hardware or software. Sustained off-resonance irradiation is accomplished by applying a rf-excitation pulse a few thousand Hertz off-resonance from the ion to be dissociated. The trajectory assumed by the kinetically excited ion is alternately in phase and out of phase with respect to the applied excitation field, thus attenuating the overall radial amplitude of the ion packet. Because numerous collisions occur during the lengthy SORI event, there is a continuous increase in ion internal energy until dissociation occurs.

Work by McLafferty and co-workers demonstrated that the SORI experiment was 92% efficient in generating sequence-specific product ions from large electrosprayed proteins (56). Sustained off-resonance irradiation and related techniques generated high-resolution, information-rich mass spectra for ubiquitin, myoglobin, and carbonic anhydrase. For example, as shown in Figure 6, SORI is performed on selected charge states of electrosprayed bovine ubiquitin to yield very different product ion spectra. Again, because the isotopic peaks in the charge envelopes are resolved, it is possible to make unambiguous assignments of product-ion charge state, which greatly simplifies the determination of sequence information. Refinements of the SORI experiment led to improved ion dissociation for large biomolecules. McLafferty and co-workers demonstrated that 74 structurally useful fragment ions could be obtained from carbonic anhydrase, including 23 amino-acid spacings (57). Consistent with earlier work, dominant fragmentation was observed at the N-terminus side of proline residues and adjacent to aspartic acid and glutamic acid residues.

3. Infrared Multiphoton Dissociation (IRMPD) in ESI/FTICR. As an alternative to collisional-dissociation techniques, McLafferty and co-workers evaluated the use of photodissociation processes. Photodissociation offers the advantages to FTICR that ion dissociation occurs on axis, thus improving detection efficiency and increasing the performance of multistage ion dissociation. Another advantage is that the analyzer chamber is not subjected to the high gas load necessary for collisional dissociation. In the past, photodissociation using high-energy photons at 193 nm has been applied to electrosprayed peptides with limited success (15); however, recent work is providing considerably more sequence information (58).

In work by McLafferty and co-workers at Cornell, infrared photodissociation was applied to multiply charged ions in the ESI/FTICR experiment (59). Efficiencies for proteins as large as carbonic anhydrase were on the order of 30–80%. Optimum irradiation times were on the order of several hundred milliseconds. Product ions were often similar to those obtained by SORI although some complementary information was obtained. Interestingly, when

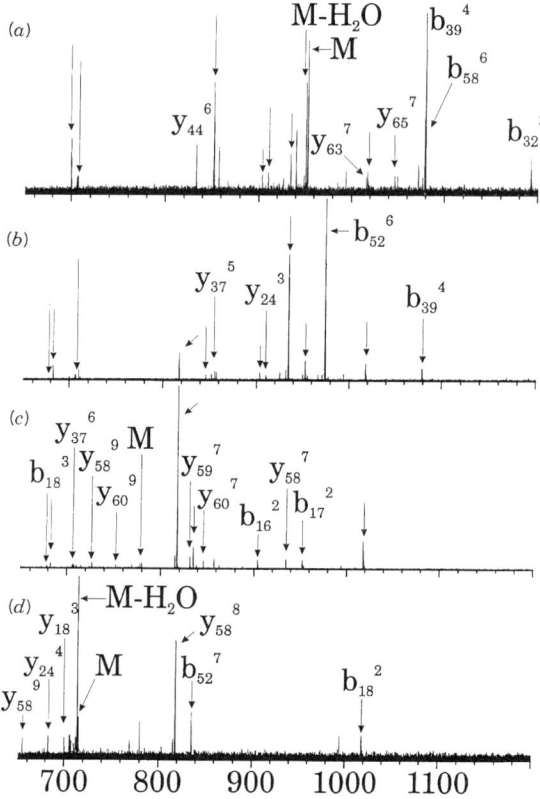

Figure 6. The SORI ESI/FTICR spectra of bovine ubiquitin for the (a) +9, (b) +10, (c) +11, and (d) +12 charge states. (Reprinted with permission from M. W. Senko et al., *Anal. Chem.*, **1994**, *66*, 2807. © 1994 American Chemical Society.)

IRMPD was applied to oligonucleotides, optimum irradiation times were an order of magnitude lower than for proteins. This greater variability in ion stability compared to collisional techniques is attributed to a photon resonance with a P–O bond. In a final experiment demonstrating the facility of photodissociation in FTICR, a second stage of dissociation was performed on a product-ion spectrum of electrosprayed ubiquitin ions to yield sequence information from 11 additional ions.

In subsequent work by McLafferty and co-workers, the IRMPD experiment was applied to a variety of proteins in the 37- to 67-kDa range (60). Sequence specific fragmentation was obtained for two different albumins, although only a small fraction of the total sequence was generated. Of particular interest, sequence errors in a Protein A sample were discerned using product spectra from an IRMPD/ESI/FTICR experiment.

4. Surface-Induced Dissociation (SID) in ESI/FTICR. The application of surface induced dissociation (SID) to ESI/FTICR was evaluated by McLafferty and co-workers (61). Surface-induced dissociation offers the advantage to FTICR that no collision gas is introduced to the mass-analyzer region. A probe-mounted SID assembly was inserted into the rear trapping electrode of an open-cell assembly. The SID surface was an electrically isolated Cu cylinder with a 60° conical collision surface. A collection of trap and SID assembly potentials and time delays were optimized for product ion efficiency. It was found that efficiencies ranged from 36% for gramicidin-S to 14% for carbonic anhydrase. Product ions were similar to those observed by SORI and IRMPD methods.

IV. CHEMICAL INFORMATION DERIVED FROM FTICR DETECTION OF ELECTROSPRAYED IONS

A. Ultrahigh-Resolution Detection in ESI/FTICR

The enormous amount of chemical information present in an ES dissociation spectrum is of limited value in a low-resolution quadrupole mass-analyzer spectrum because of the inability to readily assign charge states. In contrast, FTICR has for years demonstrated extraordinary mass-resolution capabilities. However, it is only recently in this coupling of electrospray with FTICR that an analytical application which places great demands on the mass resolution of FTICR is finally demonstrated.

The advantages of high mass resolution were demonstrated for ESI/FTICR from the beginning. Even in the first experiment at the University of Virginia, a resolving power in excess of 10^6 was shown to be more than sufficient to resolve the isotopes within charge states of gramicidin-S (5). Over the past five years, there has been a steady increase in the benchmark value for mass resolution in ESI/FTICR. For example, McLafferty and co-workers reported a mass resolution in excess of 6×10^4 for the $+17$ charge state of cytochrome c at 773 Da/z (12). With an improved cell and vacuum chamber design, McLafferty and

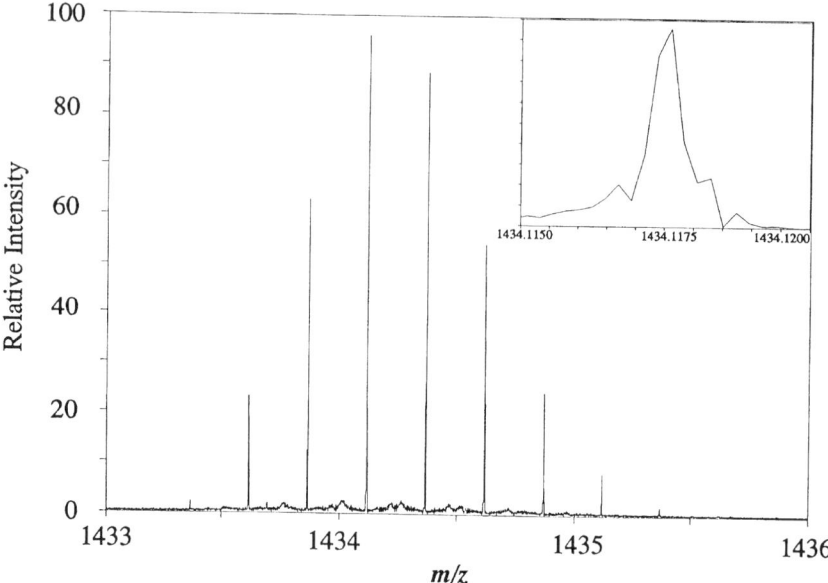

Figure 7. High-resolution FTICR spectrum of the +4 charge state of electrosprayed insulin exhibits the ^{13}C isotopic envelope. A resolving power in excess of 2.5×10^6 is achieved. Note that the full range of the Da/z scale in the inset is only 0.005 Da/z. (Reprinted with permission from Bruce et al., *Rapid Commun. Mass Spectrom.*, **1993**, *7*, 702. © 1993 John Wiley & Sons, Ltd.)

co-workers achieved a resolution for the +10 charge state of ubiquitin at 857 Da/z in excess of 2×10^6 (15). Smith and co-workers at Batelle Northwest employed a procedure to correct for the cyclotron frequency shift of very long transients and obtained a resolution on excess of 2.5×10^6 for the +4 charge state of bovine insulin at 1434 Da/z (62). This spectrum is shown in Figure 7.

Despite the ease with which high resolution could be obtained on smaller masses, there was for a time a difficulty in resolving the isotope peaks for electrosprayed ions above 30 kDa. This problem was finally overcome by McLafferty and co-workers, who were the first to resolve the charge states of an electrosprayed albumin ion (60); a resolution of 166,000 was obtained for the +43 charge state of the albumin. This spectrum was achieved by first dissociating noncovalent adducts and by minimizing coulombic effects in the trapped ion cell. Such a result by no means represents the limit to isotopic resolution of charge states in large proteins. Presently, larger magnets are being constructed at Batelle Northwest and the National High Magnetic Field Lab (NHMFL), which should readily extend the isotope resolution to proteins beyond the 100-kDa range. For example, in preliminary work on a 9.4-T magnet at the NHMFL, McLafferty and co-workers achieved a mass resolution in excess of 250,000 for electrosprayed albumin (63).

B. Accurate Mass Assignment in ESI/FTICR

The FTICR technique has long been recognized for high mass-accuracy performance—it is considered routine to obtain 1–10 ppm errors in mass assignment on samples even in the absence of calibrant ions. With greater care taken to control the size of the ion population, parts per million mass-measurement accuracy is also achievable. Extension of these FTICR performance standards to ESI applications should be routine. Not only should ESI mass accuracy be expected to benefit from the ready availability of calibration standards, but it should be possible to further reduce uncertainty in the FTICR mass assignment by averaging all of the charge states in an ES envelope.

As expected, early experiments by McLafferty suggested that 1–10 ppm error would be routine. For example, isotopically resolved peaks of myoglobin exhibited an error of only 1 ppm (12). In later work with an improved data system, isotopically resolved peaks yielded mass accuracies with subpart per million errors (15). Still later, ESI/FTICR mass accuracies were reduced below the 100 ppb level (56). For example, in the assignment of product ions for the SORI dissociation of ubiquitin, fragments in the 1000- to 7000-Da range routinely exhibited errors of between 0.01 and 0.1 ppm. The initial manuscript by Smith and co-workers verified this high mass-accuracy capability. Average error for the glycoprotein ribonuclease B were in the 1-ppm range and subsequent work is routinely reported at the subpart per million level (7).

C. Reaction Chemistry in ESI/FTICR

An expanding area of interest in the field of ESI is the study of reaction chemistry that occurs both in the electrospray and subsequently in the mass analyzer. The FTICR technique offers the previously noted advantage of high-performance analysis of the reaction products, but is also particularly attractive here because of the possibility of chemical reaction monitoring during the long storage times in the trapped ion cell.

The use of FTICR detection of chemical reactivity in the ES is well established and a few examples are given here. McLafferty and co-workers evaluated metal adduction reactions with proteins and, in particular, examined the adduction of Cu^{+2} by ubiquitin (64). The Cornell group suggested the general use of metal-ion adduction as a means to assign charge states in low resolution or contaminated mass spectra. Cassady and co-workers performed a follow-up investigation on the reactivity of Cu(I) and Cu(II) species in ES reactions with ubiquitin (65). Laude and co-workers at Texas evaluated the addition of acetone to the ES sample as a means to form Schiff base adducts with proteins. In contrast with the use of metal adducts, the covalently bound imine withstands ion dissociation and permits charge-state assignment in product ion spectra (66).

Each of the reactions just described took place in the solution phase. Alternatively, the use of ESI/FTICR to study the gas-phase chemistry of large biomolecules over extended observation times in the trapped ion cell is particularly intriguing. As will be described shortly, McLafferty and co-workers employed gas-phase deuterium exchange studies to investigate gas-phase conformation of protein ions. Laude and co-workers evaluated the competition between deprotonation and adduction in the reaction of diethylamine with electrosprayed proteins; an example of this work is presented in Figure 5 (53). More recently, Cassady and co-workers investigated the charge-state specific deprotonation of electrosprayed ubiquitin and found rate constants for the reaction to increase with number of charges (67).

D. Gas-Phase Conformation Studies in ESI/FTICR

The recognition that electrosprayed biomolecules assume a higher-order structure in the gas phase is of considerable interest to mass spectrometrists. One question that has been evaluated with great interest by the Cornell group is the extent to which solvent plays a role in protein conformation. In a first set of experiments, McLafferty and co-workers performed gas-phase H/D exchange reactions on multiply charged cytochrome c and evaluated the rate of reaction (68). They found that at least three different gas-phase conformers of the protein

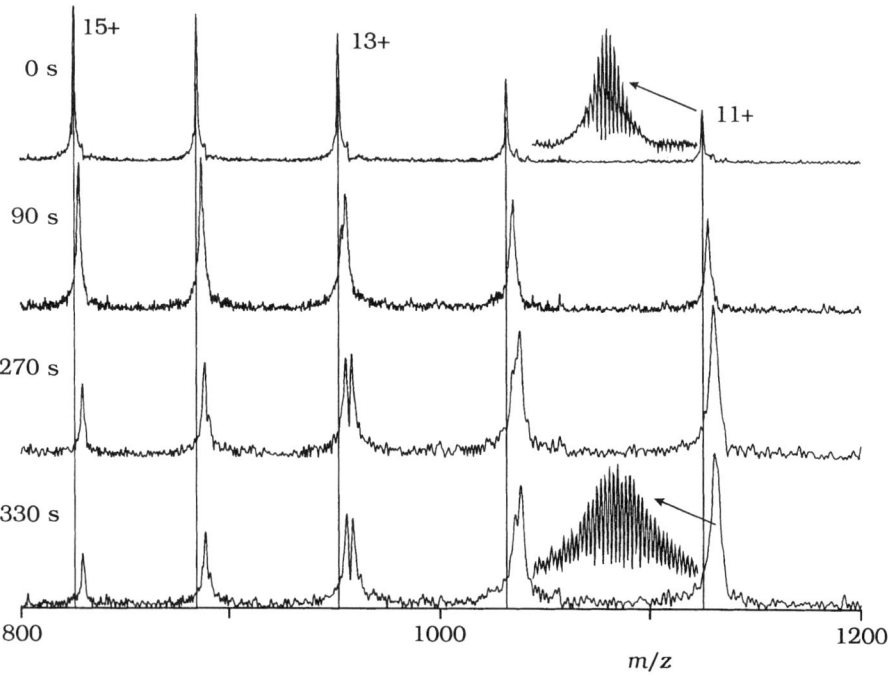

Figure 8. The FTICR spectra of electrosprayed equine cytochrome-c, allowing the ions to react for different times with ^2H_2O at 1×10^{-7} Torr before excitation and detection. Insets—Isotopic peaks at 30,000 resolving power measured in separate experiments. (Reprinted with permission from D. Suckau et al., *Proc. Natl. Acad. Sci. USA*, **1993**, *90*, 790–793. © 1993 National Academy of Sciences USA.)

exist and, in terms of compactness, each corresponds to a known solution phase conformer. Figure 8 presents a series of FTICR spectra acquired after a progressively increasing delay time, which demonstrate the H/D exchange process. In follow-up work, McLafferty and co-workers observed that at least six levels of H/D exchange occurred for electrosprayed cytochrome-c reactions (69). They found that cytochrome-c cations could undergo infrared laser-induced unfolding, as demonstrated by a new level of H/D exchange, and that charge-stripping reactions could also yield new folded and unfolded forms of the electrosprayed protein.

In independent work at the University of California at Berkeley, Gross and Williams performed H/D exchange experiments on gramicidin-S to validate a gas-phase basicity model they developed to quantify the coulomb energy in multiply charged ES ions (70).

In an alternative approach to gas-phase conformation studies, the Texas group evaluated the mobility of different gas-phase conformers in the ES expansion. Because a magnetic-field focusing source was used, a high-quality ES beam with minimal kinetic energy spread could be delivered to the trapped ion cell. This kinetic-energy resolution was sufficient to permit a distinction between gas-phase conformations of electrosprayed ions by carefully manipulating the trapping voltages. For example, Laude and co-workers were able to selectively trap open and closed forms of lysozyme and of cytochrome c in the FTICR cell. This approach to conformation selection is complementary to the H/D exchange experiments in that it provides a means to isolate individual conformers in the trapped ion cell prior to reaction or structural analysis.

E. Single-Ion Detection in ESI/FTICR

As a final example of the utility of ESI/FTICR, the work of Smith and co-workers at Batelle Northwest in detecting the presence of a single ES ion in the trapped ion cell is presented (71–74). Even by FTICR standards, this particular application is well outside the bounds of the conventional mass-spectrometry experiment. Consequently, it has not been mentioned to this point in descriptions of ESI/FTICR methodology development or performance. It should be noted, however, that this particular experiment sets mass spectrometry performance standards on numerous fronts. For example, the detection of a single ion in a routine analytical instrument is a record for detection limits. The recording of a transient lifetime in excess of 476 s would permit records to be established for mass resolution and accuracy. The observation of an ion with a mass beyond 10^8 Da is also a record. In performing the experiment, ion remeasurement is accomplished with unit efficiency and real-time monitoring of ion–molecule reactions is conducted. Clearly, this specific mass-spectrometric measurement is well ahead of its time both in terms of the capability of existing equipment to fully evaluate the data and in terms of our ability to comprehend the full range of possible applications.

A summary of the specific experiment performed at Batelle Northwest is now

314 ELECTROSPRAY IONIZATION/FTICR MASS SPECTROMETRY

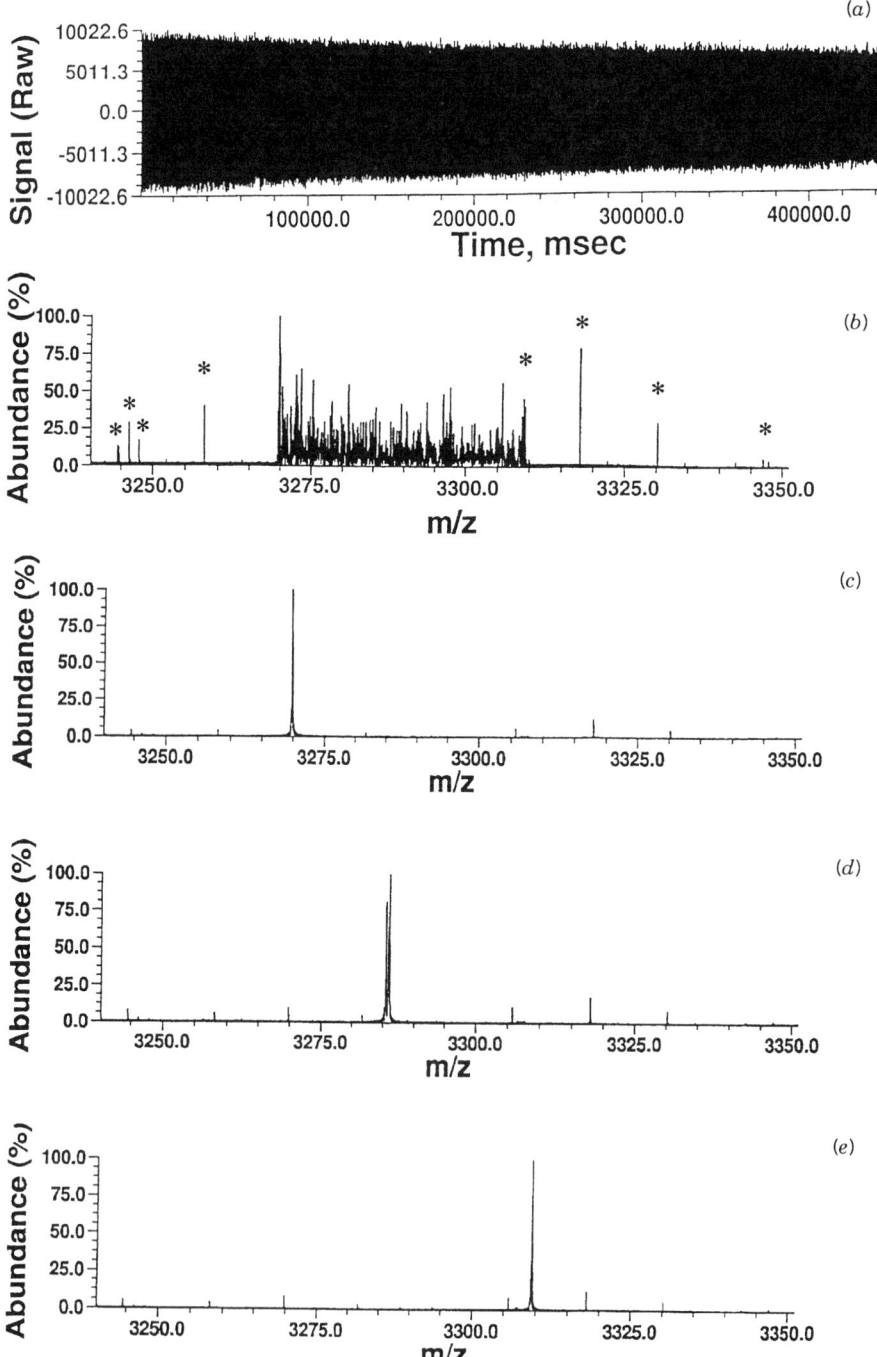

provided. It is generally accepted that as few as 100–1000 ions will generate a detectable image current in FTICR. Thus, as the number of charges on an electrosprayed molecule increases with molecule size, it is conceivable that a single highly charged ES ion would generate sufficient image current to be detected. The experiment performed at Batelle required that a collection of large-molecular-weight electrosprayed ions be introduced to the FTICR trapped ion cell and cooled. Consecutive suspended trapping events were coupled with ion remeasurement to reduce the population in the cell to as few as a single ion.

The first experiments was performed on a poly(ethyleneglycol) (PEG) sample with nominal molecular weight of 5×10^6 (71). Several thousand charges on the ions permitted a folding of the ion signal into the range around 2000 Da/z. When a single ion was isolated and allowed to react with background neutrals, a stepwise shift in ion signal corresponding to differences of a single charge state were observed. Further evaluation of the analytical utility of the single ion detection included a demonstration that the molecular weight of a single isolated PEG ion with nominal molecular weight of 5×10^6 was about 4.1×10^6 Da (72). This procedure was suggested as a potentially superior method for obtaining precise mass determination for very large molecules by examining the change in mass-to-charge ratio for the ion due to proton-transfer reactions that occurred during an extended transient-detection process (72). The FTICR spectra of individual ions were observed for a range of polymers and proteins ranging down to bovine albumin at about 66 kDa (73).

In a final paper on the single ion-detection experiment, Smith and co-workers demonstrated that a molecule, a Coliphage T4 DNA, with molecular weight exceeding 1.1×10^8 Da could be detected by ESI/FTICR (74). This molecule is, by a factor of 20, the largest-ever detected by mass spectrometry. Shown in Figure 9 is a 476-s transient for a single ion which is observed to undergo a complex series of changes in Da/z over the course of the transient lifetime.

V. CONCLUSIONS

The work presented in this chapter summarizes the efforts to develop ESI/FTICR instrumentation and methodology. An examination of the types of data generated with ESI/FTICR provides clear evidence of the extraordinary mass spectrometry performance that can be achieved in terms of mass resolution, mass accuracy, and mass range. What is equally intriguing are the collection of ES

Figure 9. (*a*) A 476-s FTICR transient acquired from an individual T4 DNA ion and (*b*) its frequency-domain spectrum. Major noise peaks are indicated by asterisks. (*c–e*) Frequency domain spectra of 5 s truncated data sets from the transient shown in *a* at 0–5 s, 230–235 s, and 471–476 s, respectively. The last three spectra confirm that the spectrum in (*b*) arises from an individual ion undergoing a complex series of changes in *b*. (Reprinted with permission from R. Chen et al., *Anal. Chem.*, **1995**, *67*, 1162. © 1995 American Chemical Society.)

mass spectrometry experiments that are currently achieved only with the FTICR as the mass analyzer. Examples include the ion-remeasurement experiment, high-resolution tandem mass spectrometry of large biomolecules, real-time monitoring of ion-molecule reaction chemistry, and gas-phase conformation studies.

The preponderance of work presented in this chapter was accomplished by the three research teams at Cornell University, Batelle Northwest, and the University of Texas at Austin. However, their efforts in generating the data presented here have prompted considerable interest in the instrumentation and methodology associated with ESI/FTICR. Electrospray ionization has clearly become the most attractive of ionization sources for FTICR and the future of ESI/FTICR for high-performance mass spectrometry of biomolecules is very bright.

ACKNOWLEDGMENTS

We gratefully acknowledge the generous financial assistance of the National Institutes of Health, the National Science Foundation, the Texas Advanced Research Program, and the Welch Foundation.

REFERENCES

1. Comisarow, M. B.; Marshall, A. G. *Chem. Phys. Lett.* **1974**, *25*, 282–283.
2. Hunt, P. F.; Shabanowitz, J.; McIver, R. T., Jr.; Hunter, R. L.; Syka, J. E. P. *Anal. Chem.* **1985**, *57*, 765–768.
3. Hunt, D. F.; Shabanowitz, J.; Yates III, J. R.; Zhu, N. Z.; Russell, D. H.; Castro, M. E. *Proc. Natl. Acad. Sci. USA* **1987**, *84*, 620–623.
4. Karas, M.; Bachmann, D.; Bahr, U.; Hillenkamp, F. *Int. J. Mass Spectrom. Ion Proc.* **1987**, *78*, 53–68.
5. Henry, K. D.; Williams, E. R.; Wang, B. H.; McLafferty, F. W.; Shabanowitz, J.; Hunt, D. F. *Proc. Natl. Acad. Sci. USA* **1989**, *86*, 9075–9078.
6. Hofstadler, S. A.; Laude, D. A. *Anal. Chem.* **1992**, *64*, 572–575.
7. Winger, B. E.; Hofstadler, S. A.; Bruce, J. E.; Udseth, H. R.; Smith, R. D. *J. Am. Soc. Mass Spectrom.* **1993**, *4*, 566–577.
8. Marshall, A. G. *Anal. Chem.* **1979**, *51*, 1710–1714.
9. Alford, J. M.; Williams, P. E.; Trevor, D. J.; Smalley, R. E. *Int. J. Mass Spectrom. Ion Processes* **1986**, *72*, 33–51.
10. Chen, F. F. *Introduction to Plasma Physics*; Plenum Press: New York, 1984; 2nd Ed., pp. 1–44.
11. Henry, K. D.; McLafferty, F. W. *Org. Mass Spectrom.* **1990**, *25*, 490–492.

12. Henry, K. D.; Quinn, J. P.; McLafferty, F. W. *J. Am. Chem. Soc.* **1991**, *113*, 5447–5449.
13. Loo, J. A.; Quinn, J. P.; Ryu, S. I.; Henry, K. D.; Senko, M. W.; McLafferty, F. W. *Proc. Natl. Acad. Sci. USA* **1992**, *89*, 286–289.
14. Beu, S. C.; Senko, M. W.; Quinn, J. P.; McLafferty, F. W. *J. Am. Soc. Mass Spectrom.* **1993**, *4*, 190–192.
15. Beu, S. C.; Senko, M. W.; Quinn, J. P.; Wampler, F. M.; McLafferty, F. W. *J. Am Soc. Mass Spectrom.* **1993**, *4*, 557–565.
16. Beu, S. C.; Laude, D. A. *Anal. Chem.* **1992**, *64*, 177–180.
17. Gauthier, J. W.; Trantman, T. R.; Jacobson, D. B. *Anal. Chim. Acta* **1991**, *246*, 211–225.
18. Baykut, G.; Watson, C. H; Weller, R. R.; Eyler, J. R. *J. Am. Chem. Soc.* **1985**, *107*, 8036–8042.
19. Watson, C. H.; Baykut, G.; Eyler, J. R. *Anal. Chem.* **1987**, *59*, 1133–1138.
20. Kofel, P.; Alleman, M.; Kellerhals, H. P.; Wanczek, K. P. *Int. J. Mass Spectrom. Ion Processes* **1985**, *65*, 97–103.
21. Stacey, C. C.; Kruppa, G. H.; Watson, C. H.; Wronka, J.; Laukien, F. H.; Banks, J. F.; Whitehouse, C. M. *Rapid Commun. Mass Spectrom.* **1994**, *8*, 513–516.
22. Winger, B. E.; Hein, R. E.; Becker, B. L.; Campana, J. E. *Rapid Commun. Mass Spectrom.* **1994**, *8*, 495–497.
23. Pasa-Tolic, L.; Marto, J. A.; Senko, M. W.; Hendrickson, C. L.; White, F. M.; Guan, S.; Marshall, A. G. *Proceedings of the 43rd ASMS Conference on Mass Spectrometry and Allied Topics*, Atlanta, GA, 1995; p. 1085.
24. Tang, L.; Buchanan, M. V.; Hettich, R. L.; Hurst, G. B. *Proceedings of the 43rd ASMS Conference on Mass Spectrometry and Allied Topics*, Atlanta, GA, 1995; p. 1083.
25. Milgram, E.; Watson, C.; Eyler, J. *Proceedings of the 43rd ASMS Conference on Mass Spectrometry and Allied Topics*, Atlanta, GA, 1995; p. 1088.
26. Guan, Z.; Campbell, V. L.; Drader, J. J.; Hendrickson, C. L.; Laude, D. A., Jr. *Rev. Sci. Instrum.* **1995**, *66*, 4507–4515.
27. Hofstadler, S. A.; Schmidt, E.; Guan, Z.; Laude, D. A., Jr. *J. Am. Soc. Mass Spectrom.* **1993**, *4*, 168–176.
28. J. R. Eyler, private communication
29. Wu, Q.; Udseth, H. R.; Anderson, G. A.; Sherman, M. G.; Van Orden, S.; Chen, R.; Hofstadler, S. A.; Michell, D. W.; Rockwood, A. L.; Smith, R. D. *Proceedings of the 43rd ASMS Conference on Mass Spectrometry and Allied Topics*, Atlanta, GA, 1995; p. 1086.
30. Hofstadler, S. A.; Wahl, J. H.; Bruce, J. E.; Smith, R. D. *J. Am. Chem. Soc.* **1993**, *115*, 6983–6984.
31. Wahl, J. H.; Hofstadler, S. A.; Smith, R. D. *Anal. Chem.* **1995**, 67, 462–465.
32. Hofstadler, S. A.; Wahl, J. H.; Bakhtiar, R.; Anderson, G. A.; Bruce, J. E.; Smith, R. D. *J. Am. Soc. Mass Spectrom.* **1994**, *5*, 894–899.
33. Hofstadler, S. A.; Swanek, F. D.; Gale, D. C.; Ewing, A. G.; Smith, R. D. *Anal. Chem.* **1995**, *67*, 1477–1480.

34. Schweikhard, L.; Guan, S.; Marshall, A. G. *Int. J. Mass Spectrom. Ion Processes* **1992**, *120*, 71–83.
35. Kelleher, N. L.; Senko, M. W.; Little, D. P.; O'Connor, P. B.; McLafferty, F.W. *J. Am. Soc. Mass Spectrom.* **1995**, *6*, 220–221.
36. Valaskovic, G. A.; Kelleher, N. L.; Little, D. P., Aaserud, D. J.; McLafferty, F. W. *Anal. Chem.* **1995**, *67*, 3802–3805.
37. Beu, S. C.; Laude, D. A. *Int. J. Mass Spectrom. Ion Processes* **1991**, *104*, 109–127.
38. Beu, S. C.; Laude, D. A. *Int. J. Mass Spectrom. Ion Processes* **1990**, *97*, 295–310.
39. Hofstadler, S. A.; Laude, D. A. *J. Am. Soc. Mass Spectrom.* **1992**, *3*, 615–623.
40. Hofstadler, S. A.; Wu, Q.; Bruce, J. E.; Chen, R.; Smith, R. D. *Int. J. Mass Spectrom. Ion Processes.* **1995**, *142*, 143–150.
41. Hofstadler, S. A.; Beu, S. C.; Laude, D. A. *Anal. Chem.* **1993**, *65*, 312–316.
42. Campbell, V. L.; Guan, Z.; Laude, D. A., Jr. *J. Am. Soc. Mass Spectrom.* **1994**, *5*, 221–229.
43. Hendrickson, C. L.; Drader, J. J.; Laude, D. A. *J. Am. Soc. Mass Spectrom.* **1995**, *6*, 76–79.
44. Hendrickson, C. L.; Hofstadler, S. A.; Beu, S. C.; Laude, D. A. *Int. J. Mass Spectrom. Ion Processes* **1993**, *123*, 49–58.
45. Bruce, J. E.; Anderson, G. A.; Hofstadler, S. A.; Van Orden, S. L.; Sherman, M. S.; Rockwood, A. L.; Smith, R. D. *Rapid Commun. Mass Spectrom.* **1993**, *7*, 914–919.
46. Bruce, J. E.; Van Orden, S. L.; Anderson, G. A.; Hofstadler, S. A.; Sherman, M. G.; Rockwood, A. L.; Smith, R. D. *J. Mass Spectrom.* **1995**, *30*, 124–133.
47. Speir, J. P.; Gorman, G. S.; Pitsenberger, C. C.; Turner, C. A.; Wang, P. P.; Amster, I. J. *Anal. Chem.* **1993**, *65*, 1746–1752.
48. Williams, E. R.; Henry, K. D.; McLafferty, F. W. *J. Am. Chem. Soc.* **1990**, *112*, 6157–6162.
49. Guan, Z.; Hofstadler, S. A.; Laude, D. A. *Anal. Chem.* **1993**, *65*, 1588–1593.
50. Campbell, V. L.; Guan, Z.; Laude, D. A. *J. Am. Soc. Mass Spectrom.* **1995**, 6, 564–570.
51. Hendrickson, C. L.; Laude, D. A. *Anal. Chem.* **1995**, *67*, 1717–1721.
52. Hendrickson, C. L.; Drader, J. J.; Laude, D. A. *J. Am. Soc. Mass Spectrom.* **1995**, *6*, 448–452.
53. Guan, Z.; Drader, J. J.; Campbell, V. L.; Laude, D. A. *Anal. Chem.* **1995**, 67, 1453–1458.
54. Loo, J. A.; Udseth, H. R.; Smith, R. D. *Rapid Commun. Mass Spectrom.* **1988**, *2*, 207–210.
55. Senko, M. W.; Beu, S. C.; McLafferty, F. W. *Anal. Chem.* **1994**, *66*, 415–417.
56. Senko, M. W.; Speir, J. P.; McLafferty, F. W. *Anal. Chem.* **1994**, *66*, 2801–2808.
57. O'Connor, P. B.; Speir, J. P.; Senko, M. W.; Little, D. P.; McLafferty, F. W. *J. Mass Spectrom.* **1995**, *30*, 88–93.
58. Guan, Z.; Kelleher, N. L.; O'Connor, P. B.; Aaserud, D. J.; Little, D. P.; McLafferty, F. W., submitted for publication in *Int. J. Mass Spectrom. Ion Processes.*
59. Little, D. P.; Speir, J. P.; Senko, M. W.; O'Connor, P. B.; McLafferty, F. W. *Anal. Chem.* **1994**, *66*, 2809–2815.

60. Speir, J. P.; Senko, M. W.; Little, D. P.; Loo, J. A.; McLafferty, F. W. *J. Mass Spectrom.* **1995**, *30*, 39–42.
61. Chorush, R. A.; Little, D. P.; Beu, S. C.; Wood, T. D.; McLafferty, F. W. *Anal. Chem.* **1995**, *67*, 1042–1046.
62. Bruce, J. E.; Anderson, G. A.; Hofstadler, S. A.; Winger, B. E.; Smith, R. D. *Rapid Commun. Mass Spectrom.* **1993**, *7*, 700–703.
63. A. G. Marshall, private communication
64. Senko, M. W.; Beu, S. C.; McLafferty, F. W. *J. Am. Soc. Mass Spectrom.* **1993**, *4*, 828–830.
65. Jiao, C. Q.; Freiser, B. S.; Carr, S. R.; Cassady, C. J. *J. Am. Soc. Mass Spectrom.* **1995**, *6*, 521–524.
66. Guan, Z.; Campbell, V. L.; Laude, D. A., Jr. *J. Mass Spectrom.* **1995**, *30*, 119-123.
67. Cassady, C. J.; Wronka, J.; Kruppa, G. H.; Laukien, F. H. *Rapid Commun. Mass Spectrom.* **1994**, *8*, 394–400.
68. Suckau, D.; Shi, Y.; Beu, S. C.; Senko, M. W.; Quinn, J. P.; Wampler, F. M., III; McLafferty, F. W. *Proc. Natl. Acad. Sci. USA*, **1993**, *90*, 790–793.
69. Wood, T. D.; Chorush, R. A.; Wampler, F. M., III; Little, D. P.; O'Connor, P. B.; McLafferty, F. W. *Proc. Natl. Acad. Sci. USA* **1995**, *92*, 2451–2454.
70. Gross, D. S.; Williams, E. R. *J. Am. Chem. Soc.* **1995**, *117*, 883–890.
71. Smith, R. D.; Cheng, X.; Bruce, J. E., Hofstadler, S. A.; Anderson, G. A. *Nature* **1994**, *369*, 137–139.
72. Chen, R.; Wu, Q.; Mitchell, D. W.; Hofstadler, S. A.; Rockwood, A. L.; Smith, R. D. *Anal. Chem.* **1994**, *66*, 3964–3969.
73. Bruce, J. E.; Cheng, X.; Bakhtiar, R.; Wu, Q.; Hofstadler, S. A.; Anderson, G. A.; Smith, R. D. *J. Am. Chem. Soc.* **1994**, *116*, 7839–7847.
74. Chen, R.; Cheng, X.; Mitchell, D. W.; Hofstadler, S. A.; Wu, Q.; Rockwood, A. L.; Sherman, M. G.; Smith, R. D. *Anal. Chem.* **1995**, *67*, 1159–1163.

PART III
INTERFACING OF SOLUTION-BASED SEPARATION TECHNIQUES TO ELECTROSPRAY

CHAPTER 9

Combining Liquid Chromatography with Electrospray Mass Spectrometry

ROBERT D. VOYKSNER

Analytical and Chemical Sciences, Research Triangle Institute, Research Triangle Park, North Carolina

	Abstract	323
I.	Introduction	324
II.	Combining LC with API–MS	324
	A. API–MS considerations	324
	B. Chromatographic considerations	327
	C. On-line LC/API–MS	331
III.	Summary of LC/API–MS	336
IV.	Future trends in LC/MS	337
	Acknowledgments	339
	References	340

Abstract

The combination of liquid chromatography with mass spectrometry (LC/MS) has evolved into a sensitive, rugged, and widely used technique since the development and commercialization of atmospheric-pressure ionization (API) MS. While on-line LC/MS can be relatively straightforward, it is still important to consider solution chemistry to achieve the best separation and API–MS response. This chapter covers the key aspects in developing a successful LC/MS analysis. Often this means a LC separation must be adjusted to be suitable for LC/MS. For example, typical LC additives such as sulfate, phosphate, and borate buffers that are not suitable for API–MS must be replaced with volatile mobile-phase additives. The API–MS sensitivity can be optimized through selection of conditions to form ions in solution, aid in nebulization, enhance desolvation, and improve ion-evaporation ionization. Conditions including solvent choice, pH, and flow rates

Electrospray Ionization Mass Spectrometry, Edited by Richard B. Cole.
ISBN 0-471-14564-5 © 1997 John Wiley & Sons, Inc.

need to be optimized for LC separation and API–MS detection to obtain the best sensitivity. Post-column techniques can be used to decouple the LC separation from API–MS detection so that optimal LC conditions can be used for the separation and be adjusted post column for optimal API–MS sensitivity.

I. INTRODUCTION

The development and commercialization of atmospheric pressure ionization–mass spectrometry (API–MS) has for the first time brought the combination of liquid chromatography (LC) with MS into the realm of a routine analytical procedure. This combination of LC/API–MS has collimated nearly 30 years of research in overcoming incompatibilities in combining LC with MS and limitations in resulting sensitivity and ruggedness. The combination of LC with the API techniques of electrospray (ES) (1) pneumatically assisted ES (2) and atmospheric-pressure chemical ionization (APCI) (3) has been extremely successful because

1. API approaches can handle volumes of liquid typically used in LC
2. API is suitable for the analysis of nonvolatile, polar, and thermally unstable compounds typically analyzed by LC
3. API–MS systems are sensitive, offering comparable or better detection limits than achieved by GC/MS
4. API systems are very rugged and relatively easy to use

For these reasons, LC/API–MS instruments are found in most commercial and government laboratories that perform research involving compounds that require LC separations.

The purpose of this chapter is not to review the hundreds of publications and presentations in the area of LC/MS, which have been the subject of several recent reviews (3–5). Rather, the chapter will provide insights in achieving good separations together with sensitive API–MS detection.

II. COMBINING LC WITH API–MS

A. API–MS Considerations

Liquid chromatography/mass spectrometry is still limited to conditions that are suitable for MS operations. There are restrictions on pH, solvent choice, solvent additives, and flow rates for LC in order to achieve optimal API–MS sensitivity (6–14). Furthermore, these restrictions are dependent on the API operation mode of ES, pneumatically assisted ES, or APCI. As a review, ES is a desorption ionization process which uses electrical fields to generate charged droplets and subsequent analyte ions by ion-evaporation ionization (15,16). Pneumatically assisted ES is the same as ES except the initial droplet formation is the result of pneumatic nebulization. Finally, APCI is a gas-phase ionization process initiated by a discharge sustained by the LC mobile phase vapor.

In general, API techniques require the use of volatile solvent additives listed in Table 1 to prevent API chamber contamination or plugging of the sampling orifice. Phosphate, sulfate, or borate additives typically used in LC are not suitable for API–MS. Electrospray operation further requires that the solvent additives do not form strong ion pairs which could result in neutralization of ions after desorption (16). Solvent additives also are important in controlling pH. This is particularly important in ES operations because the protonation or deprotonation of the analyte in solution greatly enhances ion formation. In general, for compounds with basic sites (e.g., amines) the analysis should be performed at a low pH using positive-ion detection. Components containing acidic sites (e.g., carboxylic acid) are analyzed at high pH using negative-ion detection. Neutral species can often be more effectively ionized with ES through cationization by the additions of micromolar levels of sodium or potassium acetate.

Suitable solvents for API–MS are also presented in Table 1 (18). In general, most LC solvents are compatible with APCI. However, solvents suitable for ES permit the formation of ions in solution. Also, ease in nebulization and

TABLE 1. Suitable API–MS Solvents and Additives

pH	Acetic acid, formic acid, trifluoroacetic acid (TFA) for ES positive-ion detection; ammonium hydroxide for ES negative ion-detection (typically in the 0.1–1% range)
Buffers, ion-pair reagents	Ammonium acetate, ammonium formate, triethylamine heptafluorobutyric acid (HFBA), tetraethyl or tetrabutylammonium hydroxide (TEAH or TBAH) (10–100 mM level)
Cationization reagents	Potassium or sodium acetate (20–50 μM level)
Solvents for API–MS	Methanol, ethanol, propanol,[b] isopropanol,[b] butanol,[b] acetonitrile, water, acetic acid,[a] formic acid,[a] acetone,[a] dimethylforamide,[a] dimethyl sulfoxide,[a] 2-methoxy ethanol,[b] tetrahydrofuran,[a] chloroform[a]
Solvents for APCI only (not suitable for ES)	Hydrocarbon solvents (e.g., hexane, cyclohexane, toluene), CS_2, CCl_4

[a] These solvents have been used in the 5–20% range. Higher percentages have not evaluated in our laboratory.

[b] Solvents best for negative-ion operation.

desolvation play a role in solvent suitability. For example, water easily supports the formation of ions in solution, but its surface tension and solvation energy make ion desorption more difficult than a solvent like isopropanol or acetonitrile. However, with the use of pneumatically assisted ES, the difference in response between water and acetonitrile due to differences in surface tension and heats of vaporization of the solvents are reduced (Fig. 1). Yet differences in the ability of the solvent to form and support ion in solutions or relieve charge neutralization by strong ion-pair reagents are not changed by pneumatic nebulization. For these reasons, some organics (e.g., peptides) may show better response in water, whereas other organics (e.g., oligonucleotides) may exhibit better sensitivity in organic solvents such as acetonitrile. Still, reducing the effects of surface tension, viscosity, and heat of vaporation upon ES sensitivity is an advantage when performing gradient elution in LC/MS and when operating at higher flow rates.

Negative-ion ES operation occasionally requires special solvent consideration, beyond the usual increase in the pH. Voltages applied to the API chamber to charge droplet for negative-ion ionization often lead to a corona discharge before a sufficiently strong field is generated for ion evaporation (19). The evaporation of electrons from the ES needle can be controlled by using electron-scavenging solvents or nebulization gases. For example, the use of isopropanol and water and air for nebulization (O_2 used for electron scavenging) permits operation at 1000–1500 V higher than for an acetonitrile–water

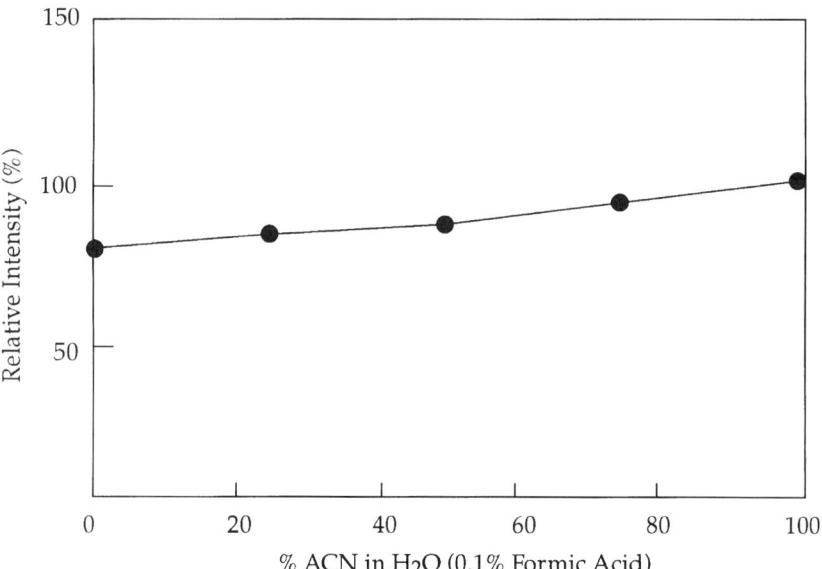

Figure 1. Pneumatically assisted ES response for the $[M + H]^+$ ion of penicillin G from 0 to 100% acetonitrile (ACN) in water containing 0.1% formic acid.

solution with nitrogen nebulization (19). The operation at the higher voltage permits droplet charging, enabling sufficient field strengths to allow ion evaporation of the analyte ions. In general, solvents like isopropanol, butanol, or 2-methoxyethanol should be substituted or added post column for methanol acetonitrile, ethanol, or tetrahydrofuran and water solutions if discharges occur (19). Also, the addition of air, oxygen, or SF_6 will further aid in preventing a discharging, enabling operation at 500–1000 V higher potential in the API chamber (19,20).

The solvent flow rate used in a LC separation ranges from 3–8 μL/min for capability LC columns (0.32 mm i.d.) to 1 mL/min for 4.6 mm i.d. columns. These ranges in flow rate are not compatible with each mode of API–MS operation. Electrospray is limited to low flow rates (5–10 μL/min is optimal for ion formation) owing to the limitations of nebulization through charging of the liquid by electrical fields. Pneumatically assisted ES optimizes at about 0.2 mL/min and can operate up to 1 mL/min without a major loss in sensitivity. The pneumatic nebulization overcomes droplet-formation limitation from electrical fields in ES. Finally, APCI is a high-flow-rate technique, optimizing at greater than 1 mL/min. Operating at flow rates lower than 0.4 mL/min is often insufficient to maintain a stable corona discharge for APCI. Figure 2 displays the response of penicillin G for different flow rates under ES, pneumatically assisted ES, and APCI conditions for a Hewlett Packard 5989 system. However, the specific optimal flow rate for each mode of operation is different, depending on the design of the API system.

B. Chromatographic Considerations

Numerous chromatographic separation techniques can be coupled to API–MS. The key to achieve separations and maintain API–MS performance in LC/MS is finding separation conditions which are compatible with API–MS. Table 2 lists the general compatibility of seven common modes of LC with ES and APCI. Such as chromatographic modes (1) reverse-phase, (2) ion-pair, (3) size-exclusion, or (4) immunoaffinity separations are compatible with ES because ions can be created in solution, the solvents can be electrosprayed, and buffers compatible with ES can be employed:

   1. Reverse phase is currently the most common method for LC and is widely used for the separation of a variety of pharmaceutical, environmental, and biologically active compounds. The mechanism of chromatographic retention is based on the partitioning of the analyte into the stationary phase. A wide range of applications exists, owing to the numerous solvents and stationary phases (e.g., C_{18}, C_8, C_2, CN, and phenol) that can be combined to obtain a selective separation. Reverse-phase chromatography is compatible with APCI if the sample exhibits sufficient volatility. Although ions in solution are optimal for ES, they often lead to poor retention in reverse-phase chromatography because there is little partitioning from the mobile phase (e.g., water) into the less-polar

stationary phase (e.g., C_{18}). Often, pH must be controlled to perform the separation on the neutral compounds, then the pH is adjusted post column to form the anion or cation of interest for ESMS. Post-column modification is very important in achieving LC/MS compatibility and is discussed in Section II.C. Larger molecules with both hydrophobic and hydrophilic groups (or ionic species) often are retained in reverse phase owing to partitioning from the hydrophobic moiety into the stationary phase. Separation of ionic species by reverse-phase methods can be performed by ion-pair chromatography if more retention is desired.

2. Ion-pair chromatography employs a cationic or anionic additive to complex with the anion or cation sample, respectively. The ion pair found is neutral, allowing for increased retention based upon partitioning into a nonpolar stationary phase. Often the ion-pair additives can be purchased in various alkyl group lengths (e.g., tetraethyl or tetrabutyl) to "fine tune" the retention and separation of compounds. Ion-pair separations are well suited for ES as long as the ion-pair additives are volatile and form weak enough ion pairs to prevent

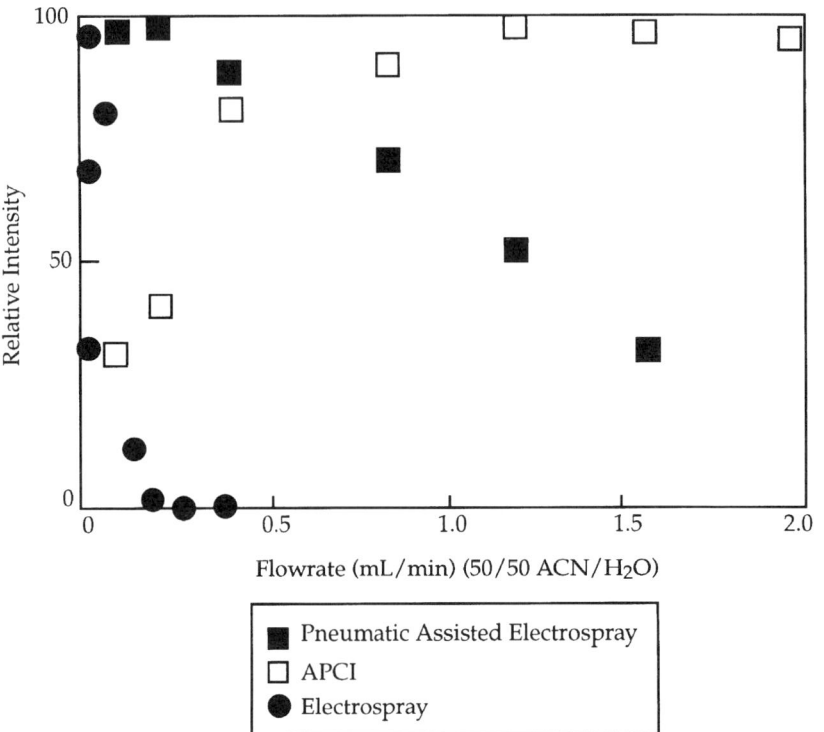

Figure 2. Response for the $[M + H]^+$ ion of penicillin G with solvent flow rate for ES, pneumatically assisted ES and APCI. The mobile phase was 50% acetonitrile in water containing 0.1% formic acid.

TABLE 2. Compatability of API–MS with Various Chromatographic Modes

Mode	ES	APCI	Comments
Reversed phase	+++	++	Formation of ions in solution is possible; usually limited sample volatility
Normal phase	+	+++	Ions in solution (nonaqueous miscibility) may be a problem; usually sample is volatile
Size exclusion	+++	+	Buffers to suppress nonexclusion mechanisms may cause problems; most likely sample is not volatile and is a high molecular weight
Ion pair	++	++	Reagent ions may compete for ion-evaporation process; volatility of mobile-phase additive
Ion exchange	+	+	High ionic strength may be a problem; limited volatility of mobile-phase additives
Hydrophobic interaction	+	+	Uses salt gradients to elute biomolecules; salt is not compatible with API–MS
Immunoaffinity	+++	+	Mobile phase often compatible with API–MS, usually nonvolatile sample

[a] A greater number of plus signs indicates a larger degree of compatibility.

charge neutralization in ES. For these reasons, ion-pair reagents such as trifluoroacetic acid, heptafluorobutric acid, or ammonium acetate (for cationic samples) and tetraethyl or tetrabutyl ammonium hydroxide (for anionic samples) are employed. An example of an ion-pair separation for β-lactam antibiotics is presented in Section II.C (Fig. 6).

3. Size-exclusion chromatography (SEC) offers a different retention mechanism compared to reverse-phase or ion-pair chromatography. The SEC columns are packed with various pore-sized polymers to separate compounds based upon molecular weight. Smaller (low molecular weight) compounds are retained while larger compounds are less retained. Size-exclusion chromatography is a powerful technique for separation of biologically active compounds in the 1–1000 K molecular weight range. The SEC columns also offer advantages over reverse-phase techniques for the separation of peptides or proteins that often stick to polar sites on the reverse-phase station phase (e.g., hydroxyl groups form silica) and provide a complimentary mode of separation. While SEC can be

used to obtain molecular weight (±10%) based upon retention, the coupling of SEC with ESMS can provide molecular-weight determination of 0.02% of the average molecular weight for biological compounds up to 100–200 K in molecular weight.

For the most part, SEC is readily compatible with ESMS because the mobile phase is usually water with some organic (e.g., acetonitrile) and an additive such as TFA to minimize sample adsorption. On-line SEC/ESMS has been demonstrated to separate neuropeptides (21) and polymers (22). Off-line approaches of SEC and SDS–PAGE have been used for the purification of oligosaccharides and peptides. Based upon SEC application to the separation of larger biologically active compounds, which are thermally liable and not volatile, APCI–MS is not suitable for their analysis.

4. Immunoaffinity chromatography (IAC) offers a specific method to bind a target compound in a solution to a solid support, permitting concentration as well as purification. The changing of the mobile phase (e.g., pH or ionic strength) releases the target compound for elution. The key for IAC is obtaining an antibody that binds the sample of interest. There are only a few commercial antibodies available; therefore, antibodies often must be grown and purified for use in IAC. Generating custom antibodies is a slow and expensive process and limits widespread usage of the technique. Once an antibody is found, it can be immobilized onto a solid support for packing into a column. Immunoaffinity chromatography is for the most part compatible with MS. Typically, a sample in a buffer or salt solution is loaded onto the IAC column (0.1–100 mL of sample). The IAC column binds the specific compound of interest while the other components are washed off the column. The buffer or salt (which is not compatible with ESMS) is washed off in water after sampling loading, then the mobile phase is adjusted to release the target compound for analysis. Often the release can be accomplished by decreasing pH with formic acid. The released target compound can be directly analyzed by ESMS or eluted to a trapping column for further concentration and followed by reverse-phase LC with ESMS detection. Both IAC/ESMS and IAC-LC/ESMS techniques have been reported for the analysis of drugs (23,24) environmental compounds (25), and biologically active compounds (26,27). The technique has been successful in concentrating samples (200-fold concentration has been demonstrated) and extremely specific, minimizing interferences in the ESMS analysis.

The remaining modes of chromatography on Table 2 are less commonly used in API–MS. Normal-phase separations are well suited for APCI because many normal-phase solvents do not substantiate ion formation in solution or can be charged for generating fields for ion evaporations. Also, many samples that are soluble in normal-phase solvents are of low polarity and have a reasonable possibility that they can be vaporized for APCI. Ion-exchange and hydrophobic-interaction chromatography are generally not compatible with API–MS owing to higher ionic strength and nonvolatile buffers. However, ion-exchange LC/MS has been reported, using a suppression column prior to the MS to remove the salts (28,29).

Another important chromatographic consideration is adapting existing LC conditions to LC/MS. Often LC/MS is employed to identify or confirm the identity of a peak in a LC chromatogram using conventional detection (e.g., UV). In adapting an existing LC method to LC/API–MS, one must consider (1) buffers, (2) pH, (3) ion pair reagents, (4) solvents, and (5) columns.

1. Buffers such as sulfates, phosphates, or borates need to be exchanged for more volatile additives such as ammonium acetate, ammonium formate, acetic acid, formic acid, trifluoroacetic acid (TFA), heptafluorobutryic acid (HFBA), or tetrabutylammonium hydroxide (TBAH). Usually ionic strength of the buffer is more important to the buffer choice, allowing the user to adapt to LC/MS compatible conditions with minimal loss in chromatographic resolution.

2. The pH of the separation should be kept the same as the existing LC separation. The pH can be adjusted using acetic acid, formic acid, TFA, or ammonium hydroxide for API–MS.

3. Ion-pair reagents used in existing LC separations often contain sulfate, phosphate, sodium, or potassium. These nonvolatile ion-pair reagents should be substituted with such reagents as HFBA, TBAH, or tetraethyl ammonium hydroxide (TEAH). It is difficult to achieve the same retention times with the ion-pair substitutes, but often the chromatographic separations can be maintained. Also the molecular weights of the ion-pair reagents should be kept to below 200 to minimize interference in the mass spectra.

4. For the most part solvents used in LC are compatible with API–MS. There is usually no need to change the mobile phase for the separation when performing LC/MS.

5. Most columns used in LC are compatible with LC/MS. However, columns used with nonvolatile ion-pair reagents for LC may exhibit contamination from alkali metal salts or anions (e.g., sulfate or phosphate) for a long period of time, suppressing ES ion current. Ion-pair columns should not be used for LC/MS.

C. On-line LC/API–MS

The schematic diagram shown in Figure 3 presents the hardware for combining LC columns ranging from 0.32 to 4.6 mm i.d. to API–MS. The system allows for gradient-elution operation for flow rates of 0.004–1 mL/min. The first splitter before the injection is used to achieve stable flow rates <200 μL/min from a typical piston pump. Furthermore, this splitter allows for rapid changes in solvent composition to reach the column, avoiding mixing dead volumes in the pump, which could add 3- to 10-min delays in a gradient separation at lower operating flow rates. This splitter is placed before the injector (vs. between the injector and column) to minimize the loss of sample. At operations above 200 μL/min, this splitter is not required.

The injection loop sizes shown in Figure 3 are average ranges for the inside diameter of the column employed. Larger injection loops can be used when

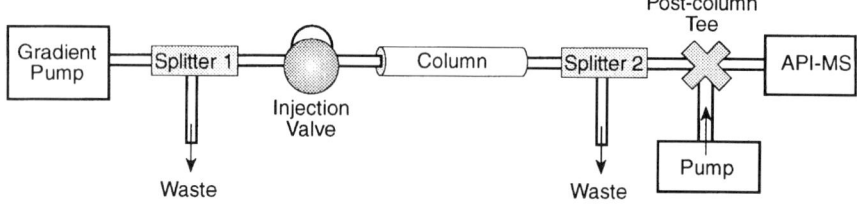

Pump mL/min	Splitter 1	Loopsize (uL)	Column i.d (mm).	Column Flow Rate (mL/min)	Post Column Splitter 2	Post-column Tee Pump (mL/min)
0.5 - 1.0	yes	0.2 - 10	0.32	0.004 - 0.008	No	0.002 - 0.005
0.5 - 1.0	yes	1 - 20	1.0	0.04 - 0.06	No	0.02 - 0.04
0.2 - 0.3	no	5 - 50	2.1	0.2 - 0.3	Yes (1)	0.1
1.0	no	20 - 100	4.6	1.0	Yes (1)	0.1 - 0.2

(1) Post column splitting for ES operation

Figure 3. Schematic diagram for hardware used to combine LC with API-MS. The table lists typical conditions for the pump, the need for splitters, and flow rates for separation using 0.32, 1.0, 2.1, and 4.6 mm i.d. LC columns.

injections are performed in a weak LC mobile phase (e.g., water in reverse-phase LC), permitting focusing and concentrating on the front of the column. After the loop is emptied, the mobile phase can be adjusted for the elution of the sample while maintaining chromatographic integrity of the separation.

The use of capillary LC columns (0.32 mm i.d.) is often a challenge, because all dead volumes must be eliminated to achieve the expected advantages of capillary LC, namely, improved peak concentration and, therefore, better ES sensitivity (30). Also 25- to 50-μm i.d. fused silica or PEEK tubing is required to minimize peak broadening. Connections after the LC column should be made using butt connections in a Teflon connector that can withstand 150 psi of pressure.

The post-column splitter (splitter 2, Fig. 3) is used to reduce LC flow into the API–MS. It is primarily used with 2.1- or 4.6-mm i.d. columns to keep the flow rate at 100–200 μL/min, which is optimal for pneumatically assisted ES operation (Fig. 2) of the Hewlett Packard 5989 instrument. However, the need to split or the split ratio is dependent on the design of the instrument and needs to be optimized for each manufacturer to achieve optimal API–MS sensitivity. For a smaller diameter column or for operation in APCI, this splitter is not required. Also, this splitter (when used) is placed prior to the post-column additions pump to minimize the amount of reagent added post column to optimize API–MS performance.

The most important aspect of the diagram in Figure 3 is the post-column addition tee and pump. The use of post-column addition is the best means to decouple the LC from the MS (31). In this way, conditions optimal for the LC separation can be used, then the solution is modified post column to achieve optimal API–MS performance. There are numerous ways that post-column additions can be employed to improve API–MS sensitivity, as itemized below:

1. The LC mobile phase pH can be adjusted to improve ES sensitivity (32). The addition of acetic acid or formic acid (e.g., 50% in isopropanol at about 5 μL/min) can lower the pH to about 3 to improve ES positive-ion detection sensitivity. Figure 4 demonstrates the use of post-column addition to acidify the mobile phase for a capillary LC/MS separation of a mixture of amines. Likewise, ammonium hydroxide could be added post column to increase the pH to improve ES negative-ion detection. Figure 5 demonstrates post-column additives to increase the pH for the negative-ion detection of bile acids.

2. Solvents that aid in desolvation or dilute low molarities of ionic buffers can be added post column to improve ES performance. Isopropanol is one of the best solvents to add post column because it aids in nebulization, desolvation, and ion evaporation. Also, additives of relative large volumes of isopropanol relative to column flow (10–100 times more isopropanol) can sufficiently dilute a mobile phase containing 1 mM phosphate buffer to achieve suitable ES signal intensities. This use of post-column additives is primarily employed for capillary LC (0.32 mm i.d.) or capillary electrophoresis.

3. Post-column addition can be used to provide a source of alkali metal ions to promote cationization of weak acidic or basic compounds such as sugars under ES operation (33,34). In this case, potassium or sodium acetate is added to achieve a final concentration of 20–50 μM at the API–MS chamber to promote cationization.

4. Post-column addition can be used to increase flow rates to achieve suitable APCI (1–2 mL/min) or pneumatically assisted ES (0.1–0.3 mL/min) operation. This is particularly important when working with 0.32- or 1.0-mm i.d. LC columns for APCI–MS operation. However, since API–MS is a concentration

---

Figure 4. Capillary LC/ES MS total-ion-current chromatogram for the separation of a mixture of aromatic amines. (Conditions: 150 × 0.32 mm C_{18} column with 5-μM particles using a gradient of 40–90% acetonitrile in 30 min at a flow rate of 6 μL/min.) Acetic acid (10% in isopropanol) was added post column at 1 μL/min to protonate the amines for ES positive-ion detection; 1 ng of each amine was injected. The identities of the peaks are as follows: (1) aniline; (2) phenol; (3) m-phenylenediamine; (4) 2-fluoroaniline; (5) 4-fluoroaniline; (6) benzidine; (7) 2,4-dinitroaniline; (8) 3,3′-dimethoxybenzidine; (9) 2-methylaniline; (10) 2-methoxyaniline; (11) 3,3′-dimethylbenzidine; (12) 4-fluoro-2-methylaniline; (13) 4-nitroaniline; (14) ethylene dianiline; (15) 2,4-dimethylaniline; (16) 4-amino-3-nitrobenzonitrile; (17) 3,3′,5,5′-tetramethylbenzidine; (18) N,N,N',N'-tetramethylbenzidine; (19) 3-methylmercaptoaniline; (20) thiochroman-4-ol; (21) 1-naphthylamine; (22) 4,4′-difluorobiphenyl; (23) 4-chloro-2-methylaniline; (24) 4,5-difluoro-2-nitroaniline; (25) 3,3′-dichlorobenzidine; (26) 2,6-dichloro-4-nitroaniline; (27) 4,4′-diaminooctafluorobiphenyl; (28) diphenylamine.

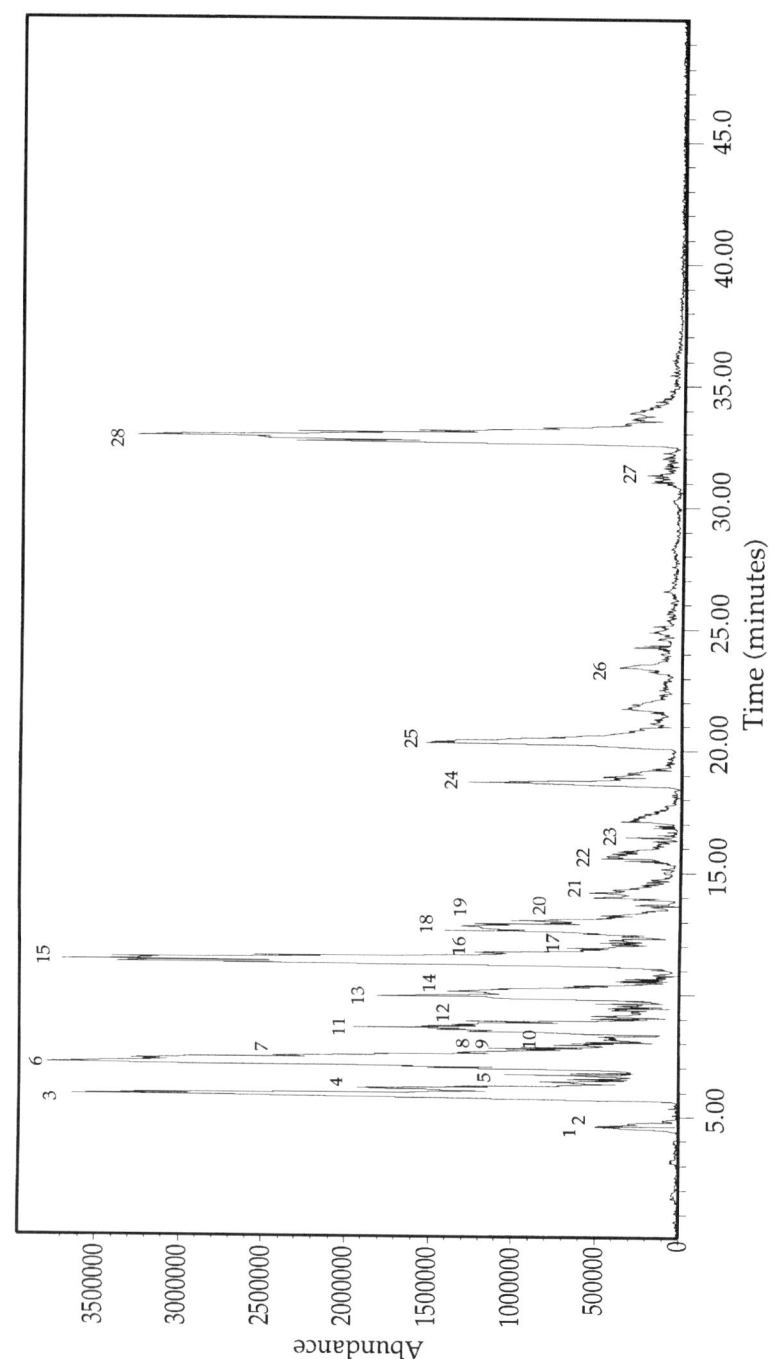

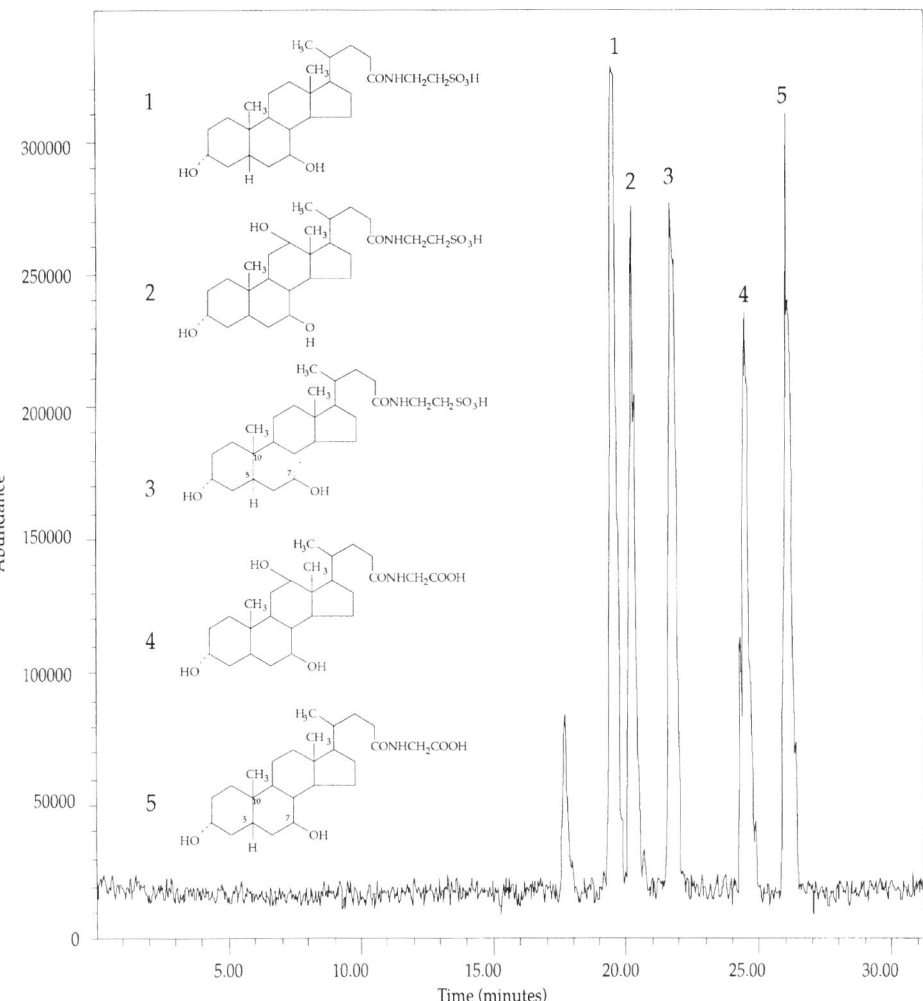

Figure 5. The LC/MS total-ion-current chromatogram for the analysis of bile acids. The separation was performed on a C_{18} (100 × 4.6 mm) column with 0–100% MeOH in 30 min at 1 mL/min. The flow rate was reduced to 200 μL/min using a post-column split. The pH was increased by post-column addition of ammonium hydroxide : water : isopropanol, 10 : 40 : 50, at 25 μL/min, to deprotonate the acids to improve ES negative-ion detection sensitivity.

detector, the addition of solvent will dilute the sample, reducing the sensitivity that might be achieved compared to ideal conditions.

5. Derivatization of a sample to improve ES sensitivity can be performed post column. For example, derivatization for a cationic species (quaternary ammonium ion) or an amine which can be protonated at low pH, will greatly

improve ES positive-ion detection sensitivity. There have been few reports of derivatization for ES/MS (35,36), and post-column derivatization has been employed in our laboratory to improve thermospray sensitivity (37,38).

6. Post-column additives can be used to displace an additive that forms stronger ion pairs with the analyte (reducing ES sensitivity) with an additive that forms weaker ion pairs (39). In the presence of stronger acids (HA) such as TFA or HFBA, Reaction 1 favors the formation of $[M + H + A]^*$:

$$\text{Reaction 1: } [M + H]^+ + [A]^- \rightleftharpoons [M + H + A]^*$$

Reaction 1 leads to charge neutralization, reducing API–MS sensitivity. However, the post-column addition of a weaker acid (RCOOH), such as propionic acid, can displace the more volatile stronger acids based on volatility (b.p. propionic acid > HFBA > TFA). With the stronger acid removed from the droplet at the time of ion evaporation, Reaction 2 favors the formation of $[M + H]^+$:

$$\text{Reaction 2: } [M + H]^+ + [RCOO]^- \rightleftharpoons [M + H + RCOO]^*$$

Reaction 2 favors the separate anion and cation species, greatly improving the API–MS sensitivity. Figure 6 demonstrates this improved sensitivity by comparing responses for the LC/MS determination of a mixture of β-lactams using 25 mM HFBA with and without post-column addition of propionic acid. A factor of almost 20 is gained in sensitivity for this determination using post-column addition of propionic acid.

This methodology of using post-column additives of propionic acid has been named the "TFA Fix" (39). However, it will work whenever the boiling point of the weak ion-pair acid (RCOOH) is greater than the boiling point of the strong acid (HA).

III. SUMMARY OF LC/API–MS

Although LC/MS has become a widespread and rugged technique, it is still important to consider solution chemistry to achieve the best separation and API–MS response. The key issues in combining LC with API–MS are the substitution of nonvolatile and ionic additives for more volatile additives. Sulfate, phosphate, and borate buffers should not be used. Solution chemistry is important in forming ions in solution, promoting solution nebulization and desolvation, and achieving ion-evaporation ionization in ES/MS. Through careful selection of pH, solvent, and flow rate, ES/MS sensitivity can be optimized within the context of the LC separation. Most important, post-column techniques can be used to decouple the LC separation from API–MS detection. In this way, optimal LC separation conditions can be adjusted post column to achieve solution conditions optimal for API–MS. The reader is also cautioned to interpret the API–MS results from changing conditions to achieve compatibility with LC and API–MS. The statements made in this chapter cover the average sample that behaves in a predictable manner. However, we all find, on occasion, the exceptions to the rule when it comes to API–MS optimization.

IV. FUTURE TRENDS IN LC/MS

Although the vast majority of LC/MS is performed by reverse-phase chromatography using the single- or triple-quadrupole API–MS system,

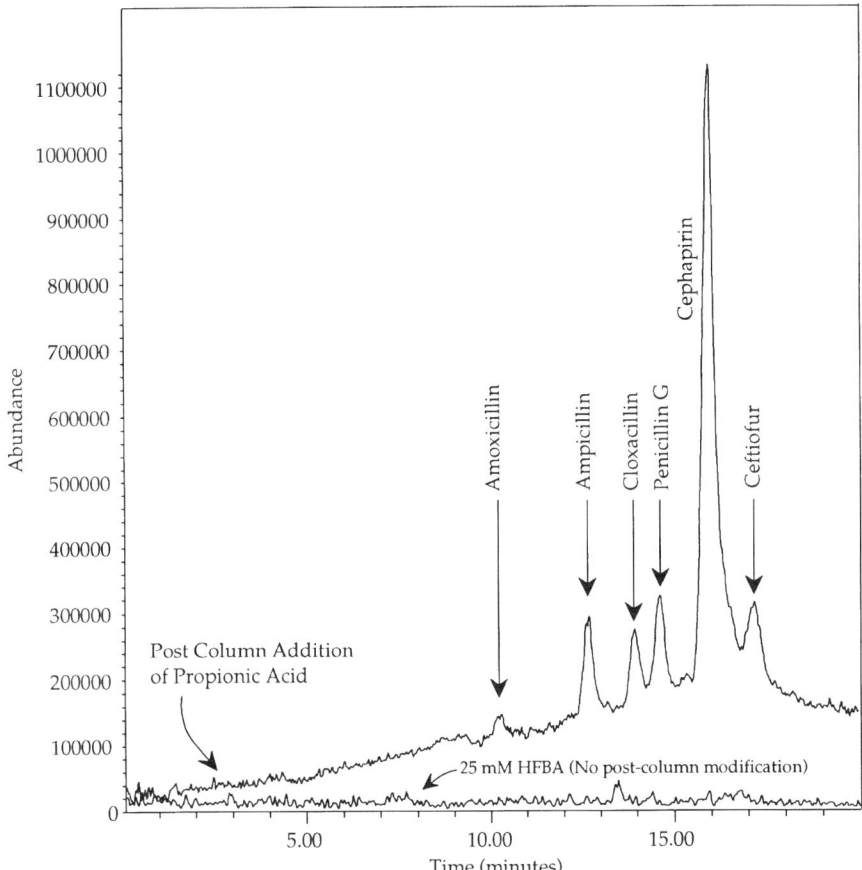

Figure 6. The LC/MS total-ion-current chromatographic separations of a mixture of β-lactam antibiotics with and without the post-column addition of propionic acid. The separation was performed on a C_{18} column (150 × 2.1 mm) using a gradient of 10–70% acetonitrile in water for 20 min at 0.25 mL/min. The ion-pair reagent HFBA was added at the 25-mM level to provide retention of the β-lactams. Propionic acid (50% acid in isopropanol) was added at 0.1 mL/min to generate the upper trace.

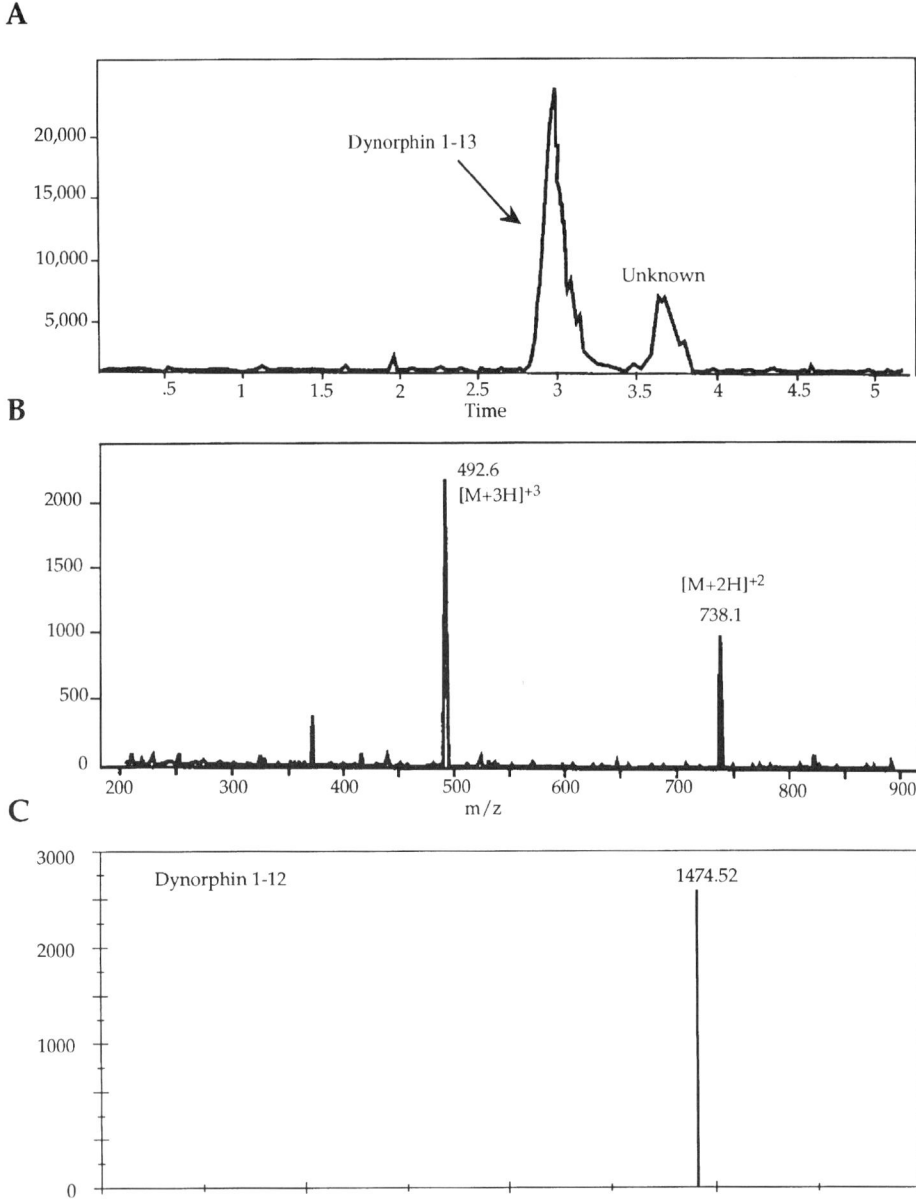

Figure 7. Perfusion LC–ES/MS analysis of dynorphin 1–13. (*A*) Total-ion-current chromatogram for a separation performed on a porous R/H (250 × 0.32 mm) column with a gradient of 90–10% water in acetonitrile (0.1% TFA) in 4 min at a flow rate of 60 μL/min. (*B*) Electrospray mass spectrometry of the unknown peak. (*C*) Deconvoluted ES/MS, indicating that the molecular weight of the unknown is 1474.5, corresponding to the hydrolysis product dynorphin 1–12.

advances in both separation techniques and mass analyzers could prove important. Immunoaffinity cleanup and concentration columns coupled with a trapping and reverse-phase analytical LC column have been demonstrated to provide on-line sample cleanup and concentration for LC/MS analysis (40). Immunoaffinity technique has been demonstrated to provide over a 200X increase in sample concentration (41,42). Another separation technique of importance is perfusion chromatography (43). Perfusion packings are porous 7–8 μM particles which have 6000–8000 Å flow through pores and 500–100 Å diffuse pores. Columns using these porous particles can be operated at 10 times higher flow rates while maintaining good mass transfer with normal operating pressure. Separation using perfusion LC can be an order of magnitude faster and gradient equilibrium times can be reduced to 1 min. Figure 7 demonstrates the rapid separation capabilities achieved from capillary perfusion LC/ES/MS determination of dynorphin 1–13 and a hydrolysis product (dynorphin 1–12). A complete analytical cycle for the analysis (gradient elution and reequilibration) can be completed in 5 min. While perfusion separations are better applied to the analysis of peptides and proteins (44,45), they can also be suitable for small-molecule separations (46).

Advances in LC/MS have also involved the mass analyzer. There are two emerging mass analyzers which offer better sensitivity and mass scan rates. Ion traps (45,47,48) and ion storage time-of-flight (TOF) (49,50) offer improved sensitivity through ion accumulation to enhance the duty cycle (i.e., the time ions are sampled relative to the total cycle time of the analysis). Secondly, these analyzers have the capability to rapidly scan the mass range of interest. This proves very important in the characterization of very narrow peaks obtained in capillary LC or capillary electrophoresis. High-resolution separation techniques that result in peak widths of 1–3 s must be adequately sampled by acquiring 10–20 points across the peak, requiring scan rates in the range of 50 to 300 ms per cycle. Scan rates in the lower-millisecond range are not achievable by quadrupole instruments.

Finally, advances in the ES spray needle have been realized, permitting the utilization of low sample flow rates (down to 10–20 nL/min) (51–53). These low-flow-rate ES or "nano ES" ion sources are well suited to capillary LC (0.18 mm i.d. columns and smaller) or capillary electrophoresis where the flow rates often do not exceed 1 μL/min. Conventional API–MS systems would require a post-column addition of solvent to reach flow rates of 3–5 μL/min to achieve stable ES ion formation. However, the post-column addition of solvent will dilute the chromatographic peak, reducing API–MS sensitivity and the advantages of these capillary chromatographic techniques.

ACKNOWLEDGMENTS

Although the work reported in this article was supported in part by The Food and Drug Administration Cooperative Agreement No. FD-U-000589 and The

Environmental Protection Agency Cooperative Agreement No. CR-819555 to Research Triangle Institute, it has not been subject to review by these agencies. Therefore, it does not necessarily reflect the views of the agencies. Any mention of the trade names of commercial products does not constitute an endorsement or recommendation.

REFERENCES

1. Whitehouse, C. M.; Dreyer, R. N.; Yamashita, M.; Fenn, J. B. *Anal. Chem.* **1985**, *57*, 675–679.
2. Huang, E. C.; Wachs, T.; Conboy, J. J.; Henion, J. D. *Anal. Chem.* **1990**, *62(13)*, 713–725.
3. Niessen, W. M. A.; Tinke, A. P. *J. Chromatogr. A* **1995**, *703*, 37–57.
4. Gelpí, E. *J. Chromatogr. A* **1995**, *703*, 59–80.
5. Slobodník, J.; van Baar, B. L. M.; Brinkman, U. A. Th. *J. Chromatogr. A* **1995**, *703*, 81–121.
6. Kebarle, P.; Tang, L. *Anal. Chem.* **1993**, *65(22)*, 972–986.
7. Zhou, S.; Hamburger, M. *Rapid Commun. Mass Spec.* **1995**, *9*, 1516–1521.
8. Wang, G.; Cole, R. B. *Anal. Chem.* **1994**, *66*, 3702–3708.
9. Kostiainen, R.; Bruins, A. P. *Rapid Commun. Mass Spectrom.* **1994**, *8*, 549–558.
10. Ikonomou, M. G.; Blades, A. T.; Kebarle, P. *Anal. Chem.* **1991**, *63*, 1989–1998.
11. Loo, J. A.; Udseth, H. R.; Smith, R. D. *Bio. Environ. Mass Spectrom.* **1988**, *17*, 411–414.
12. Ikonomou, M. G.; Blades, A. T.; Kebarle, P. *Anal. Chem.* **1990**, *62*, 957–967.
13. Sakairi, M.; Yergey, A. L.; Siu, K. W. M.; Le Blanc, J. C. Y.; Guevremont, R.; Berman, S. S. *Anal. Chem.* **1991**, *63*, 1488–1490.
14. Garcia, D. M.; Huang, S. K.; Stansbury, W. F. *J. Am. Soc. Mass Spectrom.* **1996**, *7*, 59–65.
15. Thomson, B. A.; Iribarne, J. V. *J. Chem. Phys.* **1979**, *71(11)*, 4451–4463.
16. Iribarne, J. V.; Thomson, B. A. *J. Chem. Phys.* **1976**, *64(6)*, 2287–2294.
17. Mirza, U. A.; Chait, B. T. *Anal. Chem.* **1994**, *66*, 2898–2904.
18. Hiraoka, K.; Kudaka, I. *Rapid Commun. Mass Spectrom.* **1990**, *4(12)*, 519–526.
19. Straub, R. F.; Voyksner, R. D. *J. Am. Soc. Mass Spectrom.* **1993**, *4*, 578–587.
20. Cole, R. B.; Harrata, A. K. *J. Am. Soc. Mass Spectrom.* **1993**, *4*, 546–556.
21. Nylander, I.; Tan-no, K.; Winter, A.; Silberring, J. *Life Sci.* **1995**, *57(2)*, 123–129.
22. Prokai, L.; Simonsick, Jr., W. J. *Rapid Commun. Mass Spectrom.* **1993**, *7*, 853–856.
23. Davoli, E.; Fanelli, R.; Bagnati, R. *Anal. Chem.* **1993**, *65*, 2679–2685.
24. Cai, J.; Henion, J. *Anal. Chem.* **1996**, *68*, 72–78.
25. Rule, G. S.; Mordehai, A. V.; Henion, J. *Anal. Che.* **1994**, *66*, 230–235.
26. Hsieh, Y. L. F.; Wang, H.; Elicone, C.; Mark, J.; Martin, S. A.; Regnier, F. *Anal. Chem.* **1996**, *68*, 455–462.
27. Bruce, J. E.; Anderson, G. A.; Chen, R.; Cheng, X.; Gale, D. C.; Hofstadler, S. A.; Schwartz, B. L.; Smith, R. D. *Rapid Commun. Mass Spectrom.* **1995**, *9*, 644–650.

28. Debets, A. J. J.; Mekes, T. J. L.; Ritburg, A.; Jacobs, P. L. *J. High Resolut. Chromatogr.* **1995**, *18*, 45–48.
29. Simpson, R. C.; Fenselau, C. C.; Hardy, M. R.; Townsend, R. R.; Lee, Y. C.; Cotter, R. J. *Anal. Chem.* **1990**, *66*, 248–252.
30. Hopfgartner, G.; Bean, K.; Henion, J. *J. Chromatogr.* **1993**, *647*, 51–61.
31. Voyksner, R. D.; Bursey, J. T.; Pellizzari, E. D. *Anal. Chem.* **1984**, *56*, 1507–1514.
32. Le Blanc, J. C. Y.; Guevremont, R.; Siu, K. W. M. *Int. J. Mass Spectrom. Ion Processes* **1993**, *125*, 145–153.
33. Kohler, M.; Leary, J. A. *Anal. Chem.* **1995**, *67*, 3501–3508.
34. Duffin, K. L.; Welply, J. K.; Huang, E.; Henion, J. D. *Anal. Chem.* **1992**, *64*, 1440–1448.
35. Van Berkel, G. J.; Asano, K. G. *Anal. Chem.* **1994**, *66*, 2096–2102.
36. Quirke, J. M. E.; Adams, C. L.; Van Berkel, G. *J. Anal. Chem.* **1994**, *66*, 1302–1315.
37. Voyksner, R. D.; Bush, E. D. *Bio. Environ. Mass Spectrom.* **1987**, *14*, 213–220.
38. Voyksner, R. D.; Bush, E. D.; Brent, D. *Bio. Environ. Mass Spectrom.* **1987**, *14*, 523–531.
39. Kuhlmann, F. E.; Apffel, A.; Fischer, S. M.; Goldberg, G.; Goodley, P. C. *J. Am. Soc. Mass Spectrom.* **1995**, *6*, 1221–1225.
40. Cai, J.; Henion, J. *Anal. Chem.* **1996**, *68*, 72–78.
41. Rule, G. S.; Mordehal, A. V.; Henion, J. *Anal. Chem.* **1994**, *66*, 230–235.
42. Van Ginkel, L. A. *J. Chromatogr.* **1991**, *564*, 363–384.
43. Afeyan, N. B.; Gordon, N. F.; Mazsaroff, I.; Varady, L.; Fulton, S. P.; Yang, Y. B.; Regnier, F. E. *J. Chromatogr.* **1990**, *519*, 1–29.
44. Kassel, D. B.; Shushan, B.; Sakuma, T.; Salzmann, J.-P. *Anal. Chem.* **1994**, *66*, 236–243.
45. Lin, H.-Y.; Voyksner, R. D. *Rapid Commun. Mass Spectrom.* **1994**, *8*, 333–338.
46. Voyksner, R. D. *Environ. Sci. Technol.* **1994**, *28(3)*, 118–127.
47. Henion, J.; Wachs, T.; Mordehai, A. *J. Pharm. Biomed. Anal.* **1993**, *11(11/12)*, 1049–1061.
48. McLuckey, S. A.; Van Berkel, G. J.; Goeringer, D. E.; Glish, G. L. *Anal. Chem.* **1994**, *66(14)*, 737–743.
49. Boyle, J. G.; Whitehouse, C. M. *Anal. Chem.* **1992**, *64*, 2084–2089.
50. Verentchikov, A. N.; Ens, W.; Standing, K. G. *Anal. Chem.* **1994**, *66*, 126–133.
51. Wilm, M.; Mann, M. *Anal. Chem.* **1996**, *68*, 1–8.
52. Gale, D. C.; Smith, R. D. *Rapid Commun. Mass Spectrom.* **1993**, *7*, 1017–1021.
53. Davis, M. T.; Stahl, D. C.; Hefta, S. A.; Lee, T. D. *Anal. Chem.* **1995**, *67*, 4549–4556.

CHAPTER 10

Capillary Electrophoresis—Electrospray Ionization Mass Spectrometry

JOANNE C. SEVERS AND RICHARD D. SMITH

Environmental Molecular Sciences Laboratory, Pacific Northwest National Laboratory, Richland, Washington

	Abstract	344
I.	Introduction	344
II.	Capillary electrophoresis	345
	A. Separation mechanisms	345
	1. Capillary-zone electrophoresis	347
	2. Capillary electrokinetic chromatography	347
	3. Capillary isotachophoresis	349
	4. Capillary isoelectric focusing	350
	5. Capillary gel electrophoresis	350
	B. Sample injection	350
III.	CE–MS interfaces	351
	A. MS sources employed	351
	B. On-line CE–MS interfaces based upon ESI	351
	1. Coaxial sheath-flow interface	351
	2. Liquid–junction interface	354
	3. Sheathless interface	354
IV.	Optimization of experimental methodology	358
V.	Applications	368
	A. Small molecules	368
	B. Polypeptides and enzymatic digests	371
	C. Proteins	372
	D. Nucleic acids	373
	E. Other biopolymers	375
VI.	Future prospects	376
	Acknowledgments	376
	References	376

Electrospray Ionization Mass Spectrometry, Edited by Richard B. Cole.
ISBN 0-471-14564-5 © 1997 John Wiley & Sons, Inc.

ABSTRACT

The development of electrospray ionization (ESI) sources that can produce ions directly from liquids at atmospheric pressure has provided an ideal means of detection for capillary electrophoretic separations. This chapter highlights experimental considerations, methods, and applications for capillary electrophoresis–mass spectrometry (CE–MS) based upon the ESI interface. This is currently the preferred interface for CE–MS, owing to its ease of implementation, sensitivity, and wide range of applications. A review of the techniques applied to interfacing the CE separation system to the spectrometer source is provided, and the current status of the technique is discussed, including present limitations on sensitivity and scan speeds. A brief introduction to the CE separation technique is given, with a discussion of the separation mechanisms of the different modes of CE, and ancillary methods for capillary modification and sample loading or injection.

I. INTRODUCTION

Electrophoresis is based upon the differential migration of charged species in condensed media under the influence of an electric field. It has been employed as a separation technique by chemists and biochemists for over a century. Initially, a solid support such as paper, cellulose acetate, starch, agarose, or polyacrylamide was adopted for electrophoresis, and detection was accomplished by autoradiography or staining and blotting techniques. This solid-support technique, which is still widely employed and is essential to modern biological research, has several limitations. For example, it is time consuming, often nonquantitative, and, owing to diffusion effects and solute adsorption in the gel, often does not produce high-efficiency separations (1). Although several early studies were aimed at improving efficiencies through the use of smaller volumes of the separation medium and alternative liquid separation media, it was not until 1981, when Jorgenson and Lucaks published the results of experiments carried out using capillaries with internal diameters of less than 100 μm i.d. (2,3), that the potential of this separation technique became apparent. In recent years, the use of capillary electrophoresis (CE) has expanded rapidly, driven by the development of more sensitive detectors, the introduction of "micellar electrokinetic chromatography" in 1984 (4) and automated commercial instruments in 1988.

The development of the first CE–MS was stimulated by early reports on electrospray ionization (ESI–MS) by Fenn and co-workers. Because no mass spectrometer suitable for ESI operation was available at the author's laboratory, a mass spectrometer for this purpose was developed, and resulted in the first publication on CE–MS appearing in 1987 (5). In this initial report, a metal coating on the tip of the CE capillary made contact with a metal sheath capillary to which the ESI voltage was applied. In this way, the sheath capillary acted as both the CE cathode, closing the CE electrical circuit, and the ESI source (emitter). Ideally, the interface between CE and MS should (1) maintain separation efficiency and resolution; (2) be sensitive, precise, and linear in response; (3) maintain electrical continuity across the separation capillary so as to define the CE field gradient; (4) cope with all eluents presented by the CE separation step; and (5) provide efficient ionization from low flow rates for mass analysis. Developments from several groups, especially in the last few years, have improved the technique and made it more acceptable to analytical chemists.

It is the opinion of the authors that ESI affords a near-ideal marriage between CE and MS, and will likely remain the preferred approach for the foreseeable future. Recent developments in CE separations, the more common use of small-diameter capillaries, the improvements in ESI interface efficiency (which transform into sensitivity gains), and the availability of improved mass spectrometers will almost certainly lead to increased use of CE–MS. In this chapter we summarize these developments, demonstrate the current capabilities and limitations of the technique, and propose possibilities for its future use with improvements in methodology and instrumentation.

II. CAPILLARY ELECTROPHORESIS

A. Separation Mechanisms

A schematic diagram of the basic instrumentation used in CE is presented in Figure 1. Electrophoretic separations occur in moderately conductive solutions due to differences in analyte velocity in an electric field. As represented in Figure 2, small, highly charged species have large electrophoretic mobilities, and large, minimally charged species have lower mobilities. Two ions of the same absolute

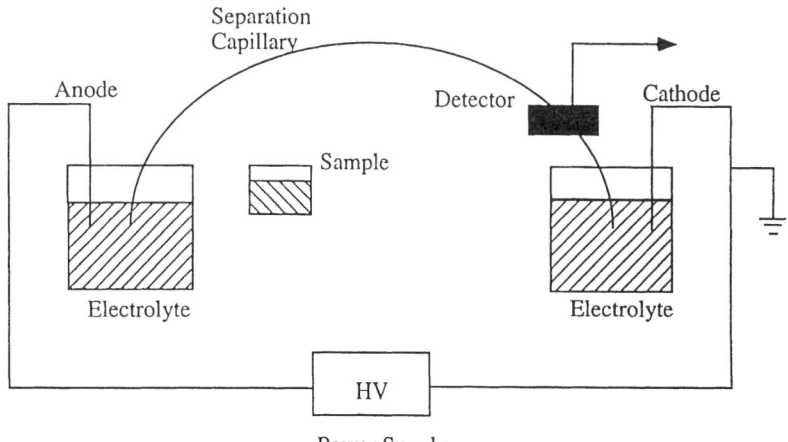

Figure 1. Schematic diagram of a CE system.

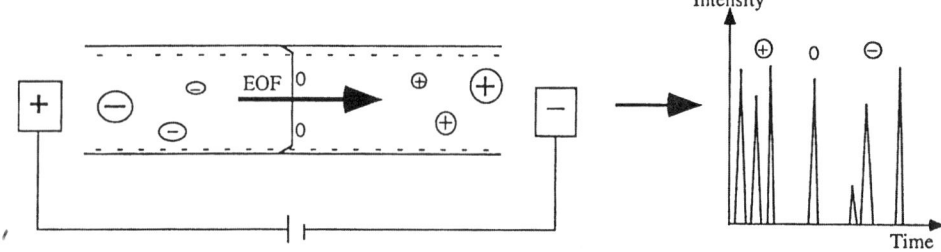

Figure 2. Schematic diagram of basic CE separation mechanism under normal polarity conditions, employing an uncoated, fused-silica capillary.

electrophoretic mobilities at a given pH can, however, be generally separated owing to differences in their pK_a values (6). The effective electrophoretic mobility of an ion through the capillary depends on both the composition and pH of the separation buffer.

Beyond the advantages of small-diameter capillaries that are specific to ESI–MS (and discussed later), are those that are key to the power of CE. Joule heating, caused by the passage of the electrical current through the capillary, can have a significant effect on separations. In small-diameter CE capillaries, the dissipation of this heat is improved, allowing higher electric fields and concentrated free-buffer solutions to be employed, thus providing faster, more efficient separations. (The effects of significant joule heating may also be minimized, however, by external cooling of the capillary.) A key property of capillaries for electrophoretic separations is the production of an electroosmotic flow (EOF) originating from the capillary walls (6). The small excess of negative charge on the inner surface of an uncoated fused-silica capillary induces a flow of the bulk solution toward the cathode. If the EOF is larger than the electrophoretic mobilities of all anions, as represented in Figure 2, all ions (and uncharged species) will migrate in the same direction. The EOF process can easily be controlled, facilitates automation of the technique, allows anions and cations to be separated and detected in the same run, and is advantageous for ESI–MS interfacing with sheathless designs (see Section III.B.3).

The EOF is generally assumed to be constant along the length of the capillary, generating a flat flow profile, as compared to the laminar flow profile of an externally pumped system such as liquid chromatography (LC). This results in more efficient separations, with qualities corresponding to plate heights of several hundred thousand being achievable. Also, because of the differences in the nature of the separation mechanisms of CE and LC, completely different analyte selectivities are achieved, and complementary information can often be obtained if both methods are employed.

Separations can be optimized by a variety of means, including variation of buffer, electric-field strength, temperature, and addition of chemical additives and organic modifiers. Another method increasingly adopted is to decrease or eliminate the EOF, by coating the inner wall of the capillary, through either physical or chemical modification. Many different coatings have been investigated, including polyacrylamide, cellulose, polyethylene glycol, and aminopropyltrimethoxysilane (1,6,7). This procedure also minimizes the problems of analytes, especially proteins, interacting (e.g., adsorption) with the charged capillary walls.

There are, in fact, three basic electrophoretic separation mechanisms: moving-boundary, zone, and steady-state electrophoresis. These mechanisms have been adapted to build a family of capillary-electrophoretic separation techniques; the choice of which to employ is based on analyte type, concentration, and volume.

1. Capillary-Zone Electrophoresis. To date capillary-zone electrophoresis (CZE) has been the most widely used CE technique, probably because it is the simplest to accomplish and the most generally applicable. The separation capillary is filled with a single buffer and a sample band (nanoliters) is introduced into one end of the capillary. Upon the application of an electric field (typically 150–300 V/cm) across the capillary, the analytes migrate at different rates through the capillary, separating into discrete sample zones, as shown in Figure 3a. Cations, anions, and neutrals can be detected in one run owing to the EOF; however, neutral analytes will not be separated from each other in CZE, and all will coelute at a time that depends only upon the EOF.

The length of the CZE sample band is generally constrained to be less than 1–2% of the total capillary length to provide good separations and to prevent overloading and degradation of peak shapes. The solute concentration (or more precisely, the solute-band conductivity) should, ideally, be 100 times less than the background electrolyte. These factors lead to the major drawback in CZE—the limited sample-loading ability and, therefore, relative to HPLC, poor concentration detection limits. Although sample concentration can be increased by the use of more conductive buffers, this approach is not advantageous for ESI–MS detection. These characteristics result in a significant demand upon ESI interface efficiency, as will be discussed later.

2. Capillary Electrokinetic Chromatography. Separation based on capillary electrokinetic chromatography (CEKC) occurs via differential partitioning of an analyte between two phases. Terabe introduced the method of micellar capillary EKC (or MEKC) in 1984 (4), and this is now one of the most important methods in CE, especially within the pharmaceutical industry. Related methods of inclusion-complex EKC, ion-exchange EKC, and microemulsion EKC have also been demonstrated and used.

Analytes, including neutrals, partition with the micellar phase at different rates, governed by electrostatic, steric, and hydrophobic properties (1). Differences in residence times in the micelles, which move through the separation capillary with different and distinctive electrokinetic mobilities compared to the

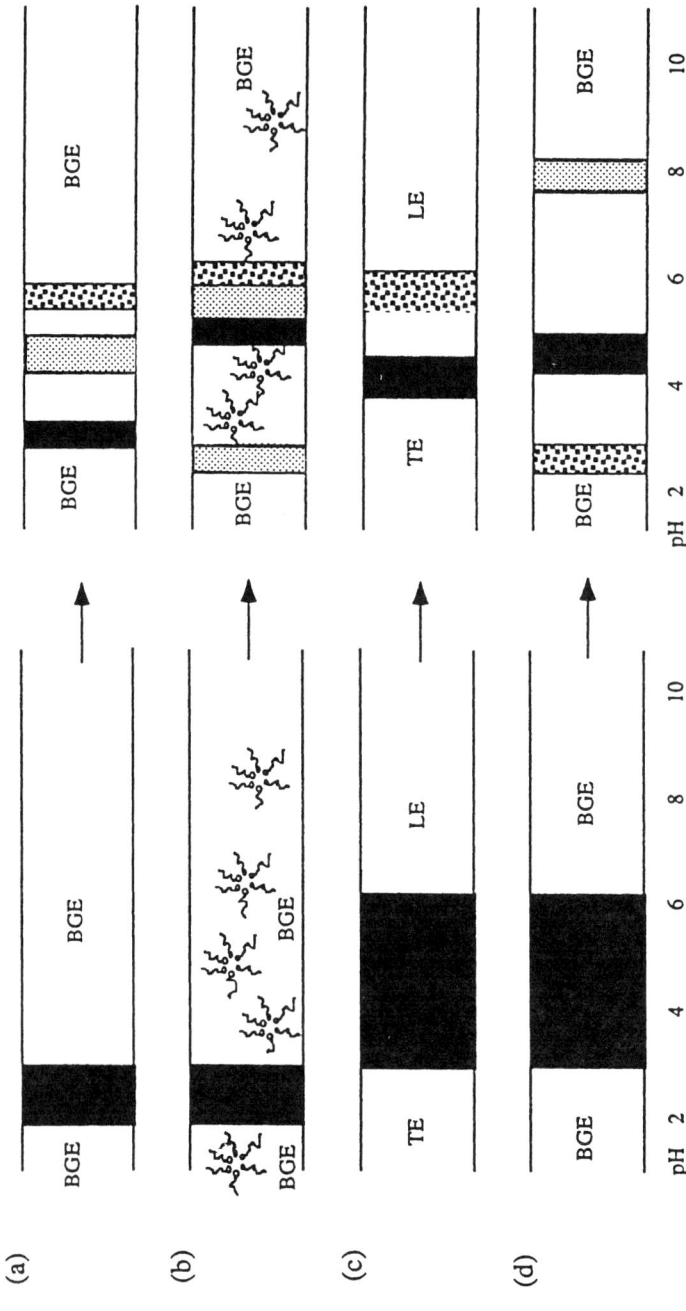

Figure 3. Schematic diagram of the different modes of CE: (*a*) CZE, (*b*) CEKC, (*c*) CITP, and (*d*) CIEF, where BGE = background electrolyte, LE = leading electrolyte, and TE = terminating electrolyte.

analytes, allow neutral analytes to be separated, as represented in Figure 3b. In a similar manner to micelles, inclusion complexes also act as a pseudostationary phase in the capillary. Complexes containing groups with chiral centers, such as cyclodextrins, have also been successfully used for obtaining enantiomeric separations (8). Affinity CE, in which separations are based upon the interactions of species with biopolymers included in the buffer, can also be viewed as a variant of CEKC.

Truly effective combinations of CEKC with ESI–MS require either the use of additives that do not significantly impact the ESI process, or a method for their removal prior to ES. Although this problem has not yet been completely solved, a recent report demonstrated that certain surfactant combinations have a less deleterious effect than others on the ESI signal (9). Also, the use of zwitterionic detergents along with a coated capillary, to reduce the electroosmotic flow, can minimize surfactant interference with the ESI signal (10). The use of two capillaries combined through a liquid junction, the first employed for MEKC and the second for CZE analyte transfer to the ESI source, has also been demonstrated (11). The processes involved in switching voltages and buffers to transfer all the analytes (with only the minimum volume of surfactant) into the second capillary were, however, rather intricate, and only charged analytes have so far been successfully analyzed. Lee and co-workers recently presented initial results demonstrating that the use of a narrow band of micelles in the capillary, through which the analytes migrate and interact, can be sufficient to carry out certain separations (12). Under carefully selected conditions, the analytes could be separated and detected before the micelle band migrated into the source. Because most analytes that benefit from the CEKC mode of operation can be addressed effectively by the interface of other separations methods (GC, SFC, or HPLC) with MS, more emphasis was usually placed upon interfacing with other CE modes.

3. Capillary Isotachophoresis. In capillary isotachophoresis (CITP), a voltage gradient is set up in the capillary by the use of two different mobility electrolytes—a terminating electrolyte and a leading electrolyte—and proper consideration of counterion selection and concentrations (1,13). The sample zone, as shown in Figure 3c, is then sandwiched between these two buffers. When the electric field is applied, analyte ions gradually separate into discrete bands according to their mobilities. These bands then move at a steady rate, as defined by the velocity of the leading electrolyte, and remain focused with sharp boundaries. The amount of each analyte present can be determined by the length of each band (6,13). This self-focusing principle has also been employed to perform transient preconcentration steps prior to CZE separation. This was initially carried out by physically interfacing two separate steps (14,15), and these principles have subsequently been extended to carry out isotachophoretic preconcentration and then separation in the same capillary (16–18). The use of CITP (and transient CITP) allows the introduction of a much larger sample volume into the capillary than CZE, greatly improving detection limits when applicable.

4. Capillary Isoelectric Focusing. In capillary isoelectric focusing (CIEF), a pH gradient is set up by placing one end of the capillary in an acid and the other in a base and by the introduction of ampholytes (a series of zwitterionic species), typically at concentrations of 1–5% in the sample solution, into the entire length of the capillary. When a voltage is applied, as represented in Figure 3d, all ions migrate to a point in the gradient corresponding to their isoelectric point, where they have zero net charge and, therefore zero mobility. Thus, this method can also be used to stack large volumes of amenable species from dilute solutions. Once the focusing has finished, the contents of the capillary are mobilized to the detector, typically either by application of pressure or the addition of a salt to one of the buffer reservoirs.

5. Capillary Gel Electrophoresis. Capillary gel electrophoresis (CGE) involves separations in which the capillary is filled with a gel, and is principally employed for the analysis of macromolecules, particularly nucleic acids. A sieving effect is achieved; the macromolecules, with approximately equivalent mass-to-charge ratios, are separated according to their size. The gel minimizes solute diffusion, and thus provides highly efficient separations (6). Although both CIEF and CGE have been interfaced to ESI–MS, initial results indicate that substantial improvements in performance and sensitivity are required to be useful for most applications. Increasingly, the use of highly cross-linked gels (often chemically bound to the capillary surface) is being displaced by the use of "pumpable" solutions of high-molecular-weight polymers. Interestingly, effective separations of larger biopolymers have been obtained with surprisingly dilute polymer solutions, suggesting their compatibility with ESI–MS.

B. Sample Injection

There are three injection methods commonly applied for CE: hydrodynamic, hydrostatic (siphoning), and electrokinetic. In contrast to LC, in which a sample plug is injected from a valve of known volume, in CE the experimental conditions (e.g., pressure difference or electric field) and the time for injection are used to calculate the injection volume (1,6). Hydrodynamic injection, in which a pressure is applied to the CE capillary while it is introduced into the sample vial, is favored at present, particularly in conjunction with ESI–MS. It can easily be automated and introduces all analytes without bias. Additionally, a system that has the ability to pressurize the capillary also allows more ready capillary conditioning, direct sample infusion to a detector, and elution of analyte bands in CITP and CIEF (where electroosmotic flow is typically reduced to zero). As previously stated, the ability to inject the maximum volume of sample onto the capillary while maintaining separation is an important sensitivity factor. Sample stacking and preconcentration techniques are now major considerations in sample-introduction techniques (19,20).

III. CE–MS INTERFACES

A. MS Sources Employed

Although this chapter is focused on the on-line interfacing of CE–MS via ESI sources, it is worthwhile to note other interfaces and ionization methods that have been employed. Several reports have been published on CE–MS conducted using continuous-flow fast-atom bombardment (cf-FAB) sources, employing either "liquid-junction" (21–25) or "coaxial sheath" (25–28) interfaces. While this is a viable approach, current results indicate that the ESI source has numerous advantages over this technique (29,30).

Pneumatically assisted ESI, or "ionspray" (IS) (31), though providing more benefits for higher flow rate LC–MS than CE–MS, has been successfully interfaced to CE (32). The use of an atmospheric-pressure chemical ionization (APCI) source, for CE–MS analysis of caffeine and some aromatic amines, has also been presented recently (33,34). The latter separation was carried out in an electrolyte system containing a high concentration of sodium dodecyl sulphate. For small-molecule CE analysis, in which micellar and inclusion complex systems are commonly used, APCI may provide a useful alternative to ESI, because the APCI process is not as greatly affected by involatile salts and additives. Capillary electrophoresis has also recently been interfaced to inductively coupled plasma MS for metal species analysis (35,36)

Plasma desorption (PD) and, more recently and importantly matrix-assisted laser desorption/ionization (MALDI) sources have been employed as off-line complements to ESI–MS, especially for larger molecules. Although fully on-line CE–MS techniques have not yet been developed for these sources, a few innovative "fraction-collection" systems have been reported (37–43). Takigiku et al. have demonstrated a porous glass joint located near the cathode end of their separation capillary, which allowed fraction collection for subsequent desorption MS (37). A coaxial sheath flow at the end of the CE capillary, to provide electrical contact and aid elution of the analytes onto either the desorption target (38) or into sample tubes (40), has been employed by other research groups. Computer-controlled collection systems have also been reported recently. One system was designed to move the end of the separation capillary to specific wells on the target (41), another to move sets of collection capillaries, in turn, to the end of the CE capillary (42). An alternative system, using a "moving-belt" design to collect separated analytes, has been described by Van Veelen et al. (43).

B. On-line CE–MS Interfaces Based upon ESI

1. Coaxial Sheath-Flow Interface. A schematic of this interface, first developed for CE–ESI–MS by Smith et al. (44), is illustrated in Figure 4*a*. The sheath liquid, with an electrolytic content, is infused into the ESI source at a

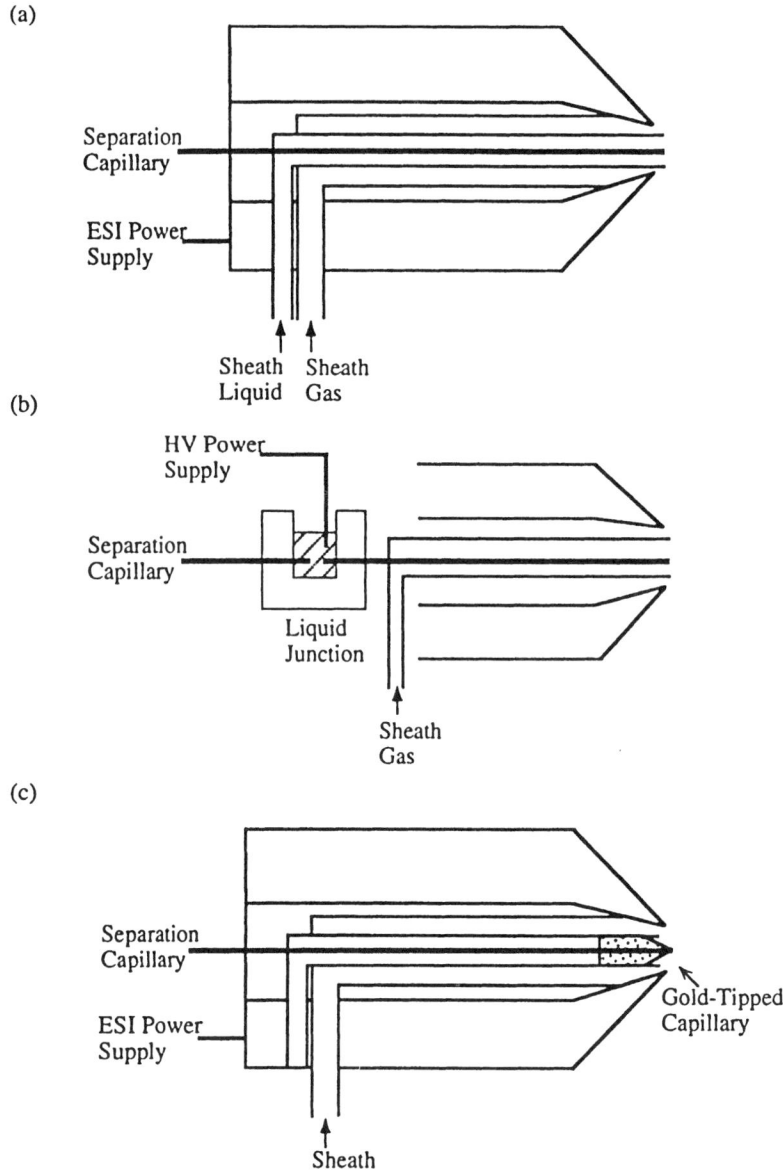

Figure 4. Schematic illustration of CE–MS interfaces to an ESI source: (*a*) a coaxial sheath-flow interface; (*b*) a liquid-junction interface; (*c*) a sheathless interface.

constant rate, through the coaxial sheath capillary which surrounds and terminates near the end of the separation capillary. This sheath liquid then mixes with the separation buffer as it elutes from the tip of the CE capillary, thus providing the necessary electrical contact between the ESI needle and the CE buffer and closing the CE circuit. Since the CE terminus and ESI source are at the same voltage, if the ESI source requires a high voltage (2–5 kV), rather than ground potential, the ESI voltage chosen also directly affects the potential difference across the separation capillary. In setting up the system, the height of the CE instrument is generally adjusted to balance the two ends of the capillary to avoid or minimize any siphoning effects. To maintain electrical contact, both the separation buffer and sheath liquid should be thoroughly degassed and filtered to avoid air bubbles, and a constant, stable flow of the sheath liquid should be verified. The magnitude of the EOF may also affect the contact. In certain cases, when the EOF is minimal, application of a small constant pressure ($\sim$ 1–200 mbar) through the CE capillary during the separation step may aid electrical contact while incurring only a minimal degradation of the separation. The ability to pressurize the CE system is of definite advantage to the mass spectrometrist, allowing sample infusion for system optimization and rapid washing and regeneration of the capillary between runs. (The application of sodium hydroxide capillary washes should be avoided once the capillary is in the ESI source, as sodium adduction to analyte ions decreases signal sensitivity by producing a wider range of ions). The use of a sheath gas is generally avoided if possible in CE–MS studies, owing to the disruptive effect on liquid flows and, therefore, electrical contact. The aspirating effect of a high sheath gas has also been reported to effect separation quality owing to a pressure-induced flow (45). Gas flows should be minimized, and both the ESI voltage (and gas, if used) should ideally be switched off during injection to avoid introduction of air at the capillary inlet.

To date the sheath–liquid interface has been the most widely used and accepted system, being the simplest to construct, with numerous published results of studies employing sheath liquids typically containing 60–80% organic solvent, modified with 1–3% acid in water, and typically introduced at flow rates of 1–4 μL/min. The composition of the sheath liquid should be optimized for the specific systems under investigation. Although negative-ion CE–MS has been demonstrated (45), most researchers have employed positive-ion ESI because of its greater stability (and lower risk of discharge). Recent reports have confirmed that the relative dimensions and positioning of the separation and sheath capillaries also influence sensitivity and stability (46), and that tapering of the CE capillary tip improves interface results (47).

Although the additional flow of an organic-containing electrolyte into the ESI source moderately extends the range of CE buffer systems that can be used, the CE buffer composition still has a dramatic effect on the ESI signal, minimizing buffer choice for best sensitivity to volatile solutions. In some cases, for reasons related to the introduction of excess ionic species to the ESI source (as discussed in the next section), particular benefits can be derived from elimination of the sheath flow. A recent report by Foret et al. has also highlighted the need for a more considered selection of sheath liquid composition (48). They presented results that showed the formation of moving ionic boundaries inside the capillary due to migration of sheath-liquid counterions into the CE capillary.

Depending on relative ionic mobilities, sharp or diffuse boundaries were created, leading to variations in migration times, resolution, and, as demonstrated in Figure 5, changes in migration order. The possibility of these effects occurring should be considered and minimized when transferring a CE method from an alternative detection system to MS. It should be noted, however, that these effects are minimized or eliminated when there is a sufficiently strong electro-osmotic or pressure-driven flow toward the CE terminus.

2. Liquid–Junction Interface. Electrical contact with this interface is established through a liquid reservoir which surrounds the junction of the separation capillary and a transfer capillary, as shown in Figure 4b. This interface has been developed and employed most extensively by Henion and co-workers, for interfacing CE to pneumatically assisted ESI interfaces (32,49). The gap between the two capillaries is adjusted, under a microscope, to approximately 10–20 μm, allowing sufficient make-up liquid from the reservoir to be drawn into the transfer capillary, while avoiding analyte loss. The flow of make-up liquid into the transfer capillary is induced by a combination of gravity and the venturi effect of the nebulizing gas at the capillary tip. It has been suggested that the high gas-flow rates used in IS could lead to extra peak broadening if the CE capillary terminated in the source rather than at a liquid junction (50). However, it has also been suggested that pulling the CE capillary back so that it is flush with the sheath tip is sufficient to avoid this problem (51).

In comparisons of coaxial sheath-flow and liquid–junction interfaces, Pleasance et al. noted that although both provided efficient coupling, the former was generally easier to operate and, for an analysis of marine toxins, provided improved S/N ratios and separations (52). One of the major disadvantages in employing the liquid–junction interface is in establishing a reproducible connection inside the tee piece. Also, the use of a transfer capillary, which has no potential difference applied across it, can lead to peak broadening, as can the pressure often applied to the separation capillary to maintain a stable flow toward the source. The advantages of this interface, however, include the possibility of combining different outer-diameter capillaries through the junction and the extra mixing time provided for the make-up liquid and CE eluant.

The problem with both interfaces described so far is that they depend upon the addition of excess electrolyte to the ESI source to maintain the circuit. This invariably decreases analyte sensitivity (although not to the extent expected based upon the relative flow rates and compositions of the sheath and CE effluents). The development of an interface that does not rely on an additional liquid flow has, therefore, attracted renewed attention.

3. Sheathless Interface. The first CE–MS interface, reported in 1987 by Olivares et al., made electrical connection between the separation buffer and the ESI needle via a metal coating on the tip of the 100-μm i.d. capillary (5), as represented in Figure 4c. Although femtomole detection limits and separation efficiencies of up to half a million theoretical plates were achieved, problems included a high dependence on the buffer system used and the need to regularly replace the metal coating on the capillary tip. These issues led directly to the development of the sheath-flow approach.

In eliminating the sheath flow, additional stability has been obtained by using metal deposited on the CE terminus that is tapered to provide an increased electric field (53). This approach was adopted for both low flow-rate infusion studies (54) and on-line "sheathless" CE–MS studies of biomolecules in purely aqueous solutions (55). In these studies, the silica capillaries were tapered by etching the tip in a 40% hydrofluoric acid solution before applying the metal coatings. The use of a micropipette puller, to taper 360-μm o.d. × 20 μm i.d. capillaries to dimensions of 40 μm o.d. × 3 μm i.d. at the tip, has also been reported recently (56). These so-called "microspray" and "nanospray" approaches have been adopted recently by several other groups, for interfacing continuous-infusion systems and LC to MS, and, as demonstrated in other chapters of this book, significant gains in sensitivity and sample usage have been observed (57,58). These gains derive from more effective ionization, as we discuss in greater detail later in this chapter.

Ramsey et al. have also used gold-coated capillary tips to carry out CE–ESI quadrupole ion-trap (QIT) MS with a sheathless interface (56,59,60). They have investigated different procedures to provide the most durable capillary tips, and have reported silanizing (59) or chromium coating (56) the tapered tips prior to the gold-coating procedure to improve stability. Gains in sensitivity of at least three- to fourfold over a sheath-flow interface were reported (56). This allowed them to attain full-scan, mid-attomole detection limits for some standard peptides in aqueous solutions.

Alternative sheathless interfaces have also been investigated briefly. One method involved forming a microhole near the end of the capillary by an electrical discharge. The hole was then covered with gold epoxy to make the connection (47). Although this was feasible, it was thought that electrochemical reactions at the joint could produce gases that would disturb the contact. Fang et al., in coupling CE to a time-of-flight (TOF) mass spectrometer with an ESI source, formed the electrical contact by placing a gold wire, attached to the source, into the outlet of the separation capillary (61). However, they were only

---

Figure 5. (pages 356–357) Analysis of a protein mixture to demonstrate the effect of liquid sheath boundaries on the separation. (*a* and *b*) 20 mM ϵ-aminocaproic acid/phosphoric acid background electrolyte (BGE), pH 4.4, 25 kV, 36(27) cm × 75 μm i.d. capillary and UV detection. (*a*) BGE in both electrode chambers, current 13 mA; (*b*) 1% acetic acid in 50% methanol/water in cathodic reservoir, decreasing current, 13–6 mA. (*c* and *d*) 67 cm × 75 μm i.d. capillary, 25 kV, and ESI–MS detection using a liquid sheath containing 1% acetic acid. (*c*) 20 mM ϵ-aminocaproic acid/phosphoric acid BGE; (*d*) 20 mM ϵ-aminocaproic acid/acetic acid BGE. Peak identification: 1, cytochrome C; 2, lysozyme; 3, aprotinin; 4, myoglobin; 5, RNase A; 6, β-lactoglobulin B; 7,8, β-lactoglobulin A + carbonic anhydrase; 9, α-chymotrypsinogen A. Adapted with permission from F. Foret et al., *Anal. Chem.*, **1994**, *66*, 4450. © 1994 American Chemical Society.

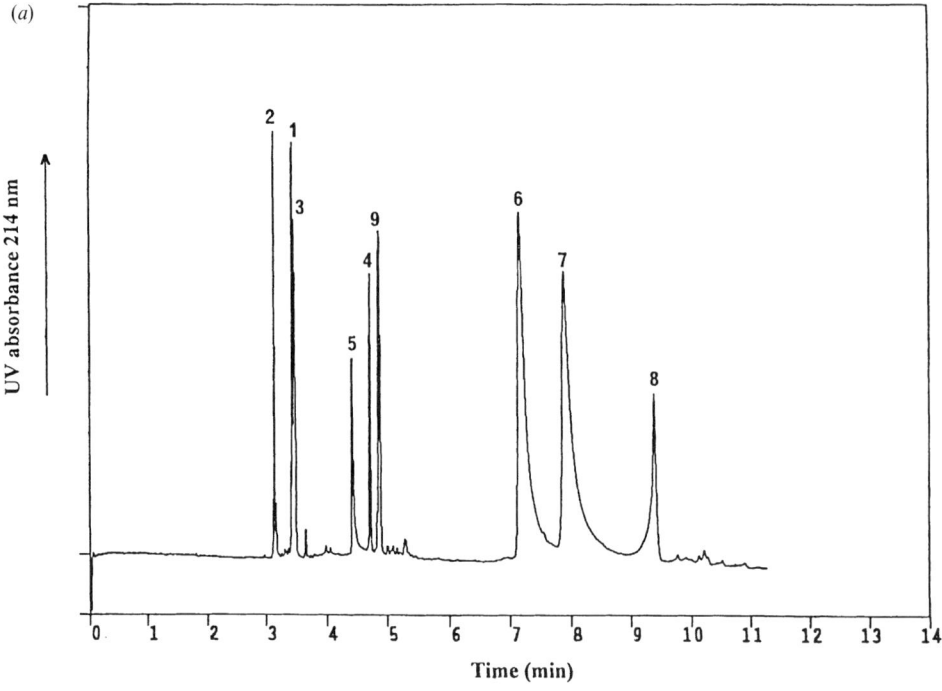

(a)

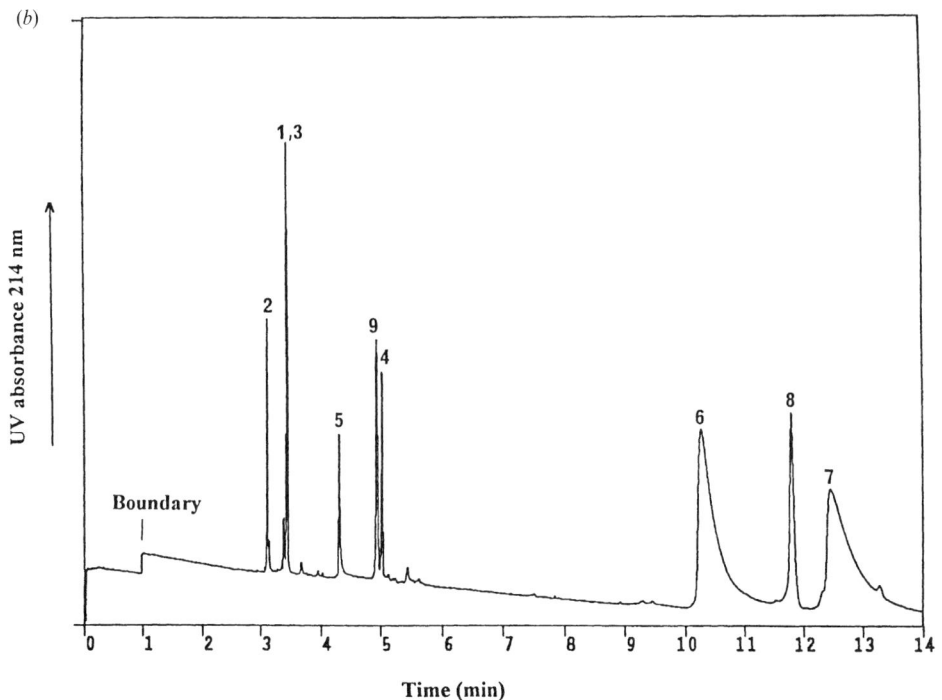

(b)

III. CE—MS INTERFACES   **357**

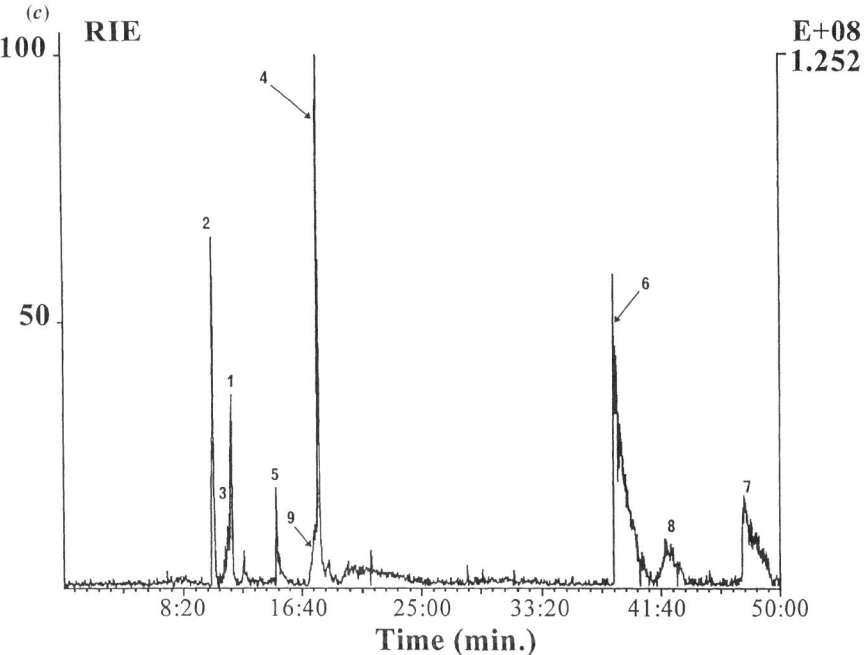

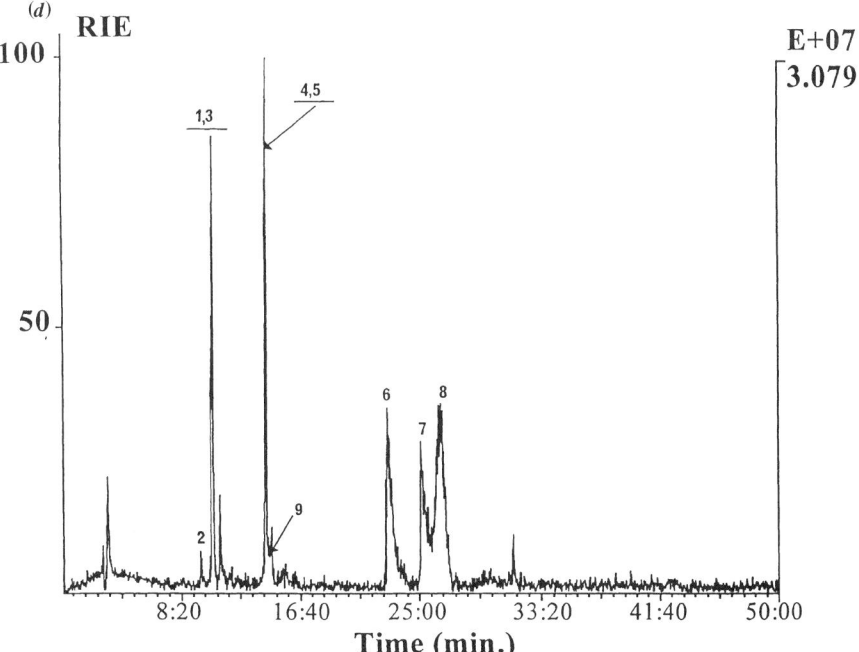

able to spray up to 90% aqueous solutions by this means, and stability was apparently less than ideal.

An interface that does not use an additional makeup flow can, as well as aiding sensitivity, also avoid such problems as charge-state distribution shifts. In addition, the ability to electrospray purely aqueous systems is often advantageous for looking at fragile biological and noncovalently bound analytes. In some cases, however, a makeup liquid may be found necessary. For example, for certain separations, the EOF may need to be minimized or eliminated in the CE capillary, and thus flow rates into the source will not be sufficiently high to maintain a stable ES. If a capillary needs to be coated to avoid analyte interaction with the capillary wall, a cationic coating, which reverses the EOF rather than eliminating it, should preferably be chosen when a sheathless system is to be employed. Also, it may be found that a makeup liquid is necessary to increase the volatility of a specific CE electrolyte system.

Another disadvantage in using the sheathless interface is the time dispensed in preparing the tapered, coated tips. Although the coatings now employed are more stable than those initially used (5), the tips do not regularly survive more than a day or two of use. This can, however, frequently be due to the tip "plugging," rather than the metal coating deteriorating. Filtering the electrolyte and analytes, and rinsing the capillary can, therefore, often prolong the capillary lifetime.

IV. OPTIMIZATION OF EXPERIMENTAL METHODOLOGY

The small solute quantities in CE require highly sensitive detection methods. The low signal intensities generally produced by ESI–MS effectively limit the maximum practical scan speeds with quadrupole mass spectrometers. Thus, depending on the desired m/z range, solute concentration, and other factors related to the nature of the solute and buffer species, maximum m/z scan speeds are often insufficient to exploit the high-quality separations feasible with CE when coupled to quadrupole or other scanning mass spectrometers.

The use of either ion-trapping methods or array detectors on sector instruments (62,63) improves sensitivity for CE–MS analyses. Ramsey et al. demonstrated that the combined use of broad-band collisional activation and one or more resonance ejection steps from an ion trap can reduce chemical background and noise (by up to 30-fold), allowing acquisition of CE–MS spectra over a wider scan range (64).

The combination of CE–ESI with Fourier-transform ion-cyclotron resonance (FTICR) MS (55,65–70), provides high-accuracy mass measurements (0.001%) and extremely high resolving powers (exceeding 10^5) for biopolymer analysis. The major disadvantage for on-line separations, at present, is the relatively slow spectral acquisition rate. Figure 6 shows an example of a separation and mass spectrum obtained from the injection of approximately 30 fmole of the protein carbonic anhydrase I, using a 20-μm i.d. capillary

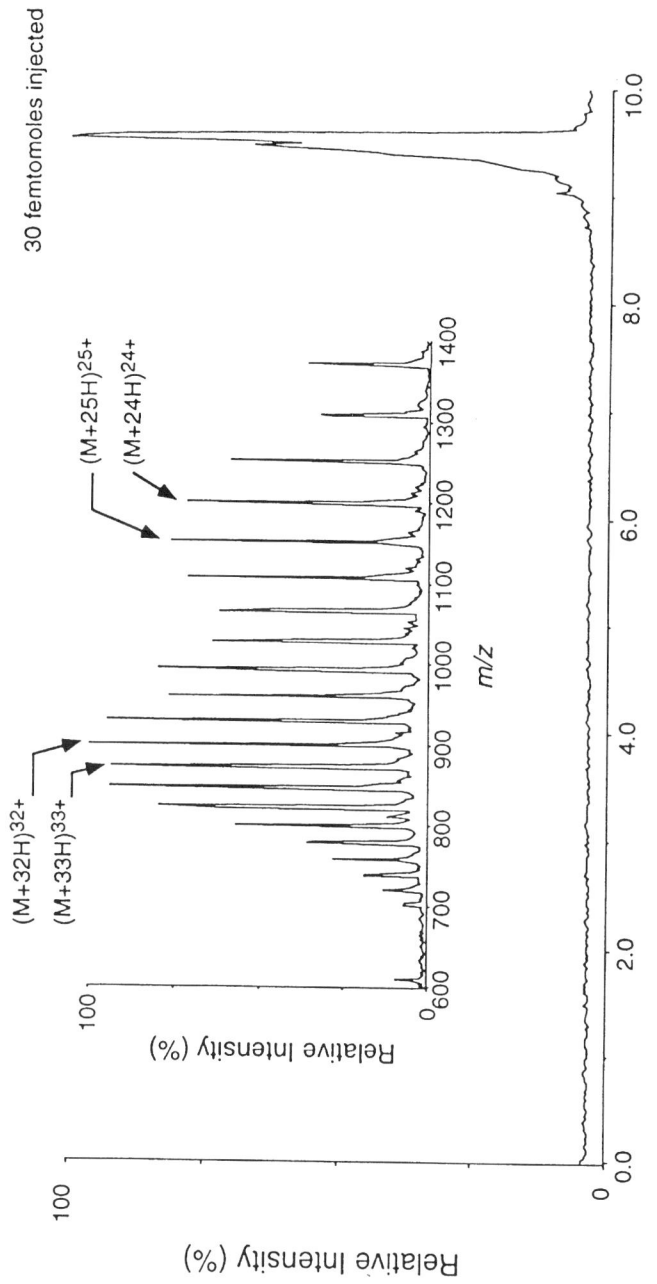

Figure 6. CE–ESI–FTICR–MS of 30 fmol of carbonic anhydrase I, injected onto a 20-μm i.d. capillary treated with 3-aminopropyltrimethoxysilane.

interfaced to FTICR. The potential for higher-order mass spectrometry, MS^n ($n \geq 3$), using FTICR or QIT MS instruments also promises exciting applications for CE–MS (71).

Alternatively, the orthogonal TOF mass spectrometer has the potential for obtaining high ion-utilization efficiency, along with rapid scan rates (72)—a potentially ideal system for on-line CE–MS analyses. Zare et al. presented the first combination of CE–TOF, using a sheathless interface (61,73) and CE–TOF results from several other groups have been presented recently (74–76).

An advantage of the mass spectrometer relative to other detectors is its high specificity. The elution times for analytes can often change somewhat with time, owing to often unavoidable capillary-surface modifications. With less-selective detectors, great care is generally necessary to establish reproducible migration times to identify eluents. In contrast, highly reproducible elution times are far less important for MS detection because of its much greater selectivity. Thus, relative migration times are usually sufficient, and one is generally more concerned with MS performance (sensitivity, resolution, scan speed and S/N). Obtaining the maximum number of theoretical plates possible with combined CE–MS is rarely required, for similar reasons, unless closely related mixture components have nearly similar molecular weights.

For samples where analyte molecular weights are known, and m/z values can be predicted, selected-ion-monitoring (SIM) detection is an obvious choice with quadrupole mass spectrometers. If sufficient sample is available, direct infusion can be utilized to produce a mass spectrum of the unseparated mixture and the results used to guide selection of m/z values for SIM detection. One of the advantages of CE–MS is that mixture components strongly discriminated against by direct infusion often show much more uniform response after CE separation.

Both the liquid–junction and sheath-flow ESI interfaces allow a range of CE buffers to be successfully electrosprayed owing to the dilution of the low-CE elution flow by a much larger volume of liquid. Probably the most widely used buffers for CE–ESI–MS are ammonium acetate, bicarbonate, and formate systems. Recently, use has been made of pure acids or bases, rather than true buffers, in CE–MS. Although this leads to variations in pH around the analyte zone, it can in many cases improve sensitivity and stability, and allow much higher concentrations of weaker electrolytes (up to 1 M) to be used for separations. Nonvolatile buffer systems may be used, but often present background "noise" difficulties. One can background subtract the spectrum acquired before the analyte peak elutes from the spectrum obtained during analyte-peak elution in such situations. We have recently applied the data-analysis technique of "sequential paired covariance" to reconstruct electropherograms obtained from CE with an orthogonal TOF mass spectrometer (77). This procedure compares adjacent spectra in the electropherogram and only generates meaningful intensities if the spectra have common features. As shown in Figure 7, the S/N is greatly enhanced by this procedure, facilitating detection of analyte peaks.

The efficiency of the ESI detection process for CE–MS can be considered in

predicted to be inversely proportional to the background electrolyte mass flow rate, $V_M(B^+)$, in this detection region:

$$I(A^+) \propto \frac{V_M(A^+)}{V_M(B^+)} \tag{1}$$

Consequently, analyte sensitivity can be improved by decreasing the background electrolyte mass flow rate.

These two detection regimes are illustrated in Figure 8, where the relative analyte signal intensity is shown as a function of the analyte mass flow rate. For curve A, the maximum and minimum analyte mass flow rate is presumed to be 10 and 0.1 times that of the background mass flow rate, respectively. The rate constants for the relative efficiencies of transfer of analyte and background electrolyte ions into the gas phase are assumed equal, and the sampling efficiency of the system, ESI current, and fraction of droplet charge converted into gas-phase ions are assumed constant. From Figure 8, a constant analyte sensitivity is then expected within the dynamic range of the detector when the mass flow rate of the analyte is small compared to the background electrolyte.

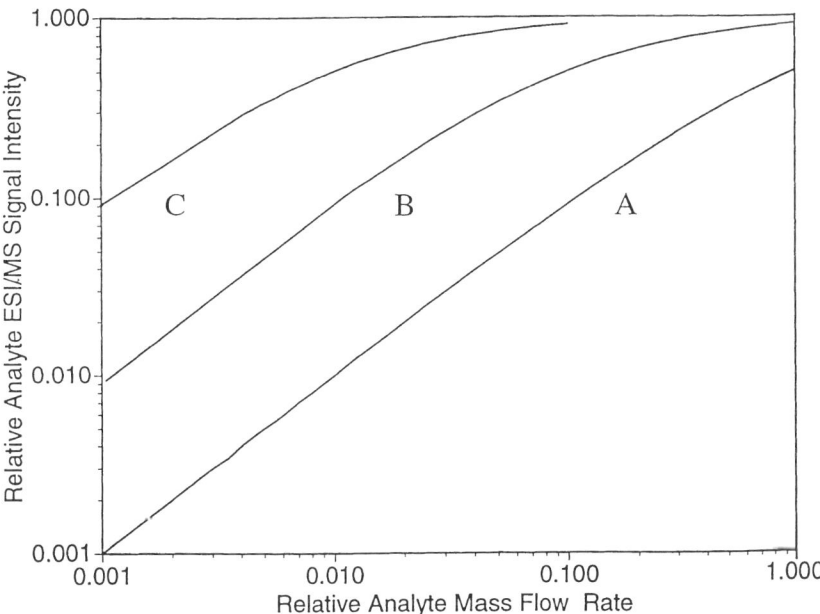

Figure 8. Electrospray analyte signal response from Eq. 1 where curve B represents the signal intensity predicted when both the analyte and background electrolyte mass flow rates are reduced by a factor of 10 relative to curve A. Curve C corresponds to a concurrent reduction of 100, that is, where the background electrolyte delivery rate to the ESI source is reduced by factors of 10 (B) and 100 (C).

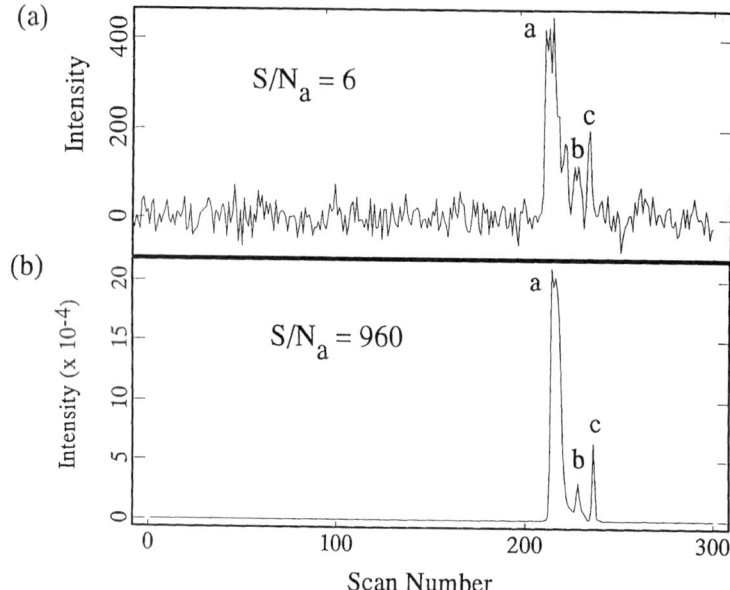

Figure 7. Electropherograms obtained by CE–ESI–TOF–MS of a mixture of a, angiotensin I, b, melittin, and c, neurotensin using (*a*) reconstructed ion current and (*b*) sequential paired covariance.

terms of the simple model of Kebarle et al. (78–80). This model proposes two disinct regimes for the ESI process. In the first, the mass flow rate of the various electrolyte species to the ESI source exceeds that capable of being transferred to the gas phase by the ESI process. In the second regime, the low mass flow rate of the various electrolytes constrains the ESI current. This second regime is most relevant for small-diameter CE capillaries with sheathless interfaces. The sheath–liquid interface, however, provides an additional flow of charge-carrying species to the ESI source, complicating any treatment of ESI efficiency. Uncertainties in the efficiency of mixing between the liquid sheath and the CE effluent prevent a simple accounting of its contributions to the ESI current. Therefore, if this interface is employed, it is probably unreasonable to expect a quantitative agreement with experimental results and this model, in which the analyte signal intensity can be expressed as a function of the relative mass flow rates of the analyte and the background electrolyte (81,82).

Two limiting detection regions are predicted (81,82). The first arises when the analyte mass flow rate is relatively high and substantially greater than the background mass flow rates. The second occurs when the background electrolyte mass flow rate is assumed constant and much greater than the analyte mass flow rate, the general case in CE–MS studies. In this high background–electrolyte region, the analyte signal intensity, $I(A^+)$, is predicted to be directly proportional to the analyte mass flow rate, $V_M(A^+)$. Moreover, the analyte sensitivity is

The predicted increase in analyte sensitivity when the mass flow rate of the background electrolyte is decreased is also illustrated in Figure 8. Curve B represents the signal intensity predicted when both the analyte and background mass-flow-rates are reduced by a factor of 10 relative to curve A, and curve C corresponds to reduction of mass flow rates by a factor of 100 for both components. When the relative analyte mass flow rate is 0.001, the analyte signal intensity is two orders of magnitude greater when both the analyte and background ions are reduced by the same magnitude (curve C) compared to when only the analyte mass flow rate is decreased (curve A). Additionally, as shown in Figure 8, this model predicts that the onset of the analyte-rich region occurs at lower mass flow rates when both the analyte and background are reduced. The increase in sensitivity arises directly from the decrease in background electrolyte, which allows a greater fraction of the analyte to be converted into gas-phase ions.

It is clear that the simple relationship described must fail at sufficiently low background-electrolyte mass flow rates to the ESI source. Consequently, the second regime of ESI detection occurs when charge-carrying species are no longer supplied to the ESI source at a rate sufficient to sustain the maximum ES current. The ionization efficiency for the "constrained" ES in this regime is maximized because there is no longer a competition between analyte and buffer for ionization. That is,

$$I(A^+) \propto V_M(A^+) \qquad (2)$$

One important distinction exists between the two regimes encountered in CE–MS and represented by Eqs. 1 and 2. For the regime represented by Eq. 1, a change in background electrolyte mass flow rate will affect analyte signal intensity, and in the course of a given CE experiment, the signal intensity will directly reflect analyte concentration. Changes in flow rate with CE are generally irrelevant, and hence the ESI–MS emulates a concentration-sensitive detector. In the case of Eq. 2, encountered at the lower limits of electrolyte mass flow rate or for very small capillary diameters, the ES current is limited by electrolyte flow to the ESI source and ionization efficiency will be optimum. Thus, in this regime, the ESI–MS detector appears to function as a mass-sensitive detector. The transition between the regimes described by Eqs. 1 and 2 should be experimentally evident when further decreases in the background electrolyte do not lead to additional gains in analyte sensitivity.

These simple considerations indicate that analyte sensitivity in CE–ESI–MS may be increased by reducing the mass flow rate of the background components. This decrease in background flow rates could be accomplished experimentally by decreasing the concentration of the supporting electrolyte; however, low-concentration buffer systems can lead to poor separation efficiencies (83). In addition, for very low conductivity buffers, the maximum analyte concentration at which linear response will be obtained will decrease, leading to a reduced dynamic range. The second regime of detection should occur when the CE current is less than the normal ESI current. The lower limits to CE current for larger capillary diameters are generally defined by the small amounts of trace ionic contaminants in aqueous solutions that are nearly impossible to eliminate; consequently, no further reduction in the relative concentration of the background electrolyte occurs. Thus, in CE the use of highly dilute buffer systems is not generally an option. The mass flow rate from the separation capillary, however, can be reduced by decreasing the electric field or employing smaller-diameter capillaries, and this predicted increase in analyte sensitivity is now well supported by experimental studies.

To reduce the elution speed of the analyte ions into the source, the electrophoretic voltage can be decreased just prior to elution of the first analyte of interest, minimizing the experimental analysis time. The reduced elution speed (RES) allows more scans to be recorded without a significant loss in ion intensity. Under conditions where the amount of solute entering the ESI source per unit of time exceeds its ionization capacity, no substantial decrease in maximum ion intensity is expected when the electric field strength is decreased. As a result, a greater fraction of the analyte ions can be transferred to the gas phase during a RES CE–MS experiment than during normal constant electric field strength CE–MS experiments.

Some of the greatest challenges for CE–MS involve analysis of complex mixtures of biopolymers. An important goal is to decrease the quantity of protein required for sequencing using methods based upon an initial enzymatic digestion of a protein. Comparison of constant electric field strength and RES CE–MS for a 40-fmole injection of peptides produced by digestion of bovine serum albumin with trypsin was carried out (84), and results indicated that only a small decrease (20%) in ion intensity was observed when the electrophoretic voltage was decreased to slow elution. The reduced complexity of individual scans aids data interpretation of complex mixtures because of the greater number of scans obtained during elution of a given component and the reduced likelihood of other components eluting during the same scan.

Reduced elution speed CE–MS provides an effective increase in the efficiency of mass-spectrometric scanning compared to conventional CE–MS methods. The prolonged analyte elution into the ES source can be exploited by increasing the m/z range scanned, increasing the number of scans recorded during migration of a given solute, and enhancing the sensitivity and signal intensity for a given solute. The method does not increase solute consumption, provides improved sensitivities for peptide and protein analyses extending into the low-femtomole regime, and incurs little loss in ion intensity, which is particularly important for tandem MS methods and their potential application to peptide sequencing.

Alternatively, as first demonstrated by Wahl et al. (82,85), the use of smaller-diameter capillaries than conventionally used for CE also increases sensitivity. An optimum CE capillary diameter should, ideally, be commercially available, amenable to alternative detection methods, and provide the necessary detector sensitivity. For the last criterion, one would expect that optimum sensitivity would be obtained for CE currents approximately equal to or less than the ESI

IV. OPTIMIZATION OF EXPERIMENTAL METHODOLOGY   **365**

current. Even for a relatively low-conductivity acetic-acid buffer system, CE capillary diameters greater than 40 μm will generally have currents that exceed that of the ESI source. A series of CE–MS separations, using a coaxial sheath-flow interface, were obtained using capillary i.d.'s of 100, 50, 20, and 10 μm, are shown in Figure 9 on the same absolute-intensity scale. For these comparisons,

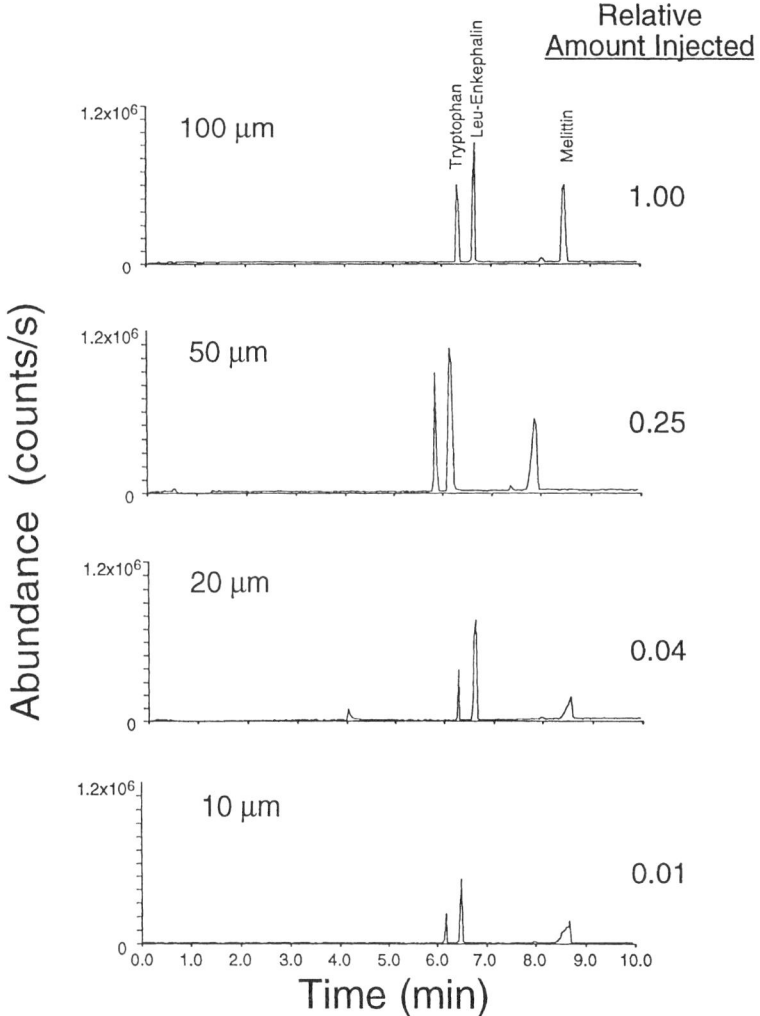

Figure 9. Comparison of the CE–MS total-ion electropherograms for separation of tryptophan, leucine-enkephalin, and melittin for four different capillary diameters. Electromigration injection and separation conditions were identical for all experiments and signal intensities are shown on the same scale. The relative amount of analyte injected is a factor of 100 lower for the 10-μm capillary compared to the 100-μm capillary. The vertical axis is the same for all separations (1.2×10^6 count/s full scale).

all the experimental conditions were the same among the separations, and duplicate separations showing good agreement were obtained for each capillary diameter. The injected amounts of sample were proportional to the capillary cross-sectional area, where, for example, the relative amount injected for the 10-μm i.d. capillary is two orders of magnitude smaller compared to the 100 μm i.d. capillary. The relative ESI–MS signal intensity, however, decreased by only approximately half for melittin in comparing the 100- and 10-μm i.d. capillaries, and the corresponding sample-injection sizes decreased from 800 to 8 fmole (82).

To examine more closely the effect of solute concentration as a function of decreasing capillary inside diameter, a series of separations were performed with dilution of the peptide mixture at ratios of 1 : 2, 1 : 5, 1 : 10, 1 : 20, and 1 : 50. The results for one solute, tryptophan, from these CE–MS separations are shown in Figure 10, where the solute zone area measured is shown as a function of the injected amount for the four different capillary inside diameter. For the lower-analyte concentrations, the observed results may be summarized as suggesting that roughly comparable sensitivities are obtained with the 10- and 20-μm i.d. capillaries.

The prediction of the simple model discussed above, that sensitivity will increase indefinitely as the mass flow rate through the CE capillary decreases, fails at a certain point. In the absence of a sheath flow, the CE current will ultimately reach a condition where the rate of analyte and background electrolyte delivered to the ESI source causes a decrease in the ESI current. The electrophoretic current is less than the ES current for both the 10- and 20-μm i.d. capillaries. As a result, the mass flow rate of ions to the ESI source from the analytical capillary may be inadequate to account for all the ions necessary for the ES process. If the contribution from the sheath liquid is small, the ionization efficiency for the analyte ions will approach its maximum, and further reduction in the CE background electrolyte delivery rate or concentration will have little effect. For the smallest-capillary diameters, the results indicate that the amount of analyte delivered to the ESI region is now the primary determinant of the mass spectral response, and little change in sensitivity is expected with further reduction in capillary diameter. The use of the sheath-flow interface will complicate this situation and probably leads to some of the deviations observed for these experimental results from those predicted. These considerations may at least partially account for variations in relative response between 10- and 20-μm i.d. capillaries for the three analytes. If the ESI current was dominated by the sheath liquid contributions, no sensitivity gain would be expected for smaller inside diameter capillaries. Similar gains in CE–MS sensitivity for protein mixtures have been reported by Wahl et al. in a study of 50- to 5-μm i.d. capillaries (85).

As mentioned previously, the most obvious approach to improving sensitivity is to modify the injection step to load more sample onto the capillary. Many of these techniques have already been applied for CE–MS. For example, CITP–MS was first demonstrated to significantly improve sensitivity in 1989 (86). A particularly useful approach involving isotachophoretic sample

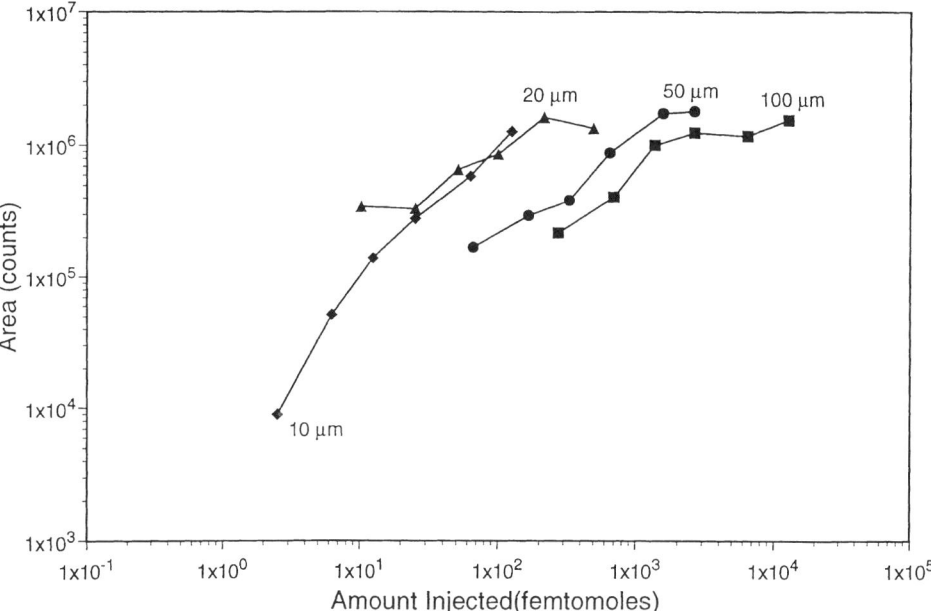

Figure 10. CE–MS sensitivity comparison for tryptophan from the mixture examined in Figure 9. The solute peak area is shown as a function of the injected amount for the four different capillary inside diameters. In general, for all solutes, the sensitivity increased with decreasing capillary diameter.

preconcentration prior to CZE separation has also been reported by Tinke et al. (14) and Karger and co-workers (18). Results obtained from this latter study are presented in Figure 11. Preconcentration of a protein mixture was obtained during the injection onto the separation capillary by consideration of the electrolyte systems used and relative ion mobilities. By dissolving the analytes in an ammonium-acetate buffer, which acted as a leading electrolyte, the proteins were focused during injection into the capillary containing the 6-aminohexanoic acid CZE separation buffer. This focusing allowed the injection of up to 750 nL of sample, as compared to < 50 nL in CZE, which, along with the stacking effect, afforded increases in sensitivity of up to two orders of magnitude. Several groups have now adopted these techniques for CE–MS, improving concentration sensitivity limits for a range of applications (87–89). The use of liquid–liquid electroextraction prior to transient CITP–MS has also been reported (90). Extraction of some β-agonists was carried out from a large volume of a methanol/ethyl-acetate solution directly into the separation capillary by careful balancing of applied voltages and low backpressures. The concentrated analytes were then focused between CITP buffers prior to separation and detection

The use of CIEF–ESI–MS has also been reported for improving protein-detection limits (91). Although the addition of ampholytes to the electrolyte system does not make this technique ideal for ESI analysis, a sensitivity increase of 100-fold using a 1% ampholyte system was reported for myoglobin and carbonic anhydrase compared to CZE–MS owing to the larger injection volume. The development of methods for obtaining immobilized ampholyte gradients in CIEF, to prevent ampholyte elution into the source, would be of great advantage to the mass spectrometrist.

Another preconcentration approach applied to CE–MS is the combination of a small plug of LC packing material (62,92,93) or, more recently and efficiently, a hydrophobic membrane (94) with the inlet of a CE capillary. Microliter volumes of the analyte solution can be preconcentrated and cleaned on this material bed before being eluted, in a concentrated solvent plug, directly into the separation capillary for CZE (or transient CITP with subsequent CZE) separations. This approach, developed by Tomlinson et al., although still under research and development, has demonstrated extremely productive initial results for both pharmaceutical analyses (62,92) and peptide-sequencing studies (94).

V. APPLICATIONS

Capillary electrophoresis–mass spectrometry has now been applied to the analysis of a wide range of species, from ions to large biomolecules (49). However, routine, quantitative applications of CE–MS are still relatively few, and most reports have aimed at evaluation of its use for specific applications. The small sample sizes injected in CE make this technique particularly useful for the biochemistry laboratory, where sample sizes are often limited, but laboratories in many fields are now realizing the potential in running an alternative, or complementary, system to LC–MS. In this section, we briefly highlight the range of CE–MS applications.

A. Small Molecules

Although its possibilities for small molecule analysis are starting to be recognized, to date relatively few groups have published work in this area. Henion and co-workers demonstrated the on-line analysis of small drug species, looking at sulphonamides, benzodiazepines, and the metabolites of flurazepam in human urine (95). They have also analyzed metabolites derived from human hepatic

---

Figure 11. (*a*) CZE–MS reconstructed-ion electropherogram (RIE) (m/z 600–2000) of a 150-nL injection of 1.2 μM each of (1) lysozyme, (2) cytochrome c, (3) ribonuclease A, (4) myoglobin, (5) β-lactoglobulin A, and (6) β-lactoglobulin B dissolved in water. (*b*) Transient CITP–MS RIE (m/z 600–1850) of a 750-nL injection of $\approx$ 500 nM of each protein in 5 mM ammonium acetate buffer. The peak marked with an asterisk is from the rear boundary of the ammonium zone. BGE: 20 mM 6-aminohexanoic acid + acetic acid, pH 4.4, 18 kV. Adapted with permission from T. J. Thompson et al., *Anal. Chem.*, **1993**, *65*, 902, 903, 905. © 1993 American Chemical Society.

V. APPLICATIONS

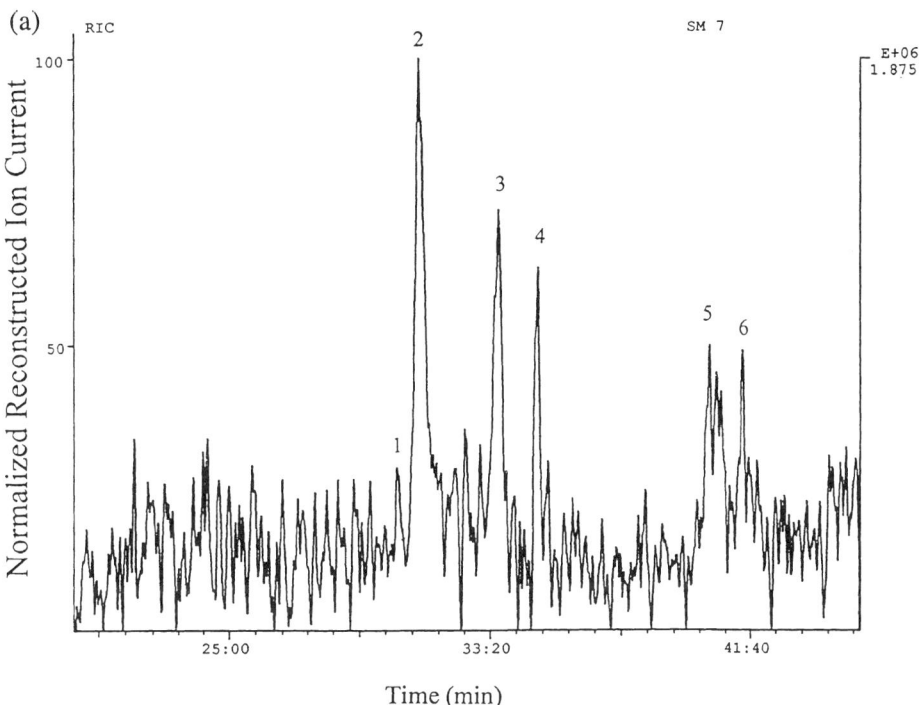

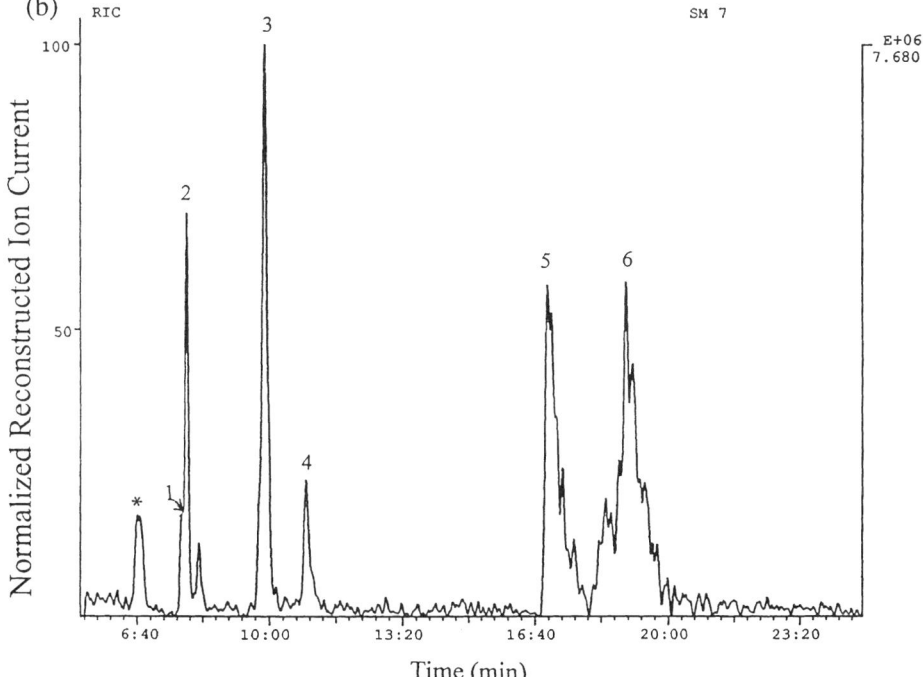

microsomal incubation of LSD (96). They have published quantitative results on some isoquinoline alkaloid natural products studied using a benchtop ion-trap system (97), and have examined the potential of identifying trace impurities at the 0.1% level in some standard peptide and alkaloid mixtures (98). Further quantitative studies and investigations into improving CE–MS reproducibility must be undertaken for the technique to become widely accepted in the analytical laboratory.

The wide use of EKC for many small-molecule applications, and the difficulty it presents for ES sources, has hindered the development of corresponding CE–MS methods. The technique, which separates species on the basis of differential interactions and hydrophobicities, is incompatible with ESI–MS because of the high concentrations of ionic additives present in the background electrolyte. As previously mentioned, the development of CE–APCI–MS may aid these studies (33,34) However, Tomlinson et al. have shown the potential for carrying out metabolism studies in free-solution CE by the addition of organic modifiers to the buffer system or the use of purely nonaqueous media; a method more readily compatible with ESI–MS (62,99). This group have also developed methods for on-line preconcentration for cleanup of metabolite samples (62,92).

The separation of sulphonamide drugs has also been studied by Perkins et al. (100), and the same group has presented a comparison of the use of nanoscale capillary LC–MS with CZE–MS for the separation of 10 macrolide antibiotics (101). The application of on-capillary transient ITP–MS, to improve concentration sensitivity, has been demonstrated for some β-agonists in calf urine, for which nanogram per milliliter detection limits were attained (87), and some β-blockers and their trace-level impurities (88). Naproxen–lysozyme conjugate mixtures (102) and noncovalently bound complexes between the immunophilin FKBP and the immunosuppressive drugs FK506 and rapamycin (103) have been studied, illustrating the ability of the analytical technique to examine noncovalent interactions of biomolecules under physiological conditions (104), an important area of research for the pharmaceutical industry. The separation and identification of chiral molecules has also become extremely important. The analysis of chiral drug mixtures by CE–MS, via the addition of a cyclodextrin to the background electrolyte, has been demonstrated by Sheppard et al. (105). Free analytes and analyte–cyclodextrin complexes were observed. This was also the case in the on-line analysis of some closely related herbicide analogs (88).

In the agrochemical and environmental fields, the separation of some sulphonylurea crop-protection agents (106), the detection of Fumonisin B1 toxin in corn samples (107) and a quantitative analysis of some quaternary ammonium herbicides (108) have been reported. Thibault et al. presented CE–MS studies on paralytic shellfish-poisoning toxins (109) and antibiotics used in the aquaculture industry (52). Other applications reported include the use of CE–MS for the analysis of textile dyes (110,111), cellulose derivatives in archaeological linen textiles (112), amino acids (113), cationic and anionic laser dyes (114), inositol phosphates (115), and warfare-agent products (51). The ability to detect inorganic ions by CE–IS–MS has also been demonstrated (116,117).

B. Polypeptides and Enzymatic Digests

The development of CE–MS techniques for analyzing small samples of complex polypeptide mixtures and digests for sequence determination is an important area of research. Standard peptide mixtures have been employed by numerous groups interested in developing this method. Ramsey et al., using a sheathless interface to combine CE with an ion trap, have obtained full scan midattomole detection limits for leucine enkephalin (56). Tomlinson et al. have shown the improvements obtained using preconcentration (PC)–CE–MS in the analysis of some MHC class 1 peptides (93) They have also demonstrated peptide detection at the low-femtomole level by the use of membrane PC–CE–MS combined with isotachophoretic elution and separation (94).

Applications demonstrated include the analysis of a 37-residue peptide from a monoclonal antibody against herpes by CE–MS, in which five impurities were identified (118). The ability to identify synthetic peptide impurities (119) and to analyze hydrophobic polypeptides, via the addition of organic solvents (38), have also been reported recently. The possible advantages of applying CE–MS and affinity-CE–MS to simplify the analysis of peptide combinatorial libraries (120,121) is a subject presently under much discussion in the pharmaceutical industry and the field of mass spectrometry. The separation and identification by CE–MS of certain ligands in a library which bind most tightly to a target receptor could provide a rapid and economical means of identifying structurally specific interactions. In addition, the ability to analyze these interactions in aqueous solutions helps minimize nonspecific interactions associated with techniques involving immobilization of one of the constituents.

Complex mixtures of peptides generated from tryptic digestion of large proteins present a difficult analytical challenge because the large number of fragments cover an extensive range of both isoelectric points and hydrophobicities. Because trypsin specifically cleaves peptide bonds on the C-terminal side of lysine and arginine residues, the resulting peptides generally form doubly as well as singly charged molecular ions by positive-ion ESI. Such doubly charged tryptic peptides generally fall within the m/z range of modern quadrupole mass spectrometers. Several examples of CE–MS applied to tryptic digest-model systems have been presented (29,104,122,123). The resolving power of the method has been demonstrated by comparing UV electropherograms obtained in conjunction with a CE–MS analysis for tryptic digests of cytochrome c (123). The individual mass electropherogram peaks indicated separation efficiencies of up to $\sim 4 \times 10^5$ theoretical plates, indicating a very small effective dead volume or residence time in the ESI interface for the particular system employed. The total ion electropherograms obtained from tryptic digests of approximately 30 fmole, before digestion, of bovine candida krusei and horse cytochrome c have also been presented by Wahl et al., employing acidic buffer conditions and a sheathless CE–MS interface (104). Mass spectrometric detection allowed each of the tryptic fragments to be identified.

Results to date suggest that nearly all tryptic fragments can be detected by

ESI–MS methods if first separated; infusion of the unseparated digest often results in dramatic discrimination against some components. Fragments observed by the UV detector are almost always detected by ESI–MS. However, detection of the very small fragments is sometimes a problem owing to difficulties in obtaining optimum ESI–MS conditions over a wide m/z range. Excessive internal excitation or large solvent-related background peaks at low m/z appear to be the origin of some of the difficulty. There are also indications that some ESI interface designs discriminate substantially against low m/z ions. In general, however, tryptic fragments not detected by CE–MS are most probably lost prior to the separation, a problem particularly evident with "nanoscale" sample handling. The further development of methods to minimize sample handling, such as the use of on-line immobilized-trypsin capillaries (124), should lead to improvements.

C. Proteins

The major difficulty in studying proteins by CE–MS is the protein interaction with the capillary surface. Generally, capillaries are coated to minimize, or prevent, wall interactions (6). Thibault et al. demonstrated that the analysis of proteins in acidic solutions using amino-based coatings in capillaries resulted in significant sensitivity improvements in CE–MS (7). Cole et al. presented the separation of a synthetic protein mixture using a hydrophilic, coated capillary (125), demonstrating separation efficiencies of up to 500,000 theoretical plates and detection limits in the low-femtomole range. Detection sensitivity for proteins is generally lower than for small peptides due to the greater number of charges per molecule and the greater number of charge states. Several means of improving protein concentration sensitivity in CE–MS have been reported (18,82,85,91,94), as discussed previously.

The potential of CE–MS for protein characterization has only begun to be explored and is expected to grow as more sensitive and powerful MS detection methods become more widely available. Some cases already reported include the characterization of recombinant somatotropins (MW $\sim$ 22 kDa) with detection of both mono- and deoxidized homologs (126), separation of glycoforms of ribonuclease B (127), and the analysis of recombinant insulin-like growth factor I variants (10). This latter separation was achieved by the addition of organic modifiers and zwitterionic detergent molecules to the system. Also, the characterization of polypeptides and small proteins in several snake venoms has been reported by Perkins et al., using both quadrupole and sector spectrometers (128,129).

The ability to inject and analyze extremely small sample volumes by CE–MS provides the possibility of confronting extreme biochemical challenges, such as the analysis of single cells, by CE–MS. This laboratory has recently reported the direct analysis of small-cell populations by CE–FTICR–MS (67,68). A small, known number of human erythrocytes was injected into the etched end of a

20-μm i.d. separation capillary with the aid of a microscope and micromanipulator. Figure 12a shows the total ion current trace from the injection of 20 cells, obtained from the ion beam impinging on the front shutter of the FTICR spectrometer (69). The first peak is due to buffer ions in the cell media and the second, broad peak represents the cell proteins. (Variations in cell lysing and injection times will lead to peak broadening, but the application of an isotachophoretic concentration step should overcome this problem.) A mass spectrum of the α chain of hemoglobin, obtained from the direct injection of 10 erythrocytes, corresponding to approximately 4.5 fmole of hemoglobin, is shown in Figure 12b; separation of the α and β chains of hemoglobin is virtually complete. The application of quadrupolar axialization and sustained off-resonance irradiation (SORI) provides partial sequence information on selected components. The MS/MS spectrum presented in Figure 12c was obtained from the selective-ion accumulation of the 18+ charge state of the α chain from the injection of 75 erythrocytes followed by SORI at -1000 Hz relative to the chain (66). The detection of hemoglobin from a single human erythrocyte has recently been achieved (corresponding to ~ 450 attomoles). With further increases in sensitivity, by improved ion generation and accumulation techniques, it would seem likely that this technique could be extended to the analysis of components in many alternative cell systems. Analysis at the single-cell level could provide insights into chemical processes occurring in the cell without the need to average over large cell populations.

The ability to study noncovalent interactions by ESI–MS and the possibility that some specific protein conformation is retained in transfer to the gas phase, has led to great excitement over the possible results of studying many biological systems and processes by MS (130). To this aim, studies into the controlling parameters involved in biological noncovalent complex analysis are under way and the application of CE–MS to this research should increase specificity and quantitation. Goodlett et al., employing CE–MS under a variety of conditions, have studied the complex formed between the 11.5-kDA protein and the peptide (2166 Da) which combine to form ribonuclease S (131). And CE–MS has also been used to analyze the noncovalent complex (59 kDa) formed between human antithrombin III, a single-chain glycoprotein, and an octasulphated pentasaccharide (132).

D. Nucleic Acids

Recent reports from our laboratory have demonstrated the use of CZE–ESI–MS, reversed-anionic CITP–MS, and CE–MS/MS techniques for the analysis of nucleotides (133) and the identification of irradiation-damaged sites (134). Damage products, caused by X-ray irradiation of d(CGTA) in N2-saturated aqueous solution, were analyzed by semi-preparative HPLC followed by CITP–MS/MS. The results, examples of which are presented in Figure 13, indicated that cytosine was modified to uracil, hydroxyuracil, and hydroxyhydrouracil; guanine was modified to cycloguanosine; and thymine was changed to dihy-

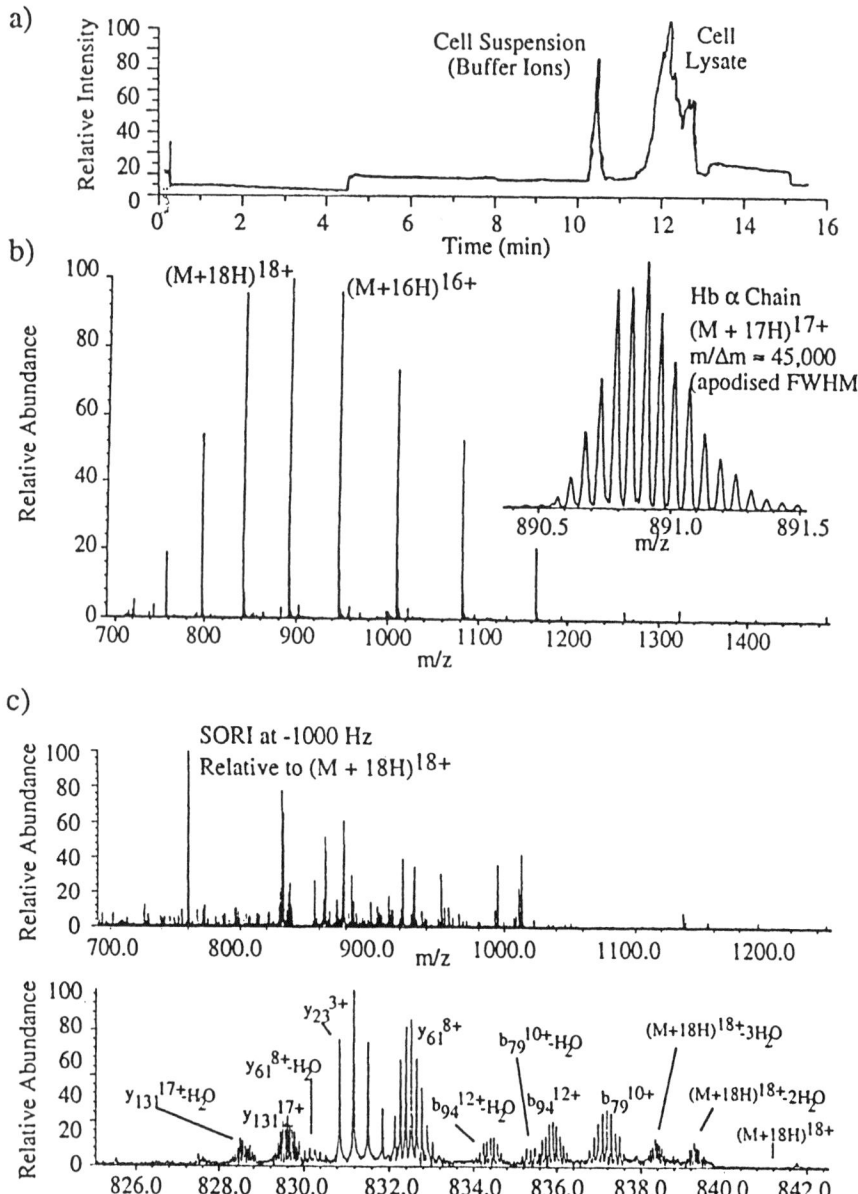

Figure 12. (*a*) Total ion current trace from the injection of 20 cells into the CE capillary obtained from the ion beam hitting the front shutter of the FTICR spectrometer. (*b*) The spectrum obtained for the hemoglobin α-chain, demonstrating the resolution obtained for the 17+ charge state. (*c*) The MS/MS spectrum, providing partial sequence information, from the selective ion accumulation of the 18+ charge state of the α-chain followed by SORI at −1000 Hz.

drothymine and hydroxyhydrothymine. The characterization of metastable intermediates of enzymatic peroxidation of nicotinamide adenine dinucleotide has also been reported (135). Further studies are currently under way into the chemistries of larger oligonucleotides. Vouros and co-workers have presented results of studies on polyaromatic hydrocarbon adducts to deoxyguanosine using CE–MS based on a CF–FAB interface (24,136), and benzo[a]pyrene metabolite adducts (137) and carcinogenic 2-amino-3-methylimidazo[4,5-f]quinoline(IQ) adducts (138) to DNA, employing negative-mode CE–MS. Janning et al. have recently presented initial results on the use of CE–MS to analyze styrene oxide and cisplatin adducts to DNA (139,140).

E. Other Biopolymers

The analysis of oligosaccharides isolated from some glycoproteins has been carried out by Kelly et al., employing isotachophoretic preconcentration and reduced-elution-speed techniques to aid mass-spectral analysis of the separated

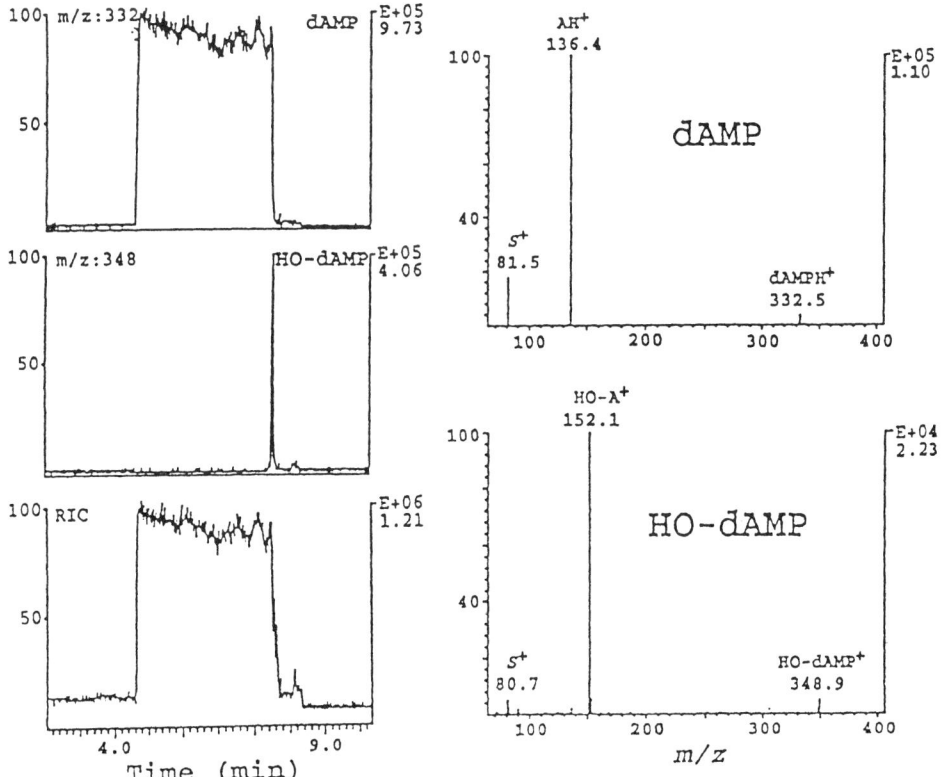

Figure 13. CITP–ESI–MS–MS of dAMP, identifying one of its irradiation-damaged products as caused by a hydroxy radical attacking the adenine base.

peaks (89). They have also presented reports on the separation and structural characterization of lipopolysaccharides from B. catarrhalis and H. influenzae type B (141), and some O-deacylated lipopolysaccharide products from the bacteria *Pseudomonas aeruginosan*, by CE–MS (142).

VI. FUTURE PROSPECTS

The combination of microscale preconcentration and cleanup techniques with CE–MS, and the resulting improvements in concentration detection limits, is likely to become an important step in the expanded use of this analytical technique. Further developments in ion trapping and TOF mass spectrometry, to achieve the predicted improvements in sensitivity, mass resolution, acquisition speeds, and tandem MS capabilities, should also accelerate implementation of the technique. The ability to provide both an electrical contact and stable ES at the capillary tip is of prime importance, and further investigations to improve interface durability, reproducibility, and applicability, while maintaining maximum sensitivity, are underway (143). However, the sensitivity already demonstrated by CE–MS systems, in combination with the minute analyte volumes sampled by this technique, suggest that routine separations and sequence determination from femtomole to attomole levels will soon become commonplace. The extension to subattomole levels is certain to follow. Future prospects would seem to include the promise of separating, sequencing, and determining the location and nature of components, modifications and noncovalent associations in, and between, systems such as proteins, single cells, and DNA segments.

ACKNOWLEDGMENTS

This research was supported by internal exploratory research and the Director, Office of Health and Environmental Research, U. S. Department of Energy. Pacific Northwest National Laboratory is operated by Battelle Memorial Institute for the U. S. Department of Energy, through contract DE-AC06-76RLO1830. We thank Dr. S. A Hofstadler for useful discussion.

REFERENCES

1. *Capillary Electrophoresis—Theory and Practice*; Camilleri, P., Ed.; CRC Press: Boca Raton, FL, 1993.
2. Jorgenson, J. W.; Lukacs, K. D. *J. Chromatogr.* **1981**, *218*, 210.
3. Jorgenson, J. W.; Lukacs, K. D. *Anal. Chem.* **1981**, *53*, 1298.
4. Terabe, S.; Otsuka, K.; Ichikawa, K.; Tsuchiya, A.; Ando T. *Anal. Chem.* **1984**, *56*, 111.

5. Olivares, J. A.; Nguyen, N. T.; Yonker, C. R.; Smith, R. D. *Anal. Chem.* **1987**, *59*, 1230.
6. Heiger, D. N. *High Performance Capillary Electrophoresis—An Introduction*; Hewlett-Packard Co., 1992; 2nd ed.
7. Thibault, P.; Paris, C.; Pleasance, S. *Rapid Commun. Mass Spectrom.* **1991**, *5*, 484.
8. Sepaniak, M. J.; Cole, R. O.; Clark, B. K. *J. Liq. Chromatogr.* **1992**, *15*, 1023.
9. Kirby, D. P; Greve, K. F.; Foret, F.; Vouros, P.; Karger, B. L.; Nashabeh, W. *Proc. of the 42nd ASMS Conference on Mass Spectrometry and Allied Topics*, 1994, p. 1014.
10. Nashabeh, W.; Greve, K. F.; Kirby, D.; Foret, F.; Karger, B. L.; Relfsnyder, D. H.; Builder, S. E. *Anal. Chem.* **1994**, *66*, 2148.
11. Lamoree, M. H.; Tjaden, U. R.; van der Greef, J. *J. Chromatogr. A* **1995**, *712*, 219.
12. Nelson, W. M.; Tang, Q.; Harrata, A. K.; Lee, C. S. *J. Chromatogr. A* **1996**, *749*, 219.
13. Stegehuis, D. S.; Irth, H.; Tjaden, U. R.; van der Greef, J. *J. Chromatogr.* **1991**, *538*, 393.
14. Tinke A. P.; Reinhoud N. J.; Niessen W. M. A., Tjaden U. R.; van der Greef J. *Rapid Commun. Mass Spectrom.* **1992**, *6*, 560.
15. Stegehuis, D.S.; Tjaden, U. R.; van der Greef, J. *J. Chromatogr.* **1992**, *591*, 341.
16. Schwer, C.: Lottspeich, F. *J. Chromatogr.* **1992**, *623*, 345.
17. Reinhoud N. J.; Tjaden U. R.; van der Greef J. *J. Chromatogr. A.* **1993**, *653*, 303.
18. Thompson T. J.; Foret F.; Vouros P.; Karger B. L. *Anal. Chem.* **1993**, *65*, 900.
19. Chien, R.-L.; Burgi, D. S. *Anal. Chem.* **1992**, *64*, 1046.
20. Albin, M.; Grossman, P.D.; Moring, S.E. *Anal. Chem.* **1993**, *65*, 489A.
21. Minard, R. D.; Chin-Fatt, D.; Curry, R., Jr.; Ewing, A. G. *Proc. of the 36th ASMS Conference on Mass Spectrometry and Allied Topics*; San Francisco, CA 1988, p. 950.
22. Caprioli, R. M.; Moore, W. T.; Martin, M.; DaGue, B. B. *J. Chromatogr.* **1989**, *480*, 247.
23. Reinhoud, N. J.; Niessen, W. M. A.; Tjaden, U. R. *Rapid Commun. Mass Spectrom.* **1989**, *3*, 348.
24. Wolf, S. M.; Vouros, P.; Norwood, C.; Jackim, E. *J. Am. Soc. Mass Spectrom.* **1992**, *3*, 757.
25. Suter, M. J.-F.; Caprioli, R. M. *J. Am. Soc. Mass Spectrom.* **1992**, *3*, 198.
26. deWit, J. S. M.; Deterding, L. J.; Moseley, M. A.; Tomer, K. B.; Jorgenson, J. W. *Rapid Commun. Mass Spectrom.* **1988**, *2*, 100.
27. Moseley, M. A.; Deterding, L. J.; Tomer, K. B.; Jorgenson, J. W. *J. Chromatogr.* **1989**, *480*, 197.
28. Moseley, M. A.; Deterding, L. J.; Tomer, K. B.; Jorgenson, J. W. *Anal. Chem.* **1991**, *63*, 109.
29. Deterding, L. J.; Parker, C. E.; Perkins, J. R.; Moseley, M. A.; Jorgenson, J. W.; Tomer, K. B. *J. Chromatogr.* **1991**, *554*, 329.
30. Nichols, W.; Zweigenbaum, J.; Garcia, F.; Johansson M.; Henion J. D. *LC-GC* **1992**, *10*, 676.
31. Bruins, A. P.; Covey, T. R.; Henion J. D. *Anal. Chem.* **1987**, *59*, 2642.
32. Lee, E. D.; Mück, W.; Henion, J. D.; Covey, T. R. *Biomed. Environ. MassSpectrom.* **1989**, *18*, 844.

33. Takada, Y.; Sakairi, M.; Koizumi H. *Rapid Commun. Mass Spectrom.* **1995**, *9*, 488.
34. Takada, Y.; Sakairi, M.; Koizumi H. *Anal. Chem.* **1995**, *67*, 1474.
35. Liu, Y.; Lopez-Avila, V.; Zhu, J. J.; Wiederen, D. R.; Beckert W. F. *Anal. Chem.* **1995**, *67*, 2020.
36. Lu, Q.; Bird, S. M.; Barnes R.M. *Anal. Chem.* **1995**, *67*, 2949.
37. Takigiku, R.; Keough, T.; Lacey, M. P.; Schneider, R. E. *Rapid Commun. Mass Spectrom.* **1990**, *4*, 24.
38. Weinmann, W.; Parker, C. E.; Baumeister, K.; Maier, C.; Tomer, K. B.; Przybylski, M. *Electrophoresis* **1994**, *15*, 228.
39. Castoro, J. A.; Chiu, R. W.; Monnig, C. A.; Wilkins C. L. *J. Am. Chem. Soc.* **1992**, *114*, 7571.
40. Tomer, K. B.; Weinmann, W.; Papac, D. I.; Palm, L.; Parker, C. E.; Deterding, L. J.; Przybylski, M.; Hoyes, J. *Proc. of the 42nd ASMS Conference on Mass Spectrometry and Allied Topics*, Chicago, IL 1994, p. 783.
41. Blakley, C. R.; Vestal, M. R.; Verentchikov, A.; Wang, Y. *Proc. of the 42nd ASMS Conference on Mass Spectrometry and Allied Topics*, Chicago, IL 1994, p. 1045.
42. Muller, O.; Foret, F.; Karger, B.L. *Anal. Chem.* **1995**, *67*, 2974.
43. Van Veelen, P. A.; Tjaden, U. R.; van der Greef, J.; Ingendoh, A.; Hillenkamp, F. *J. Chromatogr.* **1993**, *647*, 367.
44. Smith, R. D.; Olivares, J. A.; Nguyen, N.T.; Udseth H. R. *Anal. Chem.* **1988**, *60*, 436.
45. Nielen, M. W. F. *J. Chromatogr. A* **1995**, *712*, 269.
46. Tetler, L. W.; Cooper, P. A.; Powell B. *J. Chromatogr. A.* **1995**, *700*, 21.
47. Wahl, J. H.; Smith R. D. *J. Capillary Electrophoresis* **1994**, *1*, 62.
48. Foret, F.; Thompson, T. J.; Vouros, P.; Karger, B. L.; Gebauer, P.; Bocek P. *Anal. Chem.* **1994**, *66*, 4450.
49. Cai, J.; Henion, J. *J. Chromatogr. A* **1995**, *703*, 667.
50. Johansson, I. M.; Huang, E. C.; Henion, J. D.; Zweigenbaum J. *J. Chromatogr.* **1991**, *554*, 311.
51. Kostiainen, R; Bruins, A. P.; Hakkinen, V. M. A. *J. Chromatogr.* **1993**, *634*, 113.
52. Pleasance, S.; Thibault, P.; Kelly, J. *J. Chromatogr.* **1992**, *591*, 325.
53. Chowdhury, S. K.; Chait, B. T. *Anal. Chem.* **1991**, *63*, 1660.
54. Gale, D. C.; Smith R. D. *Rapid Commun. Mass Spectrom.* **1993**, *7*, 1017.
55. Wahl, J. H.; Gale, D. C.; Smith R. D. *J. Chromatogr. A.* **1994**, *659*, 217.
56. Ramsey, R. S.; McLuckey, S. A. *J. Microcol. Sep.* **1995**, *7*, 461.
57. Emmett, M. R.; Caprioli, R. M. *J. Am. Soc. Mass Spectrom.* **1994**, *5*, 605.
58. Wilm, M.; Mann, M. *Anal. Chem.* **1996**, *68*, 1.
59. Kriger, M. S.; Cook, K. D.; Ramsey R. S. *Anal. Chem.* **1995**, *67*, 385.
60. Kriger, M. S.; Cook, K. D.; Ramsey R. S. *Proc. of the 42nd ASMS Conference on Mass Spectrometry and Allied Topics*, 1994, p. 789.
61. Fang, L.; Zhang, R.; Zare, R. N. *Proc. of the 41st ASMS Conference on Mass Spectrometry and Allied Topics*, 1993, p. 755a.
62. Tomlinson, A. J.; Benson, L. M.; Naylor S. *J. Capillary Electrophoresis* **1994**, *1*, 127.

63. Reinhoud, N. J.; Schroder, E.; Tjaden; U. R.; Niessen, W. M. A.; Ten Noever de Brauw, M. C.; van der Greef, J. *J. Chromatogr.* **1990**, *516*, 147.
64. Ramsey, R. S.; Goeringer, D. E.; McLuckey, S. A. *Anal. Chem.* **1993**, *65*, 3521.
65. Hofstadler, S. A.; Wahl. J. H.; Bruce, J. E.; Smith, R. D. *J. Am. Chem. Soc.* **1993**, *15*, 6983.
66. Hofstadler, S. A.; Wahl, J. H.; Bakhtiar, R.; Anderson, G. A.; Bruce, J. E.; Smith, R.D. *J. Am. Soc. Mass Spectrom.* **1994**, *5*, 894.
67. Hofstadler, S. A.; Swanek, F. D.; Gale, D. C.; Ewing, A. G.; Smith, R. D. *Anal. Chem.*, **1995**, *67*, 1477.
68. Hofstadler, S. A.; Severs, J. C.; Smith, R. D.; Swanek, F. D.; Ewing, A. G. *J. High Resol. Chromatogr.* **1996**, *19*, 617.
69. Wahl, J. H; Hofstadler, S. A.; Smith R. D. *Anal. Chem.* **1995**, *67*, 472.
70. Johnson, P. J.; Gross, D. S.; Schnier, P. D.; Williams, E. R. *Proc. of the 42nd ASMS Conference on Mass Spectrometry and Allied Topics,* Chicago, IL 1994, p. 235.
71. Wu, Q.; van Orden, S.; Cheng, X.; Bakhtiar, R.; Smith, R. D. *Anal Chem.* **1995**, *67*, 2498.
72. Boyle, J. G.; Whitehouse, C. M. *Anal. Chem.* **1992**, *64*, 2084.
73. Zhang, R.; Fang, L.; Zare, R. N. *Proc. of the 42nd ASMS Conference on Mass Spectrometry and Allied Topics*, Chicago, IL 1994, p. 788.
74. Andrien, Jr. B.; Boyle, J.; Banks, F.; Dresch, T.; Haren, P.; Whitehouse, C. *Proc. of the 43rd ASMS Conference on Mass Spectrometry an Allied Topics*, Atlanta GA, 1995, p. 999.
75. Gao, Q.; Rockwood, A. L.; Udseth, H. R.; Follansbee J. C.; Severs J. C.; Muddiman, D. C.; Smith, R. D. *Proc. of the 43rd ASMS Conference on Mass Spectrometry and Allied Topics*, Atlanta, GA 1995, p. 125.
76. Li, X.; Wu, J.; Li, H.; Lubman, D. Proc. of the 43rd ASMS Conference on Mass Spectrometry and Allied Topics, Atlanta, GA 1995, p. 461.
77. Muddiman, D. C.; Rockwood, A. L.; Gao, Q.; Severs; J. C.; Udseth, H. R.; Smith, R. D. *Anal. Chem.*, **1995**, *67*, 4371.
78. Ikonomou, M. G.; Blades, A. T.; Kebarle P. *Anal. Chem.* **1990**, *62*, 957.
79. Ikonomou, M. G.; Blades, A. T.; Kebarle P. *Anal. Chem.* **1991**, *63*, 1989.
80. Tang L.; Kebarle P. *Anal. Chem.* **1991**, *63*, 2709.
81. *Handbook of Capillary Electrophoresis*; Landers, J. P., Ed.; CRC Press: Boca Raton, FL, 1994.
82. Wahl, J. H.; Goodlett, D. R.; Udseth, H. R.; Smith, R. D. *Electrophoresis* **1993**, *14*, 448.
83. Hjerten, S. *Electrophoresis* **1990**, *11*, 665.
84. Goodlett, D. R.; Wahl, J. H.; Udseth, H. R.; Smith R. D. *J. Microcol. Sep.* **1993**, *5*, 57.
85. Wahl, J. H.; Goodlett, D. R.; Udseth, H. R.; Smith R. D. *Anal. Chem.* **1992**, *64*, 3194.
86. Udseth, H. R.; Loo, J. A.; Smith, R. D. *Anal. Chem.* **1989**, *61*, 228.
87. Lamoree, M. H.; Reinhoud, N. J.; Tjaden, U. R.; Niessen, W. M. A.; van der Greef J. *Biol. Mass Spectrom.* **1994**, *23*, 339.

88. Severs, J. C.; Games D. E. *Proc. of the 42nd ASMS Conference on Mass Spectrometry and Allied Topics,* Chicago, IL 1994, p. 787.
89. Kelly, J. F.; Thibault, P.; Locke, S.; Ramaley, L. R. *Proc. of the 42nd ASMS Conference on Mass Spectrometry and Allied Topics,* Chicago, IL 1994, p. 1153.
90. van der Vlis, E.; Mazereeuw, M.; Tjaden, U. R.; Irth, H.; van der Greef, J. *J. Chromatogr. A.* **1995**, *712*, 227.
91. Tang, Q.; Harrata, A. K.; Lee, C. S. *Anal. Chem.* **1995**, *67*, 3515.
92. Tomlinson A. J.; Benson, L. M.; Braddock, W. D.; Oda, R. P.; Naylor, S. *J. High Res. Chromatogr.* **1994**, *17*, 729.
93. Tomlinson, A. J.; Nevala, W. K.; Braddock, W. D.; Strausbauch, M. A.; Wettstein, P. J.; Naylor, S. *Proc. of the 42nd ASMS Conference on Mass Spectrometry and Allied Topics,* Chicago, IL 1994, p. 23.
94. Tomlinson, A. J.; Naylor S. *J. Capillary Electrophoresis* **1995**, *2*, 225.
95. Johansson, J. M.; Pavelka, R.; Henion J. D. *J. Chromatogr.* **1991**, *559*, 515.
96. Cai, J.; Lim, H. K.; Henion, J. D. *Proc. of the 42nd ASMS Conference on Mass Spectrometry and Allied Topics,* **1994**, p. 358.
97. Henion, J. D.; Mordehai, A. V.; Cai, J. *Anal. Chem.* **1994**, *66*, 2103.
98. Hsieh, F. Y. L.; Cai, J.; Henion J. D. *J. Chromatogr. A.* **1994**, *679*, 206.
99. Tomlinson, A. J.; Benson, L. M.; Naylor, S. *LC-GC,* **1994**, *12*, 122.
100. Perkins, J. R.; Parker, C. E.; Tomer, K. B. *J. Am. Soc. Mass Spectrom.* **1992**, *3*, 139.
101. Parker, C. E.; Perkins, J. R.; Tomer, K. B.; Shida, Y.; O'Hara, K.; Kono M. *J. Am. Soc. Mass Spectrom.* **1992**, *3*, 563.
102. Kostiainen, R.; Franssen, E. J. F.; Bruins, A. P. *J. Chromatogr.* **1993**, *647*, 361.
103. Hsieh, Y-L.; Cai, J.; Li Y-T.; Henion, J. D.; Ganem B. *J. Am. Soc. Mass Spectrom.* **1995**, *6*, 85.
104. Smith, R. D.; Wahl, J. H.; Goodlett, D. L.; Hofstadler S. A. *Anal. Chem.* **1993**, *65*, 574A.
105. Sheppard, R.; Tong, X.; Cai, J.; Henion, J. D. *Anal. Chem.* **1995**, *67*, 2054.
106. Garcia, F.; Henion, J. D. *J. Chromatogr.* **1992**, *606*, 237.
107. Hines, H. B.; Holcomb, M.; Brueggemann, E. E.; Holder, C. L. *Proc. of the 42nd ASMS Conference on Mass Spectrometry and Allied Topics,* Chicago, IL 1994, p. 151.
108. Wycherley, D.; Rose, M. E.; Rimmer, D.; Giles, K.; Jarvis, S. A.; McDowall, M. A. *Proc. of the 41st ASMS Conference on Mass Spectrometry and Allied Topics,* San Francisco, CA 1993, p. 400a.
109. Buzy, A.; Thibault, P.; Laycock, M. V. *J. Chromatogr. A.* **1994**, *688*, 301.
110. Tetler, L. W.; Cooper, P. A.; Carr, C. M. *Rapid Commun. Mass Spectrom.* **1994**, *8*, 179.
111. Lee, E. D.; Muck, W.; Henion, J. D.; Covey, T. R. *Biomed. Environ. Mass Spectrom.* **1989**, *18*, 253.
112. Kouznetsov, D. A.; Ivanov, A. A.; Veletsky, P. R. *Anal. Chem.* **1994**, *66*, 4359.
113. Lu, W.; Yang, G.; Cole, R. B. *Electrophoresis* **1995**, *16*, 487.
114. Varghese, J.; Cole, R. B. *J. Chromatogr.* **1993**, *639*, 303.
115. Buscher, B. A. P.; van der Hoeven, R. A. M.; Tjaden, U. R.; Andersson, E.; van der Greef, J. *J. Chromatogr. A* **1995**, *712*, 235.

116. Huggins, T. G.; Henion, J. D. *Electrophoresis* **1993**, *14*, 531.
117. Corr, J. J.; Covey, T. R.; Anacleto, J. F. *Proc. of the 42nd ASMS Conference on Mass Spectrometry and Allied Topics,* Chicago, IL 1994, p. 340.
118. Kostiainen, R.; Lasonder, E.; Bloemhoff, W.; van Veelen, P. A.; Welling, G. W.; Bruins, A. P. *Biol. Mass Spectrom.* **1994**, *23*, 346.
119. Rosnack, K. J.; Stroh, J. G.; Singleton, D. H.; Guarino, B. C.; Andrews, G. C. *J. Chromatogr. A.* **1994**, *675*, 219.
120. Kirby, D. P.; Chu, Y-H.; Dunayevskiy, Y.; Karger, B. L.; Vouros P. *Presented at the 43rd ASMS Conference on Mass Spectrometry and Allied Topics,* Atlanta, GA 1995.
121. Dunayevskiy, Y. M.; Vouros, P.; Carell, T.; Wintner, E. A.; Rebek, J. Jr. *Proc. of the 43rd ASMS Conference on Mass Spectrometry and Allied Topics*, Atlanta, GA 1995, p. 493.
122. Moseley, M. A.; Shabanowitz, J.; Hunt, D. F.; Tomer, K. B.; Jorgenson, J. W. *J. Am. Soc. Mass Spectrom.* **1992**, *3*, 289.
123. Smith, R. D.; Udseth, H. R.; Barinaga, C. J.; Edmonds, C. G. *J. Chromatogr.* **1991**, *559*, 197.
124. Takigiku, R.; Keough, T.; Lacey, M. P.; Purdon, M. P.; Licklider, L.; Kuhr, W. G. *Proc. of the 42nd ASMS Conference on Mass Spectrometry and Allied Topics*, Chicago, IL 1994, p. 673.
125. Cole, R. B.; Varghese, J.; McCormick, R. M., Kadlecek, D. *J. Chromatogr, A,* **1994**, *680*, 363.
126. Tsuji, K.; Baczynskyj, L.; Bronson, G. E. *Anal. Chem.* **1992**, *64*, 1864.
127. Volk, K. J.; Liu, J.; Lee, M. S.; Klohr, S. E.; Kerns, E. H.; Rosenberg, I. E. *Proc. of the 41st ASMS Conference on Mass Spectrometry and Allied Topics*, San Francisco, CA 1993, p. 895a.
128. Perkins, J. R.; Tomer, K. B. *Anal. Chem.* **1994**, *66*, 2835.
129. Perkins, J. R.; Tomer, K. B. *J. Capillary Electrophoresis* **1994**, *1*, 231.
130. Smith, R. D.; Light-Wahl, K. J. *Biol. Mass Spectrom.* **1993**, *22*, 493.
131. Goodlett, D. R.; Wahl, J. H.; Udseth, H. R.; Ogorzalek Loo, R. R.; Loo, J. A.; Smith, R. D. *Proc. of the 40th ASMS Conference on Mass Spectrometry and Allied Topics,* Washington, DC 1992, p. 422.
132. Uzabiaga, F.; Tuong, A.; Lormeau, J. C.; Petitou, M.; Picard, C. *Proc. of the 42nd ASMS Conference on Mass Spectrometry and Allied Topics,* Chicago, IL 1994, p. 924.
133. Zhao, Z.; Wahl, J. H.; Udseth, H. R.; Hofstadler, S. A.; Fuciarelli, A. F.; Smith R. D. *Electrophoresis* **1995**, *16*, 389.
134. Severs, J. C.; Hofstadler, S. A.; Zhao, Z.; Senh, R. T.; Smith, R. D. *Electrophoresis* **1996**, *17*, in press.
135. Zhao Z., Udseth, H. R.; Smith, R. D. *Proc of the 43rd ASMS Conference on Mass Spectrometry and Allied Topics*, Atlanta, GA 1995, p. 457.
136. Wolf, S. M.; Vouros P. *Anal. Chem.* **1995**, *67*, 891.
137. Barry, J. P.; Vouros, P.; Norwood C. *Anal. Chem.* **1996**, *68*, 1432.
138. Rindgen, D.; Turesky, R. J.; Vouros P. *Proc. of the 43rd ASMS Conference on Mass Spectrometry and Allied Topics*, Atlanta, GA 1995, p. 590.
139. Janning, P.; Schrader, W.; Pesch, R.; Munster, H.; Linscheid, M. *Proc. of the 43rd*

ASMS Conference on Mass Spectrometry and Allied Topics, Atlanta, GA 1995, p. 248.

140. Janning, P.; Schrader, W.; Linscheid, M. *Rapid Commun. Mass Spectrom.* **1994**, *8*, 1035.

141. Kelly, J. F.; Thibault, P.; Masoud, H.; Perry, M. B.; Richards, J. C. *Proc. of the 41st ASMS Conference on Mass Spectrometry and Allied Topics,* San Francisco, CA 1993, p. 1079a.

142. Auriola, S.; Thibault, P.; Sadovskaya, I.; Masoud, H.; Altman, E.; Richards, J. Presented at the 43rd ASMS Conference on Mass Spectrometry and Allied Topics, Atlanta, GA 1995.

143. Severs, J. C.; Smith, R. D. *Anal. Chem.* **1997**, submitted.

PART IV
APPLICATIONS OF ELECTROSPRAY IONIZATION

CHAPTER 11

Electrospray Ionization Mass Spectrometry of Peptides and Proteins

JOSEPH A. LOO

Parke-Davis Pharmaceutical Research, Division of Warner-Lambert Company, Ann Arbor, Michigan

and

RACHEL R. OGORZALEK LOO

Department of Biological Chemistry, University of Michigan, Ann Arbor, Michigan

	Abstract	385
I.	Introduction	386
II.	Determining protein primary structure by ESI–MS	387
	A. ESI–MS detection and molecular-weight determination	392
	B. ESI–tandem mass spectrometry and peptide mapping	395
III.	Studying protein conformation by ESI–MS	399
IV.	Studying molecular interactions with ESI–MS	401
	A. Noncovalent interactions and quaternary structure	401
	B. Metal-binding proteins	403
	C. Combinatorial libraries and ligand identification	408
V.	Conclusions	411
	Acknowledgments	411
	References	411

Abstract

Electrospray ionization mass spectrometry (ESI–MS) is an important tool for the analysis of peptides and proteins. Several aspects of protein structure (primary, secondary, tertiary, and quaternary) have been studied using ESI–MS. Many examples of such applications are described, including molecular-weight determination, sequencing by tandem-mass spectrometry, determining metal-binding stoichiometry, and subunit stoichiometry of the quaternary structure.

Electrospray Ionization Mass Spectrometry, Edited by Richard B. Cole.
ISBN 0-471-14564-5 © 1997 John Wiley & Sons, Inc.

I. INTRODUCTION

"The question is no longer simply what are proteins made of, but rather how can the composition and structure of proteins be related to their specific biological functions?" Hans Neurath, the founding editor of the journal *Biochemistry* and the founding editor-in-chief of *Protein Science* wrote this in the preface of the second edition of *The Proteins* in 1963 (1). Although this statement was made over 30 years ago, recent technological advances, including mass spectrometry (MS), have made it abundantly more easy to answer this question.

Mass spectrometry has been used to solve many problems in protein chemistry, as Klaus Biemann nicely described (2). The development of electrospray ionization (ESI) and matrix-assisted laser desorption/ionization (MALDI) have facilitated the process greatly. Both MALDI (3,4) and ESI (5–7) have enabled mass spectrometers to measure molecular weights for biomolecules to greater than 150 kDa. Previously, volatilizing such large biological species without significant sample degradation had been a key limitation for gaseous-ion methods. Using the information these measurements provide is the next hurdle for the analyst; fortunately, a large number of problems exist in biochemistry and molecular biology which are amenable for study using these techniques.

Dole and colleagues originally described ESI in studies of ions from polymers of molecular weights in excess of 100 kDa (8,9). Dole conceived of using an ES process to produce intact polymeric ions from learning about electrospraying automobile paint while working as a consultant to a paint company (10). Later, Dole's experiments were extended by Fenn (5,11) and by researchers in the former Soviet Union (12). The production of molecules bearing multiple charges accesses higher molecular weights by extending the mass range for mass-to-charge (m/z) limited mass spectrometers. The mechanism of ion formation has been treated in several reports (13,14). Whatever the ultimate process of ion formation, it is clear that ESI produces multiply charged molecules from solution under mild conditions. These ions generally arise by attachment of protons, alkali cations, or ammonium ions for positive-ion formation, or with reversal of the polarity of the nebulizing electric field, negative ions are formed by proton abstraction. The multiple-charging phenomenon has been demonstrated to apply to molecules of over 150 kDa and it has permitted the measurement of relative molecular mass with precision of better than 0.05%.

Although Dole's work included proteins as the analyte and Fenn's early work with ES demonstrated small multiple charging of small peptides, it was not until the 1988 ASMS Conference on Mass Spectrometry that the explosion of biochemical applications utilizing ESI–MS began. Once researchers were accustomed to seeing many more peaks (multiply charged ions) than the number of analytes anticipated in the mass spectrum, new applications of ESI–MS for protein structure elucidation were developed.

A much earlier view of protein structure is represented by the following: "For a time the prevailing view in many circles was that these protein colloids were more or less random aggregates, not true chemical compounds, and that the attempt to investigate their properties as if they were genuine chemical substances was really a waste of time" (15). (Thankfully for protein chemists, this is not the accepted view in today's scientific arena.)

Protein structure can be classified as primary, secondary, tertiary, and quaternary. The primary structure describes the sequence in which the amino acids are strung together. Mass spectrometry provides molecular mass information, an important confirmation of primary structure. In addition, mass spectrometry and tandem mass spectrometry have made great strides in sequencing large polypeptides.

The arrangement of amino-acid sequence in three-dimensional space is described by its higher-order structure. Secondary structure refers to regular features of proteins formed by contiguous segments of the protein and defined by specific bond angles in the polypeptide backbone. Tertiary structure is the three-dimensional structure assumed by proteins in their native states and involves interactions between noncontiguous regions of the protein. Quaternary structure refers to arrangement of subunits in multi-subunit proteins. Hydrophobic, van der Waals, ionic, dipole, and hydrogen bonding are many of the forces that contribute to the formation of secondary, tertiary, and quaternary structures and govern how proteins fold and interact with other molecules to form complex structures. Many types of protein–ligand binding are examples of intermolecular noncovalent interactions, including antibody–antigen, protein–cofactor, receptor–ligand, and enzyme–substrate pairings. Intermolecular noncovalent interactions are responsible for aggregation of folded polypeptide chains into multimers which determines a protein system's quaternary structure.

II. DETERMINING PROTEIN PRIMARY STRUCTURE BY ESI–MS

Large biomolecules examined by ESI–MS typically show a distribution of multiply charged molecules and no evidence of fragmentation unless dissociation is induced during transport into the mass spectrometer by higher-energy collisions. The ESI mass spectra of protein enzymes proteinase K (29 kDa) and thermolysin (34 kDa) are shown in Figure 1. A distinctive bell-shaped distribution of charge states is typically observed in which adjacent peaks differ by one charge. A feature of ESI mass spectra for most proteins is that the average charge state increases in an approximately linear fashion with molecular weight. This is further demonstrated in the ESI mass spectra in Figure 2 for two galactosidase proteins showing over 100+ charges.

From the start, mass spectrometrists were fascinated with ES charge distributions. Where do the charges reside? How many charges can be placed on a protein? How closely can charges be spaced? Why is the most probable m/z 1000 for one protein and 2500 for another? Why are charge distributions bell shaped?

It has been suggested that protonation occurs at basic residues (Arg, Lys, NH_2-terminus, His) for positive ions and deprotonation at acidic residues (Asp,

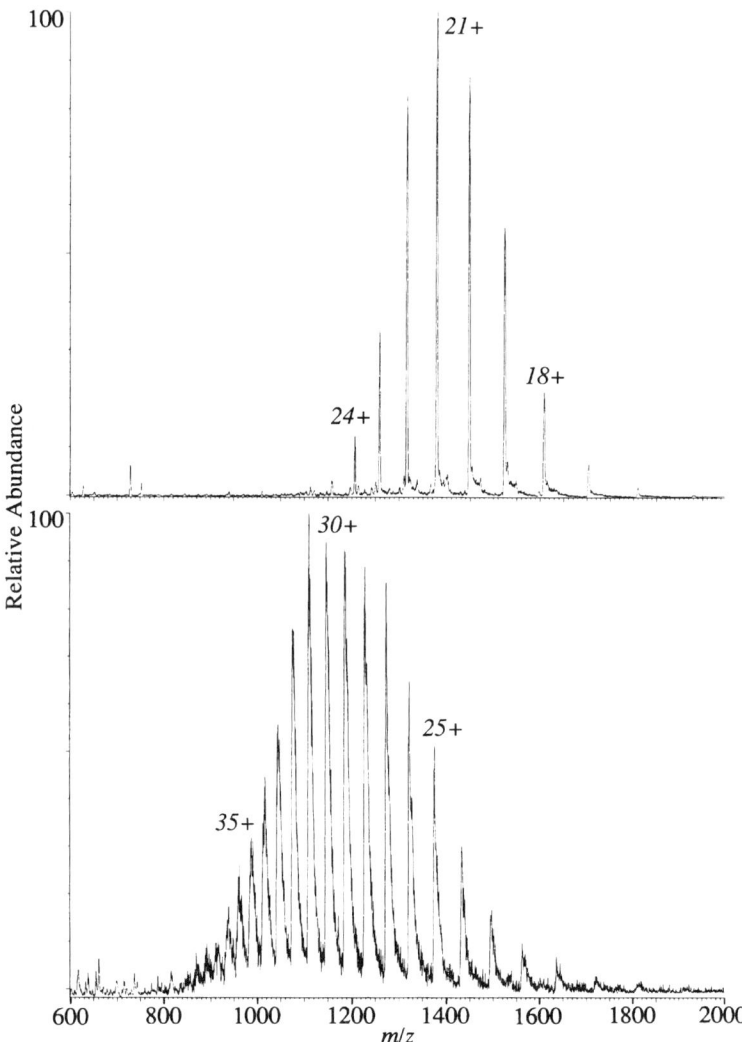

Figure 1. The ESI mass spectra of proteinase K (top) and thermolysin (bottom) in 50% acetonitrile/46% H_2O/4% acetic acid (v/v/v) obtained on a VG Fisons (Micromass) Platform single-quadrupole ESI mass spectrometer. The source temperature was 100°C.

Glu, Tyr, COOH-terminus) (16–18). Mostly, the highest charge states observed for proteins are consistent with this explanation, but a few exceptions exist. For those cases where apparent overcharging is present, it is not clear just what is charging. Overcharging has also been observed in the negative-ion mode (19), where it was suggested that the hydroxyls of serine and threonine might deprotonate. Recent work has suggested that charges can reside at the COOH terminus of a peptide during fragmentation, but what is not clear is whether that

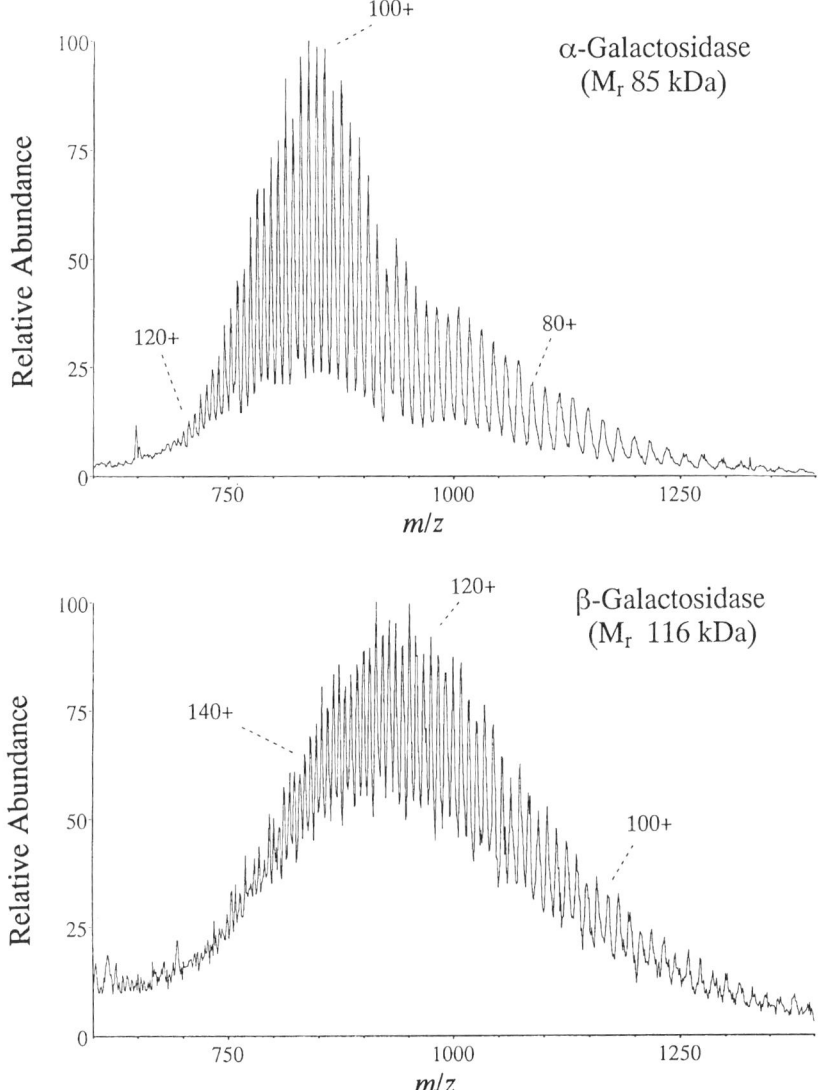

Figure 2. The ESI mass spectra of α- and β-galactosidase from 50/50 MeOH/H$_2$O with 5% acetic acid (v/v/v) obtained on a quadrupole mass spectrometer.

location represents only the transition state for dissociation. Despite the attractiveness of explaining charging by the presence of acidic or basic residues that protonate or deprotonate in solution, pK_a data does not explain the production of negative ions from very acidic peptide and protein solutions at pH levels significantly lower than the pK_a values of acidic residues (20–22). Le Blanc et al. have suggested that gas-phase contributions can explain those results (23,24).

Some effort has gone into evaluating how primary sequence and base spacing affects charging. Multiple antigenic peptides (MAPs) (routinely injected into animals to generate antibodies) were used to amplify subtle differences in positive charging (25). These peptides have a branched structure that links 2, 4, 8, or 16 copies of the same peptide. The MAPs were synthesized on a branched-core structure, where the α and ϵ amino groups were used as branch points. For immunogenic work, MAPs are frequently synthesized as octapeptide polymers, for example, a MAP-8 peptide, (antigen)$_8$-Lys$_4$-Lys$_2$-Lys-βAla-OH. Figure 3 explores the ESI multiple charging of MAP-8 core peptides containing one and two charge sites per branch, using Ac-Arg(Gly)$_5$-MAP and Ac-Arg(Gly)$_2$His(Gly)$_2$-MAP. While up to the maximum theoretically possible

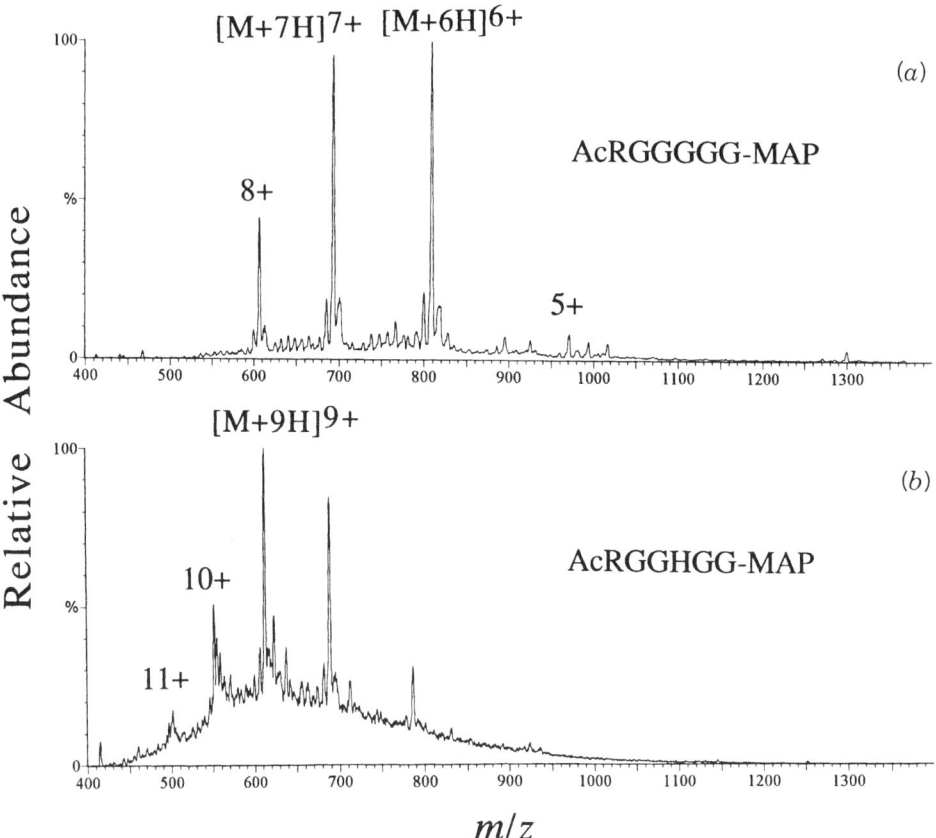

Figure 3. The ESI quadrupole mass spectra of two multiple antigenic peptides with MAP-8 cores. Samples were dissolved in 50% acetonitrile/46% H$_2$O/4% acetic acid (v/v/v).

8 charges were observed for ESI–MS of Ac-Arg(Gly)$_5$-MAP, only up to 11 charges were observed with Ac-Arg(Gly)$_2$His(Gly)$_2$-MAP. A similar observation was made for Ac-ArgLys(Gly)$_4$-MAP, while Ac-(Arg)$_2$(Gly)$_4$-MAP did not charge completely, either. Apparently, electrostatic repulsions do ultimately limit charging in MAPs. These experiments yielded considerable information about specific-sequence propensity for multiple charging.

The number of charges that can be placed on a peptide or protein is determined by other factors as well. First, it should be related to the number of acidic (negative-ion ESI) or basic (positive-ion ESI) residues, because the highest charge state observed only occasionally exceeds that value. If very basic proteins such as protamines and histones are a guide, the average charge state might be limited to 33% when the number of basic residues is not the limitation. Higher densities have been observed on short peptides (e.g., 3+ with Arg$_3$, 2+ on Arg). Chait's group has shown that anions can affect the charging observed (26).

Molecular-weight calculation for each of the observed m/z values enhances the precision of the measurement. Algorithms have been developed that deconvolute or convert the ESI m/z spectrum to a mass-domain plot for calculation of mass and provide an easier to visualize graphical representation (27). Most commercial ESI–MS systems include some form of this deconvolution software package. Some also provide a program that uses maximum entropy methods for higher-resolution deconvolution analysis (28). Relative molecular-mass measurements for proteins of greater than 100 kDa have been demonstrated and, in favorable cases, accuracy and precision of measurement may be better than ±0.05–0.01%. However, for larger molecules, more accurate measurements are required to differentiate mutations that produce a 1-Da mass difference (i.e., Asp and Asn, Leu/Ile and Asn, Glu and Gln). With higher-resolution mass spectrometers (29–32), direct charge-state assignment in even complex mass spectra is possible by measurement of the ^{12}C/^{13}C isotopic peak separation. The spacing between adjacent isotopic peaks is equal to (charge)$^{-1}$. The combination of high resolution and high accuracy may be important for unknown identification and complex mixture analysis.

Although multiple charging enables conventional mass spectrometers to measure molecular weights in excess of the mass range for singly charged molecules, the individual peaks representing each charge state need to be adequately resolved to determine an accurate mass. In general, the larger the molecule, the greater the maximum charge state and the smaller the spacing in m/z units between adjacent peaks. One method to circumvent this problem is to shift the charge-state envelope to higher m/z or lower charge, for example, by adjusting the solution pH (to higher values for positive ions), by gas-phase proton-transfer reactions with basic amines to neutralize some of the charges (33), or by collisional methods in the ESI interface (34). This increases the peak separation between adjacent charge states. Mass spectrometers with greater resolving power than conventional quadrupole analyzers, such as magnetic sector and especially Fourier transform mass spectrometers (FTMS), should be able to measure masses for larger compounds and for complex mixtures.

A. ESI–MS Detection and Molecular-Weight Determination

Determining the molecular weight of a peptide or protein is a rapid and straightforward method to confirm the identity and integrity of a sample. Peptides and proteins have been analyzed by ESI mass-spectrometric methods with high levels of success. Intact protein molecular weights in excess of 150 kDa have been measured (35). Mass spectrometry is an important tool for differentiating macromolecules of similar mass. For example, for the over 230 known variants of the β-chain human hemoglobin polypeptide, a great majority involve a single amino-acid substitution due to a single base substitution in the corresponding triplet codon of the cDNA. Mass spectrometry, in particular ESI–MS, has been used to identify many of the hemoglobin variants (36–38). Three-dimensional protein-structure analysis by NMR spectroscopy is facilitated by incorporating ^{13}C and ^{15}N in the protein. Because of the high accuracy with which molecular weights can be measured, an additional application of ESI–MS is the determination of the degree of isotope enrichment in proteins to be studied by NMR (39,40).

The sensitivity of these mass-spectrometric methods has yielded many applications in biochemical research. The combination of ESI–MS and capillary electrophoresis has been used to detect hemoglobin protein from a total of 1–10 red blood cells (41). Desiderio's group has used ESI–MS to characterize qualitatively and quantitatively neuropeptides in biological tissue and fluid samples. Endogenous methionine enkephalin (M_r 573) and β-endorphin (1-31, M_r 3463) were measured from extracts from human pituitary glands (42). A 9.8-kDa β-endorphin-containing protein from bovine pituitary was also characterized by ESI–MS (43).

It is often desirable to perform some type of separation methodology prior to ESI–MS analysis, especially proteins from biological matrices. Figure 4 shows an ESI mass spectrum of human variant hemoglobins (Hb) isolated from an isoelectric focusing gel. The analysis was performed in order to exclude putative presence of a β-globin variant comigrating with Hb F, whose molecular mass would differ from normal only by a few Daltons and thus would not be resolved from normal β-globin the ESI–MS analysis of the whole hemolysate (44).

Proteins associated with cell membranes present a challenging problem for ESI–MS analysis because of their hydrophobic nature. Obviously, analytes need

---

Figure 4. The ESI–MS spectrum of hemoglobin isolated from an isoelectric focusing gel (agarose plate, Isolab ampholines pH 6–8). The inset shows a spectrum deconvoluted with a MaxEnt® algorithm (Micromass). Hemoglobin migrating in a position of Hb F ($\alpha2\gamma2$) was eluted from the gel with water and diluted with trifluoroethanol (final concentration 50%) prior to ESI–MS analysis. The molecular masses of the detected globins are consistent with the presence of Hb F (expected average molecular masses of α, G-γ, and A-γ are 15,126.4 Da, 15,995.3 Da, and 16,009.3 Da, respectively); no significant amount of any other globin was detected (unpublished data, H. Ewa Witkowska, K. Kleman, and Cedric Shackleton, Children's Hospital Oakland).

II. DETERMINING PROTEIN PRIMARY STRUCTURE BY ESI-MS 393

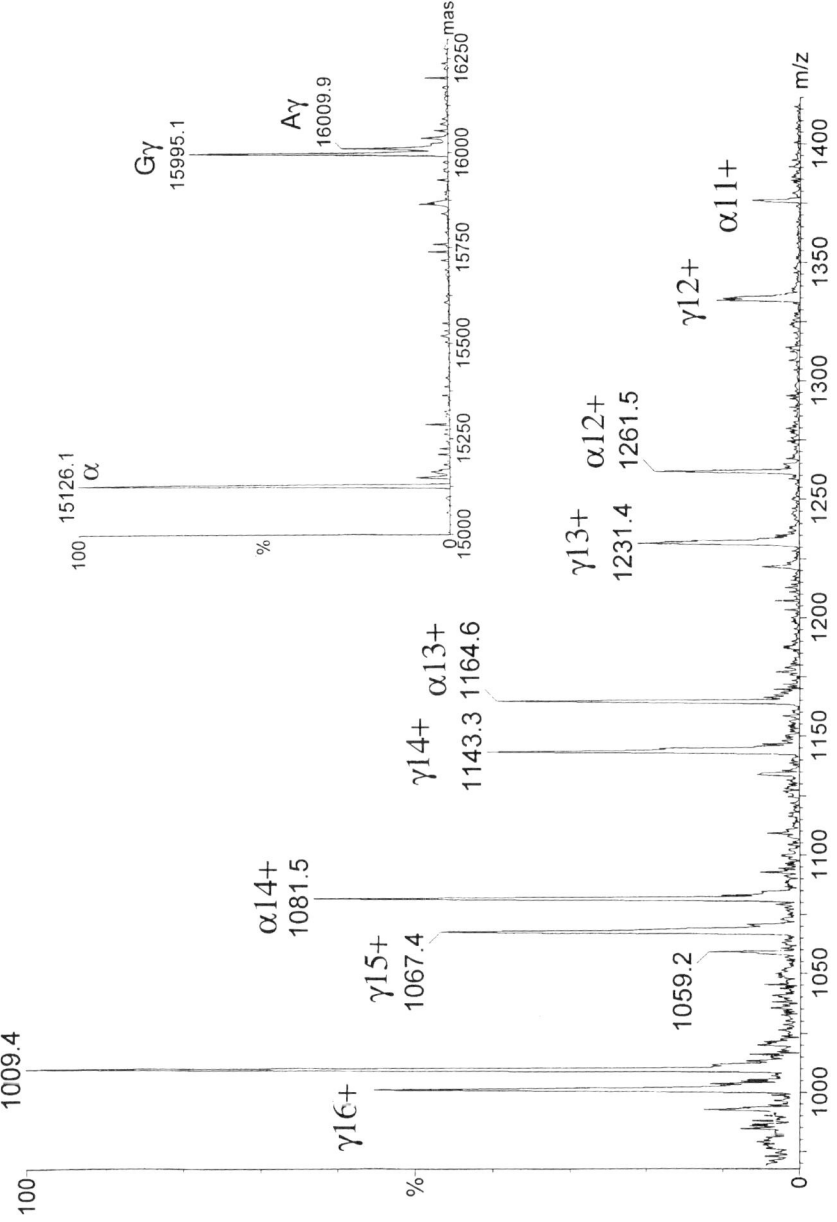

to be dissolved in the ESI solution for ion production. Membrane proteins are typically insoluble in the solutions normally used for ESI (i.e., a water/methanol or water/acetonitrile mixture with additional acetic or formic acid). However, Schindler et al. (45) found that chloroform/methanol/water mixtures work well as a carrier solution for samples dissolved in hexafluoroisopropanol (HFIP) or neat formic acid. Molecular weights were measured for hydrophobic proteins such as cytochrome c oxidase subunits and membrane protein bacterioropsin. Membrane proteins are typically studied in the presence of detergents. However, most detergents have deleterious effects for ESI–MS analysis. Ogorzalek Loo and co-workers have examined a number of potential detergents for protein analysis by ESI–MS (46), which may provide an alternative method for the analysis of membrane proteins.

The ESI–MS of large proteins such as monoclonal antibodies also presents an unusual challenge not only because of their large molecular weight (150 kDa), but also because of their relatively high mass-to-charge multiple-charge distribution (i.e., low charge). These proteins were difficult to analyze in the early days of ESI–MS because of the relatively low m/z instruments used (≤ 2000). Although multiple charging allows low m/z instruments to measure molecular weights of large biomolecules, monoclonal antibodies show charging such that the charge distribution typically shows up at m/z 2000–4000 (35,47–49). On the other hand, as discussed earlier, lower charging relaxes the requirements for resolving adjacent multiply charged peaks.

An important application of ESI–MS is to detect and identify post-translationally modified proteins. Cell regulation of protein function can be achieved through control of covalent modifications of the protein, such as glycosylation, phosphorylation, sulfation, and acylation. Mass spectrometry is a rapid and efficient method to detect these modifications. In the analysis by Roepstorff and co-workers (50) of a 45-kDa recombinant barley α-amylase, an enzyme important in malt production, at least three modifications were detected: C-terminal heterogeneous degradation, glutathionylation of a Cys residue, and O-glycosylation. ESI–MS was used to characterize legume plant lectins that commonly undergo C-terminal proteolytic processing to ragged ends, previously undetected by SDS–polyacrylamide gel electrophoresis (51).

The carbohydrate functionalities of glycoproteins are interesting because of their diverse range of vital biological activities. Characterization of the carbohydrate moiety of recombinant glycoproteins presents a significant challenge because the cDNA sequence provides little information regarding glycosylation and predicts only positions of potential sites of N-glycosylation. Moreover, although the molecular weight of most carbohydrates is relatively small compared to proteins (less than 10 kDa), the complexity from the numerous possible glycan linkages and branching patterns provides a considerable challenge for structure elucidation. Moreover, the glycosylation "signature" can vary from lot to lot or for each expression system.

Settineri and Burlingame reviewed the application of mass spectrometry for carbohydrate and glycoconjugate structure analysis (52). Reinhold and co-workers (53) have published a thorough review of the application of ESI-MS for carbohydrate structure analysis, highlighting several of the many examples from their laboratory with an emphasis on glycolipids and glycoprotein glycans.

Transferrins (Tf) are iron-binding glycoproteins of approximately 79 kDa that are widely distributed in the physiological fluids of vertebrates. Normal transferrin contains 2 Asn N-linked disialylated carbohydrate chains. ESI-MS was used to differentiate transferrin from a normal human patient and a patient with carbohydrate-deficient glycoprotein (CDG) syndrome (54). A 1.2-kDa difference between normal Tf and sialidase-treated Tf closely corresponds to the mass of four sialic acids. A 2.2-kDa difference was measured between normal Tf and Tf from a CDG patient, consistent with a Tf lacking both N-linked carbohydrate chains.

B. ESI-Tandem Mass Spectrometry and Peptide Mapping

An accurate molecular-weight measurement provides a degree of confirmation for identity of the target molecule. However, a more complete indicator is to determine the sequence of the protein. Although the speed and sensitivity of sequencing by conventional gas-phase sequencers relying on Edman chemistry has been improving, tandem mass spectrometry (MS/MS) (55,56) is also well established for peptide sequencing, especially for N-terminally blocked peptides, materials containing unnatural amino acids, and post-translationally modified proteins. In combination with chemical and proteolytic digestion, protein sequencing by MS/MS can be a sensitive and powerful methodology (57,58).

The generation of multiply charged ions by ESI has allowed the application of tandem mass spectrometry to larger biomolecules because of the higher collision cross sections and increased electrostatic forces. Collisionally activated dissociation (CAD) efficiency of *singly* charged molecules with molecular weight less than 2000 Da is generally higher for high-energy (keV) collisions versus low-energy (<500 eV) collisions. However, application of low-energy CAD for *multiply* charged molecules can be used to generate sequence ions for larger polypeptides, as demonstrated by the CAD spectra of 66-kDa albumin proteins (59,60).

The application of tandem MS ESI for protein sequencing has been demonstrated by many laboratories. The most effective method has been the combination of chemical or proteolytic digestion with MS/MS. Tandem MS of peptides derived from trypsin digestion is very useful because basic residues at the C-terminus (Arg or Lys) and the amino-terminus (if unblocked) yield a 2+ precursor ion. Dissociation of this 2+ molecule produces an easily interpretable spectrum with singly charged fragments. Further combination with chromatographic separations provides a very powerful and rapid methodology for protein sequencing.

Identification of an unknown protein can rapidly proceed from the masses of its proteolytic peptides. Several computer-based searching routines have been described in which the measured peptide masses along with any sequence information obtained by, for example, MS/MS, are entered into a program and compared to theoretical proteolytic masses of proteins in a given sequence database (61–63).

A simple, effective method to generate fragment ions is to increase the energy of the ions as they traverse the ESI atmospheric pressure/vacuum interface. This technique, originally described by Loo, Udseth, and Smith (64) and referred to by several names (e.g., in-source CAD, nozzle-skimmer dissociation), can be used to provide sequence information for peptides and proteins. For example, Figure 5 shows ESI mass spectra obtained at two interface energies (where, in this case, ΔNS refers to the voltage difference between the nozzle and skimmer elements of the ESI interface) for the undecapeptide substance P (M_r 1348). At low energy, the 2+ and 3+ molecular ions are predominant with a few low-abundant fragment ions. [The nomenclature used to identify the peptide cleavage products has been described (58). In short, fragment ions with the charge residing on the N-terminus of the peptide are denoted as a_n, b_n, and c_n product ions and fragments with the charge residing on the C-terminus are identified as x_n, y_n, and z_n ions, where n designates the residue number (counting from the N- or C-terminus, respectively).] At higher energy, nearly complete sequence information is detailed in the ESI mass spectrum.

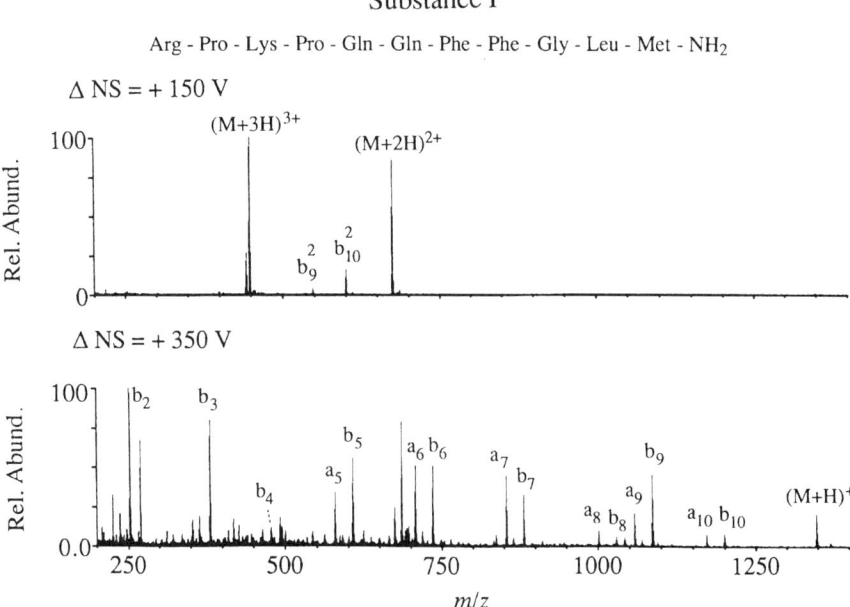

Figure 5. The ESI mass spectra of substance P [50/50 MeOH/H$_2$O, 5% acetic acid (v/v/v), 0.25 μL min^{-1}] with ΔNS (voltage difference between the nozzle and skimmer of the ESI interface of a triple quadrupole mass spectrometer) of +150 V (top) and +350 V (bottom).

Identifying the sites of post-translational modifications is an important application of MS and tandem mass spectrometry (65). Methods to detect phosphopeptides (66) or glycopeptides (67) using LC–ESI tandem mass spectrometry have been developed. Phosphopeptides containing phosphoserine, phosphothreonine, or phosphotyrosine yield fragment ions at m/z 63 (PO_2^-) and m/z 79 (PO_3^-). Glycosylated peptides are identified by a diagnostic m/z 204 (HexNAc$^+$) fragment ion. Monitoring these diagnostic ions during an LC-MS/MS experiment of a digested protein mixture can be a sensitive and time-saving procedure for identification of sites of modifications. Electrospray ionization with MS/MS can provide information on sequence, linkage, and branching. Strategies used for the ESI–MS analysis of the glycosylation patterns of glycoproteins have been treated by Carr (67–69). Carbohydrate mapping and fingerprinting for a glycoprotein can be obtained through a combination of methods for separation (liquid chromatography), digestion (glycosidases), mass analysis (ESI–MS), and further structural characterization (MS/MS).

This combination of selective proteolysis, chromatography, and ESI–MS/MS has been used successfully to locate the sites of phosphorylation of rhodopsin (70,71). Electrospray ionization and tandem mass spectrometry was used to determine the phosphorylation stoichiometry and map the complex phosphorylation sites of MAP kinase kinase (MAPKK), a key component of the MAP kinase cascade involved in signal transduction (72). Similarly, the post-translation modification of GAP-43, a 25-kDa protein involved in the regulation of neurotransmitter release and neuronal growth, was determined by LC/MS and MS/MS to be a Thr and Ser residue adjacent to a proline residue (73).

Thulin and Walsh have developed a method to identify the amino terminus of N-terminally blocked proteins (74). By proteolytically digesting a protein with trypsin, acetylating the N-terminus of each tryptic peptide, and comparing the LC/ESI–MS data to that obtained for a sample not further acetylated, the N-terminal heterogeneity of human filaggrin was determined. Three different but related blocked amino-terminal sequences were found (74). An early application of ESI–MS and MS/MS was the structural characterization of the N-terminal heterogeneity of rod transducin, a heterotrimeric GTP-binding protein. Four types of N-terminal fatty acylation were found (75).

Deamidation of asparagine and glutamine is a potential source of microheterogeneity for proteins. The combination of ESI–MS and CAD fragmentation induced in the ESI interface was used to pinpoint the sites of deamidation of a sample of recombinant hirudin to specific Asn residues (76).

There has been considerable application to ESI–MS as a tool to investigate the mechanism of enzymes. Several research groups have used ESI–MS to detect covalent intermediates for a variety of enzymes. Covalent enzyme-inhibitor complexes for human leucocyte elastase (77) and β-lactamase (78) have been observed by ESI–MS. Furthermore, tandem mass spectrometry can then be used to locate the sites of enzyme modifications. The active sites of human glucocerebrosidease (79) and bacillus subtilis xylanase (80) were identified by using tandem mass spectrometry.

Peptide sequencing can be further enhanced by developing methods to minimize sample handling. The direct incorporation of such on-line reactors with LC/ESI–MS and tandem MS has enormous potential for rapid protein sequencing. For example, low-flow reactors with on-line ESI–MS monitoring of peptides digested with carboxypeptidase P for C-terminal sequence information have been described (81). Davis et al. (82) used an immobilized trypsin column for the generation of tryptic peptides from cytochrome c and hemoglobin with subsequent ESI–MS analyses. By minimizing sample handling, subpicomole sensitivity for peptide sequencing by tandem mass spectrometry can be obtained. Microcapillary HPLC–ESI–MS/MS methods developed by Hunt and co-workers have allowed them to sequence peptide antigens associated with class I and class II MHC molecules from less than 100 fmole of material (83).

An obvious method to reduce sample handling for protein sequencing is tandem mass spectrometry of the intact protein (59,60,84–87). Multiply charged ions for the intact protein can be subjected to CAD to yield sequence-informative product ions. Sequence information for proteins as large as 66 kDa serum

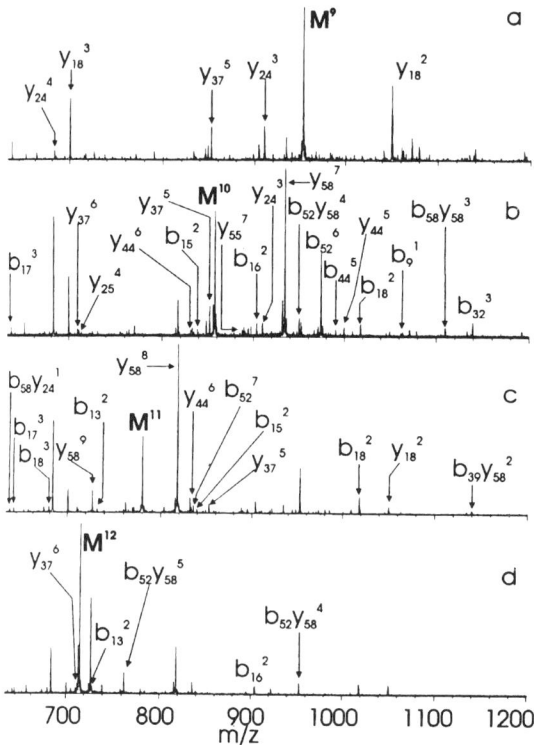

Figure 6. The IRMPD Fourier transform mass spectra of ubiquitin molecular ions: (*a*) 9+; (*b*) 10+; (*c*) 11+; (*d*) 12+ (88). The protein was dissolved in 80/20 MeOH/H_2O with 2% acetic acid (v/v/v). Reproduced with permission from D. P. Little et al., *Anal. Chem.*, **1994**, *66*, 2809–2815. © 1994 American Chemical Society.

albumin proteins from a variety of mammalian species have been obtained by direct MS/MS (59,60). The CAD mass spectra of multiply charged proteins have been compared for eight variants of the β-chain human hemoglobin polypeptide, whose modification covers various regions of the molecule (38). The application of high-resolution MS with FTMS analysis greatly simplifies the enormous data interpretation challenge because product-ion charge states can be directly identified by measuring the isotopic peak spacings.

For example, multiple charged product ions from MS/MS experiments for the protein ubiquitin (8565 Da) can be obtained (Fig. 6). McLafferty's group has demonstrated the use of infrared multiphoton dissociation (IRMPD) for generating protein fragmentation in a Fourier transform mass spectrometer (88). The results are similar to previously published spectra (85) obtained with a triple quadrupole instrument. Dissociation of the polypeptide is localized to regions around proline residues. The unusually abundant products from dissociation of the amide bond to a proline residue from multiply charged parent ions has been noted (85). Separate IRMPD experiments of the 7+ to 12+ molecular ions results in fragmentation primarily from the initial 20 residues (from the NH_2-terminal). However, an additional stage of mass spectrometry (MS^3) with IRMPD of product ions can generate additional sequence ions not produced in the MS^2 experiment (Fig. 7).

To date, dissociation of large peptides and proteins does not yield the complete sequence. Also, the location of the fragmentation events is not predictable and is highly variable [although some common themes have been found, such as dissociation of bonds N-terminally located to a proline residue (38,85,87)]. In principle, instruments with MS^n capabilities such as FTMS and ion-trap MS should be able to obtain greater amounts of sequence information through repetitive MS/MS stages.

III. STUDYING PROTEIN CONFORMATION BY ESI–MS

Knowledge of a protein's three-dimensional structure is important for understanding and predicting its function in biological systems. A great deal of interest was piqued by the discovery that higher-order protein structure affects the relative amounts of multiple charging displayed in an ESI mass spectrum (84,89–92). The higher charging almost always observed in denatured proteins (whether denatured by reduction of disulfide bonds or by addition of denaturants such as organic solvents) has been rationalized by the concept that charge sites are "more accessible." What "more accessible" really means is still open to debate; hydrophilic residues are usually at the surface of native proteins, although a few may be tied up in salt bridges and hydrogen bonds. Perhaps the charges can space themselves farther apart or there are other interactions which can make higher charging more favorable in denatured proteins. Proteins have been denatured for ESI work by changing the solution pH or by addition of organic solvents (90–92), increased temperature (23,93,94),

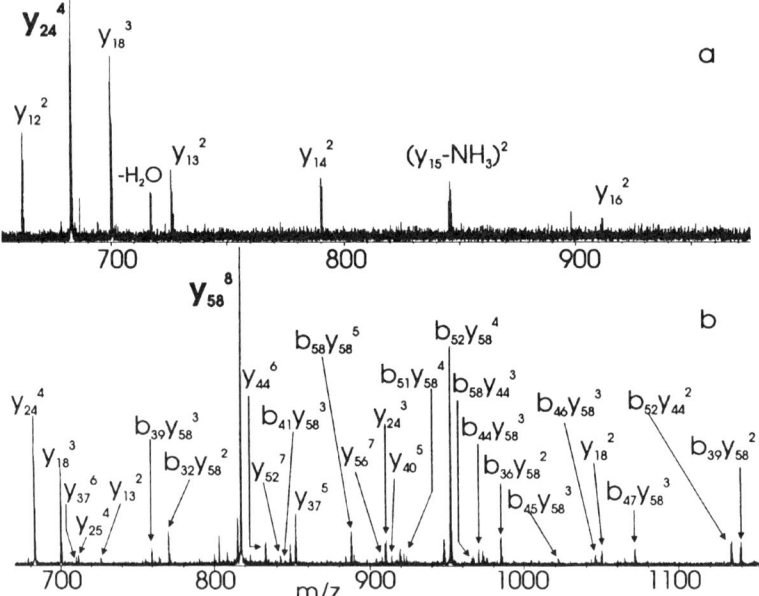

Figure 7. The MS^3 of ubiquitin of (a) $y_{24}{}^{4+}$ and (b) $y_{58}{}^{8+}$, both generated from the 11+ ion. Spectra were acquired with a Fourier transform mass spectrometer (88) (see Fig. 6). Reproduced with permission from D. P. Little et al., *Anal. Chem.*, **1994**, *66*, 2809–2815. © 1994 American Chemical Society.

or addition of detergent (46,95). While the effects of conformation on charge distributions are interesting, numerous cautions exist for relying solely on charge data to conclude if a conformational change has occurred. There are many examples where significant changes in conformation yield only subtle changes in charge distribution, easily attributable to other parameters. Moreover, mass spectrometry cannot yet provide details as to the nature of the conformational change. On the other hand, if one uses mass spectrometry for that which it was designed–to measure masses–exciting structural studies can be performed.

The hydrogen/deuterium (H/D) exchange kinetics of a protein are sensitive to its structure in solution. Hydrogen/deuterium exchange in solution is readily probed by mass spectrometry (92,96) and has found application in many problems. The H/D exchange rates for amide protons are influenced by the secondary and tertiary structure of a polypeptide. Mass spectrometry can provide important information, similar to NMR experiments, regarding protein conformation by monitoring the degree of deuterium exchange as a function of time. Internal hydrogens are less likely to be exposed to the deuterated environment; therefore no exchange will occur or exchange will occur at a much slower rate than for the more labile hydrogens on the surface of the structure.

Methods using ESI–MS have demonstrated the potential to provide information regarding a protein's higher-order structure. For example, Radford and co-workers have used ESI–MS H/D exchange experiments to study the folding rates of hen egg-white lysozyme (97). Chaperonin-assisted folding of proteins is another noteworthy example (98). Chaperonin proteins are thought to assist proteins fold to their native-like states. By starting with a deuterated form of α-lactalbumin complexed to chaperonin GroEL in solution, hydrogen exchange is initiated by introducing H_2O. The dissociated gas-phase complex is monitored by mass spectrometry. In this fashion, the hydrogen exchange rate of α-lactalbumin complexed to GroEL can be measured as a function of incubation time in H_2O. This method was shown to provide new information about the conformational properties of the bound ligand. Studies by Anderegg have shown the application of ESI–MS with H/D exchange for determining the relative amounts of α-helices and β-sheets in peptides (99,100). Tandem mass spectrometry of H/D-exchanged melittin, a 26-residue peptide, over a time-course study indicated the sites of deuterium incorporation and were in agreement with NMR secondary structure studies (101).

Gas-phase H/D exchange and ion-molecule reactions have also been explored in attempts to understand gas-phase ion structure (33,102,103). These experiments have established that disulfide-linked proteins initially react faster than their disulfide-reduced counterparts, but disulfide-reduced proteins exchange to a larger extent.

Scattering cross-section measurements of electrosprayed protein ions also provide information regarding the conformation of gas-phase ions. By using a triple-quadrupole mass spectrometer, Covey and Douglas (104) measured the loss of axial energy of an ion as it passes through a gas-filled collision cell to determine the collision cross section for a number of multiply charged ions for proteins ranging from motilin (2.7 kDa) to bovine serum albumin (66 kDa). Jarrold and co-workers (105) used ion-mobility measurements to resolve and similarly measure collision cross sections for conformers of cytochrome c (12 kDa). In both studies, cross sections were shown to increase with increasing charge, consistent for a more unfolded gas-phase molecule. For cytochrome c, cross sections ranged from 2000 to 4000 $Å^2$, depending on the charge state (104,105). Spraying solutions under denaturing conditions (i.e., high organic solvent content) yielded ions with somewhat larger cross sections, again consistent for a more denatured conformation. Hydrogen/deuterium exchange and cross-section measurement studies suggest that some elements of higher-order structure are preserved in the gas phase, a concept now embraced more strongly with the prevalence of noncovalent complex work.

IV. STUDYING MOLECULAR INTERACTIONS WITH ESI–MS

A. Noncovalent Interactions and Quaternary Structure

A more recent application of ESI–MS is the study of biochemical interactions. Many types of protein–ligand binding are examples of intermolecular noncovalent interactions, including antibody–antigen, protein–cofactor, receptor–

ligand, and enzyme–substrate pairings. Intermolecular noncovalent interactions are responsible for the aggregation of folded polypeptide chains into multimers, which determines a protein system's quaternary structure. "Protein–protein interactions are the language of cell biology" (15).

Electrospray ionization is a gentle ionization method, yielding no molecular fragmentation (unless induced in the ESI atmosphere–vacuum interface) and allowing weakly bound complexes to be detected. There have been several examples reported of ESI–MS detection of noncovalently bound complexes since the initial reports on receptor–ligand and enzyme–substrate complexes (106,107) and on the globin–heme interaction of myoglobin (108). Table 1 is a representative list of the many ESI–MS studies of noncovalent complexes reported in the literature. Evidence is building that the ESI–MS observations for these weakly bound systems reflect to some extent the nature of the interaction found in the solution phase. However, control experiments are necessary to rule out ubiquitous nonspecific interactions (i.e., nonspecific aggregation) (109). For example, for a variety of ligands of different binding strengths available for testing, the protein–ligand interaction found in the ESI–MS experiment should reflect the expected solution–phase measurements. This was found for the ribonuclease S-protein/S-peptide system (110), where the protein–peptide complex for a S-peptide analog showed a weaker attraction, as expected from solution-phase binding experiments.

Electrospray ionization mass spectrometry may potentially be used to determine the strength of these solution-phase interactions. The binding of various peptide inhibitors to Src SH2 (Src homology 2) domain protein (12.9 kDa), critical in the signal-transduction pathways of the tyrosine kinase growth-factor receptors, was examined by Loo and co-workers (111,112). From a mixture of several peptide inhibitors, where the total peptide concentration is much greater than the protein concentration (competitive binding conditions), the relative abundances of the Src SH2 protein–phosphopeptide complexes observed in the ESI mass spectrum are consistent with their measured solution-phase binding constants. Henion's group (113) recently demonstrated that data from ESI–MS experiments can be used to construct Scatchard plots for measuring the binding constants of vancomycin antibiotics with tripeptide ligands. Their gas-phase measurements were in reasonable agreement with previously reported solution-phase values. Griffey and co-workers determined solution-phase dissociation constants for oligonucleotide–serum albumin complexes by using ESI–MS titration data (114).

Multimeric proteins present an opportunity to use ESI–MS to study very large molecular-weight complexes. Many enzyme systems are composed of identical and nonidentical subunits that associate together to form the fully active species. Studying the nature of the interactions that maintain the quaternary structure of enzymes is essential for the understanding of cellular functions at the molecular level. Mass-spectrometry experiments with protein oligomers can yield information on both the stoichiometry and molecular nature of subunit interactions.

Smith and co-workers (109,115–118) and others (95) have presented ESI–MS results on tetrameric protein complexes such as avidin (64 kDa) and concanavalin A (102 kDa) (see Table 1). In these experiments, only monomer, dimer, and tetramer associations were observed. The absence of trimer and pentamer species suggests that the ESI–MS data reflects the specific solution-phase interactions known to occur.

Loo (119) reported results for alcohol dehydrogenase (ADH) proteins. Alcohol dehydrogenase is a zinc metalloenzyme responsible for the interconversion of acetaldehyde and ethanol. Yeast and mammalian ADHs are distinctly homologous, yet only 25% of all residues are conserved. Equine liver ADH is dimeric, with a monomer weight of 39.8 kDa, yet yeast alcohol dehydrogenase exists as the tetrameric complex for the active species. The ESI mass spectra of ADH in water/methanol (pH 2.4) solution show only ions for the monomer form with multiple charging to $\sim 50+$. Low pH is known to cause dissociation for a number of dehydrogenases. The ESI mass spectrum of horse ADH in water (pH ~ 6) shows multiply charged molecules for the monomer and dimer species of relatively low charge state (Fig. 8a). The relative abundance of the two forms is dependent on a number of experimental factors, including the amount of energy imparted to the ions in the ESI interface. At higher energies, CAD of the complex results in dissociation of the complex to the monomeric species. The mass spectrum of bakers yeast ADH shows only ions for the monomer and a tetramer species (Fig. 8b). No dimer yeast ADH molecules were evident, consistent with the specific solution-phase characteristics of the protein, that is, a monomer–tetramer equilibrium.

The expression of the genetic information found in nucleic acids is dependent upon the specificity of their interaction with proteins. Thus, the development of techniques to study and understand the molecular details of protein–nucleic acid interactions has broad interest. The observation of a noncovalent protein–DNA complex by ESI–MS was recently reported (120,121). Gene V single-stranded DNA binding protein (9.7 kDa) exists as a homodimer under physiological conditions; the protein dimer binds to DNA with a 1 : 1 stoichiometry for every 8 bases of DNA. Addition of a solution of a 12-mer DNA yields the exclusive formation of a protein dimer–DNA complex, consistent with NMR and gel-shift assays. Experiments with a larger 18-mer DNA sample produce the complex composed of a pair of protein dimers bound to 1 DNA molecule, again consistent with the known solution-binding stoichiometry (120,121).

B. Metal-Binding Proteins

The interaction with metal ions, essential to catalytic function and structural stability, is an integral component of many metalloenzymes. Techniques used to study the interaction between metal ions and biological materials include absorption spectroscopy, circular dichroism, and electron paramagnetic and nuclear magnetic resonance (NMR) spectroscopy. A rapidly emerging method

TABLE 1. Noncovalent Peptide–Protein Complexes Observed by ESI–MS

Peptide–Protein Complexes	Ligand Interaction	Comments and References
Peptides		
Various peptides	Multimer formation	(135,136)
Antisense peptide	Dimerization	(137)
Leucine zipper peptide	Dimer and trimer formation	(138,139)
Ribonuclease S-protein	S-peptide	(110,140)
Src SH2 domain	Phosphopeptide inhibitors	(111,112,141)
Albumin	Growth-hormone releasing factor	(142)
Polypeptide–Metal ions		
Amino acids and glutathione	Alkali metals, Hg	(143)
Angiotensin peptides	Zn	(144,145)
Rubredoxin	Fe	(146)
	Zn	(147)
	Ga	(148)
Ferredoxin	Fe	(146,149,150)
Desulforedoxin	Zn, Fe	(151)
Metallothionein	Transition metals	(122,152)
Histidine-rich glycoprotein	Cu	(153)
Estrogen receptor DNA-binding domain	Cu, Zn	Zinc finger protein (154–156)
Nucleocapsid protein (NCp7)	Zn	Zinc finger protein (125–127)
Prenisin	Zn	(126)
Ubiquitin	Cu	(157, 158)
Cytochrome c oxidase subunit	Cu	(159)
Calmodulin	Ca	EF-hand protein (123,124)
Parvalbumin	Ca	EF-hand protein (123,124)
Stromelysin catalytic domain	Zn, Ca	Matrix metalloproteinase (123, 160)
Matrilysin	Zn, Ca plus drug inhibitor	Matrix metalloproteinase (161)
Glucocorticoid receptor DNA binding domain	Zn, Cd	Zinc finger protein (128)
SPARC peptide	Cu	Extracellular matrix-binding protein (162)
Lysozyme	Cu, Zn	(163)
Carbonic anhydrase	Zn with inhibitors	(134)
Protein–Small Molecule		
Peptide–antibiotics	Vancomycin, ristocetin	(113, 164)
Myoglobin	Heme	(95,108,146,165–170)

TABLE 1. Continued

Peptide–Protein Complexes	Ligand Interaction	Comments and References
Hemoglobin	Heme	(95,106,167)
FKBP	FK506, rapamycin	Receptor ligand (106,142,169,171–173)
Albumin	FK506/FK520	(174)
Lysozyme	N-acetylglucose hexasaccharide	Enzyme substrate/enzyme product (107,169,172)
Avidin/streptavidin tetramer	Biotin	(115,116,175)
Methionine synthase	Cobalamine	(176)
Catalytic antibody (single chain)	Hapten	(177)
HIV-1 protease dimer	Inhibitor	(178)
Elastase	Peptidic substrates/ products	(179)
Antithrombin III	Heparin fragment	Glycoprotein–pentasaccharide complex (180)

Quaternary Structure Complexes

Gene V protein	Dimer (19.5 kDa)	(120, 121)
HIV-1 protease	Dimer (21.5 kDa)	(178)
4-Oxalocrotonate tautomerase	Hexamer (40.9 kDa)	(181)
gp45	Dimer (49.7 kDa)	(182)
Streptavidin	Tetramer (52 kDa)	(116,118,175)
Transthyretin	Tetramer (55 kDa)	(183)
Avidin	Tetramer (64 kDa)	(115)
Hemoglobin	Dimer/tetramer (64.5 kDa)	(95,115)
Alocohol dehydrogenase (ADH), horse	Dimer (80 kDa)	(119)
Concanavalin A	Tetramer (102 kDa)	(95,115,117)
Soybean agglutinin	Tetramer (116 kDa)	(184)
Alcohol dehydrogenase (ADH), yeast	Tetramer (148 kDa)	(119)
Pyruvate kinase	Tetramer (261 kDa)	(119)
Catalase HPII	Tetramer (339 kDa)	(185)

Nucleic Acid Complexes

Ribonuclease A	CMP	(186,187)
Aldose reductase	NADP$^+$	(188)
Ras protein	GDP	(189,190)
Adenylate kinase	AMP	(165)
bis PNA	DNA	Peptide nucleic acids (191)
Albumin	DNA	(114)
Gene V protein	DNA	Protein dimer (120,121)

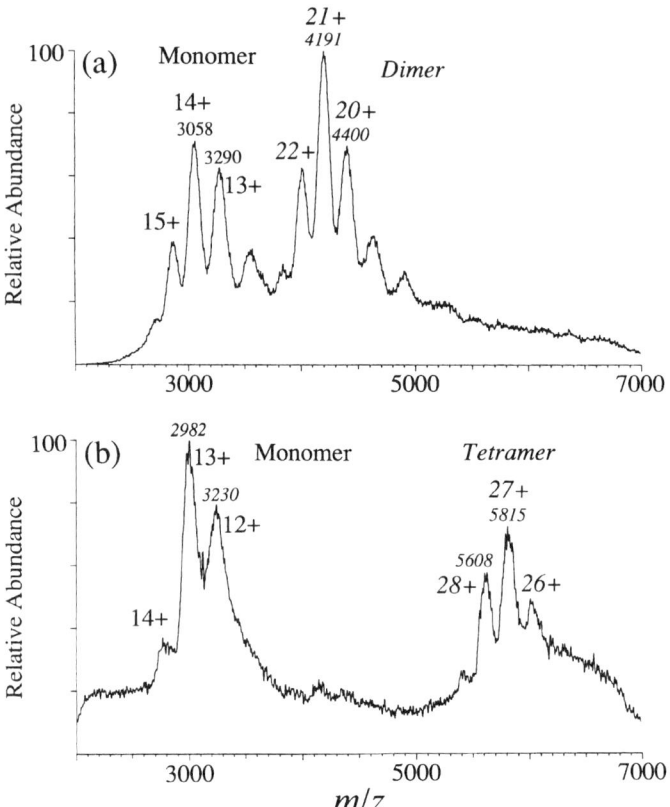

Figure 8. The ESI mass spectra of alcohol dehydrogenase from (*a*) horse liver (21.1 pmol μL^{-1} in water) and (*b*) bakers yeast (12.5 pmol μL^{-1} in 5 mM ammonium acetate) obtained on a forward-geometry magnetic-sector mass spectrometer (119). Reproduced with permission from J. A. Loo, *J. Mass Spectrom.*, **1995**, *30*, 181. © 1995 John Wiley & Sons, Ltd.

for investigating interactions between metal ions and biological molecules is mass spectrometry.

The potential of ESI–MS to determine peptide–metal ion stoichiometry in a rapid and sensitive fashion is promising. Many ESI–MS examples are provided in Table 1. Fenselau and co-workers (122) developed an ESI–MS strategy that provides analysis of metal ions in native and reconstituted metallothionein proteins. Their results suggest that ESI mass spectra reflect the *specific* association of the metal(s) to the peptide and can clearly be differentiated from nonspecific interactions due to clustering or adduct formation. Hu et al. (123,124) described the application of ESI–MS to determine the calcium-binding stoichiometry of Ca^{2+}-binding EF-hand proteins. Bovine calmodulin, bovine α-lactalbumin, and rabbit parvalbumin were found to bind specifically to 4, 1, and 2 Ca^{2+} ions, respectively, in agreement with previously reported results obtained by other physical methods.

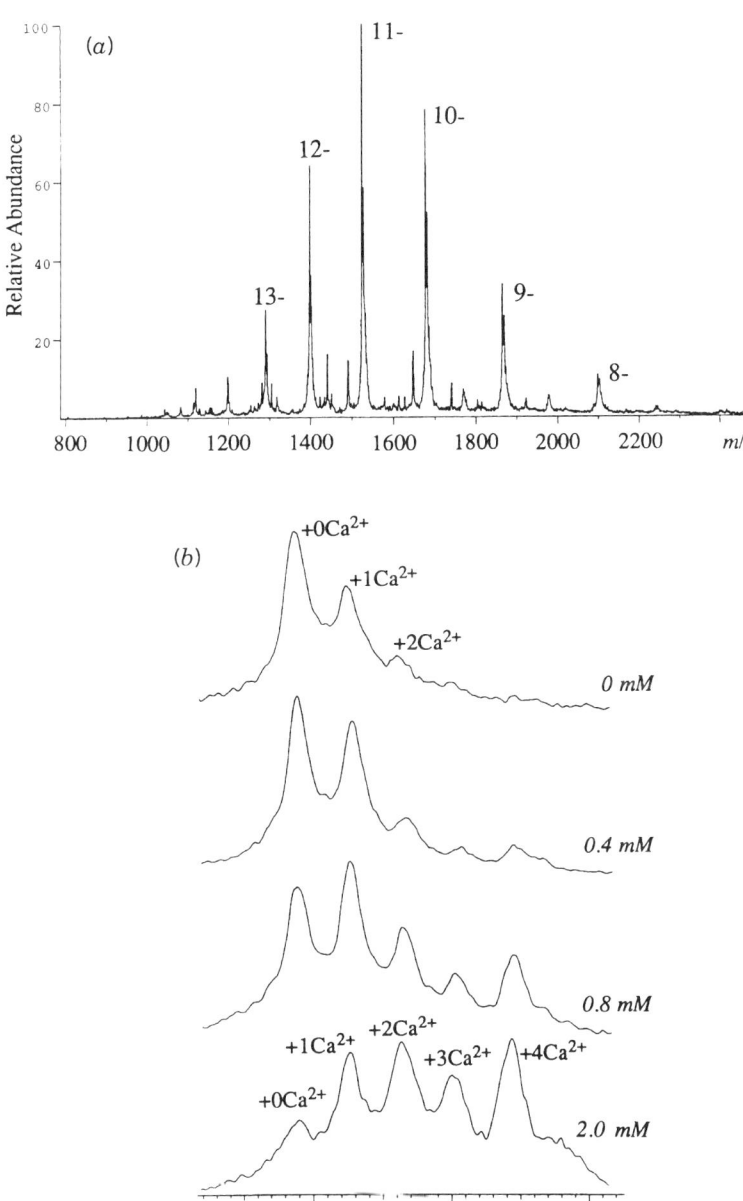

Figure 9. Titration of bovine calmodulin (M_r 16792) with calcium ion. (*a*) The full-scan negative-ion ESI–MS spectrum of calmodulin without added calcium. (*b*) The mass-converted deconvolution spectra with 0, 0.4, 0.8, and 2 mM calcium acetate. The calmodulin concentration used was ca. 60 μM. The solvent used was 15% methanol (v/v), pH 7. Spectra were acquired on a magnetic-sector mass spectrometer (124). Reproduced with permission from P. Hu et al., *J. Mass Spectrom.*, **1995**, *30*, 1080. © 1995 John Wiley & Sons, Ltd.

For calmodulin (CaM), a Ca^{2+} titration study was carried out to monitor the relative population changes of the coexisting species carrying various number of Ca^{2+} ions (124). The abundance of $CaM \cdot Ca_4$ increased and $CaM \cdot Ca_0$ decreased steadily upon increasing the calcium concentration, while the abundances of $CaM \cdot Ca_2$ and $CaM \cdot Ca_3$ lagged behind (Fig. 9). It was also observed that the $CaM \cdot Ca_4$ species was detected at $[Ca^{2+}]/[CaM]$ ratios less than 4. The observations suggest that the process of loading the last two Ca^{2+} ions are positively cooperative, as indicated by the low abundance of $CaM \cdot Ca_3$ (Fig. 9b), and that the two halves of the protein molecule interact so that the Ca^{2+} affinity of the two low-affinity sites is elevated to a level similar to the high-affinity sites upon Ca^{2+} occupation of the two high-affinity sites (124).

Zinc finger proteins contain Cys and His ligands that are believed to coordinate zinc and to participate in protein–nucleic acid interactions. Many transcription factors include zinc finger structures that appear to be suited to DNA recognition. A few zinc finger proteins have been studied using ESI–MS, as shown in Table 1. Nucleocapsid protein NCp7 contains two zinc fingers that are involved in encapsulation of genomic RNA during HIV viral assembly. Surovoy et al. (125,126) and others (127) have demonstrated that ESI–MS can determine the zinc stoichiometry for NCp7. Witkowska et al. (128) have used ESI–MS to study the zinc and cadmium binding characteristics of the DNA binding domain of the glucocorticoid receptor, which contains two zinc fingers, each with four cysteine residues coordinated to a zinc atom. The spectra in Figure 10 show the results for the glucocorticoid receptor protein-specific binding to a maximum of two zinc and two cadmium atoms.

C. Combinatorial Libraries and Ligand Identification

Traditionally, novel compounds for drug discovery in the pharmaceutical industry have been identified in receptor-based assays, screening large numbers of compounds. Using combinatorial chemistry to synthesize simultaneously greater numbers of diverse compounds for screening is becoming very popular (129). Combinatorial chemistry, by greatly increasing the molecular diversity available, has the potential to broaden the scope of the random collections of molecular structures being surveyed for biological activity. However, analytical characterization of these complex chemical libraries and identification of the active compounds from these complex mixtures are challenging tasks. Chemical libraries exceeding 10^6 unique components are commonly synthesized, with improvements being made to increase the efficiency for producing libraries several orders of magnitude larger. Approaches involving mass-spectrometric

Figure 10. The ESI mass spectra of GR DBD (glucocorticoid receptor DNA binding domain) protein containing Zn and Cd ions electrosprayed from a neutral pH aqueous solution. The corresponding spectra transformed to the mass domain are also shown (128). Reproduced with permission from H. E. Witkowska et al., *J. Am. Chem. Soc.*, **1995**, *117*, 3322. © 1995 American Chemical Society.

IV. STUDYING MOLECULAR INTERACTIONS WITH ESI–MS 409

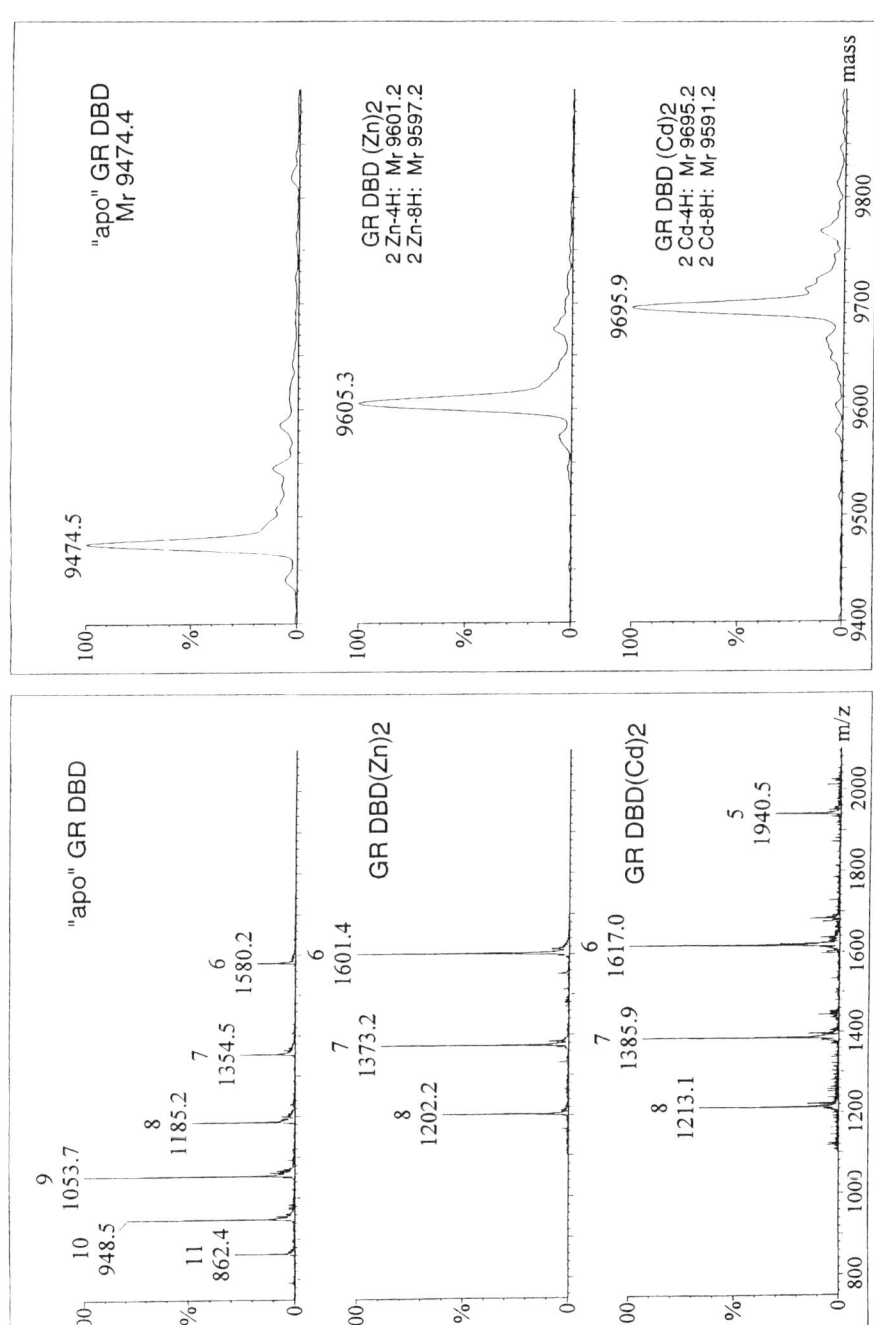

identification have been developed for the affinity selection and identification of novel protein–ligand interactions.

The analytical characterization of these complex mixtures is a challenging task. Electrospray ionization MS is a useful method to identify potential problems in the synthesis of peptide libraries (130), and in combination with MS/MS, provides a rapid and reliable method to determine the composition (131). For relatively small libraries (less than 500 unique components), the resulting ESI mass spectrum shows a distribution of peaks across the m/z scale, and in many cases, each individual component can be resolved. It can be beneficial to examine both a positive- and negative-ion spectrum to obtain a more realistic picture of the components. Some molecules have optimum ESI–MS sensitivities with one polarity over the other, depending on the functional groups present. For larger libraries, where resolution of each individual component is prohibitive, the shape and position of the unresolved distribution provides useful information regarding the success of the synthesis. One can compare a calculated theoretical mass distribution with the measured spectrum to gain insight. For example, a low-resolution spectrum can be used to validate a peptide library containing a chemical modification (e.g., phosphorylation) (130).

Ultimately, the identification of the active compounds from these complex mixtures is the final goal, but it is a difficult task. Affinity-column methods have been used to isolate selectively the active materials, which are then eluted off the column and identified by mass spectrometry. For example, an immobilized-Src SH2 domain-protein column coupled with HPLC–ESI mass spectrometry has been described by Kassel and co-workers to identify high-affinity binding phosphopeptides (132).

Affinity-capillary electrophoresis/MS (ACE/MS) is being developed as a method to identify candidate peptides from combinatorial libraries (133). The receptor is present in the capillary as part of the electrophoresis buffer. Ligands that interact tightly with the receptor will be retained as the unbound ligands travel through the column. Thus, the binding ligands will have slower electrophoretic mobilities relative to mobilities in the absence of the receptor. On-line ESI–MS is then used to identify the interacting ligand.

A method that involves the detection of noncovalent complexes and subsequent tandem mass spectrometry to identify high binding affinity ligands was described by Smith and co-workers (134). Complexes formed between a mixture of benzenesulfonamide-based inhibitors and zinc-bound carbonic anhydrase (29 kDa) were detected by ESI with FTMS detection. The relative abundances of each protein–inhibitor complex mirrored the trend in solution phase-binding constants. Further confirmation and identification of the inhibitors were accomplished by subsequent MS/MS experiments. The mass and relative abundance of each released inhibitor provided information regarding the identity and relative binding constant. A third stage of MS (MS^3) of the released inhibitor ion can be used to obtain structural information.

All of these types of advanced experiments take advantage of the strengths of

ESI: the ability to interface with chromatographic and electrophoretic separation methods, the gentleness of the ionization method to detect noncovalently bound complexes, and the ease with which tandem mass spectrometry can be used in conjunction with ESI.

V. CONCLUSIONS

In 1953, Neurath wrote that "there is no doubt that the time is drawing near when few problems will remain outside the bounds of experimental approach; and in this happy hunting ground, the cooperation of biochemists, chemists, biophysicists and physicists, biologists and many others is a prime necessity" (as described in ref. 15). It is clear that the development of ESI mass spectrometry for protein studies resulted from the needs and collaborations of many from the biochemistry, biology, chemistry, and physics communities. This continued cooperation between the various scientific disciplines and the mass spectrometry community will further advance the applicability of methods such as ESI–MS.

ACKNOWLEDGMENTS

We would not have been able to grow along with this maturing field of biological mass spectrometry without the people with whom we have interacted during our careers, in particular, Fred W. McLafferty (Cornell University), Richard D. Smith and the group at Pacific Northwest Laboratory, and Philip C. Andrews (University of Michigan).

REFERENCES

1. *The Proteins*; Neurath, H., Ed.; Academic Press: New York, 1963.
2. Biemann, K. *Protein Sci.* **1995**, *4*, 1920–1927.
3. Karas, M.; Bahr, U.; Ingendoh, A.; Hillenkamp, F. *Angew. Chem. Int. Ed. Engl.* **1989**, *28*, 760–761.
4. Hillenkamp, F.; Karas, M.; Beavis, R. C.; Chait, B. T. *Anal. Chem.* **1991**, *63*, 1193A–1203A.
5. Meng, C. K.; Mann, M.; Fenn, J. B. Z. *Phys. D–Atoms, Molecules and Clusters* **1988**, *10*, 361–368.
6. Fenn, J. B.; Mann, M.; Meng, C. K.; Wong, S. F.; Whitehouse, C. M. *Mass Spectrom. Rev.* **1990**, *9*, 37–70.
7. Fenn, J. B.; Mann, M.; Meng, C. K.; Wong, S. F.; Whitehouse, C. M. *Science* **1989**, *246*, 64–71.
8. Dole, M.; Mack, L. L.; Hines, R. L.; Mobley, R. C.; Ferguson, L. D.; Alice, M. B. *J. Chem. Phys.* **1968**, *49*, 2240–2249.

9. Dole, M.; Hines, R. L.; Mack, L. L.; Mobley, R. C.; Ferguson, L. D.; Alice, M. B. *Macromolecules* **1968**, *1*, 96–97.
10. Dole, M. *My Life in the Golden Age of America*; Vantage Press: New York, 1989.
11. Whitehouse, C. M.; Dreyer, R. N.; Yamashita, M.; Fenn, J. B. *Anal. Chem.* **1985**, *57*, 675–679.
12. Aleksandrov, M. L.; Gall', L. N.; Krasnov, N. V.; Nikolaev, V. I.; Shkurov, V. A. *Zh. Anal. Khim.* **1985**, *40*, 1570–1580.
13. Kebarle, P.; Tang, L. *Anal. Chem.* **1993**, *65*, A972–A986.
14. Fenn, J. B. *J. Am. Soc. Mass Spectrom.* **1993**, *4*, 524–535.
15. Neurath, H. *Protein Sci.* **1995**, *4*, 1939–1943.
16. Covey, T. R.; Bonner, R. F.; Shushan, B. I.; Henion, J. *Rapid Commun. Mass Spectrom.* **1988**, *2*, 249–256.
17. Smith, R. D.; Loo, J. A.; Edmonds, C. G.; Barinaga, C. J.; Udseth, H. R. *Anal. Chem.* **1990**, *62*, 882–899.
18. Smith, R. D.; Loo, J. A.; Ogorzalek Loo, R. R.; Busman, M.; Udseth, H. R. *Mass Spectrom. Rev.* **1991**, *10*, 359–451.
19. Loo, J. A.; Ogorzalek Loo, R. R.; Light, K. J.; Edmonds, C. G.; Smith, R. D. *Anal. Chem.* **1992**, *64*, 81–88.
20. Kelly, M. A.; Vestling, M. M.; Fenselau, C. C.; Smith, P. B. *Org. Mass Spectrom.* **1992**, *27*, 1143–1147.
21. Guevremont, R.; Siu, K. W. M.; LeBlanc, J. C. Y.; Berman, S. S. *J. Am. Soc. Mass Spectrom.* **1991**, *3*, 216–224.
22. Le Blanc, J. C. Y.; Guevremont, R.; Siu, K. W. M. *Int. J. Mass Spectrom. Ion Proc.* **1993**, *125*, 145–153.
23. Le Blanc, J. C. Y.; Beuchemin, D.; Siu, K. W. M.; Guevremont, R.; Berman, S. S. *Org. Mass Spectrom.* **1991**, *26*, 831–839.
24. Siu, K. W. M.; Guevremont, R.; Le Blanc, J. C. Y.; O'Brien, R. T.; Berman, S. S. *Org. Mass Spectrom.* **1993**, *28*, 579–584.
25. Ogorzalek Loo, R. R.; Andrews, P. C. In *Proceedings of the 42nd ASMS Conference on Mass Spectrometry and Allied Topics*, Chicago, IL, 1994; American Society for Mass Spectrometry: Santa Fe, NM, p. 410.
26. Mirza, U. A.; Chait, B. T. *Anal. Chem.* **1994**, *66*, 2898–2904.
27. Mann, M.; Meng, C. K.; Fenn, J. B. *Anal. Chem.* **1989**, *61*, 1702–1708.
28. Ferrige, A. G.; Seddon, M. J.; Green, B. N.; Jarvis, S. A.; Skilliing, J. *Rapid Commun. Mass Spectrom.* **1992**, *6*, 707–711.
29. Cody, R. B.; Tamura, J.; Musselman, B. D. *Anal. Chem.* **1992**, *64*, 1561–1570.
30. Dobberstein, P.; Schroeder, E. *Rapid Commun. Mass Spectrom.* **1993**, *7*, 861–864.
31. Beu, S. C.; Senko, M. W.; Quinn, J. P.; Wampler, F. M., III; McLafferty, F. W. *J. Am. Soc. Mass Spectrom.* **1993**, *4*, 557–565.
32. Winger, B. E.; Hofstadler, S. A.; Bruce, J. E.; Udseth, H. R.; Smith, R. D. *J. Am. Soc. Mass Spectrom.* **1993**, *4*, 566–577.
33. (a) Ogorzalek Loo, R. R.; Loo, J. A.; Udseth, H. R.; Fulton, J. L.; Smith, R. D. *Rapid Commun. Mass Spectrom.* **1992**, *6*, 159–165. (b) Stephenson, J. L., Jr.; McLuckey, S. A. *Anal. Chem.* **1996**, *68*, 4026–4032.

34. Smith, R. D.; Loo, J. A.; Barinaga, C. J.; Edmonds, C. G.; Udseth, H. R. *J. Am. Soc. Mass Spectrom.* **1990**, *1*, 53–65.
35. Feng, R.; Konishi, Y. *Anal. Chem.* **1992**, *64*, 2090–2095.
36. Shackleton, C. H. L.; Falick, A. M.; Green, B. N.; Witkowska, H. E. *J. Chromatogr.* **1991**, *562*, 175–190.
37. Witkowska, H. E.; Bitsch, F.; Shackleton, C. H. L. *Hemoglobin* **1993**, *17*, 227–242.
38. Light-Wahl, K. J.; Loo, J. A.; Edmonds, C. G.; Smith, R. D.; Witkowska, H. E.; Shackleton, C. H. L.; Wu, C. S. C. *Biol. Mass Spectrom.* **1993**, *22*, 112–120.
39. Chowdhury, S. K.; Vavra, K. J.; Brake, P. G.; Banks, T.; Falvo, J.; Wahl, R.; Eshraghi, J.; Gonyea, G.; Chait, B. T.; Vestal, C. H. *Rapid Commun. Mass Spectrom.* **1995**, *9*, 563–569.
40. Heath, T. G.; Thanabal, V.; Ye, Q.-Z. *Biotechnol. Tech.* **1993**, *7*, 367–372.
41. (a) Hofstadler, S. A.; Swanek, F. D.; Gale, D. C.; Ewing, A. G.; Smith, R. D. *Anal. Chem.* **1995**, *67*, 1477–1480. (b) Valaskovic, G. A.; Kelleher, N. L.; McLafferty, F. W. *Science*, **1996**, *273*, 1199–1202.
42. Dass, C.; Kusmierz, J. J.; Desiderio, D. M.; Jarvis, S. A.; Green, B. N. *J. Am. Soc. Mass Spectrom.* **1991**, *2*, 149–156.
43. Yan, L.; Tseng, J.-L.; Fridland, G. H.; Desiderio, D. M. *J. Am. Soc. Mass Spectrom.* **1994**, *5*, 377–386.
44. Witkowska, H. E.; Shackleton, C. H. L., unpublished data.
45. Schindler, P. A.; Van Dorsselaer, A.; Falick, A. M. *Anal. Biochem.* **1993**, *213*, 256–263.
46. Ogorzalek Loo, R. R.; Dales, N.; Andrews, P. C. *Protein Sci.* **1994**, *3*, 1975–1983.
47. Ashton, D. S.; Beddell, C. R.; Cooper, D. J.; Craig, S. J.; Lines, A. C.; Oliver, R. W. A.; Smith, M. A. *Anal. Chem.* **1995**, *67*, 835–842.
48. Bennett, K. L.; Hick, L. A.; Truscott, R. J. W.; Sheil, M. M.; Smith, S. V. *J. Mass Spectrom.* **1995**, *30*, 769–771.
49. Verentchikov, A. N.; Ens, W.; Standing, K. G. *Anal. Chem.* **1994**, *66*, 126–133.
50. Sogaard, M.; Andersen, J. S.; Roepstorff, P.; Svensson, B. *Biotechnology* **1993**, *11*, 1162–1165.
51. Young, N. M.; Watson, D. C.; Yaguchi, M.; Adar, R.; Arango, R.; Rodriguezarango, E.; Sharon, N.; Blay, P. K. S.; Thibault, P. *J. Biol. Chem.* **1995**, *270*, 2563–2570.
52. Settineri, C. A.; Burlingame, A. L. *J. Chromatogr. Libr.* **1995**, *58*, 447–514.
53. Reinhold, V. N.; Reinhold, B. B.; Costello, C. E. *Anal. Chem.* **1995**, *67*, 1772–1784.
54. Nakanishi, T.; Shimizu, A.; Okamoto, N.; Ingendoh, A.; Kanai, M. *J. Am. Soc. Mass Spectrom.* **1995**, *6*, 854–859.
55. *Tandem Mass Spectrometry*; McLafferty, F. W., Ed.; Wiley-Interscience: New York, 1983.
56. Busch, K. L.; Glish, G. L.; McLuckey, S. A. *Mass Spectrometry/Mass Spectrometry. Techniques and Applications of Tandem Mass Spectrometry*; VCH: New York, 1988.
57. Biemann, K.; Scoble, H. A. *Science* **1987**, *237*, 992–998.
58. Biemann, K. *Biomed. Environ. Mass Spectrom.* **1988**, *16*, 99–111.
59. Loo, J. A.; Edmonds, C. G.; Smith, R. D. *Anal. Chem.* **1991**, *63*, 2488–2499.

60. Speir, J. P.; Senko, M. W.; Little, D. P.; Loo, J. A.; McLafferty, F. W. *J. Mass Spectrom.* **1995**, *30*, 39–42.
61. Henzel, W. J.; Billeci, T. M.; Stults, J. T.; Wong, S. C.; Grimley, C.; Watanabe, C. *Proc. Natl. Acad. Sci. USA* **1993**, *90*, 5011–5015.
62. Pappin, D. J. C.; Hojrup, P.; Bleasby, A. J. *Curr. Biol.* **1993**, *3*, 327–332.
63. Yates, J. R., II; Speicher, S.; Griffin, P. R.; Hunkapiller, T. *Anal. Biochem.* **1993**, *214*, 397–408.
64. Loo, J. A.; Udseth, H. R.; Smith, R. D. *Rapid Commun. Mass Spectrom.* **1988**, *2*, 207–210.
65. Rossomando, A. J.; Wu, J.; Michel, H.; Shabanowitz, J.; Hunt, D. F.; Weber, M. J.; Sturgill, T. W. *Proc. Natl. Acad. Sci. USA* **1992**, *89*, 5779–5783.
66. Huddleston, M. J.; Annan, R. S.; Bean, M. F.; Carr, S. A. *J. Am. Soc. Mass Spectrom.* **1993**, *4*, 710–717.
67. Carr, S. A.; Huddleston, M. J.; Bean, M. F. *Protein Sci.* **1993**, *2*, 183–196.
68. Carr, S. A.; Hemling, M. E.; Bean, M. F.; Roberts, G. D. *Anal. Chem.* **1991**, *63*, 2802–2824.
69. Huddleston, M. J.; Bean, M. F.; Carr, S. A. *Anal. Chem.* **1993**, *65*, 877–884.
70. Ohguro, H.; Palczewski, K.; Ericsson, L. H.; Walsh, K. A.; Johnson, R. S. *Biochemistry* **1993**, *32*, 5718–5724.
71. Papac, D. I.; Oatis, J. E., Jr.; Crouch, R. K.; Knapp, D. R. *Biochemistry* **1993**, *32*, 5930–5934.
72. Resing, K. A.; Mansour, S. J.; Hermann, A. S.; Johnson, R. S.; Candia, J. M.; Fukasawa, K.; Vande Woude, G. F.; Ahn, N. G. *Biochemistry* **1995**, *34*, 2610–2620.
73. Taniguchi, H.; Suzuki, M.; Manenti, S.; Titani, K. *J. Biol. Chem.* **1994**, *269*, 22481–22484.
74. Thulin, C. D.; Walsh, K. A. *Biochemistry* **1995**, *34*, 8687–8692.
75. Neubert, T. A.; Johnson, R. S.; Hurley, J. B.; Walsh, K. A. *J. Biol. Chem.* **1992**, *267*, 18274–18277.
76. Bischoff, R.; Lepage, P.; Jaquinod, M.; Cauet, G.; Ackerklein, M.; Clesse, D.; Laporte, M.; Bayol, A.; Vandorsselaer, A.; Roitsch, C. *Biochemistry* **1993**, *32*, 725–734.
77. Knight, W. B.; Swiderek, K. M.; Sakuma, T.; Calaycay, J.; Shively, J. E.; Lee, T. D.; Covey, T. R.; Shushan, B.; Green, B. G.; Chabin, R.; Shah, S.; Mumford, R.; Dickinson, T. A.; Griffin, P. R. *Biochemistry* **1993**, *32*, 2031–2035.
78. Aplin, R. T.; Robinson, C. V.; Schofield, C. J.; Waley, S. G. *J. Chem. Soc., Chem. Commun.* **1993**, 121–123.
79. Miao, S. C.; McCarter, J. D.; Grace, M. E.; Grabowski, G. A.; Aebersold, R.; Withers, S. G. *J. Biol. Chem.* **1994**, *269*, 10975–10978.
80. Miao, S. C.; Ziser, L.; Aebersold, R.; Withers, S. G. *Biochemistry* **1994**, *33*, 7027–7032.
81. Rosnack, K. J.; Stroh, J. G. *Rapid Commun. Mass Spectrom.* **1992**, *6*, 637–640.
82. Davis, M. T.; Lee, T. D.; Ronk, M.; Hefta, S. A. *Anal. Biochem.* **1995**, *224*, 235–244.
83. Hunt, D. F.; Henderson, R. A.; Shabanowitz, J.; Sakaguchi, K.; Michel, H.; Sevilir, N.; Cox, A. L.; Appella, E.; Engelhard, V. H. *Science* **1992**, *255*, 1261–1263.

84. Loo, J. A.; Edmonds, C. G.; Smith, R. D. *Science* **1990**, *248*, 201–204.
85. Loo, J. A.; Edmonds, C. G.; Smith, R. D. *Anal. Chem.* **1993**, *65*, 425–438.
86. McLafferty, F. W. *Acc. Chem. Res.* **1994**, *27*, 379–386.
87. Senko, M. W.; Beu, S. C.; McLafferty, F. W. *Anal. Chem.* **1994**, *66*, 415–417.
88. Little, D. P.; Speir, J. P.; Senko, M. W.; O'Connor, P. B.; McLafferty, F. W. *Anal. Chem.* **1994**, *66*, 2809–2815.
89. Loo, J. A.; Edmonds, C. G.; Udseth, H. R.; Smith, R. D. *Anal. Chem.* **1990**, *62*, 693–698.
90. Loo, J. A.; Ogorzalek Loo, R. R.; Udseth, H. R.; Edmonds, C. G.; Smith, R. D. *Rapid Commun. Mass Spectrom.* **1991**, *5*, 101–105.
91. Chowdhury, S. K.; Katta, V.; Chait, B. T. *J. Am. Chem. Soc.* **1990**, *112*, 9012–9013.
92. Katta, V.; Chait, B. T. *Rapid Commun. Mass Spectrom.* **1991**, *5*, 214–217.
93. Allen, M. H.; Vestal, M. L. *J. Am. Soc. Mass Spectrom.* **1992**, *3*, 18–26.
94. Mirza, U. A.; Cohen, S. L.; Chait, B. T. *Anal. Chem.* **1993**, *65*, 1–6.
95. Loo, J. A.; Ogorzalek Loo, R. R.; Andrews, P. C. *Org. Mass Spectrom.* **1993**, *28*, 1640–1649.
96. Katta, V.; Chait, B. T. *J. Am. Chem. Soc.* **1993**, *115*, 6317–6321.
97. Miranker, A.; Robinson, C. V.; Radford, S. E.; Aplin, R. T.; Dobson, C. M. *Science* **1993**, *262*, 896–900.
98. Robinson, C. V.; Gross, M.; Eyles, S. J.; Ewbank, J. J.; Mayhew, M.; Hartl, F. U.; Dobson, C. M.; Radford, S. E. *Nature* **1994**, *372*, 646–651.
99. Wagner, D. S.; Anderegg, R. J. *Anal. Chem.* **1994**, *66*, 706–711.
100. Wagner, D. S.; Melton, L. G.; Yan, Y. B.; Erickson, B. W.; Anderegg, R. J. *Protein Sci.* **1994**, *3*, 1305–1314.
101. Anderegg, R. J.; Wagner, D. S.; Stevenson, C. L.; Borchardt, R. T. *J. Am. Soc. Mass Spectrom.* **1994**, *5*, 425–433.
102. Winger, B. E.; Light-Wahl, K. J.; Rockwood, A. L.; Smith, R. D. *J. Am. Chem. Soc.* **1992**, *114*, 5897–5898.
103. Suckau, D.; Shi, Y.; Beu, S. C.; Senko, M. W.; Quinn, J. P.; Wampler, F. M.; McLafferty, F. W. *Proc. Natl. Acad. Sci. USA* **1993**, *90*, 790–793.
104. Covey, T.; Douglas, D. J. *J. Am. Soc. Mass Spectrom.* **1993**, *4*, 616–623.
105. Clemmer, D. E.; Hudgins, R. R.; Jarrold, M. F. *J. Am. Chem. Soc.* **1995**, *117*, 10141–10142.
106. Ganem, B.; Li, Y.-T.; Henion, J. D. *J. Am. Chem. Soc.* **1991**, *113*, 6294–6296.
107. Ganem, B.; Li, Y.-T.; Henion, J. D. *J. Am. Chem. Soc.* **1991**, *113*, 7818–7819.
108. Katta, V.; Chait, B. T. *J. Am. Chem. Soc.* **1991**, *113*, 8534–8535.
109. Smith, R. D.; Light-Wahl, K. J. *Biol. Mass Spectrom.* **1993**, *22*, 493–501.
110. Ogorzalek Loo, R. R.; Goodlett, D. R.; Smith, R. D.; Loo, J. A. *J. Am. Chem. Soc.* **1993**, *115*, 4391–4392.
111. Loo, J. A.; Hu, P.; Thanabal, V. In *Proceedings of the 43rd ASMS Conference on Mass Spectrometry and Allied Topics*, Atlanta, GA 1995; American Society for Mass Spectrometry: Santa Fe, NM, p. 35.
112. Loo, J. A. *Bioconj. Chem.* **1995**, *6*, 644–665.

113. Lim, H.-K.; Hsieh, Y. L.; Ganem, B.; Henion, J. *J. Mass Spectrom.* **1995**, *30*, 708–714.

114. Greig, M. J.; Gaus, H.; Cummins, L. L.; Sasmor, H.; Griffey, R. H. *J. Am. Chem. Soc.* **1995**, *117*, 10765–10766.

115. Light-Wahl, K. J.; Schwartz, B. L.; Smith, R. D. *J. Am. Chem. Soc.* **1994**, *116*, 5271–5278.

116. Schwartz, B. L.; Light-Wahl, K. J.; Smith, R. D. *J. Am. Soc. Mass Spectrom.* **1994**, *5*, 201–204.

117. Light-Wahl, K. J.; Winger, B. E.; Smith, R. D. *J. Am. Chem. Soc.* **1993**, *115*, 5869–5870.

118. Schwartz, B. L.; Bruce, J. E.; Anderson, G. A.; Hofstadler, S. A.; Rockwood, A. L.; Smith, R. D.; Chilkoti, A.; Stayton, P. S. *J. Am. Soc. Mass Spectrom.* **1995**, *6*, 459–465.

119. Loo, J. A. *J. Mass Spectrom.* **1995**, *30*, 180–183.

120. Cheng, X.; Harms, A. C.; Smith, R. D.; Morin, P. E.; Goudreau, P. N.; Terwilliger, T. C. In *Proceedings of the 43rd ASMS Conference on Mass Spectrometry and Allied Topics*, Atlanta, GA, 1995; American Society for Mass Spectrometry: Santa Fe, NM, p. 1326.

121. Cheng, X.; Hofstadler, S. A.; Bruce, J. E.; Harms, A. C.; Chen, R.; Terwilliger, T. C.; Goudreau, P. N.; Smith, R. D. In *Techniques in Protein Chemistry VII*; Marshak, D. R., Ed.; Academic Press: San Diego, CA, **1996**, pp. 13–22.

122. Yu, X. L.; Wojciechowski, M.; Fenselau, C. *Anal. Chem.* **1993**, *65*, 1355–1359.

123. Hu, P. F.; Ye, Q.-Z.; Loo, J. A. *Anal. Chem.* **1994**, *66*, 4190–4194.

124. Hu, P.; Loo, J. A. *J. Mass Spectrom.* **1995**, *30*, 1076–1082.

125. Surovoy, A.; Waidelich, D.; Jung, G. *FEBS Lett.* **1992**, *311*, 259–262.

126. Surovoy, A.; Waidelich, D.; Jung, G. In *Peptides 1992, Proceedings of the 22nd European Peptides Symposium*; Schneider, C. H.; Eberle, A. N. Eds.; ESCOM Science: Leiden, Netherlands, 1993, pp. 563–564.

127. (a) Fenselau, C.; Yu, X.; Bryant, D.; Bowers, M. A.; Sowder, R. C., II; Henderson, L. E. In *Mass Spectrometry for the Characterization of Microorganisms (ACS Symposium Series)*; Fenselau, C., Ed.; American Chemical Society: Washington, DC, 1994; Vol. 541, pp. 159–172. (b) Loo, J. A.; Holler, T. P.; Sanchez, J.; Gogliotti, R.; Maloney, L.; Reily, M. D. *J. Med. Chem.* **1996**, *39*, 4313–4320.

128. Witkowska, H. E.; Shackleton, C. H. L.; Dahlman-Wright, K.; Kim, J. Y.; Gustafsson, J.-A. *J. Am. Chem. Soc.* **1995**, *117*, 3319–3324.

129. Gallop, M. A.; Barrett, R. W.; Dower, W. J.; Fodor, S. P. A.; Gordon, E. M. *J. Med. Chem.* **1994**, *37*, 1233–1251.

130. Andrews, P. C.; Boyd, J.; Ogorzalek Loo, R.; Zhao, R.; Zhu, C.-Q.; Grant, K.; Williams, S. In *Techniques in Protein Chemistry V*; Crabb, J. W., Ed.; Academic Press: San Diego, CA, 1994, pp. 485–492.

131. Metzger, J. W.; Stenanovic, S.; Brunjes, J.; Wiesmuller, K.-H.; Jung, G. *Methods (San Diego)* **1994**, *6*, 425–431.

132. Kassel, D. B.; Consler, T. G.; Shalaby, M.; Sekhri, P.; Gordon, N.; Nadler, T. In *Techniques in Protein Chemistry VI*; Crabb, J. W., Ed.; Academic Press: San Diego, CA, 1995, pp. 39–46.

133. (a) Chu, Y.-H.; Kirby, D. P.; Karger, B. L. *J. Am. Chem. Soc.* **1995**, *117*, 5419–5420.
 (b) Chu, Y.-H.; Dunayevskiy, Y. M.; Kirby, D. P., Vouros, P.; Karger, B. L. *J. Am. Chem. Soc.* **1996**, *118*, 7827–7835.
134. Cheng, X.; Chen, R.; Bruce, J. E.; Schwartz, B. L.; Anderson, G. A.; Hofstadler, S. A.; Gale, D. C.; Smith, R. D.; Gao, J.; Sigal, G. B.; Mammen, M.; Whitesides, G. M. *J. Am. Chem. Soc.* **1995**, *117*, 8859–8860.
135. Busman, M.; Knapp, D. R.; Schey, K. L. *Rapid Commun. Mass Spectrom.* **1994**, *8*, 211–216.
136. Smith, R. D.; Light-Wahl, K. J.; Winger, B. E.; Loo, J. A. *Org. Mass Spectrom.* **1992**, *27*, 811–821.
137. Loo, J. A.; Holsworth, D. D.; Root-Bernstein, R. S. *Biol. Mass Spectrom.* **1994**, *23*, 6–12.
138. Li, Y.-T.; Hsieh, Y.-L.; Henion, J. D.; Senko, M. W.; McLafferty, F. W.; Ganem, B. *J. Am. Chem. Soc.* **1993**, *115*, 8409–8413.
139. Wendt, H.; Durr, E.; Thomas, R. M.; Przybylski, M.; Bosshard, H. R. *Protein Sci.* **1995**, *4*, 1563–1570.
140. Goodlett, D. R.; Ogorzalek Loo, R. R.; Loo, J. A.; Wahl, J. H.; Udseth, H. R.; Smith, R. D. *J. Am. Soc. Mass Spectrom.* **1994**, *5*, 614–622.
141. Anderegg, R. J.; Wagner, D. S. *J. Am. Chem. Soc.* **1995**, *117*, 1374–1377.
142. Baczynskyj, L.; Bronson, G. E.; Kubiak, T. M. *Rapid Commun. Mass Spectrom.* **1994**, *8*, 280–286.
143. Canty, A. J.; Colton, R.; D'Agostino, A.; Traeger, J. C. *Inorg. Chim. Acta* **1994**, *223*, 103–107.
144. Loo, J. A.; Hu, P.; Smith, R. D. *J. Am. Soc. Mass Spectrom.* **1994**, *5*, 959–965.
145. Sullards, M. C.; Adams, J. *J. Am. Soc. Mass Spectrom.* **1995**, *6*, 608–610.
146. Jaquinod, M.; Leize, E.; Potier, N.; Albrecht, A. M.; Shanzer, A.; Van Dorsselaer, A. *Tetrahedron Lett.* **1993**, *34*, 2771–2774.
147. Petillot, Y.; Forest, E.; Mathieu, I.; Meyer, J.; Moulis, J. M. *Biochem. J.* **1993**, *296*, 657–661.
148. Kazanis, S.; Pochapsky, T. C.; Barnhart, T. M.; Penner-Hahn, J. E.; Mizra, U. A.; Chait, B. T. *J. Am. Chem. Soc.* **1995**, *117*, 6625–6626.
149. Petillot, Y.; Golinelli, M.-P.; Forest, E.; Meyer, J. *Biochem. Biophys. Res. Commun.* **1995**, *210*, 686–694.
150. Petillot, Y.; Forest, E.; Meyer, J.; Moulis, J.-M. *Anal. Biochem.* **1995**, *228*, 56–63.
151. Czaja, C.; Litwiller, R.; Tomlinson, A. J.; Naylor, S.; Tavares, P.; LeGall, J.; Moura, J. J. G.; Moura, I.; Rusnak, F. *J. Biol. Chem.* **1995**, *270*, 20273–20277.
152. Pleasance, S.; Thibault, P.; Thompson, J. In *Proceedings of the 38th ASMS Conference on Mass Spectrometry and Allied Topics*, Tucson, AZ, 1990; American Society for Mass Spectrometry: East Lansing, MI, pp. 720–721.
153. Hutchens, T. W.; Nelson, R. W.; Allen, M. H.; Li, C. M.; Yip, T.-T. *Biol. Mass Spectrom.* **1992**, *21*, 151–159.
154. Hutchens, T. W.; Allen, M. H.; Li, C. M.; Yip, T.-T. *FEBS Lett.* **1992**, *309*, 170–174.
155. Allen, M. H.; Hutchens, T. W. *Rapid Commun. Mass Spectrom.* **1992**, *6*, 308–312.
156. Hutchens, T. W.; Allen, M. H. *Rapid Commun. Mass Spectrom.* **1992**, *6*, 469–473.

157. Senko, M. W.; Beu, S. C.; McLafferty, F. W. *J. Am. Soc. Mass Spectrom.* **1993**, *4*, 828–830.

158. Jiao, C. Q.; Freiser, B. S.; Carr, S. R.; Cassady, C. J. *J. Am. Soc. Mass Spectrom.* **1995**, *6*, 521–524.

159. Kelly, M.; Lappalainen, P.; Talbo, G.; Haltia, T.; Vanderoost, J.; Saraste, M. *J. Biol. Chem.* **1993**, *268*, 16781–16787.

160. Bauer, M. D.; Sun, Y.; Anastasio, M. V.; Snider, C. E. In *Proceedings of the 43rd ASMS Conference on Mass Spectrometry and Allied Topics*, Atlanta, GA, 1995; American Society for Mass Spectrometry: Santa Fe, NM, p. 319.

161. Feng, R.; Castelhano, A. L.; Billedeau, R.; Yuan, Z. *J. Am. Soc. Mass Spectrom.* **1995**, *6*, 1105–1111.

162. Lane, T. F.; Iruelaarispe, M. L.; Johnson, R. S.; Sage, E. H. *J. Cell Biol.* **1994**, *125*, 929–943.

163. Moreau, S.; Awade, A. C.; Molle, D.; Le Graet, Y.; Brule, G. *J. Agric. Food Chem.* **1995**, *43*, 883–889.

164. Hamdan, M.; Curcuruto, O.; Di Modugno, E. *Rapid Commun. Mass Spectrom.* **1995**, *9*, 883–887.

165. Loo, J. A.; Ogorzalek Loo, R. R.; Goodlett, D. R.;; Smith, R. D.; Fuciarelli, A. F.; Springer, D. L.; Thrall, B. D.; Edmonds, C. G. In *Techniques in Protein Chemistry IV*; Angeletti, R. H., Ed.; Academic Press: San Diego, CA, 1993, pp. 23–31.

166. Loo, J. A.; Giordani, A. G.; Muenster, H. *Rapid Commun. Mass Spectrom.* **1993**, *7*, 186–189.

167. Li, Y.-T.; Hsieh, Y.-L.; Henion, J. D.; Ganem, B. *J. Am. Soc. Mass Spectrom.* **1993**, *4*, 631–637.

168. Konishi, Y.; Feng, R. *Biochemistry* **1994**, *33*, 9706–9711.

169. Ganem, B.; Henion, J. D. *Chemtracts–Org. Chem.* **1993**, *6*, 1–22.

170. McLuckey, S. A.; Ramsey, R. S. *J. Am. Soc. Mass Spectrom.* **1994**, *5*, 324–327.

171. Li, Y. T.; Hsieh, Y. L.; Henion, J. D.; Ocain, T. D.; Schiehser, G. A.; Ganem, B. *J. Am. Chem. Soc.* **1994**, *116*, 7487–7493.

172. Henion, J.; Li, Y. T.; Hsieh, Y. L.; Ganem, B. *Ther. Drug. Monit.* **1993**, *15*, 563–569.

173. Hsieh, Y.-L.; Cai, J.; Li, Y.-T.; Henion, J. D.; Ganem, B. *J. Am. Soc. Mass Spectrom.* **1995**, *6*, 85–90.

174. Bakhtiar, R.; Stearns, R. A. *Rapid Commun. Mass Spectrom.* **1995**, *9*, 240–244.

175. Eckart, K.; Spiess, J. *J. Am. Soc. Mass Spectrom.* **1995**, *6*, 912–919.

176. Drummond, J. T.; Ogorzalek Loo, R. R.; Matthews, R. G. *Biochemistry* **1993**, *32*, 9282–9289.

177. Siuzdak, G.; Krebs, J. F.; Benkovic, S. J.; Dyson, H. J. *J. Am. Chem. Soc.* **1994**, *116*, 7937–7938.

178. Baca, M.; Kent, S. B. H. *J. Am. Chem. Soc.* **1992**, *114*, 3992–3993.

179. Aplin, R. T.; Robinson, C. V.; Schofield, C. J.; Westwood, N. J. *J. Chem. Soc., Chem. Commun.* **1994**, 2415–2417.

180. Tuong, A.; Uzabiaga, F.; Petitou, M.; Lormeau, J. C.; Picard, C. *Carbohydr. Lett.* **1994**, *1*, 55–60.

181. Fitzgerald, M. C.; Chernushevich, I.; Standing, K. G.; Kent, S. B. H.; Whitman, C. P. *J. Am. Chem. Soc.* **1995**, *117*, 11075–11080.
182. Ganem, B.; Li, Y. T.; Hsieh, Y. L.; Henion, J. D.; Kaboord, B. F.; Frey, M. W.; Benkovic, S. J. *J. Am. Chem. Soc.* **1994**, *116*, 1352–1358.
183. Green, B. N.; Oliver, R. W. A. In *Proceedings of the 43rd ASMS Conference on Mass Spectrometry and Allied Topics*, Atlanta, GA, 1995; American Society for Mass Spectrometry: Santa Fe, NM, p. 1262.
184. Tang, X.-J.; Brewer, C. F.; Saha, S.; Chernushevich, I.; Ens, W.; Standing, K. G. *Rapid Commun. Mass Spectrom.* **1994**, *8*, 750–754.
185. Chernushevich, I. V.; Ens, W.; Standing, K. G.; Loewen, P. C.; Fitzgerald, M. C.; Kent, S. B. H.; Werlen, R. C.; Lankinen, M.; Tang, X.-J.; Brewer, C. F.; Saha, S. In *Proceedings of the 43rd ASMS Conference on Mass Spectrometry and Allied Topics*, Atlanta, GA, 1995; American Society for Mass Spectrometry: Santa Fe, NM, p. 1327.
186. Camilleri, P.; Haskins, N. J. *Rapid Commun. Mass Spectrom.* **1993**, *7*, 603–604.
187. Haskins, N. J.; Ashcroft, A. E.; Phillips, A.; Harrison, M. *Rapid Commun. Mass Spectrom.* **1994**, *8*, 120–125.
188. Jaquinod, M.; Potier, N.; Klarskov, K.; Reymann, J. M.; Sorokine, O.; Kieffer, S.; Barth, P.; Andriantomanga, V.; Biellmann, J. F.; Van Dorsselaer, A. *Eur. J. Biochem.* **1993**, *218*, 893–903.
189. Ganguly, A. K.; Pramanik, B. N.; Tsarbopoulos, A.; Covey, T. R.; Huang, E.; Fuhrman, S. A. *J. Am. Chem. Soc.* **1992**, *114*, 6559–6560.
190. Ganguly, A. K.; Pramanik, B. N.; Huang, E. C.; Tsarbopoulos, A.; Girijavallabhan, V. M.; Liberles, S. *Tetrahedron* **1993**, *49*, 7985–7996.
191. Griffith, M. C.; Risen, L. M.; Greig, M. J.; Lesnik, E. A.; Sprankle, K. G.; Griffey, R. H.; Kiely, J. S.; Freier, S. M. *J. Am. Chem. Soc.* **1995**, *117*, 831–832.

CHAPTER 12

Electrospray Ionization Mass Spectrometry of Nucleic Acids and their Constituents

PAMELA F. CRAIN

Department of Medicinal Chemistry, University of Utah, Salt Lake City, Utah

	Abstract	422
I.	Introduction	422
II.	Fundamental aspects of nucleic-acid structure	423
III.	Practical aspects of oligonucleotide analysis	424
	A. Sample preparation	424
	B. Solvent composition	425
	C. Additives and other enhancements	427
	D. pH effect on charge state	429
	E. Separation methods for sample introduction	429
IV.	Collision-induced dissociation of multiply charged oligonucleotides	431
	A. Quadrupole ion-trap studies	431
	B. Fourier-transform mass-spectrometry studies	434
	C. Triple-quadrupole studies	436
	D. Summary of gas-phase nucleic-acid sequencing studies	439
V.	Applications of ESI–MS to problems in nucleic-acid chemistry	439
	A. Oligonucleotide mapping/sequencing by collision-induced dissociation	440
	B. Mapping of nucleoside modification in RNA and DNA	441
	C. Sequence determination from mass measurement of exonuclease digestion products	443
	D. Probing noncovalent interactions of nucleic acids	445
VI.	Analysis of nucleic-acid monomeric constituents	449
	A. Nucleotides and nucleosides	450
	B. Nucleoside mixture analysis by LC/ESI/MS	451
VII.	Summary and prospects	452
	Acknowledgments	454
	References	454

Electrospray Ionization Mass Spectrometry, Edited by Richard B. Cole.
ISBN 0-471-14564-5   © 1997 John Wiley & Sons, Inc.

ABSTRACT

The application of electrospray ionization mass spectrometry (ESI–MS) to nucleic acids and their constituents is reviewed, with emphasis on methods that can be implemented using conventional quadrupole and trapped-ion mass spectrometers. Careful sample preparation is critical for analysis of these strongly anionic molecules, and practical aspects of sample manipulation prior to analysis are reviewed. The extent to which sequence can be derived by collision-induced dissociation (CID) of multiply charged oligonucleotide ions has been studied in both trapped-ion and triple-quadrupole mass analyzers, and results are compared for the different instruments. The use of ESI–MS to detect bimolecular noncovalent interactions in which one participant is a nucleic acid is reviewed, along with precautions for its use. There are a greater number of applications of ESI–MS to analysis of nucleic acids than to their monomeric constituents; however, the latter can easily be analyzed and with high sensitivity, particularly in combination with chromatographic sample introduction. Representative pharmaceutical and biomedical applications that illustrate the use of ESI–MS are discussed.

I. INTRODUCTION

Nucleic acids and their constituents have long represented a considerable challenge for analysis by mass-spectrometric methods, owing to their polarity and thermal lability (1). With the introduction of electrospray ionization (ESI) and matrix-assisted laser desorption ionization (MALDI), the capability for determining the masses of oligonucleotides and intact nucleic acids has been realized. These two ionization methods, as applied to nucleic acid analysis, have been compared in detail (2). Generally, MALDI yields greater sensitivity than ESI, but with lower mass accuracy and resolution as a consequence of its most common implementation with the time-of-flight (TOF) mass analyzer. The ESI process proceeds from a flowing analyte solution, and is readily suited for chromatographic and electrophoretic methods of sample introduction. Electrospray ionization can be coupled with mass analyzers that permit mass determination with an accuracy of 0.01% or better, and oligonucleotides up to 100 residues and greater have been successfully ionized and mass measured at this level of accuracy (3–5). In a striking demonstration of the ionizing capability of ESI (mass analysis using Fourier transform mass spectrometry), multiply charged ions were generated from coliphage T4 DNA (nominal M_r 1.1×10^8) (6).

Successful application of ESI–MS for analysis of oligonucleotides requires careful control of sample and instrument parameters, and these factors as they affect the quality of mass spectra are discussed in this chapter. Collision-induced dissociation (CID) of multiply charged oligonucleotide ions is discussed, as there is current interest applying the technique to determining the sequence from the fragment ions. Finally, a broad overview is presented of the types of problems to which ESI–MS has been successfully applied, both for analysis of oligonucleotides and small nucleic acids, and of their constituents, including naturally and

II. FUNDAMENTAL ASPECTS OF NUCLEIC-ACID STRUCTURE

xenobiotically modified nucleosides. Literature citations are not comprehensive, and have been chosen to emphasize useful and practical methods over exotic demonstrations. The literature cited may be consulted for leading references.

II. FUNDAMENTAL ASPECTS OF NUCLEIC-ACID STRUCTURE

The comprehensive review of the principles of nucleic acid structure by Saenger represents the single most useful overview of the chemical and physical properties of DNA and RNA (7). Figure 1 shows the structure of a tetraribonucleotide pAGUCp (7). In the analogous DNA structure, the ribose $2'$ hydroxyl groups are replaced with H (deoxyribose) and the uridine (U) nucleobase uracil is replaced with thymine (5-methyluracil). By convention, the sequence is defined in the $5' \rightarrow 3'$ chain direction, that is, the linking phosphodiester bonds are

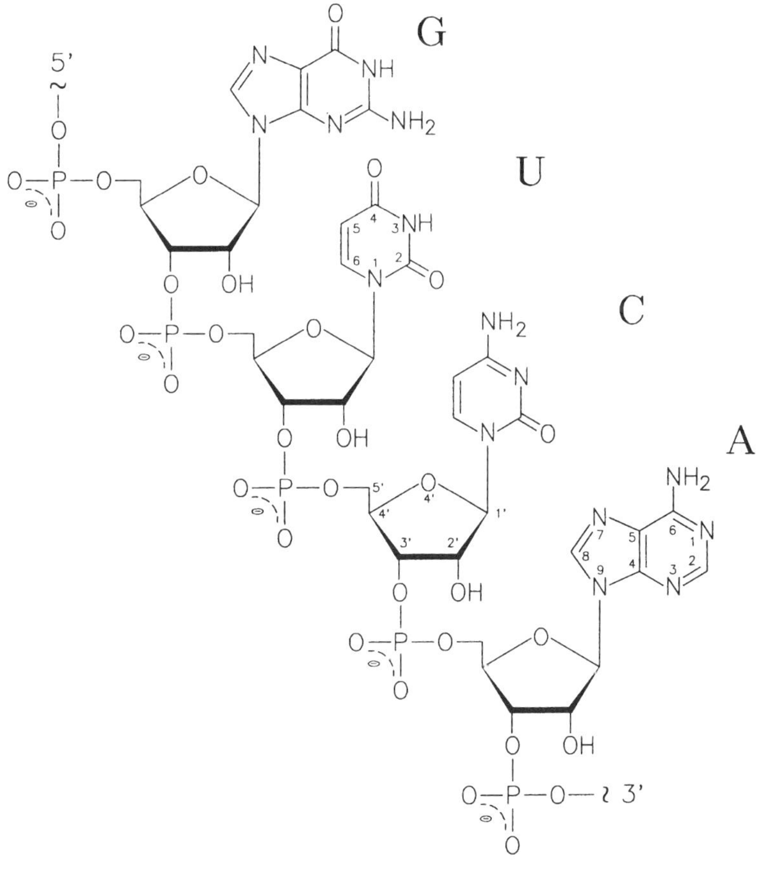

Figure 1. Structure of the tetraribonucleotide pAGUCp.

(C3' → phosphate → C5'). The bond that connects the nucleobase to the sugar is referred to as the glycosidic bond, and cleavages of this bond can be a significant factor in the mass spectra of oligonucleotides and their constituents.

The polyanionic backbone is fully dissociated in solution, and the pK_a of the internucleotide phosphates is < 1. Nucleic acids are extensively solvated: each nucleotide subunit is complexed with about 20 H_2O molecules. The nucleobases contain basic functional groups with pK_a values that range from 3.5 to 5, except for the uracil derivatives for which the pK_a (of N–3) is ≈ 10. The phosphate, sugar, and nucleobase groups all contain metal-ion binding sites. When oligonucleotides are transferred into the gas phase as multiply charged ions, the solvent is stripped off, along with counterions (depending upon their identity; discussed below). Increased coulombic repulsion from interaction between adjacent charged phosphate ions then becomes a significant issue, and it has been proposed that the negative charges can be shielded through proton sharing with the nucleobases (8).

III. PRACTICAL ASPECTS OF OLIGONUCLEOTIDE ANALYSIS

Specific conditions for analyte preparation discussed below should be approached on an individual basis, because the extent to which optimization (if any) was approached is not always clear from the relevant studies (e.g., how many organic modifiers and aqueous phase compositions were compared). Sample and ion-source tuning parameters are important considerations, and it is also evident that different instrument (ion-source) designs can give rather different spectra for the same sample (9). In many instances, conditions and protocols reviewed below were developed from samples available in large amount, and conflicting results have been reported; they should, therefore, be regarded as a starting point for method development, where sample amounts are limited. The choice of "ideal" conditions for analysis is not clear cut.

A. Sample Preparation

The presence of excessive amounts of nonvolatile cations such as Na^+, K^+, or Mg^{2+} is generally deleterious. As variable numbers of cations are complexed with the nucleic acid, the ion envelope for longer oligomers is broadened (depending on charge state and instrument resolution), making mass assignment uncertain. If instrument resolution is sufficient to resolve cationized peaks, they may be used for an additional measure of molecular weight (10), and to aid in charge-state assignment from their occurrence at $22/z$ units (e.g., for Na^+) greater than the ion in question (11). Nonetheless, and perhaps more importantly, sensitivity is reduced as a given multiply charged species (charge state) is dispersed among multiple cationized ions. Accordingly, it is highly advantageous to have the salt content as low as possible prior to the analysis, both for spectral quality and for reducing fouling of surfaces and activation of transfer lines by salt deposition. Chromatographic cleanup with volatile solvent systems is widely used and suitable for relatively short oligonucleotides. Precipitation from concentrated (e.g., 2.5 M) ammonium acetate solutions by alcohols is effective for conversion of Na^+ salts of longer oligonucleotides and small nucleic acids to ammonium salts (3,4,12), which results in displacement of metal cations (see below). Synthetic phosphorothioate oligonucleotides are more difficult to rid of cations in this fashion, and ethanol precipitation directly from 10 M ammonium acetate (13) or with increased ethanol amounts (14) have been recommended.

B. Solvent Composition

For routine analysis, oligonucleotides are commonly infused into the mass spectrometer in solutions containing an organic modifier, typically 50% or greater, with higher concentrations giving better sensitivity (15). A sufficiently great variety of "cocktails" have been successfully employed that definitive recommendations are not warranted. Few systematic studies to determine the optimal organic modifier have been reported (13,15,16). Adenosine 5′-monophosphate was used as a test compound to determine suitable solvent for negative-ion ESI–MS; isopropanol was described as the ideal choice, because it gave the highest $(M-H)^-$ ion abundance, the most stable ion current, and a high resistance to onset of corona discharge; CH_3CN was among the poorest performers (16). 2-Propanol was not included in a systematic study of solvents using longer oligonucleotides as test samples, in which it was concluded that CH_3CN was optimal, at up to 80% v/v (15). It is not clear whether differences in size of test compounds or in instrument parameters led to contrasting conclusions. Nonetheless, 2-propanol solutions (50–67%, v/v) have been successfully employed for longer oligonucleotides (e.g., 20- to 120-mers) (4,13), but acetonitrile appears to be the most widely used organic modifier (3,12,15,17,18). Methanol (90% aq.) has also been successfully used for ESI–MS of oligonucleotides (11,19,20). In general, high concentrations of organic solvent are preferable so long as oligonucleotide solubility is maintained.

The addition of volatile bases [e.g., NH_4OH, triethylamine (TEA)] to the analyte solution aids in suppression of cationization (3,4,13,15,18). Triethylamine is generally most effective in suppression of Na^+ adduction (3,13). Figure 2 shows the ESI spectra of a synthetic 20-mer in 1 : 1 mixtures of water/2-propanol, 5.0 mM piperidine/2-propanol and (2.5 mM piperidine + 2.5 mM imidazole)/2-propanol (13). The spectra are plotted to the same scale and can readily be compared. The broad, diffuse, cationized-ion clusters are converted to a single molecular ion species for each charge state following addition of piperidine. Coaddition of imidazole increases signal intensity; added alone it is much less effective in cationization suppression but still increases signal intensity. With higher amounts of imidazole, a reproducible bimodal distribution of charge states is produced (also evident in Fig. 2), and it was proposed that imidazole can induce structural alterations in the (non-complementary) oligomers that persist in the gas phase (13). In general, mass spectral response is increased at higher pH,

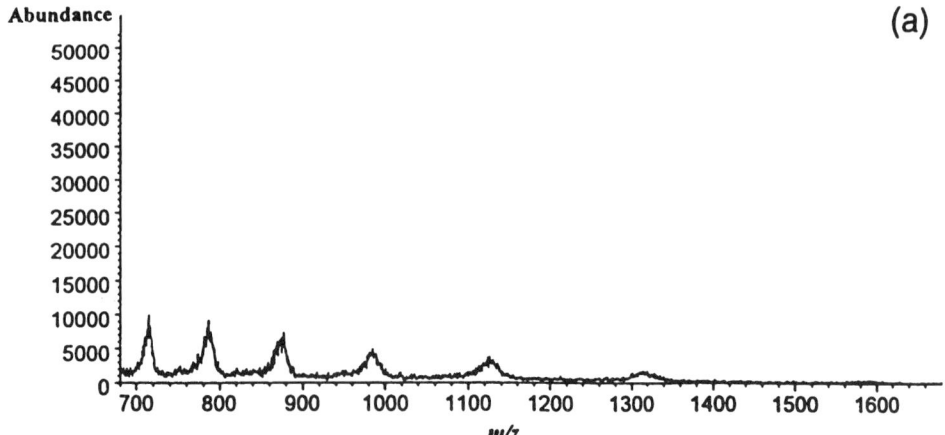

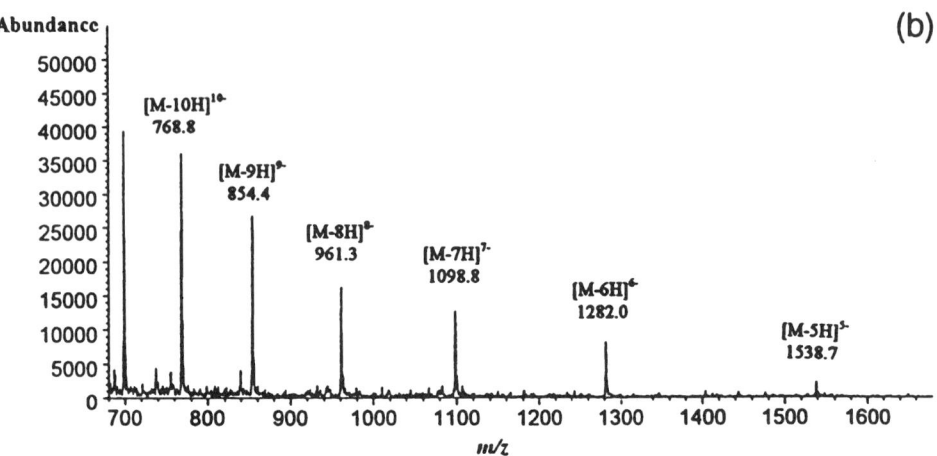

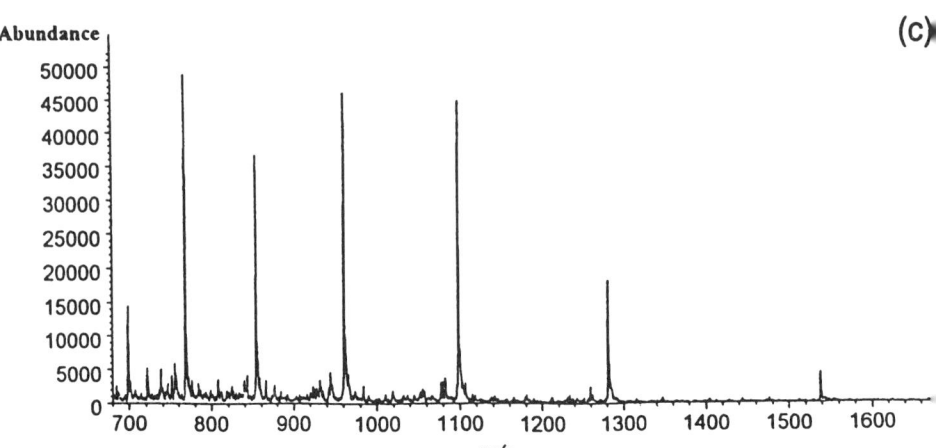

but addition of bases is generally effective only up to a certain point, after which ion signals become attenuated. The use of elevated pH is not recommended for RNA work, as oligo*ribo*nucleotides can be cleaved at alkaline pH, and also by imidazole.

Sodium ions clustered with the phosphodiester backbone are more evident in ions of lower charge states (lower z) (10). Sodium content can be decreased via sample treatment and solvent manipulations, and it is generally accepted that Na^+ is consequently displaced by NH_4^+ or $TEAH^+$. [In fact, ammonium (or $TEAH^+$) adducts can be observed directly at low (positive ions detected) (3) or high (negative ions detected) base concentrations (4,15)]. The ESI mass spectra show that the occurrence of Na^+ clusters is lessened as the volatile neutral amine is lost, effecting a net substitution of H^+ for Na^+ (4,12). The ESI spectra of oligonucleotides following treatment with nitrogenous bases in low concentrations generally reflect an increase in charge state (4,12,13), suggesting in such cases that if Na^+ is displaced by NH_4^+ or $TEAH^+$, the ion structure(s) reflects loss of the amine cation, not a neutral base.

C. Additives and Other Enhancements

Solvents, reagents, pipets, and fused-silica transfer lines used during sample synthesis, cleanup, and infusion into the mass spectrometer can contribute adventitious cations even to carefully desalted samples. To a degree, cationization can be overcome by the inclusion of volatile bases in the analyte solution. As a consequence of secondary structure or the presence of nucleoside or backbone modifications, however, nucleic acids can possess high-affinity binding sites for divalent cations, which are not removed during chromatography or ion-exchange precipitations. Figure 3 shows the ESI spectra of a mixture of ammonium salts of two transfer RNAs (a 76- and a 77-mer) (4). Addition of either TEA or the divalent-cation chelator CDTA decreased peak widths compared with untreated sample. An additive effect is evident, however, when both TEA and CDTA are present, indicating that both mono- and divalent cations were present in the desalted sample, and were effectively suppressed in this fashion. Metal chelators may be a generally useful additive for the analysis of enzymatically derived oligonucleotides, as many of the enzymes used in nucleic-acid manipulations have a divalent metal-ion requirement for activity. The crown ether 18-crown-6 was reported to enhance the abundance of duplex ions (discussed below), presumably by Na^+ displacement; amounts used were not stated (21). Replacement of the fused silica transfer line

---

Figure 2. Electrospray mass spectra of the DNA 20-mer TGAGTCAGACGCATCGT-CGTCATGG dissolved in a 1:1 mixture of buffer + isopropanol. (*a*) Ethanol precipitated from 0.5 M Na acetate without further purification; no buffer. (*b*) Dissolved in 5 mM piperidine. (*c*) Dissolved in 2.5 mM imidazole + 2.5 mM piperidine. Reprinted with permission from M. Greig and R. H. Griffey, *Rapid Commun. Mass Spectrom.*, **1995**, *9*, 101. © 1995 John Wiley & Sons, Ltd.

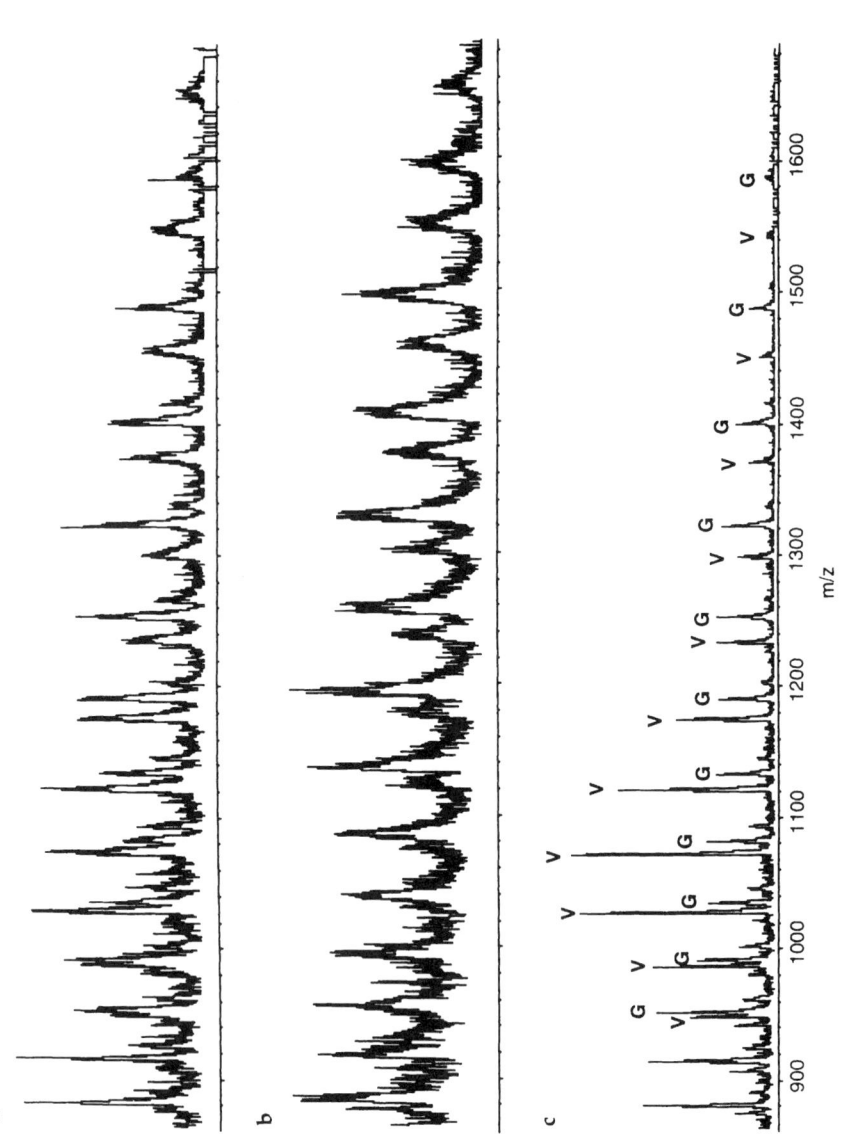

Figure 3. Electrospray mass spectra of an equimolar mixture of valine (V; 77-mer) and glycine (G; 76-mer) transfer RNAs from *E. coli* following precipitation from 2.5 M ammonium acetate. (*a*) In 0.1% triethylamine (TEA) + isopropanol, 1 : 2. (*b*) In 1 nmol ammonium 1,2-*trans*-cyclohexyl-N,N,N',N'-diamine tetraacetate (CDTA) + isopropanol, 1 : 2. (*c*) In 0.1% TEA + CDTA + isopropanol, 1 : 1 : 4. Reprinted by permission of Elsevier Science Inc. from P. A. Limbach, P. F. Crain, and J. A. McCloskey, "Measurement of intact ribonucleic acids via electrospray ionization

with PEEK tubing was reported to reduce background salt content, because it is easier to clean and is not activated by adsorption of adventitious cations carried along with analyte (4).

D. pH Effect on Charge State

Although the addition of strong bases shifts the charge states higher (discussed above), the addition of the weak base imidazole (pH of the solution was 8.0) to a 20-mer shifted the ion envelope to lower charge states (lower z) (13). N-methylimidazole, with similar pK_a, showed no such effect, nor did it suppress Na^+ adduction. It was suggested that protonated imidazole binds to two hydrogen-bond acceptor sites, and is subsequently lost as imidazole and not the imidazolium cation during desolvation. The reason for the charge-shifting effect has not been elucidated, but is under study (13).

The addition of acid to oligonucleotide solutions also leads to a reduction in charge state (22,23). Charge-state reduction was explored for ultimate application to the simplification of mixtures of oligonucleotides to facilitate their analysis by ESI–MS (22). If the charge states could be compressed into a smaller number, spectral simplification, and in principle, enhanced sensitivity could be achieved. Figure 4 shows the dramatic spectral simplification that was achieved for a test mixture of DNA homopolymers in 2 M HCO_2H, compared with water solutions (22). Fe(II) adducts of dT_{12} were a prominent feature from use of strong acid in the solvent; they could be eliminated by coaddition of 1,10-phenanthroline to the solution. Efficiency of charge-state reduction was found to be a function of the pK_a, concentration, and the identity of the acid used. Charge-state reduction was also effected by complexation of oligonucleotides with diamines to form adducts, followed by dissociation of the complex in the interface (22).

Somewhat different findings were reported from a similar study (23). A set of hexamers was analyzed by negative-ion ESI–MS over a pH range from 2.6 to 7. The extent of charge-state reduction by addition of acid to the solvent, however, was substantially less than reported in the previous study (22). For example, in Figure 4 the most abundant charge state for dT_{12} shifts from 9– to 3– upon acid treatment (22), whereas the charge state (for dT_6) was reported in the later study to be pH independent for pH values of 2.6 and 7. Charge-state reduction was found to be pH dependent but independent of the acid concentration (23). The reason(s) for this discrepancy remains to be explored.

E. Separation Methods for Sample Introduction

Although ESI is effected from a flowing liquid solution of analyte and would a priori seem ideally suited for coupling with sample-separation methods, only a few examples of the approach have been described for oligonucleotide separations (24–28). Capillary-zone electrophoretic (CZE) and capillary isotachophoretic (CITP) separations of mixtures of mononucleotides and dinucleoside monophosphates have been described (27). CZE–ESI–MS has been applied to

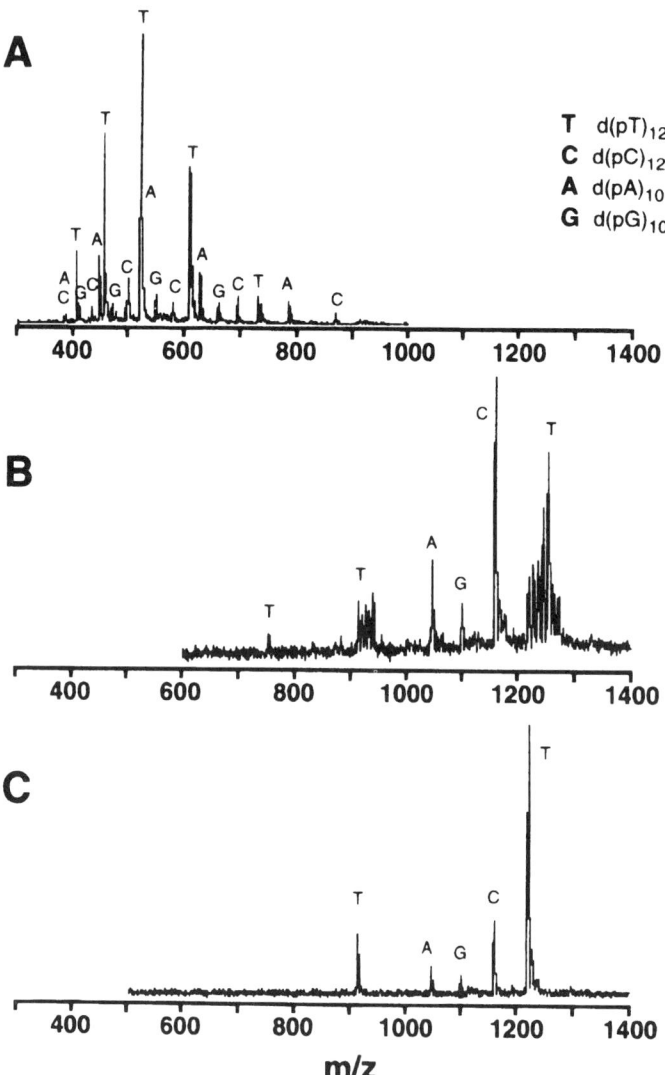

Figure 4. Electrospray mass spectra of an equimolar mixture of ammonium salts of DNA homopolymers. (*A*) Dissolved in water (pH ~7); (*B*) 2 M HCO$_2$H (pH ~1.7); (*C*) 2 M HCO$_2$H/50 mM 1,10 phenanthroline (pH ~1.7). Reprinted with permission from X. Chang et al., *Anal. Chem.*, **1995**, *67*, 591. © 1995 American Chemical Society.

the analysis of mixtures of oligothymidylic acids (25) and of modified oligonucleotides ($n = 2$–8) produced from styrene oxide-modified DNA by endonuclease digestion (discussed below) (26,28). The coupling of electrophoretic separation with ESI–MS is covered separately in this volume.

The widespread utilization of reversed-phase HPLC with volatile eluants as a general laboratory tool for oligonucleotide sample preparation and the relative ease with which these instruments can be coupled to mass spectrometers suggests that they should be more widely used than is apparent. The requirement of volatile eluants for compatibility with ESI–MS may limit the chromatographic performance, but useful gradient systems have been described (25,29,30) that can provide a starting point for methods development. Figure 5 illustrates the chromatogram from reversed-phase separation of a mixture of dA_{10}, dC_{10}, and dT_{10}, and the corresponding mass spectra (30). The gradient system employed mixtures of 10 mM ammonium acetate (pH 8.3) and acetonitrile, and was also applied for LC–ESI–MS of longer protected synthetic oligonucleotides up to a 24-mer in length (30). Relatively large samples were analyzed in this example, but the amounts could have been decreased, judging from the good quality of the spectra.

IV. COLLISION-INDUCED DISSOCIATION OF MULTIPLY CHARGED OLIGONUCLEOTIDES

There is currently great activity directed toward discerning the extent to which the sequence of an oligonucleotide can be derived from product ions following dissociation of multiply charged ions. Both ion-trapping and quadrupole mass analyzers have been employed, with somewhat different findings. A detailed review of the dissociation chemistry of multiply charged oligonucleotides is beyond the scope of this chapter; the reader may profitably consult the works cited (19,31–38). Although not discussed here, studies of the CID of dinucleotides provide insight relevant to the CID behavior of oligonucleotides (8,39–41).

A. Quadrupole Ion-Trap Studies

The detailed studies by McLuckey and colleagues have provided a significant and valuable starting point for understanding fragmentation of multiply charged oligonucleotides (19,31–33). Small ($n = 4–8$) oligomers were analyzed using an ion trap, allowing for MS/MS and MS^n measurements to be accomplished. Figure 6 is the representation of a generic oligonucleotide developed by these workers to designate the potential locations of bond cleavages (19).

The product-ion spectrum from low-energy CID of the $(M-7H)^{7-}$ ion of TGCATCGT (31) is shown in Figure 7, and illustrates some of the concepts for spectral interpretation, in uncomplicated cases, by simple inspection. Under gentle collisional activation, complementary ions arising from charge-separation reactions of the precursor are typically seen (19,31). These pairs of ions can be located in the spectrum because they appear on the mass scale at locations that reflect the appropriate m/z differences from the precursor ion. For example, in Figure 7, the precursor ion $(M-7H)^{7-}$ produces A^- and also loses A^- to give the ion designated $[-A]^{6-}$. The ratio of m/z differences between these two ions and their precursor (six and one, respectively, accounting for the partition of seven

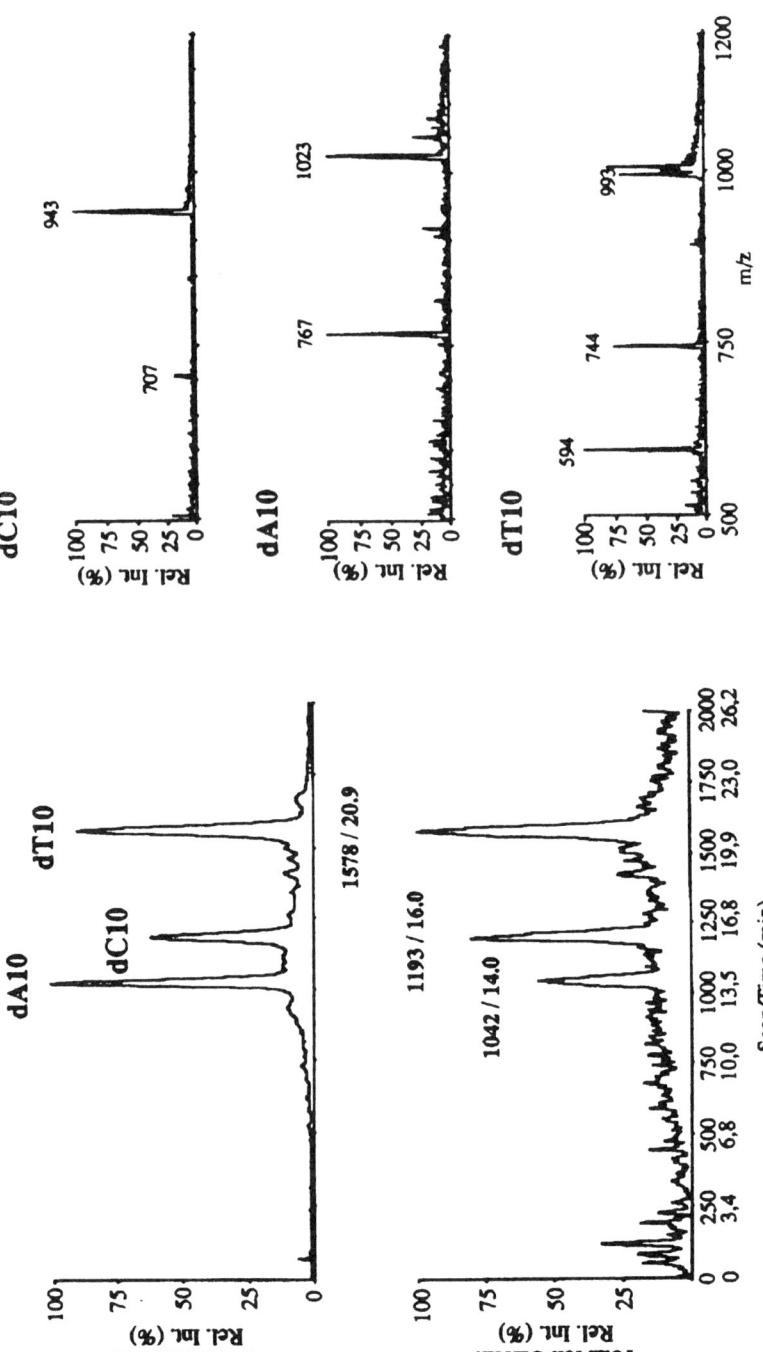

Figure 5. Reversed-phase, high-performance liquid chromatography–electrospray/mass spectrometry of a mixture of the DNA homopolymers dA10, dC10, and dT10 (5 nmol each). Left panel: chromatograms (absorbance at 260 nm) from separation in a 10-mM ammonium acetate/acetonitrile gradient. Right panel: ES mass spectra of each component. Reprinted with permission from K. Bleicher and E. Bayer, *Chromatographia*, 1994, 39, 407. © 1994 Vieweg-Publishing.

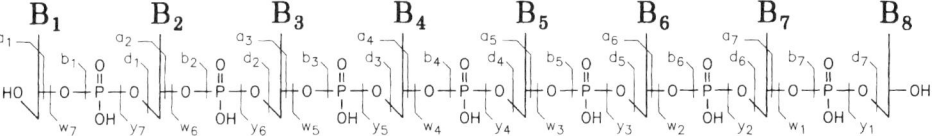

Figure 6. McLuckey nomenclature for designation of bond cleavages from CID of multiply charged oligonucleotides. The cleavages indicated as a_n represent the a_n–B_n ion series.

charges) is, therefore, six (arrows beneath the m/z scale). Likewise, it is readily apparent that the two triply charged ions $[a_4\text{–}B_4(A)]^{3-}$ and w_4^{3-} are complementary, subsequent, products of the $[-A]^{6-}$ ion and are not derived directly from the precursor ion. Ions representing the singly charged bases themselves (T^-, A^-, and G^-) are also evident in the low-mass region of the spectrum.

A notable feature of the spectrum of TGCATCGT (Fig. 7) is the dominance of fragmentation reactions adjacent to the single A residue, initiated by loss of the base; in fact, this spectrum provides essentially only the location of the A residue. Base losses are a prominent feature of the product-ion spectra: charged bases (most commonly A^- and T^-) are lost from highly charged precursor ions (33); loss of neutral bases, particularly cytosine, occurs from lower charge-state

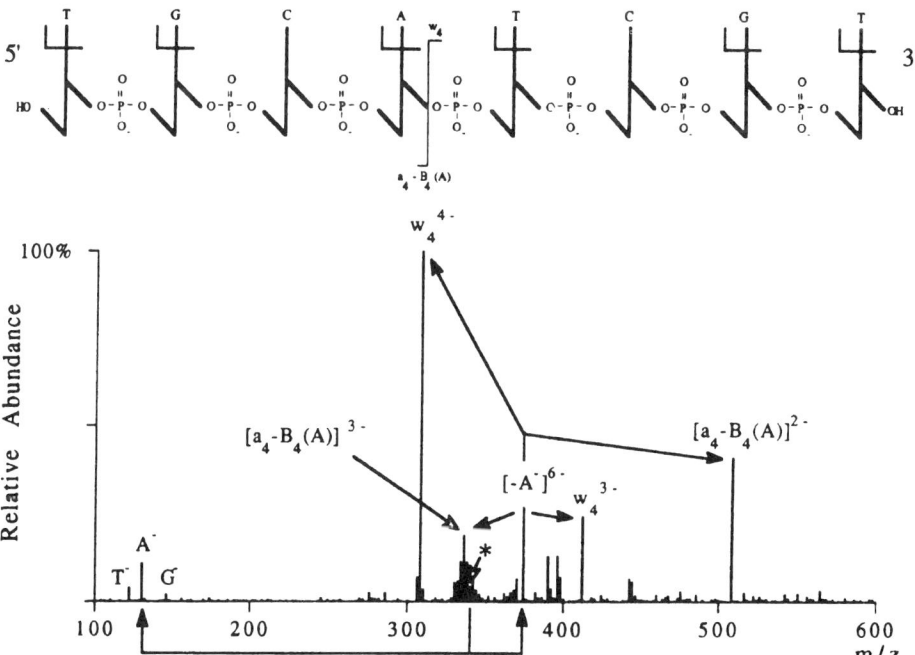

Figure 7. Product-ion spectrum from low-energy CID of $(M-7H)^{7-}$ of TGCATCGT in a quadrupole-ion trap. Reprinted with permission from S. A. McLuckey and S. Habibi-Goudarzi, *J. Am. Chem. Soc.*, **1993**, *115*, 12087. © 1993 American Chemical Society.

precursors. The capability of the ion trap for multiple stages of CID permits the remainder of the sequence to be deduced: for the $5'$ TCGT segment by dissociation of the w_4^{3-} product ion from $(M-6H)^{6-}$ and for the $3'$ segment TGC by dissociation of the $[a_4-B_4(A)]^{2-}$ product ion from $(M-5H)^{5-}$.

The fragmentation behavior of these molecules is described as remarkably consistent, and "rules of decomposition" and some observations for their fragmentation in the ion trap have been summarized (31):

- Loss of a nucleobase is the first decomposition; all bases can be lost, generally $A^- > T^- > G^-, C^-$ for highly charged species. All bases can be lost as neutrals (without particular preference) from low charge-state precursors. There is little dependence of location on probability of base loss, except for the $3'$ base, whose loss is disfavored.
- The second decomposition is cleavage of the $3'$ C–O bond of the sugar from which the base was lost to produce the complementary "w" and "a–B" type ions (Fig. 6). Although provided for in the generic fragment schematic (Fig. 6), other cleavages (e.g., "c" and "z" ions) do not occur to any significant degree.
- Loss of PO_3^- from ions with a terminal phosphate group competes with base loss.
- Complementary ions (discussed above) can be observed when ion-trapping conditions permit storage of both; their presence greatly facilitates spectral interpretation.
- Adjacent C and G residues can be difficult to sequence for two reasons. Losses of A^- and T^- and subsequent backbone cleavages at their locations dominate the product-ion spectra from highly charged precursors. At lower charge states, these processes do not dominate, but the higher m/z ions are more difficult to manipulate, and the number of stages of mass analysis is limited, owing to the lower total charge in the system.

B. Fourier-Transform Mass-Spectrometry Studies

Fourier-transform mass spectrometry (FTMS) with an ion-cyclotron resonance cell was subsequently used to expand on these earlier studies and to extend them to larger oligonucleotides (34–36). McLafferty and colleagues utilized both NS dissociation and CID with FTMS for mass analysis to investigate dissociation mechanisms for multiply charged oligonucleotides (34–36). Figure 8 illustrates the spectra from NS dissociations of the DNA 8-mer CGAGCTCG and from CID of its $(M-4H)^{4-}$ ion; their similarity is striking and both techniques were used to test the relevance of the McLuckey quadrupole ion-trap dissociation mechanisms (19, 31) to CID of multiply charged oligonucleotides in the ion-cyclotron resonance cell (34).

Because there is no mass selection in NS dissociation, a given product ion can arise via different mechanisms from molecular ions of different charges. To avoid confusion in describing and conceptualizing the product ions, all are represented as the corresponding neutral species derived from adding an H^+ for each

IV. COLLISION-INDUCED DISSOCIATION OF MULTIPLY CHARGED OLIGONUCLEOTIDES

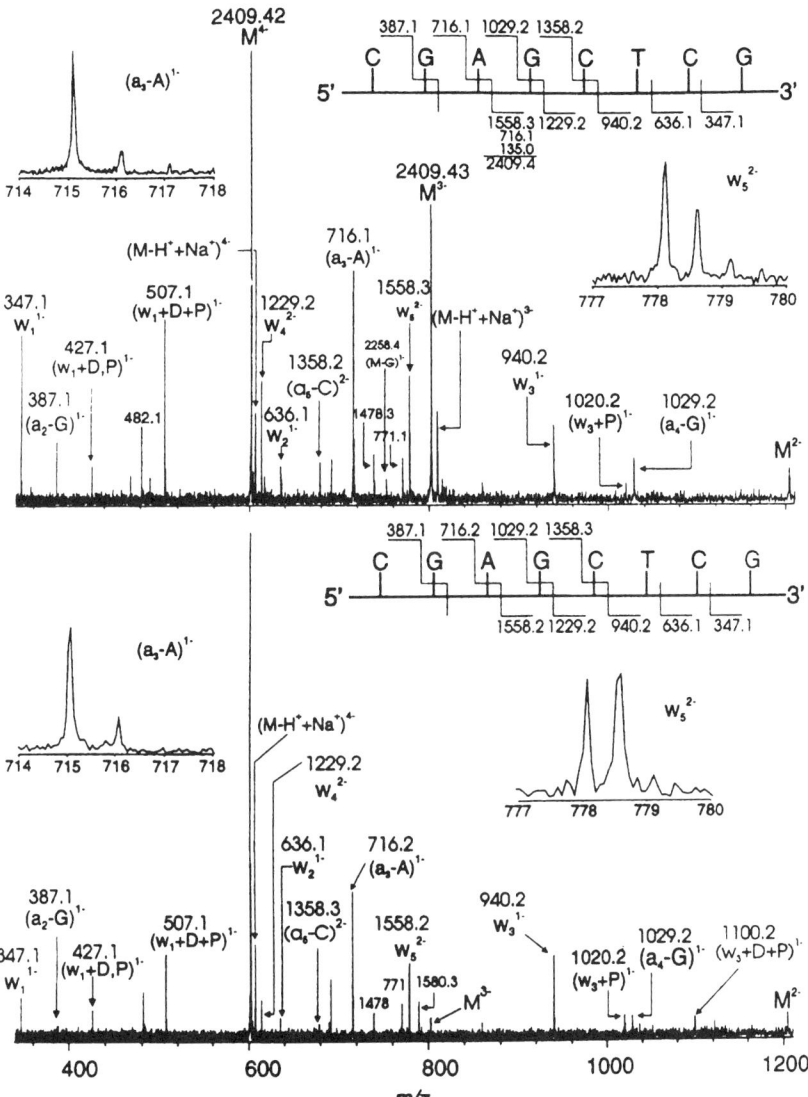

Figure 8. Spectra of CGAGCTCG from nozzle-skimmer dissociation (upper panel), and from CID of $(M-3H)^{4-}$ (lower panel) obtained by FTMS. Expanded mass scales for the $(a_3-A)^{1-}$ and w_5^{2-} ions in each panel show the resolution obtained. Reprinted with permission from D. P. Little et al., *J. Am. Chem. Soc.*, **1994**, *116*, 4895. © 1994 American Chemical Society.

negative charge. The w_n and a_n-B ion series, therefore, are referred to as $w_n + H$ and a_n-BH ions, respectively (34).

The high resolving power and consequent high accuracy of mass measurement available from FTMS allowed distinction between ion-structure candidates of the same nominal mass, but which differed in exact mass by several milliDaltons (34). For example, DNA oligomers containing CC as the 5′ terminus and GG as the 3′ terminus ($\Delta M = 80$ or phosphate) will give $w_2 + H$ and a_3-BH ions of 676 Da. The mass difference, however, is 0.014 Da, and both ions could be observed. In another example of the utility of exact mass measurement, the occurrence of ions corresponding to the addition of 80 and 160 Da to $w_n + H$ and of 80 Da to a_n-BH ions was noted. Candidates for the additional 80 Da were PO_3H (79.996 Da) and furan (80.026 Da). The m/z 427 ion (Fig. 8) was found to be a doublet, and this ion was assigned as $w_1 + H + PO_3H$, which was attributed to PO_3H transfer, and as a z_2-B_7H ion (Fig. 6), accounting for the 80-Da increment corresponding to net "addition" of furan.

Fragmentation is typically incomplete; that is, complete w and a–B ion series are rarely produced, although a complete w series was reported for a 9-mer (34). Fragmentation from either end, however, can extend sufficiently far into the interior of the chain to overlap, permitting the sequence to be derived nonetheless, at least for the few examples shown (two 8-mers and a 14-mer). As was also reported from ion-trap dissociation studies (19,31,32), certain sequences are resistant to cleavage (5,35), thus, nucleotide order within narrow regions (e.g., dimer or trimer) of the oligomer sometimes cannot be determined.

C. Triple-Quadrupole Studies

The fragmentation of a multiply charged oligonucleotide was reported by Stults (12) and resulted from use of elevated declustering voltages (nozzle-skimmer dissociation, NS). It was noted that the extent of fragmentation of the DNA 30-mer analyzed was significantly greater than for proteins of comparable size, and it was suggested that oligonucleotides were far more labile to collisional activation. The resulting spectrum was complex, and a detailed interpretation was not attempted.

Substantive studies of oligonucleotide CID in the triple-quadrupole collision cell are confined essentially to two studies (37,38), despite a far greater utilization of linear-quadrupole instruments for general oligonucleotide applications compared with ion-trap and ion-cyclotron resonance instruments. The product-ion spectrum from CID of the DNA 8-mer CGAGCTCG acquired with a triple-quadrupole instrument was extensively analyzed by Pomerantz (38) and is shown in Figure 9. This compound was selected because it is the same one studied using FTMS (34), and the spectrum shown in Figure 9 may be compared

---

Figure 9. Product-ion spectrum from CID of CGAGCTCG in a triple-quadrupole collision cell. (*A*) Product-ion spectrum from CID of $(M-4H)^{4-}$. (*B*). Edited spectrum showing only the w series ions present in the product-ion spectrum, magnified tenfold. (*C*) Edited spectrum showing only the a_n-B series ions present in the product ion spectrum, magnified tenfold. Reprinted with permission from P. F. Crain et al., *Mass Spectrometry in the Biological Sciences*, A. L. Burlingame and S. A. Carr, Ed., 1996, pp 497–517. © 1996 Humana Press.

IV. COLLISION-INDUCED DISSOCIATION OF MULTIPLY CHARGED OLIGONUCLEOTIDES 437

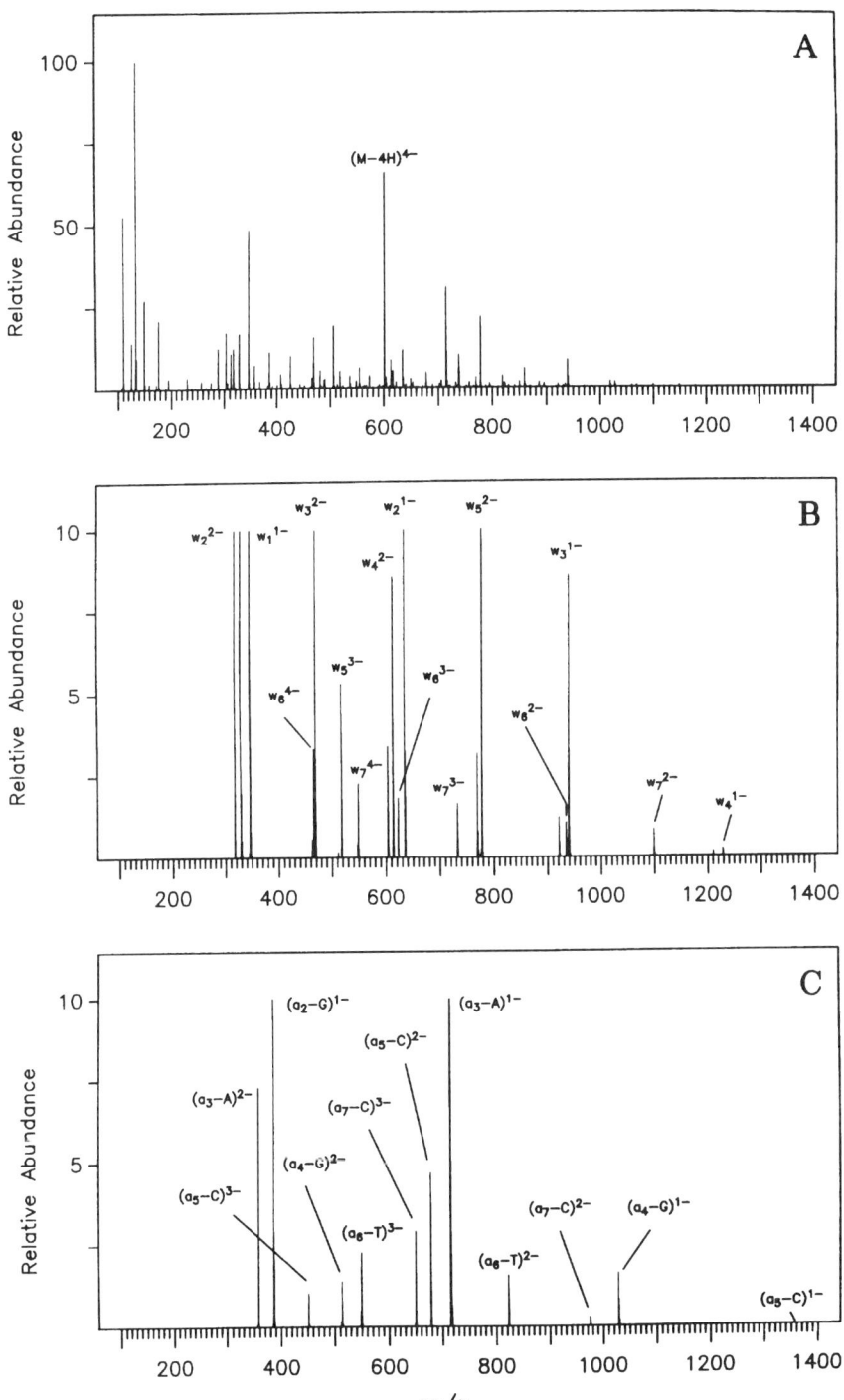

with the spectrum shown in Figure 8b as the same ion, $(M-4H)^{4-}$, was collisionally activated in both examples. It is immediately apparent (Fig. 9) that the extent of fragmentation is far greater than is generally observed from CID of trapped ions of this relatively small size (19,31,32,34). The sequence can be read in both directions because the spectrum contains complete w ($3' \rightarrow 5'$ direction) and a–B ($5' \rightarrow 3'$ direction) ion series. In addition, complete y ($3' \rightarrow 5'$) and b ($5' \rightarrow 3'$) ion series are produced (38); these ion types, along with w and d ion types, also undergo water loss to a far greater extent than with trapped-ion mass analyzers. Prior loss of a base is required for subsequent fragmentation of the chain in the ion trap analyzer to yield the b, d, and y ion series (31). In the triple-quadrupole collision cell, however, these ions can arise as well from alternative mechanisms, as was shown from the spectrum of an oligonucleotide synthesized from pseudouridine (Ψ; 5-ribosyluracil), a nucleoside with a C–C glycosidic bond, which is not cleaved in any mass-spectral method. No uracil$^-$ ion (B^-) or a_n–B ion series were produced, but b, d/y, and w ion series permitted the sequence to be read nonetheless (42). [Some a–B and w-type ions were produced by the NS CID used for charge-state reduction studies discussed above (22).]

Owing to the greater extent of fragmentation that can be produced from the triple-quadrupole collision cell, double cleavages can occur in the interior of the molecule to yield ions representing the constituent mononucleotides, along with their cyclic phosphate (from water loss) and diphosphate forms, for example, pNp (38). A subsequent study using ^{18}O-labeled d(CGAGCTGC) showed that the mononucleotides were almost exclusively 5' phosphates, and that both of the previously described mononucleotide + furan and mononucleotide + phosphate ion types (34) are likewise produced in the triple quadrupole collision cell, with pN ion types predominant (42). Transfer of a PO$_3$H group was invoked to explain the genesis of the latter ion in the FTMS studies (34), whereas in the Pomerantz study phosphate transfer would not neccessarily be required (38).

A profound effect of nucleobase identity on fragmentation was elucidated from studies of the CID of the $(M-3H)^{3-}$ ion from a set of hexamers of the structure CACGXG, where X is deoxyuridine substituted at C–5 of the base with ethyl, bromovinyl, iodo, or trifluoromethyl groups (37). Loss of X$^-$ and subsequent cleavage of the 3' C–O bond generates the $(a_5-B_5-2H)^-$ and w_1^- ions, respectively, and it was concluded that as the base modification becomes more electron withdrawing, loss of X$^-$ becomes more facile. As a consequence of enhanced base loss from this site, the abundances of other useful ion series are diminished, and spectra of the 5-iodo and 5-trifluoromethyl derivatives showed essentially complete disappearance of all w-series ions except for w_1^-. The complete sequence could be verified from further (second–generation) decomposition of the $(a_5-B_5-2H)^-$ because this ion gave m/z values unique to this ion and not to the $(M-3H)^{3-}$ precursor ion. The w, y, and c–2H ion series were present for the 5-ethyl and 5-bromovinyl derivatives.

For these hexamers, the triply charged molecular ion was chosen as precursor ion for CID because it offered the best combination of sensitivity and ease of spectral interpretation. Precursor ions with higher charges gave too many low-mass nonspecific fragment ions, while doubly and singly charged ions required higher collision energies owing to their greater stabilities, thus leading to an overall decrease in sensitivity. This paper (37) also contains a good overview and summary of (potential) mechanistic differences in ion origins from ion-trapping cells versus triple-quadrupole collision cells.

D. Summary of Gas-Phase Nucleic-Acid Sequencing Studies

The foregoing studies have employed a range of mass analyzers from the simple and relatively inexpensive ion trap to the triple quadrupole to the research-grade FT–ICR instrument. It is generally the case that the extent of fragmentation of the multiply charged precursor ion derives from the amount of energy imparted to it and is predictably greater in the triple-quadrupole instrument compared to the ion-trapping analyzers, leading to generally more complete sequence information. Sequence-context failures to cleave at every internucleotide bond, independent of oligonucleotide length, are reported for ion trap (31,32) and FTMS (5,35) studies; too few oligonucleotide sequences have been examined using triple quadrupoles to reveal whether incomplete fragmentation will likewise occur. The ion-trap and FT–ICR analyzers have the capacity for multiple stages of CID and mass analysis, which is useful for understanding the genesis of product ions. The sample amount required for such measurements is difficult to assess from these model studies where the amount of material available is relatively large, but it is undoubtedly greater than required for simple measurement of molecular weight, which is on the order of a few tens of picomoles with standard interfaces. The use of ultralow flow-rate interfaces (43–46) will lower sample requirements for sequencing to a few hundred femtomoles (45).

The size limitation is also not known, and may be sequence dependent in any event. Nearly complete sequence information for a 50-mer was reported from multiple overlapping fragments produced from multiple-dissociation regimens (36). While an impressive and useful accomplishment, gas-phase sequencing methods are presently not competitive with traditional rapid-DNA sequencing methods for long oligonucleotides (2), but can be useful for sequencing small oligonucleotides (especially RNA), for which traditional methods are more difficult to implement.

V. APPLICATIONS OF ESI–MS TO PROBLEMS IN NUCLEIC-ACID CHEMISTRY

Following are examples of the types of problems amenable to ESI–MS at the oligonucleotide level. While many of them constitute "demonstration of feasibility" as opposed to solution of an actual structure problem, they are nonetheless useful. Grouping within categories is arbitrary and designed to facilitate discussion of concepts and findings. Many questions in nucleic-acid chemistry and biochemistry can be answered by straightforward determination of the molecular weight of the targeted compound. As sample preparation conditions required to obtain high-quality mass spectra are generally understood (discussed above), accurate determination of molecular weight by ESI–MS can be considered a routine tool for nucleic-acid studies. The extent to which molecular-mass measurement of an oligonucleotide of unknown structure can be used to determine composition has been studied (47), and several interesting conclusions were reported. Measurement of molecular mass to the nearest-integer mass value permits direct determination of the chain length up to the 7-mer (DNA) or 8-mer (RNA) level. If the number of occurrences of any one nucleotide residue is known from specific enzymatic cleavage or selective modification, measurement of molecular mass to within ± 0.01% permits determination of composition for chains up to 14-mer in length, or up to 25-mer if chain length is also known. When base composition is not constrained, the number of allowable compositions increases exponentially with mass (47); the number of structure candidates for a given measurement can be reduced, however, if the mass can be measured to a greater degree of accuracy (e.g., using FTMS) (5).

Molecular-weight measurement can verify only the presumed composition, and not the sequence, of any molecule. In straightforward fashion, mass measurement of chemically (e.g., refs. 3,12–14,17,20,37,48,49) and enzymatically (11,26,28) derived oligonucleotides, and of natural nucleic acids (4,5,50), can be accomplished without interference from the presence of base or backbone modifications. In addition to verifying the composition of synthetic oligonucleotides, mass measurement reveals other important aspects of the synthesis protocol, such as successful incorporation of base or backbone modifications, complete removal of blocking groups, and the presence of "failure sequences" in the final product.

The difference between measured masses of two large oligonucleotides can also provide valuable information about the compositions of large oligonucleotides. A mutant *E. coli* initiator tRNA and its counterpart from a normal strain were mass-measured using ESI–MS; the mutant tRNA was found to be 114 Da lower in mass, consistent with the absence of 2-methylthio-N^6-isopentenyladenosine present at position 37 in the normal tRNA (50). The measurement of molecular mass provides direct evidence of the extent of modification of modified residues: partial modification would generate nucleic acids of differing masses. It was shown that convolution of the mass spectra of an oligonucleotide mixture to molecular weights gave a quantitative measure of the amounts of each component (3), and so information about the extent of modification at any one site may be obtained. This information is otherwise tedious to obtain, most commonly by HPLC with UV detection.

A. Oligonucleotide Mapping/Sequencing by Collision-Induced Dissociation

The derivation of a complete sequence by the ESI–MS and ESI–MS/MS techniques discussed above requires that every nucleotide be specfied within the sequence by its assignment to any one or more of the described ion series. To date, the largest oligonucleotide for which complete sequence information has been obtained is a 14-mer (34). Although there are a number of examples of the rationalization of fragment ions from ESI–MS/MS when the sequence is known, there are not yet any reports of the complete sequence determination of a true unknown structure.

Partial sequence information, however, provides an additional enhancement of the routine measurement of molecular weight (36,48). For example, for an oligonucleotide of known sequence, the identity and sequence location of one residue was changed; without prior knowledge of the new sequence, combination of nozzle-skimmer dissociation and CID provided a sufficient extent of overlapping sequence information to allow the placement of the new residue (36). The rapidity with which the mass spectrum can be acquired using FTMS, concomitant with high sensitivity and mass accuracy, makes this rapid-mapping technique attractive for partial sequence verification of large oligonucleotides of known sequence.

Partial sequence verification of a synthetic methylphosphonate oligonucleotide could also be obtained by CID (48). In this application, positive ions were detected because the methylphosphonate backbone is uncharged. Only a partial y ion series was observed, generated by homolytic backbone fragmentation, but the information provides greater surety of sequence correctness than molecular-weight measurement alone.

If an oligonucleotide contains a modified residue, its sequence location, in principle, can be determined from ESI–MS/MS (32,37). McLuckey has demonstrated that the N^6-methyladenine residue in the synthetic hexamer d(m^6ATGCAT) could correctly be located using CID in an ion trap (32). Figure 10 shows the CID spectra of the oligonucleotide and of the m/z 150 ion produced from MS/MS of the hexamer. The latter spectrum from the MS^3 experiment was used to show that the m/z 150 ion was consistent with the assignment of a methylated adenine structure to this ion, although isomer distinction was not possible because no ring elements were lost. The location of a known 5-methylcytosine residue in the hexamer analog d(ATGm^5CAT) could not be established owing to its failure to appear in any structurally informative ions in the MS/MS and MS^3 experiments (32). The potential of the MS^3 experiment for generating information about modified nucleobases is evident from this example; however, the approach is not general, at least with ion-trapping mass analyzers, owing to (apparent) identity and location requirements of the nucleobase.

B. Mapping of Nucleoside Modification in RNA and DNA

Protocols have been described whereby modified nucleosides can be detected in oligonucleotides derived from cellular RNA (11) and DNA (26,28) by digestion with endonucleases. In the RNA studies, the RNA is digested to nucleosides and screened by LC/MS to provide a census of modified nucleosides, which must ultimately be accounted for in the sequence. Next, the RNA is digested with

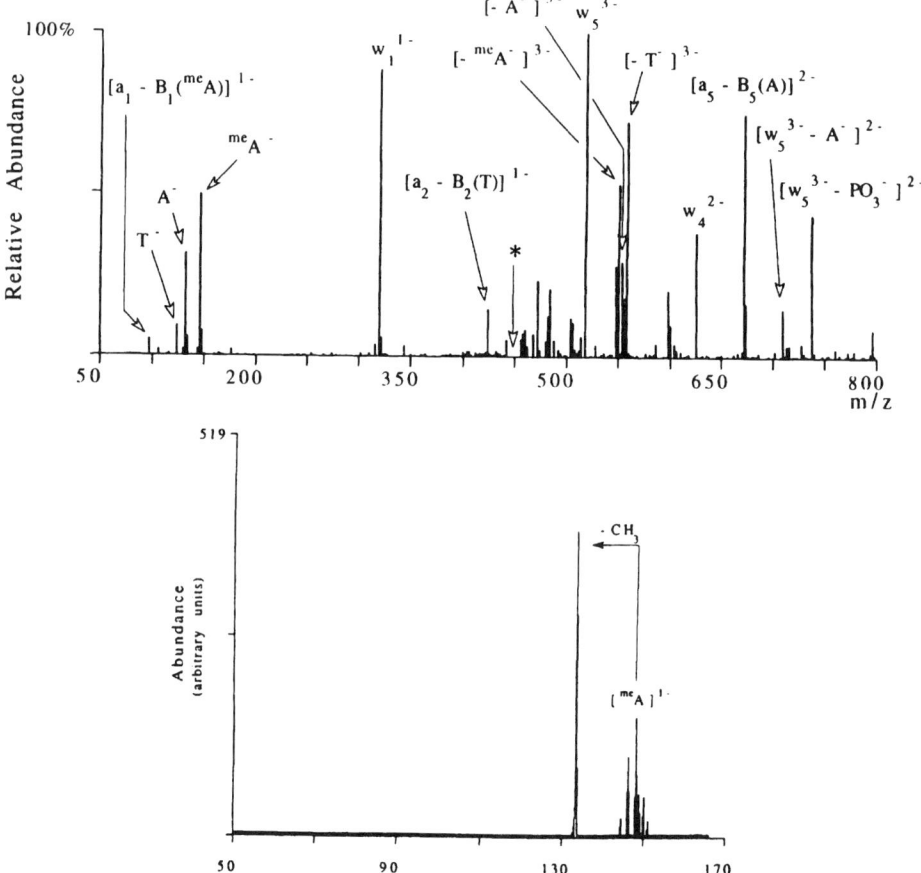

Figure 10. Tandem mass spectra of d(m⁶ATGCAT) in an ion-trap mass analyzer. Upper panel: product-ion spectrum from CID of the $(M-4H)^{4-}$ ion. Lower panel: product-ion spectrum of m/z 150 derived from CID of the $(M-4H)^{4-}$ ion. Reprinted by permission of Elsevier Science Inc. from S. A. McLuckey and S. Habibi-Goudarzi, "Ion trap tandem mass spectrometry applied to small multiply charged oligonucleotides with a modified base," *J. Am. Soc. Mass Spectrom.*, **5**, 740–747. © 1994 American Society for Mass Spectrometry.

ribonuclease T_1, which cleaves the RNA to give mixtures of oligonucleotides terminating in Gp. As discussed above, knowledge of the G content facilitates assignment of composition from molecular weight of the fragment (47). Following chromatographic fractionation, these T_1-derived oligonucleotides are then mass measured using ESI–MS to compare molecular weights of the fragments with those predicted from the gene sequence. When the M_r of the T_1-derived oligonucleotide matches a predicted value, no modification is present; oligonucleotides that contain modifications will not agree in mass with any predicted fragments and can then be singled out for further analysis. Using this approach, the thermally stable nucleoside N^4-acetyl-$2'$-O-methylcytidine was discovered, and its sequence location defined, in the 5S rRNA from *Pyrodictium occultum*, a hyperthermophilic archeum growing optimally at 105°C (51). Despite intensive analysis by classical methods over the past several decades, the 23S ribosomal RNA from *E. coli* was found by ESI–MS-based RNA mapping to contain incorrectly assigned known modifications (52) as well as a modified nucleoside not previously known in nature, 3-methylpseudouridine (53). 2-Thio-5-methyluridine (s^2T), a nucleoside whose mole percent in tRNA increases (T → s^2T) with culture of the hyperthermophilic archaeum *Pyrococcus furiosus* at progressively elevated temperatures, was mapped to the conserved –TΨCG– loop in tRNA (52).

Protocols for detecting xenobiotic nucleoside modification in DNA are being developed based on CZE/ESI–MS (26,28). Calf thymus DNA is reacted in vitro with styrene oxide, and digested to oligonucleotide fragments up to octamers by benzon nuclease, a nonspecific endonuclease. The modified products are then separated by CZE and detected by mass shifts compared with calculated masses of unmodified DNA fragments. Styrene oxide-modified guanosine monophosphate, used for CZE optimization, was shown to consist of both base- and phosphate-modified species by NS dissociation (28). The T moiety is not modified; small oligomers ($n = 2$–4) were singly modified, while pentamers were modified once or twice and hexamers generally twice. It is intended to ultimately apply this method to the detection of in vivo xenobiotic DNA modification at the oligonucleotide level (26,28), for which CZE/ESI–MS should be well suited.

C. Sequence Determination from Mass Measurement of Exonuclease Digestion Products

The use of exonucleases to determine the sequence of oligonucleotides is a time-honored technique for nucleic-acid structure determination (54). Phosphodiesterase I (from snake venom) sequentially removes $5'$-phosphorylated nucleotides from the $3'$-terminus of an oligonucleotide, and phosphodiesterase II (typically from calf spleen) sequentially removes $3'$-phosphorylated nucleotides from the $5'$-terminus. Identification of the mononucleotides as they are removed allows the sequence to be reconstructed (54). If the oligonucleotide to be sequenced contains a terminal phosphate group, it must be removed, because both enzymes are inhibited if the terminus where digestion is initiated is blocked. Matrix-assisted laser disorption ionization has been used to mass measure the residue masses after time-dependent removal of terminal nucleotides using phosphodiesterases I and II; the accuracy of mass measurement was insufficient to distinguish with certainty residues differing in mass by 1 Da (55). The generally greater accuracy of mass measurement using ESI-based methods makes them well suited for analysis of the products of exonuclease digestion, particularly for RNA-derived oligonucleotides, wherein the major pyrimidine subunits C and U (Fig. 1), in fact, differ in mass by 1 Da. Figure 11 shows the mass spectra of the

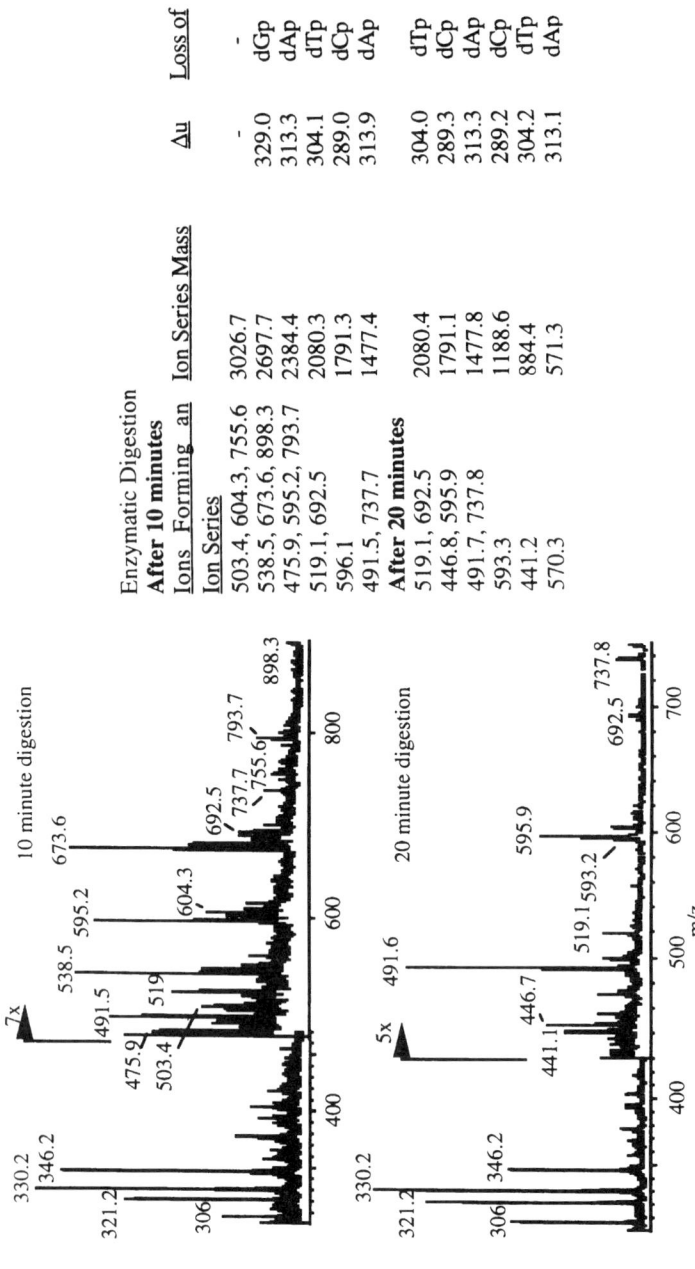

Figure 11. The ESI mass spectrum of a mixture of oligonucleotides derived from phosphodiesterase II digestion of d(GATCACTATG) at two different times. Adapted with permission from P. A. Limbach et al., *Nucleic Acids Res. Symp. Series No. 31,* **1994**, 127–128. © 1994 Oxford University Press.

phosphodiesterase II digestion products of d(GATCACTATG) at two different elapsed times, along with measured residue masses (56). The sample and enzyme are mixed and loaded into a syringe for sample delivery by infusion. Progress of the digestion is monitored in real time, and identities of both the released mononucleotide and the oligonucleotide residue can be determined, thus facilitating sequence assignment. Sample consumption was < 0.5 nmole for the example shown, which illustrates sequence determination in the $5' \rightarrow 3'$ direction. Dinucleotides are relatively resistant to phosphodiesterases, so the last two residues are not identified by digestion with analytically useful amounts of only one of the enzymes. Complete sequencing requires use of both enzymes, although only partial digestion with one of them would be required because the first two nucleotides released are the ones left undigested by the complementary enzyme. An essentially identical approach has been proposed, but with the important difference that the digestion mixture is sampled at discrete time intervals and the products are separated for off-line mass analysis by ESI–MS (57). The advantage of the latter approach is not obvious, because analysis times are longer and larger sample amounts (to compensate for losses in sample handling) will undoubtedly be required.

D. Probing Noncovalent Interactions of Nucleic Acids

Many, if not all, of the cellular processes in which nucleic acids participate involve bimolecular noncovalent interactions with other classes of molecules. Nucleic acid–nucleic acid, nucleic acid–protein, and nucleic acid–small molecule interactions are all important in cell function (and disfunction), and there are a number of reports in which ESI–MS has been used to probe different types of noncovalent interaction in which one or both of the participants is a nucleic acid (21,44,58–64). It is evident that the success of these measurements is critically dependent on both ion source (e.g., temperature, declustering voltage) and solution composition (e.g., sample concentration, electrolyte content, solvent composition) parameters. However, there are also reports that unpredicted, nonspecific, noncovalent interactions can also be detected (34,65). The perceptive overview by Smith and Light-Wahl (66) of the "promises, pitfalls, and prognosis" is required reading for investigators wishing to apply ESI–MS to detection of of noncovalent complexes.

The specific hybridization of self-complementary oligonucleotides to give the expected double-stranded forms has been observed using ESI–MS (21,58,59). The first account reported detection of hybridized 20-mers; the sequences of the two strands were designed to yield single strands of sufficiently differing mass to permit the detection of nonspecific associations (58). The strands were annealed in 10 mM ammonium acetate and analysis was carried out on this solution using an extended-mass-range quadrupole instrument, which permitted detection of charge states from 8– up to the doubly charged duplex ($\sim m/z$ 6100). The calculated mass of the duplex was higher than expected, suggesting either that desolvation was incomplete or that residual counterions remained.

A series of three 8-mer duplexes was analyzed and although all spectra showed predominantly ions corresponding to single strands, the expected duplex forms were present (59). When the spectrum of the self-complementary sequence d(C_4G_4) was compared with that of a mixture of the complementary single strands d$(A)_8$ + d$(T)_8$, a greater amount of duplex ion (3− charge state) was present from the all-CG (three base pairs/residue) duplex than from the all-AT (two base pairs/residue) duplex. This observation was concluded to reflect the greater stability of the all-CG duplex as a consequence of its greater total number of base pairs.

Correlation of duplex stability with chain length was demonstrated using ESI–MS of a series of hybrids in which one strand (a 20-mer) was invariant and the complementary strand was allowed to vary in length from 6- to 30-mer (21). The 20-mer/20-mer duplex gave the most abundant ion signals for the duplex. The 20-mer/30-mer duplex signals were less intense and were attributed to destabilization from the presence of higher charge states from charge accommodation on the overhanging ends of the 30-mer. The shorter complementary strands produced a smaller number of lower-abundance ions for the duplexes, attributed to their lower stability. Some interesting observations on sample preparation were reported. Samples were hybridized in 1 M ammonium acetate, and then exchanged with 10 mM ammonium bicarbonate eight times; ions for the duplex forms still showed broadening (attributed to residual ammonium acetate). Addition of a small amount of triethylamine increased ion signals of single-stranded ions tenfold, with little effect on duplex form ions (TEA in larger amount abolished the duplex form signals). The addition of 18-crown-6 (amount unspecified) increased duplex-form signals, but the effect (if any) on the single-stranded ions was not noted.

The analysis of a four-stranded (quadruplex) structure has also been reported and represents a particularly challenging application of ESI–MS (60). The oligomer examined, d(CGCGG_4GCG) (named dG4) is not self-complementary, but forms in solution a four-stranded complex interconnected by hydrogen bonds from the four interior guanine bases in the presence of monoatomic cations. Figure 12 shows the ESI spectra of dG4 acquired under different instrument and solvent conditions, which demonstrate that the solution stoichiometry is retained in the gas phase. It is noteworthy that the spectra were acquired from sodium phosphate buffer, which was required to permit the complex to form. At high nozzle-skimmer bias, only ions from the single strands are evident. If the nozzle-skimmer bias is lowered, ions from the quadruplex appear, while under the same instrument conditions, but in the absence of Na^+, exclusively the single strands are formed.

Figure 12. Negative-ion electrospray mass spectra of d(CGCG_4GCG) [abbreviated (dG4)]. (*A*) From 10 mM sodium phosphate (pH 7.6) + 0.1 mM EDTA; nozzle-skimmer bias −250 V. (*B*) Same solvent as in *A*; nozzle-skimmer bias −150 V. (*C*) In deionized water; nozzle-skimmer bias −150 V. Reprinted with permission from D. R. Goodlett et al., *Biol. Mass Spectrom.*, **1993**, *22*, 182. © 1993 John Wiley & Sons, Ltd.

V. APPLICATIONS OF ESI–MS TO PROBLEMS IN NUCLEIC-ACID CHEMISTRY 447

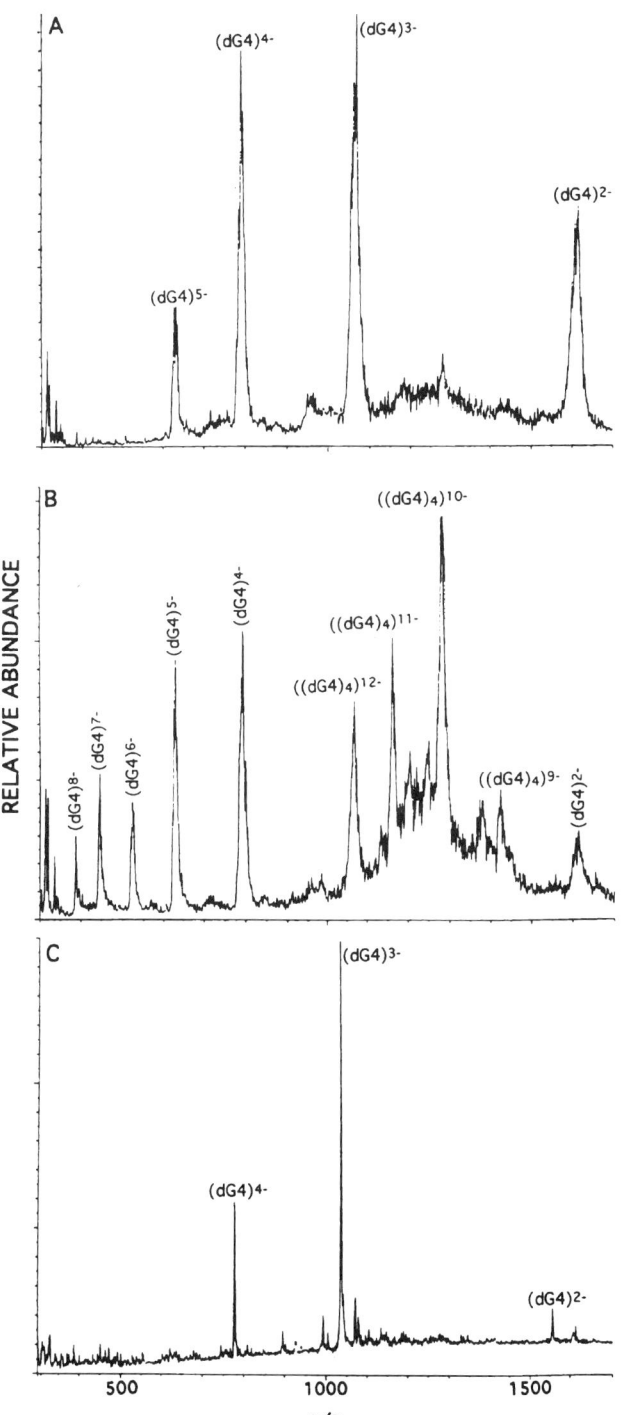

The interaction between distamycin (Dm), an antibiotic known to bind to AATT or ATTT sequences in double-stranded DNA (dsDNA), and a self-complementary 12-mer containing an –AAAATTTT– segment, has been reported (44). When dsDNA and Dm were mixed in a 2:1 molar ratio, the ES spectrum showed ions for dsDNA and the Dm·dsDNA complex in approximately equal abundance. When the dsDNA and Dm were present in solution in a 1:2 molar ratio, ions for the Dm_2·dsDNA duplex, the single-stranded DNA (ssDNA), and a Dm·ssDNA complex were observed, but neither dsDNA nor a 1:1 Dm·dsDNA complex were observed. Under the buffer and instrument conditions used, the Dm·dsDNA complexes observed in the mass spectrometer closely matched the ratios predicted from the amounts of material in solution.

Hedamycin (Hd) is an antitumor agent that both alkylates and intercalates into ds-DNA. Noncovalent interactions between the alkylated self-complementary hexamer CAC(Hd-G)TG and unmodified hexamer were examined using ESI–MS (63). When the modified oligomer was analyzed alone, no evidence for noncovalent self-association was seen in the mass spectrum, and attributed to the presence of a bulky modification in the center of the chain that would likely hinder hybridization. As increasing amounts of unmodified hexamer were added, ions from the duplex CAC(Hd-G)TG·CACGTG appeared and increased until one equivalent of unmodified hexamer was added. Only low-abundance ions were seen for modified and unmodified ss-DNAs. Under the most favorable conditions for observation of duplexes of unmodified hexamer, the duplex form represented only 25% relative to the ssDNA ions. These observations support evidence that following DNA alkylation, Hd can intercalate back into the strands, thereby increasing duplex stability.

Oligonucleotides are a new and potentially important class of "antisense" therapeutic and diagnostic agent (67). For many of their applications, interaction with cellular proteins is important, especially as it affects bioavailability. The interaction between an oligonucleotide 20-mer and serum albumin was studied using ESI–MS to compare this type of measurement against classical binding assays (64). The ESI spectra of mixtures of serum albumin and increasing amounts of 20-mer were acquired and the integrated ion abundances of the free and bound species were used to calculate dissociation constants. The values derived from the ESI–MS measurements were in close agreement with those obtained using capillary electrophoresis, suggesting the suitability of ESI–MS as a convenient and rapid technique for measuring stoichiometry and dissociation constants for protein–DNA complexes.

These studies all illustrate the (apparently) successful application of ESI–MS to detecting noncovalent interactions that are already predicted to occur in solution (21,44,58–61,63,64). Unfortunately, the detection of nonspecific duplex formation using ESI–MS has also been reported (23,34,65). Dimers of $d(A_7)$ and of $d(T_7)$ were inexplicably observed under conditions in which no duplexes were formed from any of several self-complementary DNAs of mixed composition (34). Hexameric homopolymers of the four DNA nucleotides each formed dimers at both neutral and acidic pH, although high sample concentrations were used (23).

Aspects of the specific and nonspecific formation of DNA dimers have been explored in an attempt to rationalize occurrence of the latter species (65). From analyzing a 1:1 mixture of two complementary hexamers (GAGTTC and GAACTC), the observed ion abundances gave a homodimer:heterodimer: homodimer ion ratio in reasonable agreement with the ratio predicted in the absence of specific interaction between any of the strands, calculated using binomial expansion. The melting temperature (T_m) of the duplex was calculated to be $-20.2°C$; it would not, therefore, be expected to exist in solution at room temperature or higher. A greater degree of specificity of duplex formation was evident when mixtures of longer oligomers were examined, as predicted from T_m considerations. These authors concluded that solution-based measures of duplex stability are not necessarily correlated to their gas-phase stability, and that what may be occurring is capture of transient solution complexes as gas-phase duplexes, which do not then dissociate in the absence of solvent. Inexplicably, in contrast with the notion that lower declustering voltage favors formation of noncovalent complexes (66), increasing the orifice potential actually increased duplex abundance in this study (65). The foregoing conclusions suggest that studies of noncovalent interactions should be undertaken with extreme caution, particularly regarding control measurements. If one accepts the hypothesis that solution-phase noncovalent complexes can be transferred to the gas phase, and one wishes to determine whether or not an "unknown" complex exists in solution, instrument conditions must be chosen to discriminate against artifactual aggregation. If a duplex is not predicted to occur in solution, it should not be observed in the gas phase in the context of such experiments.

VI. ANALYSIS OF NUCLEIC-ACID MONOMERIC CONSTITUENTS

The foregoing discussion has emphasized applications of ESI–MS at the oligonucleotide level; nonetheless, ESI–MS can also profitably be applied to the analysis of nucleotides and nucleosides [or bases (68–71)] derived enzymatically or chemically from nucleic acids. Two main advantages accrue for ESI–MS compared with other currently used ionization methods for nucleoside and nucleotide analysis, such as FAB (1,72) and thermospray (73). First, sensitivity is generally greater; attomole sensitivity has been demonstrated for adenosine, a well-behaved nucleoside (74), and 80-fmol detection is reported for carcinogen-modified guanine (71). Second, the extremely gentle nature of the ionization process is evident in the demonstrated ability to produce spectra from highly polar or thermally labile hypermodified nucleic acid constituents from DNA (75) and RNA (76) without degradation or excessive fragmentation. Particularly in combination with liquid-chromatographic sample introduction and tandem mass spectrometry, ESI–MS holds exceptional promise for application to nucleic-acid constituents.

A. Nucleotides and Nucleosides

Electrospray ionization produces very simple mass spectra for nucleosides, which can be detected as either positive or negative ions. Figure 13 shows the ESI mass spectra of a modified deoxyribonucleoside, 8-oxo-deoxyadenosine (77), and a stylized nucleoside structure to illustrate the fragment ion produced. A detailed study of a selection of ribo- and deoxyribonucleosides provides useful insights into nucleoside spectra and the extent to which they are affected by instrument conditions (77). Detected in the positive-ion mode (5–10 times more sensitive than negative ions), nucleosides generally produce only a protonated molecular ion (MH^+) and the protonated base (designated BH_2^+; Fig. 13). The abundance of the latter ion depends on the declustering voltage: low nozzle-skimmer potential difference favors MH^+ formation and, as the declustering voltage is increased (in-source CID), fragmentation of MH^+ to give the BH_2^+ ion is favored, and the deoxyribose ion is also produced in low abundance. Nucleosides can also form cluster ions [e.g., $(2M + H)^+$ (74, 77)], undoubtedly in a source-tuning-dependent manner. The negative-ion ESI spectra of nucleosides show generally only the $(M-H)^-$ ion; the B^- ion is not routinely observed under typical source conditions, but can be induced via in-source CID (77).

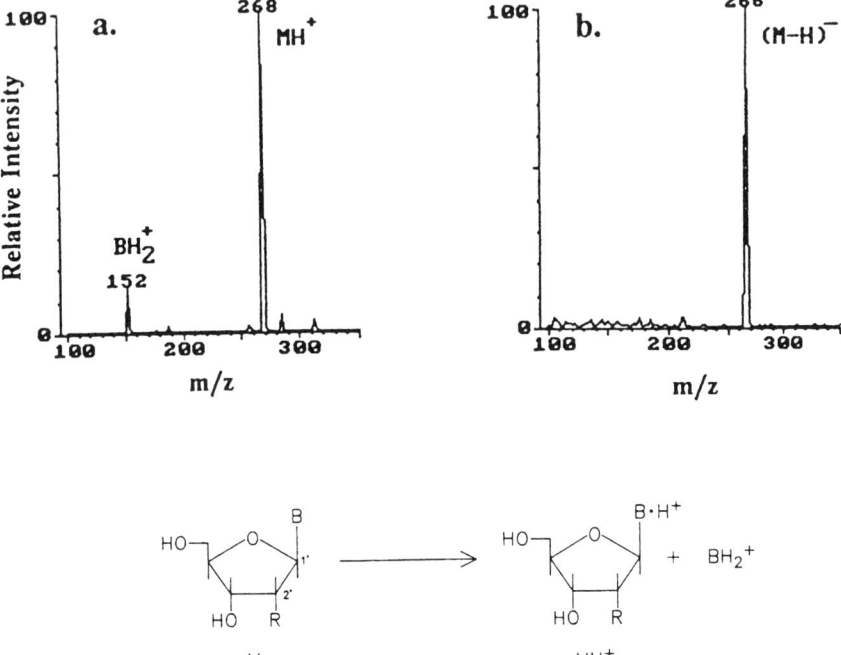

Figure 13. Electrospray mass spectra of 8-oxo-deoxyadenosine. (*a*) Positive ions detected (*b*) Negative ions detected. Adapted with permission from D. M. Reddy and C. R. Iden, *Nucleosides & Nucleotides*, **1993**, *12*, 821. © 1993 Marcel Dekker, Inc.

Electrospray ionization is particularly well suited for analysis of mononucleotides because they already exist as anions in solution. Fewer enzymatic treatments are required to produce nucleotides rather than nucleosides from nucleic acids, and analysis at the nucleotide stage may thus be especially favorable for labile constituents. The ESI mass spectra of nucleotides (40,77,78) typically consist of the $(M-H)^-$ ion, with little production of the B^- or $(M-BH)^-$ ions except at higher than optimal declustering voltages (77,78).

The ESI mass spectra of disodium thymidine 3'-monophosphate have been acquired under a variety of instrument conditions and some interesting and useful observations were reported (78). A relatively high capillary voltage (3.5 kV) produced the greatest abundance of $(M-H)^-$; lowering the voltage to 2.2–2.5 kV maximized production of $(M-2H+Na)^-$ and induced formation of $(M-2H)^{2-}$ and related water-cluster ions. Collision-induced dissociation of the $(M-H)^-$ ion gave the same fragment ions as were produced from in-source CID (78).

Because the ESI mass spectra of nucleosides and nucleotides are sensitive to ion-source conditions, tuning can be adjusted to favor one or the other major ion type [MH^+ or BH_2^+; $(M-H)^-$ or B^-]. At lower declustering voltage, production of molecular ion species is favored, thus enhancing detectability. If fragmentation is desired, for structure characterization or to produce an ion (e.g., BH_2^+) for subsequent examination by CID, it can be induced by in-source CID, thus making ESI an especially versatile technique for structure elucidation of modified nucleic-acid constituents.

B. Nucleoside Mixture Analysis by LC/ESI/MS

Directly combined LC/ESI–MS offers a particularly attractive approach to the detection and characterization of modified nucleosides directly from enzymatic digests of nucleic acids (70,74,79) or present in cell extracts (80). Reversed-phase separations can readily be accomplished utilizing volatile buffers at low ionic strengths that are compatible with ESI.

Products of the in vitro reaction between *cis*-dichlorodiamine platinum (II) (CDDP) and deoxyguanosine (dG), deoxyadenosine (dA), and DNA were characterized using microbore reversed-phase LC/ESI–MS (79). The CDDP monoadducts and dinucleosides cross linked by a CDDP group were characterized from reaction of CDDP with free nucleosides. One cross-linked dG derivative was reported (N7–N7), as were three isomers for dA (N1–N1, N1–N7, and N7–N7). Reaction of CDDP with a mixture of dA and dG yielded, in addition, mixed dA–CDDP–dG dinucleosides. Both types of dG adduct were produced from the reaction between DNA and CDDP, as well as the mixed dA–CDDP–dG dinucleoside; dA adducts were not observed. Evidence was obtained for dinucleoside monophosphates in which the nucleobases were cross linked, as a probable consequence of resistance of this product to enzyme digestion. Single-ion monitoring gave a limit of detection (S/N = 2) of 2 pmol injected, based on detection of the MH^+ ion.

The analysis of a digest of 15 pmol (about 0.4 μg) of transfer RNA by microbore LC/ESI–MS has been described (74). Spectra of the constituent nucleosides showed predominantly MH$^+$ ions, with varying amounts of BH$_2^+$ ions. A modified nucleoside present at a level of one residue per tRNA molecule corresponds to about 1.3 mole %; spectra were obtained for all but three known minor nucleosides, demonstrating the impressive sensitivity of the method. No ions could be detected for the minor nucleosides pseudouridine, 5-methyluridine, and wyeosine (a hypermodified guanosine). The failure to detect trace levels of the two modified uridines is not unexpected given their low pK_a and the water–methanol solvent system used for the separation. The missing wyeosine was not commented upon.

The combination of LC/ESI–MS and tandem mass spectrometry represents an especially powerful approach to characterization of modified nucleosides from nucleic acid digests. A useful general method is illustrated by the analysis of the major malondialdehyde–DNA adduct (designated M$_1$G-dR) in human liver (70). Liver DNA is digested to nucleosides and screened initially by LC/ESI–MS/MS using a constant neutral-loss scan (for loss of 116 Da, or deoxyribose, from the MH$^+ \to$ BH$_2^+$ decomposition). Figure 14 shows the chromatograms locating the component(s) in DNA that lose 116 Da from a precursor ion of m/z 304. The peak so identified, presumed to be M$_1$G-dR, showed m/z 326 (MNa$^+$) in addition to m/z 304, confirming that m/z 304 represented the MH$^+$ ion of M$_1$G-dR. The lower portion of Figure 14 is the product-ion spectrum from CID of the BH$_2^+$ ion (m/z 188), which was comparable to that of authentic M$_1$G-dR. Although the method was described as less sensitive than negative-ion chemical ionization GC/MS of electrophoric derivatives for detection of DNA adducts, its ease of implementation and capability to obtain structural information makes the approach an attractive one for locating and characterizing these molecules.

Cytokinins are a class of N^6-isopentenylated adenosine derivatives involved in plant growth regulation. A method has been described for their analysis in plant tissue extracts using LC/ESI–MS/MS (80). The chromatographic conditions used were a compromise between speed of analysis and nucleoside resolution; 16 different cytokinins could be determined in a 4-min analysis. Quantification was accomplished by multiple-reaction monitoring of appropriate MH$^+$ and product ions for each analyte and its corresponding deuterated analog. Response was linear over the range 1–100 pmol and a detection limit of 1 pmol injected was achieved (80).

VII. SUMMARY AND PROSPECTS

Given attention to sample preparation, molecular masses of oligonucleotides up to 120 nucleotides in length can be measured to within 0.01% or better. Exploratory investigations of noncovalent interactions in which one or more participants is a nucleic acid suggest that ESI–MS can successfully be used for this application, but experimental conditions must carefully be controlled to

VII. SUMMARY AND PROSPECTS   453

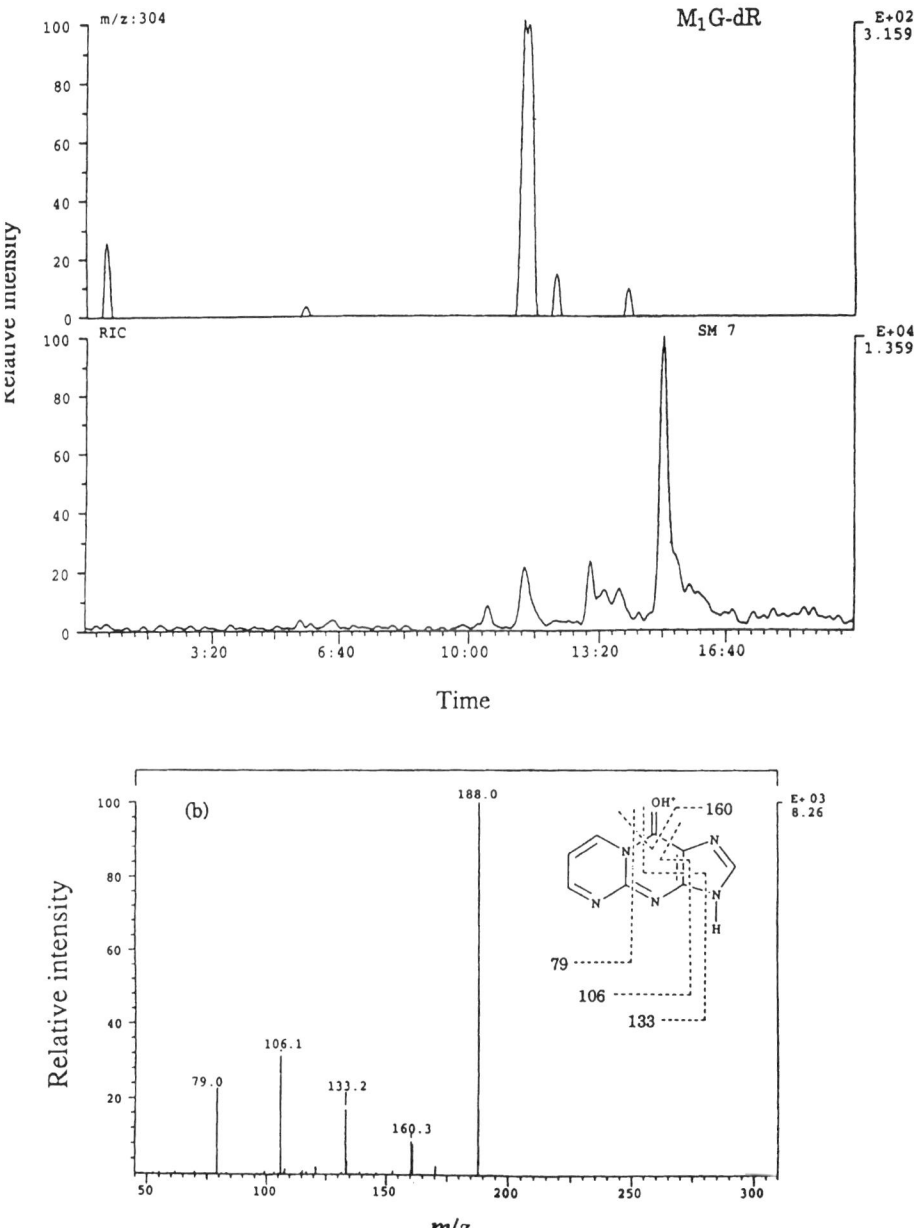

Figure 14. An LC/ESI/MS/MS analysis of an enzymatic digest of human liver DNA. Upper: chromatographic traces from MH^+ of the malondialdehyde-dG adduct M_1G-dR (m/z 304 trace) and total ion current. Lower: product-ion spectrum from m/z 188, the BH ion of M_1G-dR. Reprinted with permission from A. K. Chaudhary et al., *J. Mass Spectrom.*, **1995**, *30*, 1163–1165. © 1995 John Wiley & Sons, Ltd.

avoid artifactual aggregation where a noncovalent complex should not exist. Measurements of this type can be of great value in situations where traditional methods are difficult to apply, owing to sample limitations, for example, to measure the binding of "antisense" therapeutic agents to cellular proteins. There are presently few reports of application of ESI–MS, particularly LC/ESI–MS, for analysis of monomeric nucleic-acid constituents. There is every indication that it provides a simple and sufficiently sensitive method for screening nucleic-acid digests for the occurrence of naturally and xenobiotically modified nucleosides, despite the fact that greater sensitivity (but less specificity) is obtained compared with ^{32}P-based methods.

There is presently intense interest in direct sequencing of oligonucleotides by CID, but the number of studies reported is far too few, particularly for the widely used triple-quadrupole instruments, compared to ion-trapping mass analyzers, to assess limitations in length and composition for which full sequence information can be derived. Gas-phase sequencing is nonetheless useful, even if complete sequence information is limited to shorter oligonucleotides. Partial sequence information can always be obtained, and is clearly of greater value than measurement of molecular mass alone. Nozzle-skimmer fragmentation can even be produced in single quadrupole instruments, thus providing access to partial sequence information in this fashion.

The impact and value of the ability to accurately measure molecular weights of (formerly) intractable oligonucleotides cannot be overestimated as it becomes increasingly utilized through a range of applications, from quality assessment of synthetic oligonucleotides to structural characterization of natural nucleic acids. Although not discussed in detail herein, the recent advent of "microspray" interfaces (43–46) that reduce sample consumption to the subfemtomole level will greatly facilitate all applications of ESI–MS to nucleic acids and their constituents, most significantly sequencing efforts and oligonucleotide mass analysis at trace levels of occurrence. At present, there are no impediments to the use of ESI–MS for the routine analysis of small nucleic acids and their constituents.

ACKNOWLEDGMENTS

I thank Richard Griffey and Michael Linscheid for providing manuscript preprints, and James McCloskey and Steven Pomerantz (who also assisted with figure preparation) for helpful discussions. Preparation of this review was supported in part by NIH grant GM 21584.

REFERENCES

1. Crain, P. F. *Mass Spectrom. Rev.* **1990**, *9*, 505–554.
2. Limbach, P. A.; Crain, P. F.; McCloskey, J. A. *Curr. Opin. Biotechnol.* **1995**, *6*, 96–102.

3. Potier, N.; Van Dorsselaer, A.; Cordier, Y.; Roch, O.; Bischoff, R. *Nucleic Acids Res.* **1994**, *22*, 3895–3903.
4. Limbach, P. A.; Crain, P. F.; McCloskey, J. A. *J. Am. Soc. Mass Spectrom.* **1995**, *6*, 27–39.
5. Little, D. P.; Thannhauser, T. W.; McLafferty, F. W. *Proc. Natl. Acad. Sci. U.S.A.* **1995**, *92*, 2318–2322.
6. Chen, R.; Cheng, X.; Mitchell, D. W.; Hofstadler, S. A.; Wu, Q.; Rockwood, A. L.; Sherman, M. G.; Smith, R. D. *Anal. Chem.* **1995**, *67*, 1159–1163.
7. Saenger, W. *Principles of Nucleic Acid Structure*; Springer-Verlag: New York, 1984.
8. Phillips, D. R.; McCloskey, J. A. *Int. J. Mass Spectrom. Ion Processes* **1993**, *128*, 61–82.
9. Haskins, N. J.; Ashcroft, A. E.; Phillips, A.; Harrison, M. *Rapid Commun. Mass Spectrom.* **1994**, *8*, 120–125.
10. Smith, R. D.; Loo, J. A.; Edmonds, C. G.; Barinaga, C. J.; Udseth, H. R. *Anal. Chem.* **1990**, *62*, 882–889.
11. Kowalak, J. A.; Pomerantz, S. C.; Crain, P. F.; McCloskey, J. A. *Nucleic Acids Res.* **1993**, *21*, 4577–4585.
12. Stults, J. T.; Marsters, J. C. *Rapid Commun. Mass Spectrom.* **1991**, *5*, 359–363.
13. Greig, M.; Griffey, R. H. *Rapid Commun. Mass Spectrom.* **1995**, *9*, 97–102.
14. Fearon, K. L.; Stults, J. T.; Bergot, B. J.; Christensen, L. M.; Raible, A. M. *Nucleic Acids Res.* **1995**, *23*, 2754–2761.
15. Bleicher, K.; Bayer, E. *Biol. Mass Spectrom.* **1994**, *23*, 320–322.
16. Straub, R. F.; Voyksner, R. D. *J. Am. Soc. Mass Spectrom.* **1993**, *4*, 578–587.
17. Covey, T. R.; Bonner, R. F.; Shushan, B. I.; Henion, J. *Rapid Commun. Mass Spectrom.* **1988**, *2*, 249–256.
18. Deroussent, A.; Le Caer, J.-P.; Rossier, J.; Gouyette, A. *Rapid Commun. Mass Spectrom.* **1995**, *9*, 1–4.
19. McLuckey, S. A.; Van Berkel, G. J.; Glish, G. L. *J. Am. Soc. Mass Spectrom.* **1992**, *3*, 60–70.
20. Reddy, D. M.; Rieger, R. A.; Torres, M. C.; Iden, C. R. *Anal. Biochem.* **1994**, *220*, 200–207.
21. Bayer, E.; Bauer, T.; Schmeer, K.; Bleicher, K.; Maier, M.; Gaus, H.-J. *Anal. Chem.* **1994**, *66*, 3858–3863.
22. Cheng, X.; Gale, D. C.; Udseth, H. R.; Smith, R. D. *Anal. Chem.* **1995**, *67*, 586–593.
23. Tong, X.; Henion, J.; Ganem, B. *J. Mass Spectrom.* **1995**, *30*, 867–871.
24. Bayer, E. In *Modern Methods in Protein-Nucleic Acid Research*; Tschesche, H. Ed.; de Gruyter: Berlin, 1990; pp. 133–148.
25. Moseley, M. A.; Unger, S. E. *Proceedings of the 40th Annual Conference on Mass Spectrometry and Allied Topics*, Washington, DC, May 31–June 5, 1991, pp. 608–609.
26. Janning, P.; Schrader, W.; Linscheid, M. *Rapid Commun. Mass Spectrom.* **1994**, *8*, 1035–1040.
27. Zhao, Z.; Wahl, J. H.; Udseth, H. R.; Hofstadler, S. A.; Fuciarelli, A. F.; Smith, R. D. *Electrophoresis* **1995**, *16*, 389–395.
28. Schrader, W.; Linscheid, M. *J. Chromatogr. A* **1995**, *717*, 117–125.
29. Martin, L. B.; Mamer, C. S.; Rieger, R. A.; Iden, C. A. *Proceedings of the 41st Annual Conference on Mass Spectrometry and Allied Topics*, San Francisco, CA, May 31–June, 1992, pp. 247a–b.

30. Bleicher, K.; Bayer, E. *Chromatographia* **1994**, *39*, 405–408.
31. McLuckey, S. A.; Habibi-Goudarzi, S. *J. Am. Chem. Soc.* **1993**, *115*, 12085–12095.
32. McLuckey, S. A.; Habibi-Goudarzi, S. *J. Am. Soc. Mass Spectrom.* **1994**, *5*, 740–747.
33. McLuckey, S. A.; Gopalakrishnan, V.; Habibi-Goudarzi, S. *J. Mass Spectrom.* **1995**, *30*, 1222–1229.
34. Little, D. P.; Chorush, R. A.; Speir, J. P.; Senko, M. W.; Kelleher, N. L.; McLafferty, F. W. *J. Am. Chem. Soc.* **1994**, *116*, 4893–4897.
35. Little, D. P.; Speir, J. P.; Senko, M. W.; O'Connor, P. B.; McLafferty, F. W. *Anal. Chem.* **1994**, *66*, 2809–2815.
36. Little, D. P.; McLafferty, F. W. *J. Am. Chem. Soc.* **1995**, *117*, 6783–6784.
37. Barry, J. P.; Vouros, P.; Van Schepdael, A.; Law, S.-J. *J. Mass Spectrom.* **1995**, *30*, 993–1006.
38. Crain, P. F.; Gregson, J. M.; McCloskey, J. A.; Nelson, C. C.; Peltier, J. M.; Phillips, D. R.; Pomerantz, S. C.; Reddy, D. M. In *Mass Spectrometry in the Biological Sciences*; Burlingame, A. L.; Carr, S. A. Ed.; Humana Press: Totawa, NJ., 1996; pp. 497–517.
39. Cerny, R. L.; Gross, M. L.; Grotjahn, L. *Anal. Biochem.* **1986**, *156*, 424–435.
40. Habibi-Goudarzi, S.; McLuckey, S. A. *J. Am. Soc. Mass Spectrom.* **1995**, *6*, 102–113.
41. Rodgers, M. T.; Campbell, S.; Marzluff, E. M.; Beauchamp, J. L. *Int. J. Mass Spectrom. Ion Processes* **1994**, *137*, 121–149.
42. Pomerantz, S. C.; McCloskey, J. A. *Proceedings of the 43rd Annual Conference on Mass Spectrometry and Allied Topics*, Atlanta, GA, May 21–26, 1995, p. 600.
43. Emmett, M. R.; Caprioli, R. M. *J. Am. Soc. Mass Spectrom.* **1994**, *5*, 604–613.
44. Gale, D. C.; Goodlett, D. R.; Light-Wahl, K. J.; Smith, R. D. *J. Am. Chem. Soc.* **1994**, *116*, 6027–6028.
45. Valaskovic, G. A.; Kelleher, N. L.; Little, D. P.; Aaserud, D. J.; McLafferty, F. W. *Anal. Chem.* **1995**, *87*, 3802–3805.
46. Greig, M. J.; Gaus, H. J.; Griffey, R. H. *Rapid Commun. Mass Spectrom.* **1996**, *10*, 47–50.
47. Pomerantz, S. C.; Kowalak, J. A.; McCloskey, J. A. *J. Am. Soc. Mass Spectrom.* **1993**, *4*, 204–209.
48. Baker, T. R.; Keough, T.; Dobson, R. L. M.; Riley, T. A.; Hasselfield, J. A.; Hesselberth, P. E. *Rapid Commun. Mass Spectrom.* **1993**, *7*, 190–194.
49. Doktycz, M. J.; Hurst, G. B.; Habibi-Goudarzi, S.; McLuckey, S. A.; Tang, K.; Chen, C. H.; Uziel, M.; Jacobson, K. B.; Woychik, R. P.; Buchanan, M. V. *Anal. Biochem.* **1995**, *230*, 205–214.
50. Mangroo, D.; Limbach, P. A.; McCloskey, J. A.; RajBhandary, U. L. *J. Bacteriol.* **1995**, *177*, 2858–2862.
51. Bruenger, E.; Kowalak, J. A.; Kuchino, Y.; McCloskey, J. A.; Mizushima, H.; Stetter, K. O.; Crain, P. F. *FASEB J.* **1993**, *7*, 196–200.
52. Kowalak, J. A.; Dalluge, J. J.; McCloskey, J. A.; Stetter, K. O. *Biochemistry* **1994**, *33*, 7869–7876.
53. Kowalak, J. A.; Bruenger, E.; Hashizume, T.; Peltier, J. M.; Ofengand, J.; McCloskey, J. A. *Nucleic Acids Res.* **1996**, *24*, 688–693.
54. Holley, R. W.; Madison, J. T.; Zamir, A. *Biochem. Biophys. Res. Commun.* **1964**, *17*, 389–394.

55. Pieles, U.; Zürcher, W.; Schär, M.; Moser, H. E. *Nucleic Acids Res.* **1993**, *21*, 3191–3196.
56. Limbach, P. A.; McCloskey, J. A.; Crain, P. F. *Nucleic Acids Res. Symp. Series No. 31* **1994**, 127–128.
57. Glover, R. P.; Sweetman, G. M. A.; Farmer, P. B.; Roberts, G. C. K. *Rapid Commun. Mass Spectrom.* **1995**, *9*, 897–901.
58. Light-Wahl, K. J.; Springer, D. L.; Winger, B. E.; Edmonds, C. G.; Camp, D. G., II; Thrall, B. D.; Smith, R. D. *J. Am. Chem. Soc.* **1993**, *115*, 803–804.
59. Ganem, B.; Li, Y.-T.; Henion, J. D. *Tetrahedron Lett.* **1993**, *34*, 1445–1448.
60. Goodlett, D. R.; Camp, D. G., II; Hardin, C. C.; Corregan, M.; Smith, R. D. *Biol. Mass Spectrom.* **1993**, *22*, 181–183.
61. Griffith, M. C.; Risen, L. M.; Greig, M. J.; Lesnik, E. A.; Sprankle, K. G.; Griffey, R. H.; Kiely, J. S.; Freier, S. M. *J. Am. Chem. Soc.* **1995**, *117*, 831–832.
62. Doktycz, M. J.; Habibi-Goudarzi, S.; McLuckey, S. A. *Anal. Chem.* **1994**, *66*, 3416–3422.
63. Wickham, G.; Iannitti, P.; Boschenok, J.; Sheil, M. M. *FEBS Lett.* **1995**, *360*, 231–234.
64. Greig, M. J.; Gaus, H.; Cummins, L. L.; Sasmor, H.; Griffey, R. *J. Am. Chem. Soc.* **1995**, *117*, 10765–10766.
65. Ding, J.; Anderegg, R. J. *J. Am. Soc. Mass Spectrom.* **1995**, *6*, 159–164.
66. Smith, R. D.; Light-Wahl, K. J. *Biol. Mass Spectrom.* **1993**, *22*, 493–501.
67. Agrawal, S.; Iyer, R. P. *Curr. Opin. Biotechnol.* **1995**, *6*, 12–19.
68. Ramsey, R. S.; Van Berkel, G. J.; McLuckey, S. A.; Glish, G. L. *Biol. Mass Spectrom.* **1992**, *21*, 347–352.
69. La, D. K.; Lilly, P. D.; Anderegg, R. J.; Swenberg, J. A. *Carcinogenesis* **1995**, *16*, 1419–1424.
70. Chaudhary, A. K.; Nokubo, M.; Oglesby, T. D.; Marnett, L. J.; Blair, I. A. *J. Mass Spectrom.* **1995**, *30*, 1157–1166.
71. Rindgren, D.; Turesky, R. J.; Vouros, P. *Chem. Res. Toxicol.* **1995**, *8*, 1005–1013.
72. Slowikowski, D. L.; Schram, K. H. *Nucleosides Nucleotides* **1985**, *4*, 309–345.
73. Pomerantz, S. C.; McCloskey, J. A. *Methods Enzymol.* **1990**, *193*, 796–824.
74. Banks, J. F., Jr.; Shen, S.; Whitehouse, C. M.; Fenn, J. B. *Anal. Chem.* **1994**, *66*, 406–414.
75. Gommers-Ampt, J. H.; Van Leeuwen, F.; de Beer, A. L. J.; Vliegenthart, J. F. G.; Dizdaroglu, M.; Kowalak, J. A.; Crain, P. F.; Borst, P. *Cell* **1993**, *75*, 1129–1136.
76. Crain, P. F.; McCloskey, J. A. In *Biological Mass Spectrometry: Present and Future*; Matsuo, T., Caprioli, R. M.; Gross, M. L. Ed.; Wiley: Chichester, 1994; pp. 509–537.
77. Reddy, D. M.; Iden, C. R. *Nucleosides & Nucleotides* **1993**, *12*, 815–826.
78. Hernándes, H.; Lamb, J. H.; Farmer, P. B.; Eaton, G. *Rapid Commun. Mass Spectrom.* **1995**, *9*, 870.
79. Da Col, R.; Silvestro, L.; Baiocchi, C.; Giacosa, D.; Viano, I. *J. Chromatogr.* **1993**, *633*, 119–128.
80. Prinsen, E.; Redig, P.; Van Dongen, W.; Esmans, E. L.; Van Onckelsen, H. A. *Rapid Commun. Mass Spectrom.* **1995**, *9*, 948–953.

CHAPTER 13

Electrospray Ionization Mass Spectrometry of Carbohydrates and Lipids

YOKO OHASHI

Institute of Physical and Chemical Research (RIKEN), Wako, Saitama, Japan

	Abstract	460
I.	Introduction	460
II.	General features of ESI studies of carbohydrates and lipids	461
III.	Derivatized glycoconjugates	462
IV.	Underivatized, neutral oligosaccharides and glycoconjugates	463
V.	Acidic glycoconjugates	467
	A. Gangliosides (12)	467
	1. Conventional mass spectrometry	467
	2. CID–MS/MS (12)	469
	B. Sulfated Lewis-type oligosaccharides	473
	1. Conventional mass spectrometry of sulfated oligosaccharides (17)	473
	2. CID–MS/MS of sulfated oligosaccharides (17)	475
	C. Sulfated and sialyl Lewis-type glycosphingolipids	476
	1. Negative-ion conventional mass spectrometry of sulfated and sialyl Lewis-type glycosphingolipids (18)	476
	2. Negative-ion CID–MS/MS of Lewis-type glycosphingolipids (18)	480
	3. Positive-ion conventional mass spectrometry of Lewis-type glycosphingolipids (1)	483
	4. Positive-ion CID–MS/MS of Lewis-type glycosphingolipids (1)	483
VI.	Noncovalent binding involving sugar chains	485
VII.	Acylglycerols	488
VIII.	Lipid A	488
IX.	Glycosylphosphatidylinositol membrane anchors (GPI anchors)	491
X.	Phospholipids	492
XI.	Other lipids studied with ESI–MS	494

Electrospray Ionization Mass Spectrometry, Edited by Richard B. Cole.
ISBN 0-471-14564-5 © 1997 John Wiley & Sons, Inc.

XII. Conclusions 494
Acknowledgments 496
References 496

ABSTRACT

In the past 15 years several new mass spectrometric methods have been developed to determine the structure of extremely large biological complexes. Electrospray ionization–mass spectrometry (ESI–MS) has proven to be an excellent analytical method for structural studies of carbohydrates and lipids. Collision-induced dissociation–tandem mass spectrometry (CID–MS/MS) is an indispensable technique in ESI–MS to elucidate structures, especially because, unlike conventional mass spectrometry, CID–MS/MS can distinguish between isomers that often occur in natural compounds. Fast atom bombardment mass spectrometry (FAB–MS), which shows fragment ions even in the conventional mass spectra, is probably a method of choice if the sample is amenable to FAB ionization and has a sample size larger than a few micrograms. However, FAB often fails to ionize highly acidic compounds, and produces singly charged ions only, making it incapable of analyzing molecules whose molecular weights exceed the mass range of the analyzer. Thus ESI–MS with CID–MS/MS is currently the best analytical method for large glycoconjugates and lipids, especially if they are highly acidic.

Furthermore, ESI–MS complements or reinforces information obtained by FAB even for smaller molecules.

I. INTRODUCTION

Electrospray ionization–mass spectrometry, like other mass-spectrometric methods developed in recent years, is used primarily to mass analyze proteins which are not only difficult to ionize, but also have too large a molecular size to be analyzed with a standard mass spectrometer. Electrospray ionization is able to produce multiply charged ions, each with adequately small m/z values, thus large molecules represented by proteins are often analyzed using ESI. Additionally, daughter ions of multiply charged parent ions may be produced, which often yield structural information complementary to information obtained from daughters of singly charged ions. However, we have realized that while ESI–MS is a valuable technique for analyzing proteins, it is sometimes even more useful in structural studies of carbohydrates and lipids, even though their molecular sizes may not exceed the mass range of the analyzer. Here "carbohydrates and lipids" include oligosaccharides and their simple glycosides, neutral as well as acidic glycosphingolipids (e.g., gangliosides and sulfated glycoconjugates, glycoglycerolipids, phospholipids, lipid-A, lipopolysaccharides, GPI-anchor, and sphingomyelin). As a structural analysis technique for carbohydrates and lipids, ESI–MS is useful because processes such as desalting a sample, adding organic solvents, and applying heat to evaporate solvents, all of which are considered harsh to protein conformations, do not seriously affect the mass analysis of carbohydrates and lipids, except for binding studies, in which case conformation plays an important role.

II. GENERAL FEATURES OF ESI STUDIES OF CARBOHYDRATES AND LIPIDS

The ESI–MS studies of glycoconjugates were carried out in our lab using a triple-stage quadrupole mass spectrometer. Samples are usually dissolved in commonly used solvents such as a mixture of chloroform and methanol for conjugated general lipids, and methanol or methanol–water for oligosaccharides and polysialogangliosides. Thirty to sixty microliters of a 10- to 50-pmole/μL sample solution is mechanically driven to the ESI needle at a flow rate of 1 μL/min. A potential difference of $(-)$ or $(+)$ 3 kV is applied between the needle, which is kept at the ground potential, and the interior of the ion source. Hot nitrogen gas is used to evaporate the solvent from the charged droplets. Argon gas is used to induce collisionally activated dissociation at a pressure of 1.0–1.3 mTorr at 30–35 eV, but a lower collision energy is applied to dissociate multiply charged ions. Although ESI–MS is definitely more sensitive than FAB, it is much less sensitive than matrix-assisted laser desorption ionization (MALDI). In most cases, a 10-μg sample will be adequate for the conventional mass spectrometry of ESI in both polarities. It is safer, however, to provide 20–30 μg of a carbohydrate or lipid sample to perform CID–MS/MS studies, even when the sample is apparently pure.

Fragmentation patterns shown in ESI CID–MS/MS are much the same as in FAB CID–MS/MS, if the same instrument is used for both ionization modes. Not only ESI CID–MS/MS spectra of a singly charged ion, but even daughter spectra of a doubly charged parent have a good resemblance to FAB CID–MS/MS of the same parent moiety in the positive-ion mode (1). We assume, therefore, that once a positive charge is carried away from a doubly charged ion by a small cation such as a H^+ or Na^+, the remaining singly charged daughter ion has a similar structure and energy to that of the corresponding singly charged ion regardless of whether the FAB or ESI ionization mode was used.

Because ESI is a mild ionization method, ions observed in the conventional mass spectra of ESI are those that represent the whole molecule and not the fragments, unless skimmer CID is applied. This is usually thought to be advantageous, but it is disadvantageous if one wishes to choose an ion representing a part of the molecule as the precursor for an in-source fragmentation spectrum. Conversely, FAB provides fragment ions that can be utilized as parents of the third-generation product ions provided that the particular fragment ion has an adequate intensity, while precursor ions for ESI CID–MS/MS should be ionic forms of whole molecules of various charge states.

Samples of conjugated lipids are commonly dissolved in a mixture of chloroform and methanol in various ratios. These solvent systems, however, produce abnormal but interesting adduct ions in the negative-ion mode ESI–MS (2). That is, in addition to the normally expected molecular-weight-related negative ions such as $(M - H)^-$ or $(M' - Na)^-$ (in the case of sodium salts), negative ions produced by an addition of Cl^- or $(C_4H_9O_2)^-$ to an electrically neutral molecule are sometimes observed. In other words, more than one negative ion appears in the singly charged molecular-ion region, and a corresponding effect occurs for multiply charged ions. These abnormal adduct ions are frequently encountered in the ESI studies of glycoconjugates, because, except for derivatized or protected forms, it is more common to take ESI spectra in the negative-ion mode for saccharides. Proteins and peptides, however, are more generally measured in the positive-ion mode in water admixed with some organic solvents; therefore, these abnormal adduct ions are unlikely to be observed. A second negative charge for acidic glycosphingolipids may also be gained by an extra deprotonation. In these cases, a deconvolution program, usually supplied with the ES ionization facility, does not work. (More details will be discussed later with examples.)

Most glycoconjugates have large molecular masses, which may cause increases in observed m/z values from their nominal masses owing to the accumulated difference between the real mass of hydrogen ($=1.00782$) and its nominal mass ($=1$).

Dell et al. have clearly and concisely described the practical aspects of ESI measurements on carbohydrate-containing biopolymers (3).

III. DERIVATIZED GLYCOCONJUGATES

Derivatization of polar samples once was an essential step for electron ionization (EI) studies, but has become less important since soft-ionization techniques such as FAB, ESI, and MALDI became available. It is true that multistep derivatization processes may destroy a laboriously prepared sample of extremely small quantity, and that extraneous salt may cause fluctuations in the ES current. It is particularly risky if alkaline permethylation is applied to an unknown sample which happens to be alkali labile. However, it is also true that if the sample is amenable to derivatization, the resultant less-polar sample molecule becomes soluble in more volatile organic solvents, which in turn can provide a significant increase in detection sensitivity. Derivatization also insures that there will be clear differences in m/z values between linkage isomers after CID–MS/MS experiments (4). Gu et al. reported that reductive pyridylamination of the reducing end of oligosaccharides (PA-oligosaccharide) gave fluorescent derivatives, of which molecular weights were detectable on ESI–MS with a 1-pmole sample, and that the fragmentation pattern of those PA-oligosaccharides, obtained by applying skimmer CID, showed simple glycosidic cleavages with the positive charge retained on the PA site (5). As an improved version of PA derivatization, Okamoto et al. recommended trimethyl-(p-aminophenyl)ammonium (TMPA) derivatization (6). These derivatizations are frequently used techniques in HPLC separations.

Reinhold et al. suggested running ESI on unmodified (underivatized) glycoconjugates to see molecular profiles if the sample is present in a sulfate or a phosphate form, such as GPI anchors and lipid A (4) (ESI mass spectrometry of these anchor-type glycoconjugates is discussed in detail later). On the other hand, they recommended performing ESI measurements on derivatized glycoconjugates like glycolipids, glycoproteins, or glycans, which do not efficiently charge by protonation or deprotonation owing to weak basicity or acidity. They claim also that glycan linkage and branching are better mass separated and, consequently, better analyzed by CID–MS/MS of individual parent ions, if the sample is derivatized. According to this group, not only ionization methods, but also derivatization affect the fragmentation processes. Thus, different antennary structure as well as different linkages exhibit different mass numbers in the low-energy CID–MS/MS spectrum (see ref. 4 for details). An interesting example is found in the structural study of the exopolysaccharide of a nitrogen-fixing soil bacterium *Rhizobium meliloti*, Rm 1021. It turns out that the specific structural features of succinoglycan are needed for nodule invasion. The ESI CID–MS/MS analyses of the octasaccharide after a first methylation by CH_3I (neutral) and after a second methylation by CD_3I under alkaline conditions, revealed the total structure of the symbiotically important succinoglycan (7).

IV. UNDERIVATIZED, NEUTRAL OLIGOSACCHARIDES AND GLYCOCONJUGATES

For the linkage analysis of glucose disaccharide isomers, fragmentation rules approximately the same as those of FAB were reported in 1991. Garozzo et al. characterized fragmentational differences among glucopyranose-$\beta 1 \rightarrow$ C-2, -3, -4, and -6 of glucose by applying skimmer collision (in-source fragmentation) in negative-ion ESI–MS (8). Recently, Mulroney et al. extended this methodology to anomer analysis, which can be performed at the same time as the linkage assignment, again based on skimmer collision rather than tandem mass spectrometry (9). Although their methods depend on empirical criteria, the library can be applied to neutral disaccharides, if the reducing end, which is more readily deprotonated than other hydroxyl groups, is free.

Characterization of underivatized large neutral oligosaccharides by ESI–MS was reported in 1993 on a synthetic poly-N-acetyllactosamine-type penta-antennary pentacosasaccharide, which corresponds to the sugar chain of the glycoconjugates having blood group I activity (10). Neutral glycoconjugates are better analyzed in the positive-ion mode, especially when the polar functional groups are protected. Ii et al. measured positive-ion ESI mass spectra of

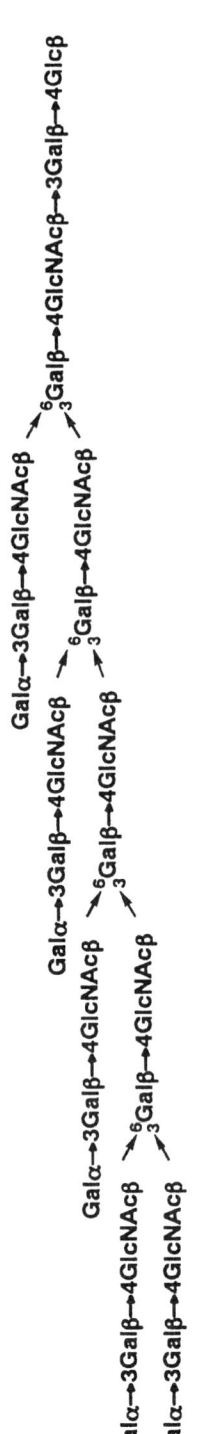

Scheme 1. Structure of the blood group I-active poly-N-acetyllactosamine-type pentacosasaccharide (MW 4441.0) Reprinted with permission from T. Li et al., *Org. Mass Spectrom.*, **1993**, *28*, 1340. © 1993 John Wiley & Sons, Ltd.

IV. UNDERIVATIZED, NEUTRAL OLIGOSACCHARIDES AND GLYCOCONJUGATES   465

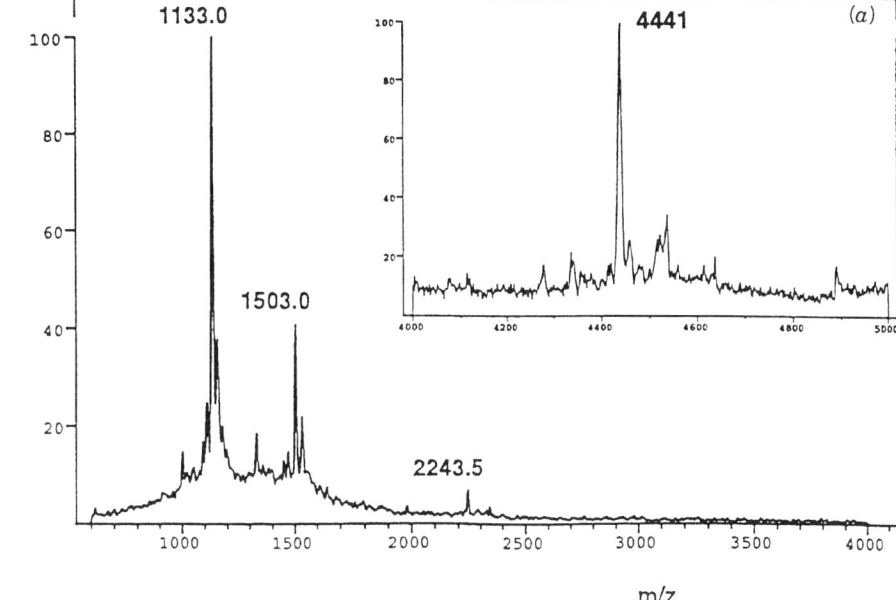

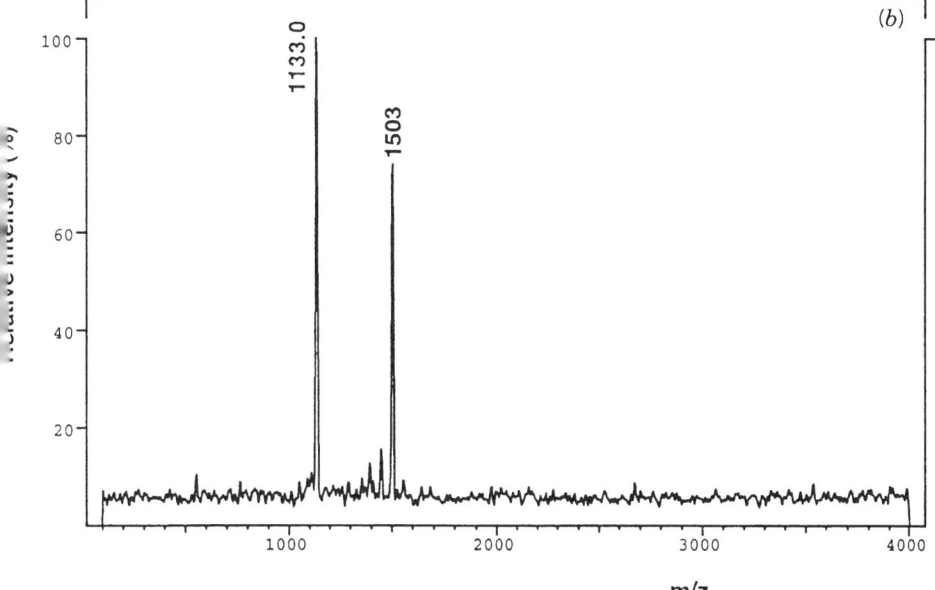

gure 1. (*a*) Positive-ion ESI mass spectrum and deconvolution spectrum (inset) of deprotected ıly-*N*-acetyllactosamine-type pentacosasaccharide (MW 4441.0). (*b*) Positive-ion ESI CID–MS/ S spectrum of this molecule having $(M + 4Na)^{4+}$ at *m/z* 1133 as the precursor ion. Reprinted with rmission from T. Ii et al., *Org. Mass Spectrom.*, **1993**, *28*, 1340. © 1993 John Wiley & Sons, Ltd.

neutral glycoconjugates both in the fully protected and deprotected intact forms. The fully protected 25-mer was observed as $(M + 3Na)^{3+}$, $(M + 4Na)^{4+}$, and $(M + 5Na)^{5+}$, and the built-in deconvolution program showed the molecular weight to be 9416, as expected from the structure. (In this chapter, molecular weights for these neutral glycoconjugates and gangliosides are represented as "M" unless otherwise stated.) As for the deprotected intact 25-mer (Scheme 1), a series of multiply charged ions, $(M + 2Na)^{2+}$, $(M + 3Na)^{3+}$, and $(M + 4Na)^{4+}$, were observed and deconvoluted to the expected molecular mass of 4441 (Fig. 1a). A CID–MS/MS analysis of the parent $(M + 4Na)^{4+}$, which has m/z of 1133, shows a daughter $(M + 3Na)^{3+}$ of m/z 1503 and also two fragment ions of m/z 1449 and 1395. These latter two m/z values correspond to two triply charged Y ions produced by the cleavage of $(Hex + Na)^+$ and another cleavage of $(Hex + Hex + Na)^+$ from the tetrasodiated precursor (Fig. 1b). Here, the term "Y ion" is borrowed from the nomenclature by Domon and Costello, and was defined originally for singly charged FAB fragment ions (11). In this chapter, "Hex" is used for convenience, to represent "anhydrohexose" or "hexose minus water," corresponding to 162 Da. Analogous representations such as HexNAc (N-acetylhexosamine – water, 203 Da) and Fuc (fucose – water, 146 Da) will be used throughout the chapter. Many daughter ions appear in higher mass ranges than the parent because their charge state is less than that of the parents. However, in lower mass ranges, the nonreducing-end trisaccharide unit, which is a repeating unit in this 25-mer structure, is also represented as a daughter ion of $(M + 4Na)^{4+}$ by $(Hex-Hex-HexNAc + Na)^+$.

Sugar chains were observed as sodiated ions rather than protonated ions in the positive-ion mode. This tendency of Na^+ addition is stronger in ESI than in positive-ion FAB for neutral sugars. As for neutral glycoconjugates, such as the ceramide decasaccharide (d18 : 1/c24 : 0), which has a sugar chain related to, but shorter than, that of the 25-mer, positive-ion ESI CID–MS/MS showed a propensity to provide the sugar-chain structure information only (10). This differs from FAB, in which the positive-ion mode tends to show fragment ions of the ceramide side if the sugar moiety is neutral.

As stated earlier, the negative-ion mode of the ESI mass spectrometry of glycoconjugates presents quite unusual features. Figure 2 shows the negative-ion ESI spectrum of N-palmitoylgalactosylceramide, of which purity was proven in the positive-ion mode both by FAB and ESI. The solvent used was chloroform/methanol = 1 : 1. In addition to the expected $(M - H)^-$ at m/z 699, $(M + Cl)^-$ and $(M + 89)^-$ were observed at m/z 735 (with isotope) and 789, respectively. The Cl^- evidently comes from the chloroform. The daughter ion spectrum of m/z 789 showed only $(M - H)^-$ at m/z 699 and an unknown negative-ion at m/z 89. A study using isotopes revealed that the composition of this negative-ion adduct is $C_4H_9O_2^-$, coming from the solvent methanol, but the ion structure has not yet been clarified (2). These abnormal adduct ions appear not only for neutral glycolipids but also for acidic oligosaccharides and other lipids in general, if chloroform/methanol, the most common solvent for lipids, is used as the solvent for negative-ion ESI measurements.

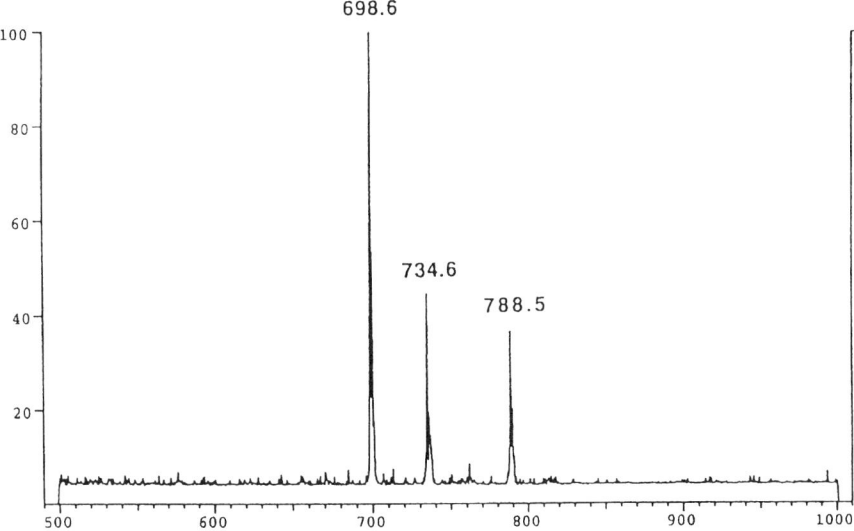

Figure 2. Negative-ion ESI mass spectrum of N-palmitoylgalactosylceramide (MW 700) in $CH_3OH:CHCl_3$ (1:1, v:v). Reprinted with permission from T. Ii et al., *Org. Mass Spectrom.*, **1993**, *28*, 927. © 1993 John Wiley & Sons, Ltd.

V. ACIDIC GLYCOCONJUGATES

A. Gangliosides (12)

1. Conventional Mass Spectrometry. Although FAB–MS of gangliosides up to GD1 (isomeric disialogangliosides, Scheme 2a) may produce molecular-weight-related ions such as $(M + Na)^+$ and $(M - H)^-$ in intensities sufficient to carry out CID–MS/MS, more highly sialylated gangliosides are measurable by FAB only in the conventional mass spectrometry. Being carboxylic acids in nature, gangliosides are not as strongly dissociative as sulfates. Therefore, molecular weights are given only in "M" (free-acid form) in this section. Sulfated glycoconjugates, discussed later in comparison with sialyl analogs, are expressed as " M' " (full-sodium salts). Sequence ions shown in the conventional FAB mass spectra sometimes fail to clearly tell positional isomers if the compound has a branched sugar chain. Electrospray ionization–mass spectrometry is desirable for highly sialylated compounds by providing ions representative of intact molecules in intensities adequate to be chosen as precursors for tandem mass spectrometry.

For instance, GQ1b, which has four sialic acids (Scheme 2b), consists of K^+- and Na^+-salt(s) of the carboxylates to various extents, the rest of the carboxylate group(s) being free acid(s). Additionally, there are so-called molecular heterogeneities in the long-chain base. Thus, the molecular-weight region is quite complex in the positive- and negative-ion mode. Fast-atom bombardment shows

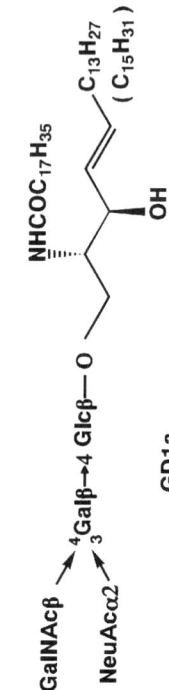
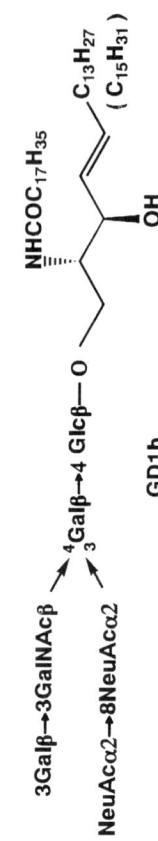
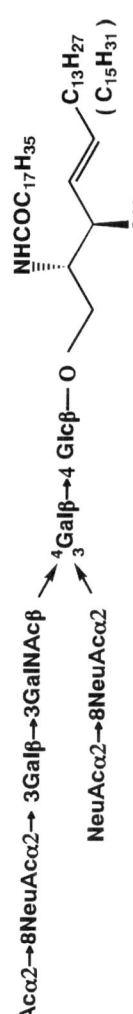

Scheme 2a

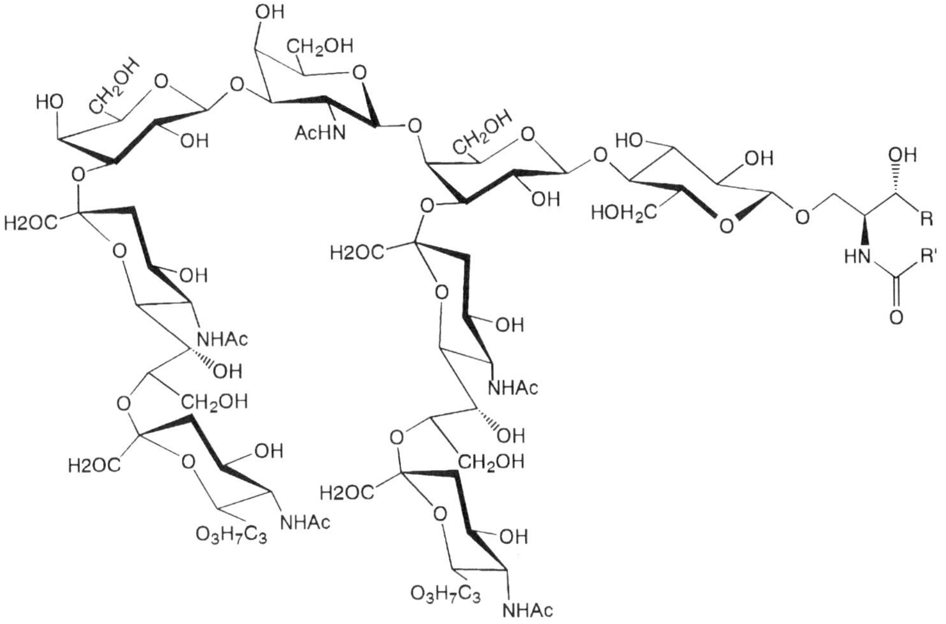

R=15:1 or 17:1
R'=17:0

GQ1b (MW 2448)
$C_{108}H_{186}N_6O_{55}$

Scheme 2b. Typical gangliosides illustrated in Svennerholm's representations and conformational formula of GQ1b.

ions representing those molecular species with some sequence ions, but they are not sufficient in intensity for CID–MS/MS. Negative-ion ESI shows doubly and triply charged ions, but not the singly nor quadruply charged ions as shown in Table 1. Table 1 also indicates that, in spite of a monosialyl property, GM1 (Scheme 2a) shows $(M - 2H)^{2-}$ in addition to the normal $(M - H)^-$, the extra charge being brought about by an additional deprotonation. Prediction of charge states in electrospray mass spectrometry is thus not straightforward and cannot be based simply on the number of the acidic functional groups. It is also true, however, that once basic rules are understood, one is not necessarily puzzled by the complex charge states of ESI spectra.

2. CID–MS/MS (12). Daughter spectra of $(M + 2Na - 4H)^{2-}$ (m/z 1243) and $(M + Na - 4H)^{3-}$ (m/z 822.5), both for d20:1/c18:0, show positions of four sialic acids in the GQ1b molecule. Assignments are given in Schemes 3a and 3b, which indicate that two sialic acids are attached to the *endo*-galactose, and the other two to the *exo*-galactose. This method was applied to distinguish GD1a from its positional isomer GD1b (Scheme 2a), the ceramide moiety for these

Table 1. Results of the negative-ion ESIMS of various gangliosides (ref. 12)

Compound		M.W. (free acid)[a]	Ions observed (m/z, assignment)					
			Singly charged ions		Doubly charged ions		Triply charged ions	
GM1	(d18:1/c18:0)	1546	1546	$[M_1-H]^-$	772	$[M_1-2H]^{2-}$		
	(d20:1/c18:0)	1574	1574	$[M_2-H]^-$	786	$[M_2-2H]^{2-}$		
GD1a	(d18:1/c18:0)	1837	1859	$[M_1+Na-2H]^-$	917.5	$[M_1-2H]^{2-}$		
	(d20:1/c18:0)	1865	1886.5	$[M_2+Na-2H]^-$	932	$[M_2-2H]^{2-}$		
GD1b	(d18:1/c18:0)	1837	1859	$[M_1+Na-2H]^-$	917.5	$[M_1-2H]^{2-}$		
	(d20:1/c18:0)	1865	1886.5	$[M_2+Na-2H]^-$	932	$[M_2-2H]^{2-}$		
GT1b	(d18:1/c18:0)	2128	2150	$[M_1+Na-2H]^-$	1075	$[M_1+Na-3H]^{2-}$	708	$[M_1-3H]^{3-}$
			2166	$[M_1+K-2H]^-$	1083	$[M_1+K-3H]^{2-}$		
	(d20:1/c18:0)	2156	2179	$[M_2+Na-2H]^-$	1089	$[M_2+Na-3H]^{2-}$	718	$[M_2-3H]^{3-}$
			2193.5	$[M_2+K-2H]^-$	1096.5	$[M_2+K-3H]^{2-}$		
GQ1b	(d18:1/c18:0)	2420			1229	$[M_1+2Na-4H]^{2-}$	813	$[M_1+Na-4H]^{3-}$
					1239	$[M_1+Na+K-4H]^{2-}$	818.5	$[M_1+K-4H]^{3-}$
	(d20:1/c18:0)	2448			1243	$[M_2+2Na-4H]^{2-}$	822.5	$[M_2+Na-4H]^{3-}$
					1253.5	$[M_2+Na+K-4H]^{2-}$	828	$[M_2+K-4H]^{3-}$

[a] M_1 and M_2 represent the molecular mass of the free acids having sphingenine (d18:1) and icosasphingenine (d20:1), respectively. Molecular masses are calculated as the exact masses.

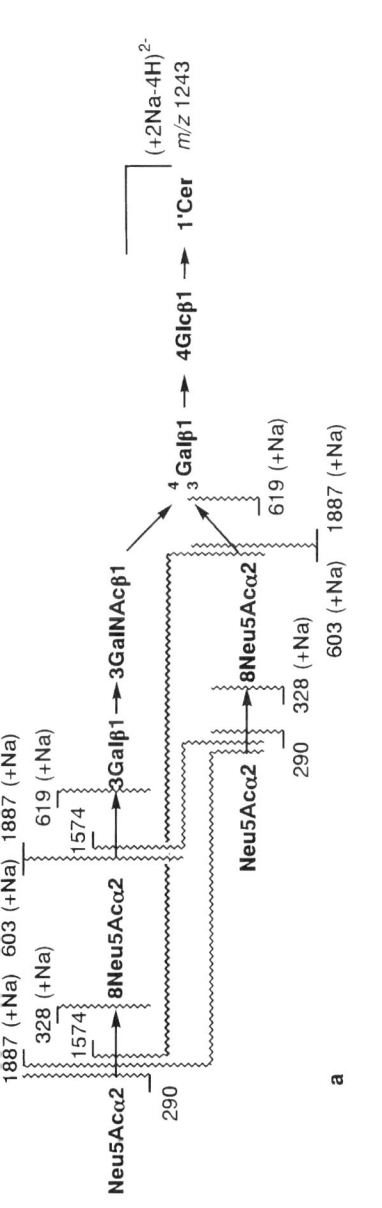
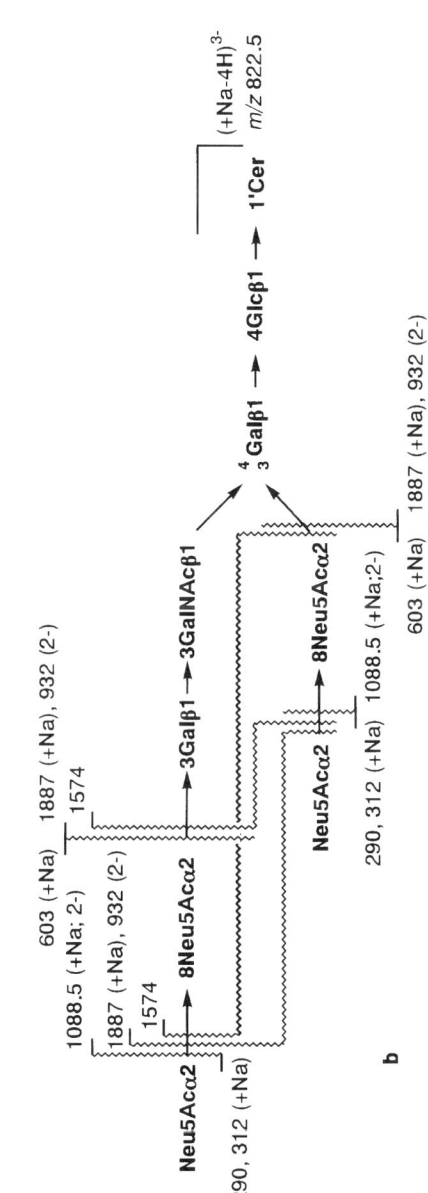

Scheme 3. Negative-ion ESI CID–MS/MS spectra of GQ1b having (a) the $(M + 2Na − 4H)^{2−}$ (m/z 1243) as the precursor ion and (b) the $(M + Na − 4H)^{3−}$ (m/z 822.5) as the precursor ion. Reprinted from T. Ii et al., "Structural elucidation of underivatized gangliosides by electrospray-ionization tandem mass spectrometry (ESIMS/MS)," *Carbohydr. Res.*, **1995**, *273*, 27, with kind permission from Elsevier Science Ltd. © 1995 Elsevier Science Ltd.

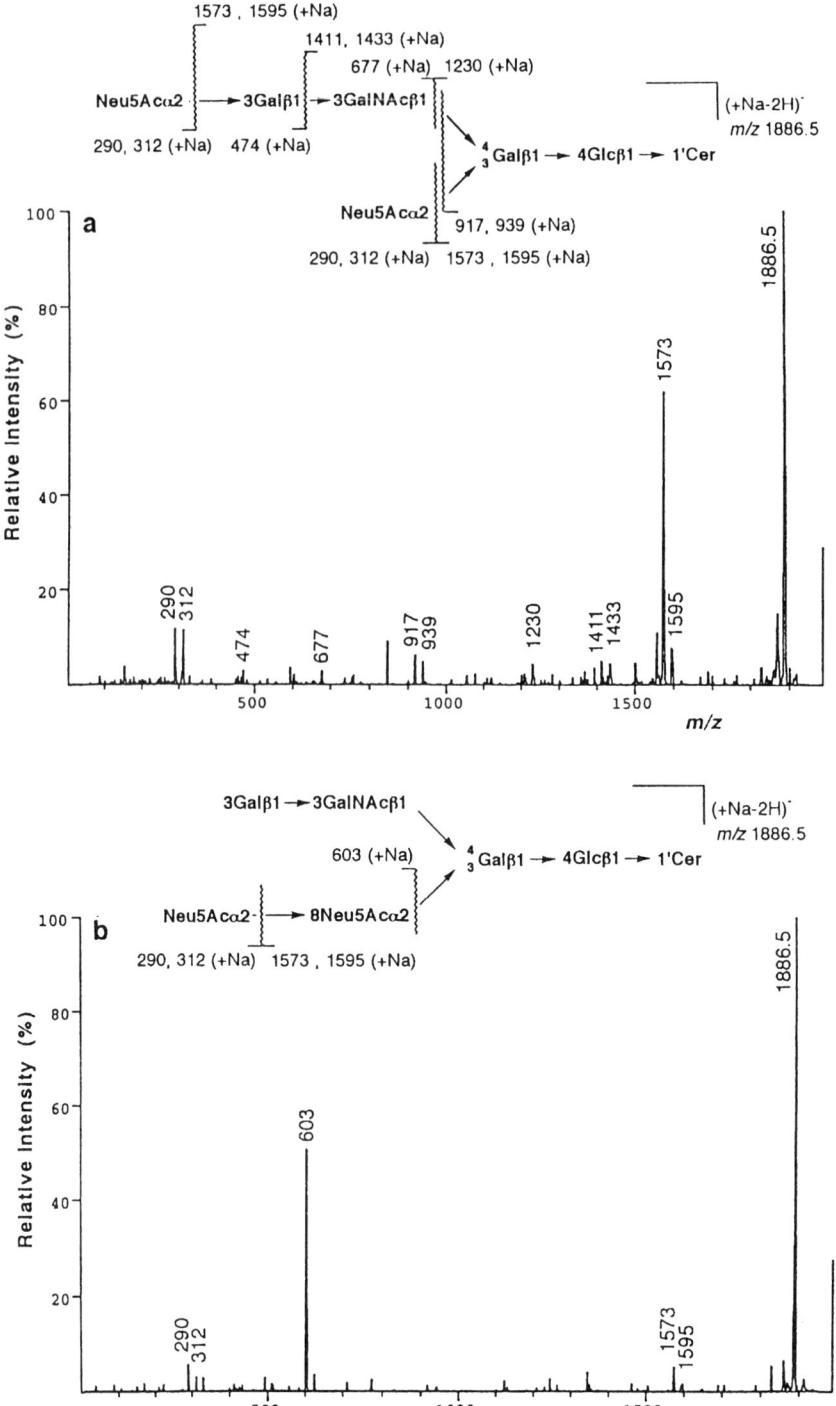

Figure 3. Negative-ion ESI CID–MS/MS spectra of isomeric disialogangliosides (MW 1865) having $(M + Na - 2H)^-$ at m/z 1886 as the precursor ion; (*a*) GD1a and (*b*) GD1b. Reprinted from T. Ii et al., "Structural elucidation of underivatized gangliosides by electrospray-ionization tandem mass spectrometry (ESIMS/MS)," *Carbohydr. Res.*, **1995**, *273*, 27, with kind permission from Elsevier Science Ltd. © 1995 Elsevier Science Ltd.

isomers being the same (d20:1/c18:0). Both singly and doubly charged ions, that is $(M + Na - 2H)^-$ (m/z 1886.5) and $(M - 2H)^{2-}$ (m/z 932), showed different fragmentations for the two isomers, as shown in Figures 3 and 4, respectively. It was also possible to distinguish positional isomers up to GD1 by positive-ion ESI CID–MS/MS. However, the peak intensities of more highly sialylated gangliosides tend to be inadequate for daughter scanning in the positive-ion mode. Thus, by performing ESI CID–MS/MS on an unknown ganglioside, it is possible to obtain not only the molecular-weight information but also the number, type (*N*-acetyl or *N*-glycolyl), and also the positions of sialic acids, all of which are significant. A-, b-, and c-paths of gangliosides are designated according to the number of sialic acids attached to the *endo*-galactose (13). Thus, gangliosides of an a-path have one sialic acid attached to the *endo*-galactose, gangliosides of a b-path have two sialic acids attached to the *endo*-galactose, and so forth. Once the positions of sialic acids are determined by ESI CID–MS/MS, the biosynthetic path of the unknown sample may be clarified.

B. Sulfated Lewis-Type Oligosaccharides

Sulfated Le-type saccharides have attracted much attention in the past several years, because they are probably ligands to E-selectins, a family of cell-adhesion proteins (14–16). Among several Le-type saccharides, two linkage isomers Le^x- and Le^a-, shown in Scheme 4, are of particular interest. In either case the acidic group, sulfate, is linked to the C3 of the Gal of the nonreducing end, but the linkages of the (Gal-C3-sulfate) and a Fuc to a GlcNAc are reversed between Le^x and Le^a, the former having 3-Fuc-4 (Gal-3-sulfate) while the latter has 4-Fuc-3-(Gal-3-sulfate).

1. Conventional Mass Spectrometry of Sulfated Oligosaccharides (17).
The principal behavior of sulfated branched trisaccharides toward ESI–MS and CID–MS/MS was studied using synthetic 3'-sulfated (**1**, molecular weight of full sodium salt, M' = 673), 2',3'-disulfated (**2**, M' = 775), and 2',3',4'-trisulfated (**3**, M' = 877) Le^x-type trisaccharide propylglycosides (Scheme 4) as typical examples. It should be remembered that molecular weights of strongly acidic samples are always given as those of full-sodium salts throughout this chapter, and special care is needed when formulas are compared to equivalent chemical species which are not salts.

In the negative-ion mode, ESI mass spectra of mono-, di-, and trisulfated Le^x-trisaccharide propylglycosides showed ions $(M' - Na)^-$ (m/z 650), $(M' - Na)^-$ (m/z 752) plus $(M' - 2Na)^{2-}$ (m/z 364.5), and $(M' - Na)^-$ (m/z 854) plus $(M' - 2Na)^{2-}$ (m/z 415.5) for mono-, di-, and trisulfated Le^x-type trisaccharides, respectively. It is very important to recall that M' here represents the full-sodium salt. A triply charged negative ion, corresponding to dissociation of all three acidic groups, was not observed for the trisulfated compound (**3**).

It was also noted that in the positive-ion mode ESI–MS, these mono-, di-, and trisulfated trisaccharides produced $(M' + Na)^+$ (m/z 696), $(M' + Na)^+$ (m/z

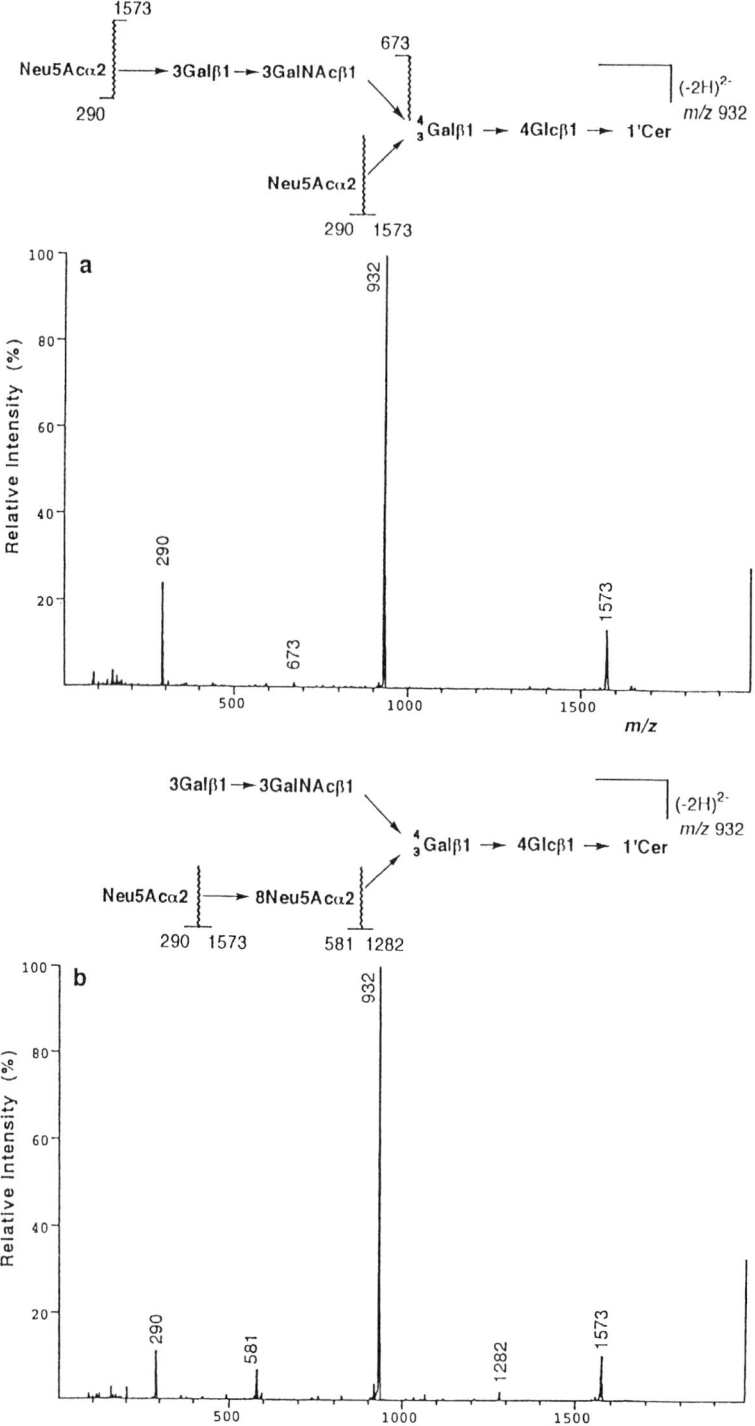

Figure 4. Negative-ion ESI CID–MS/MS spectra of isomeric disialogangliosides (MW 1865) having $(M - 2H)^{2-}$ at m/z 932 as the precursor ion; (*a*) GD1a and (*b*) GD1b. Reprinted from T. Ii et al., "Structural elucidation of underivatized gangliosides by electrospray-ionization tandem mass spectrometry (ESIMS/MS)," *Carbohydr. Res.*, **1995**, *273*, 27, with kind permission from Elsevier Science Ltd. © 1995 Elsevier Science Ltd.

798), and $(M' + Na)^+$ (m/z 900) plus $(M' + 2Na)^{2+}$ (m/z 461.5), respectively. As stated earlier, the maximum charge state cannot be predicted merely from the number of acidic functional groups. The maximum charge state may also depend on the size of the saccharide chain: Many charges within a limited space may cause destabilization due to Coulombic repulsion.

2. CID–MS/MS of Sulfated Oligosaccharides (17). Daughter spectra of sulfated Le^x- trisaccharides in negative-ion ESI having $(M' - Na)^-$ as the precursor are much like those from FAB CID–MS/MS, in which the precursor ions are singly charged and represent whole molecules. Here, Fuc is cleaved off at the nonreducing end side of the glycosidic oxygen, making a cleavage at $Y_1\beta$ (see ref. 11). Sulfate group(s) are eliminated, mostly accompanied by Na, either from $(M' - Na)^-$ or from sequence ions, the negative charge always being retained in

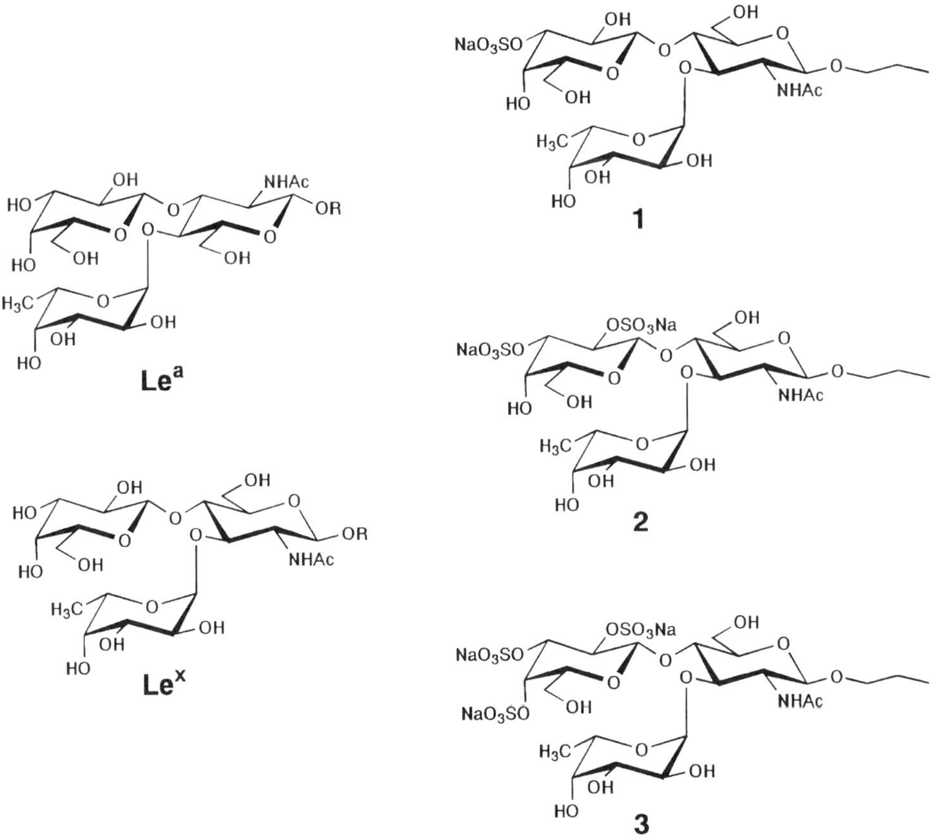

Scheme 4. Basic structures of Le^a- and Le^x-type trisaccharides (left) and mono-, di-, and trisulfated Le^x-trisaccharide propyl glycosides (**1–3**) (right).

the terminal galactose. Fragmentation patterns are shown in Scheme 5. The ESI CID–MS/MS spectra of Le^x - trisaccharides in the positive-ion mode also resemble the (+) FAB CID–MS/MS, provided that molecular-weight related, singly charged ions are chosen as the precursors.

However, ESI CID–MS/MS of such Le^x-trisaccharides presents a different set of structural information if multiply charged ions are chosen as the precursors. Figure 5 shows the negative-ion ESI CID–MS/MS spectrum employing $(M' - 2Na)^{2-}$ (m/z 364.5) of disulfated trisaccharide **2** as the precursor ion. Note that the two glycosidic cleavages and a ring-opening fragmentation of the GlcNAc each resulted in the production of doubly charged daughter ions $(Y_1\beta + H - 2Na)^{2-}$ (m/z 291.5), $(B_1\alpha - 2Na - H)^{2-}$ (m/z 160), and $(^{3,5}A_2 - 2Na)^{2-}$ (m/z 197), respectively. These doubly charged fragment ions, in turn, become singly charged after losing one negative charge with elimination of HSO_4^- or HSO_3^-. At the same time, by a removal of a neutral species SO_3 from the $(B_1\alpha - 2Na - H)^{2-}$ (m/z 160) ion with a concomitant intramolecular H^+-transfer, a negative-ion characteristic of sulfated hexose appears in high abundance at m/z 241.

In the positive-ion mode, CID–MS/MS in ESI and in FAB again gives rise to daughter ions which are much alike, if the precursor ion in ESI is a singly charged ion. Scheme 6 (**3**) shows fragmentation patterns of the trisulfated Le^x-trisaccharides in positive-ion mode ESI CID–MS/MS. On the other hand, ESI CID–MS/MS of the doubly charged parent ion $(M' + 2Na)^{2+}$ at m/z 461.5 for the trisulfated trisaccharide **3** produced abundant ions of lower charge state than the precursor, namely, $(M' + Na)^+$ at m/z 900 and $(Y_1\beta + H + Na)^+$ at m/z 754 (Fig. 6a).

Although it was not expected, ESI CID–MS/MS of a doubly charged ion $[(M' + 2Na)^{2+}, m/z\ 461.5]$ produced a fragmentation pattern very similar to that of the daughter spectrum of singly charged $(M' + Na)^+$ at m/z 900 [Scheme 6 (**3**)], which, in turn, resembles the CID–MS/MS fragmentation in the FAB mode (Fig. 6b) [compare Figs. 6a and b and Scheme 6 (**3**)].

C. Sulfated and Sialyl Lewis-Type Glycosphingolipids

Scheme 7 shows six synthetic acidic glycosphingolipids as respective full-sodium salts. Because their ceramide moiety is identical (d18:1/c24:0), these glycosphingolipids possess different chemical and physical properties attributable only to differences in oligosaccharide moieties. Charges may not be localized in acidic group(s) in the sugar chain, especially if the acidic group is not very strong, as in the case of sialyl Le-type glycosphingolipids.

1. Negative-Ion Conventional Mass Spectrometry of Sulfated and Sialyl Lewis-Type Glycosphingolipids (18).

All compounds listed in Scheme 7, except for **3**, contain only one acidic group. Nevertheless, conventional mass spectrometry of glycosphingolipids in negative-ion mode ESI–MS may provide

V. ACIDIC GLYCOCONJUGATES 477

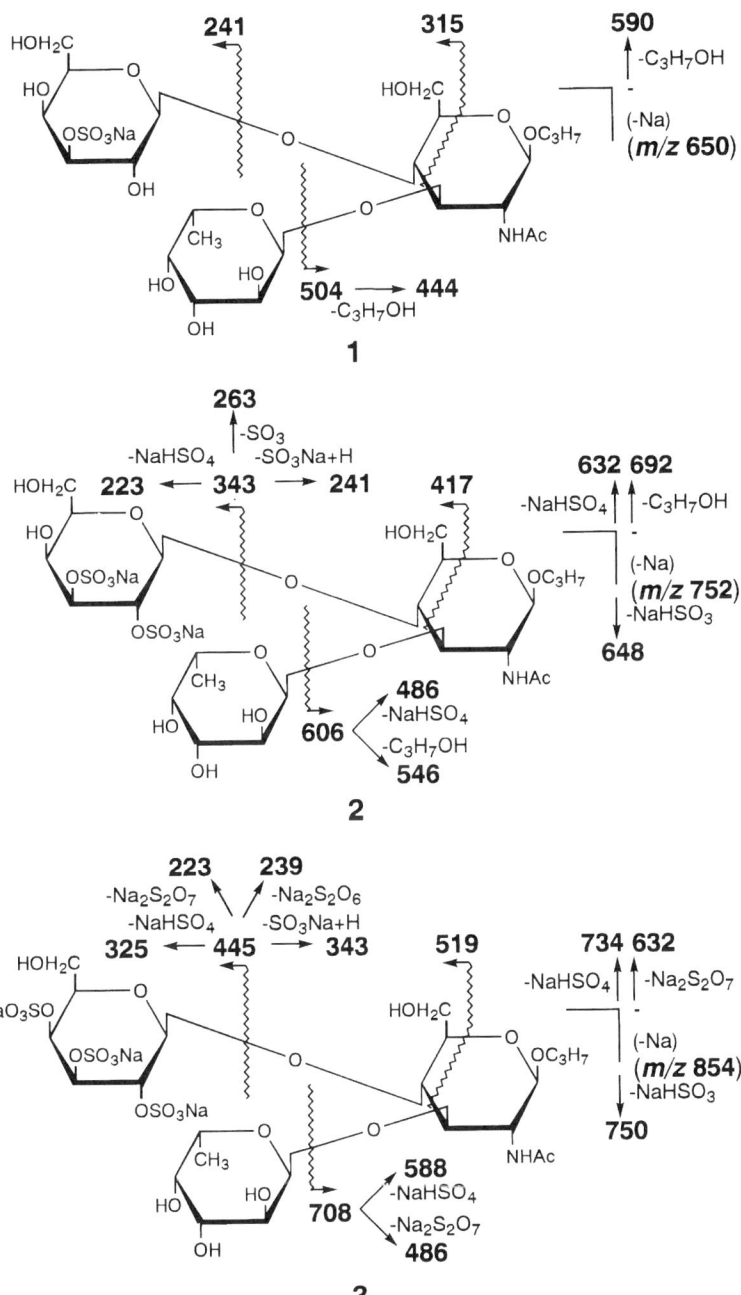

Scheme 5. Fragmentation patterns of the negative-ion ESI CID–MS/MS of sulfated Le^x-trisaccharides having the respective (M′ − Na)$^-$ as the precursor ion.

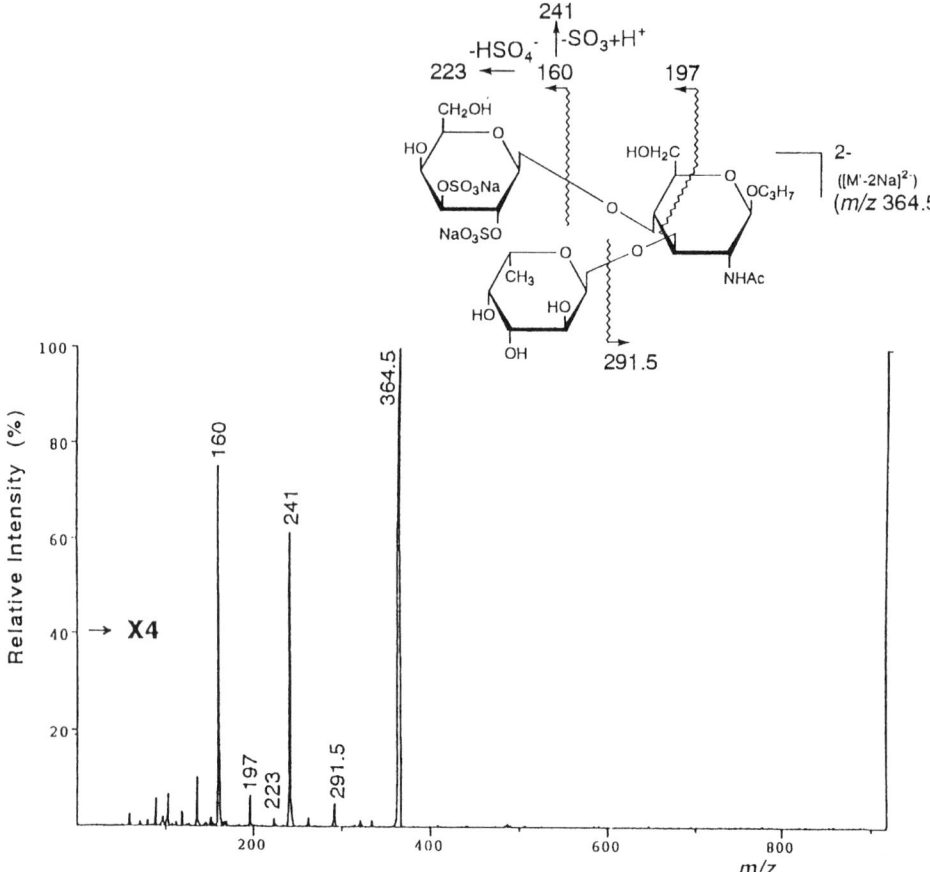

Figure 5. Negative-ion ESI CID–MS/MS spectrum of disulfated Le^x-type trisaccharide propyl-glycoside **2** (M'W 775, as the disodium salt), having $(M' - 2Na)^{2-}$ at m/z 364.5 as the precursor ion (17).

extra charge state(s) by an attachment of a small anion such as Cl^- or $C_4H_9O_2^-$ (2), or by an abnormal deprotonation, as summarized in Table 2 (18). Thus, these six compounds all yielded doubly charged ions in various forms. Compound **3** has two sulfate groups, and $(M' - 2Na)^{2-}$ is naturally produced as a result of the second dissociation. Monobasic acids **1**, **2**, and **4–6** can be endowed a second charge in unusual forms such as $(M' - Na + 89)^{2-}$, $(M' - Na + Cl)^{2-}$, or $(M' - Na - H)^{2-}$, as shown in Table 2. The unexpected deprotonation, which generates $(M' - H)^-$ in addition to the normal $(M' - Na)^-$ for **1**, **2**, and **4**, and generation of the above stated $(M' - Na - H)^{2-}$ for the five monobasic acids, takes place at a nonacidic polar group in each molecule. No difference was observed between ESI mass spectra of linkage isomers **1** and **2**.

V. ACIDIC GLYCOCONJUGATES 479

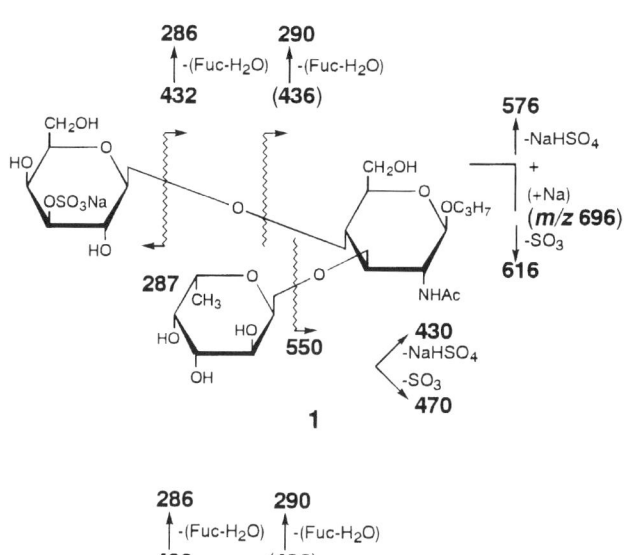

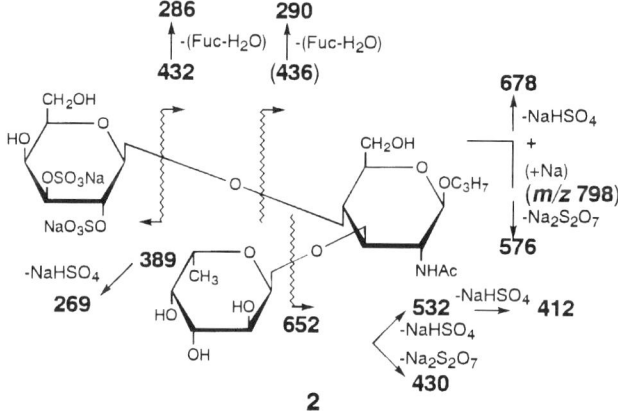

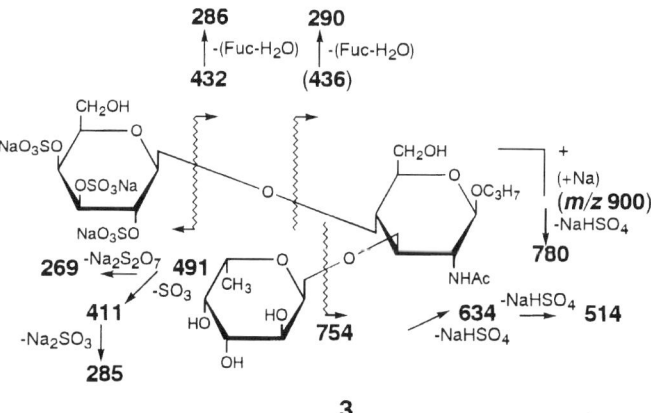

Scheme 6. Fragmentation patterns of the positive-ion ESI CID–MS/MS of sulfated Le^x-trisaccharides having the respective (M′ + Na)$^+$ as the precursor ion.

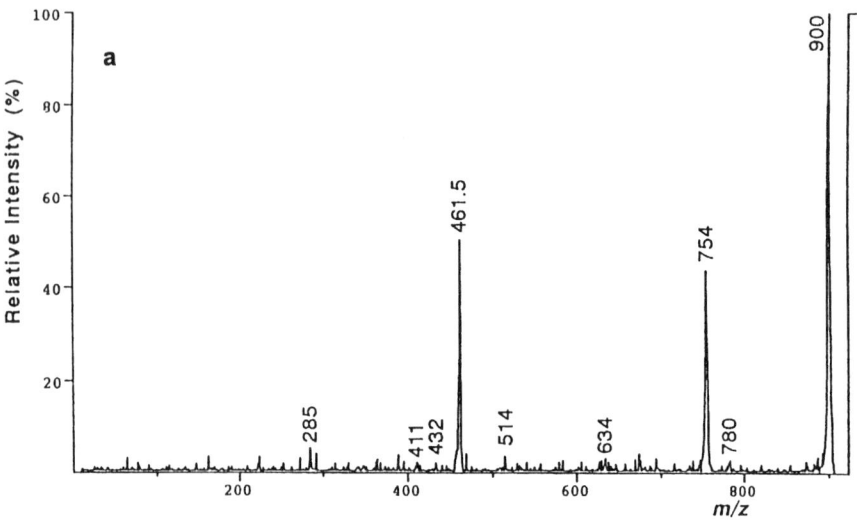

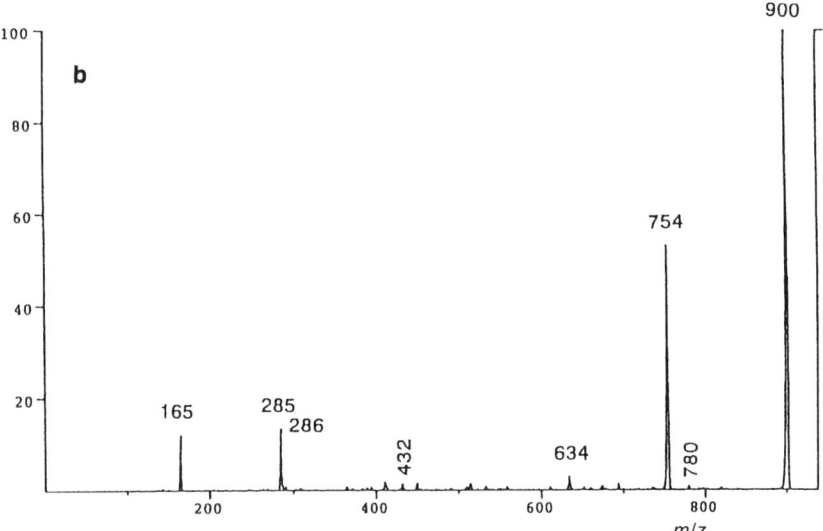

Figure 6. (a) Positive-ion ESI CID–MS/MS spectrum of trisulfated Le^x-type trisaccharide propylglycoside 3 (M'W 877, as the trisodium salt) having the $(M' + 2Na)^{2+}$ at m/z 461.5 as the precursor ion. (b) Positive-ion FAB CID–MS/MS spectrum of the same sample having the $(M' + Na)^+$ at m/z 900 as the precursor ion (17).

2. Negative-Ion CID–MS/MS of Lewis-Type Glycosphingolipids (18). The game of structure elucidation using ESI mass spectrometry begins only when CID tandem mass spectrometry has entered. In the ESI mode, negative-ion daughter spectra of singly charged $(M' - Na)^-$ of glycosphingolipids look

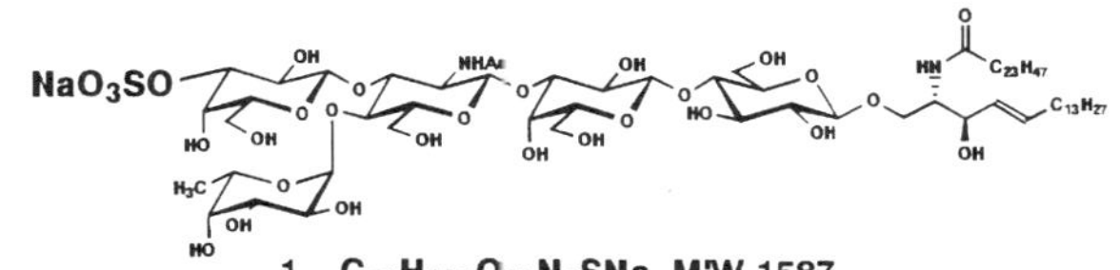

V. ACIDIC GLYCOCONJUGATES 481

Scheme 7. Structures of sulfated and sialyl Lewis-type glycosphingolipids samples.

ceramide: d18:1 / c24:0

1 $C_{74}H_{135}O_{30}N_2SNa$ M'W 1587 sulfated Le^a-pentaosylceramide
2 $C_{74}H_{135}O_{30}N_2SNa$ M'W 1587 sulfated Le^x-pentaosylceramide 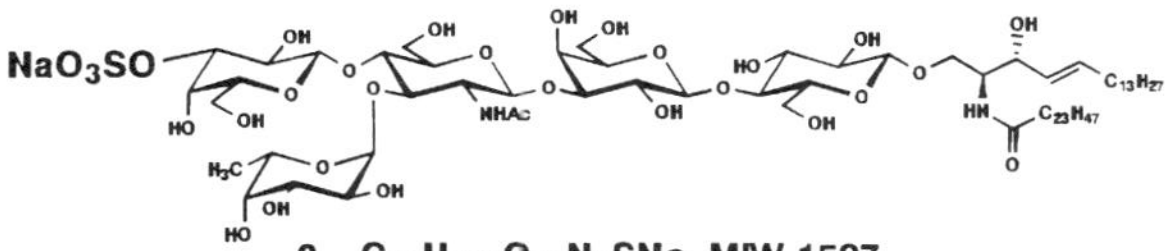
3 $C_{74}H_{134}O_{33}N_2S_2Na_2$ M'W 1689 disulfated Le^x-pentaosylceramide 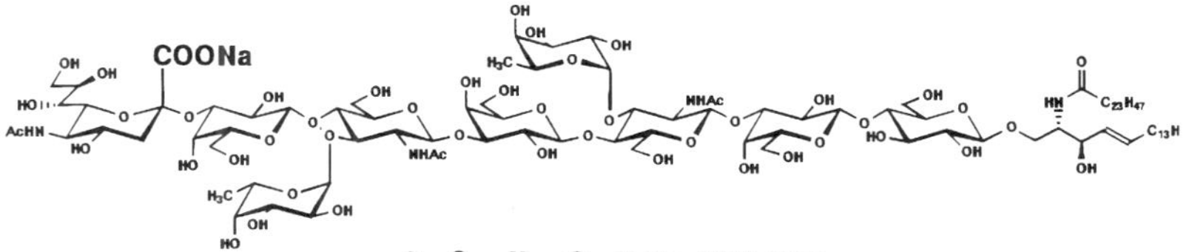
4 $C_{85}H_{152}O_{35}N_3Na$ M'W 1798 sialyl Le^a-pentaosylceramide
5 $C_{99}H_{175}O_{45}N_4Na$ M'W 2163 sialyl Le^x-heptaosylceramide
6 $C_{105}H_{185}O_{49}N_4Na$ M'W 2309 sialyl Le^x-Le^x-octaosylceramide

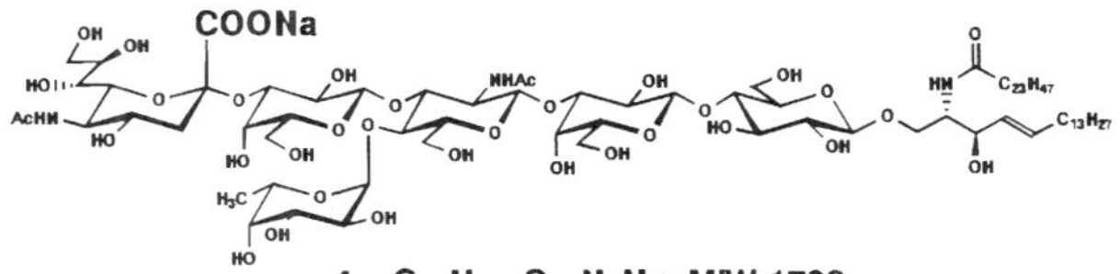

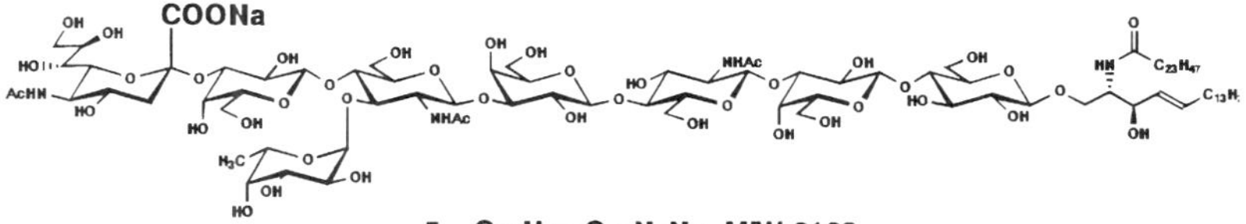

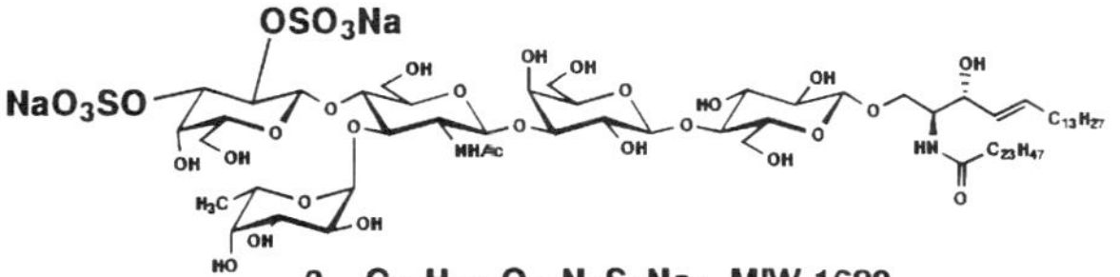

Table 2. Results of the negative-ion ESIMS of sulfated and sialyl Le-type glycosphingolipids (ref. 18).

Compound [a]	M'[b]	singly charged ion		observed ions m/z (relative intensity %) doubly charged ion			
		[M' - H]⁻	[M' - Na]⁻	[M' - Na + 89]²⁻	[M' - Na + Cl]²⁻	[M' - Na - H]²⁻	[M' - 2Na]²⁻
1	1587	1586 (6)	1564 (22)	826.5 (17)	799.5 (100)	781.5 (77)	
2	1587	1586 (5)	1564 (52)	826.5 (12)	799.5 (26)	781.5 (100)	
3	1689	—[c]	1666 (14)	—	—	—	821.5 (100)
4	1798	1797 (7)	1775 (31)	932 (13)	905 (35)	887 (100)	
5	2163	—	2140 (55)	—	1087.5 (14)	1069.5 (100)	
6	2309	—	—	1187.5 (19)	1160.5 (100)	1142.5 (43)	

[a] See Scheme 7 for structures.
[b] M' represents the full sodium salt.
[c] Intensities (—) are below 2%.

similar to the corresponding daughter spectra in the FAB mode except for the very low mass regions (<90 Da), and a complete sugar-sequence information, but not linkage information, is obtainable (Scheme 8a). Daughter spectra of doubly charged parents in the form of $(M' - Na - H)^{2-}$ of a monosialyl compound such as **4** do not add much sequence information to that obtainable from daughter spectra of singly charged ions, as shown in Scheme 8b. However, in the case of CID–MS/MS of $(M' - 2Na)^{2-}$ of a dibasic acid such as **3** (shown in Scheme 8c), the sugar sequence is deducible from both doubly and singly charged daughter ions, the latter being produced after an elimination of SO_3 with a concomitant intramolecular transfer of a proton. In the ESI, sulfate group(s) can behave more as a neutral leaving moiety than as a charge-retaining group that produces an intense signal at m/z 97(HSO_4)$^-$, which by contrast is prominent in FAB–MS or FAB CID–MS/MS in the negative-ion mode.

Unless collision is applied, ESI CID–MS/MS is able to make use of an ion representing exclusively the whole molecule as a precursor, as stated in Section II. In other words, there is no way to choose a part of the molecule as the precursor without the help of collision. This is a disadvantage of ESI CID–MS/MS. It has been reported that isomeric sulfated glycosphingolipids **1** (Le^a-) and **2** (Le^x-) can be distinguished by the negative-ion FAB CID–MS/MS if $(C_2 - Na + H)^-$, which represents the nonreducing end trisaccharide moiety, is chosen as the precursor ion (18). However, conventional mass spectrometry of ESI–MS does not exhibit such a fragment ion unless collision is applied.

3. Positive-Ion Conventional Mass Spectrometry of Lewis-Type Glycosphingolipids (1).
Underivatized acidic glycoconjugate samples are more frequently analyzed in the negative-ion mode than in the counterpolarity (positive) mode using either FAB or ESI. However, in some laboratories, the mass spectrometer is operated almost exclusively in the positive-ion mode. In fact, not only neutral glycoconjugates (10), but also sulfated and sialyl branched glycosphingolipids are amenable to ESI–MS and ESI CID–MS/MS analyses in the positive-ion mode.

Electrospray ionization–mass spectrometry in the positive-ion conventional mass spectrometry for acidic glycosphingolipids such as **1–5** produces $(M' + Na)^+$ and $(M' + 2Na)^{2+}$, rather than protonated species. A triply charged ion $(M' + 3Na)^{3+}$ was observed only for the sialyl Le^x-type heptaosylceramide **5**, but not for the disulfated pentaosylceramide **3** (compound **6** was not measured in the positive-ion mode). Again, the maximum charge state is difficult to predict, and the number of sugar units may heavily influence the maximum charge state (Table 3). Acidic glycoconjugates from a biological origin often show a singly charged positive-ion peak at 16, and/or 32 Da higher than $(M' + Na)^+$. In those cases, participation of the potassium salt or an attachment of K^+ in place of Na^+ is indicated (see Section V.A.1 for GQ1b).

4. Positive-Ion CID–MS/MS of Lewis-Type Glycosphingolipids (1).
As in the case of negative-ion ESI CID–MS/MS, daughter spectra of acidic glyco-

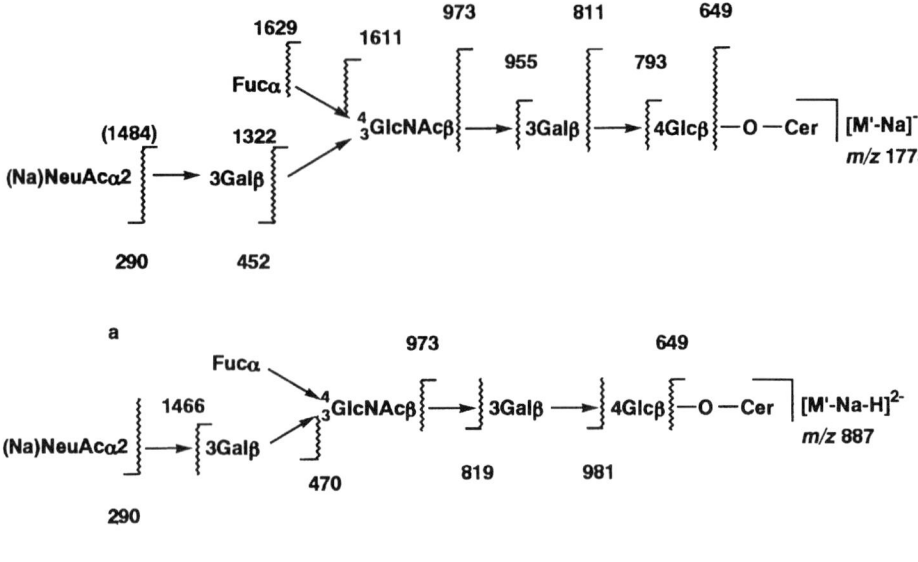

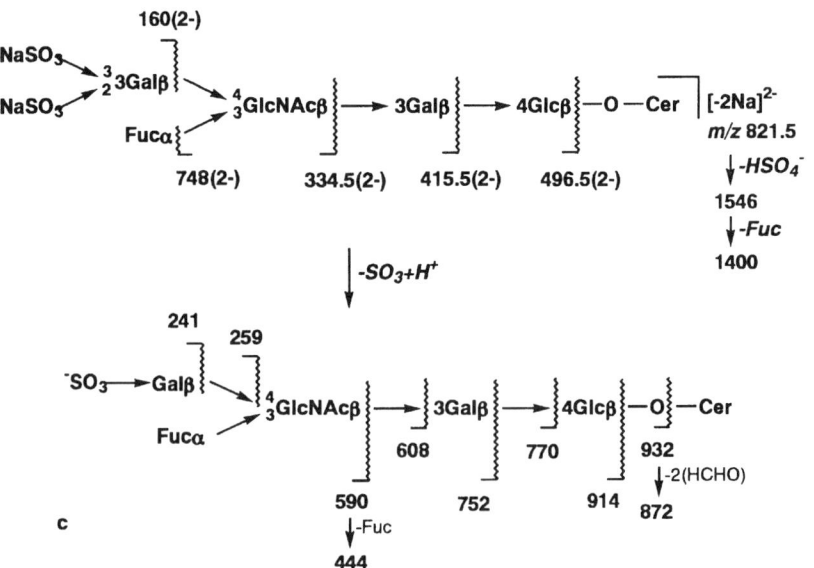

Scheme 8. Fragmentation patterns of the negative-ion ESI CID–MS/MS of sialyl Le[a]-pentaosylceramide (**4**), having $(M' - Na)^-$ (m/z 1775) as the parent in a, and $(M' - Na - H)^{2-}$ (m/z 887) as the parent in b. Scheme 8c shows the fragmentation pattern of the negative-ion ESI CID–MS/MS of disulfated Le[x]-pentaosylceramide (**3**), having $(M' - 2Na)^{2-}$ (m/z 821.5) as the parent. Note that, in this case, doubley charged and singly charged fragment ions are shown in one spectrum.

Table 3. Results of positive-ion ESIMS of Le-type sulfated and sialyl glycosphingolipids

Compound[a]	M'[b]	observed ions	m/z (relative intensity, %)	
		[M' + Na]⁺	[M' + 2Na]²⁺	[M' + 3Na]³⁺
1	1587	1610 (9)	816.5 (100)	—[c]
2	1587	1610 (10)	816.5 (100)	—
3	1689	1712 (47)	867.5 (100)	—
4	1798	1821 (9)	922 (100)	—
5	2163	2186 (13)	1104.5 (100)	744 (86)

[a] See Scheme 7 for structures. [b] M' represents the full sodium salt. [c] Not detected.

conjugates in the positive-ion mode strongly resemble the corresponding FAB CID–MS/MS spectra, provided that singly charged ions are selected as the ESI precursors. This again means that $(M' + Na)^+$ in the ESI mode has both an ion structure and also an internal energy similar to those of the corresponding ion produced in FAB. Additionally, CID–MS/MS of doubly charged $(M' + 2Na)^{2+}$ at m/z 867.5 for **3** produces a daughter spectrum (Fig. 7b), which is much like that of $(M' + Na)^+$ in the ESI mode, as shown in Figure 7a. In other words, in the positive-ion mode, ESI daughters of $(M' + 2Na)^{2+}$, ESI daughters of $(M' + Na)^+$, and FAB daughters of $(M' + Na)^+$ all resemble each other. The implication is, therefore, that after a facile loss of Na^+ from the doubly sodiated ESI parent ion, the resulting singly charged ion is much like the $(M' + Na)^+$ produced in either ionization mode. This interpretation is analogous to that proposed for the negative-ion CID–MS/MS mode.

VI. NONCOVALENT BINDING INVOLVING SUGAR CHAINS

The molecular weight of a noncovalently bound associate (complex, supermolecule, or supramolecule) can be determined by ESI–MS with little difficulty (19). Just considering the chemistry, it is already impressive that such a weakly associated (an order of magnitude lower than typical covalent bond energies) supermolecule, such as a protein–sugar complex, survives the harsh conditons of heat and spray energy encountered in ESI mass spectrometry. However, biologists question the significance of detected noncovalent binding if it fails to conclusively prove to be specific binding representing the same supermolecule as the physiological entity. A large molecule like a protein comprises many single

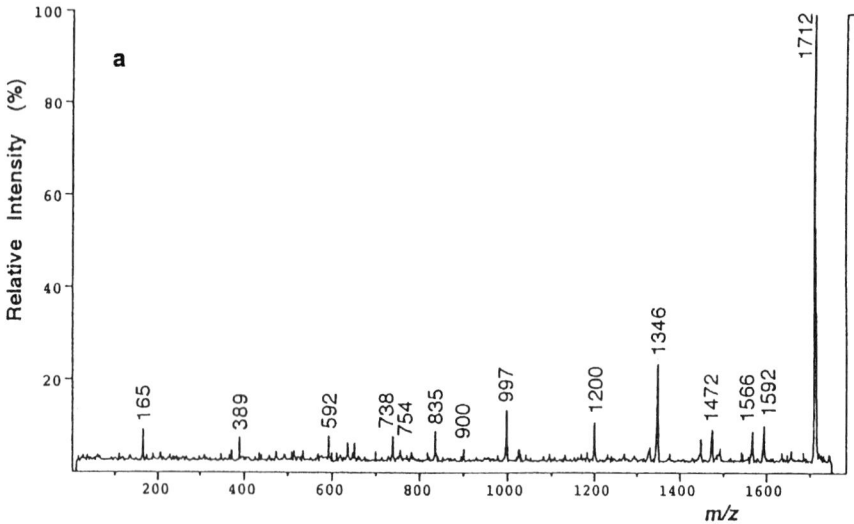

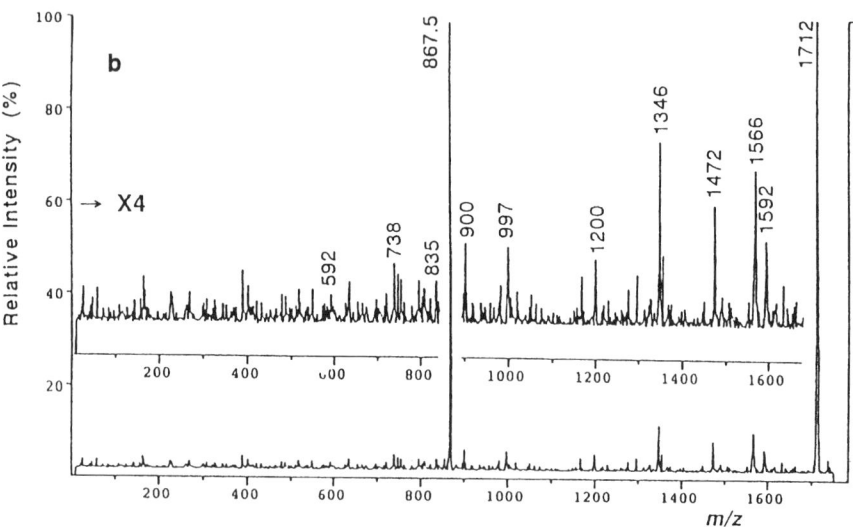

Figure 7. Positive-ion ESI CID–MS/MS spectra of disulfated Le^x-pentaosylceramide **3** (M′W 1689, as the disodium salt), having (M + Na)$^+$ at m/z 1712 as the precursor ion for **a**, and (b) (M + 2Na)$^{2+}$ at m/z 867.5 as the precursor ion for **b**. Reprinted with permission from T. Ii et al., *J. Mass Spectrom. Soc. Japan*, **1996**, *44*, 183–195. © 1996 Mass Spectrom. Soc. Japan.

bonds and thus many conformational degrees of freedom–only one of which is strictly required to form a definite supermolecule of physiological shape with a specific receptor, such as a glycoconjugate. Noncovalent binding through a hydrogen bond (8–30 kJ/mol) is much weaker than noncovalent ionic bonding.

On the other hand, many experimental conditions favor dissociation of weakly bound sample molecules prior to or at the instant of ionization or desorption. Once dissociated, monomers may form random clusters, which can then be detected as noncovalently bound complexes by the detector. Therefore, additional experimental evidence is required to claim the validity of mass spectrometric analyses of specifically bound biological complexes.

In 1991, using a pneumatically assisted electrospray (ion spray), Ganem et al. reported the epoch-making landmark of detection of noncovalent bindings between an enzyme substrate, via an enzyme intermediate, and finally, an enzyme product, where the substrate was a hexasaccharide of GlcNAc (20). Hen-egg white lysozyme (HEWL) hydrolyzes the sixmer of GlcNAc, producing mainly one tetramer and one dimer of the GlcNAc. They monitored the time course of an optimum buffer solution of (HEWL and $GlcNAc_6$) mixture at room temperature. Immediately after mixing, the enzyme–substrate complex $(HEWL + GlcNAc_6)^{8+}$ was detected in addition to the ions corresponding to unreacted $(HEWL + 8H)^{8+}$ and $(HEWL + 7H)^{7+}$ Ions of $(HEWL + GlcNAc_4)^{8+}$ and $(HEWL + GlcNAc_3)^{8+}$ began to appear in increasing abundance while $(HEWL + GlcNAc_2)^{8+}$ was not detected, owing to the even weaker binding. The presence of an equimolar known competitive inhibitor, N-acetylchitotetraose δ-lactone, reduced the ion intensity of the enzyme–substrate complex, indicating that the detected enzyme–substrate complex was not a random aggregate on the exterior surface of the enzyme protein. Many related studies followed, including one by Siuzdak et al. who reported on Ca^{2+}-dependent carbohydrate association detected by pneumatically assisted ESI (ion spray) (21). They examined interactions of the cell–surface carbohydrates, that is, Le^x and Le^x-lactosylceramide (Le^x–LacCer) in the presence of divalent cations, Ca^{2+}, Mn^{2+}, and Mg^{2+}, under ion spray conditions. The associated complexes were detected as doubly charged ions, $(nLe^x + \text{divalent cation})^{2+}$ and $(nLe^x\text{–LacCer} + \text{divalent cation})^{2+}$, where $n = 1$, 2, or 3. The dimer ($n = 2$) showed greater stability than the trimer ($n = 3$). No such association was detected when monovalent cations, such as Li^+, Na^+, and K^+, were present to replace the divalent cations. Furthermore, it was shown by CID–MS/MS that the Ca^{2+} complexation site was within the Le^x moiety, and that both fucose functionalities, of Le^x and of Le^x–LacCer, may be exposed on the dimer of Le^x and Le^x–LacCer. Le^x–LacCer also associates with LacCer, GalCer, and even Cer itself in the presence of divalent cations. The binding affinity between homodimeric Le^x–LacCer molecules was found to be greater than the binding between Le^x–LacCer and LacCer, but the binding affinity of Le^x–LacCer decreases even further with GalCer and still further with Cer. In addition, they proposed a model in which Ca^{2+} is complexed through a crown ether-like cavity within the monomer, complemented by a study using space-filling models as well as energy-minimum calculations. The consequent dimer model is in accord with cellular interaction between opposing cells which proceed through the surface carbohydrates in opposite configurations.

VII. ACYLGLYCEROLS

It is well known that acylglycerides, that is, tri-, di-, and monoglycerides are important in nutrition, cosmetics, and biochemistry, but characterization by conventional analytical methods is not satisfactory. Essentially being fat, acylglycerols do not dissolve in aqueous solvents, which are the most common solvents used in ESI measurements. Duffin et al. used a nonpolar solvent, chloroform–methanol, containing some salt or acid, for ion spray tandem mass spectrometry of an unknown lipid mixture recovered from a mammalian-cell culture reactor over 3 months (22). They preferred to use pneumatically assisted electrospray (ion spray) rather than the standard type ESI ion source. The two ionization methods are similar, but the former uses coaxial nebulizing gas to aid droplet formation. In the positive-ion daughter spectra primarily loss of fatty acids and acylinium ions of the fatty acids were observed, but at 100 eV and higher collision energy, C–C bonds in the fatty acid chains were also shown to be cleaved. Nevertheless, locations of double bonds could not be determined. As for sensitivity among these three types of acylglycerols, ion-current response decreased as the analyte polarity decreased. On the other hand, absolute ion abundance within the same acylglycerol varied linearly with its concentration in solution over a 10^4 sample concentration range. It should be noted that when $(M + NH_4)^+$ was chosen as the precursor for CID–MS/MS, the most facile loss was that of NH_3, and all other product ions were detected in protonated forms.

VIII. LIPID A

Lipopolysaccharides (LPSs) are another type of glycoconjugates located on the bacterial cell surface, and usually found together with phospholipids and proteins. The LPS consists of three structural regions, that is, lipid A, core oligosaccharide, and O-chain polysaccharide. Among them, lipid A is the most responsible for the pathogenisis and manifestation of infection of Gram-negative bacteria. The basic structure of Lipid A is a dimer of 2-amino-2-deoxy-glucose (GlcNβ1-6GlcN), in which the two amino groups at C2 and C$'$2, as well as several OH-groups, at least at C3 and C$'$3, are acylated. The non-reducing end C$'$4 is, in most cases, phospholylated (as shown in Scheme 9), but in some other cases, the reducing end C1 may be also phosphorylated. Heterogeneity within monophosphoryl lipid A (MLA) is found mostly in the alkane chain length and variations in the acylation.

Thus, structural determination of lipid A has been a very challenging subject in mass spectroscopy, especially since the emergence of the soft-ionization methods. Pioneering experiments were performed using FAB–MS (23), a laser microprobe instrument (24), LDMS (25), plasma desorption mass spectrometry (PDMS) (26), and more recently ESI–MS. In 1993, Harrata et al. applied ESI–MS, using a single-stage quadrupole instrument with CID, to the analysis

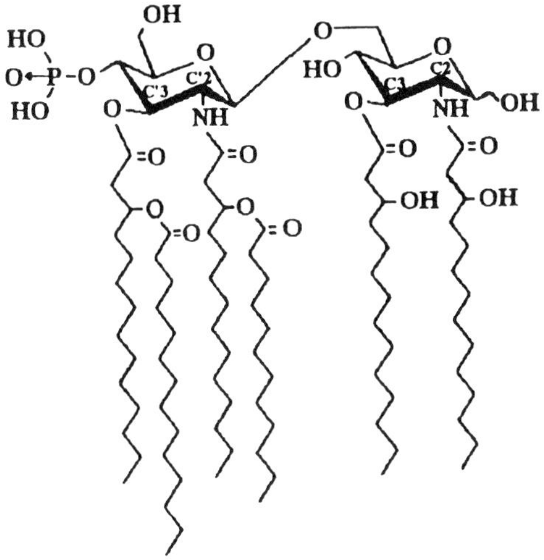

Scheme 9. A typical structure of lipid A. Reprinted with permission from V. N. Reinhold et al., *Anal. Chem.*, **1995**, *67*, 1772. © 1995 American Chemical Society.

of underivatized lipid A, which is suspected to be the etiological agent of byssinosis caused by *Enterobacter agglomerans* that colonizes field cotton (27). The spectra were compared with those of the standard lipid A isolated from *Salmonella minnesota*. Features of the collision spectra were much dependent upon the energy of the skimmer collision and the pressure in certain regions. Especially in the negative-ion CID spectra, the negative charge on all fragment ions is retained on the deprotonated phosphate, thus a series of deacylations are considered to be "charge-remote fragmentations" (28). In this way, the usefulness of skimmer-collision CID was demonstrated.

In addition to chemical reactions, NMR, FAB and EI studies, Bhat et al. utilized ESI–MS to elucidate a new structure of lipid A, which is a component of LPSs of Gram-negative *Rhizobiaceae* family (29). Bacteria belonging to this family are able to form nitrogen-fixing symbiotic relationships with legume plants. The backbone structure, which had been previously unknown, was a trisaccharide consisting of GalAα1-4GlcNβ1-6GlcN-onate, where GalA stands for anhydrogalacturonic acid. Unlike enteric lipid As, this family lacks 3-acyloxyacyl groups. It is also quite unusual that this lipid A does not contain a phosphate group, but still is as hydrophilic as common diphosphorylated lipid A, because both the reducing and nonreducing ends carry dissociative acid groups. Besides, the C5–OH is esterified with a 28:0 fatty acid, in which C27 is hydroxylated. The position of the OH group in the 28:0 fatty acid, as well as the linkage of the GlcN-onate, is still to be investigated, but the proposed structure is shown in Scheme 10. Ion-spray mass spectrometry was performed with skimmer CID mainly to confirm the molecular weights of de-O-acyl lipid As.

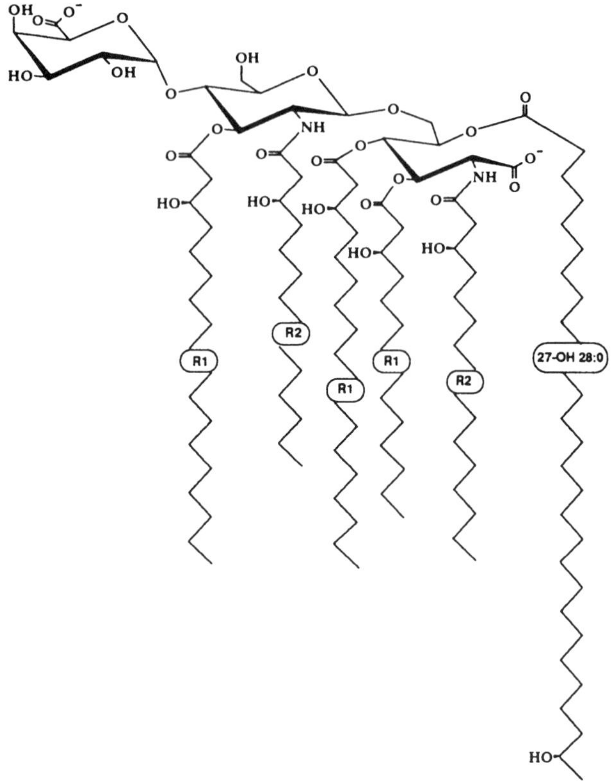

Scheme 10. Proposed structure of lipid A isolated from *Rhizobium leguminosarum*. Reprinted with permission from U. R. Bhat et al., *J. Biol. Chem.*, **1994**, *269*, 14402. © 1994 American Society for Biochemistry & Molecular Biology.

More detailed structural studies of monophosphoryl lipid A were performed by Chan and Reinhold, who applied CID–MS/MS in the ESI mode (30). The MLA profiles were compared between *Salmonella minnesota* Re595 and *Shigella flexneri*. It was found without separation that each sample is composed of more than 20 structures related to each other as acyl analogs, through alkane heterogeniety or as monosaccharides, and that the difference between the two samples mostly lay in the relative abundance of each component. The CID–MS/MS studies revealed that acyloxyesters fragment more readily than amide or glycosidic cleavages. The CID analyses of selected species also correlated MLA of *S. flexneri* with that of *E. coli* J-5, rather than that of *Shigella sonnei*. In this case, no diphosphoryl nor C1-monophosphoryl species were detected. According to these authors, FAB, LSIMS, LD, and PDMS all involve a ballistic ionization mechanism, thus giving excess internal energies to the sample molecules to cause fragmentations. This is why earlier lipid A analyses using particle-induced desorption yielded complicated spectra compared to recent ESI–MS.

IX. GLYCOSYLPHOSPHATIDYLINOSITOL MEMBRANE ANCHORS (GPI ANCHORS)

The GPI anchors are found among eukaryotic cells ranging from protozoa to mammalians. GPI anchoring is one of the common modes of post-translational modification of proteins by lipids. It has a well conserved core structure of peptide-ethanolamine-PO_4-Manα1-2Manα1-6Manα1-4GlcN1-6-inositol-1-PO_4-lipid motif. Here, Man designates anhydromannose or (mannose – H_2O). However GPI anchors may have many different types of additional residues such as ethanolaminephosphate(s) or hexose(s) pendant to the core structure. Using positive-ion ESI–MS, Redman et al. determined the pendant galactosyl residues of GPI-peptide from *Trypanosoma brucei* surface glycoprotein (31). The mass profile of the (4+) region of the peptide-GPI showed seven ions 162/4 u apart from each other, indicating that heterogeneity in the galactose residue ranges from zero to six (Scheme 11). Reinhold et al. presumed that sensitivity of this peptide-conjugated GPI anchor was enhanced by the charge-carrying ability of the peptide (4). The CID–MS/MS of the $(M + 4H)^{4+}$ of two GPI-peptides showed fragment ions produced by cleavages occurring preferentially in the GPI-anchor portion, thus providing easy interpretation of the GPI structures.

Taguchi et al. applied ESI–MS to the analysis of the GPI-anchor of 5′-nucleotidase isolated from bovine liver (32). Purified 5′-nucleotidase was cleaved by CNBr, and the inositolphosphoglycan (IPG)-containing C-terminal peptides in methanol/water was subjected to ESI–MS on a sector-type analyzer (JEOL

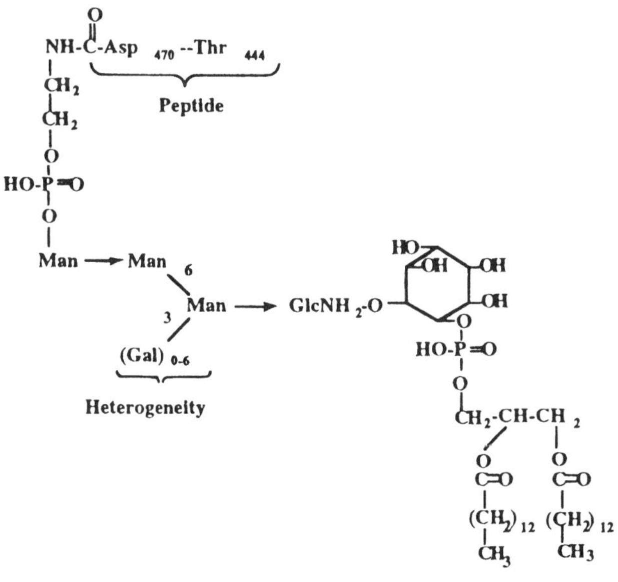

Scheme 11. An example of a peptide and its pendant GPI anchor. Reprinted with permission from V. N. Reinhold et al., *Anal. Chem.*, **1995**, *67*, 1772. © 1995 American Chemical Society.

JMS-LX 2000 Tokyo, Japan). On the other hand, some fractions of CNBr-cleaved IPG-peptides were treated with trypsin, and analyzed by ESI–MS using, in this case, a reversed-geometry sector instrument (JEOL JMS-SX102A Tokyo, Japan). Thus, they determined five species of molecular microheterogeneities in residues pendant to the core structure, such as extra phosphoethanolamine, Man, HexNAc, or their combinations.

X. PHOSPHOLIPIDS

Phospholipids (Scheme 12) maintain the lipid bilayer structures of cell membranes, and are thus important for membranal functions such as transport, endocytosis, or various physiopathological changes. They are also considered important to the activity of membrane-bound enzymes and they serve as the source of polyunsaturated fatty acids. Phospholipids have been amenable to mass spectrometry without derivatization because so-called soft ionization, such as FAB–MS and SIMS, became available. Characterization of the polar head group, namely a phosphodiester group, and the nonpolar tail(s), that is, glycerolipid or sphingolipid, by conventional mass spectrometry (33, 34) and more elaborately by CID–MS/MS analyses has a history of well over 10 years (35–38). However, FAB–MS is not sufficiently sensitive to analyze diminutive amounts of phospholipids; thus, it is not easy to see the profile of complex phospholipids quantitatively without the aid of liquid chromatography. Moreover, differential volatilization among chemically distinct phospholipid classes may also affect quantification by FAB–MS. In 1993, Silvestro et al. analyzed platelet-activating factor (PAF), which chemically corresponds to 1-alkyl-2-acetylglycero-3-phosphocholine and lyso-PAF from human polymorphonuclear neutrophils (PMN) using pneumatically assisted ESI (ion spray) as the interface for reversed-phase HPLC (39). Reconstructed-ion chromatograms indicated an adequate separation, and the SIM mode showed a detection limit as low as 0.3 ng. The CID–MS/MS of individual PAF-related protonated molecules $(M + H)^+$ as precursors all showed the phosphoryl-choline positive-ion at m/z 184 (for the ion structure, see ref. 33).

Using a 0.5% ammonium hydroxide/aqueous-methanol-hexane mixture as the solvent for HPLC/ESI–MS, Kim et al. reported that phospholipids showed a linear response over a 10^2 range with total 0.5 pmol complex samples (40). The relative response was more dependent on the type of head group than the composition of the nonpolar group(s) within a phospholipid class, with phosphatidylcholine being the most sensitive and phosphatidylserine about 20 times less in the positive-ion mode. This means that quantitative analysis of complex phospholipid mixture is not straight forward. Although ESI–MS had been applied previously to analyze some classes of phospholipids (41), Kim et al. improved the quantitation precision by employing an HPLC with a splitter before the ESI (40). Han and Gross, on the other hand, applied positive- and negative-ion mode ESI–MS to a chloroform/methanol extract of human

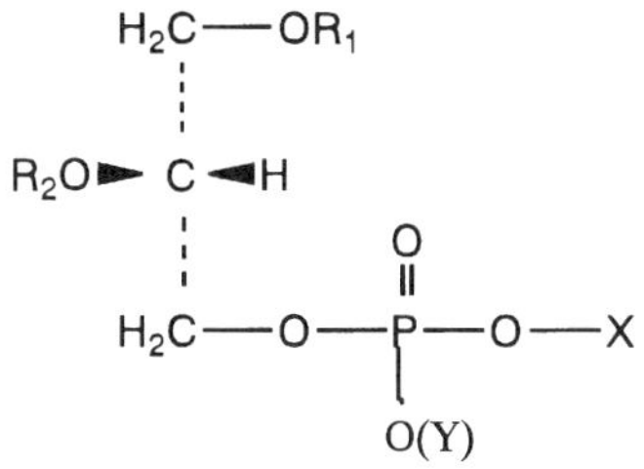

X. PHOSPHOLIPIDS

$$\begin{array}{c} H_2C-OR_1 \\ R_2O \blacktriangleright C \blacktriangleleft H \\ H_2C-O-\overset{\overset{O}{\|}}{\underset{O(Y)}{P}}-O-X \end{array}$$

*Phosphatidylcholine (PC)
R_1, R_2: acyl
$X = (CH_2)_2\overset{+}{N}(CH_3)_3$

*Phosphatidylinonsitol (PI)
R_1, R_2: acyl
$X =$ (typically myo-) inositol
$Y = H$

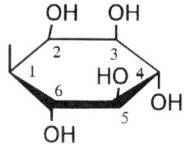

*Phosphatidylethanolamine (PE)
R_1, R_2: acyl
$X = (CH_2)_2\overset{+}{N}H_3$

*Phosphatidylserine (PS)
R_1, R_2: acyl
$X =$ serine

$$(-CH_2-\underset{\underset{NH_3}{|}}{CH}-CO_2^-)$$

*Platelet activating factor (PAF)
R_1: alkyl or alkenyl
$R_2 =$ acetyl
$X = (CH_2)_2\overset{+}{N}(CH_3)_3$

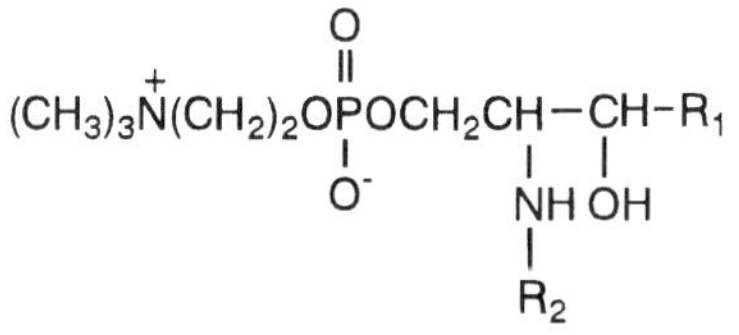

$$(CH_3)_3\overset{+}{N}(CH_2)_2O\underset{\underset{O^-}{|}}{\overset{\overset{O}{\|}}{P}}OCH_2\underset{\underset{NH}{|}}{CH}-\underset{\underset{OH}{|}}{CH}-R_1$$
$$\underset{R_2}{|}$$

Scheme 12. Phospholipids.

erythrocyte plasma membrane from less than 1 µL whole blood without prior chromatography (42). They reported that phosphatidylethanolamine in that membranal source was mostly composed of plasmalogens (18:0/20:4-plasmenyl-ethanolamine). The CID–MS/MS spectrum in the negative-ion mode carried out in the presence of slight excess of NaOH showed carboxylate ions, by which ester species were assigned. After this, the regiospecificity of the two ester groups was determined from the observed lyso-type daughter ions. These molecular species were also elucidated by conventional techniques. According to these authors, ESI is less sensitive to small differences in surface activities of phopholipid classes or subclasses when compared with FAB, hence ESI has advantages for quantification of mixtures. However, the response factors are still widely different among different head groups; phosphatidylcholine being the most sensitive, showing $(M + Na)^+$ in the positive-ion mode, and phosphatidylethanolemine showing the most intense $(M - H)^-$ in the negative-ion mode when equimolar mixture was analyzed by ESI. Thus, profiling or quantitation of phospholipids having different head groups by ESI is not still straight forward unless stable isotopes are used as internal standards. ESI–MS can be used to fortify structural information obtained through other analytical methods. Hechtberger et al. characterized three classes of PI-containing sphingolipids of unique structures, namely, inositolphosphorylceramide (Ins$P$Cer), mannosylinositolphosphorylceramide (ManIns$P$Cer), and mannosyldiinositolphosphorylceramide (Man$P$ Ins$_2P$Cer), all of which were isolated from yeast subcellular membranes mainly by the use of conventional analytical methods, but used ESI–MS to determine molecular weights of the proposed structures (43).

XI. OTHER LIPIDS STUDIED WITH ESI–MS

Interesting applications of ESI–MS in structural studies of other types of lipids have also been reported. For example, Bitsch et al. elucidated structures of taxol and related diterpenoids using HPLC–ESI–MS and the CID–MS/MS (44). Wilson et al. used ESI–MS for the study of vitamin-D derivatives (45), and Yeung et al. made use of HPLC–ESI–MS to characterize rat renal metabolites of 1,25-dihydroxy-16-ene vitamin D-3 (46). Roda et al. used ESI–MS as the detector for HPLC analysis of bile acids (47).

XII. CONCLUSIONS

The joy of establishing a new fragmentation, after the hard work of extracting structures of ions from known glycoconjugate or lipid samples, is indeed enormous. The results are then readily applicable to real samples, which may be deeply related to human diseases or normal development. Even this latter process is like a cross-word puzzle, where all the knowledge and information available are put together to yield a rational structure and verify conformity of information from analytical data.

Significance of mass spectrometry in the research of carbohydrates and lipids lies, first, in the sensitivity, and second, in the structural information, which complements or sometimes exceeds information obtained by NMR. Mass spectrometer sensitivity is becoming increasingly critical as minute quantities of important, biological specimens become the focus of interest. Mass spectrometry usually does not require radioactive experiments, yet the sensitivity competes with radioimmunological assays. Although compound dependent, ESI–MS is probably next to the most sensitive ionization method, which is MALDI-MS, among several others. As for structural information, FAB may still be better for smaller molecules, but FAB is incapable of ionizing molecules larger than several thousand daltons, especially if the molecule is highly acidic, besides FAB requires much larger sample amounts.

For structural analysis utilizing ESI, CID–MS/MS is an absolute requirement, because unless CID is applied, ESI does not produce fragment ions in the conventional mass spectrometry. One of the advantages of CID–MS/MS in the ESI mode is that multiply charged ions can be chosen as the parent. However, a phenomenon of interest is that daughter spectra having $(M - H)^-$ or $(M' - Na)^-$ as the parent are similar to each other in ESI and in FAB when the charge is localized. This phenomenon is even more remarkable in the positive-ion mode. That is, the daughter spectra of $(M' + 2Na)^{2+}$ and $(M' + Na)^+$ in ESI, and that of $(M' + Na)^+$ in FAB, resemble each other. A multiply charged parent ion generally requires less energy to cleave bonds, and provides additional information on the structure.

In the future, microseparation techniques such as HPLC or capillary-zone electrophoresis using ESI–MS as the interface will become essential to see the profile of ultratrace amounts of biological samples. Mass spectrometry in the field of carbohydrates and lipids has several important aspects, most notably, structural studies in the interest of pure science and characterization of biological materials, including their profiles. As for the first, ESI–MS, especially its CID–MS/MS mode, now provides much more information than generally credited. One of the future problems will probably be to obtain steric information, such as the determination of sugar epimers. The second, and continually ongoing quest is to reduce the sample amount required for mass spectrometric analyses. This will likely become increasingly important in the near future. Lastly, profiling of known glycoconjugates and lipids *in vivo* is important, if we intend to be of greater service to the medical sciences.

As for the detection of specific, noncovalent weak bindings, pneumatically assisted electrospray mass spectrometry seems to be more advantageous than standard ESI–MS. Yet, one problem that remains is how to modulate ESI conditions in general so as to detect specific bindings selectively. The author thinks that, since specific bindings should be energetically more favorable than nonspecific bindings, they are more likely to survive ESI measurement conditions than otherwise bound random clusters, if appropriate methodology can be established.

The author regrets that it is impossible to cover the total range of quickly growing important ESI applications in carbohydrate and lipid research areas. Subjects described here are only a few examples, which will hopefully be of some service to the readers.

ACKNOWLEDGMENTS

The author acknowledges Professor Y.-T. Li of Tulane University Medical Center for supporting my involvement in this book. Research the author was involved in was carried out at Frontier, RIKEN under the supervision of her director, Professor Y. Nagai, and in collaboration with Dr. S. Kurono, Dr. T. Ii (presently at Soda Aromatic Co. Ltd.), and Mr. M. Kubota. Most samples, except for commercial chemicals, were generously given by Professor T. Ogawa of RIKEN and Dr. S. Nunomura of Nissin Food Products Corp. Mr. H. Ishii, a student at the University of Electro-communications, helped in organizing figures and schemes.

The author is especially indebted to Dr. J. Sokoloff of Frontier, RIKEN who kindly read through and edited this chapter.

REFERENCES

1. Ii, T.; Ohashi, Y.; Ogawa, T.; Nagai, Y. *J. Mass Spectrom. Soc. Japan* **1996**, *44*, 183–195.
2. Ii, T.; Ohashi, Y; Nagai, Y. *Org. Mass Spectrom.* **1993**, *28*, 927–928.
3. Dell, A.; Reason, A. J.; Khoo, K.-H.; Panico, M.; McDowell, R. A.; Morris, H. R. In *Mass Spectrometry of Carbohydrate-Containing Biopolymers*; Academic Press: San Diego, CA, 1994, Vol. 230, pp. 108–132.
4. Reinhold, V. N.; Reinhold, B. B.; Costello, C. E. *Anal. Chem.* **1995**, *67*, 1772–1784.
5. Gu, J.; Hiraga, T.; Wada, Y. *Biol. Mass Spectrom.* **1994**, *23*, 212–217.
6. Okamoto, M.; Takahshi, K.; Doi, T. *Rapid Commun. Mass Spectrom.* **1995**, *9*, 641–643.
7. Reinhold, B. B.; Chan, S. Y.; Reuber, T. L.; Marra, A.; Walker, G. C.; Reinhold, V. N. *J. Bacteriology* **1994**, *176*, 1997–2002.
8. Garozzo, D.; Impallomeni, G.; Spina, E.; Green, B. N.; Hutton, T. *Carbohydr. Res.* **1991**, *221*, 253–257.
9. Mulroney, B.; Traeger, J. C.; Stone, B. A. *J. Mass Spectrom.* **1995**, *30*, 1277–1283.
10. Ii, T.; Ohashi, Y; Matsuzaki, Y.; Ogawa, T.; Nagai, Y. *Org. Mass Spectrom.* **1993**, *28*, 1340–1344.
11. Domon, B; Costello, C. E., *Glycoconjugate J.* **1988**, *5*, 397–409.
12. Ii, T.; Ohashi, Y; Nagai, Y. *Carbohydr. Res.* **1995**, *273*, 27–40.

13. Svennerholm, L. *J. Neurochemistry*, **1963**, *10*, 613–623.
14. Green, P. J.; Tamatani, T.; Watanabe, T.; Miyasaka, M.; Hasegawa, A.; Kiso, M.; Yuen, C.-T.; Stoll, M. S.; Feizi, T. *Biochem. Biophys. Res. Commun.* **1992**, *188*, 244–251.
15. Yuen, C.-T.; Bezouska, K.; O'Brien, J.; Stoll, M.; Lemoine, R.; Lubineau, A.; Kiso, M.; Hasegawa, A.; Bockovich, N. J.; Nicolaou, K. C.; Feizi, T. *J. Biol. Chem.* **1994**, *269*, 1595–1598.
16. Erbe, D. V.; Wolitzky, B. A.; Presta, L. G.; Norton, C. R.; Ramos, R. J.; Burns, D. K.; Rumberger, J. M.; Rao, B. N. N.; Foxall, C.; Brandley, B. K.; Lasky, L. A. *J. Cell Biol.* **1992**, *119*, 215–227.
17. Ii, T.; Ohashi, Y; Nunomura, S.; Ogawa, T.; Nagai, Y. *J. Biochem.* **1995**, *118*, 526–533.
18. Ii, T.; Ohashi, Y.; Ogawa, T.; Nagai, Y. *Glycoconjugate J.* **1996**, *13*, 273–283.
19. Ohashi, Y.; Hirabayashi, Y.; Kanai, M.; Nagai, Y. in *ESI—A Novel Probe for Protein-Sugar Chain Bindings*; Matsuo, T. Ed.; San-ei Publishing: Kyoto, Japan, 1992, pp. 168–169.
20. Ganem, B.; Li, Y.-T.; Henion, J. D. *J. Am. Chem. Soc.* **1991**, *113*, 7818–7819.
21. Siuzdak, G.; Ichikawa, Y.; Caulfield, T. J.; Munoz, B.; Wong, C.-H.; Nicolau, K. C. *J. Am. Chem. Soc.* **1993**, *115*, 2877–2881.
22. Duffin, K. L.; Henion, J. D.; Shieh, J. J. *Anal. Chem.* **1991**, *63*, 1781–1788.
23. Qureshi, N.; Takayama, K.; Heller, D.; Fenselau, C. *J. Biol. Chem.* **1983**, *258*, 12947–12951.
24. Seydel. U.; Lindner, B.; Wollenweber, H.-W.; Rietschel, E. T. *Eur. J. Biochem.* **1984**, *145*, 505–509.
25. Qureshi, N.; Honovich, J. P.; Hara, H.; Cotter, R. J.; Takayama, K. *J. Biol. Chem.* **1988**, *263*, 5502–5504.
26. Masoud, H.; Weintraub, S. T.; Wang, R.; Cotter, R.; Holt, S. C. *Eur. J. Biochem.* **1991**, *200*, 775–779.
27. Harrata, A. K.; Domelsmith, L. N.; Cole, R. B. *Biol. Mass Spectrom.* **1993**, *22*, 59–67.
28a. Adams, J.; *Mass Spectrom. Rev.* **1990**, *9*, 141–186.
28b. Contado, M. J.; Adams, J.; Jensen, N. J.; Gross, M. L.; *Am. Soc. Mass Spectrom.* **1991**, *2*, 180–183.
28c. Gross, M. L.; *Int. J. Mass Spectrom. Ion Processes* **1992**, *118/119*, 137–165.
29. Bhat, U. R.; Forsberg, L. S.; Carlson, R. W. *J. Biol. Chem.* **1994**, *269*, 14402–14410.
30. Chan, S.; Reinhold, V. N. *Anal. Biochem.* **1994**, *218*, 63–73.
31. Redman, C. A.; Green, B. N.; Thomas-Oates, J. E.; Reinhold, V. N.; Ferguson, M. A. J. *Glycoconjugate J.* **1994**, *11*, 187–193.
32. Taguchi, R.; Hamakawa, N.; Harada-Nishida, M.; Fukui, T.; Nojima, K.; Ikezawa, H. *Biochemistry* **1994**, *33*, 1017–1022.
33. Ohashi, Y. *Biomed. Mass Spectrom.* **1984**, *11*, 383–385.
34. Weintraub, S. T.; Ludwig, J. C.; Mott, G. E.; McManus, L. M.; Lear, C.; Pinckard, R. N. *Biochem. Biophys. Res. Commun.* **1985**, *129*, 868–876
35. Jensen, N. J.; Tomer, K. B.; Gross, M. L. *Lipids* **1987**, *22*, 480–489.

36. Jensen, N. J.; Gross, M. L. *Mass Spectrom. Rev.* **1988**, *7*, 41–69.
37. Munster, H.; Budzikiewicz, H. *Biol. Chem. Hoppe-Seyler* **1988**, *369*, 303–308.
38. Haroldsen, P. E.; Gaskell, S. J. *Biomed. Environ. Mass Spectrom.* **1989**, *18*, 439–444.
39. Silvestro, L.; Col, R. D.; Scappaticci, E.; Libertucci, D.; Biancone, L.; Camussi, G. *J. Chromatogr.* **1993**, *647*, 261–269.
40. Kim, H.-Y.; Wang, T.-C. L.; Ma, Y.-C. *Anal. Chem.*, **1994**, *66*, 3977–3982.
41. Weintraub, S. T.; Pinckard, R. N.; Hail, M. *Rapid Commun.* **1991**, *5*, 309–311.
42. Han, X.; Gross. R. W. *Proc. Natl. Acad. Sci.* **1994**, *91*, 10635–10639.
43. Hechtberger, P.; Zinser, E.; Saf, R.; Hummel, K.; Paltauf, F.; Daum, G. *Eur. J. Biochem.* **1994**, *225*, 641–649.
44. Bitsch, F.; Ma, W.; Macdonald, F.; Nieder, M.; Shackleton, C. H. L. *J. Chromatogr.* **1993**, *615*, 273–280.
45. Wilson, S. R.; Tulchinsky, M. L; Wu, Y. H. *Bioorg. Med. Chem. Lett.* **1993**, *3*, 1805–1808.
46. Yeung, B.; Vouros, P.; Siucaldera, M. L.; Reddy, G. S. *Biochem. Pharmacol.* **1995**, *49*, 1099–1110.
47. Roda, A.; Gioacchini, A. M.; Cerre, C.; Baraldini, M. *J. Chromatogr. B—Biomed. Appl.* **1995**, *665*, 281–294.

CHAPTER 14

Drug Metabolism and Pharmacokinetics

GRACE K. POON

Drug Metabolism and Pharmacokinetics, SmithKline Beecham Pharmaceuticals, The Frythe, Welwyn, United Kingdom

	Abstract	499
I.	Introduction	500
II.	Identification of metabolites by LC–ESI–MS (qualitative analysis)	502
	A. Identification of phase I metabolites	502
	B. Identification of phase II metabolites	508
III.	Pharmacokinetics studies (quantitative analysis)	511
IV.	Capillary electrophoresis–electrospray ionization–mass spectrometry (CE–ESI–MS)	515
V.	The LC–ESI–Ion trap (IT) in drug-metabolism and pharmacokinetics	516
VI.	The LC–ESI–Magnetic-sector mass spectrometer in drug metabolism	517
VII.	Metabolic studies of anticancer reagents by LC–ESI–MS	517
VIII.	Conclusion	520
	Acknowledgments	521
	References	521

ABSTRACT

The process of drug metabolism involves biotransformation of the drug to metabolites that are chemically different from the parent drug. These metabolites can (1) have little or no pharmacological effect, (2) be more pharmacologically active than the parent drug, or (3) generate metabolites that elicit additional biological responses unrelated to the pharmacological properties of the parent compound. The ability to metabolize a particular drug is of prime importance in determining the efficacy, duration of action, and toxicity of the drug. Moreover, an understanding of the impact of drug metabolism on activity and toxicity can lead to the development of new drugs with enhanced pharmacological activities. Conse-

---

Electrospray Ionization Mass Spectrometry, Edited by Richard B. Cole.
ISBN 0-471-14564-5 © 1997 John Wiley & Sons, Inc.

quently, it is necessary to characterize and quantitate both the parent drugs and the metabolites in the body after administration. The recent literature suggests that the best technology currently available to fulfill these criteria is on-line liquid chromatography–mass spectrometry.

I. INTRODUCTION

The major criteria for the selection of a drug candidate are good bioavailability and appropriate biological half-life. Pharmacokinetics is the study of the uptake, distribution, and excretion of drugs with respect to time. It is an extremely important subject area from a clinical view, as the intensity and duration of action of a drug are related to what concentration of drug is present at the active site. Several ways of measuring the concentrations of drugs in various tissues and body fluids over a period of time are available, including radioimmunoassay (RIA), gas chromatography with mass spectrometric detection (GC–MS), liquid chromatography–ultraviolet detection (LC–UV), LC with radioactivity detection, or LC with mass spectrometric detection (LC–MS). The first of these methods (RIA) has an inherent disadvantage in its lack of specificity when applied to human samples, which arises from the cross reactivity of different compounds (having similar structures) with the antibodies. As a consequence, results obtained from RIA analysis have sometimes been difficult to interpret (1–3). Both LC–UV and LC–radioactive-detection methods are less specific, and require that the analytes possess UV chromophores or radioactivity, respectively.

Traditionally, GC–MS has been used to study drug metabolism. More recently, LC–MS has replaced GC–MS because the LC technique does not require sample derivatization for polar metabolites. Some attractive features of LC–MS are its good sensitivity, reliability, and specificity for a wide variety of compounds, with minimal sample handling. Because of these advantages, LC–MS has become widely adopted for studies of drug metabolism and pharmacokinetics. The inherent value of the correlation of retention times for a reference compound and the sample is essential in the support of sample identification. A series of ionization techniques compatible with LC is available, which include thermospray ionization (TSP), electron impact (EI) or chemical ionization (CI) with particle beam interface (PB), continuous-flow fast-atom bombardment (CFFAB), atmospheric pressure ionization (API) using a heated nebulizer interface (APCI), API with electrospray interface (ESI), and the closely related API with pneumatically assisted electrospray ionization (ionspray, IS). Most of these methods have their limitations: TSP and APCI require critical control of the vaporizer temperature during analysis, and thermal degradation of labile molecules may occur; PB lacks sensitivity; CFFAB can only accept low flow rates, and requires the presence of a matrix to assist ionization, thereby affecting the ion-current stability and making the method susceptible to interference from background ions. However, LC–ESI/IS–MS is now free of most major technical problems and has become the method of choice for drug metabolism and pharmacokinetic research, as reflected in the impressive number of applications published in the last few years.

Both ESI and IS are soft-ionization techniques based on the application of a large electric field to nebulize the liquid (ESI), where nebulization can be assisted by pneumatics and/or by heat (IS). Thus LC–ESI–MS and LC–IS–MS have gained widespread recognition as powerful analytical tools, because they permit the separation and ionization of polar nonvolatile or thermally labile compounds, such as drugs, conjugated metabolites, peptides, or DNA adducts. In addition, these techniques facilitate qualitative and quantitative studies on the parent compounds and their polar metabolites.

Electrospray ionization is typically interfaced with a single- or triple-quadrupole mass spectrometer, which can tolerate high LC flow rates. Electrospray ionization interfaced with magnetic-sector (4,5), Fourier-transform (6,7), or ion-trap mass spectrometers (8,9) have also been reported. The ESI method usually generates the pseudomolecular ion $[M + H]^+$ or sometimes NH_4^+, Na^+, or K^+ adduct ions for the labile drugs with little fragmentation. Collision-induced dissociation (CID) is sometimes performed to enhance fragmentation of the analytes either at the ES source (in-source CID) or in conjunction with tandem mass spectrometry (MS–MS) (10). Depending on the design of the source, in-source CID can either be induced by increasing the entrance orifice voltage so that fragmentation occurs in the atmosphere–vacuum interface, or by increasing the collision energy at the desolvation capillary. This technique is applicable to single- or triple-quadrupole mass spectrometers. Additionally, a triple-quadrupole mass spectrometer is capable of performing MS–MS such as neutral-loss scan, parent-ion scan, or product-ion scan. For tandem MS, the first and third quadrupoles are set up to transmit only the targeted MH^+ ion and its selected fragment ion, thus providing optimum sensitivity and selectivity. It is useful in generating fragment ions characteristic of the analytes, but there is a trade-off in sensitivity due to the loss of ions during transmission across the quadrupole filters. Tandem MS and in-source CID both gave very similar product-ion mass spectra (11).

Several stages are involved in the study of drug metabolism and pharmacokinetics. The initial steps involve in vitro studies in which the drug (radioactive or nonradioactive) is incubated with isolated animal–human hepatocytes, microsomes, or human liver slices. Radioactive profiles will confirm that the components observed are drug related, and mass spectrometry can then provide molecular-weight information on these components. This approach can generate relatively large quantities of metabolites for structural characterization, and so assist in establishing the metabolic pathways. This first phase is followed by in vivo experiments involving the dosing of a wide variety of animals, such as rats, mice, dogs, or even primates, to extrapolate biotransformation data and risk assessment from animals to humans. Further qualitative and quantitative investigations will be carried out before the drug enters human clinical trials. Such experiments generate information which enables us to correlate in vivo pharmacokinetic and metabolic data with in vitro metabolic data, and subsequently to design a dosing schedule for the human clinical trials.

The route of drug excretion is largely influenced by its molecular weight. Drugs having a molecular weight of approximately 300 Da are largely eliminated in urine, whereas drugs with higher molecular weights are mainly excreted in the bile, and thereafter by the intestine. Once in the intestine, a glucuronide conjugate can either be excreted in the feces or it can be hydrolyzed back to the parent drug (e.g., by β-glucuronidase). The free drug can then be absorbed through the gut wall and undergo enterohepatic circulation. In vivo metabolic biotransformation information will be obtained by examining biological fluids such as urine, bile, plasma, or fecal extracts. Phase I metabolism involves modification or addition of polar functional groups on lipophilic drugs (e.g., hydroxylation, reduction, oxidation, and dealkylation). Phase II metabolism occurs when a functional group reacts with an endogenous substrate to yield a conjugated metabolite which is readily excreted from the body. These conjugates consist of O- or N-glucuronides, O-sulphates, glutathiones, cysteine, or glycine. TSP/APCI studies have proven useful for generating quasi-molecular ions ($[M + H]^+$ or $[M - H]^-$) for the phase I metabolites, but not necessarily for the phase II conjugated metabolites. Thermal degradation is likely to occur, forming only the molecular ion of the aglycone. Conjugate metabolites tend to form protonated or deprotonated molecular ions using ESI–MS positive or negative ionization modes respectively. Metabolites with an addition of 176, 80, or 305 mass units suggests conjugation with glucuronide, sulphate, or glutathione, respectively. In tandem mass spectrometry, it is feasible to monitor the loss of a neutral moiety of mass 176 Da, corresponding to a glucuronide moiety (12); a neutral loss of 80 Da, corresponding to the sulphoconjugate (13); and a loss of m/z 129, corresponding to the γ-glutamyl residue from the glutathione conjugate (14,15).

II. IDENTIFICATION OF METABOLITES BY LC–ESI–MS (QUALITATIVE ANALYSIS)

Searching for trace amount of metabolites in biological fluids can be a challenge. The dominant state-of-the-art analytical approach applicable to drug metabolism research is LC–ESI–MS with quadrupole mass spectrometers.

A. Identification of Phase I Metabolites

Electrospray ionization–mass spectrometry usually produces pseudomolecular ions for the analytes, which can easily be confirmed by radioactive signals if radioactive parent drugs are used in the study. A 1 : 1 mixture of omeprazole and [^{34}S] omeprazole (Fig. 1) was given to rats, and their urine samples were analyzed by LC–IS–MS (16). More than 40 drug-related compounds were observed.

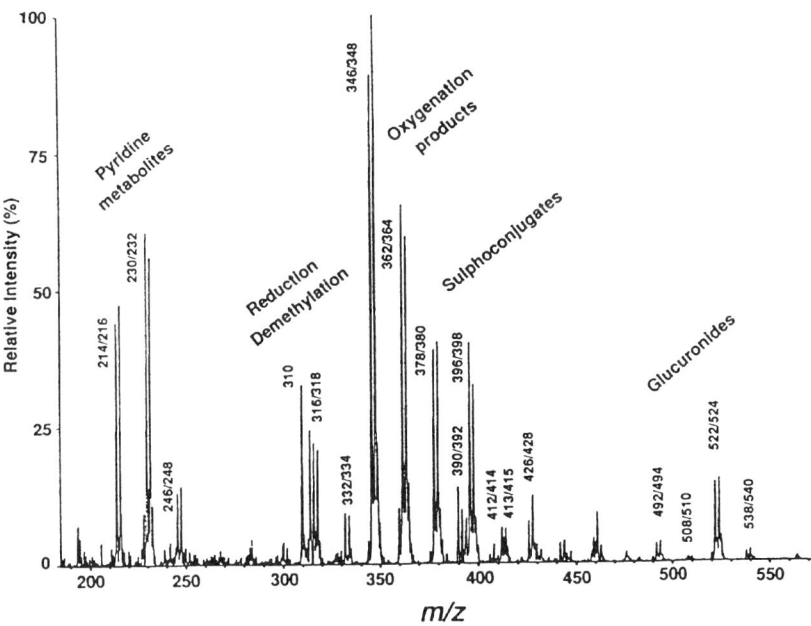

Figure 1. (*a*) Schematic diagram to show the metabolic pathways of omeprazole; (*b*) the summed, background-subtracted spectrum of all ions acquired during the analysis and labelled with the corresponding metabolites. Reprinted with permission from L. Weidolf and T. Covey, *Rapid Commun. Mass Spectrom.*, **1992**, *6*, 195. © 1992 John Wiley & Sons, Ltd.

Because each produced a 1 : 1 ratio of MH^+ and $[MH + 2]^+$ cluster ions, the authors were able to generate a summed, background-subtracted mass spectrum of the entire TIC trace. This "metabolite mass profile" clearly displayed all the compounds containing the molecular-ion 1 : 1 clusters in the sample, which represent the metabolites derived from oxygenation, reduction, and sulphate, glucuronide, or glutathione conjugation of the parent drug (Fig. 1).

Structural elucidation by tandem mass spectrometry reveals partial structures of the metabolites and definitive confirmation is best carried out by NMR spectroscopy or by comparison with authentic reference compounds. Numerous studies have utilized this technique, which has led to the discovery of metabolic pathways or identification of novel metabolites. For example, a novel metabolic pathway for leukotriene B_4 (LTB_4) with a human-derived hepatoma cell line, was established using negative-ionization ESI–MS. Further characterization of these metabolites was achieved by GC–MS following studies with the derivatized drug. The investigation revealed that β-oxidation from the carboxyl terminus was the major metabolic pathway and 20-carboxy-LTB_4 was a minor metabolic route for this endogenous compound (17). Investigation of the C_{24} oxidation pathway of 1,25-dihydroxy-16-ene vitamin D_3 in rat kidney perfusate was achieved by on-line capillary LC–ESI–MS after derivatization with PTAD (4-phenyl-1,2,4-triazoline-3,5-dione) (18). A novel metabolite, 1,23,25-trihydroxy-24-oxo-16-ene vitamin D_3, was observed. The disposition of the substrate and its metabolite in liver perfusates was monitored by tandem mass spectrometry based on the ion transition from MH^+ to m/z 314, a facile fragment ion characteristic of vitamin D. Other studies include an inhibitor of the cytochrome P450 monooxygenases, thiazopyr (containing a carboxylic acid ester group), which was biotransformed to its corresponding carboxylic acid in rat microsomes. This new metabolite was identified by LC–ESI–MS and the suggested cleavage of the ester group proceeds by an oxidative rather than a hydrolytic mechanism (19).

A comprehensive metabolic study of iloperidone (a dopamine D2/serotonin–$5HT_2$ antagonist, Fig. 2a) in human, dog, and rat urine, plasma, and bile extracts using LC–IS–MS–MS supported by LC–NMR was described by Mutlib et al. (20). The compound is extensively metabolized in all species, including the unusual formation of a benzisoxazole ring-opened analog of iloperidone. Quantitation of the parent drug and its principal circulating metabolite (methoxy analog) in human plasma was described, and improved sensitivity was achieved when using SIM–ES–MS. The detection limit was 250 pg/mL when the plasma was spiked with the two standards (21).

Novel peptide drugs are emerging as therapeutic agents (22), but the range of bioassays available for their determination is rather limited. However, LC–ESI–MS has widely been used to provide structural information on peptides and proteins at the picomole level, and the technique has been extended to analyze peptide drugs and their metabolites in biological fluids. In vivo conjugation of the anticancer agent melphalan to the ubiquitous cytosolic protein metallothionein, followed by tryptic digestion into a mixture of peptides, was examined by LC–MS–MS (23). A mixture of known peptides such as neurotensin, buccalin, and substance P (fragment 4-11) was used for mass calibration and for tuning the analyzer. Mass spectral evidence was presented to show that the compound generates two metabolites, formed by replacement of a chlorine atom of the parent drug with two different modified peptide fragments. The limit of detection was 10 pmol for each peptide. Further confirmation of the site of alkylation within each tryptic peptide is expected from tandem mass spectrometry studies.

Andren and Caprioli (24) developed an on-line in vivo microdialysis micro-LC–ESI–MS technique capable of monitoring the metabolism of substance P (an 11 amino-acid-residue polypeptide) in rat brain and providing qualitative and quantitative information about the metabolites. Substance P (1 pmol/μL) was infused into the rat brain, at 0.3 μl/min (approximately 320 fmol/min entered the brain). A microdialysis probe was inserted into the rat brain, and regular samples were withdrawn and transferred to a valco 10-port injection valve. The samples were loaded onto a capillary C_{18} column and eluted directly into the mass spectrometer. Substance P is metabolized in vivo by a proteolytic enzyme in the extracellular fluid within the rat striatum. The metabolites identified include the N-terminal fragments 1-9, 1-8, and 1-7 and C-terminal fragments 3-11, 5-11, 6-11, 7-11, and 8-11. From the reconstructed ion chromatograms, the relative concentrations of these metabolites were calculated from their peak areas. The authors emphasized that, unlike RIA where cross reactivity has been observed (25), LC–MS generates molecular-weight information on these peptides unambiguously. Similar studies by Takada et al. (26) demonstrated the effectiveness of on-line microdialysis CE–MS to monitor γ-aminobutyric acid (GABA) in rat brain, using an electrophoresis buffer containing 1% formic acid in methanol:water (50:50). Good CE–MS sensitivities were achieved for the compounds studied: 10^{-4} M for catecholamines; 10^{-5} M for amino acids and GABA; 10^{-6}–10^{-7} M for brain peptides and hormones.

Traditionally, structural elucidation of carcinogen DNA adducts has been carried out by GC–MS. Malondialdehyde (MDA) is an endogenous genotoxic agent, and it readily forms a pyrimido[1,2-α] purine(10/3H)-one (M_1G-dR) adduct with calf thymus DNA (10) (Fig. 3a). M_1G-dR has been detected in rat liver (27) and disease-free liver (28) after derivatization with pentafluorobenzyl bromide (PFB) and analysis by GC–MS. The mass spectrum gave a peak corresponding to the molecular ion $[(M) - PFB]^-$. M_1G-dR was isolated by LC and analyzed by ESI–MS, producing the ion $[M + H]^+$ at m/z 304, and a prominent fragment ion at m/z 188, $[BH_2]^+$ (a base adduct ion with two hydrogen atoms attached), which corresponds to the loss of the 2′ deoxyribose moiety (Fig. 3b). When the mass spectrometer was set up to perform in-source CID, followed by MS–MS analysis of m/z 188, several structurally informative ions appeared at m/z 160, 133, 106, and 79 which demonstrated unequivocally that M_1G-dR is an endogenous adduct in human liver DNA (Fig. 3c). The detection limit, approximately 10 pg (30 fmol), is an order of magnitude lower than that for GC–MS (27,28), but the simpler sample preparation procedure makes LC–ESI–MS a more attractive alternative. Another carcinogen, 1,2,3-trichloropropane (TCP), used industrially as a solvent, also forms DNA adducts in vivo in rats. The ESI–MS–MS analysis and confirmation with authentic

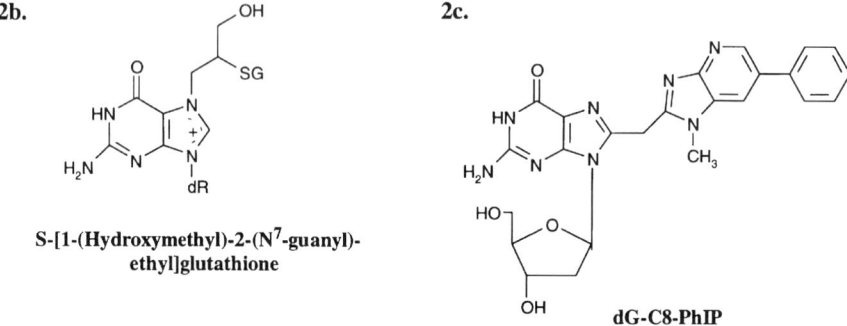

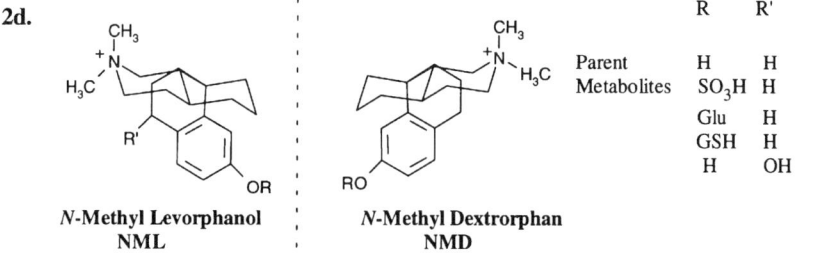

Figure 2. Schematic diagrams to show the metabolic pathways of (*a*) iloperidone; (*b*) structure of *S*-[1-hydroxymethyl)-2-(N^7-guanyl)-ethyl]glutathione; (*c*) *N*-(deoxy-guanosin-8-yl)-PhIP; (*d*) NML and NMD; (*e*) cisatracurium; (*f*) BCNU; (*g*) disulfiram; and (*h*) *N*-formylamphatamine.

II. IDENTIFICATION OF METABOLITES BY LC–ESI–MS (QUALITATIVE ANALYSIS)   507

2e.

Cleavage (Tetrahydropapaverine)
Cleavage (Laudanosine)
Cleavage (Quarternary acid)
Cleavage (Quarternary alcohol)

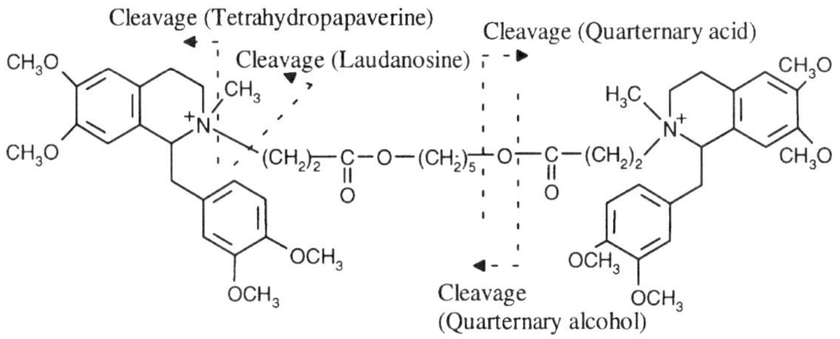

Cisatracurium

2f.

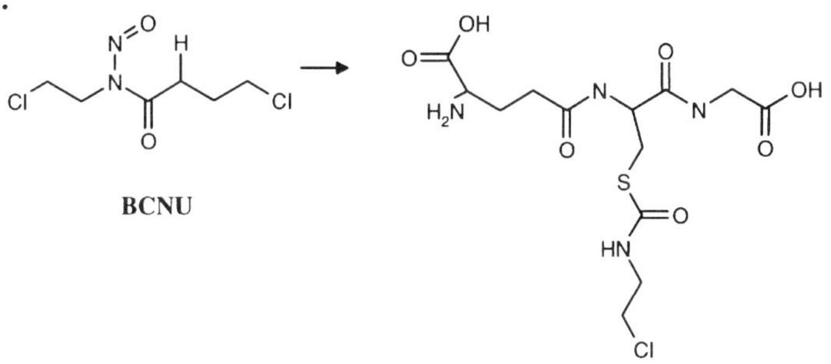

BCNU

**S-[N-(2-Chloroethyl)carbamoyl]glutathione
SCG**

2g.

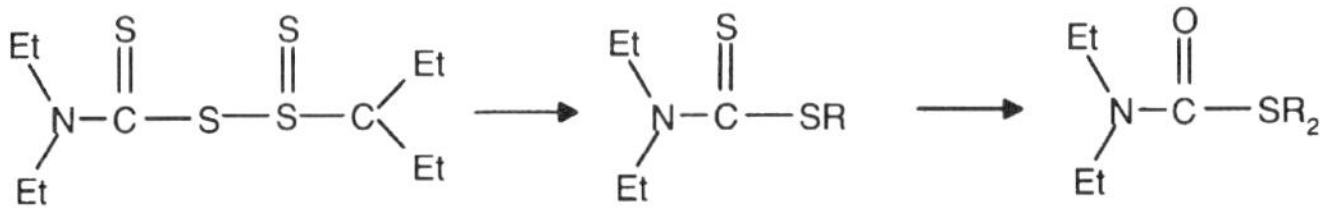

Disulfiram

DDTC
R = H, Me or glutathione

DETC
R_2 = H, Me, (O)Me, (O)$_2$Me
or glutathione

2h.

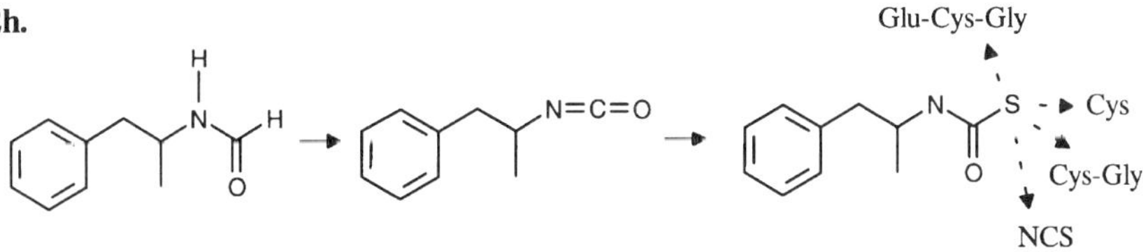

Glu-Cys-Gly
Cys
Cys-Gly
NCS

N-Formylamphatamine

Figure 2.(*cont.*)

compounds revealed the adduct to be S-[1-(hydroxymethyl)-2-(N^7-guanyl)-ethyl]glutathione (29) (Fig. 2b). Using capillary LC–ESI–MS–MS in the selective multiple-reaction monitoring (MRM) scanning mode, the DNA adduct of a food mutagen 2-amino-1-methyl-6-phenylimidazo(4,5-b)pyridine (PhIP) was determined (Fig. 2c). In addition, an unexpected product in the in vitro incubation mixture was observed; the ESI product-ion mass spectrum suggested a ring-opened product, presumably resulting from cleavage through the guanine base (30).

To avoid sample preparation, Henion's laboratory (31,32) designed an on-line "cleanup method" for incorporation into their analytical method for LSD in human urine. They utilized a high-performance protein G immunoaffinity column, an internal-surface reversed-phase (ISRP) trapping column, and a switching valve coupled with a reversed-phase analytical column to study two basic drugs in diluted urine samples. Detection of the analytes was achieved by either UV or IS–MS detection, operated in the selected-ion monitoring (SIM) mode. For each analysis, a small amount of antibody was used to extract either propranolol or lysergic acid diethylamide (LSD) from human urine. The detection limit of LC–IS–MS was 2.5 ng/mL and 500 pg/mL for propranolol and LSD respectively, compared to 250 pg/mL using UV detection.

B. Identification of Phase II Metabolites

The metabolism of an analgesic compound S12813 (P) was examined by incubating S12813 with rat liver slices and analyzing metabolites by LC–ESI–MS (33). Phase I metabolites include 4-OHP, phenyl piperazine, hydroxyethyl P, and possibly a carboxymethyl cleavage product AMCP. LC–MS–MS operated in the neutral-loss mode (masses 176 and 80) confirmed the presence of 4-OHP glucuronide and sulphate conjugates, ring-opened P glucuronide and sulphate, and 4-OH ring-opened P glucuronide and sulphate.

Balani et al. (34) incubated [^{3}H] L-735524 (a potent and specific inhibitor of the human immunodeficiency virus type I protease) with human liver slices to

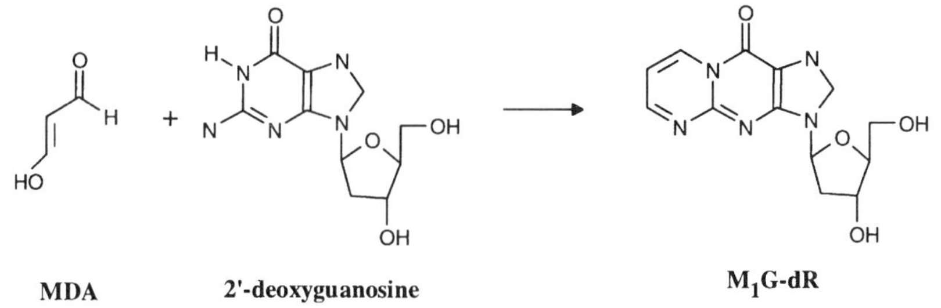

Figure 3. (a) Formation of M_1G-dR from MDA; (b) ESI mass spectrum of M_1G-dR; (c) production of BH_2^+ ion (m/z 188) by in-source CID, followed by tandem MS–MS on BH_2^+ ion in Q_2. Reprinted with permission from A. K. Chaudhary et al., *J. Mass Spectrom.*, **1995**, *30*, 1159–1160. © 1995 John Wiley & Sons, Ltd.

II. IDENTIFICATION OF METABOLITES BY LC–ESI-MS (QUALITATIVE ANALYSIS) 509

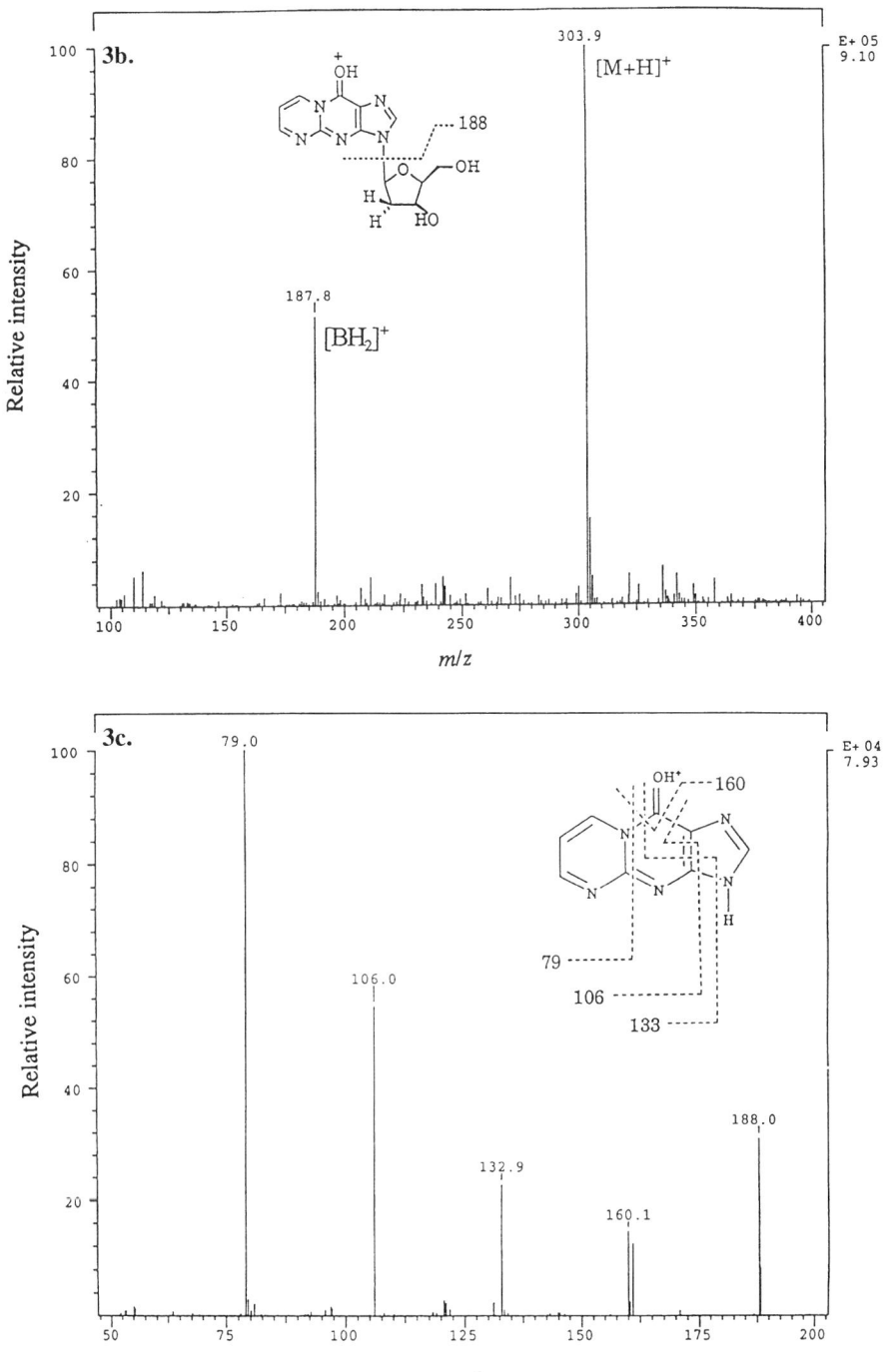

Figure 3.(*cont.*)

establish the metabolic profile, which was compared with metabolites present in urine samples of healthy male volunteers. Metabolites were isolated by LC, characterized by off-line IS-MS–MS, and confirmed by NMR spectroscopy. There were six metabolites observed in the human liver slices and urine extracts. The metabolic pathways involved N-glucuronidation, N-oxidation, p-hydroxylation of the phenylmethyl group, and N-depyridomethylation. The product-ion mass spectra of these metabolites generate diagnostic fragments, and thus any structural modifications of the metabolites in different parts of the molecule are easily recognized from examination of the fragment ions.

Swart et al. (35,36) compared the metabolism of the potent dopamine D2 agonist (against Parkinson's disease) by incubating [^{3}H] 2-(N-propyl-N-2-thienylethylamino)-5-hydroxytetralin in rat, human, and monkey liver microsomes. The resulting metabolites were examined to establish possible differences in metabolism for these species. The radioactive profiling indicated that there were five metabolites in all three incubations. Metabolic pathways include N-depropylation of the propyl side chain, hydroxylation to form a catechol, and cleavage of the thienylethyl chain. Further studies of rat-liver bile by IS–MS revealed several novel metabolites. These include glucuronide and sulphate conjugates of the parent drug, N-depropylation followed by glucuronidation, N-dealkylation of the thienylethyl group, followed by glucuronidation and sulphation, and low levels of a glucuronide conjugate of the catechol-desthienylethyl analog.

The metabolic fate of morphinan analogs [^{2}H$_3$]-N-methyl dextrorphan (NMD) and N-methyl levorphanol (NML, Fig. 2d) in rat bile and liver perfusates has been targeted in two studies. Both NMD and NML are quaternary ammonium compounds and require ion-pairing reagents or buffers in the LC mobile phase compatible with MS, and the separation was achieved with 0.025% trifluoroacetic acid (37). Both compounds formed glucuronide and glutathione conjugates, and in the case of NML the sulphate conjugate was also observed. The structures of these conjugates have been unequivocally assigned by LC–MS–MS, thus confirming that glucuronidation and sulphation are common metabolic pathways of opiates (38).

Urinary and biliary conjugated metabolites of the neuromuscular blocker [^{14}C]-cisatracurium were characterized by LC–ESI–MS–MS (Fig. 2e). Apart from the unusual phase I metabolites (laudanosine, tetrahydropapaverine, quaternary monoacrylate, and quaternary acid), several O-glucuronide acid conjugates of monodesmethyl laudansoine and monodesmethyl tetrahydropapaverine were identified, in addition to a sulphate conjugate of monodesmethyl laudanosine, as determined in cat bile (39). After dosing rats with the anti-inflammatory agent diflunisal, the labile 3-hydroxy diflunisal monosulphate conjugate was detected in rat urine samples and confirmed by off-line LC–ESI–MS in glucuronosyltransferase-deficient rats (40).

In vivo studies indicated that a trace amount of S-(p-chlorophenoxy-2-methylpropanoyl) glutathione (CA-SG) was excreted in bile after an intravenous infusion of clofibric acid to rats. When 1-O-clofibryl glucuronide (1-O-CAG, a known metabolite of clofibrate) was incubated with glutathione, positive ESI–MS allowed assignment of the structure of the product as a CA–GS adduct (41), and hence confirmed the hypothesis that 1-*O*-CAG is a substrate for glutathionetransferase (42).

Another example of the investigation of glutathione conjugates by ESI–MS was given by Davis and Baillie (43). The anticancer drug BCNU (Fig. 2*f*) is a selective inhibitor of the enzyme glutathione reductase. Owing to its high lipophilicity, it is capable of crossing the blood–brain barrier, and has been used for the treatment of malignant brain tumors. It produces reactive intermediates in vitro or in vivo in rats which subsequently form cross linkages with DNA. The authors examined the nature of the reactive electrophilic intermediate liberated during decomposition of BCNU. Analysis of rat bile by LC–IS–MS–MS using constant neutral loss of 129 Da (which corresponds to the loss of pyroglutamic acid) and precursor ion scanning on m/z 179 revealed the presence of four GSH conjugates.

Applying the same strategy, similar LC–IS–MS–MS experiments have been reported on the identification of novel GSH adducts as biliary metabolites of disulfiram and diethyldithiocarbamate (DDTC, Fig. 2*g*) (44). These reactive metabolites are responsible for the parent drug's aldehyde dehydrogenase inhibitory effects in animals and humans. The bioactivation of *N*-formyl-amphetamine (NFA, Fig. 2*h*) to its corresponding isothiocyanate in rats was investigated by ESI–MS assisted by the stable-isotope methodology (45). Glutathione, cysteinylglycine, and cysteine conjugates were observed in rat bile. The biliary excretion of NFA via the mercapturate pathways is novel for formamide metabolism. A breakdown product of a new fluorinated anesthetic, servoflurane, which can cause nephrotoxicity in rats, was investigated and IS tandem mass spectrometry provided evidence for the formation of glutathione and cysteine conjugates in rat bile and urine (46). Cleavage of these conjugates by β-lyase resulted in the formation of reactive intermediates which bind covalently to cellular macromoleucles and cause organ-selective nephrotoxicity.

III. PHARMACOKINETICS STUDIES (QUANTITATIVE ANALYSIS)

The recent demand for monitoring trace levels of drugs and metabolites in animals or humans to support drug discovery has increased dramatically. Although other LC–MS techniques show wide variations in response of the analytes and prove difficult in quantitation research, LC–API–MS–MS plays a significant role in supporting these studies, because of its ease of operation, high sample throughput, and reliability.

Quantitative LC–MS involves selecting an internal standard, ideally the stable isotopically labeled analog of the parent drug or another compound with a structure closely resembling the parent drug, so that they all have similar response on the mass spectrometer. The stable isotope analogs are usually labelled with ^{2}H, ^{13}C, ^{15}N, or ^{18}O. Even though the difference in masses may be distinguishable by mass spectrometry, care must be taken to ensure the mass peaks monitored are not common to both compounds. The MW of the analyte and its isotope analog should preferably be at least 3 mass units apart. The mass-spectrometric responses of the analyte and the internal standard are measured using either selected-ion monitoring (SIM), selected-reaction monitoring (SRM), where the mass spectrometer is tuned to one set of ions, or multiple-reaction monitoring (MRM), where the mass spectrometer is tuned to several sets of ions. The SIM technique provides lower detection limits at a greater speed than the full-scan mode because only a few specific ions are monitored, and it is useful for detecting small quantities of target compounds in a complex mixture. In the SRM–MRM method, a limited set or sets of precursor-ion/product-ion pairs are monitored. Detection is targeted only on compounds with the chosen precursor ions, and each precursor ion should also fragment to form a product ion of the chosen mass-to-charge ratio. SRM–MRM is performed on tandem mass spectrometers, and the specificity is much better than that obtained with SIM. However, in the SRM–MRM method, the analytes must be capable of forming precursor-ion/product-ion pairs consistently and with good intensity. Once the ratio of analyte peak area to internal standard (i.s.) peak area is obtained, for each analyte, a calibration curve is constructed. The quantitation criteria, such as detection limits and quantification limits, were addressed by McLaughlin et al. (47). The exploration of product ions for quantitative purposes was carried out using isotopically labeled parent drugs as internal standards. The authors demonstrated that quantitative data can be generated by measuring the ratio of peak areas for product ions to that of the internal standard. The linearity of detection is supported by calibration plots of samples spiked with known amounts of analytes and internal standards.

Covey et al. (48) first demonstrated the capability of high-speed LC–APCI–MS–MS and applied it to crude equine urine and to extracted plasma using SIM. Promazine is metabolized extensively to give 3-OH, 3-OH-N-oxide, and N-oxide/sulfoxide analogs. These metabolites all give product-ion signals at m/z 58 and 86. By performing parent-ion scans on these ions, the potential metabolites were detected by the mass spectrometer. Similarly, using the product-ion signals of phenylbutazone at m/z 77, 93, 120, and 190, the feasibility for carrying out a 60-sample, 48-h pharmacokinetic study of phenylbutazone and its oxidative metabolites in horse urine and plasma samples was demonstrated (Fig. 4). Since then numerous assays have been described in the literature, such as LC–APCI–MS–MS analysis of a new cholecystokinin receptor antagonist in clinical plasma samples which takes less than 2 min per sample. Remarkably, 1 pg of the drug could be detected with a S/N ratio of 10 (2,49).

Figure 4. (*a*) Structure of phenylbutazone and its CID fragment ions; (*b*) LC–MS–MS analysis of 60 samples from a phenylbutazone pharmacokinetic study in the horse in 1 h. The ion current from SRM of four product ions of phenylbutazone is shown. Reprinted with permission from T. R. Covey et al., *Anal. Chem.*, **1986**, *58*, 2459. © 1986 American Chemical Society.

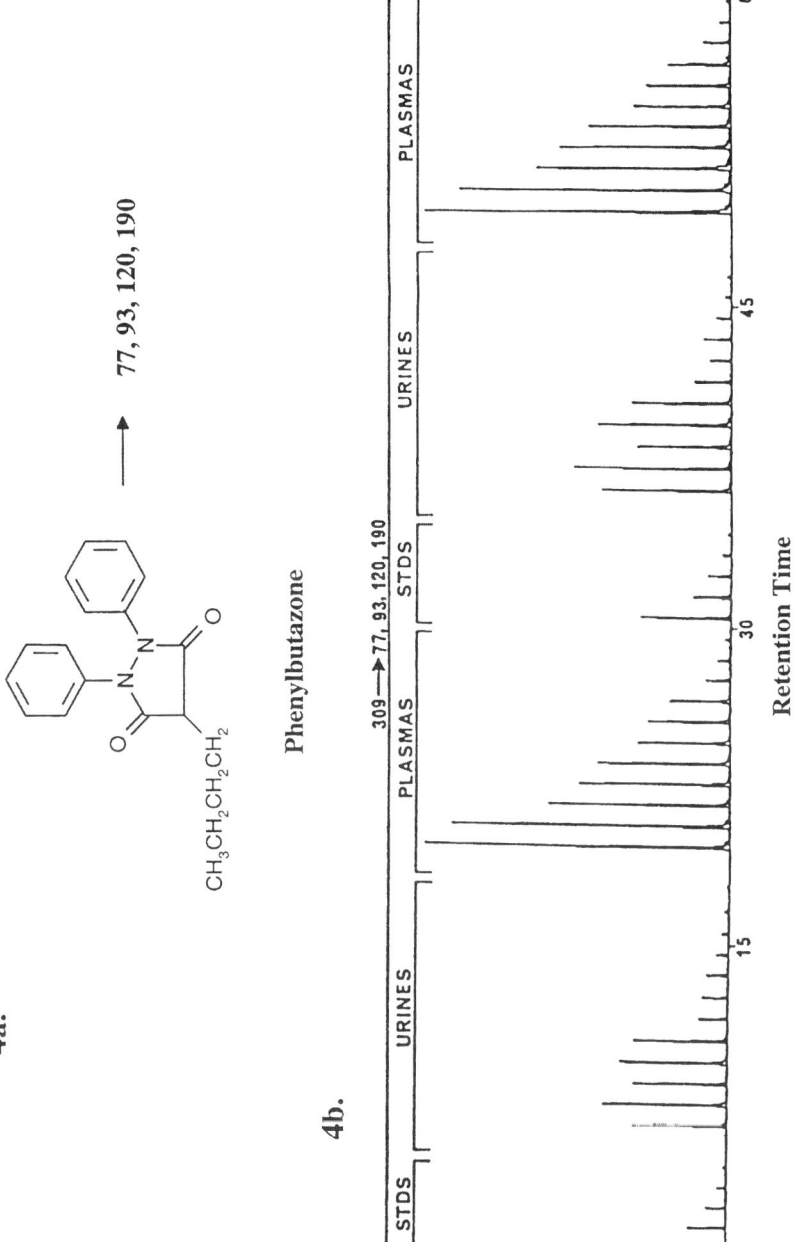

Other examples include the determination of the antiarrhythmic agent MK-0499 and finasteride in human plasma (50,51); methandrostenolone in equine urine (52); nicergoline (an ergot alkaloid with vasodilating and α-receptor blocking activity) in human urine and plasma (53); quantitation of a new antiinflammatory agent tenidap in serum (54); and detection of two muscarinic antagonists, atropine and pirenzepine, in human plasma (55). Generally, for rapidly performed analyses, great care must be taken to avoid bond cleavages of the phase II metabolites which could convert to the aglycones and contribute to the parent drug measurements. This problem is less likely to occur when the LC–ESI–MS technique is used.

Applying the same strategy, LC–IS–MS–MS has been used to determine a collagenase inhibitor and its major metabolite in plasma and urine samples obtained from clinical trials (56). Quantitation of the calcium antagonist nimodipine in human plasma was achieved by Muck (57). The racemic drug was given orally to healthy volunteers, and the enantiomers were separated by LC using a chiral column. In the MS profile, both optical isomers formed fragment ions at m/z 343 and 301, which were selected for MS–MS treatment. The resulting distribution curve of the two enantiomers in human plasma compared favorably with the results obtained by GC–ECD analysis. However, the shorter time required for sample preparation and analysis rendered LC–MS–MS the method of choice. Since the optimum flow rate of LC–ESI–MS was ~ 5 μL/min (58), these early reports had to rely on either microbore LC (1 mm i.d.), packed-capillary columns (<320 μm i.d.) (13,18), or narrow-bore columns (2 mm i.d.) and a LC eluent split. Post-column splitting of the LC eluent means that a high proportion of the analyte is directed to waste.

Recently, ESI–IS sources have been modified and they are now capable of accepting high LC flow rates from 150 μL/min (turbo ionspray) to 1 mL/min (heated capillary), making LC–ESI–MS an equally attractive quantitation technique for determining parent drugs and their corresponding metabolites. The sensitivity remained constant when identical amounts of sample were injected into the turbo ionspray interface at higher flow rates, where better sensitivity was achieved compared to normal ionspray (59). Recent examples include an LC–MS–MS quantitation method for the determination of an endothelin receptor antagonist (Bosentan) in human plasma (60). An LC–MS–MS assay was developed by manipulating an intense product ion at m/z 202, operated in the SRM mode. The response was linear in the range of 0.5–200 ng/mL, with much improved sensitivity and selectivity over LC–UV detection. A sensitive LC–ESI–MS assay was developed to monitor iloperidone and its principal circulating metabolite in human plasma at the 250 pg/mL level (21). Quantitation of the major GSH metabolites of DDTC and disulfiram in rat bile (44), pharmacokinetic studies of a muscarinic agent LY297802 in rat, rabbit, and monkey plasma by LC–ESI–MS–MS (61), quantitative assay of the potent immunosuppressant drug tacrolimus in blood samples (3), and an LC–ESI–MS assay developed to assess α-hydroxytamoxifen level in human microsomes after incubation with tamoxifen (62) are all examples which illustrate the power of LC–MS techniques. These investigations have proved LC–MS–MS to be a convenient, specific, and highly successful technique for quantitation of drugs in body fluids, and the methodology is approved by the U.S. Food and Drug Administration.

Other interesting studies include that of cyclosporine and its metabolites (desmethyl and hydroxyl analogs) which were observed in needle-biopsy samples, isolated from the liver or kidney of patients following transplantation. The HPLC–MS response is linear from 5 to 500 pg of cyclosporine, and the limit of detection is 500 fg (~415 amol) using SIM (63). The authors reported that samples contained 100 pg/mL of cyclosporine in whole blood, equivalent to ~5 pg (5 fmol) introduced into the mass spectrometer, and thus demonstrated the sensitivity of ESI–MS. Pacifici et al. (64) developed an LC–MS assay to monitor morphine, codeine, naltrexone, and morphine 3- and 6-glucuronide simultaneously. The limit of detection ranges from 10 ng/mL morphine to 100 ng/mL morphine-3-glucuronide. The methodology was successfully applied to evaluation of these chemicals in the sera of mice, patients, and drug addicts. It is necessary to quantify these metabolites in plasma because morphine 6-glucuronide is a more potent opioid agonist than morphine, whereas morphine 3-glucuronide has no opioid effects. After oral administration of morphine to cancer patients, there is evidence of a significant passage of the glucuronides to the cerebrospinal fluid, in the concentration range where they may have an influence on analgesia (65).

IV. CAPILLARY ELECTROPHORESIS–ELECTROSPRAY IONIZATION–MASS SPECTROMETRY (CE–ESI–MS)

The coupling of CE with ESI–MS to analyze pharmaceutical compounds and xenobiotic metabolites has been pursued enthusiastically. The CE–MS technique combines the benefits of the highly efficient separation of analytes, short analysis times and low sample consumption of CE, with the wide scope of MS detection. Capillary electrophoresis relies on the separation of analytes within a capillary under the influence of a high electrical field. The limit of optical detection for CE is 1–2 orders of magnitude higher than in HPLC because of the low detection path length in CE. However, the absolute detectable amount of analytes is low because of the small injection volume. The first report by Johansson et al. (66) demonstrated that the feasibility of studying small drug molecules by CE–ESI–MS using 15–20 mM aqueous ammonium acetate with 15–20% methanol. The capillary end from the CE system was coupled via a liquid junction to a pneumatically-assisted electrospray interface (67). Separation of six sulfonamides and four benzodiazepines was achieved with only 600 pg (2 fmol) sulfamethazine required to produce a satisfactory product-ion mass spectrum. Analysis of flurazepam, its major metabolite (N-1-hydroxyethyl-flurazepam), and two minor desmethyl metabolites in human urine has also been described.

Following Johansson's pioneering investigation, several other CE–MS studies have been reported. Notable examples are the analysis of sulfonamide analogs and indole alkaloids having antitumor properties (68), a comparison of CE–MS with nanoscale packed-capillary LC–MS for the analysis of macrolide antibiotics (69), the separation of chiral drugs (70), the synthetic coupling of an antiinflammatory drug Naproxen to lysozyme (a drug-delivery carrier) employing CE–MS monitoring (71), and the on-line in vivo microdialysis CE–MS studies of GABA in rat brain, as described earlier (26).

Benson et al. (72) investigated the metabolism of pyragoloacridine in rat urine using preconcentration by isotachophoresis and a nonaqueous separation buffer. The objective of introducing organic modifiers was to improve the resolution of small molecules and to enhance the solubility of the hydrophobic analytes. Thereafter, the authors applied a similar CE–ESI–MS technique to establish the in vitro metabolic profiles of haloperidal (the neuroleptic agent used clinically to treat schizophrenia), with limits of detection of these components at approximately 100 fmol on a magnetic-sector instrument (73). Other examples of the analysis of drugs and metabolites by CE–ESI–MS include the H2 antagonist mifentidine (74) and the breast-cancer drug tamoxifen (75).

The CE-ESI–MS technique, used in the negative-ionization mode, was employed to analyze ibuprofen, flurbiprofen, and aspirin in healthy human urine extracts, using a total aqueous CE buffer of ammonium acetate (pH 9) (76). The metabolites include ibuprofen, hydroxyibuprofen, and their corresponding glucuronides, as well as the carboxylated glucuronides.

V. THE LC–ESI–ION TRAP (IT) IN DRUG-METABOLISM AND PHARMACOKINETICS

A recent novel development concerns the implementation of LC to a quadrupole-ion trap which allows highly specific structural analysis by multistage tandem mass spectrometry (MS^6) (with up to six-stage tandem mass spectrometry). The instrument has been designed to use external ion sources (API) to produce analyte ions which are then injected into the ion trap. The pharmaceutical industry is eager to see the outcome of this approach and the feasibility of applying the technique to drug-metabolism and pharmacokinetics. Using a specially constructed ionspray source, Henion et al. (77) first demonstrated that it was feasible to study small drug molecules using the ESI–IT mass spectrometer. ESI–IT–MS is capable of producing intense fragment ions during SRM operation, an extremely useful feature for quantitative drug analysis. Quantitation of isoquinoline alkaloids was achieved, a standard curve was prepared, and the method was validated over the range of 1.7 to 45 fmol with a correlation coefficient of 0.998 (78). However, the instrument is not capable of performing neutral-loss scans or parent-ion scans. The obvious question is whether the performance of LC–ESI–IT can challenge the performance of a triple-quadrupole mass spectrometer at a lower cost.

VI. THE LC–ESI–MAGNETIC-SECTOR MASS SPECTROMETER IN DRUG METABOLISM

In addition to reports by Naylor's group (72–74) describing the interfacing of CE with a magnetic-sector mass spectrometer to characterize metabolites, other examples include the metabolic profile of the anticholinergic agent bentropine in rat, using off-line LC–MS with electron-impact ionization (EI) and on-line LC–ES–MS (79). Electron ionization provides diagnostic fragment ions useful for characterizing the metabolites, while microcolumn LC–ES–MS generates pseudomolecular ions for both phase I and II metabolites. Another example is the study of platinum anticancer drugs incubated with human plasma ultrafiltrate (HPUF). The assessment of the production of platinum complexes in vivo has been a significant challenge, because these complexes are labile and can readily undergo ligand-exchange reactions with the LC mobile phase. However, recent studies on platinum complexes have demonstrated that ESI–MS may have adequate sensitivity for identification of metabolites of these compounds (80,81). Simultaneous on-line LC–ESI–MS with accurate mass measurement has been utilized to confirm the presence of several metabolites of an anticancer drug JM216 in HPUF. This technique provided the elemental compositions of these metabolites, which was required for verification of their structures (82).

VII. METABOLIC STUDIES OF ANTICANCER REAGENTS BY LC–ESI–MS

In the field of cancer chemotherapy, the determination of metabolites by ESI–MS contributes to an understanding of the metabolic pathways of a variety of anticancer drugs and thereby makes a valuable contribution toward the design of such drugs. Compound S9788 is a triazine with potential adjunct therapy in cases of multiple-drug resistance. When the compound was examined by ESI–MS, it exhibited the loss of a neutral fragment of 233 Da, which represents the metabolically stable di(4-fluorophenyl)ethyl amino moiety (83). Consequently, rat bile and urine samples were analyzed using flow injection, and the metabolites were monitored by neutral loss of 233 Da. The mass spectra clearly displayed molecular ions derived from mono- or bis-hydroxylation of the propylene side chain, mono- or bisdihydroxylation of the propylene side chain, or complete cleavage of the propylene side chain.

The metabolism of ifosfamide (a clinically useful isomer of the alkylating antitumor agent cyclophosphamide, Fig. 5a) was investigated by CF–FAB, but the mass spectrum of the metabolite, hydroperoxy-ifosfamide was masked by the FAB matrix-cluster ions. However, a "clean" LC–ESI–MS spectrum was obtained for this unusual metabolite, giving $[M + H]^+$ as the dominant ion (84). The application of ESI–MS to determine the drug and its metabolites simultaneously is promising.

Two DNA-intercalating agents, CI-937 and CI-941, tamoxifen (an anti-

5a.

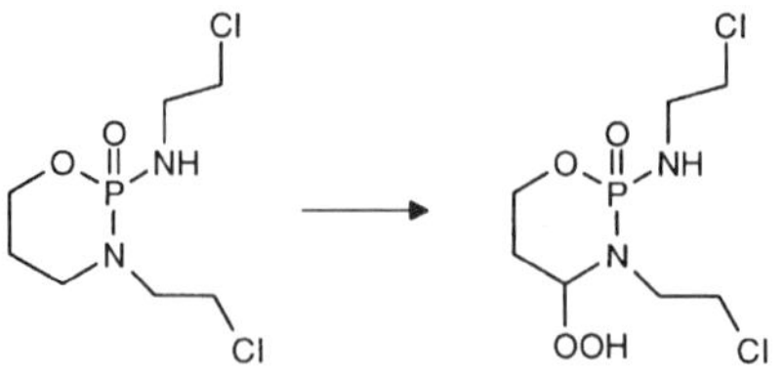

Ifosfamide

5b.

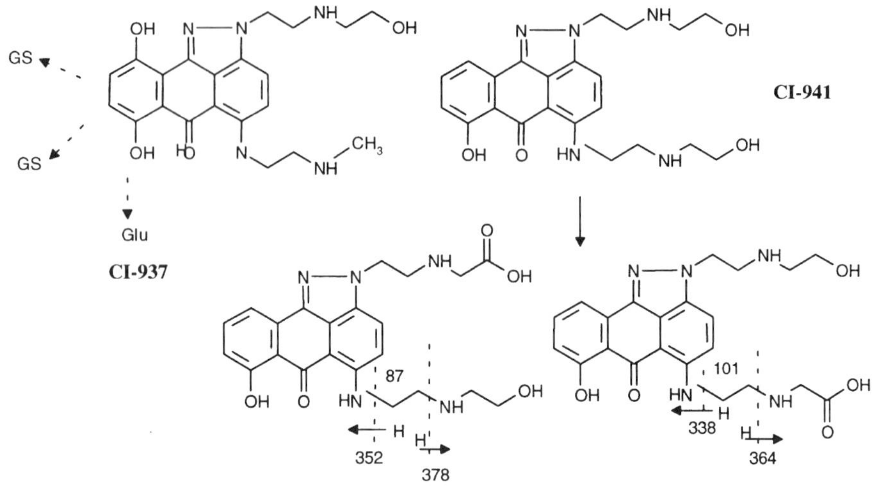

5c.

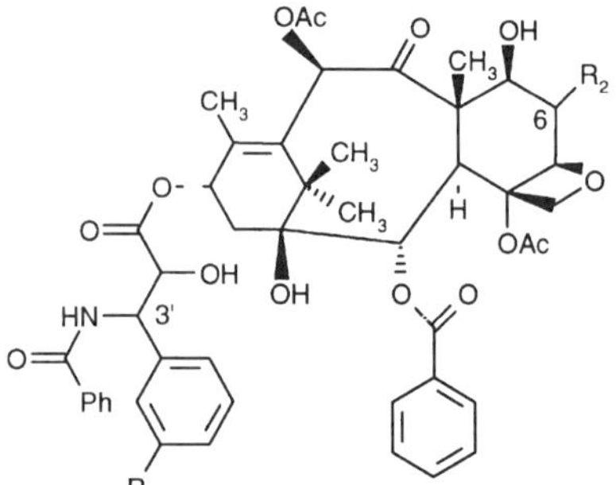

Taxol

	R_1	R_2
Taxol	H	H
6-Hydroxy	H	OH
3'-Hydroxy	OH	H
6,3'-Dihydroxy	OH	OH

Figure 5. Schematic diagrams to show the metabolic pathways of the anticancer drugs (*a*) ifosfamide, (*b*) CI-737 and CI-941, and (*c*) taxol.

estrogen), taxol (an antimitotic agent), and 4-hydroxyandrostenedione (4OHA, an aromatase inhibitor) are all currently undergoing clinical trials for the treatment of breast cancer (85). The metabolic profiling of these anticancer agents has recently been carried out by LC–ESI–MS.

Incubation of CI-937 with rat hepatocytes resulted in the formation of mono- and bis-glutathione conjugates and a glucuronic-acid conjugate (Fig. 5b). Mass spectra of these metabolites gave strong $[M + H]^+$ signals. Further structural confirmation of the metabolites was achieved by NMR using authentic reference compounds for comparison (86). Similarly LC–IS–MS studies of the antrapyrazole derivative biantrazole CI-941 (Fig. 5b) in human urine showed two polar metabolites, 14 and 28 mass units higher than the parent drug. The CID spectra of these metabolites showed (1) an intense fragment ion characteristic of aminoalkylamino-substituted anthrapyrazoles and (2) loss of one or two hydroxyethyl aminoethyl side chains (m/z 88 and 102, respectively), thus confirming that oxidation occurs at the anilino side chains (87). Further studies by Richards and Sun (88) characterized the monocarboxylic metabolites unequivocally. Careful examination of the CID mass spectra of the two metabolites showed that each exhibited a unique fragmentation pattern which enabled the authors to establish the regioisomers of the monocarboxylic acids of biantrazole (Fig. 5b).

Tamoxifen has had considerable early success in the treatment of breast cancer, encouraging its use as a preventive agent for breast cancer. However long-term risks associated with tamoxifen are still being evaluated, particularly with cancers occurring at sites other than the breast (endometrial or ovarian). The question of the safety of tamoxifen as a chemopreventive agent for women at high risk for breast cancer has prompted the study of the metabolism of tamoxifen in patients. Tamoxifen is subjected to extensive biotransformation in humans and laboratory animals. Novel phase I and II metabolites have been detected by LC–ESI–MS in rat hepatocytes (89,90) and human hepatocytes (91) and in the plasma and urine samples of patients treated with tamoxifen (92). The limits of detection by LC–ESI–MS in the SIM mode are 0.5 and 1 ng/mL of plasma for tamoxifen and for the metabolites, respectively. Quantitation by LC–ESI–MS of α-hydroxytamoxifen in rat, mouse, and human hepatocytes, where cells had been incubated with tamoxifen, showed that human liver cells produced less α-hydroxytamoxifen than rat liver cells.

Taxol (paclitaxel, Fig. 5c) is a natural product isolated from the Pacific yew tree *Taxus canadensis* or *Taxus brevifolia*. It is a potent inhibitor of cell replication during the late G2-mitotic phase of the cell cycle, presumably by stabilizing the microtubule cytoskeleton. The drug is under clinical investigation for the treatment of breast and platinum-resistant ovarian cancers. There are several LC–IS–MS–MS reports of the analysis of taxol and its analogs, either from cell cultures (93) or from plant extracts (94,95). The detection limit for a full-scan mass spectrum was 50 ng/mL for extracts and 100 pg for the authentic compound when operated in the SIM mode. Most of the chemical assays applied to the study of the metabolism of taxol in human plasma and urine and rat bile extracts have been performed either by FAB–MS (96–98), or by APCI–MS (99). These studies indicated that 6-OH,3′-OH-epitaxol and 3′-OH,6-OH-taxol were the major metabolites present in the biological samples.

The metabolic profile of the aromatase inhibitor 4-OHA was examined by LC–IS–MS–MS methods, and several glucuronide and sulphate conjugates were detected for the first time in urine samples of patients (100). The biotransformation of the [^{14}C] zanoterone (an androgen receptor antagonist currently under clinical trial for benign prostate cancer) in human urine was examined utilizing MS–MS. By monitoring the neutral loss of 78 mass units (formed by the loss of the methylsulfonyl group with a hydrogen-atom transfer to the pyrazole ring), better sensitivity and improved mass spectra were obtained for the sulphate and glucuronide conjugates (101).

VIII. CONCLUSION

Even though most of the studies reported in this chapter were published within the last five years, the application of LC–ESI–MS methodologies in drug-metabolism pharmacokinetics is well established. The technique provides reliable data in terms of structural confirmation of metabolites and their quantitation at trace levels, as well as high sample throughput. The advent of ESI–MS has greatly increased the range of compounds amenable to mass-spectral analysis, and future application extends to screening for active compounds in combinatorial libraries. A combinatorial library consists of an automated system which is designed to generate large numbers of compounds. Since hundreds and thousands of compounds are prepared in one experiment, the successful development of this technique depends on reliable, fast, analytical screening assays to search for the pharmacologically active candidates in the complex mixtures, with subsequent identification of enzyme substrates and inhibitors, and receptor agonists or antagonists. ESI–MS–MS (102), MALDI–MS (103), and imaging time-of-flight secondary ion mass spectrometry (TOF–SIMS) (104) have been evaluated for analysis of nonpeptide, bead-bound, small-molecule libraries directly, or for investigation of the angiotensin II receptor antagonist synthesized on beads (105). Comparison of the three techniques indicates that the high mass resolution associated with TOF–SIMS provides more accurate fragment-ion masses, while MALDI–MS and ESI–MS–MS provide a greater number of structurally informative fragments. The ESI–MS–MS method can easily be set up for automation, hence offering the capacity to handle a large number of samples.

In summary, LC–ESI–MS is a state-of-the-art analytical technique which plays an important role in drug metabolism and pharmacokinetics research. The continued development of these technologies will provide valuable information to aid in the development of more effective drugs in the future.

ACKNOWLEDGMENTS

The author would like to thank Dr. H. P. Bennetto and Mr. P. B. East for helpful discussions in preparing the manuscript.

REFERENCES

1. Olah, T. V.; Gilbert, J. D.; Barrish, A.; Greber, T. F.; McLoughlin, D. A. *J. Pharm. Biomed. Anal.* **1994**, *12*, 705–712.
2. Gilbert, J. D.; Olah, T. V.; Barrish, A.; Greber, T. F. *Biol. Mass Spectrom.* **1992**, *21*, 341–346.
3. Taylor, P. J.; Jones, A.; Balderson, G. A.; Lynch, S. V.; Norris, R. L. G.; Pond, S. M. *Clin. Chem.* **1996**, *42*, 279–285.
4. Loo, J.; Pesch, R. *Anal. Chem.* **1994**, *66*, 3659–3663.
5. Dobberstein, P.; Muenster, H. *J. Chromatogr. A* **1995**, *712*, 3–15.
6. Winger, B. E.; Hein, R. E.; Becker, B. L.; Campana, J. E. *Rapid Commun. Mass Spectrom.* **1994**, *8*, 495–497.
7. Haskins, N. J.; Eckers, C.; Organ, A. J.; Dunk, M. F.; Winger B. E. *Rapid Commun. Mass Spectrom.* **1995**, *9*, 1027–1030.
8. Van Berkel, G. J.; Glish, G. L.; McLuckey, S. A. *Anal. Chem.* **1990**, *62*, 1284–1295.
9. Henion, J. D.; Mordehai, A. V.; Cai, J. *Anal. Chem.* **1994**, *66*, 2103–2109.
10. Chaudhary, A. K.; Nokubo, M.; Oglesby, T. D.; Marnett, J.; Blair, I. A. *J. Mass Spectrom.* **1995**, *30*, 1157–1166.
11. Rozman, E.; Galceran, M. T.; Albet, C. *Rapid Commun. Mass Spectrom.* **1995**, *9*, 1492–1498.
12. Rudewicz, P. J.; Straub, K. M. *Anal. Chem.* **1986**, *58*, 2928–2934.
13. Weidolf, L. O. G.; Lee, E. D.; Henion, J. D. *Biol. Environ. Mass Spectrom.* **1988**, *15*, 283–289.
14. Baillie, T. A.; Davies, M. R. *Biol. Mass Spectrom.* **1993**, *22*, 319–325.
15. Murphy, A. L.; Fenselau, C.; Gutierrez, P. L. *J. Am. Soc. Mass Spectrom.* **1992**, *3*, 815–822.
16. Weidolf, L.; Covey, T. *Rapid Commun. Mass Spectrom.* **1992**, *6*, 192–196.
17. Wheelan, P.; Murphy, R. C. *Arch. of Biochem. Biophys.* **1995**, *321*, 381–389.
18. Yeung, B.; Vouros, P.; Siu-Caldera, M.-L.; Reddy, G. S. *Biochem. Pharmacol.* **1995**, *49*, 1099–1110.
19. Feng, P. C. C.; Solsten, R. T. *Xenobiotica* **1994**, *24*, 729–734.
20. Mutlib, A. E.; Strupczewski, J. T.; Chesson, S. M. *Drug Metab. Dispos.* **1995**, *23*, 951–964.
21. Mutlib, A. E.; Strupczewski, J. T. *J. Chromatogr. B* **1995**, *669*, 237–246.
22. Taylor, M. D.; Amidon, G. Eds. *Peptide Based Drug Design: Controlling Transport and Metabolism*; American Chemical Society: Washington, DC, 1995.
23. Kaltashov, I. A.; Yu, X.; Fenselau, C. *J. Pharm. Biomed. Anal.* **1995**, *13*, 279–284.
24. Andren, P. E.; Caprioli, R. M. *J. Mass Spectrom.* **1995**, *30*, 817–824.

25. Toreson, G.; Brodin, E.; Wahlstrom, A.; Bertilsson, L. *J. Neurochem.* **1988**, *50*, 1701–1705.
26. Takada, Y.; Yoshida, M.; Sakairi, M.; Koizumi, H. *Rapid Commun. Mass Spectrom.* **1995**, *9*, 895–896.
27. Chaudhary, A. K.; Nokubo, M.; Marnett, L. J.; Blair, I. A. *Biol. Mass Spectrom.* **1994**, *23*, 457–464.
28. Chaudhary, A. K.; Nokubo, M.; Reddy, G. R.; Yeola, S. N.; Morrow, J. D.; Blair, I. A.; Marnett, L. J. *Science* **1994**, *265*, 1580–1582.
29. La, D. K.; Lilly, P. D.; Anderegg, R. J.; Swenberg, J. A. *Carcinogenesis* **1995**, *16*, 1419–1424.
30. Rindgen, D.; Turesky, R. J.; Vouros, P. *Chem. Res. Toxicol.* **1995**, *8*, 1005–1013.
31. Rule, G. S.; Henion, J. D. *J. Chromatogr.* **1992**, *582*, 103–112.
32. Cai, J.; Henion, J. *Anal. Chem.* **1996**, *68*, 72–78.
33. Brownsill, R.; Combal, J.-P.; Taylor, A.; Bertrand, M.; Luijten, W.; Walther, B. *Rapid Commun. Mass Spectrom.* **1994**, *8*, 361–365.
34. Balani, S. K.; Arison, B. H.; Mathai, L.; Kauffman, L. R.; Miller, R. R.; Stearns, R. A.; Chen, I.-Wu; Lin, J. H. *Drug Metab. Dispos.* **1995**, *23*, 266–230.
35. Swart, P. J.; Bronner, G. M.; Bruins, A. P.; Ensing, K.; Tepper, P. G.; De Zeeuw, R. A. *Toxicol. Methods* **1993**, *3*, 279–290.
36. Swart, P. J.; Oelen, W. E. M.; Bruins, A. P.; Tepper, P. G.; De Zeeuw, R. A. *J. Anal. Toxicol.* **1994**, *18*, 71–77.
37. Lanting, A. B.; Bruins, A. P.; Drenth, B. F. H.; de Jonge, K.; Ensing, K.; de Zeeuw, R. A.; Meijer, D. F. K. *Biol. Mass Spectrom.* **1993**, *22*, 226–234.
38. Misra, A. L. In *Factors Affecting the Action of Narcotics*; Adler, M. L.; Manara, L.; Samanin, R. Eds.; Raven Press: New York, 1978; p. 297.
39. Dear, G. J.; Harelson, J. C.; Jones, A. E.; Johnson, T. E.; Pleasance, S. *Rapid Commun. Mass Spectrom.* **1995**, *9*, 1457–1464.
40. Dickson, R. G.; King, A. R.; Kelly, M. A.; Kaltashov, I. A.; Fenselau, C. *J. Pharm. Biomed. Anal.* **1994**, *12*, 1075–1078.
41. Shore, L. J.; Fenselau, C.; King, A. R.; Dickinson, R. G. *Drug Metab. Dispos.* **1995**, *23*, 119–123.
42. Stogniew, M.; Fenselau, C. *Drug Metab. Dispos.* **1982**, *10*, 609–613.
43. Davis, M. R.; Baillie, T. A. *J. Mass Spectrom.* **1995**, *30*, 57–68.
44. Jin, L.; Davies, M. R.; Hu, P.; Baillie, T. A. *Chem. Res. Toxicol.* **1994**, *7*, 526–533.
45. Borel, A. G.; Abbott, F. S. *Chem. Res. Toxicol.* **1995**, *8*, 891–899.
46. Jin, L.; Baillie, T. A.; Davies, M. R.; Kharasch, E. D. *Biochem. Biophys. Res. Commun.* **1995**, *210*, 498–506.
47. McLaughlin, L. G.; Henion, J. D.; Kijak, P. J. *Biol. Mass Spectrom.* **1994**, *23*, 417–429.
48. Covey, T. R.; Lee, E. D.; Henion, J. D. *Anal. Chem.* **1986**, *58*, 2453.
49. Gilbert, J. D.; Hand, E. L.; Yuan, A. S.; Olah, T. V.; Covey, T. R. *Biol. Mass Spectrom.* **1992**, *21*, 63–68.
50. Gilbert, J. D.; Greber, T. F.; Ellis, J. D.; Barrish, A.; Olah, T. V.; Fernandez-Metzler, C.; Yuan, A. S.; Burke, C. J. *J. Pharm. Biomed. Anal.* **1995**, *13*, 937–950.

51. Constanzer, M. L.; Chavez, C. M.; Matuszewski. *J. Chromatogr. B* **1994**, *658*, 281–287.
52. Edlund, P. O.; Bowers, L.; Henion, J.; Covey, T. R. *J. Chromatogr.* **1989**, *497*, 49–57.
53. Banno, K.; Horimoto, S.; Mabuchi, M. *J. Chromatogr.* **1991**, *568*, 375–384.
54. Avery, M. J.; Mitchell, D. Y.; Falkner, F. C.; Fouda, H. G. *Biol. Mass Spectrom.* **1992**, *21*, 353–357.
55. Chavez, C. M.; Constanzer, M. L.; Matuszewski, B. K. *J. Pharm. Biomed. Anal.* **1995**, *13*, 1179–1184.
56. Knebel, N. G.; Sharp, S. R.; Midgan, M. J. *J. Chromatogr. A* **1995**, *673*, 213–222.
57. Muck, W. M. *J. Chromatogr. A* **1995**, *712*, 45–53.
58. Bruins, A. P. *Mass Spectrom. Rev.* **1991**, *100*, 53–77.
59. Oda, Y.; Mamo, N.; Asakawa, N. *J. Mass Spectrom.* **1995**, *30*, 1671–1678.
60. Lausecker, B.; Hopfgartner, G. *J. Chromatogr. A* **1995**, *712*, 75–83.
61. Cornpropst, J. D.; Gillespie, T. A.; Shipley, L. A. *J. Chromatogr. B* **1995**, *673*, 67–74.
62. Phillips, D. H.; Carmichael, P. L.; Hewer, A.; Cole, K. J.; Hardcastle, I. R.; Poon, G. K., Keogh, A.; Strain, A. J. *Carcinogenesis* **1996**, *17*, 89–94.
63. Whitman, D. A.; Abbott, V.; Fregien, K.; Bowers, L. D. *Ther. Drug Monit.* **1993**, *15*, 552–556.
64. Pacifici, R.; Pichini, S.; Altieri, I.; Caronna, A.; Passa, A. R.; Zuccaro, P. *J. Chromatogr. B* **1995**, *664*, 329–334.
65. Wolff, T.; Samuelsson, H.; Hedner, T. *Pain J.* **1995**, *62*, 147–154.
66. Johansson, I. M.; Pavelka, R.; Henion, J. D. *J. Chromatogr.* **1991**, *559*, 515–528.
67. Bruins, A. P.; Covey, T. R.; Henion, J. D. *Anal. Chem.* **1987**, *59*, 2642–2646.
68. Mylchreest, I. *Proceedings of the 40th Conference of the American Society of Mass Spectrometry*, Washington, DC 1992, p. 616.
69. Parker, C. E.; Perkins, J. R.; Tomer, K. B.; Shida, Y.; O'Hara, K.; Kono, M. *J. Mass Spectrom.* **1992**, *3*, 563–574.
70. Sheppard, R.; Tong, X.; Cai, J.; Henion, J. *Proceedings of the 42nd Conference of the American Society of Mass Spectrometry*, Chicago, IC 1994, p. 601.
71. Kostianinen, R.; Franssen, E. J.; Bruins, A. P. *J. Chromatogr.* **1993**, *647*, 361.
72. Benson, L. M.; Tomlinson, A. J.; Reid, J. M.; Walker, D. L.; Ames, M. M.; Naylor, S. *J. High Resolut. Chromatogr.* **1993**, 324–326.
73. Tomlinson, A. J.; Benson, L. M.; Johnson, K. L.; Naylor S. *J. Chromatogr.* **1993**, *621*, 239–248.
74. Tomlinson, A. J.; Benson, L. M.; Gorrod, J. W.; Naylor, S. *J. Chromatogr. B* **1994**, 373–381.
75. Lu, W.; Poon, G. K.; Carmichael, P.; Cole, R. B. *Anal. Chem.* **1996**, *68*, 668–674.
76. Ashcroft, A. E.; Major, H. J.; Lowes, S.; Wilson, I. D. *Anal. Proc.* **1995**, *32*, 459–462.
77. Henion, J. D.; Mordehai, A. V.; Cai, J. *Anal. Chem.* **1994**, *66*, 2103–2109.
78. Taylor, L. C. E.; Singh, R.; Chang, S. Y.; Johnson, R. L.; Schwartz, J. *Rapid Commun. Mass Spectrom.* **1995**, *9*, 902–910.
79. He, H.; McKay, G.; Midha, K. K. *Xenobiotica* **1995**, *25*, 857–872.

80. Poon, G. K.; Mistry, P.; Lewis, S. *Biol. Mass Spectrom.* **1991**, *20*, 687–692.
81. Poon, G. K.; Bisset, G. M. F.; Mistry, P. *J. Am. Soc. Mass Spectrom.* **1993**, *4*, 588–595.
82. Poon, G. K.; Raynaud, F. I.; Mistry, P.; Odell, D. E.; Kelland, L. R.; Harrap, K. R.; Barnard, C. F. J.; Murrer, B. A. *J. Chromatogr.* **1995**, *61*–66.
83. Jackson, P. J.; Brownsill, R. D.; Taylor, A. R.; Walther, B. *J. Mass Spectrom.* **1995**, *30*, 446–451.
84. Siethoff, C.; Nigge, W.; Linscheid, M. *Fresenius' J. Anal. Chem.* **1995**, *352*, 801–805.
85. Lønning, P. E.; Lien, E. A.; Lundgren, S.; Kvinnsland, S. *Clin. Pharmacokinet.* **1992**, *22*, 327–358.
86. Renner, U.; Joachim, B.; Freund, S.; Waidelich, D.; Ehninger, G.; Zeller, K.-P. *Drug Metab. Dispos.* **1995**, *23*, 94–101.
87. Blanz, J.; Renner, U.; Schmeer, K.; Ehninger, G.; Zeller, K.-P. *Drug Metab. Dispos.* **1993**, *21*, 955–961.
88. Richards, L. E.; Sun, J.-H. *Drug Metab. Dispos.* **1995**, *23*, 600–602.
89. Jarman, M.; Poon, G. K.; Rowlands, M. G.; Grimshaw, R. M.; Horton, M. N.; Potter, G. A.; McCague, R. *Carcinogenesis* **1995**, *16*, 683–688.
90. Phillips, D. H.; Carmichael, P. L.; Hewer, A.; Cole, K. J.; Poon, G. K. *Cancer Res.* **1994**, *54*, 5518–5522.
91. Poon, G. K.; Walter, B.; Lønning, P. E.; Horton, M. N.; McCague, R. *Drug Metab. Dispos.* **1995**, *23*, 377–382.
92. Poon, G. K.; Chui, Y. C.; McCague, R.; Lønning, P. E.; Feng, R.; Rowlands, M. G.; Jarman M. *Drug Metab. Dispos.* **1993**, *20*, 1119–1124.
93. Bitsch, F.; Ma, W.; Macdonald, F.; Nieder, M.; Shackleton, C. H. L. *J. Chromatogr.* **1993**, *615*, 273–280.
94. Blay, P. K. S.; Thibault, P.; Thiberge, N.; Kiecken, B.; Lebrun, A.; Mercure, C. *Rapid Commun. Mass Spectrom.* **1993**, *7*, 626–634.
95. Kerns, E. H.; Volk, K. J.; Hill, S. E. *J. Nat. Prod.* **1994**, *57*, 1391–1403.
96. Sparreboom, A.; Huizing, M. T.; Boesen, J. J. B.; Nooijen, W. J.; Van Tellingen, O.; Beijnen, J. H. *Cancer Chemother. Pharmacol.* **1995**, *36*, 299–304.
97. Walle, T.; Walle, U. K.; Kumar, G. N.; Bhalla, K. N. *Drug Metab. Dispos.* **1995**, *23*, 506–512.
98. Monsarrat, B.; Alvinerie, P.; Gares, M.; Wright, M.; Dubois, J.; Gueritte-Voegelein, F.; Guenard, D.; Donehower, R.; Rowinsky, E. *Cell. Pharmacol.* **1993**, Suppl. 1, S77–S81.
99. Royer, I.; Alvinerie, P.; Armand, J. P.; Ho, L. K.; Wright, M.; Monsarrat, B. *Rapid Commun. Mass Spectrom.* **1995**, *9*, 495–502.
100. Poon, G. K.; Chui, Y. C.; Jarman, M.; Rowlands, M. G.; Kokkonen, P. S.; Niessen, W. M. A.; van der Greef, J. *Drug Metab. Dispos.* **1992**, *20*, 941–947.
101. Stack, R. F.; Rudewicz, P. J. *J. Mass Spectrom.* **1995**, *30*, 857–866.
102. Dunayevskiy, V.; Vouros, P.; Carell, T.; Winter, E. A.; Rebek, J. *Anal. Chem.* **1995**, *67*, 2906–2915.
103. Youngquist, R. S.; Fuentes, G. R.; Lacey, M. P.; Keough, T. *Rapid Commun. Mass Spectrom.* **1994**, *8*, 77–81.

104. Brummel, C. L.; Lee, I. N. W.; Zhou, Y.; Benkovic, S. J.; Winograd, N. *Science* **1994**, *264*, 399–402.
105. Brummel, C. L.; Vickerman, J. C.; Carr, S. A.; Hemling, M. E.; Roberts, G. D.; Johnson, W.; Weinstock, J.; Gaitanopoulos, D.; Benkovic, S. J.; Winograd, N. *Anal. Chem.* **1996**, *68*, 237–242.

CHAPTER 15

Electrospray Ionization of Inorganic and Organometallic Complexes

CHRISTINE L. GATLIN AND FRANTIŠEK TUREČEK

Department of Chemistry, University of Washington, Seattle, Washington

	Abstract	527
I.	Introduction	528
II.	Formation of gas-phase metal complexes by electrospray	528
III.	Electrospray ionization in inorganic and elemental analysis	532
IV.	Organic complexes with alkali and alkaline earth cations	536
	A. Coronands and cryptands	537
	B. Carbohydrates and oligonucleotides	539
	C. Polyether antibiotics and toxins	541
V.	Electrospray ionization of nonpolar molecules via metal complexes	543
VI.	Metal complexes with proteins, peptides, and amino acids	543
	A. Binary protein complexes	545
	B. Binary peptide complexes	548
	C. Binary amino-acid complexes	550
	D. Ternary complexes	550
VII.	Supramolecular metal complexes	558
VIII.	Coordination compounds and miscellaneous applications	560
IX.	Conclusions and perspectives	563
	Acknowledgments	564
	References	564

ABSTRACT

Inorganic and organometallic complexes have been studied extensively by electrospray ionization–mass spectrometry, as summarized in this chapter. The developments of the last five years have brought new insights into coordination chemistry in the gas-phase and solution. Alkali-metal complexes of a variety of organic

Electrospray Ionization Mass Spectrometry, Edited by Richard B. Cole.
ISBN 0-471-14564-5   © 1997 John Wiley & Sons, Inc.

ligands, including polyethylene glycols, coronands, cryptands, carbohydrates, ionophore antibiotics, peptides, and nucleotides, have been obtained by electrospray ionization and their gas-phase properties have been correlated with their chemistries in solution. A variety of coordination compounds have been analyzed by electrospray ionization–mass spectrometry, which provided a wealth of information regarding molecular weight and structure, and often represented the analytical method of choice. Transition-metal complexes of biological molecules, amino acids, peptides, proteins, and carbohydrates are readily transferred to the gas-phase by electrospray ionization of their solutions, which allows their gas-phase and solution properties to be correlated. In addition, new charge-transfer and dissociation reactions have been found to take place in the gas-phase complexes that contribute to the analytical potential of mass spectrometry of biomolecules.

I. INTRODUCTION

The potential of electrospray ionization (ESI) for obtaining and studying gas-phase metal complexes was realized soon after the method had been introduced by Fenn and co-workers as a soft-ionization technique for polar involatile molecules (1–3). The first reports by Kebarle and co-workers (4–7) and Chait and co-workers (8) were the harbingers of numerous further studies of metal complexes, in both the fundamental and applied areas of research. In this chapter, we cover and discuss several selected topics relevant to organometallic and metal–ion chemistry. We begin by discussing the formation of gas-phase complexes in relation to their solution counterparts, and the redox processes that accompany the ES process. Inorganic complexes relevant to elemental analysis are briefly surveyed. Transition-metal complexes with large organic ligands are treated in several sections, including ligand dissociations triggered by the metal-ion and attachment of metal ions to chemically modified nonpolar ligands. A survey of analytical applications aimed at the detection of organometallics by electrospray ionization–mass spectrometry (ESI–MS) is also included, although the buoyant development in this area is likely to make this section rapidly obsolete.

II. FORMATION OF GAS-PHASE METAL COMPLEXES BY ELECTROSPRAY

Electrospraying aqueous-organic solutions of metal salts typically results in the formation of metal-solvent cluster ions, $M(S)_n^{y+}$. Bare alkali metal ions ($n = 0$, Li^+ through Cs^+) are also obtained abundantly. The number of solvent molecules in the gas-phase complexes depends on the nature of both the metal-ion and the solvent, and on the ESI conditions, such as pressure, temperature, and electrostatic potentials in the ESI interface. Kebarle and co-workers studied the formation of gas-phase $M(H_2O)_n^{y+}$ clusters in a special interface chamber in which the metal-ions formed by ESI were allowed to equilibrate with water vapor seeded in nitrogen gas according to Eq. 1 (7):

$$M(H_2O)_{n-1}^{y+} + H_2O(g) \rightleftarrows M(H_2O)_n^{y+} \quad (1)$$

Under these conditions, alkali metal-ions formed several $M(H_2O)_n^+$ clusters whose intensities peaked at different values of n, which decreased from $n = 4$ for Li^+ to $n = 1$ for Rb^+. The ΔG values for the gas-phase complexes have been deduced from the cluster-ion relative abundances (7). Alkaline earth ions, Mn^{2+}, and Co^{2+} gave relatively broad distributions of $M(H_2O)_n^{2+}$, $n = 7\text{–}13$, which peaked at $n = 9\text{–}10$. The estimated water binding energies in the gas-phase ions were found to roughly increase with the decreasing ion-crystal radii (9).

Ligand–ligand proton transfer and metal–ligand electron transfer have been studied for simple binary complexes of dications (M^{2+}) and trications (M^{3+}). The thermochemistry of the first type of reactions [e.g., for aquo-complexes, $M(H_2O)_n^{y+} \rightarrow MOH(H_2O)_{n-2}^{(y-1)+} + H_3O^+$] can be expressed by the enthalpy terms for complete complex dissociation ($\Delta H_{n,0}$), electron transfer from the ligand to metal-ion, $\Delta IE = IE(H_2O) - IE(M^{(y-1)+})$, self-CI of the ionized ligand, $H_2O^{+\cdot} + H_2O \rightarrow H_3O^+ + OH^\cdot$, ΔH_{PT}, formation of the $M^{(y-1)+}$–OH bond, –D(M–OH), hydration of the $M^{(y-1)+}$OH ion, $\Delta H_{n-2,0}$(M–OH), and coulombic repulsion (E_{coul}), as shown in Eq. 2:

$$\Delta H_r = \Delta H_{n,0}(M^{y+}) + IE(H_2O) - IE(M^{(y-1)+}) + \Delta H_{PT}$$
$$- D(M-OH) - \Delta H_{n-2,0}(M-OH)^{(y-1)+} + E_{coul} \quad (2)$$

Equation 2 could potentially be useful for predicting thermodynamic stabilities of gas-phase complexes, provided all the enthalpy terms are known or subject to reliable estimates. Although this is not the case at present for the first and last terms in Eq. 2, bond-dissociation energies in complexes produced by ESI may soon become available from threshold energies measured by collision-activated dissociation (CAD) (10). The enthalpies of the alternative charge transfer, $M(L)_n^{y+} \rightarrow M(L)_{n-1}^{(y-1)+} L^{+\cdot}$, and ligand loss reactions, $M(L)_n^{y+} \rightarrow M(L)_{n-1}^{y+} + L$, are expressed by Eqs. 3 and 4, respectively:

$$\Delta H_r = \Delta H_{n,0}(M^{y+}) + IE(L) - IE(M^{(y-1)+}) - \Delta H_{n-1,0}(M^{(y-1)+}) \quad (3)$$
$$\Delta H_r = \Delta H_{n,n-1} \quad (4)$$

Kebarle and co-workers compared the processes in Eqs. 2 and 3 and concluded that the former will be favored by large D(M–OH) and $-\Delta H_{PT}$ values for aquo complexes and by the relatively high ionization energy of water (5). By contrast, organic ligands, which typically have ionization lower than that of water, favor metal–ligand electron transfer or ligand elimination (5,7).

The aforementioned charge transfer reactions result in coulombic repulsion between the positively charged metal complex and the departing ligand cation.

The potential energy due to the coulombic repulsion is released as kinetic energy of the separating fragments and thus contributes to the activation energy for the overall dissociation. Hence complexes, which are thermochemically unstable according to Eq. 2 or 3, may be stable kinetically. For example, $Cu(bpy)_2^{2+}$ is produced by ESI as a stable species in the gas-phase in spite of the large difference in the IE(bpy) and IE_2(Cu), which makes the charge transfer 11.94 eV exothermic and cannot be compensated for by the bpy-Cu binding energy (~ 6.5 eV) according to Eq. 3. The coulombic repulsion energy depends on the charge distribution in the transition state, and hence the reaction kinetics is likely to depend on the polarizability of the ligands. Kinetic studies of metal–ligand charge-transfer reactions in gas-phase organometallic ions and applications of the Marcus theory (11) would be extremely interesting in this respect.

A different reaction occurs in ternary carboxylate complexes of transition-metal dications, which undergo intramolecular electron transfer accompanied by ligand fragmentation, as shown for decarboxylation (Eq. 5) (12). Such reactions resemble reductive eliminations known in solution organometallic chemistry, in which metal–ligand bonds are cleaved homolytically (13).

$$[M^{2+}(R-COO)^-(L)]^+ \rightarrow [M(R^{\cdot})(L)]^+ + CO_2 \qquad (5)$$

A distinct feature of the reactions in Eq. 5 is retention of the charge state in the product ion.

Unusual electron-transfer reactions have been observed recently for gas-phase $Cu^{(I)}$ complexes with dipeptide alkali metal salts prepared by ESI (14). The peptide residue, which is coordinated to $Cu(I)bpy^+$, is oxidized with the alkali metal cation on CAD to eliminate a neutral alkali metal atom (Scheme 1). This ligand oxidation is a dominant CAD process for sodium salts, while potassium and rubidium salts show both ligand oxidation, resulting in losses of neutral K and Rb, and formation of K^+ or Rb^+, respectively. However, Cs^+ does not oxidize the ligand and is eliminated as a Cs^+ cation (14).

Redox reactions have been found to occur upon ESI of organometallic complexes, in particular those with low ionization energies. The $Mg^{(II)}$, $Ni^{(II)}$, $Cu^{(II)}$, and $Zn^{(II)}$ complexes of octaethylporphyrin (OEP), which are neutral in solution, gave the corresponding cation radicals by ESI (15). The complexes have low ionization energies ($IE = 6.19-6.38$ eV) and low half-wave potentials ($E_{1/2} = 0.63-0.79$ V), which favor their electrochemical oxidation at the metal–solution interface. The formation of $[M^{(II)}(OEP)]^{+\cdot}$ was enhanced in the presence of trifluoroacetic acid, which has been known to stabilize cation radicals by solvation in solution (16).

Metallocenes, Cp_2Fe, Cp_2Ru, and Cp_2Os ($Cp = \eta^5$-cyclopentadienyl) and their derivatives provide several ion species by ESI of their acetonitrile solutions (17). Ferrocene Cp_2Fe forms chiefly Cp_2Fe^+ and $CpFe^+$, which result, respectively, from electrochemical oxidation of Cp_2Fe followed by CAD in the ESI interface. Cp_2Ru forms acetonitrile complexes, $Cp_2Ru(CH_3CN)_{0-3}^+$, which also result by one-electron oxidation of neutral ruthenocene. A correlation

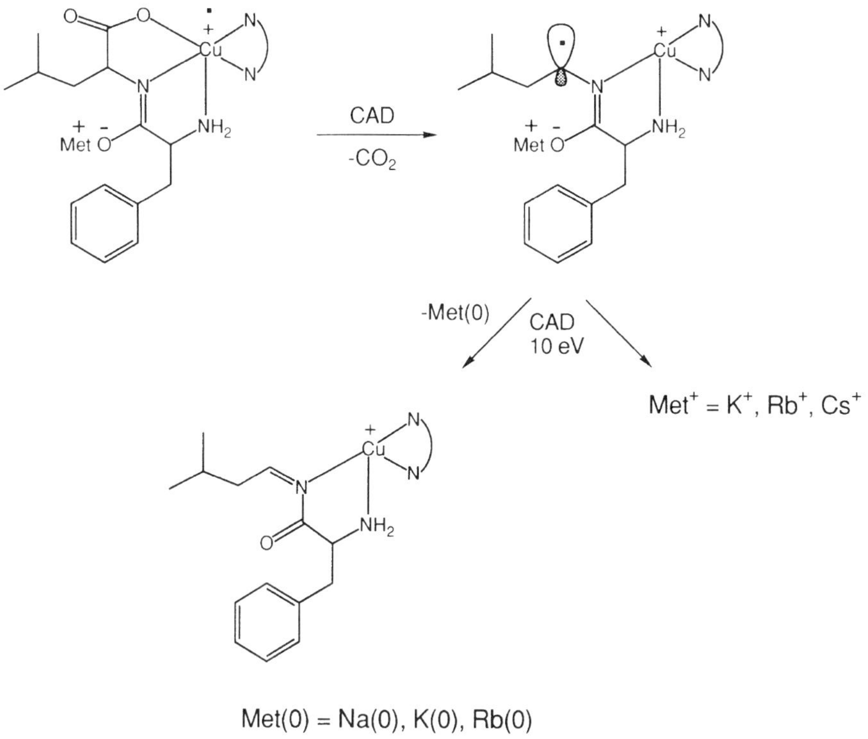

Scheme 1

Met(0) = Na(0), K(0), Rb(0)

Scheme 1

was drawn between the onset potential applied on the electrospray needle and the half-wave potential of the metallocene, which supported the electrochemical mechanism for the formation of Cp_2M^+ (17). Substantial solvent effects on the production of Cp_2M^+ were found, which, however, differed for Fe, Ru, and Os and do not appear to be well understood (17). In contrast to the case of OEP complexes (15), trifluoroacetic acid had a negligible effect on the formation of Cp_2M^+, whereas protonation to give $[Cp_2M + H]^+$ was observed in the presence of CF_3COOH. The authors noted that different optimum ESI conditions existed for protonation and electrochemical oxidation, which appear to occur in competition (17).

The role of electrochemical processes in the formation of charged droplets in ESI has been documented by investigations of metal–ion redox systems. Kebarle and co-workers have shown that electrospraying methanol solutions containing dimethylsulfoxide (DMSO) from a Zn-tipped capillary at conventional positive potentials (3.5–4 kV) resulted in the formation of Zn^{2+} ions that were detected as $Zn(DMSO)_3^{2+}$ complexes in the gas phase (18). Likewise, stainless steel and silver were oxidized by ESI to provide metal ions, which were detected as $Fe(DMSO)_3^{2+}$ and $Ag(DMSO)_{1-2}^+$, respectively.

The formation of H_3O^+ in ESI was studied by the present authors with the use of a pH-sensitive organometallic system (19). Both Ni^{2+} and Fe^{2+} form stable complexes with 2,2′-bipyridyl (bpy), which show large formation constants, $\beta_3(Fe) = 1.6 \times 10^{17}$ and $\beta_3(Ni) = 1.4 \times 10^{20}$ (20). The complexes dissociate negligibly in solution at neutral pH (Eq. 6) even at low concentrations used in ESI. Since bpy is a weak base ($K_{BH+} = 1.5 \times 10^{-4}$), the

$$Fe^{2+}(bpy)_3 \rightleftarrows Fe^{2+}(bpy)_{3-n} + n bpy \qquad (n = 1-3) \qquad (6)$$

equilibria in Eq. 6 are coupled with bpy protonation, which is pH dependent. The organometallic complex thus provides a pool of bpy, which is released depending on the pH of the solution. Measurements of $[bpy + H]^+$ ion currents from electrospraying nominally pH-neutral $Fe^{2+}(bpy)_3$ and $Ni^{2+}(bpy)_3$ solutions showed a substantially increased dissociation of the complexes on ESI. This dissociation was attributed to an increase in the $[H_3O]^+$ concentration in the microdroplets formed by ESI, which was equivalent to pH 2.5–3.3 in the electroneutral bulk solution. These measurements have shed some light on the mechanism of protonation of basic solutes on ESI. Estimates of coulombic repulsion and Rayleigh stability limits for microdroplets of aqueous methanol (21) indicate that the droplets could not hold the positive charge if their entire volume was filled with 3×10^{-3} M H_3O^+ corresponding to pH 2.5. Hence, it was concluded that the H_3O^+ ions must be confined within a smaller volume, such as a thin surface layer in which coulomb repulsion will be minimized. The layer thickness was estimated to be 5–27 nm to accommodate the positive charge due to H_3O^+ at pH 2.5–3.3 (19).

Electrochemical redox reactions in ESI have also been observed for copper complexes with organic ligands that stabilize $Cu^{(I)}$ or $Cu^{(II)}$ oxidation states (12). These redox reactions depend strongly on the solvent and can be related to the electrochemical potentials of the $Cu^{(II)}/Cu^{(I)}$ redox couple in solution (22,23).

On-line electrochemical oxidation of metal complexes followed by ESI–MS analysis has been reported (24). Neutral N,N-diethyldithiocarbamate (Et_2dtc) complexes $M(Et_2dtc)_2$, M = Cu, Ni, and Co, were oxidized to $[M(Et_2dtc)_2]^+$ in a flow-electrochemical cell in the presence of 0.001 M $(n\text{-}C_4H_9)_4NPF_6$ as a support electrolyte, and the cations were detected by ESI–MS (24). The method holds promise for the detection of neutral organometallic complexes that are not protonated well under standard ESI conditions (12).

III. ELECTROSPRAY IONIZATION IN INORGANIC AND ELEMENTAL ANALYSIS

Compared to solution-phase inorganic chemistry, the field of gas-phase inorganic chemistry is relatively new (25). Mass spectrometric methods for the analysis of inorganic solids and liquids were recently reviewed by Duckworth et al. (26). By far, ICP–MS has been the technique of choice for the analysis of inorganic solutions, based on its superior sensitivity and a wide linear dynamic range. However, ICP and related techniques do not provide information on the analyte as to its valence state, molecular form, and counteranions. In contrast, ESI–MS can, in principle, provide this information.

Since the reintroduction of ESI–MS in 1984 by Yamashita and Fenn (1), the majority of ESI applications has concerned organic and bioorganic molecules (27–29), whereas the analysis of inorganic ions has lagged far behind. The early studies of Douglas (30) and Kebarle and co-workers (4–7) showed that inorganic ions other than alkali metals could be analyzed by ESI–MS. Earlier attempts to observe uncomplexed divalent and trivalent metals by electrospray were unsuccessful (1,31). To circumvent the problem of multiply charged metal-ions, metals in their higher oxidation states can be detected as overall +1 ions by the addition of negatively charged ligands. For example, +3 lanthanides can be detected as single-charged β-diketone complexes, $[La(diketone)_3 + H]^+$ (32,33). For monovalent ions such as alkali metals, declustering (solvent or ligand loss) is a relatively simple process (34). Multiply charged "naked" metal ions are more difficult to produce in the gas phase, because ligand ionization [charge reduction (6)] starts to compete with ligand loss when the solvation sphere can no longer stabilize the charge state of the metal. The ionization potential of the metal (Met^+) and ligand (L) are indicative parameters. As a rule of thumb, when $IP(Met^+) > IP(L)$, charge-reduced species will form.

Beside the detection of ions from simple alkali and alkaline earth-metal salt solutions, there are numerous reports on the analysis of uncomplexed transition metals (35), including metal oxides (36), metal oxo ions (37), polyoxoanions (38,39), organoarsenics (40,41), and lanthanide metal-ions (42). Given the ability to obtain a wide variety of bare metal-ions by electrospray under relatively mild declustering conditions, the preliminary results by Douglas (43), Cheng et al. (44), and Van Berkel et al. (45) indicated the potential of ESI–MS as a technique for elemental inorganic analysis. This potential has been explored in detail by Agnes and Horlick and described in several papers (42,46–49).

For example, ESI mass spectra were attained for several elements in the lanthanide series. Under harsh declustering conditions of +3 lanthanides, bare +1 and +2 metal-ions dominate and, as well as metal oxides (MO^+), were obtained, as shown in Figure 1 (42). The spectrum for the mixture of Pr, Gd, and Lu is fairly complex with a number of isobaric overlaps due to the presence of oxides. For instance, the peak of $^{141}PrO^+$ overlaps with that of the bare $^{157}Gd^+$ isotope, and the peak of $^{158}GdO^+$ overlaps with the peak of $^{176}Lu^+$. This kind of overlap is common to ICP–MS as well, but it is less of a problem because oxides and hydroxides are present at lower levels. The authors point out that this problem could be circumvented if gentle source and interface conditions are used when running the mixture (42). Under gentle declustering conditions, the same types of clusters are formed from Gd, Pr, and Lu, such that mass overlap can be minimized (Fig. 2) (42). As yet no attempt has been made to try and quantify unknown mixtures under either gentle or harsh declustering conditions.

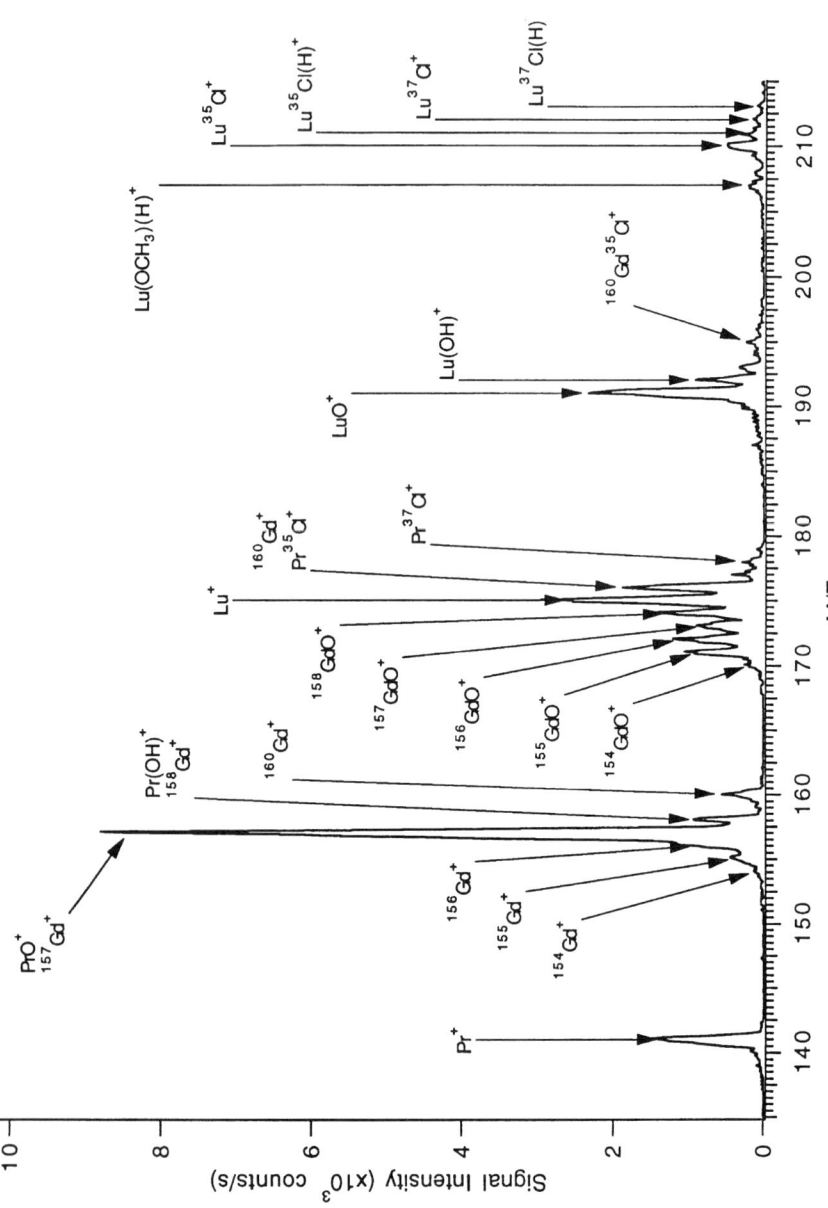

Figure 1. The ESI mass spectrum of $Pr(NO_3)_3$, $GdCl_3$, and $LuCl_3$ (2×10^{-4} M each) acquired under harsh declustering conditions. Reproduced with permission from I. I. Stewart and G. Horlick, *Anal. Chem.*, **1994**, *66*, 3983–3993. © 1994 American Chemical Society.

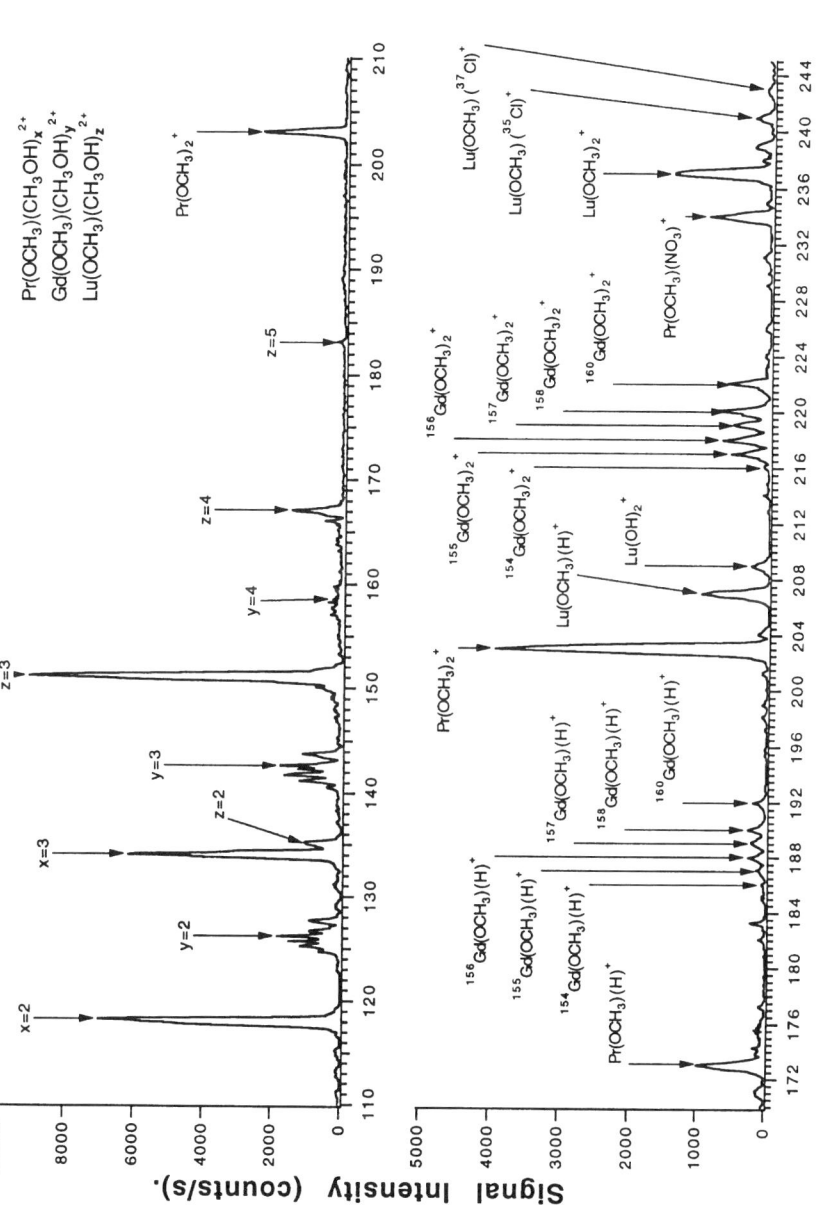

Figure 2. The ESI mass spectrum of $Pr(NO_3)_3$, $GdCl_3$, and $LuCl_3$ (2×10^{-4} M each) acquired under gentle declustering conditions. Reproduced with permission from I. I. Stewart and G. Horlick, *Anal. Chem.*, **1994**, *66*, 3983–3993. © 1994 American Chemical Society.

Investigations of mixtures containing more than one type of metal ion (e.g., alkali, alkaline earth, transition metal, or lanthanide) must still be carried out.

IV. ORGANIC COMPLEXES OF ALKALI AND ALKALINE EARTH METAL IONS

The ability of alkali metal-ions to form stable gas-phase complexes with polar molecules has been long known from FAB (50) and MALDI–MS (51) studies. Sodium complexes of polyethylene glycols were obtained early in the development of ESI–MS (3). The number of sodium ions attached to the polyethylene glycol molecule shows a distribution that depends on the PEG molecular weight and hence on the number of ether linkages. PEG 400 forms mostly monosodium complexes, whereas PEG 3350 gives a distribution ranging from $(PEG + 2Na)^{2+}$ to $(PEG + 6Na)^{6+}$. Fenn and co-workers discussed the question of the maximum number of Na^+ ions that can be accommodated by the PEG molecule in a gas-phase ion (52). They presumed that the Na^+ ions bind to the ether oxygen atoms at equidistant intervals, and the gas-phase ion was presumed to adopt a stretched-out all-trans conformation to minimize coulombic repulsion. The potential energy due to coulombic repulsion is highest at the centermost charge. The number of Na^+ ions reaches a maximum when the potential energy of the centermost charge equals the binding energy between Na^+ and the oxygen atom. The latter was taken to be equal to the Na^+ binding energy in 1,2-dimethoxyethane (2.05 eV). This simple model somewhat exaggerates the number of charges in PEG; for instance, it predicts PEG 200-mer and 400-mer to accommodate 18 and 31 charges, respectively, compared to the 12 and 23 charges found experimentally (52). A similar discrepancy has been found for multiply protonated proteins, in which the number of protons is limited kinetically by gas-phase deprotonation with solvent molecules in the high-pressure region of the ESI interface (53). From a practical point of view, Na^+–PEG complexes are easily made by ESI, and the variety of ions produced makes them convenient internal standards for mass-scale calibration in ESI–MS.

Complexation of Na^+ and K^+ with dibenzylether, bis-(4-methoxybenzyl)-ether, bis-(4-nitrobenzyl)ether, and diferrocenylether has been studied by ESI–MS (54). Both cations form monomer and dimer complexes, for example, $(PhCH_2)_2ONa^+$ and $[(PhCH_2)_2O]_2Na^+$, respectively (54).

Formation of potassium complexes in ESI has been reported for some functionalized hydrocarbons (e.g., **1** and **2**) (55). An interesting aspect of this work is the use of aprotic solvents dimethylformamide (DMF) and tetrahydrofuran (THF). This may broaden the scope of organometallic compounds amenable to analysis by ESI–MS to include very labile complexes that do not survive in protic solvents. Electrospraying DMF–THF solutions apparently requires relatively high needle potentials of 7.5 kV (55), as compared with the standard 6.1–6.3 kV used on the same instrument for spraying aqueous–organic solutions (19). The sensitivity for organometallic ion detection from aprotic ESI has not been addressed, as the reported spectra were obtained with $1-3 \times 10^{-4}$ M solutions.

A. Coronands and Cryptands

Complexes of alkali and alkaline earth metals with crown ethers, macrocyclic polyamines, polysulfides, and cryptands have been studied for over two decades (56,57). The ESI of such complexes has been reported recently by Colton (58), Enke (59), and their co-workers. Alkali-metal cations form mixtures of gaseous monomers (3) and dimers (4) with 12-crown-4-ether and dibenzo-18-crown-6-ether. In contrast, complexes with 18-crown-6-ether appear as monomers following ESI. These variations in the metal ion–crown ether stoichiometry have been interpreted by considering the cation radius and the space provided by the crown cavity. The cavity of 18-crown-6-ether is large enough to accommodate alkali-metal cations such that the metal-ion lies close to the plane of the donor oxygen atoms for optimum bonding (58). In the smaller 12-crown-4-ether, the metal-ion does not fit well in the cavity and remains coordinationally unsaturated in a 1 : 1 complex. This allows coordination of a second crown molecule to give a dimer complex. Dibenzo-18-crown-6-ether is an intermediate case, in which the flexibility of the polyether ring is diminished by the ortho-condensed benzene rings, resulting in a cavity of a smaller effective size. Small alkali cations (Li^+, Na^+) apparently fit well into this crown-ether cavity and give rise to monomer complexes in the gas phase. The larger K^+, Rb^+, and Cs^+ do not fit the cavity, which, because of its rigidity, cannot adjust to the cation size. Consequently, both monomer and dimer complexes are detected by ESI–MS (58).

The facile detection by ESI–MS of crown-ether complexes with alkali-metal ions has been utilized for the analysis of macropolycycles prepared by con-

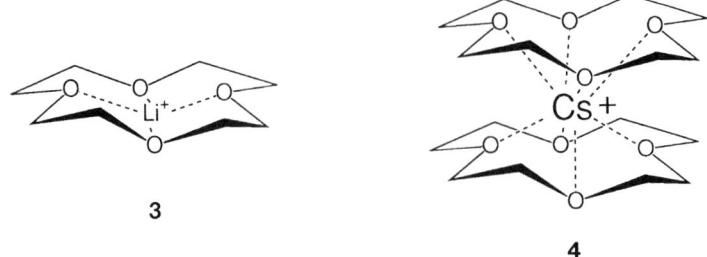

densation of 1,10-diaza-18-crown-6-ether with isophthaloyl dichloride. The reaction affords a mixture of oligomers (**5**), which are readily detected by ESI as the corresponding singly charged Na^+ complexes whose m/z should differ by the mass of the monomer (392 u, Scheme 2). This was indeed observed, although some of the m/z values were incorrectly assigned in the published ESI spectrum (60).

A preliminary report on the CAD of alkali-metal ion complexes with 18-crown-6-ether showed formation of bare metal ions from K^+ complexes with 18-crown-6-ether, 1-aza-18-crown-6, and 1,10-diaza-18-crown-6 (59). The $(M + 18\text{-}C\text{-}6)^+/M^+$ ion abundance ratios were found to depend strongly on the ion-collision energy and collision-gas pressure. At higher laboratory collision energies, fragmentations within the ligand were observed for Ag^+ complexes, which depended on the macrocycle type. In particular, the azacrown complexes underwent facile dehydrogenation by losing one H_2 molecule per a secondary amino group (59).

n = 1, m/z 415
n = 2, m/z 807
n = 3, m/z 1199
n = 4, m/z 1591
n = 5, m/z 1983

Scheme 2

The ESI of crown ethers with salts of alkaline earth dications gives several types of ions, which depend on the cation radius and crown-ether cavity size, as discussed for the alkali-metal cations (58).

B. Carbohydrates and Oligonucleotides

By far, the majority of interest in ESI–MS has been in the area of metal-ion interaction with biomolecules, which included carbohydrates, acylglycerols, polyether antibiotics and toxins, oligonucleotides, amino acids, peptides, and proteins. This section deals with highly oxygenated biomolecules, which are particularly amenable to ESI–MS analysis of their adducts with alkali or alkaline earth-metal cations.

Leary and co-workers have studied metal-ion interactions with carbohydrate molecules, in particular with oligosaccharides. Most of this work has involved FAB–MS analysis of metal–oligosaccharide species; however, recent reports have utilized ESI–MS. Carbohydrates show affinity to alkali and alkaline earth-metal ions and give rise to cationized adducts on ESI. Fura and Leary investigated the formation of Ca^{2+} and Mg^{2+} complexes of isomeric trisaccharides, which differed in the linkage of a fucose unit (61), for example, β-D-Gal-[1 → 4]-β-D-Glc-[3 → 1]-β-L-Fuc (3-FuLa), α-D-Gal-[1 → 3]-β-D-Gal-[2 → 1]-β-L-Fuc (B), and β-L-Fuc-[1 → 2]-β-D-Gal-[1 → 4]-β-D-Glc (2'-FuLa). These fucosyl derivatives (Scheme 3) form several complexes with Ca^{2+} in acetonitrile–water solutions at $1.2-10 \times 10^{-5}$ M concentrations. ESI produces $[Ca^{2+} + FuLa + CH_3CN]^{2+}$, $[Ca + (FuLa)_n]^{2+}$ ($n = 1-6$), and $[CaCl + FuLa]^+$ ions, whose relative abundances were found to depend on the concentration ratio of the metal and the trisaccharide and on the skimmer potential (61). Structure information about the hexose-unit linkage was obtained from the CAD spectra of the singly charged $[CaCl + FuLa]^+$ ions, which are formed preferentially by ESI of solutions at high metal-to-sugar ratios (61). All three fucosyl complexes undergo elimination of HCl, which is thought to trigger the subsequent skeletal fragmentations (61). 3-FuLa fragments by cleavage of the reducing lactose ring and loss of the fucose unit to give the ion at m/z 321 (Scheme 4). 2'-FuLa retains the fucose unit, while the reducing galactose ring is cleaved to form a disaccharide complex at m/z 407. Trisaccharide B eliminates the fucose unit to give another disaccharide ion at m/z 381 (Scheme 4). The Mg^{2+} and Li^+ trisaccharide complexes have also been investigated, but found to be less useful for structure analysis (61).

Carbohydrate analysis by LC–ESI–MS of metal complexes has been addressed in a recent communication (62). An effluent from an LC column was mixed with a metal salt solution in a triaxial capillary arrangement and the mixture was electrosprayed to provide gas-phase metal complexes. The Li^+ complexes of cellobiose, di-N-acetylchitobiose, and Lewis x trisaccharide are formed readily by post-column mixing and give rise to abundant (analyte + Li)$^+$ ions by ESI. Sensitivity enhancement by a factor of 5–50 has been achieved against ESI of free carbohydrates based on (M + H)$^+$ ions.

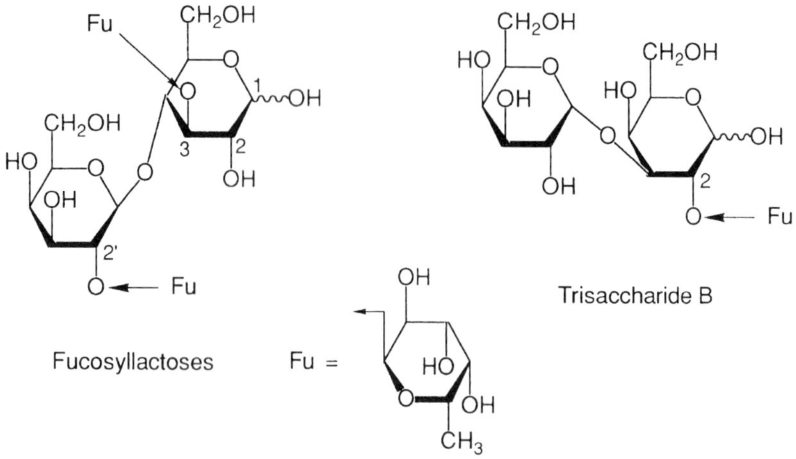

Scheme 3

Kinetic-energy release (KER) measurements for the distinction of structural and chiral isomers of metal-oligosaccharides have been reported by the same group (63). In this case, cobalt complexes of two different isomers, laminaribiose and nigerose, were differentiated by their different KER values. Whether this methodology proves to be useful has yet to be resolved.

Complex glycosphingolipids, such as β-D-Gal-$[1 \rightarrow 4]$-(β-L-Fuc-$[1 \rightarrow 3]$)-β-D-GlcNAc-$[1 \rightarrow 3]$-β-D-Gal-$[1 \rightarrow 4]$-β-D-Glc-ceramide (Le^x-LaCer), form homodimers in the presence of divalent cations (Mg^{2+}, Ca^{2+}, and Mn^{2+}) that

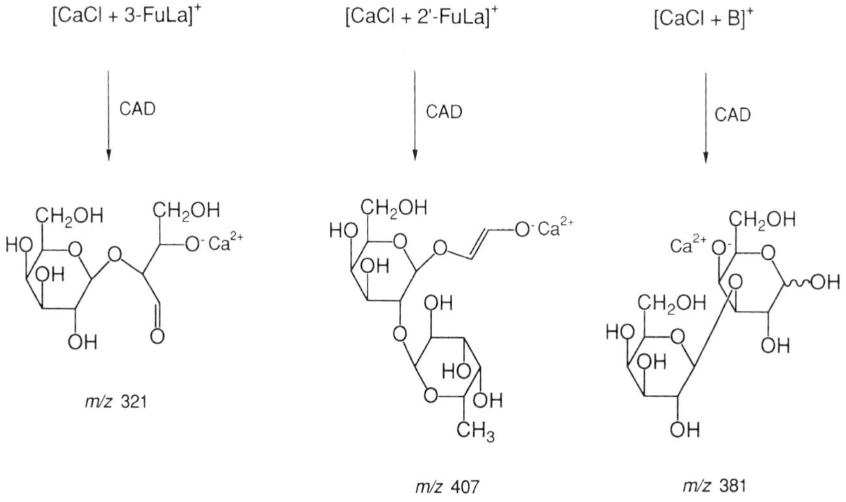

Scheme 4

are detected as doubly charged ions by ESI–MS (64). It has been suggested that interactions of glycosphingolipids occur during cellular adhesion prior to protein–carbohydrate and protein–protein interactions. The existence of these complexes as stable species provides supporting evidence as to the function of Ca^{2+} in interactions of cell-surface carbohydrates. The gas-phase complexes eliminate an entire LacCer unit on CAD, but also show a few sequence-specific fragment ions due to glycosidic bond cleavages (Scheme 5) (64). ESI–MS has been applied to the analysis of acylglycerol mixtures (65), which utilized formation of Na^+ complexes.

There have been few analyses of metal–ion interactions with oligonucleotides by ESI–MS. Instead, a great deal of effort has been applied to desalting biological samples of DNA/RNA prior to ESI–MS analysis. This is a serious problem, because the phosphate anions in oligonucleotides readily form ion pairs with alkali-metal cations, which decreases the overall negative charge in ESI. This in turn results in increased ion m/z values and diminished detection efficiency. Besides desalting, other factors such as pH value and choice of solvent composition can improve signal intensity and allow determination of correct molecular mass (66). Cheng and co-workers (67) have probed some metal–oligonucleotide structures using CAD in a FTICR–MS. Upon CAD of the multiply charged metal–oligonucleotide anionic complexes, the Na^+, Mg^{2+}, and UO_2^{2+} counterions were found to affect the dissociations of the oligonucleotide anions differently. This suggests a different nature of interaction for each of the metals. The dissociation patterns of the metal–oligonucleotide anions were also indicative of locations of metal cations and higher-order structures. Thus, there exists a potential to probe metal-binding site and higher-order structures of metal-bound oligonucleotides by CAD of their electrospray-generated ions.

C. Polyether Antibiotics and Toxins

Complexation of alkali and alkaline earth-metal ions by natural ionophores has been utilized for the analysis by ESI–MS. For example, coccidiostatic polyether

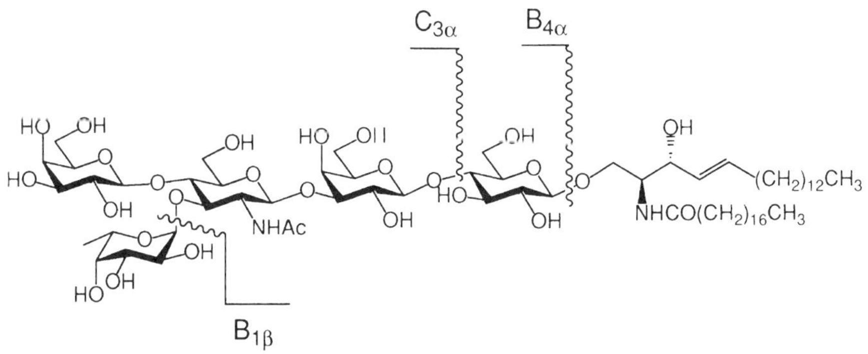

Scheme 5

ionophores semduramicin (Sem, **6**) lasalocid (Las, **7**), salinomycin (Sal), and maduramicin (Mad) form abundant gas-phase adducts with Na^+, K^+, and Cs^+ by electrospray of their $CH_3CN/THF/H_2O$ solutions containing alkali metal salts (68). CAD at $E_{LAB} = 49$ eV of the sodium adducts $(Sem + Na)^+$ and $(Mad + Na)^+$ resulted in decarboxylation followed by elimination of water, but did not reveal the structural differences between the ionophores. An important result of this study was an increase by up to two orders of magnitude of the ESI ion currents due to sodium-ion adducts, allowing detection of 50 pg of **6**.

Brevetoxins, produced by the marine dinoflagellate *Gymnodinium breve*, cause massive poisoning of fish and shellfish during the blooming periods known as "red tides." Upon ESI, the brevetoxins form several complexes with spurious Na^+ and K^+ present in solution (69). Besides 1:1 adducts with Na^+ and K^+, which are the predominating ion species, the ESI spectra also show larger complexes with brevetoxin:metal-ion stoichiometries of 3:2 and 2:1. Such complexes have not been observed for some other ionophore polyether toxins such as okadaic acid and dinophysistoxin-1 (70, 71). Detection limits in the $10^{-7}-10^{-6}$ M range have been achieved for the 1:1 sodium adducts (69).

In another recent ESI–MS study, alkali-metal complexes of the peptide ionophore valinomycin have been investigated (72).

6
Semduramicin, $(M + Na)^+$, *m/z* 895

7
Lasalocid, $(M + Na)^+$, *m/z* 613

V. ELECTROSPRAY IONIZATION OF NONPOLAR MOLECULES VIA METAL COMPLEXES

Complexation with metal ions in solution, followed by efficient transport of charged complexes to the gas phase by ESI, represents a promising method for ionization of a variety of analytes, including those that are not directly amenable to ESI. Wilson and co-workers have developed a methodology for ESI–MS analysis of nonpolar compounds, such as fullerenes and some steroids. The nonpolar molecule is first covalently tethered to an auxiliary ligand that is capable of coordinating a metal cation. For example, the vitamin D metabolite calcitriol (**8**) is treated with N-(benzo-15-crown-5)-1,2,4-triazole-3,5-dione to give a Diels-Alder adduct (**9**, Scheme 6). The crown-ether ring in the latter coordinates Na^+ in solution to give rise to an abundant ion by ESI. Detection limits approaching 10^{-6} M of adduct **9** have been achieved, which could potentially lead to detection of vitamin D and its metabolites at physiological levels (73).

A similar strategy was used for detection by ESI of functionalized fullerenes (74). A benzo-18-crown-6-ether ring was attached to C_{60} by carbene addition and the product **10** was analyzed as a potassium complex by ESI–MS (Scheme 7). Since the functionalized fullerenes undergo reactions similar to those of the parent C_{60}, products of their oxidation (74), methoxylation (75), nitrene addition (76), cyclopropanation (76), and [2 + 2] photolytic cycloaddition with enones (77) can readily be monitored by ESI. This provides molecular-weight information, which is invaluable for the determination of the number of reactive molecules attached to the fullerene core (76). Attachment to C_{60} of 4-(1,3-butadienyl)benzo-18-crown-6-ether by Diels–Alder reaction has also been achieved (78), although the yield of the adduct (25%) is lower than in the carbene addition (42%) (74).

Binding the analyte in a charged-metal complex can be beneficial even for molecules that are amenable to analysis by ESI–MS under standard conditions of protonation or deprotonation (e.g., amino acids and peptides) (79, 80), as discussed later for Cu complexes.

VI. METAL COMPLEXES WITH PROTEINS, PEPTIDES, AND AMINO ACIDS

Metal ions play an essential role in the ways many proteins function. Of fundamental interest is the structure and stability of the metal–protein complex itself. Among the more traditional methods of metal–protein structural analysis, such as X-ray crystallography, EXAFS, NMR, EPR, absorption spectroscopy, and circular dichroism, mass spectrometry is the newest addition, and its use has been growing rapidly. Gross (81–83), Adams (84–87), and Leary (88,89) have reported extensively on singly charged binary metal–peptide complexes produced by FAB ionization. Assuming that the metal ion induces binding site-

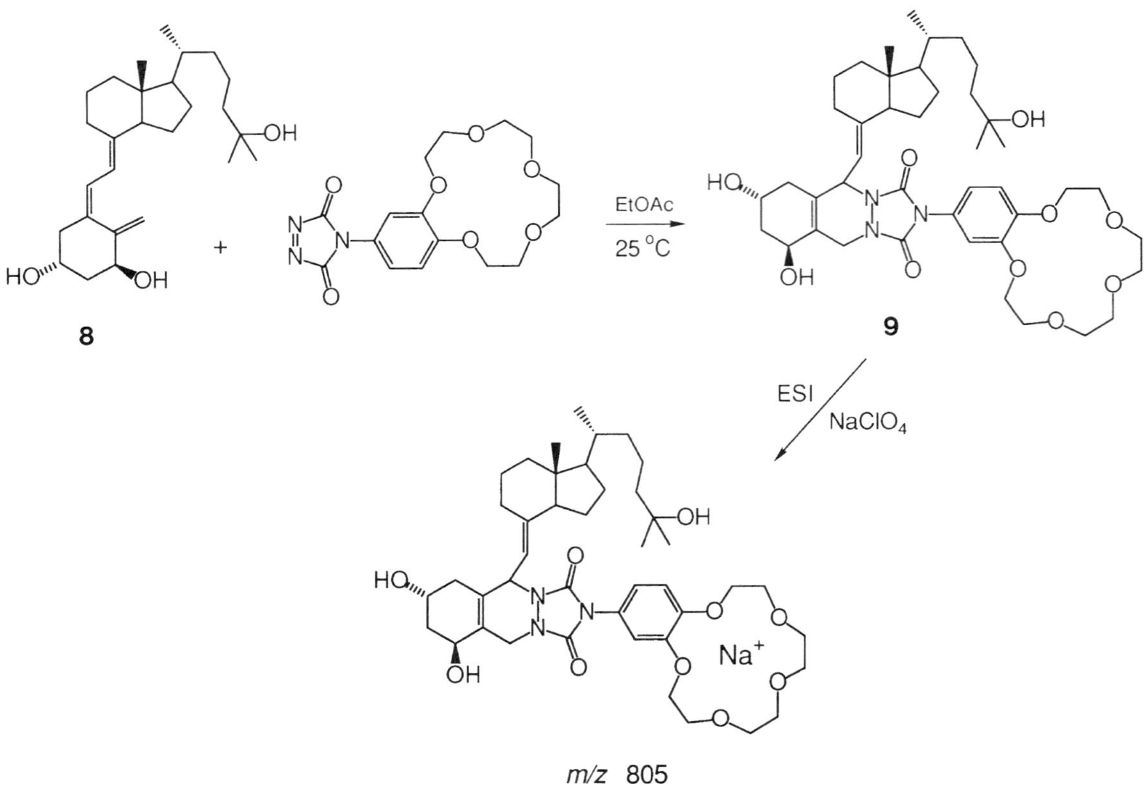

Scheme 6

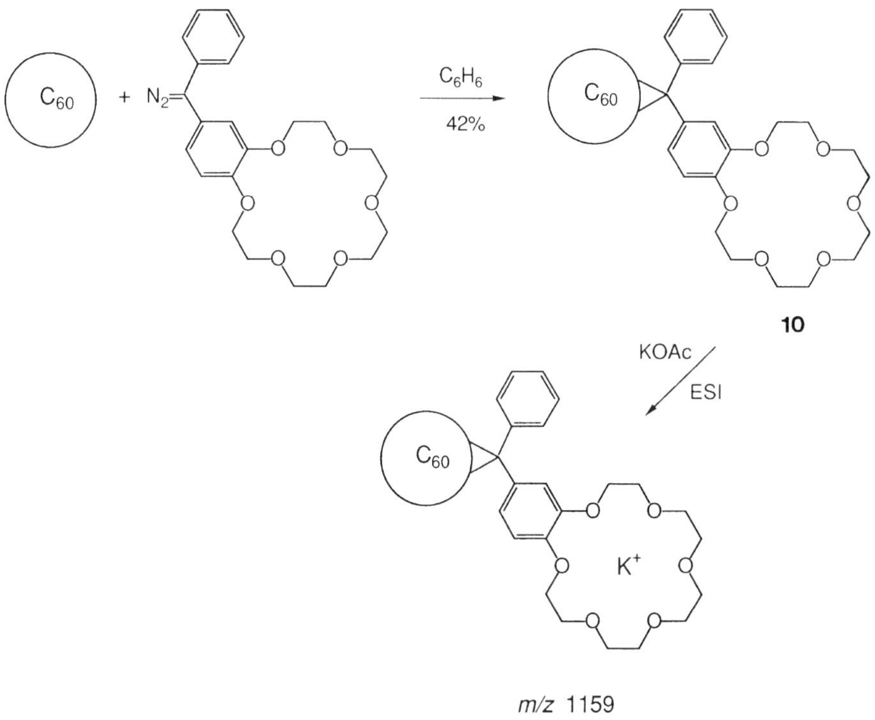

Scheme 7

specific fragmentation, information about the intrinsic metal–peptide interactions have been extracted from the dissociation chemistry. Using ESI–MS allows one to study metal–protein interactions in solution and also brings a substantial flexibility into the preparation of metal complexes with peptides and amino acids. This is especially true of gas-phase ternary metal–amino acid and metal–peptide complexes, which are not produced by FAB or laser desorption, and which are of interest because of their low detection limits and useful fragmentations.

A. Binary Protein Complexes

Although there have been a number of studies of metal-ion interactions with peptides and proteins, most have focused on the detection of metalloproteins and stoichiometry of metal binding. Hutchens and co-workers (90) have reported numerous examples of ESI–MS detection of metal-ion binding to peptides/proteins with clear determination of stoichiometry. Examples include $Cu^{(II)}$ and $Zn^{(II)}$ binding to Cys-rich synthetic peptides (91), His-rich synthetic peptides (92), and Zn-finger proteins (93,94). Other literature examples of detection by ESI–MS of metalloproteins include Fe-S proteins (95–97), Zn-finger proteins (98,99), $Cu^{(II)}$-ubiquitin (100), $Ca^{(II)}$ proteins (101–103), and metallothionein (104). In all cases, the metal-ion stoichiometry was established, but not necessarily the metal-ion oxidation state.

An example, in which the metal oxidation state has been determined, is given in Figure 3, which shows a portion of the high-resolution FT–ICR spectrum of a +12 charged Cu–ubiquitin complex (100). The assignment of the Cu oxidation state is based on the 1/12 u difference in the m/z for [ubiquitin + 11H + Cu$^{(I)}$]$^{12+}$ and [ubiquitin + 10H + Cu$^{(II)}$]$^{12+}$, and on the frame shift in the distribution of ^{13}C, ^{63}Cu, and ^{65}Cu isotopomers. It is obvious that the spectrum is difficult to interpret unambiguously, and the assignment has been made on the basis of peak relative intensities, which showed a somewhat better fit for the theoretical isotope distribution in [ubiquitin + 10H + Cu$^{(II)}$]$^{12+}$ (100).

Reactivity studies involving interactions of metalloproteins with organic substrates is another area of possible ESI–MS application. A recent report by Sam and co-workers (105) is the first of its kind. Electrospray MS and EPR

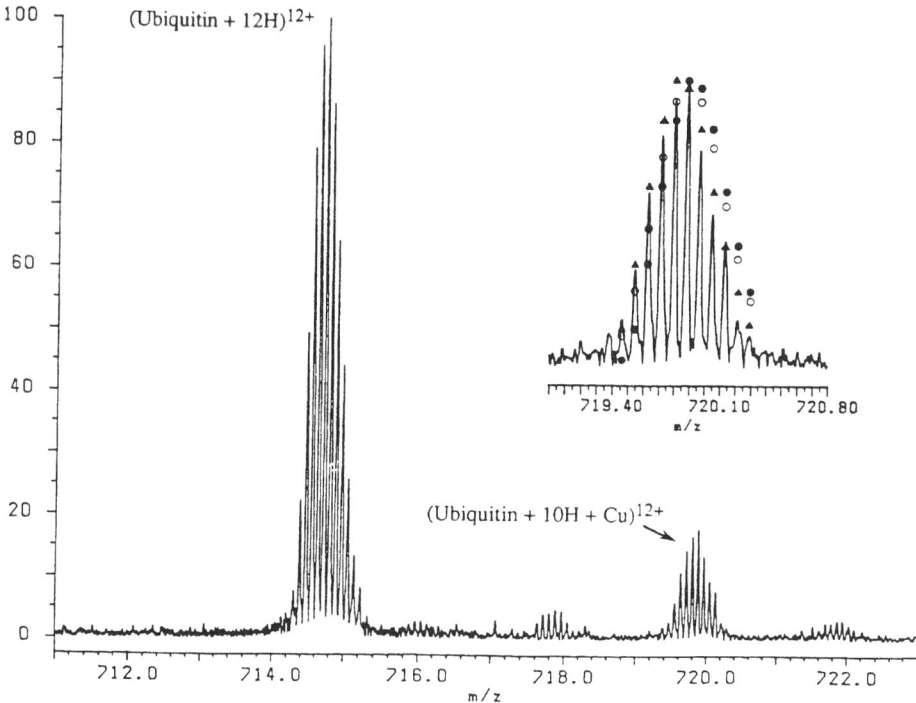

Figure 3. High-resolution ESI mass spectrum of a mixture of ubiquitin and CuBr$_2$ in methanol. The inset is an enlarged portion showing adduct isotope peaks along with the calculated data for [ubiquitin + 10H + Cu]$^{12+}$ (filled triangles), [ubiquitin + 11H + Cu]$^{12+}$ (filled circles), and [ubiquitin + 10H + Cu]$^{12+}$ with 25% of [ubiquitin + 10H + Na + K]$^{12+}$ (open circles). Reprinted by permission of Elsevier Science Inc. from "An Electrospray Ionization Mass Spectrometry Study of Copper Adducts of Protonated Ubiquitin," by C. Q. Jiao et al., *J. Am. Soc. Mass Spectrom.*, **1995**, *6*, 522. © 1995 American Society for Mass Spectrometry.

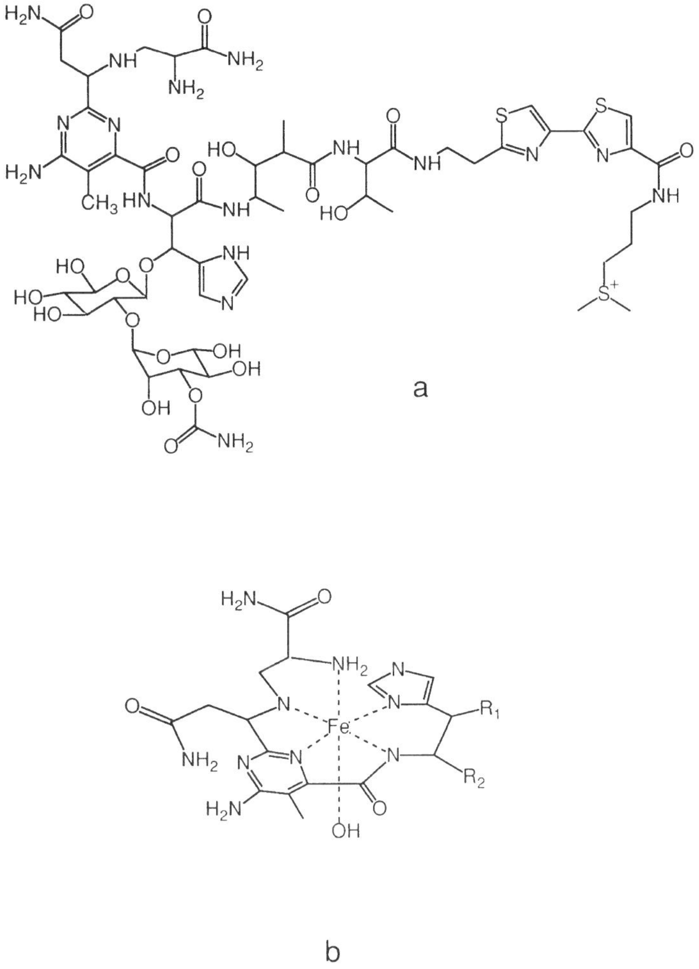

VI. METAL COMPLEXES WITH PROTEINS, PEPTIDES, AND AMINO ACIDS   547

Figure 4. Structures of (*a*) bleomycin and (*b*) the iron-binding complex.

spectroscopy were used to investigate the reaction of ferric bleomycin (Fe^{3+}BLM) and iodosobenzene (PhIO). Bleomycin A2 (Fig. 4*a*) is a glycopeptide antibiotic, which, after binding iron or cobalt (Fig. 4*b*), activates O_2 and cleaves DNA and RNA by attacking the sugar moiety. When reacting $Fe^{(III)}$BLM with PhIO, several product ions were observed in the ESI mass spectrum, which were identified by their masses, but HOO-$Fe^{(III)}$BLM was not detected. Instead, the intermediates observed by ESI–MS suggested a mechanism in which the metal-ion was not required for the oxidation of BLM by PhIO (105). This example shows that ES–MS has great potential to provide supporting or refuting evidence, as well as mechanistic insights, into reactivity studies involving metal-containing biomolecules.

B. Binary Peptide Complexes

Several authors have investigated the use of ESI–MS to probe metal–peptide interactions and possibly to obtain peptide sequence information. Loo and co-workers (106) have utilized ESI–MS to probe the interaction of angiotensin

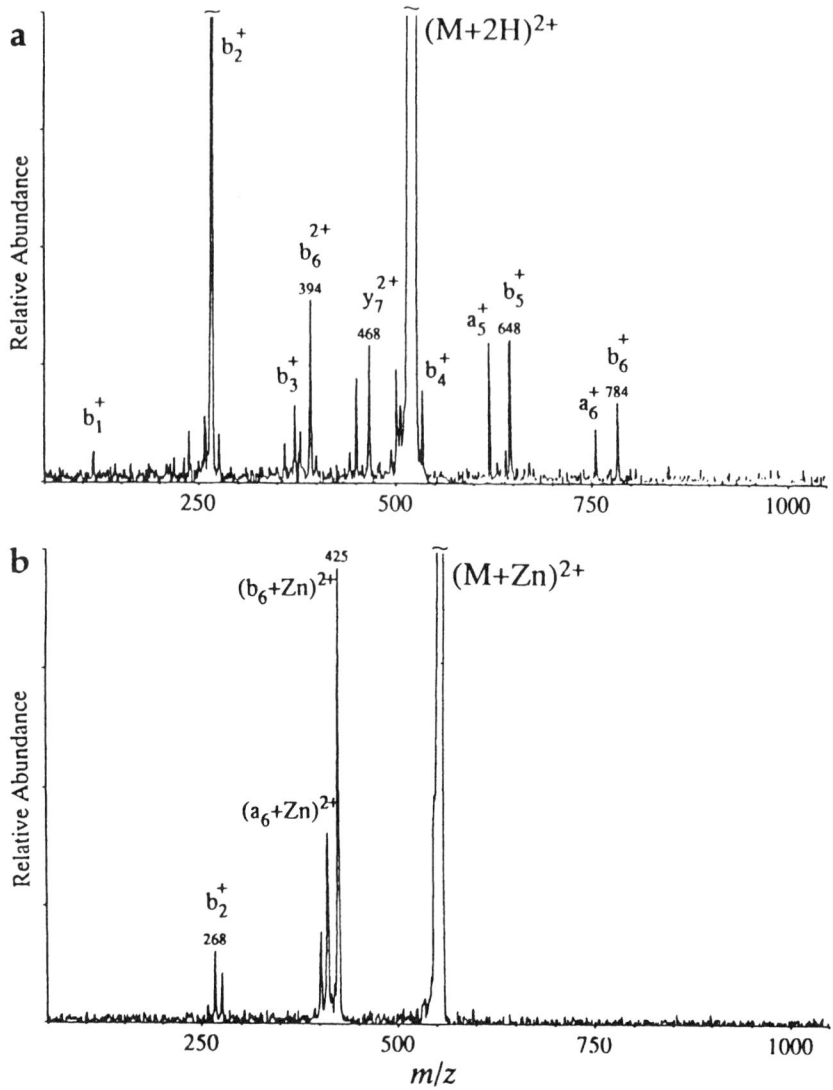

Figure 5. The ESI–CAD mass spectra of the (*a*) [M + Zn]$^{2+}$ and (*b*) [M + 2H]$^{2+}$ ions from angiotensin II, acquired on a tandem quadrupole instrument. Reprinted by permission of Elsevier Science Inc., from "Interaction of Angiotensin Peptides and Zinc Metal Ions Probed by Electrospray Ionization Mass Spectrometry," by J. A. Loo et al., *J. Am. Soc. Mass Spectrom.*, **1994**, *5*, 962. © 1994 American Society for Mass Spectrometry.

peptides and Zn^{2+} ions. In the presence of zinc, histidine-containing angiotensin II (DRVYIHPF) and angiotensin I (DRVYIHPFHL) show abundant multiply charged zinc ions in the ESI mass spectra. The CAD of the $[M + Zn]^{2+}$ ion for angiotensin II shows an abundant $[b_6 + Zn]^{2+}$ ion (Fig. 5b), which indicates that Zn^{2+} binds in the vicinity of the His_6 residue. Furthermore, abundant $[b_6 + Zn]^{2+}$ and $[b_9 + Zn]^{2+}$ CAD fragment ions suggest both His_6 and His_9 are involved in zinc coordination. This is not surprising because His is most frequently found in the coordination sphere of Zn-containing proteins. For comparison, Figure 5a shows the ESI–CAD mass spectrum of the angiotensin II $[M + 2H]^{2+}$ ion, which reveals a complete series of b ions. It is clear that metal complexation by Zn^{2+} dramatically reduces the number of CAD fragmentation channels for the peptide. In contrast, when zinc is added to a nonhistidine-containing peptide, substance P, the major fragmentation channels for [substance $P + Zn]^{2+}$ are the same as for [substance $P + 2H]^{2+}$ (107), although the fragmentation efficiency for the Zn^{2+} complex is dramatically reduced. So far, obtaining sequence information from protonated peptides is far superior than from metal-cationized peptides. The ESI–MS spectra of (Cu-peptide)$^{2+}$ binary complexes were reported recently by Hu and Loo (108).

Siu and co-workers (109,110) have reported on alkali metal-ion interactions with tripeptides. The most abundant ions in the ESI mass spectra are $[M - H + 2X]^+$ and $[M - 2H + 3X]^+$ (X = Li, Na, K) for nine tripeptide solutions adjusted to pH 12 with alkali hydroxides. Based on CAD results of the $[M - H + 2X]^+$ and $[M - 2H + 3X]^+$ ions, their proposed gas-phase precursor ion structures are, respectively, **11** and **12** (Scheme 8). The core structure of **11** and **12**, where the metal is coordinated with the amino and carboxylate terminus and the amide nitrogens, is similar to the structures proposed by Gross (81–83) and Adams (84–87); the difference being that the amide nitrogens in the anionic complexes are deprotonated. The authors present examples in which they were able to distinguish tripeptide isomers by the relative abundance of y_1 and C_1 fragments. The peptide examples all contained amino acids with alkyl side chains for simplicity, and all were sequenced with assistance of the alkali metal. Therefore, metal ions can assist in providing peptide sequence information.

11   **12**

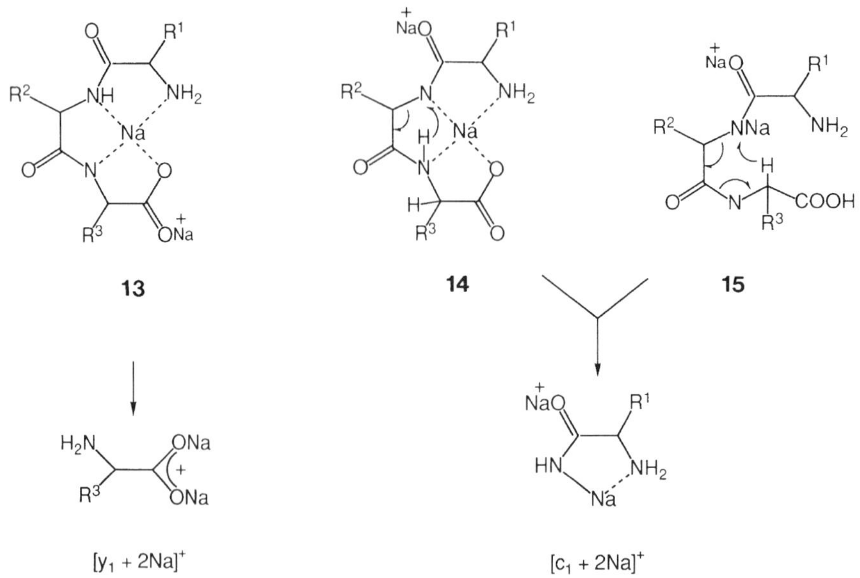

Scheme 8

C. Binary Amino-Acid Complexes

In contrast to the numerous reports of metal–peptide and metal–protein interactions, amino-acid interactions with metal ions have received much less attention. Hoppilliard and co-workers have reported formation and fragmentation of metal complexes involving aliphatic α-amino acids and transition-metal cations, which were produced by ^{252}Cf plasma desorption (111,112). Amster and co-workers have used laser desorption FTICR–MS to investigate the reactions of Cu^+ and Fe^+ with the 20 common amino acids to determine the intrinsic chemistry of metal ions with peptide building blocks (113). Wilson and Tureček used ESI–MS to investigate metal–amino acid complexes, which included 2,2'-bipyridyl (bpy) as a coligand (79,114–118). Wilson and co-workers were interested in binary $Cu^{(I)}$ complexes with amino acids chemically bonded to the bpy moiety, which they detected by ESI–MS (114). Gatlin and Tureček have investigated ternary complexes of $Ni^{(II)}$, $Co^{(II)}$, $Cu^{(II)}$, and $Zn^{(II)}$ containing both diimine and amino-acid ligands, as discussed in the following section (79,115–118).

D. Ternary Complexes

Ternary metal complexes of both amino acids (79,115–118) and peptides (119–121) have recently been obtained by ESI–MS and showed some interesting chemistry. Electrospraying CH_3OH/H_2O (50/50) solutions of

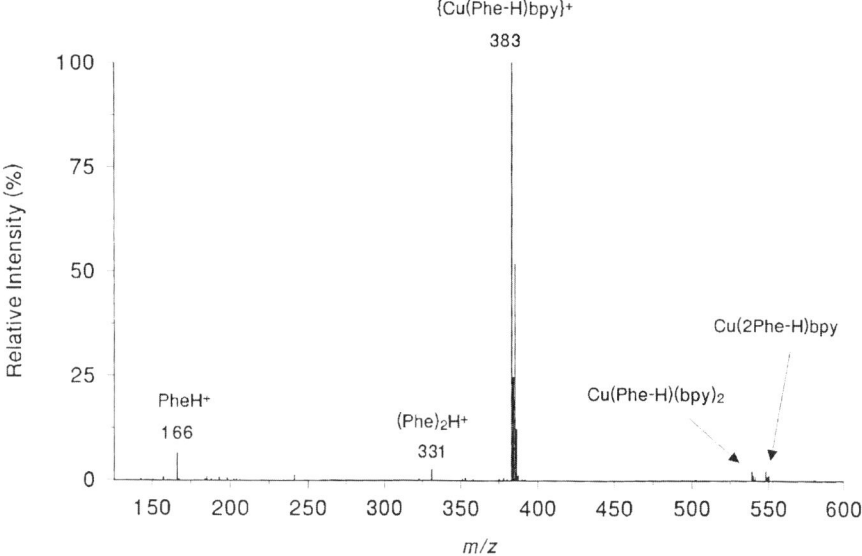

Figure 6. The ESI mass spectrum of $[Cu^{(II)}(Phe - H)bpy]^+$ in 50/50 aqueous methanol acquired on a sector instrument at a low skimmer potential.

amino acids, $CuSO_4 \cdot 5H_2O$, and 2,2'-bipyridine produced abundant gas-phase cations of the form $[Cu^{(II)}(RCOO)bpy]^+$, as shown for phenylalanine in Figure 6. Analogous complexes were obtained with phenanthroline (phen) and 2,9-dimethylphenanthroline. These ternary complexes are known to have large stability constants ($\beta > 10^{16}$) in solution (122). Contrasting this, the binary complex $[Cu^{(II)}(RCOO)RCOOH]^+$, which is formed in solution in the absence of the diimine coligand, gives rise to gas-phase ions that are 100–1000 times less abundant than the tertiary complexes $[Cu^{(II)}(RCOO)bpy]^+$ or $[Cu^{(II)}(RCOO)phen]^+$ obtained at the same amino acid concentration. All 20 common α-amino acids form abundant singly charged ions with $Cu^{(II)}$bpy and $Cu^{(II)}$phen. Cysteine is an exception because it is readily oxidized to cystine in the presence of $Cu^{(II)}$ to form $[Cu^{(II)}(cystine)bpy]^+$ (79). The basic amino acids lysine and arginine form doubly charged ions, $[Cu^{(II)}(RCOO + H)bpy]^{2+}$, owing to protonation of the basic amino-acid side chain. Double charging provides facile distinction of lysine from the isobaric glutamine which forms singly charged ions only.

The proposed structure of the complex involves coordination to $Cu^{(II)}$ of the amino and deprotonated carboxyl termini, as shown in Scheme 9. Methyl esters, in which carboxylic group deprotonation is blocked, do not form stable complexes with $Cu^{(II)}$bpy to be detected by ESI (117,118). Other carboxylic acids (formic, acetic, propionic, trifluoroacetic, lactic, γ-aminobutyric, β-alanine) also form singly charged complexes of the $[Cu^{(II)}(RCOO)bpy]^+$ type by ESI. In collisional activation of the $[Cu^{(II)}(RCOO)bpy]^+$ complexes, the first

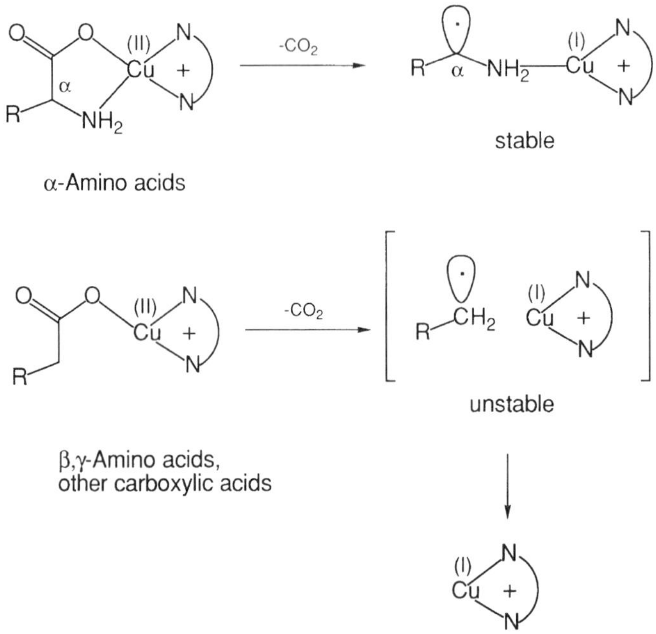

Scheme 9

α-Amino acids

β,γ-Amino acids, other carboxylic acids

step is decarboxylation to give $[Cu^{(I)}(R^{\cdot})bpy]^+$ ions (Scheme 10, m/z 305). For aliphatic amino acids, further dissociations of the α-amino alkyl radical complex $[Cu^{(I)}(R^{\cdot})bpy]^+$ involve fission of the C_β–C_γ bond of the alkyl chain which gives information on its branching. This β-fission allows one to distinguish isomeric amino acids leucine and isoleucine. In Scheme 10, leucine readily loses a $C_3H_7^{\cdot}$ radical, while β-fission products of isoleucine result in losses of $C_2H_5^{\cdot}$ and $CH_3^{\cdot}$. This unambiguous distinction of the two isomers occurs at collision energies as low as 5 eV (79,117).

In the ternary complexes, the auxiliary neutral ligand plays two major roles. First, it occupies two coordination sites of $Cu^{(II)}$ and thus blocks coordination of a second amino-acid carboxylate that would form an electroneutral complex. Tri- and tetradentate neutral ligands, such as 2,2′,6′,2″-terpyridyl and 1,5,8,12-tetraazacyclotetradecane, respectively, also form singly charged complexes of the $[Cu(II)(RCOO)(L)]^+$ by ESI. This is consistent with the coordination flexibility of Cu(II), which can accommodate 4–6 ligands in various geometries about the metal atom (22).

The other important role of the auxiliary ligand is to function as an electron donor for $Cu^{(II)}$ in the ternary complexes and for $Cu^{(I)}$ in the decarboxylated ions and their dissociation products. Electron densities obtained from ab initio calculations of $[Cu^{(I)}bpy]^+$ and several $[Cu^{(I)}(L)bpy]^+$ complexes, where L is an amino-acid fragment, showed the Cu atom to be essentially electroneutral, whereas most of the positive charge was delocalized within the bpy molecule

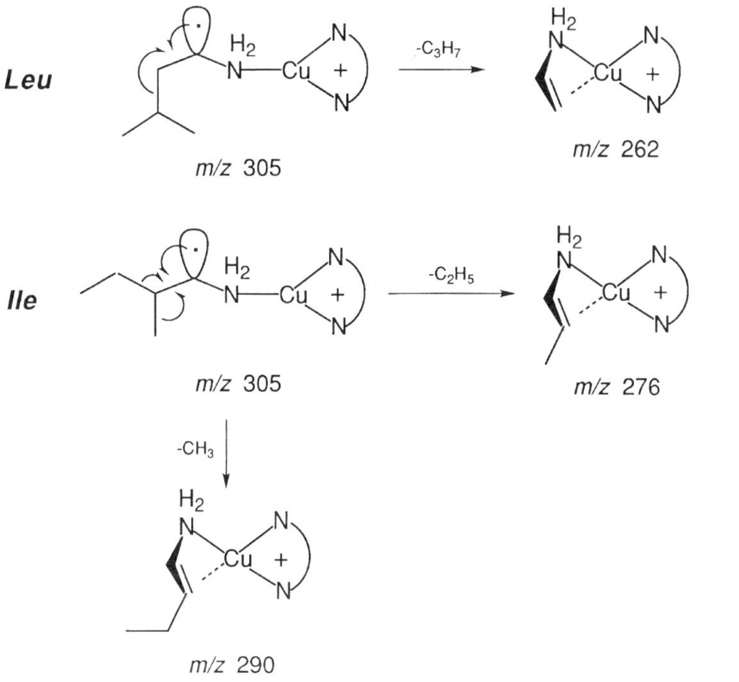

VI. METAL COMPLEXES WITH PROTEINS, PEPTIDES, AND AMINO ACIDS 553

Leu

m/z 305 → (−C$_3$H$_7$) → m/z 262

Ile

m/z 305 → (−C$_2$H$_5$) → m/z 276

↓ −CH$_3$

m/z 290

Scheme 10

(116). This makes the Cu$^{(I)}$ atom in [Cu$^{(I)}$bpy]$^+$ a very weak Lewis acid, which can form stable tricoordinated or tetracoordinated complexes. Hence, amino acid or peptide residues containing one or more donor ligands in the form of NH$_2$, SR, OR, imidazole, guanidine, or peptide groups can bind to [Cu$^{(I)}$bpy]$^+$ in stable gas-phase complexes. Since Cu(I) has a closed d^{10} valence shell, reactions typical for high-spin transition metals (e.g., C–H and C–C insertions and dehydrogenation) do not occur.

An interesting reaction occurs within the amino-acid ligand for the OH-containing serine and threonine. The Cu$^{(II)}$bpy complexes of these two amino acids do not readily decarboxylate, but instead lose an aldehyde molecule by aldol retrogression (Scheme 11). The CAD chemistry of all 20 α-amino acid

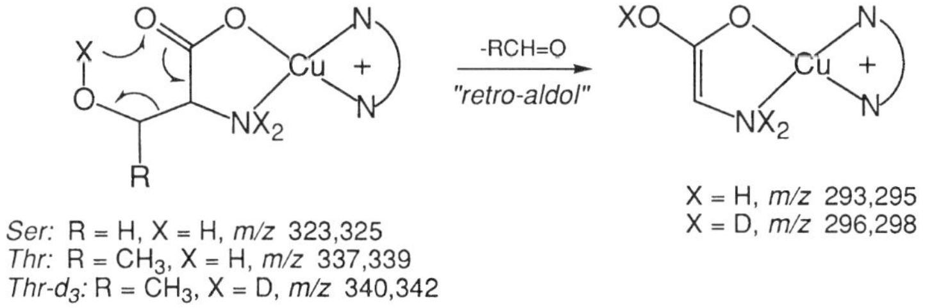

−RCH=O
"retro-aldol"

Ser: R = H, X = H, m/z 323,325
Thr: R = CH$_3$, X = H, m/z 337,339
Thr-d$_3$: R = CH$_3$, X = D, m/z 340,342

X = H, m/z 293,295
X = D, m/z 296,298

Scheme 11

complexes with $Cu^{(II)}$bpy has been documented, and the suggested ion structures have been supported by deuterium labeling of the active hydrogens and by CAD analysis of the intermediates (117,118).

Singly and doubly charged peptide complexes with $Cu^{(II)}$bpy are readily formed in the gas phase by electrospraying their methanol–water solutions (119–121). Collision-activated dissociation at low kinetic energies of the singly charged complexes, $[Cu^{(II)}(peptide - H)bpy]^+$, provides information on the amino-acid sequence for dipeptides such as *Phe-Leu, Leu-Phe, Phe-Pro, Pro-Phe, Phe-Met, Met-Phe, Ser-Phe, Asp-Phe*, and *His-Phe*. Dissociations of doubly charged complexes $[Cu^{(II)}(peptide)bpy]^{2+}$ also allow identification of the C- and N-terminal amino-acid residues (121). Two examples of distinguishing dipeptide isomers will be presented.

The $Cu^{(II)}$bpy complexes of *Leu-Phe* and *Phe-Leu* (*LF* and *FL*) can readily be distinguished based on their CAD spectra (121). The proposed CAD mechanisms are shown in Schemes 12 and 13. For $[Cu^{(II)}(LF - H)bpy]^+$ (Scheme 12), the fragmentations support a coordination of the dipeptide to the copper center in an analogous fashion as with the amino acids, that is, through the carboxylate terminus and the amide or amine nitrogen. Both *LF* and *FL* methyl esters, in which the C-terminal carboxylic groups are blocked, do not form complexes with $Cu^{(II)}$bpy (121). The *LF* complex readily decarboxylates (m/z 452) as do all the $Cu^{(II)}$–bpy–dipeptide complexes. An abundant sequence ion at m/z 304 is formed, which corresponds to a Cu(bpy)–imine complex from the leucine residue. This dissociation is analogous to an *a*-type cleavage with concomitant transfer of the Cu(bpy) residue to the N-terminus with reverse transfer of one hydrogen atom onto the neutral fragment, which is lost. In the CAD spectrum of the other isomer, $[Cu^{(II)}(FL - H)bpy]^+$, the mass of the *a*-type sequence ion (m/z 338) is indicative of the N-terminal Phe (Scheme 13). Loss of the benzyl side chain ($C_7H_7^·$) of *Phe* only occurs when *Phe* is at the N-terminus (m/z 405). The C-terminal *Leu* is confirmed by the peak at m/z 409, which results from β-fission and loss of the $C_3H_7^·$ radical from the *Leu* side chain after decarboxylation.

Isomers *Ala-Leu* and *Ala-Ile* (*AL* and *AI*) both form singly (m/z 420) and doubly (m/z 210.5) charged ions with $Cu^{(II)}$bpy. While the ESI spectra of the

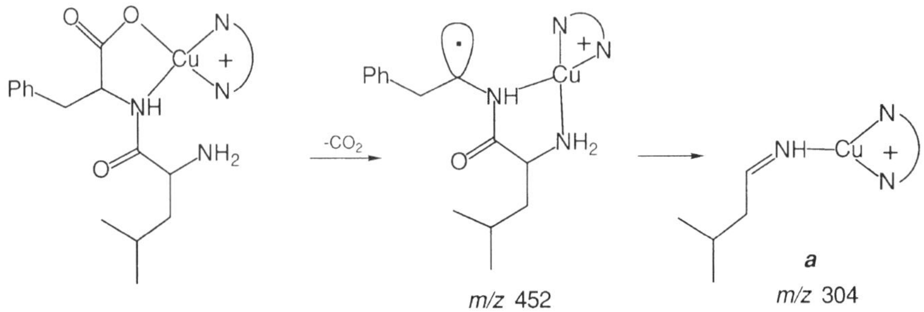

m/z 452 m/z 304

Scheme 12

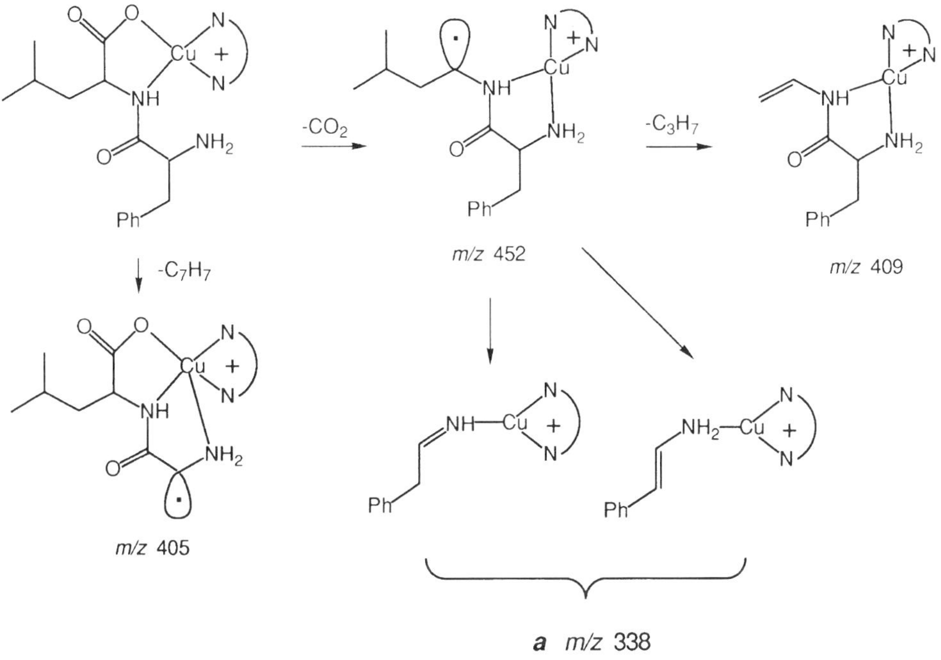

Scheme 13

dipeptide complexes are very similar, the CAD spectra of the m/z 420 ions from $[\text{Cu}^{(II)}(AL - H)\text{bpy}]^+$ and $[\text{Cu}^{(II)}(AI - H)\text{bpy}]^+$ differ. Following CO_2 loss, the Ala-Leu complex eliminates the $C_3H_7^{\cdot}$ radical from the leucine residue to form the ion at m/z 333 (Fig. 7a). The a-type sequence ion, corresponding to N-terminal *Ala*, appears at m/z 262. The *Ala-Ile* complex shows the same a-type ion for *Ala* at m/z 262 (Fig. 7b). In addition, the decarboxylated *Ala-Ile* complex (m/z 376) undergoes β-fission in the *Ile* residue, which results in losses of C_2H_5 (m/z 347) and CH_3 (m/z 361). These side-chain fragmentations in the singly charged ions clearly distinguish C-terminal leucine and isoleucine in the dipeptides (119,121), and similar results have been obtained for a series of tripeptides (120). The chemistry of gas-phase $\text{Cu}^{(II)}$bpy complexes with larger peptides is currently being explored by the same authors to establish the usefulness of $\text{Cu}^{(II)}$bpy ions of peptides in providing sequence information.

In summarizing this section, the examples presented show clearly that ESI–MS has much to offer to the study of metal-ion interaction with biomolecules. Because of its "soft" nature and high molecular-weight capability, ESI–MS has allowed for detection, stoichiometry determination, structural assignments, and reactivity studies of metal-biomolecule complexes. Metal-assisted sequencing of small peptides has also been attained. It is expected that studies of metal–biomolecule interactions by ESI–MS will continue to flourish and support (or refute) findings from the more traditional techniques including

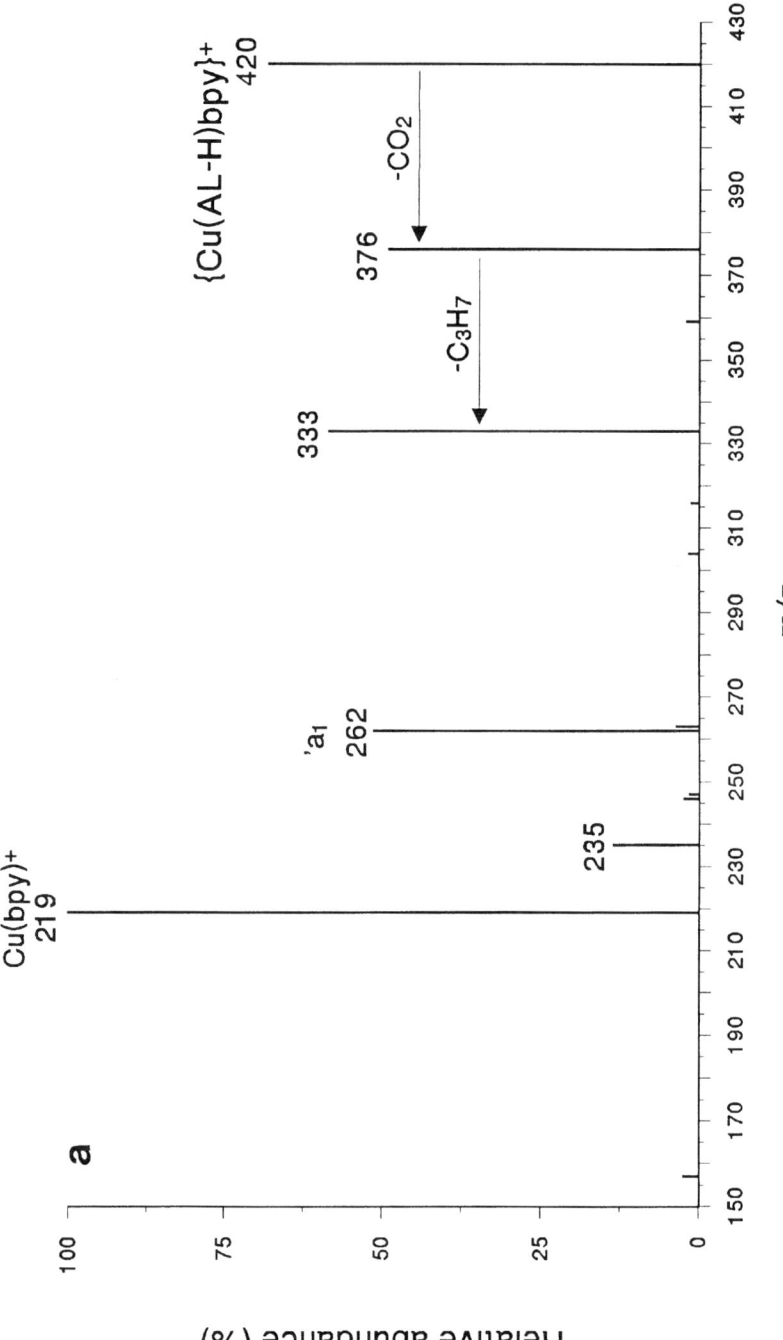

Figure 7a

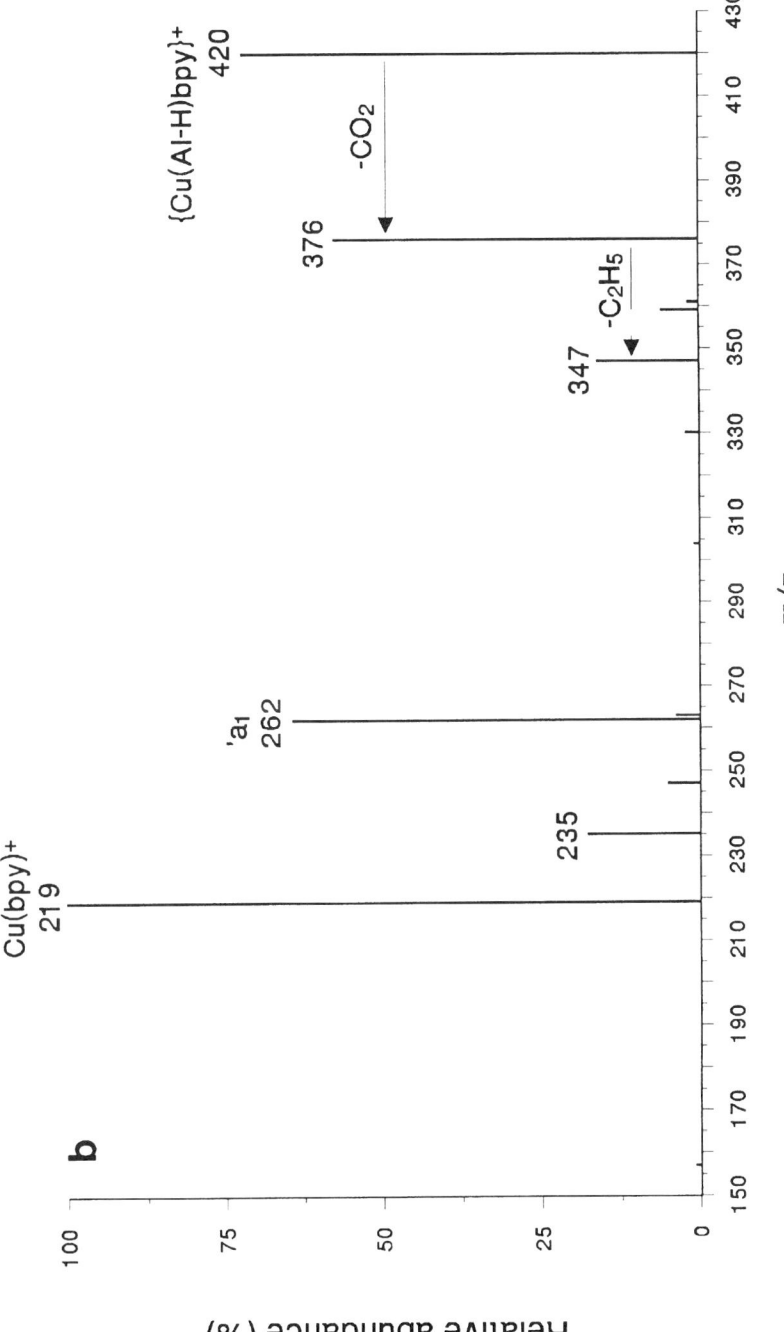

Figure 7b. The CAD spectra at $E_{LAB} = 15$ eV of mass-selected (a) $[^{63}Cu^{(II)}(Ala\text{-}Leu - H)bpy]^+$ and (b) $[^{63}Cu^{(II)}(Ala\text{-}Ile - H)bpy]^+$.

NMR, EPR, absorption, spectroscopy, circular dichroism, EXAFS, and X-ray crystallography.

VII. SUPRAMOLECULAR METAL COMPLEXES

Owing to the extremely wide span of binding energies of metals to carbon, nitrogen, oxygen, sulfur, and phosphorus ligands (123), gas-phase metal complexes represent a more or less smooth transition between the bonding extremes of covalent molecules on the one hand and noncovalent complexes on the other. Noncovalent complexes with suitable haptenes of large molecules, such as proteins and nucleic acids, have been obtained by ESI in the gas-phase (124,125), indicating that the conditions for ion transfer from electrospray droplets to the gas-phase are mild enough to preserve weakly bound species. The organometallic analogs of supramolecular complexes range from loosely bound complexes with "hard" ligands up to large clusters in which coordination to transition metal-ions of polydentate ligands allows for assembly of monomer units.

Weak complexes of small molecules with Na^+, Cu^{2+}, Ni^{2+}, and Co^{2+} have been observed by ESI under special conditions. For example, $Na^+(H_2O)_x(N_2)_y$ complexes are formed by expansion of $Na^+(H_2O)_x$ clusters in nitrogen gas (7). $Cu(bpy)^+$, $Ni(bpy)^+$, and $Co(bpy)^+$ ions, which are formed by CAD of ternary $[Metal(L)bpy]^+$ complexes, bind N_2 at 10–15 Torr in a high-pressure ESI interface to give rise to $[Metal(N_2)bpy]^+$ ions that dominate the ESI mass spectra (115). Similar coordination in the ESI interface is observed for $Cu(bpy)^+$ with CO_2 and O_2 (126). To stabilize such small, weak complexes of binding energies in the 50–100 kJ/mol range (115,127,128), collisional cooling in the expanding bath gas is necessary to lower the internal energy of the gas-phase ions (7,115).

Polydentate heterocyclic ligands (e.g., L^3, L^7, L^8) (129) form self-assembled binuclear complexes with transition-metal ions in acetonitrile solution (130). The ESI of these complexes gives rise to multiply charged ions, as shown for $[Co_2(L)_3]^{4+}$ (Fig. 8). The structure of the gas-phase $[Co_2(L^7)_3]^{4+}$ is thought to correspond to the known crystal structure, which consists of three tetradentate heterocycle molecules bundled in a triple-helical arrangement to provide six ligand sites for each octahedral Co^{2+} ion (129). A CAD spectrum of $[Co_2(L^7)_3]^{4+}$ has been reported (131).

Binuclear double-helical and trinuclear-toroidal complexes with polydentate ligands (L^7, L^8) of Cu^+, Co^{2+}, Fe^{2+}, Ni^{2+}, La^{3+}, Eu^{3+}, Gd^{3+}, Tb^{3+}, and Lu^{3+} ions have also been obtained by ESI of acetonitrile solutions, and collision-activated dissociations were studied for some of these complexes under high-pressure conditions in the ESI interface (131).

Other types of fascinating supramolecular organometallic species that have been characterized by ESI–MS include $Pd^{(II)}$[2]catenanes (132), $Fe^{(II)}$–$Ag^{(I)}$ heteronuclear helical complexes (133), capped $Cu^{(I)}$ trinuclear complexes (134), and multiring catenates held together by $Cu^{(I)}$–diimine coordination (135).

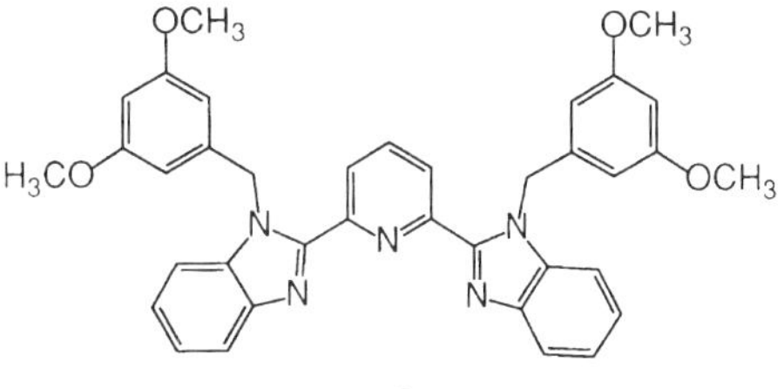

L³

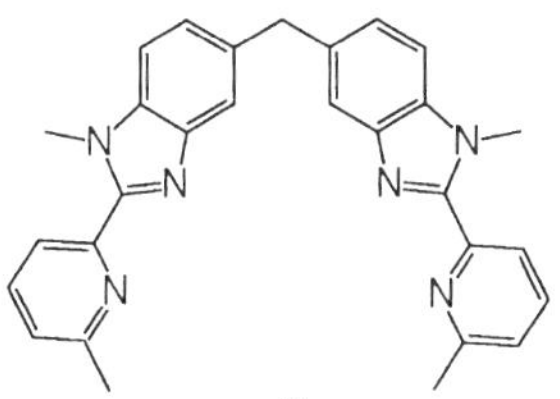

L⁷

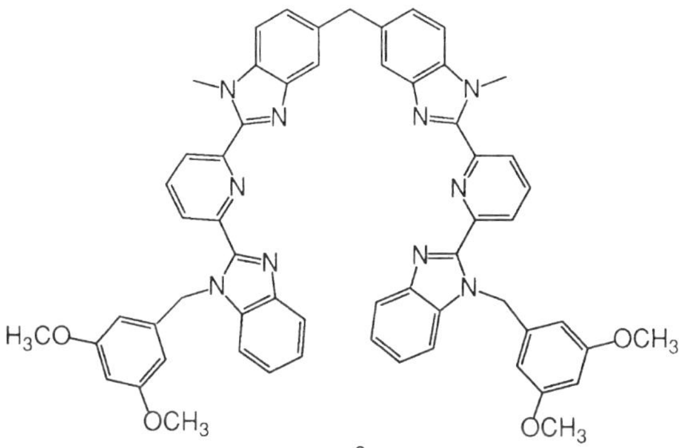

L⁸

A significant advantage of ESI when compared with other mass-spectrometric ionization techniques (FAB, laser desorption, etc.) is that the metal complexes can be made and sampled in solvents in which they are stable and their properties can be studied by other spectroscopic methods, such as NMR or UV–VIS. This minimizes matrix interferences, metal reduction, or demetallation of the complexes, and lends credibility to the proposed structures of gas-phase ions.

Binuclear complexes are formed by ESI of solutions containing ternary complexes of amino-acids or peptides with Cu^{2+} and bpy (117). For example, a binuclear complex is obtained by ESI of aspartyl-phenylalanine (*Asp-Phe*), in which the C-terminal carboxylate and the *Asp* carboxylate each coordinate one $Cu^{(II)}$bpy unit to give a doubly charged ion, $[(Cu^{(II)}bpy)_2(Asp\text{-}Phe-2H)]^{2+}$. Although a detailed structure of the gas-phase complex is unknown, structural studies showed that free carboxylates in the peptide are necessary for coordination to Cu(bpy) (121).

Preliminary results have been reported on the ESI–MS detection of $Ru^{(II)}$–diimine complexes associated with 12-mer and 15-mer oligonucleotide duplexes (136,137). Both single- and double-strand nucleotides form noncovalent complexes with two equivalents of $[Ru^{(II)}(phen')_2dppz]^{2+}$, which are detected in the gas-phase as multiply charged negative ions. The interactions with nucleotides of Ru–diimine complexes in solution have been studied in detail (138), and different

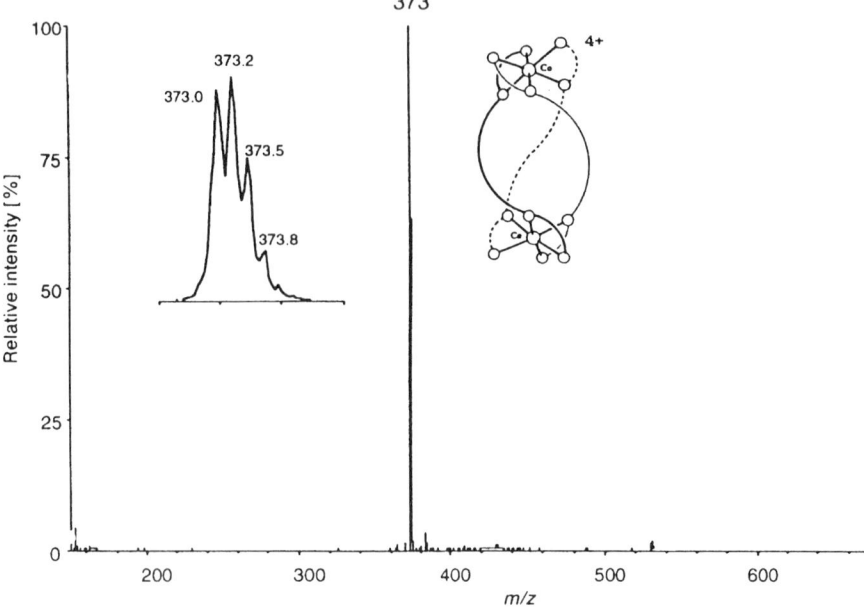

Figure 8. The SI mass spectrum of $[Co_2(L^7)_3]^{4+}$ in acetonitrile. Reproduced by permission of Helvetica Chimica Acta from *Helv. Chim. Acta*, **1993**, *76*, 1761. © 1993 Verlag Helvetica Chimica Acta AG.

binding affinities have been found for $Ru^{(II)}(bpy)_3$ (**16**), $Ru^{(II)}(phen)_3$, and $[Ru^{(II)}(phen')_2dppz]$ (**17**).

VIII. COORDINATION COMPOUNDS AND MISCELLANEOUS APPLICATIONS

Since the report of the first coordination complex detected by ESI–MS in 1990 (8), coordination chemistry studies by ESI–MS have developed substantially. The majority of the work in this area has been carried out by Colton and Traeger, whose focus has been mainly on the organophosphorus and organosulfur chemistry of second- and third-row transition metals. Most of the studies have involved simple detection of metal complexes and extensive listings of the gas-phase ions present, and their relationship to expected ions in solution. The observations of ESI mass spectra of simple coordination complexes have extensively added to the notion that gas-phase ions formed by electrospray are closely related to ions present in solution. In all the ensuing examples, the solution chemistry of the metal complexes is well known (unless otherwise stated), and ESI mass spectra display ions which are highly reflective of the ions existing in solution.

16
[Ru(bpy)₃]²⁺ — $[Ru(bpy)_3]^{2+}$

17
[Ru(phen')₂(dppz)]²⁺ — $[Ru(phen')_2(dppz)]^{2+}$

Colton and co-workers have performed extensive studies on metal phosphine (PR_3) complexes (139–145). For example, a number of $Au^{(I)}$ phosphine derivatives in dichloromethane/methanol were observed by ESI–MS as gas-phase ions of the type $[Au(PR_3)_2]^+$ and $[Au(PR_3)_3]^+$. In solution, rapid ligand exchange on the NMR time scale at room temperature is known to occur for mixed-ligand $Au^{(I)}$ complexes (139). However, ESI–MS is capable of identifying the individual components in such mixtures. To illustrate, when tri-*p*-chlorophenylphosphine (Ppt) ligand is added to a solution of $ClAu(PPh_3)$, a mixture of gas-phase ion products results consisting of $[Au(PPh_3)_x]^+$, $[Au(PPh_3)(Ppt)]^+$, $[Au(PPh_3)_x(Ppt)_{3-x}]^+$, and $[Au(Ppt)_x]^+$. Thus, ESI–MS at room temperature can reveal individual components of a system in which ligand exchange of phosphines is fast on the time scale of NMR. This occurs because ligand exchange stops immediately when solution ions enter the gas phase, so each ion is "frozen" into the original identity at the moment of transfer to the gas phase. In this case, ESI–MS is a method of choice, since room-temperature NMR cannot distinguish these components. Hence in this particular system, ESI–MS is more informative than NMR for characterization.

Other examples include phosphine complexes of Hg(II) (140,141) and Cd (141, 142) and dithiocarbamate complexes of Fe, Co, Ni, Cu (143), Hg (142), and Ag (144). Complexes containing ligands such as CO (145,146), alkyl groups (147–151), alkenes (146,152), arenes (146), and cyclopentadienyls (146,152,153) have been analyzed by ESI–MS. In these studies, ESI–MS of organometallic compounds was able to provide information on structure (145-153), insight into ligand lability (146,147), and qualitative mixture analysis (153). Organometallic reactions have also been monitored by ESI–MS, as in reductive elimination of methyls (146), C–S bond activation (152), and phosphine methylation (145).

Transition-metal complexes with nitrogen-containing ligands have also been detected by ESI–MS, although studies to date have only addressed simple molecular identification and ligand lability. Amine complexes such as cisplatin and its derivatives, $Pt(NR_3)_2Cl_2$, have been detected as protonated and sodiated ions (154). Whether ESI–MS proves to be a sensitive method for analysis of these anticancer drugs in biological samples has yet to be addressed. A number of researchers have studied transition metals containing bidentate diimine ligands such as 2,2′-bipyridine and phenanthroline, of which $[Ru(bpy)_3]^{2+}$ and $[Ru(phen)_3]^{2+}$ were the first metal complexes detected by ESI–MS by Chait and co-workers (8). Other diimine complexes studied include multiply charged ions of $Ru^{(II)}$, $Rh^{(III)}$, and $Co^{(III)}$ (155) and $Cu^{(I)}$ phosphine complexes (156).

The facility with which stable metal–ion complexes are transferred in the gas-phase by ESI has been utilized in numerous analytical applications to various chemical problems. For example, Raney Nickel-catalyzed coupling of 6-methyl-

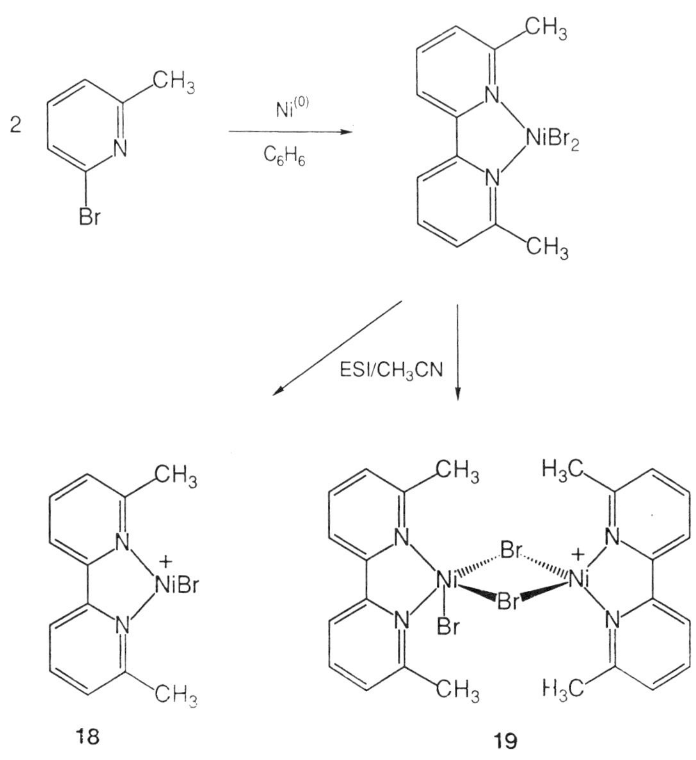

Scheme 14

2-bromopyridine gives a Ni-complex of the dimer 6,6′-dimethyl-2,2′-bipyridine (dmbp, Scheme 14) (157). The ESI mass spectrum revealed the presence of two Ni complexes–a monomer, which manifests itself as a $[Ni^{(II)}(dmbp)Br]^+$ ion (**18**), and a dimer, which gives rise to a $[(Ni^{(II)}dmbp)_2Br_3]^+$ ion (**19**). Loss of the dmbp ligand is the major dissociation of the monomer ion on CAD. In a footnote to this short study, the authors raise the interesting question of whether the complexes detected by ESI–MS correspond to species present in bulk solution, or whether they are formed in the ionization process (157). For another recent application see ref. 158.

Other transition metal–diimine complexes studied by ESI include cyano-bridged trinuclear Ru and Os complexes of the $[(NH_3)_5Ru\text{-}NC\text{-}Ru(bpy)_2\text{-}CN\text{-}Ru(NH_3)_5(PF_6)_5]^+$ type (159), tetradentate Schiff-base complexes of alkali metal-ions, $Cu^{(II)}$, $Ni^{(II)}$, and $Zn^{(II)}$ (160), and Schiff-base complexes of $Co^{(II)}$ (161). Metalloporphyrins have been analyzed successfully by ESI–MS for both synthetic complexes with $Zn^{(II)}$, (162) and $Ni^{(II)}$ or VO^{2+} containing geoporphyrins isolated from Gilsonite bitumen (163). The type of gas-phase ions produced by ESI depended on both the porphyrin nucleus and the metal-ion (163).

In another application of ESI–MS, Gatlin et al. (12) detected the presence of Dupont Metal Deactivator (DMD) fuel additive as a protonated $Cu^{(II)}$ ion in Air Force JP8 jet fuel. Colton and co-workers have been able to characterize by ESI–MS an engine oil inhibitor and antiwear additives, which contain zinc dithiophosphates with alkyl and aryl substituents (164).

IX. CONCLUSIONS AND PERSPECTIVES

To summarize, we have attempted to show through examples that ESI–MS in its present stage of development is a powerful technique for structural analysis of coordination complexes. Further applications are likely to follow, in particular for compounds that are at least moderately stable in protic solvents, which tend to give the highest ESI ion yields. The use of acetonitrile or tetrahydrofuran may further extend the utility of ESI–MS to the analysis of labile and air- and water-sensitive organometallics. The very mild ionization conditions under which ions are transferred from the solution to the gas-phase can make ESI–MS the method of choice for the characterization of metal complexes. In this area, ESI–MS can complement solution NMR, and in a few special cases it has been found to be superior. ESI–MS is the technique of choice when kinetically labile, paramagnetic, or sparingly soluble metal complexes are involved. It presents some advantages for qualitative mixture analysis and for identifying reaction products. The potential to provide correct solution-phase structures, which may differ from those inferred from solid-phase analysis, has been demonstrated by ESI–MS. Lastly, another potential use of ESI–MS is the analysis of small metal complexes in biological and environmental samples, a few examples of which have been presented.

ACKNOWLEDGMENTS

The authors thank the University of Washington Royalty Research Fund for partial suport of this work. C. L. G. thanks Chevron Research for a Graduate Fellowship. We enjoyed enlightening discussions with Professors James M. Mayer, Tomikazu Sasaki, John B. Yates, III, Ruedi Aebersold, and Laszlo Prokai and with Dr. Vladimir Kral, which contributed to this work and are gratefully acknowledged. Special thanks are due to Dr. Tomas Vaisar for continuing collaboration.

REFERENCES

1. Yamashita, M.; Fenn, J. B. *J. Phys. Chem.* **1984**, *88*, 4451, 4671.
2. Whitehouse, C. M.; Dreyer, R. N.; Yamashita, M.; Fenn, J. B. *Anal. Chem.* **1985**, *57*, 675.
3. Wong, S. F.; Menz, C. K.; Fenn, J. B. *J. Phys. Chem.* **1988**, *92*, 546.
4. Jayaweera, P.; Blades, A. T.; Ikonomou, M. G.; Kebarle, P. *J. Am. Chem. Soc.* **1990**, *112*, 2452–2453.
5. Blades, A. T.; Jayaweera, P.; Ikonomou, M. G.; Kebarle, P. *Int. J. Mass Spectrom. Ion Processes* **1990**, *101*, 325–336.
6. Blades, A. T.; Jayaweera, P.; Ikonomou, M. G.; Kebarle, P. *Int. J. Mass Spectrom. Ion Processes* **1990**, *102*, 251.
7. Blades, A. T.; Jayaweera, P.; Ikonomou, M. G.; Kebarle, P. *J. Chem. Phys.* **1990**, *92*, 5900–5906.
8. Katta, V.; Chowdhury, S. K.; Chait, B. T. *J. Am. Chem. Soc.* **1990**, *112*, 5348–5349.
9. Dzidic, I.; Kebarle, P. *J. Phys. Chem.* **1970**, *74*, 1466.
10. Anderson, S. G.; Blades, A. T.; Klassen, J.; Kebarle, P. *Int. J. Mass Spectrom. Ion Processes* **1995**, *141*, 217–228.
11. Marcus, S. A.; Sutin, N. *Biochim. Biophys. Acta* **1985**, *811*, 265.
12. Gatlin, C. L.; Tureček, F.; Vaisar, T. *Anal. Chem.* **1994**, *66*, 3950–3958.
13. Collman, J. P. *Acc. Chem. Res.* **1968**, *1*, 136.
14. Vaisar, T.; Tureček, F.; Gatlin, C. L. *J. Am. Chem. Soc.* **1996**, *118*, 11321–11322.
15. Van Berkel, G. J.; McLuckey, S. A.; Glish, G. L. *Anal. Chem.* **1992**, *64*, 1586–1593.
16. Bechgaard, K.; Parker, V. D. *J. Am. Chem. Soc.* **1972**, *94*, 4749.
17. Xu, X.; Nolan, S. P.; Cole, R. B. *Anal. Chem.* **1994**, *66*, 119–125.
18. Blades, A. T.; Ikonomou, M. G.; Kebarle, P. *Anal. Chem.* **1991**, *63*, 2109–2114.
19. Gatlin, C. L.; Tureček, F. *Anal. Chem.* **1994**, *66*, 712–718.
20. Martell, A. E.; Smith, R. M. *Critical Stability Constants*; Plenum: New York, 1979; Vol. 2, p. 235.
21. Taflin, D. C.; Ward, T. L.; Davis, E. J. *Langmuir* **1989**, *5*, 376–384.
22. Hathaway, B. J. In *Comprehensive Coordination Chemistry*; Wilkinson, G., Ed.; Perganon: Oxford, 1977; Vol. 5, pp. 533–765.

23. Ochiai, E.; *Bioinorganic Chemistry, An Introduction*; Allyn and Bacon: Boston, 1977.
24. Bond, A. M.; Colton, R.; D'Agostino, A.; Downard, A. J.; Traeger, J. C. *Anal. Chem.* **1995**, *67*, 1691–1695.
25. *Gas Phase Inorganic Chemistry*; Russell, D. H., Ed.; Plenum Press: New York, 1989.
26. Colodner, D.; Salters, V.; Duckworth, D. C. *Anal. Chem.* **1994**, *66*, 1079A–1089A.
27. Smith, R. D.; Loo, J. A.; Edmonds, C. G.; Baringa, C. J.; Udseth, H. R. *Anal. Chem.* **1990**, *62*, 882–899.
28. Smith, R. D.; Loo, J. A.; Ogorzalek Loo, R. R.; Busman, M.; Udseth, H. R. *Mass Spectrom. Rev.* **1991**, *10*, 359–451.
29. Mann, M. *Org. Mass Spectrom.* **1990**, *25*, 575–587.
30. Douglas, D. J. *Presented at the Winter Conference on Plasma Spectrochemistry*, St. Petersburg, FL, January 1990; Paper IL-11.
31. Thomson, B. A.; Iribarne, J. V. *J. Chem. Phys.* **1979**, *71*, 4451.
32. Curtis, J. M.; Derrick, P. J.; Schnell, A.; Constantin, E.; Gallagher, R. T.; Chapman, J. R. *Org. Mass Spectrom.* **1992**, *27*, 1176–1180.
33. Curtis, J. M.; Derrick, P. J.; Schnell, A.; Constantin, E.; Gallagher, R. T.; Chapman, J. R. *Inorg. Chim. Acta* **1992**, *201*, 197–201.
34. *Gas-Phase Ion Chemistry*; Bowers, M. T., Ed.; Academic Press: New York, 1979.
35. Cheng, Z. L.; Siu, K. W. M.; Guevremont, R.; Berman, S. S. *J. Am. Soc. Mass Spectrom.* **1992**, *3*, 281–288.
36. Cheng, Z. L.; Siu, K. W. M.; Guevremont, R.; Berman, S. S. *Org. Mass Spectrom.* **1992**, *27*, 1370–1376.
37. Lau, T.-C.; Wang, J.; Siu, K. W. M.; Guevremont, R. *J. Chem. Soc. Chem. Commun.* **1994**, 1487–1588.
38. Le Quan Toui, J.; Muller, E. *Rapid Commun. Mass Spectrom.* **1994**, *8*, 692–694.
39. Summerfield, S.; Howarth, O.; Jennings, K. R. *Presented at the 43rd ASMS Conference of Mass Spectrometry and Allied Topics*, Atlanta, GA, 1995; Comm. No. WPD 091.
40. Siu, K. W. M.; Gardner, G. J.; Berman, S. S. *Rapid Commun. Mass Spectrom.* **1988**, *2*, 69–71.
41. Jones, T. L.; Betowski, L. D. *Rapid Commun. Mass Spectrom.* **1993**, *7*, 1003.
42. Stewart, I. I.; Horlick, G. *Anal. Chem.* **1994**, *66*, 3983–3993.
43. Douglas, D. J. In *Inductively Coupled Plasmas in Analytical Atomic Spectroscopy*, 2nd ed.; Montaser, A.; Golightly, D. W. Eds.; VCH Publishers: New York, 1992.
44. Cheng, Z. L.; Siu, K. W. M.; Guevremont, R.; Berman, S. S. *Proceedings of the 39th ASMS Conference on Mass Spectrommetry and Allied Topics*, San Francisco, CA, 1993; p. 1255.
45. Van Berkel, G. J.; McLuckey, S. A.; Glish, G. A. *Proceedings of the 39th ASMS Conference on Mass Spectrommetry and Allied Topics*, San Francisco, CA, 1993; p. 292.
46. Agnes, G. R.; Horlick, G. *Appl. Spectrosc.* **1992**, *46*, 401–406.
47. Agnes, G. R.; Horlick, G. *Appl. Spectrosc.* **1994**, *48*, 649–654.
48. Agnes, G. R.; Horlick, G. *Appl. Spectrosc.* **1994**, *48*, 655–661.
49. Agnes, G. R.; Horlick, G. *Appl. Spectrosc.* **1995**, *49*, 324–334.

50. Fenselau, C.; Cotter, R. J. *Chem. Rev.* **1987**, *87*, 501–512.
51. Karas, M.; Hillenkamp, F. *Anal. Chem.* **1988**, *60*, 2299–2301.
52. Fenn, J. B.; Mann, M.; Meng, C. K.; Wong, S. F.; Whitehouse, C. M. *Mass Spectrom. Rev.* **1990**, *9*, 37–70.
53. Gross, D. S.; Williams, E. R. *J. Am. Chem. Soc.* **1995**, *117*, 883.
54. Wang, K.; Han, X.; Gross, R. W.; Gokel, G. W. *J. Am. Chem. Soc.* **1995**, *117*, 7680–7686.
55. Saf, R.; Mirtl, C.; Hummel, K. *Tetrahedron Lett.* **1994**, *35*, 6653–6656.
56. Pedersen, C. J. *J. Am. Chem. Soc.* **1967**, *89*, 7017.
57. Gokel, G. W. *Crown Ethers and Cryptands*; Royal Society of Chemistry: London, 1991.
58. Colton, R.; Mitchell, S.; Traeger, J. C. *Inorg. Chim. Acta* **1995**, *231*, 87–93.
59. Znamirovschi, C, G.; Frutos, D.; Enke, C. G. *Presented at the 43rd ASMS Conference on Mass Spectrometry and Allied Topics*, Atlanta, GA, 1995; Comm. No. WPD 092.
60. Wilson, S. R.; Tulchinsky, M. L. *J. Org. Chem.* **1993**, *58*, 1407–1408.
61. Fura, A.; Leary, J. A. *Anal. Chem.* **1993**, *65*, 2805–2811.
62. Kohler, M.; Leary, J. A. *Presented at the 43rd ASMS Conference on Mass Spectrometry and Allied Topics*, Atlanta, GA, 1995; Comm. No. ThPE 183.
63. Smith, G.; Bott, G; Leary, J. A. *Presented at the 43rd ASMS Conference of Mass Spectrometry and Allied Topics*, Atlanta, 1995; Comm. No. ThPE 184.
64. Siuzdak, G.; Ichikawa, Y.; Caulfield, T. J.; Munoz, B.; Wong, C.-H.; Nicolaou, K. C. *J. Am. Chem. Soc.* **1993**, *115*, 2877–2881.
65. Duffin, K. L.; Henion, J. D.; Shieh, J. J. *Anal. Chem.* **1991**, *63*, 1781–1788.
66. Bleicher, K.; Bayer, E. *Biol. Mass Spectrom.* **1994**, *23*, 320–322.
67. Cheng, X.; Wu, Q.; Gao, Q.; Hofstadler, S. A.; Smith, R. D. *Presented at the 43rd ASMS Conference of Mass Spectrometry and Allied Topics*, Atlanta, GA, 1995; Comm. No. ThPF 221.
68. Schneider, R. P.; Lynch, M. J.; Ericson, J. F.; Fouda, H. G. *Anal. Chem.* **1991**, *63*, 1789–1794.
69. Hua, Y.; Lu, W.; Henry, M. S.; Pierce, R. H.; Cole, R. B. *Anal. Chem.* **1995**, *34*, 1815–1823.
70. Pleasance, S.; Quilliam, M. A.; de Freitas, A. S. W.; Marr, J. C.; Cembella, A. D. *Rapid. Commun. Mass Spectrom.* **1990**, *4*, 206–213.
71. Quilliam, M. A.; Thompson, B. A.; Scott, G. J.; Siu, K. W. M. *Rapid. Commun. Mass Spectrom.* **1989**, *3*, 145–150.
72. Wilson, S. R.; Wu, Y. *Supramol. Chem.* **1994**, *3*, 273–277.
73. Wilson, S. R.; Lu, Q.; Tulchinsky, M. L.; Wu, Y. *J. Chem. Soc. Chem. Commun.* **1993**, 664–665.
74. Wilson, S. R.; Wu, Y. *J. Chem. Soc. Chem. Commun.* **1993**, 784–786.
75. Wilson, S. R.; Wu, Y. *J. Am. Chem. Soc.* **1993**, *115*, 10334–10337.
76. Wilson, S. R.; Wu, Y. *J. Am. Soc. Mass Spectrom.* **1993**, *4*, 596–603.
77. Wilson, S. R.; Kaprinidis, N.; Wu, Y.; Schuster, D. I. *J. Am. Chem. Soc.* **1993**, *115*, 8495–8496.

78. Wilson, S. R.; Lu, Q. *Tetrahedron Lett.* **1993**, *34*, 8043–8046.
79. Gatlin, C. L.; Tureček, F.; Vaisar, T. *J. Am. Chem. Soc.* **1995**, *117*, 3637–3638.
80. Vaisar, T.; Gatlin, C. L.; Tureček, F. *Int. J. Mass Spectrom. Ion Processes* **1997**, in press.
81. Hu, P.; Gross, M. L. *J. Am. Chem. Soc.* **1993**, *115*, 8821–8828.
82. Hu, P.; Gross, M. L. *J. Am. Chem. Soc.* **1992**, *114*, 9153–9160.
83. Hu, P.; Gross, M. L. *J. Am. Chem. Soc.* **1992**, *114*, 9161–9169.
84. Reiter, A.; Adams. J.; Zhao, H. *J. Am. Chem. Soc.* **1994**, *116*, 7827–7838.
85. Zhao, J.; Reiter, A.; Teesch, L. M.; Adams, J. *J. Am. Chem. Soc.* **1993**, *115*, 2854–2863.
86. Teesch, L. M.; Orlando, R. C.; Adams, J. *J. Am. Chem. Soc.* **1991**, *113*, 3668–3675.
87. Teesch, L. M.; Adams, J. *J. Am. Chem. Soc.* **1991**, *113*, 812–820.
88. Leary, J. A.; Zhou, Z.; Ogden, S. A.; Williams, T. D. *J. Am. Soc. Mass Spectrom.* **1990**, *1*, 473–480.
89. Leary, J. A.; Williams, T. D.; Bott, G. *Rapid Commun. Mass Spectrom.* **1989**, *3*, 192–196.
90. Hutchens, T. W.; Yip, T.-T. *Methods: A Companion to Methods in Enzymology* **1992**, *4*, 79–96.
91. Allen, M. H.; Hutchens, T. W. *Rapid Commun. Mass Spectrom.* **1992**, *6*, 308–312.
92. Hutchens, T. W.; Nelson, R. W.; Allen, M. H.; Li, C. M.; Yip, T.-T. *Biol. Mass Spectrom.* **1992**, *21*, 151–159.
93. Allen, M. H.; Hutchens, T. W. *Rapid Commun. Mass Spectrom.* **1992**, *6*, 469–473.
94. Hutchens, T. W.; Allen, M. H.; Li, C. M.; Yip, T.-T. *FEBS Lett.* **1992**, *309*, 170–174.
95. Kazanis, S.; Pochapsky, T. C.; Barnhart, T. M.; Penner-Hahn, J. E.; Mirza, U. A.; Chait, B. T. *J. Am. Chem. Soc.* **1995**, *117*, 6625–6626.
96. Ptillot, Y.; Forest, E.; Meyer, J.; Moulis, J.-M. *Presented at the 43rd ASMS Conference of Mass Spectrometry and Allied Topics*, Atlanta, GA, 1995; Comm. No. MPF 278.
97. Jaquinod, M.; Leize, E.; Potier, N.; Albrecht, A.-M.; Shanzer, A.; Dorsselaer, A. V. *Tetrahedron Lett.* **1993**, *34*, 2771–2774.
98. Witkowska, H. E.; Shackleton, C. H. L.; Dahlman-Wright, K.; Kim, J. Y.; Gustafsson, J.-A. *J. Am. Chem. Soc.* **1995**, *117*, 3319–3324.
99. Surovoy, A.; Waidelich, D.; Jung, G. *FEBS Lett.* **1992**, *311*, 259–262.
100. Jiao, C. Q.; Freiser, B. S.; Carr, S. R.; Cassady, C. J. *J. Am. Soc. Mass Spectrom.* **1995**, *6*, 521–524.
101. Hu, P.; Ye, Q.-Z.; Loo, J. A. *Anal. Chem.* **1994**, *66*, 4190–4194.
102. Dell'Angelica, E. C.; Schleicher, C. H.; Santome, J. A. *J. Biol. Chem.* **1994**, *269*, 28929–28936.
103. Hofmann, F.; James, P.; Vorherr, T.; Carafoli, E. *J. Biol. Chem.* **1994**, *268*, 10252–10259.
104. Yu, X.; Wojciechowski, M.; Fenselau, C. *Anal. Chem.* **1993**, *65*, 1355–1359.
105. Sam, J. W.; Tang, X.-J.; Magliozzo, R. S.; Peisach, J. *J. Am. Chem. Soc.* **1995**, *117*, 1012–1018.

106. Loo, J. A.; Hu, P.; Smith, R. D. *J. Am. Soc. Mass Spectrom.* **1994**, *5*, 959–965.
107. Loo, J. A.; Ogorzalek Loo, R. R.; Andrews, P. C. *Org. Mass Spectrom.* **1993**, *28*, 1640–1649.
108. Hu, P.; Loo, J. A. *J. Am. Chem. Soc.* **1995**, *117*, 11314–11319.
109. Wang, J.; Guevremont, R.; Siu, K. W. M. *Eur. Mass Spectrom.* **1995**, *1*, 171–181.
110. Le Blanc, J. C. Y.; Wang, J.; Siu, K. W. M.; Guevremont, R. *Proceedings of the 42nd ASMS Conference of Mass Spectrometry and Allied Topics*, Chicago, IL, 1994; p. 416.
111. Bouchonnet, S.; Hoppilliard, Y.; Ohanessian, G. *J. Mass Spectrom.* **1995**, *30*, 172–179.
112. Bouchonnet, S.; Hoppilliard, Y. *Org. Mass Spectrom.* **1992**, *27*, 71–76.
113. Lei, Q. P.; Amster, I. J. *Presented at the 43rd ASMS Conference of Mass Spectrometry and Allied Topics*, Atlanta, GA, 1995; Comm. No. FOA 11:10
114. Wilson, S. R.; Yasmin, A.; Wu, Y. *J. Org. Chem.* **1992**, *57*, 6941–6945.
115. Gatlin, C. L.; Tureček, F.; Vaisar, T. *J. Mass Spectrom.* **1995**, *30*, 775–777.
116. Gatlin, C. L.; Tureček, F. *J. Mass Spectrom.* **1995**, *30*, 1605–1616.
117. Gatlin, C. L.; Tureček, F.; Vaisar, T. *J. Mass Spectrom.* **1995**, *30*, 1617–1627.
118. Gatlin, C. L.; Tureček, F.; Vaisar, T. *J. Mass Spectrom.* **1995**, *30*, 1636–1637.
119. Gatlin, C. L.; Tureček, F.; Vaisar, T. *Presented at the 43rd ASMS Conference of Mass Spectrometry and Allied Topics*, Atlanta, GA, 1995; Comm. No. MPE 231.
120. Gatlin, C. L.; Rao, R. D.; Tureček, F.; Vaisar, T. *Presented at the 43rd ASMS Conference of Mass Spectrometry and Allied Topics*, Atlanta, GA, 1995; Comm. No. MPE 232.
121. Gatlin, C. L.; Rao, R. D.; Tureček, F.; Vaisar, T. *Anal. Chem.* **1996**, *68*, 263–270.
122. Yamauchi, O.; Odani, A. *J. Am. Chem. Soc.* **1985**, *107*, 5938.
123. Martinho-Simoes, J. A.; Beauchamp, J. L. *Chem. Rev.* **1990**, *90*, 629–688.
124. Ganem, B.; Li, Y. T.; Henion, J. D. *J. Am. Chem. Soc.* **1991**, *113*, 6294–6296.
125. Smith, D. L.; Zhang, Z. *Mass Spectrom. Rev.* **1994**, *13*, 411–429.
126. Gatlin, C. L.; Tureček, F. unpublished results.
127. Rincon, L.; Ruette, F.; Hernandez, A. *J. Mol. Struct. (THEOCHEM)* **1992**, *254*, 395–403.
128. Schwarz, J.; Heinemaa, C.; Schwarz, H. *J. Phys. Chem.* **1995**, *99*, 11405–11411.
129. Piguet, C.; Bernardinelli, G.; Bocquet, B.; Quattropani, A.; Williams, A. F. *J. Am. Chem. Soc.* **1992**, *114*, 7440.
130. Hopfgartner, G.; Piguet, C.; Henion, J. D.; Williams, A. F. *Helv. Chim. Acta* **1993**, *76*, 1759–1766.
131. Hopfgartner, G.; Piguet, C.; Henion, J. D. *J. Am. Soc. Mass Spectrom.* **1994**, *5*, 748–756.
132. Fujita, M.; Ibukuro, F.; Hagihara, H.; Ogura, K. *Nature* **1994**, *367*, 720–722.
133. Piguet, C.; Hopfgartner, G.; Bocquet, B.; Schaad, O.; Williams, A. F. *J. Am. Chem. Soc.* **1994**, *116*, 9092–9102.
134. Leize, E.; Van Dorsselaer, A.; Kramer, R.; Lehn, J.-M. *J. Chem. Soc. Chem. Commun.* **1993**, 990–993.

135. Bitsch, F.; Dietrich-Buchecker, C. O.; Khemiss, A.-K.; Sauvage, J.-P.; Van Dorsselaer, A. *J. Am. Chem. Soc.* **1991**, *113*, 4023–4025.
136. Light-Wahl, K. J.; Delinger, S. L.; Gale, D. C.; Hams, A. C.; Smith, R. D.; Jenkins, Y.; Barton, J. K. *Proceedings of the 42nd ASMS Conference on Mass Spectrometry and Allied Topics*, Chicago, IL, 1994; p. 916.
137. Harms, A. C.; Smith, R. D.; Jenkins, Y.; Barton, J. K. *Presented at the 43rd ASMS Conference on Mass Spectrometry and Allied Topics*, Atlanta, GA, 1995; Comm. TPF 233.
138. Murphy, C. J.; Barton, J. K. In *Methods in Enzymology*, 1993; Vol. 226, p. 576.
139. Colton, R.; Harrison, K. L.; Mah, Y. A.; Traeger, J. C. *Inorg. Chim. Acta* **1995**, *231*, 65–71.
140. Colton, R.; Dakternieks, D. *Inorg. Chim. Acta* **1993**, *208*, 173–177.
141. Colton, R.; Tedesco, V.; Traeger, J. C. *Inorg. Chem.* **1992**, *31*, 3865–3866.
142. Bond, A. M.; Colton, R.; Trager, J. C.; Harvey, J. *Inorg. Chim. Acta* **1993**, *212*, 233–239.
143. Bond, A. M.; Colton, R.; D'Agostino, A.; Harvey, J.; Traeger, J. C. *Inorg. Chem.* **1993**, *32*, 3952–3956.
144. Bond, A. M.; Colton, R.; Mah, Y. A.; Traeger, J. C. *Inorg. Chem.* **1994**, *33*, 2548–2554.
145. Ahmed, I.; Bond, A. M.; Colton, R.; Jurcevic, M.; Traeger, J. C.; Walter, J. N. *J. Organomet. Chem.* **1993**, *447*, 59–65.
146. Kane-Maguire, L. A. P.; Kanitz, R.; Sheil, M. M. *J. Organometal. Chem.* **1995**, *486*, 243–248.
147. Canty, A. J.; Colton, R. *Inorg. Chim. Acta* **1994**, *215*, 179–184.
148. Canty, A. J.; Traill, P. R.; Colton, R.; Thomas, I. M. *Inorg. Chim. Acta* **1993**, *210*, 91–97.
149. Canty, A. J.; Colton, R.; Thomas, I. M. *J. Organometal. Chem.* **1993**, *445*, 283–289.
150. Jones, T. L.; Betowski, L. D. *Rapid Commun. Mass Spectrom.* **1993**, *7*, 1003–1008.
151. Corr, J. J.; Siu, K. W. M. *Presented at the 41st ASMS Conference of Mass Spectrometry and Allied Topics*, Atlanta, GA, 1995; Comm. No. WPF 164.
152. Aplin, R. T.; Robinson, C. V.; Smith, V. C. M. *Proceedings of the 41st ASMS Conference of Mass Spectrometry and Allied Topics*, San Francisco, CA, 1993; p. 1071.
153. Colton, R.; Klaui, W. *Inorg. Chim. Acta* **1993**, *211*, 235–242.
154. Poon, G. K.; Mistry, P.; Lewis, S. *Biol. Mass Spectrom.* **1991**, *20*, 687–692.
155. Arakawa, R.; Matsuo, T.; Ito, H.; Katakuse, I.; Nozaki, K.; Ohno, T.; Haga, M. *Org. Mass Spectrom.* **1994**, *29*, 289–294.
156. Colton, R.; James, B. D.; Potter, I. D.; Traeger, J. C. *Inorg. Chem.* **1993**, *32*, 2626–2629.
157. Wilson, S. R.; Wu, Y. *Organometallics* **1993**, *12*, 1478–1480.
158. Aliprantis, A. O.; Canary, J. W. *J. Am. Chem. Soc.* **1994**, *116*, 6985–6986.
159. Hamdan, M.; Curcuruto, O.; Bortolini, O.; Bignozzi, C. A. *Presented at the 43rd ASMS Conference on Mass Spectrometry and Allied Topics*, Atlanta, GA, 1995; Comm. WPD 090.

160. Cardwell, T. J.; Colton, R.; Mitchell, S.; Trager, J. C. *Inorg. Chim. Acta*, **1994**, *216*, 75–81.
161. Raffaelli, A.; Pucci, S.; Di Sacco, S.; Isola, M. *Presented at the 43rd ASMS Conference on Mass Spectrometry and Allied Topics*, Atlanta, GA, 1995; Comm. MPC 125.
162. Anderson, S.; Anderson, H. L.; Sanders, J. K. M. *Angew. Chem. Int. Ed. Engl.* **1992**, *31*, 907–910.
163. Van Berkel, G. J.; Quinones, M. A.; Quirke, J. M. E. *Energy Fuels*, **1993**, *7*, 411–419.
164. Cardwell, T. J.; Colton, R.; Lambropoulos, N.; Traeger, J. C.; Marriott, P. J. *Anal. Chim. Acta* **1993**, *208*, 239–244.

INDEX

Accumulated trapping, 301
Accurate mass measurement (*see* mass accuracy)
Acetonitrile complexes, 530
Acylglycerols, 488, 541
Additives for LC-MS (*see* volatile additives)
Adduct ions, 230, 306, 311, 462
Aerosol, 108, 109, 116–117, 129
Aerosol, charged, 108, 116
Affinity capillary electrophoresis, 410
Affinity chromatography, 410
Agrochemicals, 370
Albumin, 395, 398, 401, 402
Alcohol dehydrogenase, 403, 406
Alkali metal atom, 530
Alkali metal ions, 529, 533
Alkaline earth ions, 529, 533
Alkaloids, 516
Amines, 115, 325, 333
Amino acids, 370, 550–558
Ampholytes, 350, 368
Anomer analysis, 463, 495
Antibodies, 218, 330, 387, 394, 401, 500
APCI-MS (*see* atmospheric pressure chemical ionization)
API (*see* atmospheric pressure ionization)
Aquo complexes, 529, 558
Array detectors, 358
Atmospheric pressure chemical ionization (APCI), 114, 197, 324, 351, 370, 500, 512
Atmospheric pressure ionization (API), 118, 120, 123–127, 131, 324
Au complexes (*see* gold complexes)

Background electrolyte, 347, 360–367
Background subtraction, 360
Bile acids, 333, 335, 494
Brevetoxins, 542
Buffers (*see* volatile additives)

CAD (*see* collision-induced dissociation)
Calibration (*see* mass calibration)
Cancer, 517–520

Capillary coatings, 347, 358, 372
Capillary diameters in capillary electrophoresis, 363–366
Capillary electrokinetic chromatography (CEKC), 347–349, 370
Capillary electrophoresis (CE) 108, 110, 111, 344–350 (*see also* capillary electrophoresis-mass spectrometry)
Capillary electrophoresis-mass spectrometry, 111–113, 128, 131, 191, 205, 219–221, 277–279, 297, 299–300, 333, 339, 343–376, 392, 429, 443, 505, 515–516
Capillary gel electrophoresis (CGE), 350
Capillary isoelectric focusing (CIEF), 348, 350, 367–368
Capillary isotachophoresis (CITP), 348–350, 367, 370, 373, 375, 429, 516
Capillary LC (*see* liquid chromatography)
Capillary zone electrophoresis (CZE), 347 (*see also* capillary electrophoresis)
Carbohydrate, 394–395, 459–496, 539–541 (*see also* oligosaccharides)
CE (*see* capillary electrophoresis; capillary electrophoresis-mass spectrometry)
CE-ESI-MS (*see* capillary electrophoresis-mass spectrometry)
Cell analysis, 372
CE-MS (*see* capillary electrophoresis-mass spectrometry)
Ceramide, 476, 483–484, 486–487, 540–541
CF-FAB (*see* continuous flow-fast atom bombardment)
Charge balance, 68, 70, 71, 72, 73
Charge of droplets, 15–17
Charge excess in droplets, 151–153
Charge-remote fragmentations, 489
Charge state determination, 258, 273–277
Charge state distribution, 137–171, 387, 425, 427
  Effect of analyte concentration, 148–153
  Effect of analyte structures/conformations, 156–161
  Effect of counterion, 151, 153–156

571

Charge state distribution (*continued*)
  Effect of instrumental conditions, 161–165
  Effect of solution pH, 7, 142–146, 158, 429
  Effect of solvent, 146–148, 358
  Gas phase modifications, 165–170
Charge stripping, 306, 313
Charged aerosol (*see* aerosol, charged)
Charged residue model, 25–26, 31–33, 138
Chelation, 427
Chemical ionization, 115, 120, 123, 275 (*see also* atmospheric pressure chemical ionization)
Chiral molecules, separation of, 370, 514
Chromatography (*see* gas chromatography; liquid chromatography)
CI (*see* chemical ionization)
CID (*see* collision-induced dissociation)
Cisplatin, 562
Cleaning, 128
Cluster ion, 118, 120–122, 132, 165–167, 230, 246, 273, 528–529
Coaxial sheath-flow interface (*see* sheath-flow interface)
Collision-induced dissociation (CID), 122, 130–132, 153, 161, 164, 206, 227–229, 258–262, 267, 273, 281, 306–307, 395–399, 431, 433–442, 462–486, 490–494, 501, 529, 530, 538–541, 548–549, 553–558 (*see also* in-source collision-induced dissociation)
Collisional dampening, 237, 244
Collisionally activated dissociation (*see* collision-induced dissociation)
Combinatorial libraries, 197, 371, 408, 410, 520
Complex, noncovalent (*see* noncovalent interactions)
Concentration of analyte (*see also* dynamic range), 39–48, 148–153
Concentration of electrolytes, 17–20
Conductivity of solution, 17–19
Cone CID (*see* in-source collision-induced dissociation)
Cone-jet mode, 10, 11, 47
Conformational charges in solution, 142
Conjugates, 508, 510–511, 519–520
Continuous-flow fast atom bombardment (CF-FAB), 351, 375, 500, 517
Controlled current electrolytic (CCE) cell, 77–86, 88, 89
Controlled potential electrolytic (CPE) cell, 78
Copper complexes, 530–532, 545–546, 549–559, 561–563
Corona discharge (*see* electrical gas discharge)
Corrosion, 70, 81, 84, 87
Coulomb energy, 313
Coulomb fission of droplets (*see* droplet fission)
Coulombic repulsion, 145–146, 158, 167, 168–170, 475, 529, 532
Counter current gas, 162–164
Counter electrode, 67, 68, 78
Cross-section measurements, 401
Crown ethers, 427, 446, 487, 537–539, 543–545
Cryptands, 537–539
C-terminal sequence determination, 398
Cu complexes (*see* copper complexes)
Current due to charged droplets, 15–17, 133
Cyclodextrins, 349, 370
Cyclopentadienyl ligands, 530–531, 561

Damping (*see* collisional dampening)
Declustering, 122, 210, 230, 533–535
Deconvolution of multiply charged ion peaks, 183–186, 267–268, 391, 462, 465–466
Deprotonation reactions, 312
Derivatization, 335, 462–463, 505
Desorption rate constant (*see* ion evaporation rate constants)
Detection limits, 277–279
Detergents (*see* surfactants)
Deuterium labeling, 554 (*see also* H/D exchange)
Dielectric of solvent, 147–148, 170
Digests, enzymatic (*see* enzymatic digests)
Digests, tryptic (*see* enzymatic digests)
Diquaternary ammonium ions, 155
Discharge suppression (*see* electrical gas discharge suppression)
Disulfide proteins, 158–160, 168–169, 399, 401
DNA, 267–277, 375, 403, 408–409, 422–454, 505, 508, 511, 541, 547 (*see also* nucleic acids)
Dole, charged residue mechanism (*see* charged residue mechanism)
Double-strand nucleotides (*see* nucleotides)
Droplet electrospray device (DES), 50–53, 141
Droplet fission, 21–25, 53, 54
Duplex oligonucleotides, 272, 445–449
Duty cycle, 205, 206, 210, 254, 261, 279, 301, 339
Dynamic range, 108, 133–135, 215–216, 241, 514, 516

EH (*see* electrohydrodynamic ionization)
EI (*see* electron ionization)
Electric field at ESI capillary tip, 9
Electric field for onset of ESI, 13–14, 147
Electric field on charged residues, 11
Electrical gas discharge suppression, 14, 115, 326–327
Electrical gas discharges, 12, 13–14, 114–115, 127, 132, 326–327, 425

INDEX   **573**

Electrochemical cell (*see* electrolytic cell)
Electrochemical ionization, 11, 88, 94, 96, 102
Electrochemical oxidation, 11, 68, 71, 73, 530–531
Electrode reactions in ESI, 11
Electrohydrodynamic ionization, 14, 68–71, 74, 87
Electrokinetic injection in capillary electrophoresis, 350
Electrolysis, 11, 67, 79–86, 89–94, 97, 99
Electrolytic cell, 11, 70, 73, 78, 532
Electron ionization, 248, 517
Electron transfer, 529–530
Electroosmotic flow, 346–347
Electrophoresis, capillary (*see* capillary electrophoresis)
Electrophoresis, gel, 281
Electrophoretic mechanism of charge separation, 10, 12–13, 67, 70, 148
Electrophoretic mobility (velocity), 113, 345–347
Electrospray (*see* specific subtopic)
Electrostatic focusing (*see* ion optics)
Electrostatic lens (*see* ion optics)
Electrostatic mirror, 204, 208, 211
Electrostatic sprayer, 66, 68
Elemental analysis, 532–536
Enzymatic digests, 371–372
  trypsin digestion, 188, 194, 205, 213, 280–281, 364, 371–372, 395, 397, 398, 492, 504–505
Enzyme mechanism, 397
Enzymes, 224–225, 387–389, 402, 440, 443, 487, 492
E-Selectins, 473
ESI (*see* specific subtopic)
ESI-MS-MS (*see* MS/MS)
Evaporation of solvent from droplets, 37–39, 162
External source (*see* ion source, external)

Faradaic current, 68, 72, 80, 82–84, 92
Faraday's law, 72–73, 87, 92, 94
Fast atom bombardment-mass spectrometry, 7, 520, 543, 545 (*see also* continuous-flow FAB)
  -in carbohydrate and lipid analysis, 460–463, 466–467, 475–476, 480, 483, 485, 492, 495
FAB-MS (*see* fast atom bombardment-mass spectrometry)
Fe complexes (*see* iron complexes)
Fish and shellfish poisoning, 541–542
Flow electrochemical cell (*see* electrolytic cell)
Flow injection analysis, 256, 279
Flow rates for LC-MS, 327
Focusing optics (*see* ion optics)
Formation constants, 532

Fourier transform, 250–256
Fourier transform ion cyclotron resonance (FTICR) mass spectrometry, 258, 273, 291–316, 358, 372, 374, 391, 399, 434–436, 541, 550
Fraction collection, 351
Fragmentation, 122, 130–132, 466 (*see also* collision-induced dissociation; in-source collision-induced dissociation)
Free jet expansion, 118–120, 124–125, 129, 132
FTMS (*see* Fourier transform ion cyclotron resonance mass spectrometry)
Fucose and Fucosyl complexes, 539–540
Fullerenes, 543

Galactose, 469, 473, 539
Gangliosides, 466–473
Gas chromatography-mass spectrometry (GC-MS), 324, 500, 504–505
Gas phase basicity, 167, 169, 313
Gas phase reactions, 165–170
Gated trapping, 301
GC-MS (*see* gas chromatography-mass spectrometry)
Gel electrophoresis (*see* electrophoresis, gel)
Glucopyranose, 463
Glucuronide, 502–504, 508, 510, 516, 519–520
Glutathione, 502–504, 511, 516
Glycans (*see* oligosaccharides)
Glycerides (*see* acylglycerols)
Glycoconjugates, 462–485
Glycolipids, 463 (*see also* glycosphingolipids)
Glycoproteins, 226, 394, 395, 397, 463, 491
Glycosides, 475–480
Glycosphingolipids, 480–485, 540–541
Glycosylation, 397
Gold complexes, 561
Gram-negative bacteria, 488–490

H/D exchange reactions, 295, 312–313, 400–401
Heated capillary, 246
Hendricks equation, 79, 89
Hepatocytes, 501, 519
Herbicides, 370
High performance liquid chromatography (HPLC) (*see* liquid chromatography)
High resolution mass spectrometry, 186–188, 243, 253–254, 258, 262–265, 295, 309–310, 391, 399, 546
High voltage, 110–113, 124, 131
Higher order protein structure (*see* protein conformation)
History of electrospray, 178–179 (*see* also Foreword)

Hydrodynamic injection in capillary electrophoresis, 350
Hydrogen/deuterium exchange (*see* H/D exchange reactions)
Hydrophobic interaction chromatography, 330
Hydrostatic injection in capillary electrophoresis, 350

ICP-MS (*see* inductively coupled plasma-mass spectrometry)
ICR (*see* Fourier transform ion cyclotron resonance)
Immunoaffinity separations, 327, 329, 330, 508
Inclusion complexes, 349, 351
Inductively coupled plasma mass spectrometry (ICP-MS), 296, 351, 533
Infrared multiphoton dissociation (IRMPD), 307–309, 398–399
Injection of sample (*see* sample introduction)
Inorganic analysis, 351, 370, 527–563
Inositol phosphates, 491–494
In-source collision-induced dissociation, 130–132, 153, 161, 164–165, 188–190, 253, 278, 306, 396, 434–436, 438, 450, 461–463, 483, 489, 501, 505, 508, 539
In-source fragmentation (*see* in-source collision-induced dissociation)
Intramolecular coulombic repulsion (*see* coulombic repulsion)
Introduction of sample (see sample introduction)
Ion charge state (*see* charge state determination; charge state distribution)
Ion counting, 208, 216–218
Ion cyclotron resonance, 261 (*see also* Fourier transform ion cycloton resonance)
Ion dissociation (*see* collision-induced dissociation)
Ion ejection, 238, 243, 249–250, 303 (*see also* resonance ejection)
Ion evaporation model and theory, 25–31, 34–39, 138, 150
Ion evaporation rate constants, 26–33, 141, 156
Ion exchange LC, 330
Ion focusing (*see* ion optics)
Ion guide
  quadrupole, 123, 125, 129, 211, 294–296
  octapole, 247–248, 296
Ion injection, 238, 241, 244, 248, 250, 254, 302
Ion/ion reaction, 273–277
Ion isolation, 237, 242, 254–256
Ion mobility spectrometry, 120
Ion/molecule reactions, 248–249, 258, 273
Ion optics, 114, 116, 122–127, 129, 131, 180, 294–297

Ion pair LC (*see* liquid chromatography, ion pair)
Ion pairing, 151–155
Ion solvation (*see* cluster ion)
Ion source, 111, 113, 115–118, 121–122, 124, 131, 207, 351
Ion source, external, 295–296
Ion spray (*see* pneumatically-assisted electrospray)
Ion storage (*see* quadrupole ion trap; Fourier transform ion cyclotron resonance)
Ion transport, 124–125, 127, 133, 246–247
Ion trap (*see* quadrupole ion trap)
Ion trapping in FT-ICR, 301–303
Ionization efficiency, 115
Ionization energies, 530, 533
Ionophores, 541–542
Iribarne, Thomson ion evaporation theory (*see* ion evaporation model and theory)
Iron complexes, 530–532
IS-MS (*see* pneumatically-assisted electrospray)
Isoelectric focusing (*see* capillary isoelectric focusing)
Isotachophoresis (*see* capillary isotachophoresis)
Isotope-enriched proteins, 392

Joule heating, 346

Kinetic energy release, 540

Lanthanides, 533
LC (*see* liquid chromatography)
LC-ESI-MS (*see* liquid chromatography-mass spectrometry)
LC-MS/MS, 397–398, 508–520
LC-MS (*see* liquid chromatography-mass spectrometry)
Ligand exchange, 561
Ligand identification, 408, 410
Ligand loss reactions, 529
Ligands, 552, 558
Linkage isomers, 462, 478
Lipid A, 488–490
Lipids, 459–496
Lipopolysaccharides (LPS), 376, 488
Lithium complexes, 539
Liquid chromatography (LC), 116, 346, 350, 500, 514 (*see also* liquid chromatography-mass spectrometry)
Liquid chromatography, ion pair, 326–329, 331, 336
Liquid chromatography, normal phase, 330
Liquid chromatography, perfusion, 338–339
Liquid chromatography, reverse phase, 327–329, 492

Liquid chromatography, size exclusion, 198, 327, 329–330
Liquid chromatography-mass spectrometry, 128, 131, 190–198, 205, 219, 221, 297, 299, 301, 324, 351, 370, 397, 398, 410, 431, 451–452, 500–520
Liquid junction interface, 351, 354, 360
Low-flow electrospray, 108, 110, 180–181, 191, 208, 229, 339, 355

Mach disk, 118–120, 122, 209, 246
Macro ions, ESI mechanism, 53–54
Magnetic field focusing, 296–297
Magnetic field gradient, 294
Magnetic mirror effect, 294
Magnetic sector mass spectrometer, 111, 113, 114, 124, 127, 129–131, 177–179, 181–190, 358, 517
Magnetron motion, 303
MALDI (*see* matrix-assisted laser desorption/ionization)
MAP peptide (*see* multiple antigenic peptide)
Mass accuracy, 186–188, 214–215, 221, 239, 258, 264, 267–268, 295, 310–311, 391, 436, 443
Mass calibration, 132, 214–215, 243, 250
Mass-flow rates, 361–366
Mass measurement accuracy (*see* mass accuracy)
Mass range (*see* mass-to-charge range)
Mass resolution, 182, 183, 186–188, 204, 212–214, 243, 250–254, 258, 295, 309–310
Mass-to-charge (m/z) range, 132, 204, 218, 243, 249
Mathieu equation, 239–241
Matrix-assisted laser desorption/ionization (MALDI), 195, 198, 204, 237, 292, 351, 386, 422, 433, 461, 520
Maximum entropy, 391
Mechanism of ion formation, 3–60, 67–77, 140–161
Membrane anchors, 491–492
Metabolism studies, 368, 370, 494, 499–520
Metal-binding proteins, 403, 406–408
Metal-coated ESI tips, 354–355
Metal complexes, 528–532
Metal oxo ions, 533
Metal-solvent cluster ions, 528–529
Metallocenes, 530–531
metalloenzymes, 403
Metalloporphyrins (*see* porphyrins)
Metalloproteins, 543, 545–547
Methylation by methyl iodide, 463
MHC peptides, 398

Micellar electrokinetic chromatography (MEKC), 344, 347–349
Microdialysis, 505
Microelectrospray (*see* low-flow electrospray)
Microscan, 243
Microspray (*see* low-flow electrospray)
Molecular beam, 118–120, 124, 126
Molecular weight determination, 182–188, 392
Monoclonal antibodies (*see* antibodies)
Moving ionic boundaries, 353
MRM (*see* selected reaction monitoring)
MS^n (*see* multiple stage MS)
MS/MS, 110, 131, 194, 306, 395–399, 431–442, 461–494
Multiple antigenic peptide (MAP peptide), 390–391, 397
Multiple stage MS (MS^n), 237, 258–261, 265, 267, 273, 277, 360, 399–400, 410, 434, 441, 516
Multiply charged ions (*see* charge state determination; charge state distribution)

N/N_o ratio, 150–153
N-terminus determination, 397
Nanoelectrospray (*see* low-flow electrospray)
Nanospray (*see* low-flow electrospray)
Narrow bore (*see* low-flow electrospray)
Natural ionophores (*see* ionophores)
Nebulization, 108, 110
Nebulizer, 109
Negative ion ESI, 147–148, 326, 388, 425, 462, 466–467, 469–478, 489
Neutral-loss scan, 502, 508, 516, 517
Nickel complexes, 530–532, 550, 561–563
Noncovalent associations (*see* noncovalent interactions)
Noncovalent complexes (*see* noncovalent interactions)
Noncovalent interactions, 197, 222–230, 304, 358, 370, 373, 387, 401–405, 410. 445–449, 485–487, 558
Normal phase LC (*see* liquid chromatography, normal phase)
Nozzle-skimmer dissociation (*see* in-source collision-induced dissociation)
Nucleic acids, 350, 373, 375, 403, 421–454 (*see also* DNA, RNA)
Nucleobases, 424, 434, 438, 449
Nucleosides, 449–452
Nucleotides, 267–272, 431–451, 541, 559 (*see also* duplex oligonucleotides)

Octapole guide (*see* ion guide)
Oligonucleotide sequencing, 431, 438–445

Oligonucleotides (*see* nucleotides)
Oligosaccharides, 375, 462–463, 475–476, 488, 539–541
Onset potential (*see* electric field for onset of ESI)
Organic modifiers (see volatile additives)
Organoarsenics, 533
Organometallics, 527–563
Orthogonal injection time-of-flight, 205–211, 360

Particle beam interface, 237, 500
Peptide sequencing, 198, 265, 555
Peptides, 110, 144–146, 262–265, 365–367, 371, 385–411, 491–492, 504, 548–558
Perfusion chromatography (*see* liquid chromatography, perfusion)
Pesticides, 108
pH control for LC-MS, 325, 331
pH effects, 39–42, 142–146, 158, 427, 532
Pharmaceutical analysis, 131, 368, 499–520
Pharmacokinetics, 448, 499–520
Phosphine (PR_3) complexes, 561
Phospholipids, 492–494
Photodissociation, 261–262, 295
Plasma desorption (PD), 7, 351
Pneumatically-assisted electrospray, 109–110, 115–117, 180, 190, 324, 327, 332, 351, 354, 487–489, 492
Polyether toxins, 541–542
Polyethylene glycols, 132, 142, 146, 170, 198, 536
Polymers, 197–198
Polyoxoanions, 533
Polysaccharides (*see* oligosaccharides)
Porphyrins, 74–77, 81–86, 92–94, 530, 563
Post-column addition in LC, 328, 332, 539
Post-translational modifications, 394, 395, 397, 491
Potassium complexes, 536–538
Potential at electrode/solution interface, 79–86, 89
Pre-analyzer collision induced dissociation (*see* in-source collision-induced dissociation)
Pressure gradient, 293–294
Protein conformation, 142–144, 156–161, 312–313, 399–401 (*see also* protein structure)
Protein denaturation (*see* protein comformation)
Protein sequencing, 395–396
Protein structure, 385–411
  primary, 387
  secondary, 387
  tertiary, 387
  quaternary, 387, 401–402

Proteins, 117, 131–132, 264–267, 281, 358–359, 364–366, 371–374, 385–411, 485, 543, 545–547
Proton transfer, 167–168, 161, 273–277, 315, 391

Quadrupolar excitation (QE), 303–306
Quadrupole ion trap (QIT), 206, 235–281, 339, 355, 358, 370, 371, 399, 431, 433–434, 439, 516
  stability diagram, 240, 249, 254, 274
Quadrupole mass spectrometer, 111, 114, 119, 125–127, 129–130, 177–181, 204, 358
Quadrupole, rf-only ion guide (*see* ion guide)
Quantitation, 133–134, 501, 504–505, 511–516

Radioimmunological assays, 500
Radius of charged droplets, 15–17, 34–37
Rayleigh stability limit, 20, 532
Redox potential, 86, 99
Redox reactions, 70, 71–77, 79, 86, 87, 99–102, 530, 532
Reduced elution speeds in capillary electrophoresis, 364
Remeasurement experiment, 303–306
Resolution (*see* mass resolution; high resolution mass spectrometry)
Resonance ejection, 241–243, 249–250, 254–256, 358
Resonance excitation, 241–243, 246, 248, 258–261, 267, 272
Reverse phase LC (*see* liquid chromatography, reverse phase)
RF-only quadrupole ion guide (*see* ion guide)
RIA (*see* radioimmunological assays)
Ring opening, 476
RNA (*see also* nucleic acids), 273–274, 423, 427, 439–443, 449–450, 452, 541, 547
Ruthenium complexes, 530–531, 559–560, 562–563

Sample handling, 372
Sample introduction in capillary electrophoresis, 347, 350
Sample preconcentration in capillary electrophoresis, 349, 350, 366, 370, 371
Sampling orifice, 115–118, 120–129
Scan function, 241–243
Scan modes, 250–256
Scan rate, 243, 249, 265
Schiff-base complexes, 563
Sector mass spectrometers (*see* magnetic sector mass spectrometer)
Selected ion monitoring (SIM), 253–254, 256, 360, 508, 512

INDEX **577**

Selected reaction monitoring (SRM), 258–261, 508, 512, 516
Selective ion accumulation (SIA), 304
Separation efficiencies, 346, 354, 363, 371, 372
Separations (*see* liquid chromatography; gas chromatography; capillary electrophoresis)
Sequencing (*see* peptide sequencing; protein sequencing; oligonucleotide sequencing)
Sheath-flow interface, 351–357, 360–361, 366
Sheath gas (*see* electrical gas discharge suppression)
Sheathless interface, 352, 354–358, 360–361, 371
Signal dependence on concentration, 42–46 (*see also* concentration of analyte; dynamic range)
Signal suppression, 42–50, 80, 92
SIM (*see* selected ion monitoring)
Single ion counting (*see* ion counting)
Single ion detection 313, 315
Single ion in droplet theory (*see* charged residue mechanism)
Single ion monitoring (*see* selected ion monitoring)
Size exclusion chromatography (*see* liquid chromatography, size exclusion)
Skimmer, 120, 122, 124–125, 128–130, 133
Skimmer CID (*see* in-source collision-induced dissociation)
Small molecule analysis, 368–370, 499–520, 527–563
Solution-phase equilibria, 140–146
Solvent effect on charge state distribution (*see* charge state distribution, effect of solvent)
Solvent evaporation from droplets, 20–22, 115, 161–164
Solvents for LC-MS, 325
Space charge, 122, 243–244, 254–258
Spray current (*see* current due to charged droplets)
SRM (*see* selected reaction monitoring)
Stability diagram, quadrupole ion trap (*see* quadrupole ion trap)
Steroids, 543

Supramolecular metal complexes, 558–560
Surface activity, 32–33, 49
Surface-induced dissociation (SID), 309
Surfactants and detergents, 349, 394, 400
Sustained off-resonance irradiation (SORI), 307, 373–374
SWIFT, 254–256

Tailored waveform (TWF), 250–251, 254–256
Tandem mass spectrometry (*see* MS/MS; MS^n)
Tapered capillary tips, 353, 355, 358
Taxol, 520
Taylor cone, 10, 11, 13
Ternary metal complexes, 550–559
TFA-fix, 336
Thermospray, 500
Time-of-flight MS, 203–231, 355
Trace analysis, 194
Transfer capillary, 112–113, 116
Transferrins, 395
Transition metal oxides, 533
Triple quadrupole mass spectrometer, 241, 261, 436–439, 461, 501, 516
t-RNA (*see* RNA)
Trypsin digestion (*see* enzymatic digests)
TSP (*see* thermospray)
Tube lens, 125, 129, 180, 246, 250, 253
Tube lens CID/CAD (*see* in-source collision-induced dissociation)

Ultrahigh resolution (*see* high resolution)
Ultrasonic nebulization, 109, 110, 116
Ultraspray (*see* ultrasonic nebulization)
Up-front CID (*see* in-source collision-induced dissociation)

Volatile additives, 325, 352–354, 360, 370, 425, 427
Voltage (*see* high voltage)

Water clusters (*see* aquo complexes)
Working electrode, 67, 71, 72, 78, 79, 86

Zinc finger proteins, 408, 545–546